AF606561

SOFT COMPUTING AND INTELLIGENT DATA ANALYSIS IN OIL EXPLORATION

Volumes 1-7, 9-18, 19a, 21-27, 29, 31 are out of print.

8 *Fundamentals of Reservoir Engineering*
19b *Surface Operations in Petroleum Production, II*
20 *Geology in Petroleum Production*
28 *Well Cementing*
30 *Carbonate Reservoir Characterization: A Geologic-Engineering Analysis, Part I*
32 *Fluid Mechanics for Petroleum Engineers*
33 *Petroleum Related Rock Mechanics*
34 *A Practical Companion to Reservoir Stimulation*
35 *Hydrocarbon Migration Systems Analysis*
36 *The Practice of Reservoir Engineering (Revised Edition)*
37 *Thermal Properties and Temperature Related Behavior of Rock/Fluid Systems*
38 *Studies in Abnormal Pressures*
39 *Microbial Enhancement of Oil Recovery – Recent Advances – Proceedings of the 1992 International Conference on Microbial Enhanced Oil Recovery*
40a *Asphaltenes and Asphalts, I*
40b *Asphaltenes and Asphalts, II*
41 *Subsidence due to Fluid Withdrawal*
42 *Casing Design – Theory and Practice*
43 *Tracers in the Oil Field*
44 *Carbonate Reservoir Characterization: A Geologic-Engineering Analysis, Part II*
45 *Thermal Modeling of Petroleum Generation: Theory and Applications*
46 *Hydrocarbon Exploration and Production*
47 *PVT and Phase Behaviour of Petroleum Reservoir Fluids*
48 *Applied Geothermics for Petroleum Engineers*
49 *Integrated Flow Modeling*
50 *Origin and Prediction of Abnormal Formation Pressures*
51 *Soft Computing and Intelligent Data Analysis in Oil Exploration*

SOFT COMPUTING AND INTELLIGENT DATA ANALYSIS IN OIL EXPLORATION

Edited by
M. NIKRAVESH
F. AMINZADEH
L.A. ZADEH

2003

ELSEVIER
Amsterdam – Boston – London – New York – Oxford – Paris
San Diego – San Francisco – Singapore – Sydney – Tokyo

First edition 2003

Library of Congress Cataloging in Publication Data
A catalog record from the Library of Congress has been applied for.

British Library Cataloguing in Publication Data
A catalogue record from the British Library has been applied for.

ISBN: 0-444-50685-3
ISSN: 0376-7361 (Series)

∞ The paper used in this publication meets the requirements of ANSI/NISO Z39.48-1992 (Permanence of Paper).
Printed in The Netherlands.

Dedicated to

Laura and Nikolas Nikravesh
Kathleen, Sara, David and Diana Aminzadeh
Fay, Norman, and Estella Zadeh

FOREWORD

In his foreword to "Soft Computing for Reservoir Characterization and Modeling", Dr. Bertrand Braunschweig posed the question: "What's next?". In this companion volume, the authors have provided an imaginative and comprehensive answer. They have extended the application of soft computing techniques to methodologies used for oil exploration in general and have indicated how these increasingly popular methodologies can be integrated with our more traditional industry techniques. In this collection of articles you will find contributions from largely diverse disciplines, ranging from geostatistics and time lapse seismic to biostratigraphy and core analysis.

It has been nearly 50 years since John McCarthy first coined the term 'artificial intelligence' and 40 years since Lotfi A. Zadeh first coined the term 'fuzzy logic' the disciplines that might have been arrayed under that banner have grown many fold. Now soft computing like evolutionary algorithms, machine reasoning, fuzzy logic, neural systems, etc., crowd the computational landscape and new techniques are being developed every day.

What is 'soft computing'? Lotfi Zadeh, one of the editors of this volume, who originally coined the term, defined it as follows: "*Soft computing differs from conventional (hard) computing in that, unlike hard computing, it is tolerant of imprecision, uncertainty, and partial truth.*" He further declared that these techniques provide the opportunity to achieve robust, tractable solutions whilst, at the same time, offering low solution cost.

This book comes at a very opportune time for the oil and gas industry. Knowledge, and the processes whereby that knowledge is managed, are clearly important assets of any organization. Value of information, quantification of risk, and uncertainty assessment are becoming increasingly important to the industry, as a whole, as it seeks to better understand the factors that influence its ability to make better, more informed, decisions. Decisions that lead to improved efficiency, in resource exploitation and utilization, increased profitability and enhanced shareholder return. It is fortunate that such decisions can be made even when the supporting data is uncertain and imprecise, if that uncertainty is accounted for in a rigorous and consistent fashion.

Another distinguishing feature of soft computing is the concept of incorporating heuristic information, in the form of expert knowledge, in the problem solving processes. This capability is sure to become increasingly relevant to the oil industry. The demographics of the energy business are well known, and many major oil companies face a potential problem in the next several years, as large numbers of experienced personnel become eligible to retire from the industry. It is imperative, that procedures be developed that enable some retention of this expertise as a component of our problem solving capability.

In the oil industry today we are facing somewhat of a data explosion. We have seen a proliferation of pre-stack analysis of 3D seismic data coupled with increased acceptance of time-lapse or 4D seismic data. Both of these activities are data intensive, however, even this data volume could be dwarfed by the advent of the so-called instrumented oilfield or electric oilfield (E-field). These installations, with their permanently in-place sensors, on the ocean floor and in the borehole, together with the use of smart wells with the ability to intelligently interact with the oil field, will produce real time data at an unprecedented rate.

The purpose of these advanced data generation and acquisition facilities is to enable timely decisions affecting the production and development of the asset. Increased speed and computing power alone will not be enough to enable us to get the most out of this valuable data resource. The ability of soft computing techniques to extract rules or patterns hidden in the data, to allow seamless incorporation of additional data into highly complex systems and to do it with mostly 'white box' methods makes these methodologies attractive additions to conventional techniques.

The subject matter of this book has intrigued me ever since I first came in to contact with soft computing and machine reasoning at the IEEE World Congress on Computational Intelligence in Orlando, Florida in 1994. Even though I was a relative latecomer to these areas, it was obvious to me that our industry could only benefit from the incorporation of these emerging techniques in to the mainstream approaches of geology, geophysics and engineering. In addition, it is clear that, while the interest in this material has been steadily growing within the energy industry, as well as a number of other industries, there is still some way to go before these methods fulfill their undoubted promise.

The editors of this volume, Drs. Lotfi Zadeh, Fred Aminzadeh and Masoud Nikravesh, have a long and distinguished history in the development and practical application of soft computing techniques and, in particular, their application within the oil industry. I heartily congratulate them on the outstanding job they have done in putting this book together. I feel certain that we are just scratching the surface when it comes to unlocking the potential inherent in these approaches, and that as far as the oil business is concerned, the uses herein described are just the thin end of a very large wedge.

David A. Wilkinson
Research Scientist
ChevronTexaco

PREFACE

Integration, handling data involving uncertainty and risk management are among key issues in geoscience and oil industry applications. In recent years there has been tremendous efforts to find new methods to address theses issues. As we approach the dawn of the next millennium, and as our problems become too complex to rely only on one discipline to solve them more effectively, and the cost associated with poor predictions (such as dry holes) increases, the need for proper integration of disciplines, data fusion, risk reduction and uncertainty management, and multi-disciplinary approaches in the petroleum industry become more important and of a necessity than professional curiosity. We will be forced to bring down the walls we have built around classical disciplines such as petroleum engineering, geology, geophysics and geochemistry, or at the very least make them more permeable. Our data, methodologies and approaches to tackle problems will have to cut across various disciplines. As a result, today's "integration" which is based on integration of results will have to give way to a new form of integration, that is, integration of disciplines. In addition, to solve our complex problem one needs to go beyond standard techniques and silicon hardware. The model needs to use several emerging methodologies and soft computing techniques. Soft Computing is consortium of computing methodologies (Fuzzy Logic (FL), Neuro-Computing (NC), Genetic Computing (GC), and Probabilistic Reasoning (PR) including; Genetic Algorithms (GA), Chaotic Systems (CS), Belief Networks (BN), Learning Theory (LT)) which collectively provide a foundation for the Conception, Design and Deployment of Intelligent Systems. The role model for Soft Computing is the Human Mind. Soft computing differs from conventional (hard) computing in that, unlike hard computing, it is tolerant of imprecision, uncertainty, and partial truth. Soft Computing is also tractable, robust, efficient and inexpensive. In this volume, we reveal (explore) the role of Soft Computing techniques for intelligent reservoir characterization and exploration. The major constituent of soft computing is fuzzy logic, which was first introduced by Prof. Lotfi Zadeh back in 1965. In 1991, Prof. Zadeh introduced the Berkeley Initiative in Soft Computing (BISC) at the University of California, Berkeley. In 1994, a new BISC special interest group in Earth Sciences was formed. Broadly, Earth Sciences subsumes but is not limited to Geophysics (seismology, gravity, and electromagnetic), Geology, Hydrology, Borehole wireline log evaluation, Geochemistry, Geostatistics, Reservoir Engineering, Mineral Prospecting, Environmental Risk Assessment (nuclear waste, geohazard, hydrocarbon seepage/spill) and Earthquake Seismology.

Soft Computing methods such as neural networks, fuzzy logic, perception-based logic, genetic algorithms and other evolutionary computing approaches offer an excellent opportunity to address different challenging practical problems. Those to focus on in this volume are the following issues:

- Integrating information from various sources with varying degrees of uncertainty;
- Establishing relationships between measurements and reservoir properties; and
- Assigning risk factors or error bars to predictions.

Deterministic model building and interpretation are increasingly replaced by stochastic and soft computing-based methods. The diversity of soft computing applications in oil field problems and prevalence of their acceptance are manifested by the overwhelming increasing interest among the earth scientist and engineers.

The present volume starts with an introductory article written by the editors explaining the basic concepts of soft computing and the past/present/future trends of soft computing applications in reservoir characterization and modelling. It provides a collection of thirty (30) articles containing: (1) Introduction to Soft Computing and Geostatistics (6 articles in Part 1), (2) Seismic Interpretation (4 articles in Part 2), (3) Geology (6 articles in Part 3), (4) Reservoir and Production Engineering (5 articles in Part 4), (5) Integrated and Field Studies (5 articles in Part 5), and (6) General Applications (4 articles in Part 6). Excellent contributions on applications of neural network fuzzy logic, evolutionary techniques, and development of hybrid models are included in this book.

We would like to take this opportunity to thank all the contributors and reviewers of the articles. We also wish to acknowledge our colleagues who have contributed to the areas directly or indirectly related to the contents of this book.

Masoud Nikravesh
Fred Aminzadeh
Lotfi A. Zadeh
Berkeley

ABOUT THE EDITORS

Masoud Nikravesh received his BS from Abadan Institute of Technology, MS and PhD in Chemical Engineering from the University of South Carolina. He is BISC (Berkeley Initiative in Soft Computing) Associate Director and BT Senior Research Fellow in the Computer Science Division, Department of Electrical Engineering and Computer Sciences at the University of California, Berkeley and Visiting Scientist in the Imaging and Collaborative Computing Group at the Lawrence Berkeley National Laboratory. In addition, he is serving as Associate Director (Co-founder) of Zadeh Institute for Information Technology (Information Technology and Chairs of BISC-Earth Sciences, BISC-Fuzzy Logic and Internet, and BISC-Recognition Technology Groups). He has over 10 years research and industrial experience in soft computing and artificial intelligence. He worked as a consultant to over 10 major companies and funded several key projects in the area of soft computing, data mining and fusion, control, and earth sciences through US government and major oil companies. He published and presented over 100 articles on diverse topics and served as SPE Technical Editor and several national and international technical committees and technical chairs. He served as member of IEEE, SPE, AICHE, SEG, AGU, and ACS. His credentials have led to front-page news at Lawrence Berkeley National Laboratory News and headline news at the Electronics Engineering Times.

Fred Aminzadeh received his BSEE from University of Tehran, MSEE and PhD from the University of Southern California. He is the President of dGB-USA and FACT (Houston, Texas, faz@dgbusa.com) since 1999. He held various technical and management positions at Unocal and Bell Laboratories. Fred also had different academic positions at University of Tabriz, USC, and Rice. Among areas of his technical contributions are: elastic seismic modeling, seismic attribute analysis, reservoir characterization, signal processing, artificial intelligence, Kalman filtering and soft computing applications. He has published over 100 articles and 7 books. He is the co-inventor of three US patents on AVO modeling, seismic while drilling and hybrid reservoir characterization. He served as the chairman of the SEG Research Committee 1994–1996 and vice president of SEG, 2001–2002. He has served as a member of the National Research Council's Committee on Seismology, Foreign Member of Russian Academy of Natural Sciences, *an honorary member of* Azerbaijan Oil Academy, and the Scientific Advisory Board of the Center for Engineering Systems Advanced Research of Oak Ridge National Laboratory. He is a Fellow of IEEE for his contributions to the application of modeling, signal processing, pattern recognition and expert systems in the analysis of seismic and acoustic data. He is Associate Editor of various journals.

Lotfi A. Zadeh is a Professor in the Graduate School, Computer Science Division, Department of EECS, University of California, Berkeley. In addition, he is serving as the Director of BISC (Berkeley Initiative in Soft Computing). Lotfi Zadeh is an alumnus of the University of Teheran, MIT and Columbia University. He held visiting appointments at the Institute for Advanced Study, Princeton, NJ; MIT; IBM Research Laboratory, San Jose, CA; SRI International, Menlo Park, CA; and the Center for the Study of Language and Information, Stanford University. His earlier work was concerned in the main with systems analysis, decision analysis and information systems. His current research is focused on fuzzy logic, computing with words and soft computing, which is a coalition of fuzzy logic, neurocomputing, evolutionary computing, probabilistic computing and parts of machine learning. The guiding principle of soft computing is that, in general, better solutions can be obtained by employing the constituent methodologies of soft computing in combination rather than in stand-alone mode. Lotfi Zadeh is a Fellow of the IEEE, AAAS, ACM, AAAI, and IFSA. He is a member of the National Academy of Engineering and a Foreign Member of the Russian Academy of Natural Sciences and an honorary member of Azerbaijan Oil Academy. He is a recipient of the IEEE Education Medal, the IEEE Richard W. Hamming Medal, the IEEE Medal of Honor, the ASME Rufus Oldenburger Medal, the B. Bolzano Medal of the Czech Academy of Sciences, the Kampe de Feriet Medal, the AACC Richard E. Bellman Central Heritage Award, the Grigore Moisil Prize, the Honda Prize, the Okawa Prize, the AIM Information Science Award, the IEEE-SMC J.P. Wohl Career Achievement Award, the SOFT Scientific Contribution Memorial Award of the Japan Society for Fuzzy Theory, the IEEE Millennium Medal, the ACM 2000 Allen Newell Award, and other awards and honorary doctorates. He has published extensively on a wide variety of subjects relating to the conception, design and analysis of information/intelligent systems, and is serving on the editorial boards of over fifty journals.

LIST OF CONTRIBUTORS

R.D. ADAMS	Energy and Geoscience Institute, University of Utah, Salt Lake City, Utah 84108, USA
F. AMINZADEH	dGB-USA, Houston, TX, USA and Fact Incorporated, 14019 SW Freeway, Suite 301-225, Sugar Land, TX 77478, USA
R. BELOHLAVEK	Institute for Research and Applications of Fuzzy Modeling, University of Ostrava, Brafova 7, 70103, Czech Republic
J. CAERS	Department of Petroleum Engineering, Stanford University, Stanford, CA 94305-2220, USA
F. CAILLY	Beicip Franlab, 232 Avenue Napoléon Bonaparte, 92500 Rueil Malmaison, France
J.N. CARTER	Department of Earth Science and Engineering, Imperial College of Science Technology and Medicine, South Kensington, London, SW7 2BP, UK
A. CHAWATHÉ	New Mexico Petroleum Recovery Research Center
R.J. COOK	BG Group, 100 Thames Valley Park Drive, Reading RG6 1PT, UK
R.V. DEMICCO	Department of Geological Sciences and Environmental Studies, Binghamton University, Binghamton, NY 13902-6000, USA
P. DIGRANES	Statoil Gullfaks Production, 5021 Bergen, Norway
E.B. EDWARDS	Pacific Operators Offshore Inc., Santa Barbara, CA, USA
A.M. ELSHARKAWY	Petroleum Engineering Department, Kuwait University, P.O. Box 5969, Safat 13060, Kuwait
I. ERSHAGHI	University of Southern California, Los Angeles, CA 90007, USA
A. FARAJ	Institut Français du Pétrole, 1–4 Avenue de Bois-Préau, 92500 Rueil Malmaison, France
T.D. GEDEON	School of Information Technology, Murdoch University, Perth, Australia
M. HASSIBI	Fact Incorporated, 14019 SW Freeway, Suite 301-225, Sugar Land, TX 77478, USA

E. HILDE	Statoil Research Centre, Postuttak 7005 Trondheim, Norway
H. JACKSON	BG Group, 100 Thames Valley Park Drive, Reading RG6 1PT, UK
V.M. JOHNSON	Lawrence Livermore National Laboratory, Livermore, CA 94551, USA
G.J. KLIR	Center for Intelligent Systems, Watson School of Engineering and Applied Science, Binghamton University, Binghamton, NY 13902-6000, USA
M. LANDRØ	Statoil Research Centre, Postuttak 7005 Trondheim, Norway and Department of Petroleum Engineering and Applied Geophysics, NTNU, 7491 Trondheim, Norway
R.A. LEVEY	Energy and Geoscience Institute, University of Utah, Salt Lake City, Utah 84108, USA
J.H. LIGTENBERG	dGB Earth Sciences, Boulevard-1945 24, 7511 AE, Enschede, The Netherlands
J.-S. LIM	Division of Ocean Development Engineering, Korea Maritime University, Dongsam-Dong, Yeongdo-Gu, Puasn, 606-791, Republic of Korea
T. LIN	Mathematical and Information Sciences, CSIRO, Canberra, Australia
S.D. MOHAGHEGH	West Virginia University, 345E Mineral Resources Building, Morgantown, WV 26506, USA
M. NIKRAVESH	Berkeley Initiative in Soft Computing (BISC) and Computer Science Division – Department of Electrical Engineering and Computer Sciences, University of California, Berkeley, CA 94720, USA
A. OUENES	Reservoir Characterization, Research & Consulting (RC)2, a subsidiary of Veritas DGC, 13 rue Pierre Loti, 92340 Bourg-La-Reine, France
R.J. PAWAR	Los Alamos National Laboratories, Los Alamos, NM, USA
L.L. ROGERS	Lawrence Livermore National Laboratory, Livermore, CA 94551, USA
C. ROMERO	PDVSA Intevep, P.O. Box 76343, Caracas 1070-A, Venezuela
M.S. ROSENBAUM	Civil Engineering Division, The Nottingham Trent University, Newton Building, Burton Street, Nottingham NG1 4BU, UK

S.A.R. SHIBLI	Landmark Graphics (M) Snd. Bhd., Menara Tan and Tan, 55100 Kuala Lumpur, Malaysia
E.A. SHYLLON	Department of Geomatics, University of Melbourne, Parkville, Victoria 3010, Aurtralia
O.A. SOLHEIM	Statoil Research Centre, Postuttak 7005 Trondheim, Norway
S. SRINIVASAN	University of Calgary, Department of Chemical and Petroleum Engineering, 2500 University Drive, N.W., Calgary, AB T2N 1N4, Canada
L.K. STRØNEN	Statoil Gullfaks Production, 5021 Bergen, Norway
D. TAMHANE	School of Petroleum Engineering, University of New South Wales, Sydney, Australia
P. THOMPSON	BG Group, 100 Thames Valley Park Drive, Reading RG6 1PT, UK
K.M. TINGDAHL	Department of Earth Sciences – Marine Geology, Göteborg University, Box 460, SE-405 30 Göteborg, Sweden
M.I. WAKEFIELD	BG Group, 100 Thames Valley Park Drive, Reading RG6 1PT, UK
A.G. WANSINK	dGB Earth Sciences, Boulevard-1945 24, 7511 AE, Enschede, The Netherlands
E.M. WHITNEY	Pacific Operators Offshore Inc., Santa Barbara, CA, USA
P.M. WONG	School of Petroleum Engineering, University of New South Wales, Sydney NSW 2052, Australia
Y. YANG	Civil Engineering Division, The Nottingham Trent University, Newton Building, Burton Street, Nottingham NG1 4BU, UK
M. YE	Equator Technologies Inv.
A.M. ZELLOU	Reservoir Characterization, Research & Consulting (RC)2, a subsidiary of Veritas DGC, 13 rue Pierre Loti, 92340 Bourg-La-Reine, France

CONTENTS

Foreword . . . vii
Preface . . . ix
About the Editors . . . xi
List of Contributors . . . xiii

Part 1 Introduction: Fundamentals of Soft Computing

Chapter 1 SOFT COMPUTING FOR INTELLIGENT RESERVOIR CHARACTERIZATION AND MODELING
M. Nikravesh and F. Aminzadeh . . . 3
Abstract . . . 3
1. Introduction . . . 3
2. The role of soft computing techniques for intelligent reservoir characterization and exploration . . . 4
   2.1. Mining and fusion of data . . . 5
   2.2. Intelligent interpretation and data analysis . . . 7
   2.3. Pattern recognition . . . 9
   2.4. Clustering . . . 9
   2.5. Data integration and reservoir property estimation . . . 10
   2.6. Quantification of data uncertainty and prediction error and confidence interval . . . 11
3. Artificial neural network and geoscience applications of artificial neural networks for exploration 13
   3.1. Data processing . . . 13
      3.1.1. First-arrival picking . . . 13
      3.1.2. Noise elimination . . . 14
   3.2. Identification and prediction . . . 14
4. Fuzzy logic . . . 15
   4.1. Geoscience applications of fuzzy logic . . . 16
5. Genetics algorithms . . . 17
   5.1. Geoscience applications of genetic algorithms . . . 17
6. Principal component analysis and wavelet . . . 18
7. Intelligent reservoir characterization . . . 18
8. Fractured reservoir characterization . . . 20
9. Future trends and conclusions . . . 22
Appendix A. A basic primer on neural network and fuzzy logic terminology . . . 22
Appendix B. Neural networks . . . 24
Appendix C. Modified Levenberge–Marquardt technique . . . 26
Appendix D. Neuro-fuzzy models . . . 26
Appendix E. K-means clustering . . . 27
Appendix F. Fuzzy c-means clustering . . . 28
Appendix G. Neural network clustering . . . 28
References . . . 29

Chapter 2 FUZZY LOGIC
G.J. Klir . . . 33
Abstract . . . 33
1. Fuzzy sets . . . 33

2. Operations on fuzzy sets . . . 36
3. Arithmetic of fuzzy intervals . . . 38
4. Fuzzy relations . . . 40
5. Fuzzy systems . . . 42
6. Fuzzy propositions . . . 43
7. Approximate reasoning . . . 46
8. Suggestions for further study . . . 48
References . . . 48

Chapter 3 INTRODUCTION TO USING GENETIC ALGORITHMS
J.N. Carter . . . 51
1. Introduction . . . 51
2. Background to Genetic Algorithms . . . 51
 2.1. Advantages and Disadvantages . . . 52
 2.2. Review of Genetic Algorithms Literature . . . 53
3. Design of a Genetic Algorithm . . . 53
 3.1. Terminology . . . 53
  3.1.1. Example of the various data structures . . . 56
 3.2. Basic Structure . . . 57
 3.3. Structure of the genome . . . 60
 3.4. Crossover operators . . . 65
  3.4.1. k-point crossover operator . . . 65
 3.5. Crossover operators for real valued genomes . . . 66
  3.5.1. k-point crossover for real valued strings . . . 67
  3.5.2. The BLX-a operator . . . 67
  3.5.3. UNDX operator . . . 67
  3.5.4. The SBX crossover operator . . . 68
  3.5.5. Comparison of the three crossover operators . . . 68
 3.6. Combining k-point and gene-based crossover operators . . . 70
 3.7. Crossover operator for multi-dimensional chromosomes . . . 70
 3.8. Selection of parents . . . 71
 3.9. Construction of new populations . . . 72
 3.10. Mutation operators . . . 73
 3.11. Population size . . . 73
 3.12. Generation of the initial population . . . 74
 3.13. General parameter settings . . . 74
4. Conclusions . . . 75
References . . . 75

Chapter 4 HEURISTIC APPROACHES TO COMBINATORIAL OPTIMIZATION
V.M. Johnson . . . 77
1. Introduction . . . 77
2. Decision variables . . . 77
3. Properties of the objective function . . . 79
4. Heuristic techniques . . . 80
References . . . 83

Chapter 5 INTRODUCTION TO GEOSTATISTICS
R.J. Pawar . . . 85
1. Introduction . . . 85
2. Random variables . . . 86
3. Covariance and spatial variability . . . 87
4. Kriging . . . 90
5. Stochastic simulations . . . 93

References 95

Chapter 6 GEOSTATISTICS: FROM PATTERN RECOGNITION TO PATTERN REPRODUCTION
J. Caers 97
1. Introduction 97
2. The decision of stationarity 98
3. The multi-Gaussian approach to spatial estimation and simulation 99
 3.1. Quantifying spatial correlation with the variogram 99
4. Spatial interpolation with kriging 101
 4.1. Stochastic simulation 102
 4.2. Sequential simulation 102
 4.3. Sequential Gaussian simulation 105
 4.4. Accounting for secondary attributes 106
 4.5. Secondary data as trend information 106
 4.6. Full co-kriging 106
 4.7. Accounting for scale of data sources 107
5. Beyond two-point models: multiple-point geostatistics 109
 5.1. Accounting for geological realism 109
 5.2. From variogram to training image to multiple stochastic models 110
 5.3. Data integration 111
6. Conclusions 113
7. Glossary 113
References 115

Part 2 Geophysical Analysis and Interpretation

Chapter 7 MINING AND FUSION OF PETROLEUM DATA WITH FUZZY LOGIC AND NEURAL NETWORK AGENTS
M. Nikravesh and F. Aminzadeh 119
Abstract 119
1. Introduction 119
2. Neural network and nonlinear mapping 120
 2.1. Travel time (DT) prediction based on SP and resistivity (RILD) logs 121
 2.2. Gamma ray (GR) prediction based on SP and resistivity (RILD) logs 123
 2.3. Density (RHOB) prediction based on sp and resistivity (RILD) logs 123
 2.4. Travel time (DT) prediction based on resistivity (RILD) 126
 2.5. Resistivity (RILD) prediction based on travel time (DT) 126
3. Neuro-fuzzy model for rule extraction 126
 3.1. Prediction of permeability based on porosity, grain size, clay content, P-wave velocity, and P-wave attenuation. 129
4. Conclusion 135
Appendix A. Basic primer on neural network and fuzzy logic terminology 137
Appendix B. Neural networks 138
Appendix C. Modified Levenberge–Marquardt technique 140
Appendix D. Neuro-fuzzy models 140
Appendix E. K-means clustering 141
References 141

Chapter 8 TIME LAPSE SEISMIC AS A COMPLEMENTARY TOOL FOR IN-FILL DRILLING
M. Landrø, L.K. Strønen, P. Digranes, O.A. Solheim and E. Hilde 143
Abstract 143
1. Introduction 143

2. Feasibility study . . . 144
3. 3D seismic data sets . . . 145
4. 4D seismic analysis approach . . . 145
5. Seismic modeling of various flow scenarios . . . 146
6. 4D seismic for detecting fluid movement . . . 147
7. 4D seismic for detecting pore pressure changes . . . 150
8. 4D seismic and interaction with the drilling program . . . 153
9. Conclusions . . . 154
Acknowledgements . . . 155
References . . . 155

Chapter 9 IMPROVING SEISMIC CHIMNEY DETECTION USING DIRECTIONAL ATTRIBUTES
K.M. Tingdahl . . . 157
1. Introduction . . . 157
1.1. Introduction to seismic chimney detection . . . 158
1.2. Introduction to dip calculations . . . 159
2. Dip calculations . . . 161
3. Dip steering . . . 165
4. Chimney detection . . . 165
4.1. Similarity . . . 166
4.2. Energy . . . 167
5. Dip-related attributes . . . 167
5.1. Dip-steered similarity . . . 167
5.2. Dip variance . . . 168
6. Processing and results . . . 169
7. Discussion and conclusions . . . 171
Acknowledgements . . . 172
References . . . 172

Chapter 10 MODELING A FLUVIAL RESERVOIR WITH MULTIPOINT STATISTICS AND PRINCIPAL COMPONENTS
P.M. Wong and S.A.R. Shibli . . . 175
Abstract . . . 175
1. Introduction . . . 175
2. Neural networks revisited . . . 177
3. Case study . . . 179
3.1. Procedures . . . 180
3.2. Results . . . 180
3.3. Discussion . . . 181
4. Conclusions . . . 184
References . . . 184

Part 3 Computational Geology

Chapter 11 THE ROLE OF FUZZY LOGIC IN SEDIMENTOLOGY AND STRATIGRAPHIC MODELS
R.V. Demicco, G.J. Klir and R. Belohlavek . . . 189
Abstract . . . 189
1. Introduction . . . 189
2. Basic principles of fuzzy logic . . . 192
2.1. Fuzzy sets . . . 192
2.2. Fuzzy logic systems . . . 194
2.3. Application of ‘if–then’ rules to coral reef growth . . . 195

2.4. Application of multi-part 'if–then' rules to a hypothetical delta model . . . 199
3. Fuzzy inference systems and stratigraphic modeling . . . 204
3.1. Production of carbonate sediment on the Great Bahama Bank . . . 204
3.2. Death Valley, California . . . 209
4. Summary and conclusions . . . 213
Acknowledgements . . . 215
References . . . 215

Chapter 12 SPATIAL CONTIGUITY ANALYSIS. A METHOD FOR DESCRIBING SPATIAL STRUCTURES OF SEISMIC DATA
A. Faraj and F. Cailly . . . 219
1. Introduction . . . 219
2. State-of-the-art . . . 220
3. Local variance and covariance between statistics and geostatistics . . . 221
3.1. Variogram–crossed covariogram . . . 221
3.2. Local variance and covariance . . . 222
4. Spatial proximity analysis: a particular SCA . . . 223
4.1. Statistical and spatial properties of SCA components . . . 224
5. SCA result interpretation aid tools . . . 226
6. Application to seismic image description and filtering . . . 228
6.1. Seismic images . . . 228
6.2. Analyzed data . . . 229
6.3. Descriptive preliminary geostatistical analysis of initial variables . . . 231
6.4. SCA results in the anisotropic case for $h = 1$ m . . . 235
6.5. SCA results in the E–W direction for $h = 15$ m . . . 235
6.6. Optimum extraction of large-scale structures and random noise from the spatial components obtained from the two analyzes . . . 237
7. Conclusion . . . 241
References . . . 244

Chapter 13 LITHO-SEISMIC DATA HANDLING FOR HYDROCARBON RESERVOIR ESTIMATE: FUZZY SYSTEM MODELING APPROACH
E.A. Shyllon . . . 247
Abstract . . . 247
1. Introduction . . . 247
2. Uncertainties in hydrocarbon reservoir estimate . . . 248
2.1. Types of uncertainties in hydrocarbon reservoir estimate . . . 248
2.1.1. Uncertainty in data acquisition . . . 248
2.1.2. Uncertainty in model formulation . . . 248
2.1.3. Uncertainty due to linguistic imprecision . . . 249
2.1.4. Uncertainty due to resolution limit of the equipment . . . 249
2.1.5. Uncertainty due to incomplete information . . . 250
2.2. Magnitude of errors and uncertainty . . . 251
3. Litho-seismic data handling . . . 251
3.1. Seismic data . . . 251
3.2. Well lithology . . . 251
3.3. Litho-seismic data restructuring . . . 252
3.3.1. Acreage . . . 253
3.3.2. Most likely porosity . . . 254
3.3.3. Saturation hydrocarbon . . . 255
3.3.4. Formation volume factor . . . 255
3.3.5. Net thickness . . . 255
3.4. Training data set . . . 255
4. Fuzzy system modeling approach . . . 256

4.1. Fuzzy system . . . 256
4.1.1. Fuzzification of hydrocarbon reservoir parameters . . . 256
4.1.2. Operation on fuzzy subsets . . . 260
4.1.3. Defuzzification of the result . . . 264
5. Interpretation of result . . . 265
5.1. Most likely estimate . . . 265
5.2. Optimal estimate – good estimate . . . 265
5.3. Very good estimate . . . 265
5.4. Slightly good estimate . . . 266
5.5. Rule-based estimate . . . 267
6. Conclusion . . . 267
7. C codes to compute the estimates . . . 267
References . . . 271

Chapter 14 NEURAL VECTOR QUANTIZATION FOR GEOBODY DETECTION AND STATIC MULTIVARIATE UPSCALING
A. Chawathé and M. Ye . . . 273
Abstract . . . 273
1. Introduction . . . 274
2. Concepts in neural vector quantization . . . 276
2.1. The HSC algorithm . . . 279
2.2. Cluster delineation . . . 279
2.3. Neuron accumulation/merging . . . 279
3. HSC performance . . . 281
3.1. Application 1 . . . 281
3.2. Application 2 . . . 284
4. Conclusions . . . 285
References . . . 287

Chapter 15 HIGH RESOLUTION RESERVOIR HETEROGENEITY CHARACTERIZATION USING RECOGNITION TECHNOLOGY
M. Hassibi, I. Ershaghi and F. Aminzadeh . . . 289
Abstract . . . 289
1. Introduction . . . 289
2. Complex sedimentary environments . . . 294
3. Pattern classification techniques . . . 295
3.1. Vector quantization . . . 295
4. Essential pre-processes . . . 296
5. Reservoir compartmentalization and continuity correlation . . . 298
6. Synthetic and real field data . . . 299
6.1. Synthetic data . . . 299
6.2. Real field data and results . . . 302
7. Conclusions . . . 306
Acknowledgements . . . 307
References . . . 307

Chapter 16 EXTENDING THE USE OF LINGUISTIC PETROGRAPHICAL DESCRIPTIONS TO CHARACTERISE CORE POROSITY
T.D. Gedeon, P.M. Wong, D. Tamhane and T. Lin . . . 309
Abstract . . . 309
1. Introduction . . . 309
2. Lithological descriptions . . . 310
3. Data descriptions . . . 310
4. Experiments . . . 312

5. Expert system . . . 312
6. Supervised clustering . . . 313
7. Neural networks . . . 314
8. Results . . . 315
9. Extensions . . . 316
10. Conclusions . . . 318
References . . . 319

Part 4 Reservoir and Production Engineering

Chapter 17 USING GENETIC ALGORITHMS FOR RESERVOIR CHARACTERISATION
C. Romero and J.N. Carter . . . 323
1. Introduction . . . 323
2. Reservoir Modelling . . . 323
3. Survey of previous work . . . 325
4. Methodologies for reservoir modelling . . . 331
4.1. Geostatistical simulation . . . 332
4.2. Fault properties . . . 334
4.3. Well skin factors . . . 337
4.4. Summary of reservoir description . . . 337
5. Reservoir Description . . . 338
5.1. PUNQ complex model . . . 338
5.2. Reservoir model . . . 338
5.3. Production plan and well measurements . . . 340
6. Design of the Genetic Algorithm . . . 341
6.1. General parameters . . . 342
6.2. Design of the genome . . . 343
6.2.1. Chromosome for reservoir property fields . . . 344
6.3. Crossover operators . . . 344
6.3.1. Crossover for three dimensional chromosomes . . . 344
6.4. Mutation operators . . . 345
6.4.1. Jump mutation . . . 345
6.4.2. Creep mutation . . . 345
6.4.3. Shift mutation . . . 346
6.5. Function evaluation . . . 347
6.6. Generation of the initial population . . . 348
7. Results . . . 349
7.1. Progression of the optimisation . . . 349
7.2. Analysis of results for each well . . . 350
7.3. Comparison with other optimisation schemes . . . 351
7.3.1. Simulated Annealing and random search . . . 355
7.3.2. Hill-climber . . . 357
8. Conclusions . . . 358
8.1. Suggestions for further work . . . 360
References . . . 361

Chapter 18 APPLYING SOFT COMPUTING METHODS TO IMPROVE THE COMPUTATIONAL TRACTABILITY OF A SUBSURFACE SIMULATION–OPTIMIZATION PROBLEM
V.M. Johnson and L.L. Rogers . . . 365
Abstract . . . 365
1. Introduction . . . 366
1.1. Statement of the problem . . . 366
1.2. ANN–GA/SA approach to optimization . . . 367

1.3. Design optimization in petroleum engineering . . . 368
2. Reservoir description . . . 370
3. Management question . . . 373
3.1. Assumptions and constraints . . . 374
3.2. Cost estimates . . . 374
3.3. Performance measure (objective function) . . . 375
4. Application of the ANN–GA/SA methodology . . . 376
4.1. Create a knowledge base of simulations . . . 376
4.1.1. Define the problem scope . . . 376
4.1.2. Select the candidate pool of well locations . . . 377
4.1.3. Sample over the decision variables . . . 377
4.1.4. Carry out the simulations . . . 378
4.2. Train ANNs to predict reservoir performance . . . 378
4.3. Search for optimal well combinations . . . 382
4.3.1. Genetic algorithm . . . 382
4.3.2. Simulated annealing . . . 384
4.3.3. Procedures common to both GA and SA searches . . . 386
4.4. Verify optimal combinations with the simulator . . . 387
5. Search results . . . 387
5.1. Context scenarios . . . 387
5.2. Best in knowledge base . . . 388
5.3. ANN–GA search results . . . 388
5.4. ANN–SA search results . . . 390
5.5. VIP®–SA search results . . . 390
6. Summary and conclusions . . . 391
6.1. Outstanding issues . . . 392
6.1.1. Substantive interpretation of results . . . 392
6.1.2. ANN accuracy issues . . . 393
6.1.3. Uncertainties in the underlying model . . . 393
Acknowledgements . . . 394
References . . . 394

Chapter 19 NEURAL NETWORK PREDICTION OF PERMEABILITY IN THE EL GARIA FORMATION, ASHTART OILFIELD, OFFSHORE TUNISIA
J.H. Ligtenberg and A.G. Wansink . . . 397
Abstract . . . 397
1. Introduction . . . 397
2. Geological setting . . . 398
3. Neural networks . . . 399
4. Available data . . . 400
5. Data analysis . . . 401
5.1. Dunham classification . . . 401
5.2. Core porosity . . . 402
5.3. Core permeability . . . 405
6. Conclusions . . . 410
Acknowledgements . . . 411
References . . . 411

Chapter 20 USING RBF NETWORK TO MODEL THE RESERVOIR FLUID BEHAVIOR OF BLACK OIL SYSTEMS
A.M. Elsharkawy . . . 413
Abstract . . . 413
1. Introduction . . . 413
2. Present study . . . 416

2.1. Development of the RBFNM . . . 416
2.2. Training the model . . . 417
2.3. Testing the model . . . 418
3. Accuracy of the model . . . 418
3.1. Solution gas–oil ratio . . . 418
3.2. Oil formation volume factor . . . 419
3.3. Oil viscosity . . . 422
3.4. Oil density . . . 422
3.5. Undersaturated oil compressibility . . . 423
3.6. Gas gravity . . . 426
4. Behavior of the model . . . 429
5. Conclusion . . . 429
6. Notation . . . 429
7. SI metric conversion factors . . . 431
Appendix A. Radial basis functions . . . 431
Appendix B. Solution gas oil ratio correlations . . . 432
Appendix C. Oil formation value factor correlations . . . 438
Appendix D. Oil viscosity correlations . . . 439
Appendix E. Saturated oil density correlations . . . 440
Appendix F. Undersaturated oil compressibility correlations . . . 441
Appendix G. Evolved gas gravity correlations . . . 441
References . . . 442

Chapter 21 ENHANCING GAS STORAGE WELLS DELIVERABILITY USING INTELLIGENT SYSTEMS
S.D. Mohaghegh . . . 445
1. Introduction . . . 445
2. Methodology . . . 447
3. Results and discussion . . . 450
3.1. Genetic optimization . . . 455
3.2. Procedure . . . 456
3.2.1. Stage 1: Screening . . . 456
3.2.2. Stage 2: Optimization . . . 457
3.3. Chemical treatments . . . 460
4. Application to other fields . . . 464
5. Conclusions . . . 466
Acknowledgements . . . 466
References . . . 466

Part 5 Integrated Field Studies

Chapter 22 SOFT COMPUTING: TOOLS FOR INTELLIGENT RESERVOIR CHARACTERIZATION AND OPTIMUM WELL PLACEMENT
M. Nikravesh, R.D. Adams and R.A. Levey . . . 471
Abstract . . . 471
1. Introduction . . . 471
1.1. Neural networks . . . 472
1.2. Fuzzy logic . . . 473
1.3. Pattern recognition . . . 474
1.4. Clustering . . . 474
2. Reservoir characterization . . . 475
2.1. Examples . . . 475
2.1.1. Area 1 . . . 476
3. Conclusions . . . 490

4. Potential research opportunities in the future . . . 491
4.1. Quantitative 3D reconstruction of well logs and prediction of pay zone thickness . . . 491
4.2. IRESC model . . . 493
4.3. Neuro-fuzzy techniques . . . 493
Appendix A . . . 494
A.1. K-means clustering . . . 494
A.2. Fuzzy c-means clustering . . . 494
A.3. Neural network clustering . . . 495
References . . . 496

Chapter 23 COMBINING GEOLOGICAL INFORMATION WITH SEISMIC AND PRODUCTION DATA
J. Caers and S. Srinivasan . . . 499
Abstract . . . 499
1. Introduction . . . 499
2. A demonstration . . . 500
3. Borrowing structures from training images . . . 501
3.1. Pattern extraction . . . 502
3.2. Pattern recognition . . . 503
3.3. Pattern reproduction . . . 505
4. Conditioning to indirect data . . . 506
4.1. Pattern extraction . . . 508
4.2. Pattern recognition . . . 508
4.3. Pattern reproduction . . . 510
5. Production data integration . . . 510
5.1. Information in production data . . . 513
5.2. Integrating production based data into reservoir models . . . 518
5.3. Results and discussion . . . 519
6. Discussion and conclusions . . . 521
Appendix A . . . 524
References . . . 525

Chapter 24 INTERPRETING BIOSTRATIGRAPHICAL DATA USING FUZZY LOGIC: THE IDENTIFICATION OF REGIONAL MUDSTONES WITHIN THE FLEMING FIELD, UK NORTH SEA
M.I. Wakefield, R.J. Cook, H. Jackson and P. Thompson . . . 527
Abstract . . . 527
1. Introduction . . . 528
2. The fundamentals of fuzzy logic . . . 528
2.1. Linguistic variables . . . 530
2.2. Membership functions . . . 530
2.3. Fuzzy logic rules . . . 530
2.4. Defuzzification . . . 531
2.5. Previous application of fuzzy logic to palaeontological data analysis . . . 531
3. Application of fuzzy logic modelling in the Fleming field . . . 532
3.1. Geological setting . . . 532
3.2. Stratigraphical data . . . 532
3.3. Graphic correlation of bioevent data . . . 534
3.4. Agglutinated foraminiferal community structure and mudstone continuity . . . 535
3.5. Calibration of the fuzzy model . . . 537
3.6. Data handling . . . 539
3.7. Results of the fuzzy logic modelling . . . 541
3.8. Integration of graphic correlation and mudstone continuity modelling . . . 542
4. The use of biostratigraphical correlation in reservoir modelling . . . 543

4.1. The ten-layer model . . . 545
4.2. The hybrid model . . . 546
4.3. Parameter grids . . . 546
4.4. History matching results . . . 547
4.5. Discussion . . . 548
5. Conclusions . . . 548
Acknowledgements . . . 550
References . . . 550

Chapter 25 GEOSTATISTICAL CHARACTERIZATION OF THE CARPINTERIA FIELD, CALIFORNIA
R.J. Pawar, E.B. Edwards and E.M. Whitney . . . 553
Abstract . . . 553
1. Introduction . . . 553
2. Reservoir geology and geologic structure modeling . . . 554
3. Available data . . . 555
4. Porosity distribution . . . 557
4.1. Semivariogram for porosity . . . 557
4.2. Porosity realizations . . . 559
5. Shale volume fraction realization . . . 563
5.1. Spatial correlation . . . 563
5.2. Realizations of shale fraction . . . 565
6. Permeability distribution . . . 568
6.1. Input data . . . 572
6.2. Results . . . 572
7. Uncertainty analysis . . . 574
7.1. Pore volume . . . 574
7.2. Uncertainty in porosity and shale fraction . . . 575
7.3. Variation in productive volume around wells . . . 577
8. Discussion of results . . . 579
9. Conclusions . . . 580
10. List of symbols . . . 581
Acknowledgements . . . 581
References . . . 581

Chapter 26 INTEGRATED FRACTURED RESERVOIR CHARACTERIZATION USING NEURAL NETWORKS AND FUZZY LOGIC: THREE CASE STUDIES
A.M. Zellou and A. Ouenes . . . 583
Abstract . . . 583
1. Introduction . . . 583
2. Fractured reservoir modeling using AI tools . . . 584
2.1. Ranking the drivers . . . 584
2.2. Training and testing the models . . . 585
2.3. Simulation process . . . 585
2.4. Transforming fractured models into 3D effective permeabilities . . . 585
3. Case study 1: Faulted limestone reservoir, North Africa . . . 586
3.1. Field geology . . . 586
3.2. Factors affecting fracturing at this field . . . 586
3.3. Application of the fractured reservoir modeling using AI tools . . . 589
3.3.1. Ranking the fracturing drivers . . . 589
3.3.2. Training and testing . . . 589
3.3.3. Simulation results . . . 589
3.4. Conclusions and recommendations . . . 589
4. Case study 2: Slope carbonate oil reservoir, SE New Mexico . . . 590

4.1. Background . . . 590
4.2. Results and discussions . . . 592
4.3. Conclusions . . . 592
5. Case study 3: A sandstone gas reservoir, NW New Mexico . . . 593
5.1. Dakota production and geology . . . 593
5.2. Factors affecting fracturing in the Dakota . . . 594
5.3. Building a geologic model . . . 598
5.3.1. Ranking fracturing factors . . . 598
5.3.2. Neural network analysis . . . 599
5.4. Conclusions . . . 599
6. Conclusions . . . 601
Acknowledgements . . . 601
References . . . 601

Part 6 General Applications

Chapter 27 VIRTUAL MAGNETIC RESONANCE LOGS, A LOW COST RESERVOIR DESCRIPTION TOOL
S.D. Mohaghegh . . . 605
Abstract . . . 605
1. Introduction . . . 605
2. Methodology . . . 607
2.1. Wells from different formations . . . 607
2.2. Wells from the same formation . . . 609
2.3. Synthetic conventional logs . . . 611
3. Results and discussion . . . 611
4. Conclusions . . . 631
Acknowledgements . . . 631
References . . . 632

Chapter 28 ARTIFICIAL NEURAL NETWORKS LINKED TO GIS
Y. Yang and M.S. Rosenbaum . . . 633
Abstract . . . 633
1. Introduction . . . 633
2. Geographical information systems and the overlay operation . . . 634
3. Artificial neural networks . . . 635
4. Relative strength of effect . . . 638
5. Integration of ANN with GIS . . . 642
6. Application of NRSE to environmental sedimentology . . . 644
7. Conclusions . . . 649
Acknowledgements . . . 649
References . . . 649

Chapter 29 INTELLIGENT COMPUTING TECHNIQUES FOR COMPLEX SYSTEMS
M. Nikravesh . . . 651
Abstract . . . 651
1. Introduction . . . 651
2. Neuro-statistical method . . . 652
3. Hybrid neural network–alternative conditional expectation (HNACE/ACE neural network) technique . . . 657
4. Application of a neuro-statistical method for synthetic data sets . . . 657
5. Application of neuro-statistical method for a metal-contaminated fill at Alameda County . . . 659
6. Conclusion . . . 664

Appendix A. Robust algorithm for training the neural network models (non-linear model for imprecise data) 669
A.1. Current methods 670
A.1.1. Gauss method or the Gausee–Newton method 670
A.1.2. Gradient methods 670
A.1.3. Levenberg–Marquardt/Marquardt–Levenberg 670
A.2. Proposed method 671
A.3. Neural network models 671
References 671

Chapter 30 MULTIVARIATE STATISTICAL TECHNIQUES INCLUDING PCA AND RULE BASED SYSTEMS FOR WELL LOG CORRELATION
J.-S. Lim 673
Abstract 673
1. Introduction 673
2. Multivariate statistical analysis 674
2.1. Principal component analysis (PCA) 674
2.2. Electrofacies determination 676
3. Rule-based correlation system 677
3.1. Database 677
3.2. Rule base 678
3.3. Inference program 679
4. Applications 679
4.1. Verification 679
4.2. Comparison of methods 680
4.3. Field examples 682
5. Conclusions 683
References 687

Author Index 689

Subject Index 701

PART 1.

INTRODUCTION: FUNDAMENTALS OF SOFT COMPUTING

Developments in Petroleum Science, 51
Editors: M. Nikravesh, F. Aminzadeh and L.A. Zadeh

Chapter 1

SOFT COMPUTING FOR INTELLIGENT RESERVOIR CHARACTERIZATION AND MODELING

MASOUD NIKRAVESH [a,1] and F. AMINZADEH [b,2]

[a] *Computer Science Division — Department of EECS University of California, Berkeley, CA 94720, USA*
[b] *dGB-USA Houston, TX, USA*

ABSTRACT

As our problems become too complex to rely only on one discipline and as we find ourselves at the midst of an information explosion, multi-disciplinary analysis methods and data mining approaches in the petroleum industry become more of a necessity than professional curiosity. To tackle difficult problems ahead of us, we need to bring down the walls we have built around traditional disciplines such as petroleum engineering, geology, geophysics and geochemistry, and embark on true multi-disciplinary solutions. Our data, methodologies and workflow will have to cut across different disciplines. As a result, today's 'integration' which is based on integration of results will have to give way to a new form of integration, that is, discipline integration. In addition, to solve our complex problems we need to go beyond standard mathematical techniques. Instead, we need to complement the conventional analysis methods with a number of emerging methodologies and soft computing techniques such as expert systems, artificial intelligence, neural network, fuzzy logic, genetic algorithm, probabilistic reasoning, and parallel processing techniques. Soft computing differs from conventional (hard) computing in that, unlike hard computing, it is tolerant of imprecision, uncertainty, and partial truth. Soft computing is also tractable, robust, efficient and inexpensive. In this overview paper, we highlight role of soft computing techniques for intelligent reservoir characterization and exploration.

1. INTRODUCTION

The last decade has witnessed significant advances in transforming geosciences and well data into drillable prospects, generating accurate structural models and creating reservoir models with associated properties. This has been made possible through improvements in data integration, quantification of uncertainties, effective use of geophysical modeling for better describing the relationship between input data and reservoir properties, and use of unconventional statistical methods. Soft computing techniques such as neural networks and fuzzy logic and their appropriate

[1] E-mail: nikravesh@cs.berkeley.edu; URL: www.cs.berkeley.edu/~nikraves/
[2] E-mail: faz@dgbusa.com, URL: www.dgbusa.com

usage in many geophysical and geological problems has played a key role in the progress made in recent years. However there is a consensus of opinion that we have only begun the scratch the surface in realizing full benefits of soft computing technology. Many challenges remain when we are facing with characterization of reservoirs with substantial heterogeneity and fracturing, exploring in the areas with thin-bedded stacked reservoirs and regions with poor data quality or limited well control and seismic coverage and quantifying uncertainty and confidence interval of the estimates. Among the inherent problems we need to overcome are: inadequate and uneven well data sampling, non-uniqueness in cause and effect in subsurface properties versus geosciences data response, different scales of seismic, log and core data and finally how to handle changes in the reservoir as the characterization is in progress.

This paper reviews the recent geosciences applications of soft computing (SC) with special emphasis on exploration. The role of soft computing as an effective method of data fusion will be highlighted. SC is consortium of computing methodologies [fuzzy logic (FL), neuro-computing (NC), genetic computing (GC), and probabilistic reasoning (PR) including genetic algorithms (GA), chaotic systems (CS), belief networks (BN), learning theory (LT)] which collectively provide a foundation for the conception, design and deployment of intelligent systems. The role model for soft computing is the human mind. Among main components of soft computing, the artificial neural networks, fuzzy logic and the genetic algorithms in the 'exploration domain' will be examined. Specifically, the earth exploration applications of SC in various aspects will be discussed.

These applications are divided into two broad categories. One has to do with improving the efficiency in various tasks that are necessary for the processing and manipulation and fusion of different types of data used in exploration. Among these applications are: first-arrival picking, noise elimination, structural mapping, horizon picking, event tracking and integration of data from different sources. The other application area is pattern recognition, identification and prediction of different rock properties under the surface. This is usually accomplished by training the system from known rock properties using a number of attributes derived from the properly fused input data (e.g., 2D and 3D seismic, gravity, well log and core data, ground penetrating radar and synthetic aperture radar and other types remote sensing data). Then a similarity measure with certain threshold level is used to determine the properties where no direct measurement is available.

2. THE ROLE OF SOFT COMPUTING TECHNIQUES FOR INTELLIGENT RESERVOIR CHARACTERIZATION AND EXPLORATION

Soft computing is bound to play a key role in the earth sciences. This is in part due to subject nature of the rules governing many physical phenomena in the earth sciences. The uncertainty associated with the data, the immense size of the data to deal with and the diversity of the data type and the associated scales are important factors to rely on unconventional mathematical tools such as soft computing. Many of these issues are addressed in a recent book, Wong et al. (2001).

Recent applications of soft computing techniques have already begun to enhance our ability in discovering new reserves and assist in improved reservoir management and production optimization. This technology has also been proven useful in production from low permeability and fractured reservoirs such as fractured shale, fractured tight gas reservoirs and reservoirs in deep water or below salt which contain major portions of future oil and gas resources. Through new technology and data acquisition to processing and interpretation the rate of success in exploration has risen to 40% in 1990 from 30% in the 1980s. In some major oil companies the overall, gas and oil well drilling success rates have risen to an average of 47% in 1996 from 3 to 30% in the early 1990s. For example, in the US only, by year 2010, these innovative techniques are expected to contribute over 2 trillion cubic feet (Tcf)/year of additional gas production and 100 million barrels per year of additional oil. This cumulative will be over 30 Tcf of gas reserves and 1.2 billion barrels in oil reserve and will add over $8 billion to revenue in 2010 (Nikravesh, 2000; NPC, 1992; US Geological Survey, 1995).

Intelligent techniques such as neural computing, fuzzy reasoning, and evolutionary computing for data analysis and interpretation are an increasingly powerful tool for making breakthroughs in the science and engineering fields by transforming the data into information and information into knowledge.

In the oil and gas industry, these intelligent techniques can be used for uncertainty analysis, risk assessment, data fusion and mining, data analysis and interpretation, and knowledge discovery, from diverse data such as 3D seismic, geological data, well log, and production data. It is important to mention that during 1997, the US industry spent over $3 billion on seismic acquisition, processing and interpretation. In addition, these techniques can be a key to cost effectively locating and producing our remaining oil and gas reserves. Techniques can be used as a tool for:

(1) Lowering exploration risk
(2) Reducing exploration and production cost
(3) Improving recovery through more efficient production
(4) Extending the life of producing wells.

In what follows we will address data processing/fusion/mining, first. Then, we will discuss interpretation, pattern recognition and intelligent data analysis.

2.1. Mining and fusion of data

In the past, classical data processing tools and physical models solved many real-world problems. However, with the advances in information processing we are able to further extend the boundaries and complexities of the problems we tackle. This is necessitated by the fact that, increasingly, we are faced with multitude of challenges: On the one hand we are confronted with more unpredictable and complex real-world, imprecise, chaotic, multi-dimensional and multi-domain problems with many interconnected parameters in situations where small variability in parameters can change the solution completely. On the other hand, we are faced with profusion and complexity of computer-generated data. Making sense of large amounts of imprecise and chaotic data, very common in earth sciences applications, is beyond the scope of human ability and understanding. What this implies is that the classical data processing tools and physical models that have addressed many problems in the past may not be sufficient to deal effectively with present and future needs.

In recent years in the oil industry we have witnessed massive explosion in the data volume we have to deal with. As outlined in Aminzadeh (1996) this is caused by increased sampling rate, larger offset and longer record acquisition, multi-component surveys, 4D seismic and, most recently, the possibility of continuous recording in 'instrumented oil fields'. Thus we need efficient techniques to process such large data volumes. Automated techniques to refine the data (trace editing and filtering), selecting the desired event types (first-break picking) or automated interpretation (horizon tracking) are needed for large data volumes. Fuzzy logic and neural networks have been proven to be effective tools for such applications. To make use of large volumes of the field data and multitude of associated data volumes (e.g. different attribute volumes or partial stack or angle gathers), effective data compression methods will be of increasing significance, both for fast data transmission efficient processing, analysis and visualization and economical data storage. Most likely, the biggest impact of advances in data compression techniques will be realized when geoscientists have the ability to fully process and analyze data in the compressed domain. This will make it possible to carry out computer-intensive processing of large volumes of data in a fraction of the time, resulting in tremendous cost reductions. Data mining is another alternative that helps identify the most information rich part of the large volumes of data. Again in many recent reports, it has been demonstrated that neural networks and fuzzy logic, in combination of some of the more conventional methods such as eigenvalue or principal component analysis are very useful.

Fig. 1 shows the relationship between intelligent technology and data fusion/data mining. Tables 1 and 2 show the list of the data fusion and data mining techniques. Fig. 2 and Table 3 show the reservoir data mining and reservoir data fusion concepts and techniques. Table 4 shows the comparison between geostatistical and intelligent techniques. In Sections 2.2, 2.3 and 2.4 we will highlight some of the recent applications of these methods in various earth sciences disciplines.

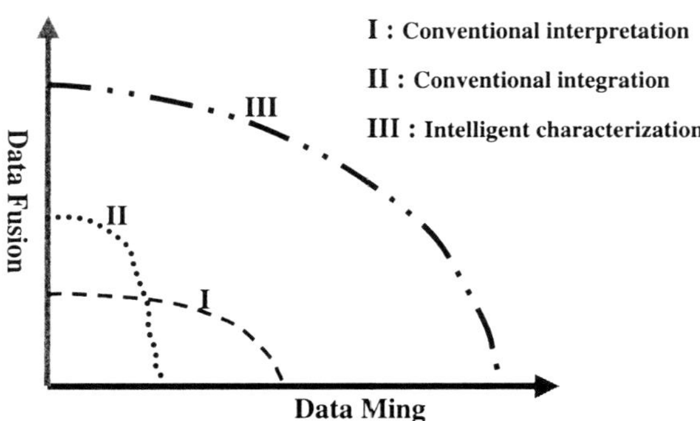

Fig. 1. Intelligent technology.

TABLE 1

Data mining techniques

- Deductive database client
- Inductive learning
- Clustering
- Case-based reasoning
- Visualization
- Statistical package

TABLE 2

Data fusion techniques

- Deterministic
  - Transform based (*projections, . . .*)
  - Functional evaluation based (*vector quantization, . . .*)
  - Correlation based (*pattern match, if/then productions*)
  - Optimization based (*gradient-based, feedback, LDP, . . .*)
- Non-deterministic
  - Hypothesis testing (*classification, . . .*)
  - Statistical estimation (*Maximum likelihood, . . .*)
  - Discrimination function (*linear aggregation, . . .*)
  - Neural network (*supervised learning, clustering, . . .*)
  - Fuzzy Logic (*Fuzzy c-Mean Clustering, . . .*)
- Hybrid (*Genetic algorithms, Bayesian network, . . .*)

2.2. Intelligent interpretation and data analysis

Once all the pertinent data is properly integrated (fused) one has to extract the relevant information from the data and draw the necessary conclusions. This can be done either true reliance on human expert or an intelligent system that has the capability to learn and modify its knowledge base as new information become available. For detailed review of various applications of soft computing in intelligent interpretation, data analysis and pattern recognition see Aminzadeh (1989a, 1991) and Aminzadeh and Jamshidi (1995).

Although seismic signal processing has advanced tremendously over the last four decades, the fundamental assumption of a 'convolution model' is violated in many practical settings. Sven Treitel, in Aminzadeh and Jamshidi (1995) was quoted to pose the question: "*What if, mother earth refuses to convolve?*" Among such situations are: highly heterogeneous environments, very absorptive media (such as unconsolidated sand and young sediments), fractured reservoirs, and mud volcano, karst and gas chimneys. In such cases we must consider non-linear processing and interpretation methods. Neural networks fractals, fuzzy logic, genetic algorithms, chaos and complexity theory are among such non-linear processing and analysis techniques that have been proven to be effective. The highly heterogeneous earth model that geophysics attempts to quantify is

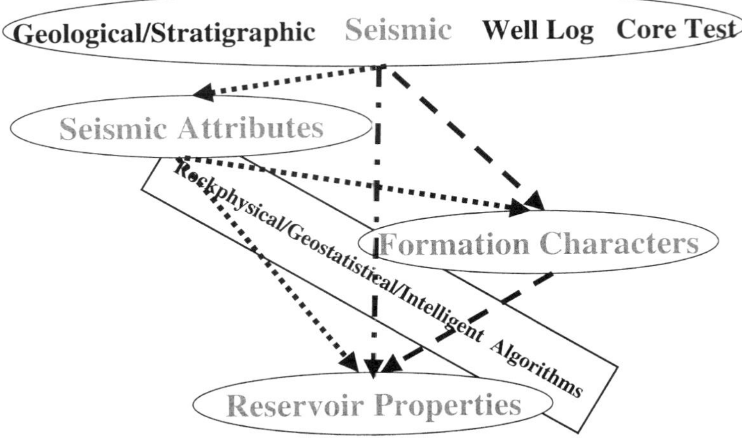

Fig. 2. Reservoir data mining.

TABLE 3

Reservoir data fusion

- Rockphysical
  - Transform seismic data to attributes and reservoir properties
  - Formulate seismic/log/core data to reservoir properties
- Geostatistical
  - Transform seismic attributes to formation characters
  - Transform seismic attributes to reservoir properties
  - Simulate the 2D/3D distribution of seismic and log attributes
- Intelligent
  - Clustering anomalies in seismic/log data and attributes
  - ANN layers for seismic attribute and formation characters
  - Supervised training model to predict unknown from existing
  - Hybrid such as GA and SA for complicated reservoirs

TABLE 4

Geostatistical vs. intelligent

- Geostatistical
  - Data assumption: a certain probability distribution
  - Model: weight functions come from variogram trend, stratigraphic facies, and probability constraints
  - Simulation: Stochastic, not optimized
- Intelligent
  - Data automatic clustering and expert-guided segmentation
  - Classification of relationship between data and targets
  - Model: weight functions come from supervised training based on geological and stratigraphic information
  - Simulation: optimized by GA, SA, ANN, and BN

an ideal place for applying these concepts. The subsurface lives in a hyper-dimensional space (the properties can be considered as the additional space dimension), but its actual response to external stimuli initiates an internal coarse-grain and self-organization that results in a low-dimensional structured behavior. Fuzzy logic and other non-linear methods can describe shapes and structures generated by chaos. These techniques will push the boundaries of seismic resolution, allowing smaller-scale anomalies to be characterized.

2.3. Pattern recognition

In the 1960s and 1970s, pattern recognition techniques were used only by statisticians and were based on statistical theories. Due to recent advances in computer systems and technology, artificial neural networks and fuzzy logic models have been used in many pattern recognition applications ranging from simple character recognition, interpolation, and extrapolation between specific patterns to the most sophisticated robotic applications. To recognize a pattern, one can use the standard multi-layer perception with a back-propagation learning algorithm or simpler models such as self-organizing networks (Kohonen, 1997) or fuzzy c-means techniques (Bezdek, 1981; Jang and Gulley, 1995). Self-organizing networks and fuzzy c-means techniques can easily learn to recognize the topology, patterns, or seismic objects and their distribution in a specific set of information. Much of the early applications of pattern recognition in the oil industry were highlighted in Aminzadeh (1989a).

2.4. Clustering

Cluster analysis encompasses a number of different classification algorithms that can be used to organize observed data into meaningful structures. For example, k-means is an algorithm to assign a specific number of centers, k, to represent the clustering of N points ($k < N$). These points are iteratively adjusted so that each point is assigned to one cluster, and the centroid of each cluster is the mean of its assigned points. In general, the k-means technique will produce exactly k different clusters of the greatest possible distinction.

Alternatively, fuzzy techniques can be used as a method for clustering. Fuzzy clustering partitions a data set into fuzzy clusters such that each data point can belong to multiple clusters. Fuzzy c-means (FCM) is a well-known fuzzy clustering technique that generalizes the classical (hard) c-means algorithm and can be used where it is unclear how many clusters there should be for a given set of data.

Subtractive clustering is a fast, one-pass algorithm for estimating the number of clusters and the cluster centers in a set of data. The cluster estimates obtained from subtractive clustering can be used to initialize iterative optimization-based clustering methods and model identification methods.

In addition, the self-organizing map technique known as Kohonen's self-organizing feature map (Kohonen, 1997) can be used as an alternative for clustering purposes. This technique converts patterns of arbitrary dimensionality (the pattern space) into the response of one- or two-dimensional arrays of neurons (the feature space). This unsupervised learning model can discover any relationship of interest such as patterns, features, correlations, or regularities in the input data, and translate the discovered relationship into outputs. The first application of clustering techniques to combine different seismic attributes was introduced in the mid eighties (Aminzadeh and Chatterjee, 1984/1985).

2.5. Data integration and reservoir property estimation

Historically, the link between reservoir properties and seismic and log data have been established either through 'statistics-based' or 'physics-based' approaches. The latter, also known as model based approaches attempt to exploit the changes in seismic character or seismic attribute to a given reservoir property, based on physical phenomena. Here, the key issues are sensitivity and uniqueness. Statistics based methods attempt to establish a heuristic relationship between seismic measurements and prediction values from examination of data only. It can be argued that a hybrid method, combining the strength of statistics and physics based method would be most effective. Fig. 3, taken from Aminzadeh et al. (1999), shows the concepts schematically.

Many geophysical analysis methods and consequently seismic attributes are based on physical phenomena. That is, based on certain theoretical physics (wave propagation, Biot–Gassman equation, Zoeppritz equation, tuning thickness, shear wave splitting, etc.) certain attributes may be more sensitive to changes in certain reservoir properties. In the absence of a theory, using experimental physics (for example, rock property measurements in a laboratory environment such as the one described in the last section of this paper) and/or numerical modeling, one can identify or validate suspected relationships. Although physics-based methods and direct measurements (the ground truth) is the ideal and reliable way to establish such correlations, for various reasons it is not always practical. Those reasons range from lack of known theories, difference between the laboratory environment and field environment (noise, scale, etc.) and the cost for conducting elaborate physical experiments.

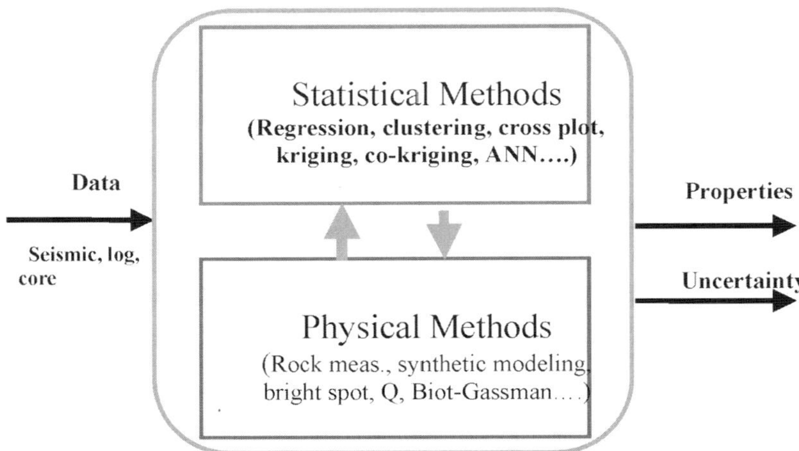

Fig. 3. A schematic description of physics-based (blue), statistics-based (red) and hybrid method (green).

Statistics-based methods aim at deriving an explicit or implicit heuristic relationship between measured values and properties to be predicted. Neural networks and fuzzy-neural networks-based methods are ideally suitable to establish such implicit relationships through proper training. We all attempt to establish a relationship between different seismic attributes, petrophysical measurements, laboratory measurements and different reservoir properties. In such statistics-based method one has keep in mind the impact of noise on the data, data population used for statistical analysis, scale, geologic environment, scale and the correlation between different attributes when performing clustering or regressions. The statistics-based conclusions have to be reexamined and their physical significance explored.

2.6. Quantification of data uncertainty and prediction error and confidence interval

One of the main problems we face is to handle non-uniqueness issue and quantify uncertainty and confidence intervals in our analysis. We also need to understand the incremental improvements in prediction error and confidence range from introduction of new data or a new analysis scheme. Methods such as evidential reasoning and fuzzy logic are most suited for this purpose. Fig. 4 shows the distinction between conventional probability and theses techniques. ‘Point probability’, describes the probability of an event, for example, having a commercial reservoir. The implication is we know exactly what this probability is. Evidential reasoning, provides an upper bound (plausibility) and lower bound (credibility) for the event the difference between the two bounds is considered as the ignorance range. Our objective is to reduce this range through use of all the new information. Given the fact that in real life we may have non-rigid boundaries for the upper and lower bounds and we ramp up or ramp down our confidence for an event at some point, we introduce fuzzy logic to handle and we refer to it as ‘membership grade’. Next-generation earth modeling will incorporate quantitative representations of geological processes and stratigraphic/structural variability. Uncertainty will be quantified and built into the models.

On the issue of non-uniqueness, the more sensitive the particular seismic character to a given change to reservoir property, the easier to predict it. The more unique influence of the change in seismic character to changes in a specific reservoir property, the higher

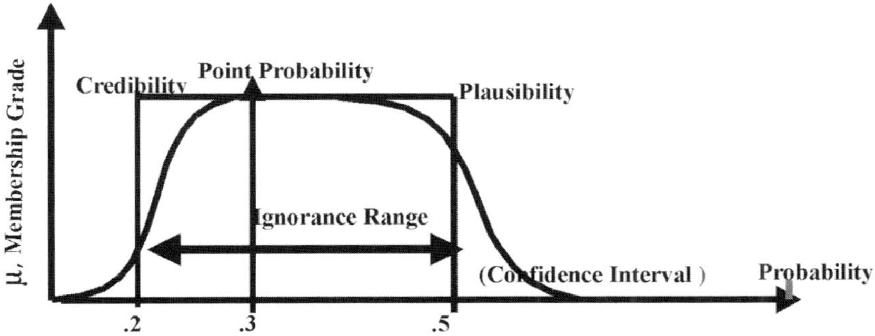

Fig. 4. Point probability, evidential reasoning and fuzzy logic.

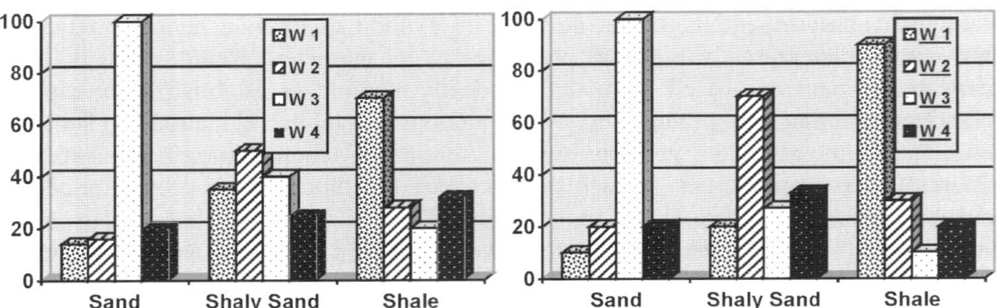

Fig. 5. Statistical distribution of different wavelet types versus lithologies. (a) Pre-stack data; (b) stacked data.

the confidence level in such predictions. Fuzzy logic can handle subtle changes in the impact of different reservoir properties on the wavelet response. Moreover, comparison of multitude of wavelet responses (for example near, mid and far offset wavelets) is easier through use of neural networks. As discussed in Aminzadeh and de Groot, 2001, let us assume a seismic pattern for three different lithologies (sand, shaly sand and shale) are compared from different well information and seismic response (both model and field data) and the respective seismic character within the time window or the reservoir interval with four 'classes' of wavelets (w1, w2, w3, and w4). These 4 wavelets (basis wavelets) serve as a segmentation vehicle. The histograms in Fig. 5a show what classes of wavelets that are likely to be present for given lithologies. In the extreme positive (EP) case we would have one wavelet uniquely representing one lithology. In the extreme negative case (EN) we would have a uniform distribution of all wavelets for all lithologies. In most cases unfortunately we are closer to NP than to EP. The question is how best we can get these distributions move from the EN side to EP side thus improving our prediction capability and increasing confidence level. The common sense is to add enhance information content of the input data.

How about if we use wavelet vectors comprised of pre-stack data (in the simple case, mid, near far offset data) as the input to a neural network to perform the classification? Intuitively, this should lead to a better separation of different lithologies (or other reservoir properties). Likewise, including three component data as the input to the classification process would further improve the confidence level. Naturally, this requires introduction of a new 'metric' measuring 'the similarity' of these 'wavelet vectors'. This can be done using the new basis wavelet vectors as input to a neural network applying different weights to mid, near and far offset traces. This is demonstrated conceptually, in Fig. 5 to predict lithology. Compare the sharper histograms of the vector wavelet classification (in this case, mid, near, and far offset gathers) in Fig. 5b, against those of Fig. 5a based on scalar wavelet classification.

3. ARTIFICIAL NEURAL NETWORK AND GEOSCIENCE APPLICATIONS OF ARTIFICIAL NEURAL NETWORKS FOR EXPLORATION

Although artificial neural networks (ANN) were introduced in the late fifties (Rosenblatt, 1962), the interests in them have been increasingly growing in recent years. This has been in part due to new applications fields in the academia and industry. Also, advances in computer technology (both hardware and software) have made it possible to develop ANN capable of tackling practically meaningful problems with a reasonable response time.

Simply put, neural networks are computer models that attempt to simulate specific functions of human nervous system. This is accomplished through some parallel structures comprised of non-linear processing nodes that are connected by fixed (Lippmann, 1987), variable (Barhen et al., 1989) or fuzzy (Gupta and Ding, 1994) weights. These weights establish a relationship between the inputs and output of each 'neuron' in the ANN. Usually ANN have several 'hidden' layers each layer comprised of several neurons. If the feed-forward (FF) network (FF or concurrent networks are those with unidirectional data flow). If the FF network is trained by back propagation (BP) algorithms, they are called BP. Other types of ANN are supervised (self-organizing) and auto (hetero) associative networks.

In what follows we will review the geoscience applications in these broad areas: data processing and prediction. We will not address other geoscience applications such as: classification of multi-source remote sensing data (Benediktsson et al., 1990), earthquake prediction (Aminzadeh et al., 1994), and ground water remediation, (Johnson and Rogers, 1995).

3.1. Data processing

Various types of geoscience data are used in the oil industry to ultimately locate the most prospective locations for oil and gas reservoirs. These data sets go through extensive amount of processing and manipulation before they are analyzed and interpreted. The processing step is very time consuming yet a very important one. ANN have been utilized to help improve the efficiency of operation in this step. Under this application area we will examine: First seismic arrival (FSA) picking, and noise elimination problems. Also, see Aminzadeh (1991), McCormack (1991), Zadeh and Aminzadeh (1995) and Aminzadeh et al. (1999) for other related applications.

3.1.1. First-arrival picking

Seismic data are the response of the earth to any disturbance (compressional waves or shear waves). The seismic source can be generated either artificially (petroleum seismology, PS) or, naturally, (earthquake seismology, ES). The recorded seismic data are then processed and analyzed to make an assessment of the subsurface (both the geological structures and rock properties) in PS and the nature of the source (location or epicenter and magnitude, for example, in Richter scale) in ES. Conventional PS relies heavily on compressional (P-wave) data while ES is essentially based on the shear (S-wave) data.

The first arrivals of P and S waves on a seismic record contain useful information both in PS and ES. However one should make sure that the arrival is truly associated with a seismically generated event and not a noise generated due to various factors. Since we usually deal with thousands of seismic records, their visual inspection for distinguishing FSA from noise, even if reliable, could be quite time consuming.

One of the first geoscience applications of ANN has been to streamline the operation of identifying the FSA in an efficient and reliable manner. Among the recent publications in this area are: McCormack (1990) and Veezhinathan et al. (1991). Key elements of the latter (V91) are outlined below:

Here, the FSA picking is treated as a pattern recognition problem. Each event is classified either as an FSA or non-FSA. A segment of the data within a window is used to obtain four 'Hilbert' attributes of the seismic signal. The Hilbert attributes of seismic data were introduced by Taner et al. (1979). In V91, these attributes are derived from seismic signal using a sliding time window. Those attributes are: (1) maximum amplitude; (2) mean power level, MPL; (3) power ratios; and (4) envelop slope peak. These types of attributes have been used by Aminzadeh and Chatterjee (1984/1985) for predicting gas sands using clustering and discernment analysis technique.

In V91, the network processes three adjacent peaks at a time to decide whether the center peak is an FSA or a non-FSA. A BPN (Backpropagation Neural Network) with five hidden layers combined with a post-processing scheme accomplished correct picks of 97%. Adding a fifth attribute, the distance from travel time curve, generated satisfactory results without the need for the post-processing step.

McCormack (1990) created a binary image from the data and used it to train the network to move up and down across the seismic record to identify the FSA. This image-based approach captures space–time information in the data but requires a large number of input units, thus necessitating a large network. Some empirical schemes are used to ensure its stability.

3.1.2. Noise elimination

A related problem to FSA is editing noise from the seismic record. The objective here is to identify events with non-seismic origin (the reverse of FSA) and then remove them from the original data in order to increase the signal to noise ratio. Liu et al. (1989), McCormack (1990) and Zhang and Li (1995) are some of the publications in this area.

Zhang and Li (1995) handled the simpler problem, to edit out the whole noisy trace from the record. They initiate the network in the 'learning' phase by 'scanning' over the whole data set. The weights are adapted in the learning phase either with some human input as the distinguishing factors between 'good' and 'bad' traces or during an unsupervised learning phase. Then in the 'recognizing' phase the data are scanned again and depending upon whether the output of the network is less than or greater than a threshold level the trace is either left alone or edited out as a bad trace.

3.2. Identification and prediction

Another major application area for ANN in the oil industry is to predict various reservoir properties. This ultimately is used a decision tool for exploration and development drilling and redevelopment or extension of the existing fields. The input data to this prediction problem is usually processed and interpreted seismic and log data and/or a set of attributes derived from the original data set. Historically, many 'hydrocarbon indicators' have been proposed to make such predictions. Among them are: the bright spot analysis (Sheriff and Geldart, 1982, 1983), amplitude versus offset analysis (Ostrander, 1982), seismic clustering analysis (Aminzadeh and Chatterjee, 1984/1985), fuzzy pattern recognition (Griffiths, 1987) and other analytical methods (Agterberg and Griffiths, 1991). Many of the ANN developed for this purpose are built around the earlier techniques either for establishing a relationship between the raw data and physical properties of the reservoirs and/or to train the network using the previously established relationships.

Huang and Williamson (1994) have developed a general regression neural network (GRNN) to predict rock's total organic carbon (TOC) using well log data. First, they model the relationship between the resistivity log and TOC with a GRNN, using published data. After training the ANN in two different modes, the GRNN found optimum values of sigma. Sigma is an important smoothing parameter used in GRNN. They have established the superiority of GRNN over BP-ANN in determining the architecture of the network. After completing the training phase a predictive equation for determining TOC was derived. Various seismic attributes from partial stacks (mid, near and far offsets) as an input to ANN. The network was calibrated using synthetic (theoretical) data with pre stack seismic response of known lithologies and saturation from the well log data. The output of the network was a set of classes of lithologies and saturations.

4. FUZZY LOGIC

In recent years, it has been shown that uncertainty may be due to fuzziness (Aminzadeh, 1991) rather than chance. Fuzzy logic is considered to be appropriate to deal with the nature of uncertainty in system and human error, which are not included in current reliability theories. The basic theory of fuzzy sets was first introduced by Zadeh (1965). Unlike classical logic which is based on crisp sets of 'true and false', fuzzy logic views problems as a degree of 'truth', or 'fuzzy sets of true and false' (Zadeh, 1965). Despite the meaning of the word 'fuzzy', fuzzy set theory is not one that permits vagueness. It is a methodology that was developed to obtain an approximate solution where the problems are subject to vague description. In addition, it can help engineers and researchers to tackle uncertainty, and to handle imprecise information in a complex situation. During the past several years, the successful application of fuzzy logic for solving complex problems subject to uncertainty has greatly increased and today fuzzy logic plays an important role in various engineering disciplines (Adams et al., 1999a,b; Aminzadeh, 1989b; Aminzadeh and Jamshidi, 1995; Aminzadeh and Chatterjee, 1984/1985). In recent years, considerable attention has been devoted to the use of hybrid neural network-fuzzy logic approaches (Adams et al., 1999a,b; Aminzadeh, 1989a,b; Aminzadeh and Chatterjee, 1984/1985) as an alternative for pattern recognition, clustering, and statistical and mathematical modeling. It has been shown that neural network models can be use to construct internal models that capture the presence of fuzzy rules. However, determination of the input structure and number of membership functions for the inputs has been one of the most important issues of fuzzy modeling.

4.1. Geoscience applications of fuzzy logic

The uncertain, fuzzy, and linguistic nature of geophysical and geological data makes it a good candidate for interpretation through fuzzy set theory. The main advantage of this technique is in combining the quantitative data and qualitative information and subjective observation. The imprecise nature of the information available for interpretation (such as seismic data, wireline logs, geological and lithological data) makes fuzzy sets theory an appropriate tool to utilize. For example, Chappaz (1977) and Bois (1983, 1984) proposed to use fuzzy sets theory in the interpretation of seismic sections. Bois used fuzzy logic as pattern recognition tool for seismic interpretation and reservoir analysis. He concluded that fuzzy set theory, in particular, can be used for interpretation of seismic data which are imprecise, uncertain, and include human error. He maintained these type of error and fuzziness cannot be taken into consideration by conventional mathematics. However, they are perfectly seized by fuzzy set theory. He also concluded that using fuzzy set theory one can determine the geological information using seismic data. Therefore, one can predict the boundary of reservoir in which hydrocarbon exists. Baygun et al. (1985) used fuzzy logic as classifier for delineation of geological objects in a mature hydrocarbon reservoir with many wells. Baygun et al. have shown that fuzzy logic can be used to extract dimensions and orientation of geological bodies and the geologist can use such a technique for reservoir characterization in a very quick way through bypassing several tedious steps. Chen et al. (1995) in their study used the fuzzy set theory as fuzzy regression analysis for extraction of the parameter for the Archie equation. Bezdek (1981) also reported a series of the applications of fuzzy sets theory in geostatistical analysis. Tamhane et al. (2002) show how to integrate linguistic descriptions in petroleum reservoirs using fuzzy logic.

Many of our geophysical analysis techniques such as migration, DMO, wave equation modeling as well as the potential methods (gravity, magnetic, electrical methods) use conventional partial differential wave equations (PDEs) with deterministic coefficients. The same is true for the partial differential equations used in reservoir simulation. For many practical and physical reasons deterministic parameters for the coefficients of these PDEs leads unrealistic (for example, medium velocities for seismic wave propagation or fluid flow for Darcy equation). Stochastic parameters in theses cases can provide us with a more practical characterization. Fuzzy coefficients for PDEs can prove to be even more realistic and easy to parameterize. Today's deterministic processing and interpretation ideas will give way to stochastic methods, even if the industry has to rewrite the book on geophysics. That is, using wave equations with random and fuzzy coefficients to describe subsurface velocities and densities in statistical and membership grade terms, thereby enabling a better description of wave propagation in the subsurface – particularly when a substantial amount of heterogeneity is present. More generalized applications of geostatistical techniques will emerge, making it possible to introduce risk and uncertainty at the early stages of the seismic data processing and interpretation loop.

5. GENETICS ALGORITHMS

Genetic algorithm (GA) is one of the stochastic optimization methods which is simulating the process of natural evolution. GA follows the same principles as those in nature (survival of the fittest, Charles Darwin). GA first was presented by John Holland as an academic research. However, today GA turn out to be one of the most promising approaches for dealing with complex systems which at first nobody could imagine that from a relative modest technique. GA is applicable to multi-objectives optimization and can handle conflicts among objectives. Therefore, it is robust where multiple solution exist. In addition, it is highly efficient and it is easy to use.

Another important feature of GA is its ability to extract knowledge in terms of fuzzy rules. GA is now widely used and applied to discovery of fuzzy rules. However, when the data sets are very large, it is not easy to extract the rules. To overcome such a limitation, a new coding technique has been presented recently. The new coding method is based on biological DNA. The DNA coding method and the mechanism of development from artificial DNA are suitable for knowledge extraction from large data set. The DNA can have many redundant parts which is important for extraction of knowledge. In addition, this technique allows overlapped representation of genes and it has no constraint on crossover points. Also, the same type of mutation can be applied to every locus. In this technique, the length of chromosome is variable and it is easy to insert and/or delete any part of DNA.

Today, genetic algorithm can be used in a hierarchical fuzzy model for pattern extraction and to reduce the complexity of the neuro-fuzzy models. In addition, GA can be use to extract the number of the membership functions required for each parameter and input variables, and for robust optimization along the multidimensional, highly non-linear and non-convex search hyper-surfaces.

5.1. Geoscience applications of genetic algorithms

Most of the applications of the GA in the area of petroleum reservoir or in the area of geoscience are limited to inversion techniques or used as optimization technique. While in other filed, GA is used as a powerful tool for extraction of knowledge, fuzzy rules, fuzzy membership, and in combination with neural network and fuzzy-logic. Recently, Nikravesh et al. (1999a,b) proposed to use a neuro-fuzzy-genetic model for data mining and fusion in the area of geoscience and petroleum reservoirs. In addition, it has been proposed to use neuro-fuzzy DNA model for extraction of knowledge from seismic data and mapping the wireline logs into seismic data and reconstruction of porosity (and permeability if reliable data exist for permeability) based on multi-attributes seismic mapping. Seismic inversion was accomplished using genetic algorithms by Mallick (1999). Potter et al. (1999) used GA for stratigraphic analysis. For an overview of GA in exploration problems see McCormack et al. (1999).

6. PRINCIPAL COMPONENT ANALYSIS AND WAVELET

Some of the data fusion and data mining methods used in exploration applications are as follows.

First we need to reduce the space to make the data size more manageable as well as reducing the time required for data processing. We can use principal component analysis. Using the eigenvalue and vectors, we can reduce the space domain. We choose the eigenvector corresponding to the largest eigenvalues. Then in the eigenvector space we use fuzzy k-mean or fuzzy c-mean technique. For details of fuzzy c-means algorithm see Cannon et al. (1986). Also, see Lashgari (1991), Aminzadeh (1989b) and Aminzadeh (1994) for the application of fuzzy logic and fuzzy k-means algorithm in several earth exploration problems.

We can also use wavelet and extract the patterns and wavelets describing different geological settings and the respective rock properties. Using the wavelet and neural network, we can fuse the data for non-linear modeling. For clustering purposes, we can use the output from wavelet and use fuzzy c-mean or fuzzy k-mean. To use uncertainty and see the effect of the uncertainty, it is easy to add the distribution to each point or some weight for importance of the data points. Once we assign some weight to each point, then we can correspond each weight to number of points in a volume around each point.

Of course the techniques based on principal component analysis has certain limitations. One of the limitations is when SNR (Signal to Noise Ratio) is negative or zero causing the technique to fail. The reason for this is the singularity of the variance and covariance matrices. Therefore, an important step is to use KF (Kalman Filtering) or some sort of fuzzy set theory for noise reduction and extraction of signal.

7. INTELLIGENT RESERVOIR CHARACTERIZATION

In reservoir engineering, it is important to characterize how 3D seismic information is related to production, lithology, geology, and logs (e.g. porosity, density, gamma ray, etc.) (Aminzadeh and Chatterjee, 1984/1985; Yoshioka et al., 1996; Boadu, 1997; Chawathe et al., 1997; Monson and Pita, 1997; Schuelke et al., 1997; Nikravesh, 1998a,b; Nikravesh et al., 1998). Knowledge of 3D seismic data will help to reconstruct the 3D volume of relevant reservoir information away from the well bore. However, data from well logs and 3D seismic attributes are often difficult to analyze because of their complexity and our limited ability to understand and use the intensive information content of these data. Unfortunately, only linear and simple non-linear information can be extracted from these data by standard statistical methods such as ordinary least squares, partial least squares, and non-linear quadratic partial least squares. However, if *a priori* information regarding non-linear input–output mapping is available, these methods become more useful.

Simple mathematical models may become inaccurate because several assumptions are made to simplify the models in order to solve the problem. On the other hand, complex models may become inaccurate if additional equations, involving a more or less approximate description of phenomena, are included. In most cases, these models require a number of parameters that are not physically measurable. Neural networks (Hecht-Nielsen, 1989) and fuzzy logic (Zadeh, 1965) offer a third alternative and have the potential to establish a model from non-linear, complex, and multi-dimensional data. They have found wide application in analyzing experimental, industrial, and field data (Baldwin et al., 1989, 1990; Rogers et al., 1992; Wong et al., 1995a,b; Nikravesh et al., 1996; Pezeshk et al., 1996; Nikravesh and Aminzadeh, 1998). In recent years, the utility of neural network and fuzzy logic analysis has stimulated growing interest among reservoir engineers, geologists, and geophysicists (Klimentos and McCann, 1990; Aminzadeh et al., 1994; Yoshioka et al., 1996; Boadu, 1997; Chawathe et al., 1997; Monson and Pita, 1997; Schuelke et al., 1997; Nikravesh, 1998a,b; Nikravesh and Aminzadeh, 1998; Nikravesh et al., 1998). Boadu (1997) and Nikravesh et al. (1998) applied artificial neural networks and neuro-fuzzy successfully to find relationships between seismic data and rock properties of sandstone. In a recent study, Nikravesh and Aminzadeh (2001) used an artificial neural network to further analyze data published by Klimentos and McCann (1990) and analyzed by Boadu (1997). It was concluded that to find non-linear relationships, a neural network model provides better performance than does a multiple linear regression model. Neural network, neuro-fuzzy, and knowledge-based models have been successfully used to model rock properties based on well log databases (Nikravesh, 1998b).

Monson and Pita (1997), Chawathe et al. (1997) and Nikravesh (1998b) applied artificial neural networks and neuro-fuzzy techniques successfully to find the relationships between 3D seismic attributes and well logs and to extrapolate mapping away from the well bore to reconstruct log responses.

Adams et al. (1999a,b), Levey et al. (1999), Nikravesh et al. (1999a) and Nikravesh et al. (1999b) showed schematically the flow of information and techniques to be used for intelligent reservoir characterization (IRESC) (Fig. 6). The main goal will be to integrate soft data such as geological data with hard data such as 3D seismic, production data, etc.

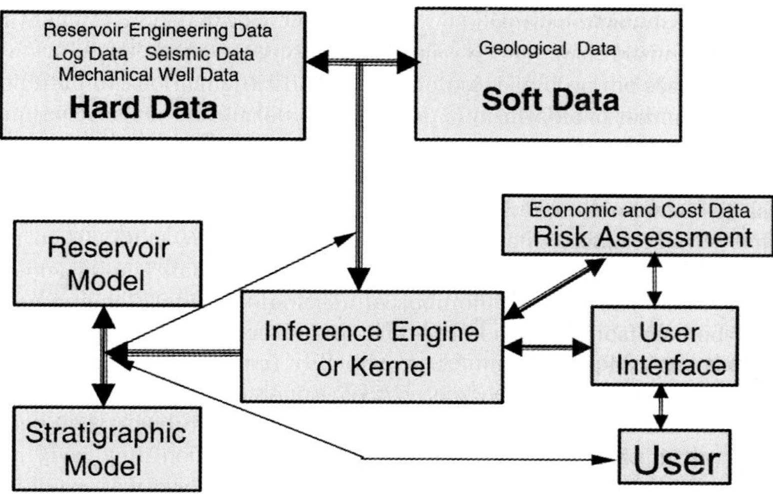

Fig. 6. Intelligent-integrated reservoir characterization (IRESC).

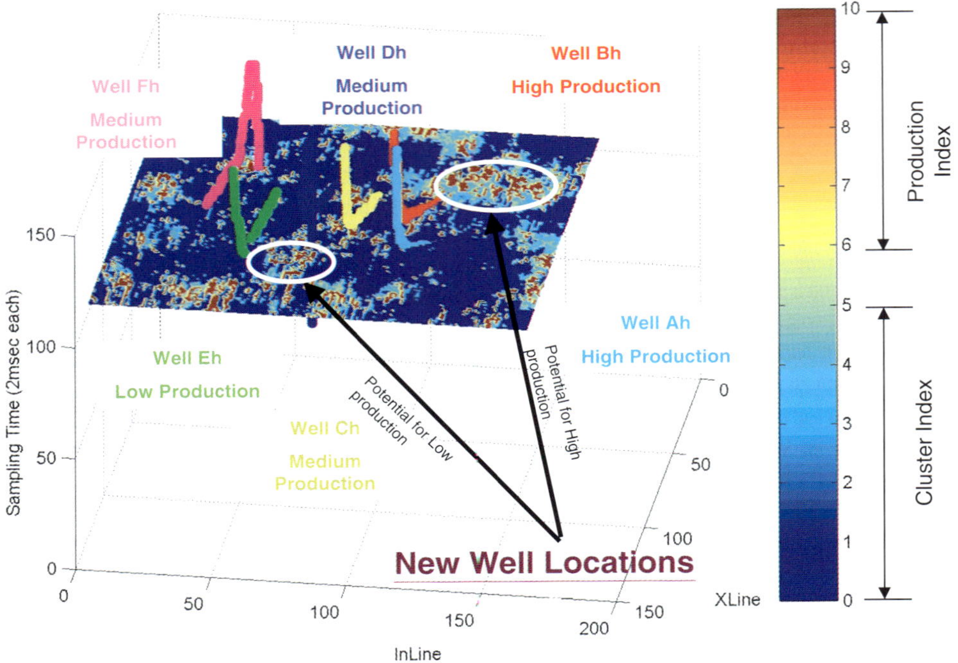

Fig. 7. Optimal well placement (Nikravesh et al., 1999a,b).

to build a reservoir and stratigraphic model. Nikravesh et al. (1999a,b) were developed a new integrated methodology to identify a non-linear relationship and mapping between 3D seismic data and production-log data and the technique was applied to a producing field. This advanced data analysis and interpretation methodology for 3D seismic and production-log data uses conventional statistical techniques combined with modern soft-computing techniques. It can be used to predict: (1) mapping between production-log data and seismic data, (2) reservoir connectivity based on multi-attribute analysis, (3) pay zone recognition, and (4) optimum well placement (Fig. 7). Three criteria have been used to select potential locations for infill drilling or recompletion (Nikravesh et al., 1999a,b): (1) continuity of the selected cluster, (2) size and shape of the cluster, and (3) existence of high production-index values inside a selected cluster with high cluster-index values. Based on these criteria, locations of the new wells were selected, one with high continuity and potential for high production and one with low continuity and potential for low production. The neighboring wells that are already in production confirmed such a prediction (Fig. 7).

Although these methodologies have limitations, the usefulness of the techniques will be for fast screening of production zones with reasonable accuracy. This new methodology, combined with techniques presented by Nikravesh (1998a,b), Nikravesh et al. (1998), and Nikravesh and Aminzadeh (2001) can be used to reconstruct well logs such as DT, porosity, density, resistivity, etc. away from the well bore. By doing so, net-pay-zone thickness, reservoir models, and geological representations will be accurately identified. Accurate reservoir characterization through data integration is an essential step in reservoir modeling, management, and production optimization.

8. FRACTURED RESERVOIR CHARACTERIZATION

In particular when we faced with fractured reservoir characterization, an efficient method of data entry, compiling, and preparation becomes important. Not only the initial model requires considerable amount of data preparation, but also subsequent stages of model updating will require a convenient way to input the new data to the existing data stream. Well logs suites provided by the operator will be supplied to the project team. We anticipate a spectrum of resistivity, image logs, cutting and core where available. A carefully designed data collection phase will provide the necessary input to develop a 3D model of the reservoir. An optimum number of test wells and training wells needs to be identified. In addition, a new technique needs to be developed to optimize the location and the orientation of each new well to be drilled based on data gathered from previous wells. If possible, we want to prevent clustering of too many wells at some locations and under-sampling in other locations thus maintaining a level of randomness in data acquisition. The data to be collected will be dependent on the type of fractured reservoir.

The data collected will also provide the statistics to establish the trends, variograms, shape, and distribution of the fractures in order to develop a non-linear and non-parametric statistical model and various possible realizations of this model. For example, one can use stochastic models techniques and alternative conditional expectation (ACE) model developed by Breiman and Friedman (1985) for initial reservoir model prediction This provides crucial information on the variability of the estimated models. Significant changes from one realization to the other indicate a high level of uncertainty, thus the need for additional data to reduce the standard deviation. In addition, one can use our neuro-fuzzy approach to better quantify and perhaps reduce the uncertainties in the characterization of the reservoir.

Samples from well cuttings (commonly available) and cores (where available) from the focus area can also be analyzed semi-quantitatively by XRD analysis of clay mineralogy to determine vertical variability. Calibration to image logs needs to be performed to correlate fracture density to conventional log signature and mineralogical analysis.

Based on the data obtained and the statistical representation of the data, an initial 3D model of the boundaries of the fractures and its distribution can be developed. The model is represented by a multi-valued parameter, which reflects different subsurface properties to be characterized. This parameter is derived through integration of all the input data using a number of conventional statistical approaches.

A novel 'neuro-fuzzy' based algorithm that combines the training and learning capabilities of the conventional neural networks with the capabilities of fuzzy logic to incorporate subjective and imprecise information can be refined for this application. Nikravesh (1998a,b) showed the significant superiority of the neuro-fuzzy approach for data integration over the conventional methods for characterizing the boundaries.

Similar method with minor modifications can be implemented and tested for fractured reservoirs.

Based on this information, an initial estimate for distribution of reservoir properties including fracture shape and distribution in 2D and 3D spaces can be predicted. Finally, the reservoir model is used as an input to this step to develop an optimum strategy for management of the reservoir. As data collection continues in the observation wells, using new data the model parameters will be updated. These models are then continually evaluated and visualized to assess the effectiveness of the production strategy. The wells chosen in the data collection phase will be designed and operated through a combination of an intelligent advisor.

9. FUTURE TRENDS AND CONCLUSIONS

We have discussed the main areas where soft computing can make a major impact in geophysical, geological and reservoir engineering applications in the oil industry. These areas include facilitation of automation in data editing and data mining. We also pointed out applications in non-linear signal (geophysical and log data) processing. And better parameterization of wave equations with random or fuzzy coefficients both in seismic and other geophysical wave propagation equations and those used in reservoir simulation. Of significant importance is their use in data integration and reservoir property estimation. Finally, quantification and reduction of uncertainty and confidence interval is possible by more comprehensive use of fuzzy logic and neural networks.

Given the level of interest and the number of useful networks developed for the earth science applications and specially oil industry, it is expected soft computing techniques will play a key role in this field. Many commercial packages based on soft computing are emerging. The challenge is how to explain or 'sell' the concepts and foundations of soft computing to the practising explorationist and convince them of the value of the validity, relevance and reliability of results based on the intelligent systems using soft computing methods.

APPENDIX A. A BASIC PRIMER ON NEURAL NETWORK AND FUZZY LOGIC TERMINOLOGY

Neural networks. Neural networks are systems that ". . . use a number of simple computational units called 'neurons'. . . " and each neuron ". . . processes the incoming inputs to an output. The output is then linked to other neurons" (von Altrock, 1995). Neurons are also called 'processing elements'.

Weight. When used in reference to neural networks, 'weight' defines the robustness or importance of the connection (also known as a link or synapse) between any two neurons. Medsker (1994) notes that weights ". . . express the relative strengths (or mathematical value) of the various connections that transfer data from layer to layer".

Backpropagation learning algorithm. In the simplest neural networks, information (inputs and outputs) flows only one way. In more complex neural networks, information can flow in two directions, a 'feedforward' direction and a 'feedback' direction. The feedback process is known as 'backpropagation'. The technique known as a 'backpropagation learning algorithm' is most often used to train a neural network towards a desired outcome by running a 'training set' of data with known patterns through the network. Feedback from the training data is used to adjust weights until the correct patterns appear. Hecht-Nielsen (1990) and Medsker (1994) provide additional information.

Perception. There are two definitions of this term (Hecht-Nielsen, 1990). The 'perception' is a classical neural network architecture. In addition, processing elements (neurons) have been called 'perceptrons'.

Fuzziness and fuzzy. It is perhaps best to introduce the concept of 'fuzziness' using Zadeh's original definition of fuzzy sets (Zadeh, 1965): "A fuzzy set is a class of objects with a continuum of grades of membership. Such a set is characterized by a membership (characteristic) function which assigns to each object a grade of membership ranging between zero and one". Zadeh (1973) further elaborates that fuzzy sets are "... classes of objects in which the transition from membership to non-membership is gradual rather than abrupt". Fuzzy logic is then defined as the "... use of fuzzy sets defined by membership functions in logical expressions" (von Altrock, 1995). Fuzziness and fuzzy can then be defined as having the characteristics of a fuzzy set.

Neuro-fuzzy. This is a noun that looks like an adjective. Unfortunately, 'neuro-fuzzy' is also used as an adjective, e.g. 'neuro-fuzzy logic' or 'neuro-fuzzy systems'. Given this confusing situation, a useful definition to keep in mind is: "The combination of fuzzy logic and neural net technology is called 'NeuroFuzzy' and combines the advantages of the two technologies" (von Altrock, 1995). In addition, a neuro-fuzzy system is a neural network system that is self-training, but uses fuzzy logic for knowledge representation, the rules for behavior of the system, and for training the system.

Crisp sets and fuzzy sets. "Conventional (or crisp) sets contain objects that satisfy *precise properties* required for membership" (Bezdek and Pal, 1992). Compare this to their definition that 'fuzzy sets' ". . . contain objects that satisfy *imprecise* properties to varying degrees. . . ". Each member of a crisp set is either 'true' or is 'false', whereas each member of a fuzzy set may have a certain degree of truth or a certain degree of falseness or may have of some degree of each!

APPENDIX B. NEURAL NETWORKS

Details of neural networks are available in the literature (Kohonen, 1987, 1997; Cybenko, 1989; Hecht-Nielsen, 1989; Widrow and Lehr, 1990; and Lin and Lee, 1996) and therefore only the most important characteristics of neural networks will be mentioned. The typical neural network (Fig. B.1) has an input layer, an output layer, and at least one hidden layer. Each layer is in communication with the succeeding layer via a set of connections of various weights, i.e. strengths. In a neural network, non-linear elements are called various names, including nodes, neurons, or processing elements (Fig. B.2).

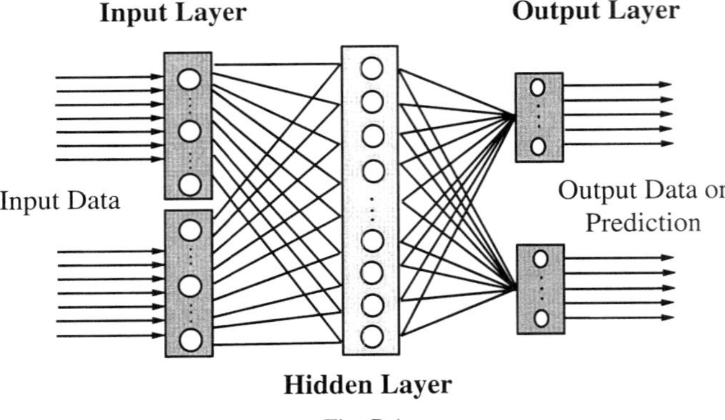

Fig. B.1

A biological neuron is a nerve cell that receives, processes, and passes on information. Artificial neurons are simple first-order approximations of biological neurons.

Consider a single artificial neuron (Fig. B.2) with a transfer function ($y1^{(i)} = f(z^{(i)})$), connection weights, w_j, and a node threshold, θ. For each pattern i,

$$z^{(i)} = x_1^{(i)} w_1 + x_2^{(i)} w_2 + \ldots + x_N^{(i)} w_N + \theta \quad \text{for} \quad i = 1, \ldots, P. \tag{B.1}$$

All patterns may be represented in matrix notation as,

$$\begin{bmatrix} z^{(1)} \\ z^{(2)} \\ \cdot \\ \cdot \\ \cdot \\ z^{(P)} \end{bmatrix} = \begin{bmatrix} x_1^{(1)} & x_2^{(1)} & \cdot & \cdot & \cdot & x_N^{(1)} & 1 \\ x_1^{(2)} & x_2^{(2)} & \cdot & \cdot & \cdot & x_N^{(2)} & 1 \\ \cdot & \cdot & \cdot & \cdot & \cdot & \cdot & \cdot \\ \cdot & \cdot & \cdot & \cdot & \cdot & \cdot & \cdot \\ \cdot & \cdot & \cdot & \cdot & \cdot & \cdot & \cdot \\ x_1^{(P)} & x_2^{(P)} & \cdot & \cdot & \cdot & x_N^{(P)} & 1 \end{bmatrix} \begin{bmatrix} w_1 \\ w_2 \\ \cdot \\ \cdot \\ \cdot \\ w_N \\ \theta \end{bmatrix} \tag{B.2}$$

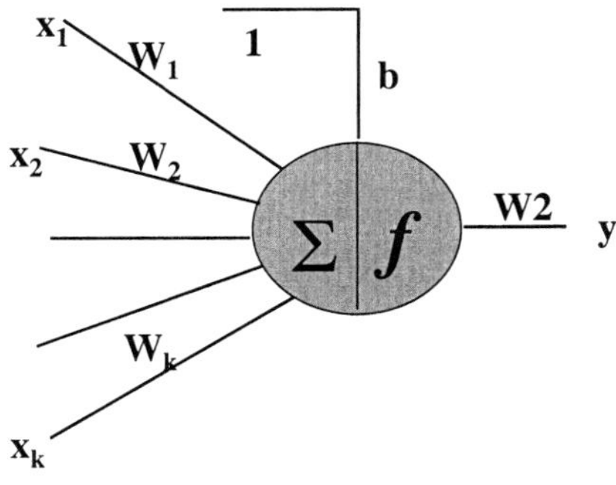

$$\mathbf{y = f\,[\,b + w_1 x_1 + w_2 x_2 + \ldots + w_k x_k\,]\ .\ w_2}$$

Fig. B.2

and

$$\underline{y1} = f(\underline{z}). \tag{B.3}$$

The transfer function, f, is typically defined by a sigmoid function such as the hyperbolic tangent function,

$$f(z) = \frac{e^z - e^{-z}}{e^z + e^{-z}}.$$

In more compact notation,

$$\underline{z} = \underline{\underline{X}}_1 \times \underline{w}_\theta = \underline{\underline{X}} \times \underline{w} + \underline{\theta} \tag{B.4}$$

where

$$\underline{w}_\theta = \left[\underline{w}^{\mathrm{T}} \mid \theta\right]^{\mathrm{T}} \tag{B.5}$$

$$\underline{\underline{X}}_1 = \left[\underline{\underline{X}} \mid \underline{1}\right] \tag{B.6}$$

and, $\underline{1}$ = column vector of ones with P rows; $\underline{\underline{X}} = P \times N$ matrix with N input and P pattern; $\underline{\theta}$ = bias vector, vector with P rows of $\overline{\theta}$; $\underline{w}$ = weights, vector with N rows.

During learning, the information is propagated back through the network and used to update the connection weights (back-propagation algorithm). The objective function for the training algorithm is usually set up as a squared error sum,

$$E = \frac{1}{2} \sum_{i=1}^{P} (y^i_{(\text{observed})} - y^i_{(\text{prediction})})^2. \tag{B.7}$$

This objective function defines the error for the observed value at the output layer, which is propagated back through the network. During training, the weights are adjusted to minimize this sum of squared errors.

APPENDIX C. MODIFIED LEVENBERGE–MARQUARDT TECHNIQUE

Several techniques have been proposed for training the neural network models. The most common technique is the backpropagation approach. The objective of the learning process is to minimize the global error in the output nodes by adjusting the weights. This minimization is usually set up as an optimization problem. Here, we use the Levenberg–Marquardt algorithm, which is faster and more robust than conventional algorithms, but it requires more memory. Using non-linear statistical techniques, the conventional Levenberge–Marquardt algorithm (optimization algorithm for training the neural network) is modified. In this situation, the final global error in the output at each sampling time is related to the network parameters and a modified version of learning coefficient is defined. The following equations briefly show the difference between the conventional and the modified technique as used in this study.

For the conventional technique:

$$\Delta W = \left(\underline{J}^{\mathrm{T}} J + \mu^2 \underline{I}\right)^{-1} \underline{J}^{\mathrm{T}} e \tag{C.1}$$

whereas in the modified technique

$$\Delta W = \left(\underline{J}^{\mathrm{T}}\underline{\Delta}^{\mathrm{T}}\underline{\Delta}J + \Gamma^{\mathrm{T}}\Gamma\right)^{-1}\underline{J}^{\mathrm{T}}\underline{\Delta}^{\mathrm{T}}\underline{\Delta}e \tag{C.2}$$

where

$$\underline{\Delta}^{\mathrm{T}}\underline{\Delta} = \underline{\hat{V}}^{-1} \tag{C.3}$$

$$\hat{V}_{ij} = \frac{1}{2m+1}\sum_{k=-m}^{m}\hat{e}_{i+k}\hat{e}_{j+k} \tag{C.4}$$

$$\underline{\hat{V}} = \hat{\sigma}^2\underline{I} \tag{C.5}$$

$$\hat{W} = W \pm k\hat{\sigma}. \tag{C.6}$$

APPENDIX D. NEURO-FUZZY MODELS

In recent years, considerable attention has been devoted to the use of hybrid neural network-fuzzy logic approaches (Jang, 1991, 1992) as an alternative for pattern recognition, clustering, and statistical and mathematical modeling. It has been shown that neural network models can be used to construct internal models that capture the presence of fuzzy rules.

Neuro-fuzzy modeling is a technique of describing the behavior of a system using fuzzy inference rules using a neural network structure. The model has a unique feature in which it can express linguistically the characteristics of the complex non-linear system. In this study, we will use the neuro-fuzzy model originally presented by Sugeno and Yasukawa (1993). The neuro-fuzzy model is characterized by a set of rules. The rules are expressed as follows:

$$R^i: \text{ if } x_1 \text{ is } A_1^i \text{ and } x_2 \text{ is } A_2^i \ \dots \text{ and } x_n \text{ is } A_n^i \text{ (Antecedent)} \tag{D.1}$$

$$\text{then } y^* = f_i(x_1, x_2, \dots, x_n) \text{ (Consequent)}$$

where $f_i(x_1, x_2, \dots, x_n)$ can be constant, linear, or fuzzy set. For the linear case

$$f_i(x_1, x_2, \dots, x_n) = a_{i0} + a_{i1}x_1 + a_{i2}x_2 + \dots + a_{in}x_n. \tag{D.2}$$

Therefore, the predicted value for output y is given by:

$$y = \sum_i \mu_i f_i(x_1, x_2, \dots, x_n) \Big/ \sum \mu_i \tag{D.3}$$

with

$$\mu_i = \prod_j A_j^i(x_j) \tag{D.4}$$

where R_i is the ith rule, x_j are input variables, y is output, A_j^i are fuzzy membership functions (fuzzy variables), and a_{ij} are constant values.

In this study, we will use the adaptive neuro-fuzzy inference system (ANFIS) technique (Jang and Gulley, 1995; The Math Works™, 1995). The model uses neuro-adaptive learning techniques. This learning method is similar to that of neural networks. Given an input/output data set, the ANFIS can construct a fuzzy inference system (FIS) whose membership function parameters are adjusted using the backpropagation algorithm or similar optimization techniques. This allows fuzzy systems to learn from the data they are modeling.

APPENDIX E. K-MEANS CLUSTERING

An early paper on k-means clustering was written by MacQueen (1967). K-means is an algorithm to assign a specific number of centers, k, to represent the clustering of N points ($k < N$). These points are iteratively adjusted so that each point is assigned to one cluster, and the centroid of each cluster is the mean of its assigned points. In general, the k-means technique will produce exactly k different clusters of the greatest possible distinction.

The algorithm is summarized in the following:

(1) Consider each cluster consisting of a set of M samples that are similar to each other: $x_1, x_2, x_3, \ldots, x_m$
(2) Choose a set of clusters $\{y_1, y_2, y_3, \ldots, y_k\}$
(3) Assign the M samples to the clusters using the minimum Euclidean distance rule
(4) Compute a new cluster so as to minimize the cost function
(5) If any cluster changes, return to step 3; otherwise stop.
(6) End

APPENDIX F. FUZZY C-MEANS CLUSTERING

Bezdek (1981) presents comprehensive coverage of the use of fuzzy logic in pattern recognition. Fuzzy techniques can be used as an alternative method for clustering. Fuzzy clustering partitions a data set into fuzzy clusters such that each data point can belong to multiple clusters. Fuzzy c-means (FCM) is a well-known fuzzy clustering technique that generalizes the classical (hard) c-means algorithm, and can be used where it is unclear how many clusters there should be for a given set of data.

Subtractive clustering is a fast, one-pass algorithm for estimating the number of clusters and the cluster centers in a set of data. The cluster estimates obtained from subtractive clustering can be used to initialize iterative optimization-based clustering methods and model identification methods.

The algorithm is summarized in the following:

(1) Consider a finite set of elements $X = x_1, x_2, x_3, \ldots, x_n$ or $x_j,\ j = 1, 2, \ldots, n$
(2) Select a number of clusters c
(3) Choose an initial partition matrix, $U^{(0)}$

- $U = [u_{ij}]_{i=1,2,\ldots,c;\, j=1,2,\ldots,n}$
- where u_{ij} express the degree to which the element of x_j belongs to the ith cluster
- $\sum u_{ij} = 1$ for all $j = 1,2,\ldots,n$
- $0 < \sum u_{ij} < 1$ for all $i = 1,2,\ldots,c$

(4) Choose a termination criteria, ε
(5) Set the iteration index l to 0
(6) Calculate the fuzzy cluster centers using $U^{(l)}$ and an objective function
(7) Calculate the new partition matrix $U^{(l+1)}$ using an objective function
(8) Calculate $\Delta = \|U^{(l+1)} - U^{(l)}\| = \max_{i,j}\left|u_{ij}^{(l+1)} - u_{ij}^{(l)}\right|$
(9) If $\Delta > \varepsilon$, then set $l = l + 1$ return to step 6, otherwise stop.
(10) End

APPENDIX G. NEURAL NETWORK CLUSTERING

Kohonen (1987, 1997) wrote two fundamental books on neural network clustering. The self-organizing map technique known as Kohonen's self-organizing feature map (Kohonen, 1997) can be used as an alternative for clustering purposes (Fig. G.1). This technique converts patterns of arbitrary dimensionality (the pattern space) into the response of one- or two-dimensional arrays of neurons (the feature space). This

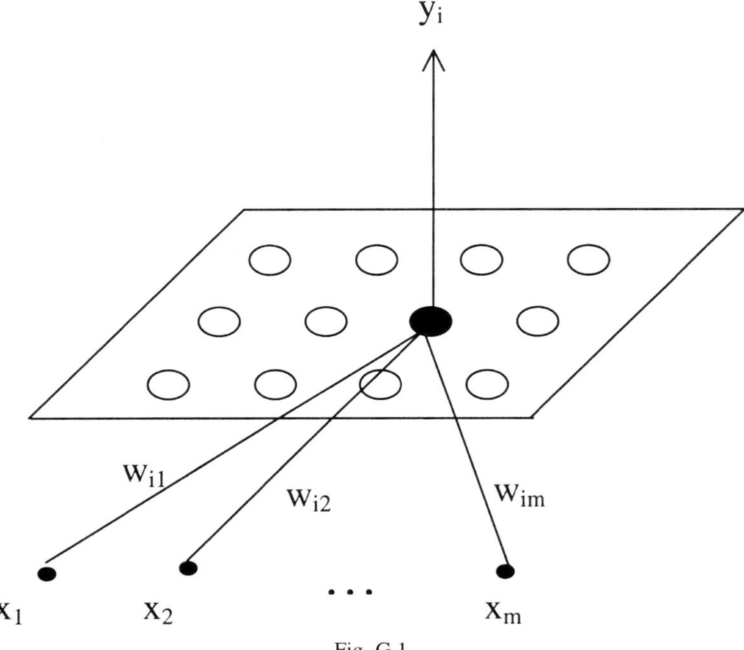

Fig. G.1

unsupervised learning model can discover any relationship of interest such as patterns, features, correlations, or regularities in the input data, and translate the discovered relationship into outputs.

The algorithm is summarized in the following:

(1) Consider the network structure as shown in Fig. G.1.

(2) The learning rule is defined as:

- The similarity match is defined as: $\|\boldsymbol{x} - \boldsymbol{w}_{i^*}\| = \mathrm{Min}_j \|\boldsymbol{x} - \boldsymbol{w}_j\|$
- The learning rule is defined as: $w_{ij}^{(k+1)} = \begin{cases} w_{ij}^{(k)} + \alpha^{(k)}\left[x_j^{(k)} - w_{ij}^{(k)}\right], \text{ in the neighborhood set of the winner node } i^* \text{ at the time step } k. \\ w_{ij}^{(k)}, \text{ otherwise.} \end{cases}$

  with α defined as a learning constant.

REFERENCES

Adams, R.D., Collister, J.W., Ekart, D.D., Levey, R.A. and Nikravesh, M., 1999a. Evaluation of gas reservoirs in collapsed paleocave systems. Ellenburger Group, Permian Basin, Texas, AAPG Annual Meeting, San Antonio, TX, 11–14 April.

Adams, R.D., Nikravesh, M., Ekart, D.D., Collister, J.W., Levey, R.A. and Siegfried, R.W., 1999b. Optimization of well locations in complex carbonate reservoirs. Summer 1999, GasTIPS, GRI 1999, Vol. 5, No. 2.

Agterberg, F.P., Griffiths, C.M., 1991. Computer applications in stratigraphy. *Computers and Geosciences* (Special review issue, Ed. Cubitt, J.M.), 17(8): 1105–1118.

Aminzadeh, F., 1989a. Pattern Recognition and Image Processing. Elsevier Science, Amsterdam.

Aminzadeh, F., 1989b. Future of expert systems in the oil industry. Proc. of HARC Conf. on Geophysics and Computers, A Look at the Future.

Aminzadeh, F., 1991. Where are we now and where are we going? In: F. Aminzadeh and M. Simaan (Eds.), Expert Systems in Exploration. SEG Publications, Tulsa, OK, pp. 3–32.

Aminzadeh, F., 1994. Applications of fuzzy expert systems in integrated oil exploration. *Computer Electr. Eng.*, 2: 89–97.

Aminzadeh, F., 1996. Future geoscience technology trends. In: Stratigraphic Analysis, Utilizing Advanced Geophysical, Wireline, and Borehole Technology For Petroleum Exploration and Production. GCSEPFM, pp. 1–6.

Aminzadeh, F. and Chatterjee, S., 1984/1985. Applications of clustering in exploration seismology. *Geoexploration*, 23: 147–159.

Aminzadeh, F. and de Groot, P., 2001. Seismic characters and seismic attributes to predict reservoir properties. Proc. of SEG-GSH Spring Symp.

Aminzadeh, F. and Jamshidi, M. (Eds.), 1995. Soft Computing. PTR Prentice Hall.

Aminzadeh, F., Katz, S. and Aki, K., 1994. Adaptive neural network for generation of artificial earthquake precursors. *IEEE Trans. Geosci. Remote Sensing*, 32(6).

Aminzadeh, F., Barhen, J., Glover, C.W. and Toomanian, N.B., 1999. Estimation of reservoir parameters using a hybrid neural network. *J. Sci. Pet. Eng.*, 24: 49–56.

Aminzadeh, F., Barhen, J., Glover, C.W. and Toomanian, N.B., 2000. Reservoir parameter estimation using hybrid neural network. *Computers Geosci.*, 26: 869–875.

Baldwin, J.L., Otte, D.N. and Wheatley, C.L., 1989. Computer emulation of human mental process: application of neural network simulations to problems in well log interpretation. *Soc. Pet. Eng.*, SPE Paper #19619: 481.

Baldwin, J.L., Bateman, A.R.M. and Wheatley, C.L., 1990. Application of neural network to the problem of mineral identification from well logs. *Log Analysis*, 3: 279.

Barhen, J., Zak, M., Gulati, S., 1989. Fast neural learning algorithms using networks with non-Lipschitzian dynamics. In: J.C. Rault (ed.), *Neuro-Nimes '89*, EC2 Press, Paris, pp. 55-68.

Baygun et al., 1985. Applications of neural networks and fuzzy logic in geological modeling of a mature hydrocarbon reservoir. Rep. #ISD-005-95-13a, Schlumberger-Doll Research.

Benediktsson, J.A., Swain, P.H. and Erson, O.K., 1990. Neural networks approaches versus statistical methods in classification of multisource remote sensing data. *IEEE Geosci. Remote Sensing, Trans.*, 28(4): 540–552.

Bezdek, J.C., 1981. Pattern Recognition with Fuzzy Objective Function Algorithm. Plenum Press, New York, NY.

Bezdek, J.C. and Pal, S.K. (Eds.), 1992. Fuzzy Models for Pattern Recognition. IEEE Press, 539 pp.

Boadu, F.K., 1997. Rock properties and seismic attenuation: neural network analysis. *Pure Appl. Geophys.*, 149: 507–524.

Bois, P., 1983. Some applications of pattern recognition to oil and gas exploration. *IEEE Trans. Geosci. Remote Sensing*, GE-21(4): 687–701.

Bois, P., 1984. Fuzzy seismic interpretation. *IEEE Trans. Geosci. Remote Sensing*, GE-22(6): 692–697.

Breiman, L. and Friedman, J.H., 1985. *J. Am. Stat. Assoc.*, 580.

Cannon, R.L., Dave, J.V. and Bezdek, J.C., 1986. Efficient implementation of fuzzy c-means algorithms. *IEEE Trans. Pattern Anal. Machine Intelligence*, PAMI 8:(2).

Chappaz, R.J., 1977. Application of the fuzzy sets theory to the interpretation of seismic. Abstract of the 47th SEG meeting, Calgary, Paper R-9.

Chawathe, A., Quenes, A. and Weiss, W.W., 1997. Interwell property mapping using crosswell seismic attributes. SPE 38747. SPE Annu. Tech. Conf. and Exhibition, San Antonio, TX, 5–8 Oct.

Chen, H.C., Fang, J.H., Kortright, M.E., Chen, D.C., 1995. Novel approaches to the determination of Archie parameters, II: fuzzy regression analysis. SPE 26288-P.

Cybenko, G., 1989. Approximation by superposition of a sigmoidal function. *Math. Control Sig. System*, 2: 303.

Griffiths, C.M., 1987. An example of the use of fuzzy-set based pattern recognition approach to the problem of strata recognition from drilling response. In: Aminzadeh, F. (Ed.), Handbook of Geophysical Exploration, Section I: Seismic Prospecting. Pattern Recognition and Image Processing, Vol. 20. Geophysical Press, London/Amsterdam, pp. 504–538.

Gupta, M.M. and Ding, H., 1994. Foundations of fuzzy neural computations. In: F. Aminzadeh and M. Jamshidi (Eds.), Soft Computing. PTR Prentice Hall, pp. 165–199.

Hecht-Nielsen, R., 1989. Theory of backpropagation neural networks. Presented at IEEE Proc., Int. Conf. Neural Network, Washington DC.

Hecht-Nielsen, R., 1990. *Neurocomputing*. Addison-Wesley, 433 pp.

Horikawa, S., Furuhashi, T., Uchikawa, Y. and Tagawa, T., 1991. A study on fuzzy modeling using fuzzy neural networks. Fuzzy Engineering toward Human Friendly Systems, IFES'91.

Horikawa, S., Furuhashi, T. and Uchikawa, Y., 1992. On fuzzy modeling using fuzzy neural networks with the backpropagation algorithm. *IEEE Trans. Neural Networks*, 3(5).

Horikawa, S., Furuhashi, T., Kuromiya, A., Yamaoka, M. and Uchikawa, Y., 1996. Determination of antecedent structure for fuzzy modeling using genetic algorithm. Proc. ICEC'96, IEEE Int. Conf. on Evolutionary Computation, Nagoya, Japan, May 20–22.

Huang, Z. and Williamson, M.A., 1994. Geological pattern recognition and modeling with a general regression neural network. *Can. J. Expl. Geophys.*, 30(1): 60–68.

Jang, J.S.R., 1991. Fuzzy modeling using generalized neural networks and Kalman filter algorithm. Proc. of the Ninth Natl. Conf. on Artificial Intelligence, pp. 762–767.

Jang, J.S.R., 1992. Self-learning fuzzy controllers based on temporal backpropagation. *IEEE Trans. Neural Networks*, 3(5).

Jang, J.S.R. and Gulley, N., 1995. Fuzzy Logic Toolbox. The Math Works Inc., Natick, MA.

Johnson, V.M. and Rogers, L.L., 1995. Location analysis in ground water remediation using artificial neural networks. Lawrence Livermore Natl. Lab. Rep. UCRL-JC-17289.

Kohonen, T., 1987. Self-Organization and Associate Memory, 2nd edn. Springer Verlag, Berlin.

Kohonen, T., 1997. Self-Organizing Maps, 2nd edn. Springer, Berlin.

Klimentos, T. and McCann, C., 1990. Relationship among compressional wave attenuation, porosity, clay content and permeability in sandstones. *Geophysics*, 55: 991014.

Lashgari, B., 1991. Fuzzy Classification with Applications. In: F. Aminzadeh and M. Simaan (Eds.), Expert Systems in Exploration. SEG Publications, Tulsa, OK, pp. 3–32.

Levey, R., Nikravesh, M., Adams, R., Ekart, D., Livnat, Y., Snelgrove, S. and Collister, J. 1999. Evaluation of fractured and paleocave carbonate reservoirs. AAPG Annu. Meet., San Antonio, TX, 11–14 April.

Lin, C.T. and Lee, C.S.G., 1996. Neural Fuzzy Systems. Prentice Hall, Englewood Cliffs, NJ.

Lippmann, R.P., 1987. An introduction to computing with neural networks. *ASSP Mag.*, April, 4–22.

Liu, X., Xue, P. and Li, Y., 1989. Neural network method for tracing seismic events. 59th Annu. SEG Meet., Expanded Abstracts, pp. 716–718.

MacQueen, J., 1967. Some methods for classification and analysis of multivariate observation. In: L.M. LeCun and J. Neyman (Eds.), The Proceedings of the 5th Berkeley Symposium on Mathematical Statistics and Probability. Univ. of California Press, 1, pp. 281–297.

Mallick, S., 1999. Some practical aspects of prestack waveform inversion using a genetic algorithm: an example from the east Texas Woodbine gas sand. *Geophysics, Soc. Explor. Geophys.*, 64: 326–336.

Matsushita, S., Kuromiya, A., Yamaoka, M., Furuhashi, T. and Uchikawa, Y., 1983. A study on fuzzy GMDH with comprehensible fuzzy rules. 1994 IEEE Symp. on Emerging Technologies and Factory Automation. Gaussian Control Problem, Mathematics of Operations Research, 8(1).

McCormack, M.D., 1990. Neural computing in geophysics. *Geophys. The Leading Edge Explor.*, 10(1): 11–15.

McCormack, M.D., 1991. Seismic trace editing and first break piking using neural networks, 60th Annu. SEG Meet. Expanded Abstracts, pp. 321–324.

McCormack, M.D., Stoisits, R.F., Macallister, D.J. and Crawford, K.D., 1999. Applications of genetic algorithms in exploration and production. *The Leading Edge*, 18(6): 716–718.

Medsker, L.R., 1994. Hybrid Neural Network and Expert Systems. Kluwer Academic Publishers, Dordrecht, p. 240.

Monson, G.D. and Pita, J.A., 1997. Neural network prediction of pseudo-logs for net pay and reservoir property interpretation: Greater Zafiro field area, Equatorial Guinea. SEG 1997 Meet., Dallas, TX.

Nikravesh, M., 1998a. Mining and fusion of petroleum data with fuzzy logic and neural network agents. CopyRight© Report, LBNL-DOE, ORNL-DOE, and DeepLook Industry Consortium.

Nikravesh, M., 1998b. Neural network knowledge-based modeling of rock properties based on well log databases. SPE 46206, 1998 SPE Western Reg. Meet., Bakersfield, CA, 10–13 May.

Nikravesh, M., 2000. Intelligent Computing Techniques for Reservoir Characterization of Fractured Networks in the Mancos Shale. Internal Proposal, EGI, University of Utah.

Nikravesh, M. and Aminzadeh, F., 1997. Knowledge discovery from data bases. Intelligent data mining. FACT Inc. and LBNL Proposal, submitted to SBIR-NASA.

Nikravesh, M. and Aminzadeh, F., 2001. Mining and fusion of petroleum data with fuzzy logic and neural network agents. *J. Pet. Sci. Eng.*, 29: 221–238.

Nikravesh, M., Farell, A.E. and Stanford, T.G., 1996. Model identification of nonlinear time-variant processes via artificial neural network. *Computers Chem. Eng.*, 20(11): 1277.

Nikravesh, M., Novak, B. and Aminzadeh, F., 1998. Data mining and fusion with integrated neuro-fuzzy agents. Rock properties and seismic attenuation. JCIS 1998. Fourth Joint Conf. on Information Sciences, NC, USA, October 23–28.

Nikravesh, M., Levey, R.A. and Ekart, D.D., 1999a. Soft computing: tools for reservoir characterization (IRESC) and optimal well placement (OWP). To be published in Special Issue, Computer and Geosc. J., 1999–2000.

Nikravesh, M., Levey, R.A. and Ekart, D.D., 1999b. Soft computing: tools for reservoir characterization (IRESC). To be presented at 1999 SEG Annu. Meet.

NPC (National Petroleum Council), 1992. The potential for Natural Gas in the United States. The National Petroleum Council, Washington, DC, USA.

Ostrander, W.J., 1982. Plane wave reflection coefficients for gas sands at nonnormal angles of incidence. 52nd Ann. Intern. Mtg.: Soc. of Expl. Geophys., Session: S16.4.

Pezeshk, S., Camp, C.C. and Karprapu, S., 1996. Geophysical log interpretation using neural network. *J. Comput. Civil Eng.*, 10: 136.

Potter, D., Wright, J. and Corbett, P., 1999. A genetic petrophysics approach to facies and permeability prediction in a Pegasus well,.61st Meet., Europ. Assoc. Geosci. Eng., Session 2053.

Rogers, S.J., Fang, J.H., Karr, C.L. and Stanley, D.A., 1992. Determination of lithology, from well logs using a neural network. *AAPG Bull.*, 76: 731.

Rosenblatt, F., 1962. Principal of Neurodynamics. Spartan Books.

Schuelke, J.S., Quirein, J.A., Sarg, J.F., Altany, D.A. and Hunt, P.E., 1997. Reservoir architecture and porosity distribution, Pegasus field, West Texas – an integrated sequence stratigraphic-seismic attribute study using neural networks. SEG 1997 Meet., Dallas, TX.

Selva, C., Aminzadeh, F., Diaz, B. and Porras, J.M., 2001. Using geostatistical techniques for mapping a reservoir in eastern Venezuela. Proc. of 7th Int. Congr. of the Brazilian Geophysical Society, October 2001, Salvador, Brazil.

Sheriff, R.E. and Geldart, L.P., 1982. Exploration Seismology, Vol. 1. Cambridge.

Sheriff, R.E. and Geldart, L.P., 1983. Exploration Seismology, Vol. 2, Cambridge.

Sugeno, M. and Yasukawa, T., 1993. A fuzzy-logic-based approach to qualitative modeling. *IEEE Trans. Fuzzy Syst.*, 1.

Taner, M.T., Koehler, F. and Sherrif, R.E., 1979. Complex seismic trace analysis. *Geophysics*, 44: 1196–1212.

Tamhane, D., Wong, P.M. and Aminzadeh, F., 2002. Integrating linguistic descriptions and digital signals in petroleum reservoirs. *Int. J. Fuzzy Syst.*, 4(1): 586–591.

The Math Works™, 1995. The Math Works, Inc., Natick, MA.

Tura, A. and Aminzadeh, F., 1999. Dynamic reservoir characterization and seismically constrained production optimization. Extended Abstracts of Soc. of Explor. Geophys., Houston.

U.S. Geological Survey, National Oil and Gas Resource Assessment Team, 1995. National assessment of U.S. oil and gas resources. USGS Circular 1118, 20 pp.

Veezhinathan, J., Wagner, D. and Ehlers, J., 1991. First break picking using neural network. In: F. Aminzadeh and M. Simaan (Eds.), Expert Systems in Exploration. SEG, Tulsa, OK, pp. 179–202.

von Altrock, C., 1995. Fuzzy logic and neuro-fuzzy applications explained. Prentice Hall PTR, 350 pp.

Widrow, B. and Lehr, M.A., 1990. 30 years of adaptive neural networks: perception, madaline, and backpropagation. *Proc. IEEE*, 78(9): 1414.

Wong, P.M., Jiang, F.X. and Taggart, I.J., 1995a. A critical comparison of neural networks and discrimination analysis in lithofacies, porosity and permeability prediction. *J. Pet. Geol.*, 18: 191.

Wong, P.M., Gedeon, T.D. and Taggart, I.J., 1995b. An improved technique in prediction: a neural network approach. *IEEE Trans. Geosci. Remote Sensing*, 33: 971.

Wong, P.M., Aminzadeh, F. and Nikravesh, M., 2001. Soft computing for reservoir characterization. In: Studies in Fuzziness. Physica Verlag, Germany.

Yoshioka, K., Shimada, N. and Ishii, Y., 1996. Application of neural networks and co-kriging for predicting reservoir porosity-thickness. *GeoArabia*, 1(3).

Zadeh, L.A., 1965. Fuzzy sets. *Information and Control*, 8: 33353.

Zadeh, L.A., 1973. Outline of a new approach to the analysis of complex systems and decision processes. *IEEE Trans. Syst., Man, Cybernet.*, SMC-3: 244.

Zadeh, L.A., 1992. The calculus of fuzzy if-then-rules. *AI Expert*, 7(3).

Zadeh, L.A., 1996. The roles of fuzzy logic and soft computing in the concept, design and deployment of intelligent systems. *BT Technol. J.*, 14(4).

Zadeh, L.A. and Aminzadeh, F., 1995. Soft computing in integrated exploration. Proc. of IUGG/SEG Symp. on AI in Geophysics, Denver, CO, July 12.

Zadeh, L.A. and Yager, R.R., 1991. Uncertainty in Knowledge Bases. Springer-Verlag, Berlin.

Zhang, X. and Li, Y., 1995. The application of artificial neural network with SOM — a neural network approach. SEG 65th Annual Meeting, Houston, USA, Oct. 1995.

Developments in Petroleum Science, 51
Editors: M. Nikravesh, F. Aminzadeh and L.A. Zadeh

Chapter 2

FUZZY LOGIC

GEORGE J. KLIR

Center for Intelligent Systems, Watson School of Engineering and Applied Science, Binghamton University, Binghamton, NY 13902-6000, USA

ABSTRACT

Fuzzy logic is generally recognized as one of the ingredients of soft computing. In this context, a broad view of fuzzy logic is employed: fuzzy logic is viewed as a formalized language, based on fuzzy set theory, that is sufficiently expressive to deal with modes of reasoning that are approximate rather than exact. The principal role of fuzzy logic, viewed in this broad sense, is to represent, in a realistic way, knowledge expressed by statements in natural language, and to formalize reasoning based upon this representation.

1. FUZZY SETS

The nucleus of fuzzy logic is the concept of a *fuzzy set*. Contrary to classical sets, fuzzy sets are not required to have sharp boundaries that distinguish their members from other objects. The membership in a fuzzy set is not matter of affirmation or denial, as it is in any classical set, but it is a matter of degree.

Degree of membership of objects in fuzzy sets are most commonly expressed by real numbers in the unit interval $[0, 1]$. Fuzzy sets in which membership degree are expressed in this way are predominant in the literature and are called *standard fuzzy sets*. Due to the limited space, this chapter is restricted to this type of fuzzy sets. Only a reference to other fuzzy sets is made at the end of this chapter.

Each standard fuzzy set, say A, is formally defined by a function μ_A of the form

$$\mu_{\mathrm{A}}: \quad X \rightarrow [0, 1],$$

where X denotes the set (classical, nonfuzzy) of all objects that are relevant in the context of a particular application. For each application, set X forms the universe of discourse of that application and it is usually referred to as the *universal set*.

Function μ_{A} is called a *membership function* of fuzzy set A. For each $x \in X$, $\mu_{\mathrm{A}}(x)$ specifies the *degree of membership* of object (element) x in fuzzy set A. Alternatively, $\mu_{\mathrm{A}}(x)$ may be viewed as the *degree of compatibility* of x with the concept represented by the fuzzy set.

A degenerate fuzzy set A in which either $\mu_{\mathrm{A}}(x) = 0$ or $\mu_{\mathrm{A}}(x) = 1$ for all $x \in X$ is called a *crisp set*. This alternative view of classical sets makes a significant difference. Classical sets are often defined in terms of *characteristic functions* by which objects that belong to a given set are labeled by 1 and those that do not belong to it are labeled by 0. However, these numbers, 1 and 0, play in classical set theory a purely *symbolic role*. They are convenient, but may be as well replaced by any other pair of symbols (e.g., m for members and n for nonmembers, $\in$ for members and $\notin$ for nonmembers, etc.). Contrary to this symbolic role of numbers 1 and 0 in characteristic functions, numbers assigned to objects by membership functions of fuzzy sets have clearly a numerical significance. This significance is preserved when classical sets are viewed (from the standpoint of fuzzy set theory) as special fuzzy sets – crisp sets.

Two distinct notations are most commonly employed in the literature to denote membership functions. In one of them, which is already introduced in previous paragraphs of this section, the membership function of a fuzzy set A is denoted by μ_{A}. That is, the symbol of the fuzzy set is distinguished from the symbol of its membership function. According to the second notation, this distinction is not made. That is, the same symbol, say A, is used to denote a fuzzy set as well as the membership function of this fuzzy set. This is acceptable as a notational convention since no ambiguity results from this double use of the same symbol: each fuzzy set is uniquely defined by one particular membership function. In this section, the second notation is adopted since it is simpler and, by and large, more popular in current literature on fuzzy set theory.

Examples of fuzzy sets (membership functions) are shown graphically in Fig. 1. These fuzzy sets are defined on the set of real numbers, which is the universal set in this case. They may represent the concept 'around 1' (or 'close to 1'). Which of these or other possible membership functions is actually an appropriate representation of the concept must be determined in the context of each particular applications.

Membership functions in Fig. 1 are quite typical. Each of them is defined in piece-meal fashion by a set of formulas. For example,

$$A(x) = \begin{cases} x & \text{when } x \in [0,1] \\ 2-x & \text{when } x \in [1,2] \\ 0 & \text{otherwise} \end{cases}$$

$$C(x) = \begin{cases} 2x - x^2 & \text{when } x \in [0,2] \\ 0 & \text{otherwise} \end{cases}$$

$$F(x) = \begin{cases} x^2 & \text{when } x \in [0,1] \\ (2-x)^2 & \text{when } x \in [1,2] \\ 0 & \text{otherwise} \end{cases}$$

Given a fuzzy set A defined on X and a real number $\alpha \in [0,1]$, the *crisp set* that contains those elements x in X for which $A(x) \geq \alpha$ is called an α-cut, ${}^{\alpha}A$, of A. Formally,

$${}^{\alpha}A = \{x \in X \mid A(x) \geq \alpha\}.$$

It follows directly from this definition that by increasing α the next α-cut is always contained in the previous one. Hence, the set of all α-cuts of any given fuzzy set always

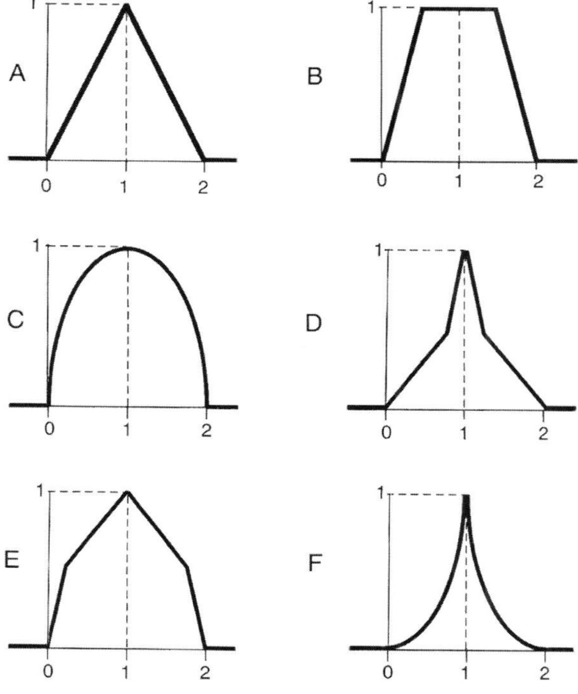

Fig. 1. Examples of fuzzy sets.

forms a nested family of sets, which uniquely represents the fuzzy set. Observe, for example, that the α-cuts of all fuzzy sets in Fig. 1 are closed intervals of real numbers. We can obtain them in this case by determining the inverse functions corresponding to the endpoints of the intervals. For example,

$$
\begin{aligned}
{}^{\alpha}A &= [\alpha, 2-\alpha] \\
{}^{\alpha}C &= [1-\sqrt{1-\alpha}, 1+\sqrt{1-\alpha}] \\
{}^{\alpha}F &= [\sqrt{\alpha}, 2-\sqrt{\alpha}]
\end{aligned}
$$

The significance of the α-cut representation of fuzzy sets is that it connects fuzzy sets with crisp sets. While each crisp set is a collection of objects that are conceived as a whole, each fuzzy set is a collection of nested crisp sets that are also conceived as a whole. Fuzzy sets are thus wholes of a higher category.

The α-cut representation of fuzzy sets allows us to extend the various properties of crisp sets, established in classical set theory, into their fuzzy counterparts. This is accomplished by requiring that the classical property be satisfied by all α-cuts of the fuzzy set concerned. Any property that is extended in this way from classical set theory into the domain of fuzzy set theory is called *cutworthy property*. For example, when convexity of fuzzy sets is defined by the requirement that all α-cuts of a fuzzy convex set be convex in the classical sense, this conception of fuzzy convexity is cutworthy. Other important examples are cutworthy definitions of fuzzy equivalence, fuzzy compatibility, and various kinds of fuzzy orderings.

It is important to realize that many (perhaps most) properties of fuzzy sets, perfectly meaningful and useful, are not cutworthy. These properties do not have any counterparts in classical set theory.

2. OPERATIONS ON FUZZY SETS

As is well known, each of the three basic operations on fuzzy sets – complement, intersection, and union – is unique in classical set theory. However, the counterparts of these operations in fuzzy set theory are not unique. Each of them consists of a class of functions that satisfy certain properties. In this section, only some notable feature of these operations are described. A more comprehensive coverage can be found, for example, in chapter 3 of the text by Klir and Yuan, 1995.

Complementation operations on fuzzy sets are functions of the form

$$c\colon \quad [0,1] \to [0,1]$$

that are order reversing and such that $c(0) = 1$ and $c(1) = 0$. Moreover, they are usually required to possess the property

$$c\big(c(a)\big) = a$$

for all $a \in [0,1]$; this property is called an *involution*. Given a fuzzy set A and a particular complementation function c, the complement of A with respect to c, ${}^{c}A$, is defined for all $x \in X$ by the formula

$${}^{c}A(x) = c\big(A(x)\big).$$

An example of a practical class of involutive complementation functions, c_λ, is defined for each $a \in [0,1]$ by the formula

$$c_\lambda(a) = (1 - a^\lambda)^{1/\lambda}, \qquad \lambda > 0, \tag{1}$$

where λ is a parameter whose values specify individual complements in this class. When $\lambda = 1$, the resulting complement is usually called a *standard complement*.

Complements stand for logical negations of concepts represented by fuzzy sets. Which of possible complements to choose is basically an experimental question. The choice is determined in the context of each particular application by eliciting the meaning of negating a given concept. This can be done by employing a suitable parameterized class of complementation functions, such as the one defined by Eq. (1), and determining a proper value of the parameter.

Intersections and unions of fuzzy sets are defined in terms of functions i and u, respectively, which assign to each pair of numbers in the unit interval $[0,1]$ a single number in $[0,1]$. These functions, i and u, are communitative, associative, monotone nondecreasing, and consistent with the characteristic functions of classical intersections and unions, respectively. They are often referred to in the literature as *triangular norms* (or *t-norms*) and *triangular conorms* (or *t-conorms*). It is well know that the inequalities

$$\begin{aligned} i_{\min}(a,b) \le i(a,b) \le \min(a,b), \\ \max(a,b) \le u(a,b) \le u_{max}(a,b) \end{aligned} \tag{2}$$

are satisfied for all $a,b,\in[0,1]$, where

$$i_{\min}(a,b) = \begin{cases} \min(a,b) & \text{when } \max(a,b)=1 \\ 0 & \text{otherwise,} \end{cases}$$

$$u_{\max}(a,b) = \begin{cases} \max(a,b) & \text{when } \min(a,b)=0 \\ 1 & \text{otherwise.} \end{cases}$$

Operations expressed by min and max are usually called *standard operations*; those expressed by $i_{\min}$ and $u_{\max}$ are called *drastic operations*.

Given fuzzy sets A and B, defined on the same universal set X, their intersections and unions with respect to i and u, $A \cap_i B$ and $A \cup_i B$, are defined for each $x \in X$ by the formulas

$$\begin{aligned} (A \cap_i B)(x) = i[A(x), B(x)], \\ (A \cup_u B)(x) = u[A(x), B(x)]. \end{aligned}$$

It is useful to capture the full range of intersections and unions, as expressed by (2), by suitable classes of functions. For example, functions

$$\begin{aligned} i_\lambda(a,b) = 1 - \min\{1, [(1-a)^\lambda + (1-b)^\lambda]^{1/\lambda}\}, \\ u_\lambda(a,b) = \min\{1, (a^\lambda + b^\lambda)^{1/\lambda}\}, \end{aligned}$$

where $\lambda > 0$, qualify as intersection and unions, and cover the whole range by varying the parameter λ. The standard operations are obtained in the limit for $\lambda \to \infty$, while the drastic operations are obtained in the limit for $\lambda \to 0$. In each particular application, a fitting operation is determined by selecting an appropriate values of the parameter λ by knowledge acquisition techniques.

Two additional types of operations are applicable to fuzzy sets, but they have no counterparts in classical set theory. They are called modifiers and averaging operations.

Modifiers are unary operations that are order preserving. Their purpose is to modify fuzzy sets to account for *linguistic hedges* such *very*, *fairly*, *extremely*, *more of less*, etc. The most common modifiers either increase or decrease all values of a given membership function. A convenient class of functions, m_λ, that qualify as increasing or decreasing modifiers is defined for each $a \in [0,1]$ by the simple formula

$$m_\lambda(a) = a^\lambda,$$

where $\lambda > 0$ is a parameter whose value determines which way and how strongly m_λ modifies a given membership function. Clearly, $m_\lambda(a) > a$ when $\lambda \in (0,1)$, $m_\lambda(a) < a$ when $\lambda \in (1,\infty)$, and $m_\lambda(a) = a$ when $\lambda = 1$. The farther the value of λ from 1, the stronger the modifier m_λ.

Averaging operations are monotone nondecreasing and idempotent, but are not associative. Due to the lack of associativity, they must be defined as functions of n

arguments for any $n \geq 2$. It is well known that any averaging operation, h, satisfies the inequalities

$$\min(a_1,a_2,\ldots,a_n) \leq h(a_1,a_2,\ldots,a_n) \leq \max(a_1,a_2,\ldots,a_n)$$

for any n-tuple $(a_1,a_2,\ldots,a_n) \in [0,1]^n$. This means that the averaging operations fill the gap between intersections (t-norms) and unions (t-conorms).

One class of averaging operations, h_λ, which covers the entire interval between min and max operations, is defined for each n-tuple $(a_1,a_2,\ldots,a_n)$ in $[0,1]^n$ by the formula

$$h_\lambda(a_1,a_2,\ldots,a_n) = \left(\frac{a_1^\lambda + a_2^\lambda + \ldots + a_n^\lambda}{n}\right)^{1/\lambda},$$

where λ is a parameter whose range is the set of all real numbers except 0. For $\lambda = 0$, function h_λ is defined by the limit

$$\lim_{\lambda\to 0} h_\lambda(a_1,a_2,\ldots,a_n) = (a_1,a_2,\ldots,a_n)^{1/n},$$

which is the well-known geometric mean. Moreover

$$\lim_{\lambda\to -\infty} h_\lambda(a_1,a_2,\ldots,a_n) = \min(a_1,a_2,\ldots,a_n),$$
$$\lim_{\lambda\to \infty} h_\lambda(a_1,a_2,\ldots,a_n) = \max(a_1,a_2,\ldots,a_n).$$

This indicates that the standard operations of intersection and union may also be viewed as extreme opposites in the range of averaging operations.

Other classes of averaging operations are now available, some of which use weighting factors to express relative importance of the individual fuzzy sets involved. For example, the function

$$h(a_i, w_i \,|i = 1,2,\ldots,n) = \sum_{i=1}^{n} w_i a_i,$$

where the weighting factors w_i take usually values in the unit interval $[0,1]$ and

$$\sum_{i=1}^{n} w_i = 1,$$

expresses for each choice of values w_i the corresponding weighted average of values a_i $(i = 1,2,\ldots,n)$. Again, the choice is an experimental issue.

3. ARITHMETIC OF FUZZY INTERVALS

Fuzzy sets that are defined on the set of real numbers, $\mathbb{R}$, have a special significance in fuzzy set theory. Among them, the most important are cutworthy *fuzzy intervals*. They are defined by requiring that each α-cut be a single closed and bounded interval of real numbers for all $\alpha \in (0,1]$. A fuzzy interval, A, may conveniently by represented for each $x \in \mathbb{R}$ by the canonical form

$$A(x) = \begin{cases} f_A(x) & \text{when } x \in [a,b) \\ 1 & \text{when } x \in [b,c] \\ g_A(x) & \text{when } x \in (c,d] \\ 0 & \text{otherwise,} \end{cases}$$

where a, b, c, d are specific real numbers such that $a \leq b \leq c \leq d$, f_A is a real-valued function that is increasing, and g_A is a real-valued function that is decreasing. In most applications, functions f_A and g_A are continuous, but, in general, they may be only semi-continuous from the right and left, respectively. When $A(x) = 1$ for exactly one $x \in \mathbb{R}$ (i.e., $b = c$ in the canonical representation), A is called a *fuzzy number*.

For any fuzzy interval A expressed in the canonical form, the α-cuts of A are expressed for all $\alpha \in (0,1]$ by the formula

$${}^{\alpha}A = \begin{cases} [f_A^{-1}(\alpha), g_A^{-1}(\alpha)] & \text{when } \alpha \in (0,\ 1), \\ [b,c] & \text{when } \ \alpha = 1, \end{cases}$$

where f_A^{-1} and g_A^{-1} are the inverse functions of f_A and g_A, respectively.

Consider fuzzy intervals A and B whose α-cut representations are

$$\begin{aligned} {}^{\alpha}A &= {}^{\alpha}[\underline{a}, \overline{a}], \\ {}^{\alpha}B &= {}^{\alpha}[\underline{b}, \overline{b}], \end{aligned}$$

where $\underline{a}$, $\overline{a}$, $\underline{b}$, $\overline{b}$ are functions of α. Then, the individual arithmetic operations on these fuzzy intervals are defined for all $\alpha \in (0,1]$ in terms of the well-established arithmetic operations on closed intervals of real numbers (Moore, 1966; Neumaier, 1990):

$${}^{\alpha}(A * B) = {}^{\alpha}A * {}^{\alpha}B,$$

where $*$ denotes any of the four basic arithmetic operations (addition, subtraction, multiplication, and division). That is, arithmetic operations on fuzzy intervals are cutworthy. According to interval arithmetic,

$${}^{\alpha}A * {}^{\alpha}B = \{a * b \mid a \in {}^{\alpha}A,\ b \in {}^{\alpha}B\}, \tag{3}$$

with the requirement that $0 \notin {}^{\alpha}B$ for all $\alpha \in (0,1]$ when $*$ stands for division. The sets ${}^{\alpha}A * {}^{\alpha}B$ (closed intervals of real numbers) defined by (3) can be obtained for the individual operations from the endpoints of ${}^{\alpha}A$ and ${}^{\alpha}B$ in the following ways:

$$\begin{aligned} {}^{\alpha}A + {}^{\alpha}B &= {}^{\alpha}[\underline{a} + \underline{b}, \overline{a} + \overline{b}], \\ {}^{\alpha}A - {}^{\alpha}B &= {}^{\alpha}[\underline{a} - \overline{b}, \overline{a} - \underline{b}]; \\ {}^{\alpha}A \cdot {}^{\alpha}B &= {}^{\alpha}[\min(\underline{a}\underline{b}, \underline{a}\overline{b}, \overline{a}\underline{b}, \overline{a}\overline{b}), \max(\underline{a}\underline{b}, \underline{a}\overline{b}, \overline{a}\underline{b}, \overline{a}\overline{b})]; \\ {}^{\alpha}A / {}^{\alpha}B &= {}^{\alpha}[\underline{a}, \overline{a}] \cdot {}^{\alpha}[1/\overline{b}, 1/\underline{b}]. \end{aligned}$$

The operation of division, ${}^{\alpha}A / {}^{\alpha}B$, requires that $0 \notin {}^{\alpha}[\underline{b}, \overline{b}]$ for all $\alpha \in (0,1]$.

Fuzzy intervals together with these operations are usually referred to as *standard fuzzy arithmetic*. Algebraic expressions in which values of variables are fuzzy intervals are evaluated by standard fuzzy arithmetic in the same way as classical algebraic expressions are evaluated by classical arithmetic on real numbers.

It turns out that standard fuzzy arithmetic does not take into account constraints among fuzzy intervals employed in algebraic expressions. For example, it does not distinguished between ${}^{\alpha}A * {}^{\alpha}B$, where ${}^{\alpha}B$ happens to be equal to ${}^{\alpha}A$, and ${}^{\alpha}A * {}^{\alpha}A$. In the latter case, both fuzzy intervals represent a value of the same variable, and, consequently, they are strongly constrained: whatever value we select in the first interval, the value in the second interval must be the same. Taking this equality constraint into account, (3) must be replaced with

$${}^{\alpha}A * {}^{\alpha}A = \{a * a \mid a \in {}^{\alpha}A\},$$

Thus, for example, ${}^{\alpha}A - {}^{\alpha}A = [0,0] = 0$ under the equality constraint, while

$${}^{\alpha}A - {}^{\alpha}A = {}^{\alpha}[\underline{a} - \overline{a}, \overline{a} - \underline{a}]$$

under standard fuzzy arithmetic.

By ignoring constraints among fuzzy intervals involved, standard fuzzy arithmetic produces results that are, in general, deficient of information, even though they are principally correct. To avoid this information deficiency, all known constraints among fuzzy intervals in each application must be taken into account. This is a subject of recently emerging *constrained fuzzy arithmetic* (Klir, 1997; Klir and Pan, 1998).

Constrained fuzzy arithmetic is essential for evaluating algebraic equations and for solving algebraic equations in which values of variables are fuzzy intervals. It is also a basis for developing fuzzy calculus and other areas of mathematics that involve numbers. All these developments are now quite advanced, but they are beyond the scope of this brief overview (Dubois and Prade, 1999, chapter 11).

4. FUZZY RELATIONS

When fuzzy sets are defined on universal sets that are Cartesian products of two or more sets, we refer to them as *fuzzy relations*. Individual sets in the Cartesian product of a fuzzy relation are called *dimensions* of the relation. When n-sets are involved in the Cartesian product, we call the relation n-dimensional ($n \geq 2$). Fuzzy sets may be viewed as degenerate, one-dimensional relations.

All concepts and operations applicable to fuzzy sets are applicable to fuzzy relations as well. However, fuzzy relation involve additional concepts and operations due to their multi-dimensionality. Among the additional operations, two of them are applicable to any n-dimensional fuzzy relations ($n \geq 2$). They are called projections and cylindric extensions. For the sake of simplicity, they are discussed here in terms of three-dimensional relations; a generalization to higher dimensions is quite obvious.

Let R denote a three-dimensional (ternary) fuzzy relation on $X \times Y \times Z$. A *projection* of R is an operation that converts R into a lower-dimensional fuzzy relation, which in this case is either a two-dimensional or one-dimensional (degenerate) relation. In each projection, some dimensions are suppressed (not recognized) and the remaining dimensions are consistent with R in the sense that each α-cut of the projection is a projection of α-cut of R in the sense of classical set theory. Formally, the three two-dimensional projection of R on $X \times Y$, $X \times Z$, and $Y \times Z$, R_{XY}, R_{XZ}, and R_{YZ}, are defined for all $x \in X$, $y \in Y$, $z \in Z$ by the following formulas:

$$R_{XY}(x,y) = \max_{z \in Z} R(x,y,z),$$
$$R_{XZ}(x,z) = \max_{y \in Y} R(x,y,z),$$
$$R_{YZ}(y,z) = \max_{x \in X} R(x,y,z).$$

Moreover, the three one-dimensional projections of R on X, Y, and Z, R_X, R_Y, and R_Z, can be then obtained by similar formulas from the two-dimensional projections:

$$\begin{aligned} R_X(x) &= \max_{y \in Y} R_{XY}(x,y) \\ &= \max_{z \in Z} R_{XZ}(x,z) \\ R_Y(y) &= \max_{x \in X} R_{XY}(x,y) \\ &= \max_{z \in Z} R_{YZ}(y,z) \\ R_Z(z) &= \max_{x \in X} R_{XZ}(x,z) \\ &= \max_{y \in Y} R_{YZ}(y,z) \end{aligned}$$

Any relation on $X \times Y \times Z$ that is consistent with a given projection of R is called an *extension* of R. The largest among the extensions is called a *cylindric extension*. Let R_{EXY} and R_{EX} denote the cylindric extensions of projections R_{XY} and R_X, respectively. Then, R_{EXY} and R_{EX} are defined for all triples $\langle x,y,z \rangle \in X \times Y \times Z$ by the formula

$$R_{EXY}(x,y,z) = R_{XY}(x,y),$$
$$R_{EX}(x,y,z) = R_X(x).$$

Cylindric extensions of the other two-dimensional and one-dimensional projections are defined in a similar way. This definition of cylindric extension for fuzzy relations is a cutworthy generalization of the classical concept of cylindric extension.

Given any set of projections of a given relation R, their standard fuzzy intersection (expressed by the minimum operator) is called a *cylindric closure* of the projections. This is again a cutworthy concept. Regardless of the given projections, it is guaranteed that their cylindric closure contains the fuzzy relation R.

Projections, cylindric extensions, and cylindric closures are the main operations for dealing with n-dimensional relations. For dealing with binary relations, an additional important operation is a *relational composition*.

Consider two binary fuzzy relations P and Q that are defined on set $X \times Y$ and $Y \times Z$, respectively. Any such relations, which are connected via the common set Y, can be composed to yield a relation on $Y \times Z$. The standard composition of these relations, which is denoted by $P \circ Q$, produces a relation R on $X \times Z$ defined by the formula

$$R(x,z) = (P \circ Q)(x,z) = \max_{y \in Y} \min[P(x,y), Q(y,z)] \tag{4}$$

for all pairs $\langle x,z \rangle \in X \times Z$.

Other definitions of a composition of fuzzy relations, in which the min and max operations are replaced with t-norms and t-conorms, respectively, are possible and useful in some applications. All compositions are associative:

$$(P \circ Q) \circ R = P \circ (Q \circ R).$$

However, the standard fuzzy composition is the only one that is cutworthy.

Equations (4), which describe $R = P \circ Q$ are called *fuzzy relation equations*. Normally, it is assumed that P and Q are given and R is determined by (4). However, two *inverse problems* play important roles in many applications. In one of them R and P are given and Q is to be determined; in the other one, R and Q are given and P is to be determined. Various methods for solving these problems exactly as well as approximately have been developed (Di Nola et al., 1989; Dubois and Prade, 1999, chapter 6).

It should also be mentioned that cutworthy fuzzy counterparts of the various classical binary relations on $X \times X$, such as equivalence relations, and the various ordering relations, have been extensively investigated. However, many types of fuzzy relations on $X \times X$ that are not cutworthy have been investigated as well and found useful in many applications (Dubois and Prade, 1999, chapter 5).

5. FUZZY SYSTEMS

In general, each classical system is ultimately a set of variables together with a relation among states (or values) of the variables. When states of variables are fuzzy sets, the system is called a *fuzzy system*. In most typical fuzzy systems, the states are fuzzy intervals that represent linguistic terms such as *very small*, *small*, *medium*, *large*, etc., as interpreted in the context of a particular application. If they do, the variables are called *linguistic variables*.

Each linguistic variable is defined in terms of a base variable, whose values are usually real numbers within a specific range. A base variable is a variable in the classical sense, as exemplified by any physical variable. Some examples of base variables relevant to geology are tidal range, grain size, temperature, distance from source, water depth, and rainfall.

Linguistic terms involved in a linguistic variable are used for approximating the actual values of the associated base variable. Their meanings are captured, in the context of each particular application, by appropriate fuzzy intervals. That is, each linguistic variable consists of:

- a *name*, which should reflect the meaning of the base variable involved;
- a *base variable* with its *range of values* (a closed interval of real numbers);
- a set a *linguistic terms* that refer to values of the base variable;
- a set of *semantic rules*, which assign to each linguistic term its meaning in terms of an appropriate fuzzy interval defined on the range of the base variable.

An example of a linguistic variable is shown in Fig. 2. Its name 'growth rate' captures the meaning of the associated base variable – a variable that expresses the coral reef growth rate in millimeters per year. The range of the base variable is $[0, 15]$. Five linguistic states are distinguished by the linguistic terms *very slow*, *slow*, *medium*, *fast*,

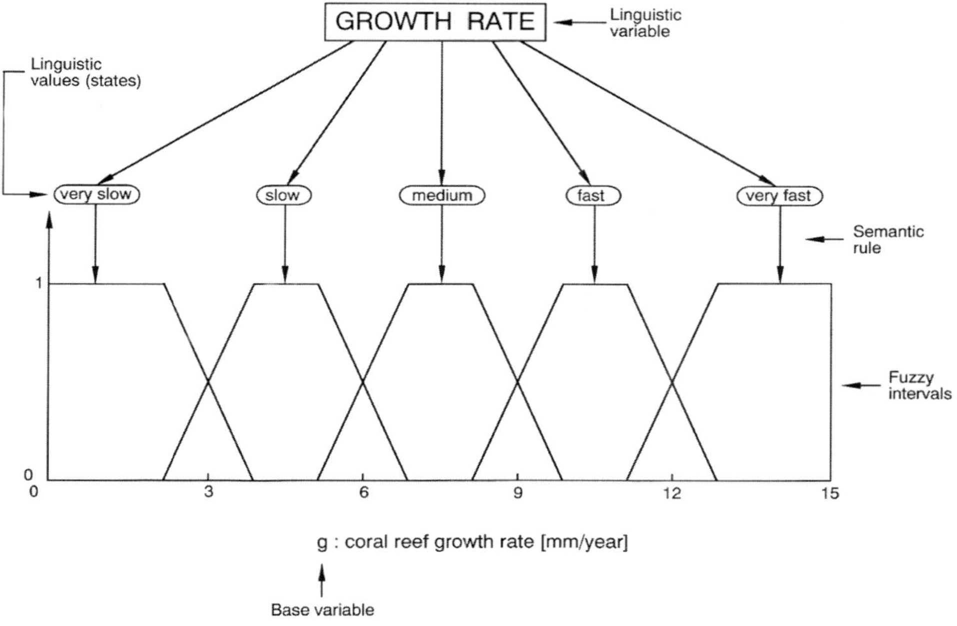

Fig. 2. An example of a linguistic variable.

and *very fast*. The meaning of each of these terms is represented by a trapezoid-shape fuzzy interval, as shown in Fig. 2.

In principle, fuzzy systems can be knowledge-based, model-based, or hybrid. In *knowledge-based fuzzy systems*, relationships between variables are described by a collections of *fuzzy if–then rules* (conditional fuzzy propositional forms). These rules attempt to capture knowledge of a human expert, expressed often in natural language. *Model-based fuzzy systems* are based on traditional systems modelling, but they employ appropriate areas of fuzzy mathematics (fuzzy analysis, fuzzy geometry, etc.). *Hybrid fuzzy systems* are combinations of knowledge-based and model-based fuzzy systems. At this time, knowledge-based fuzzy systems are more developed than model-based or hybrid fuzzy systems.

As already mentioned, the relationship between input and output linguistic variables in each knowledge-based fuzzy system is expressed in terms of a set of fuzzy if–then rules. From these rules and any fact describing actual states of input variables, the actual states of output variables are derived by appropriate rules of fuzzy inference. Before discussing these rules, we need to clarify the meaning of fuzzy propositions.

6. FUZZY PROPOSITIONS

To establish a connection between fuzzy set theory and fuzzy logic, it is essential to connect degrees of membership in fuzzy sets with degrees of truth of fuzzy propositions.

This can only be done when the degrees of membership and the degrees of truth refer to the same objects. Let us consider first the simplest connection, in which only one fuzzy set is involved.

Given a fuzzy set A, its membership degree $A(x)$ for any x in the underlying universal set X may be interpreted as the degree of truth of the associated fuzzy proposition 'x is a member of A.' Conversely, given an arbitrary proposition of the simple form 'x is A,' where x is from X and A is a fuzzy set that represent an inherently vague linguistic term (such as *low*, *high*, *near*, *fast*, etc.), its degree of truth may be interpreted as the membership degree of x in A. That is, the degree of truth of the proposition is equal to the degree with which x belongs to A.

This simple correspondence between membership degrees and degrees of truth, which conforms well to our intuition, forms a basis for determining degrees of truth of more complex propositions. Moreover, negations, conjunctions, and disjunctions of fuzzy propositions are defined under this correspondence in exactly the same way as complement, intersections, and unions of fuzzy sets, respectively.

Let us examine now basic propositional forms of fuzzy propositions. To do that, let us introduced a convenient notation. Let X, Y denote base variables whose states (values) are in sets X, Y, respectively, and let A, B denote fuzzy sets on X, Y, respectively, which represent specific linguistic states (*slow*, *fast*, *shallow*, *deep*, etc.) of linguistic variables associated with X, Y.

Using this notation, the simplest fuzzy proposition (introduced already in this section) can always be expressed in the following canonical propositional form:

$$p_A: \quad \mathsf{X} \text{ is } A$$

Given this propositional form, a fuzzy proposition, $p_A(x)$, is obtained when a particular object (value) from X is substituted for variable X in the propositional form. That is,

$$p_A(x): \quad x \text{ is } A,$$

where $x \in X$, is a particular fuzzy proposition of propositional form p_A. For simplicity, let $p_A(x)$ denote also the degree of truth of the proposition 'x is A.' This means that the symbol p_A denotes a propositional form as well as a function by which degrees of truth are assigned to fuzzy propositions based on the form. This double use of the symbol p_A does not create any ambiguity since there is only one function for each propositional form that assigns degrees of truth to individual propositions subsumed under the form. In this case, the function is defined for all $x \in X$ by the simple equation

$$p_A(x) = A(x).$$

The propositional form p_A may be modified by qualifying the claims for the degree of truth of the associated fuzzy propositions. Two types of qualified propositional forms are recognized:

- *truth-qualified propositional form*

  $$p_{T(A)}: \quad \mathsf{X} \text{ is } A \text{ is } T,$$

  where T is a fuzzy set defined on $[0,1]$, called a *fuzzy truth qualifier*, which represents a linguistic term (such as *very true*, *fairly true*, *false*, *fairly false*, etc.) that qualifies the meaning of degrees of truth of fuzzy propositions associated with given propositional form.

- *probability-qualified propositional form*

$$p_{P(A)}:\quad \text{Pro}\{\mathsf{X} \text{ is } A\} \text{ is } P,$$

where Pro{X is A} denotes the probability of the fuzzy event 'X is A,' and P is a fuzzy set defined on $[0,1]$, called a *probability qualifier*, which represents a linguistic term (such as *likely*, *very likely*, *extremely unlikely*, etc.) that qualifies the claims of individual propositions associated with the propositional form; the probability Pro{X is A} is determined for finite X by the formula

$$\text{Pro}\{\mathsf{X} \text{ is } A\} = \sum_{x \in X} A(x) f(x),$$

where f is given (known) classical probability distribution function, and by formula

$$\text{Pro}\{\mathsf{X} \text{ is } A\} = \int_{x} A(x) f(x)\, \mathrm{d}x,$$

where f is a given probability density function, when X is an interval of real numbers (Zadeh, 1968).

To obtain the degree of truth of a qualified proposition of either type, we need to compose A with the respective qualifier. That is, for all $x \in X$,

$$p_{TA}(x) = T\big(A(x)\big),$$
$$p_{PA}(x) = P\big(A(x)\big)$$

and, if both qualifiers are involved,

$$p_{TPA}(x) = T\big(P\big(A(x)\big)\big).$$

An important type of fuzzy propositions, which are essential for knowledge-based fuzzy systems, are *conditional fuzzy propositions*. They are based on the propositional form

$$p_{B|A}:\quad \text{If } \mathsf{X} \text{ is } A, \text{ then } \mathsf{Y} \text{ is } B.$$

These propositions may also be expressed in an alternative, but equivalent form

$$p_{B|A}:\quad (\mathsf{X}, \mathsf{Y}) \text{ is } R,$$

where R is a fuzzy relation on $X \times Y$. It is assumed here that R is determined for each $x \in X$ and each $y \in Y$ by the formula

$$R(x, y) = I\big(A(x), B(x)\big),$$

where the symbol I stands for a binary operation on $[0,1]$ that represents in the given application context an appropriate fuzzy implication. Clearly,

$$p_{B|A}(x, y) = R(x, y)$$

for all $(x, y) \in X \times Y$. Moreover, if a truth qualification or a probability qualification is employed, R must be composed with the respective qualifier to obtain for each $(x, y) \in X \times Y$ the degree of truth of the conditional and qualified proposition.

As is well known, operations that qualify as fuzzy implications form a class of binary operations on $[0,1]$, similarly as fuzzy intersections and fuzzy unions (Klir and Yuan, 1995, chapter 11). An important class of fuzzy implication, referred to as *Lukasiewicz implications*, is defined for each $a \in [0,1]$ and each $b \in [0,1]$ by the formula

$$I(a,b) = \min[1, 1 - a^{\lambda} + b^{\lambda})^{1/\lambda},$$

where $\lambda > 0$ is a parameter by which individual implications are distinguished from one another.

Fuzzy propositions of any of the introduced types may also be quantified. In general, fuzzy quantifiers are fuzzy intervals. This subject is beyond the scope of this introduction; basic ideas are summarized in the text by Klir and Yuan (1995).

7. APPROXIMATE REASONING

Reasoning based on fuzzy propositions of the various types is usually referred to as *approximate reasoning*. The most fundamental components of approximate reasoning are conditional fuzzy propositions, which may also be truth qualified, probability qualified, qualified, or any combination of these. Special procedures are needed for each of these types of fuzzy propositions. This great variety of fuzzy propositions make approximate reasoning methodologically rather intricate. This reflects the richness of natural language and the many intricacies of common-sense reasoning, which approximate reasoning based upon fuzzy set theory attempts to model.

To illustrate the essence of approximate reasoning, let us characterize the fuzzy-logic generalization of one of the most common inference rules of classical logic: modus ponents. The *generalized modus ponents* is expressed by the following schema:

Fuzzy rule:	If X is A, then Y is B
Fuzzy fact:	X is F
Fuzzy conclusion:	Y is C

Clearly, A and F in this schema are fuzzy sets defined on X, while B and C are fuzzy sets defined on Y. Assuming that the fuzzy rule is already converted to the alternative form

$$(\mathsf{X},\mathsf{Y}) \text{ is } R,$$

where R represents the fuzzy implication employed, the fuzzy conclusion C is obtained by composing F with R. That is

$$B = F \circ R$$

or, more specifically,

$$B(y) = \max_{x \in X} \min[F(x), R(x,y)]$$

for all $y \in Y$. This way of obtaining the conclusion according to the generalized modus ponens schema is called a *compositional rule of inference*.

To use the compositional rule of inference, we need to choose a fitting fuzzy implication in each application context and express it in terms of a fuzzy relation R. There are several ways in which this can be done. One way is to derive from the application context (by observing or expert's judgements) pairs F, C of fuzzy sets that are supposed to be inferentially connected (facts and conclusions). Relation R, which represents a fuzzy implication, is then determined by solving the inverse problem of fuzzy relation equations. This and other issues regarding fuzzy implications in approximate reasoning are discussed fairly thoroughly in the text by Klir and Yuan (1995).

In knowledge-based fuzzy systems, the relation between a set of input variables and a set of output variables is expressed in terms of a set of fuzzy if–then rules (conditional propositional forms) such as

If X_1 is A_1, X_2 is $A_2, \ldots,$ and X_n is A_n, then Y_1 is B_1, Y_2 is $B_2, \ldots,$and Y_m is B_m.

States of input variables as well as output variables in each rule are combined by an operation of fuzzy intersection. The rules, which are usually interpreted as disjunctive, are combined by an operation of fuzzy union. It is convenient to convert the rules into their relational forms (as illustrated in the case of generalized modus ponents). Given an input state in the form

X_1 is F_1 and X_2 is $F_2, \ldots,$ and X_n is F_n,

the output state of the form

Y_1 is C_1 and Y_2 is $C_2, \ldots,$ and Y_m is C_m

is then derived by composing the input state with relation representing the rules.

The result of each fuzzy inference that involves numerical variables is a fuzzy set defined on the set of real numbers. If needed, this fuzzy set is converted to a single real number by a *defuzzification method*. The number, $\mathrm{d}(A)$, obtained by any defuzzification method should be the best representation, in the context of each application, of the given fuzzy set A. The most common defuzzification method, which is called a *centroid method*, is defined by the formula

$$\mathrm{d}(A) = \frac{\displaystyle\int_R xA(x)\,\mathrm{d}x}{\displaystyle\int_R A(x)\,\mathrm{d}x}$$

or, when A is defined on a finite universal set $X = \{x_1, x_2, \ldots, x_n\}$, by the formula

$$\mathrm{d}(A) = \frac{\displaystyle\sum_{i=1}^{n} x_i A(x_i)}{\displaystyle\sum_{i=1}^{n} A(x_i)}.$$

A good overview of various other defuzzification methods was prepared by Van Leekwijck and Kerre (1999).

8. SUGGESTIONS FOR FURTHER STUDY

As is well known, the idea of fuzzy set theory was introduced in the mid 1960s by Lotfi Zadeh (1965). Since the publication of this seminal paper, Zadeh has originated most of the key ideas that advanced the theory and has conceived of many of its applications. Fortunately, his crucial role in the development of fuzzy set theory and fuzzy logic is now well documented by two volumes of his selected papers in the period 1965–1995 (Yager et al., 1987; Klir and Yuan, 1996). These volumes are indispensable for thorough understanding of the field and its development.

It is also fortunate that several broad textbooks on fuzzy set theory and fuzzy logic are now available, including an undergraduate textbook (Klir et al., 1997) and several graduate textbooks (Kandel, 1986; Novák, 1986; Klir and Yuan, 1995; Zimmermann, 1996; Nguyen and Walker, 1997; Pedrycz and Gomide, 1998). Two comprehensive and thorough encyclopedic resources are now available: (i) *The Handbooks of Fuzzy Sets Series* published by Kluwer, which consists now of seven volumes; and (ii) a large *Handbook of Fuzzy Computation* edited by Ruspini et al. (1998).

There are many books on knowledge-based fuzzy systems, but most of them are oriented to fuzzy control. Three excellent books with emphasis on issues of systems modeling were written by Babuška (1998), Hellendoorn and Driankov, 1997 (and Yager and Filev (1994)). The recent book by Mendel (2001) contains the first comprehensive treatment of fuzzy systems based on second-order fuzzy sets – sets in which degrees of membership are expressed by fuzzy intervals. A specialized book on fuzzy logic for geologists was put together by Demicco and Klir (2003).

The prime journal in the field is *Fuzzy Sets and Systems*, which is sponsored by the International Fuzzy Systems Association.

REFERENCES

Babuška, R., 1998. *Fuzzy Modeling for Control*. Kluwer, Boston, MA.

Bezdek, J.C., Dubois, D. and Prade, H. (Eds.), 1999. *Fuzzy Sets in Approximate Reasoning and Information Systems. Handbooks of Fuzzy Sets, Vol. 3*. Kluwer, Boston, MA.

Demicco, R.V. and Klir, G.J., 2003. *Fuzzy Logic in Geology*. Academic Press, San Diego, CA.

Di Nola, A., Sessa, S., Pedrycz, W. and Sanches, E., 1989. *Fuzzy Relation Equations and Their Applications to Knowledge Engineering*. Kluwer, Boston, MA.

Dubois, D. and Prade, H. (Eds.), 1999. *Fundamentals of Fuzzy Sets. Handbooks of Fuzzy Sets, Vol. 1*. Kluwer, Boston, MA.

Hellendoorn, H. and Driankov, D. (Eds.), 1997. *Fuzzy Model Identification: Selected Approaches*. Springer-Verlag, New York, NY.

Kandel, A., 1986. *Fuzzy Mathematical Techniques with Applications*. Addison-Wesley, Reading, MA.

Klir, G.J., 1997. Fuzzy arithmetic with requisite constraints. *Fuzzy Sets Syst.*, 91(2): 165–175.

Klir, G.J., 1999. On fuzzy-set interpretation of possibility theory. *Fuzzy Sets Syst.*, 108(3): 263–273.

Klir, G.J., 2001. Foundations of fuzzy set theory and fuzzy logic: a historical overview. *Int. J. General Syst.*, 30(2): 91–134.

Klir, G.J. and Pan, Y., 1998. Constrained fuzzy arithmetic: basic questions and some answers. *Soft Comput.*, 2(2): 100–108.

Klir, G.J. and Yuan, B. (Eds.), 1995. *Fuzzy Sets and Fuzzy Logic: Theory and Applications*. Prentice-Hall, PTR, Upper Saddle River, NJ.

Klir, G.J. and Yuan, B. (Eds.), 1996. *Fuzzy Sets, Fuzzy Logic, and Fuzzy Systems: Selected Papers by Lotfi A. Zadeh.* World Scientific, Singapore.

Klir, G.J., St. Clair, U.H. and Yuan, B., 1997. *Fuzzy Set Theory: Foundations and Applications.* Prentice-Hall PTR, Upper Saddle River, NJ.

Kosko, B., 1993. *Fuzzy Thinking: The New Science of Fuzzy Logic.* Hyperion, New York, NY.

Mendel, J.M., 2001. *Uncertain Rule-Based Fuzzy Logic Systems.* Prentice Hall PTR, Upper Saddle River, NJ.

Moore, R.E., 1966. *Interval Analysis.* Prentice-Hall, Englewood Cliffs, NJ.

Neumaier, A., 1990. *Interval Methods for Systems of Equations.* Cambridge Univ. Press, Cambridge, UK.

Nguyen, H.T. and Walker, E.A., 1997. *A First Course in Fuzzy Logic.* CRC Press, Boca Raton, FL.

Nguyen, H.T. and Sugeno, M. 2000. *Fuzzy Systems: Modeling and Control. Handbooks of Fuzzy Sets, Vol. 7.* Kluwer, Boston, MA.

Novák, V., 1986. *Fuzzy Sets and Their Applications.* Adam Hilger, Philadelphia.

Pedrycz, W. and Gomide, F., 1998. *An Introduction to Fuzzy Sets: Analysis and Design.* MIT Press, Cambridge, MA.

Ruspini, E.H., Bonissone, P.P. and Pedrycz, W. (Eds.), 1998. *Handbook of Fuzzy Computation.* Institute of Physics Publ., Bristol (UK) and Philadelphia, PA.

Van Leekwijck, W. and Kerre, E.E., 1999. Defuzzification: criteria and classification. *Fuzzy Sets Syst.*, 108(2): 159–178.

Yager, R.R. and Filev, D.P., 1994. *Essentials of Fuzzy Modeling and Control.* John Wiley, New York, NY.

Yager, R.R., Ovchinnikov, S., Tong, R.M. and Nguyen, H.T. (Eds.), 1987. *Fuzzy Sets and Applications – Selected Papers by L.A. Zadeh.* John Wiley, New York.

Zadeh, L.A., 1965. Fuzzy Sets. *Inf. Control*, 8(3): 338–353.

Zadeh, L.A., 1968. Probability measures of fuzzy events. *J. Math. Anal. Appl.*, 23: 421–427.

Zadeh, L.A., 1975. The concept of a linguistic variable and its application to approximate reasoning. *Inf. Sci.*, 8: 199–249; 301–357; 9: 43–80.

Zimmermann, H.J., 1996. *Fuzzy Set Theory and Its Applications* (3rd edition). Kluwer, Boston, MA.

Developments in Petroleum Science, 51
Editors: M. Nikravesh, F. Aminzadeh and L.A. Zadeh

Chapter 3

INTRODUCTION TO USING GENETIC ALGORITHMS

J.N. CARTER

Department of Earth Science and Engineering, Imperial College of Science Technology and Medicine, South Kensington, London, SW7 2BP, UK

1. INTRODUCTION

> What exactly is a Genetic Algorithm, what sort of problems can it solve, or is it just another over-hyped algorithm?

These are the three questions that are most often asked of users of Genetic Algorithms, about Genetic Algorithms. In this chapter I will attempt to give brief answers to the three questions. By the end of the chapter it is hoped that the reader will: know when the algorithm might be used to solve problems in earth sciences; known how to set up a Genetic Algorithm (GA) and be aware of the design issues involved in its use.

In the next section I will attempt to answer the second and third of the questions and review the general background of the Genetic Algorithm. This is followed by two sections that cover the first of the questions and describes the structure and design of a Genetic Algorithm. Finally there will be some conclusions and references.

What follows is a personal view of how to go about using a GA. It should all be viewed as a starting point, rather than a definitive statement. In writing this chapter, I have attempted to write down the starting point I would give to a research student with no experience of GAs. I have not attempted to give an exhaustive set of references, as these are available elsewhere. In many places I make suggestions as to how things should be done, without justifying these suggestions with references to the published literature, or with examples within the text. My answer to the obvious criticism that this is unscientific is that this is how I tackle problems currently based on a decade of using GAs on a variety of problems. Whenever I find something in the literature that appears to offer a benefit, it gets tested, modified if I think necessary, and if it proves useful it forms part of my arsenal to tackle new problems.

2. BACKGROUND TO GENETIC ALGORITHMS

Let me start with the third question first:

> . . . are genetic algorithms just another over-hyped algorithm?

In my opinion, GAs have suffered the same fate as many other new techniques, e.g. artificial neural networks. The early practitioners found that the method worked well on certain problems, provided that it had been properly designed. This was followed by a huge increase in its use with generally good results. The next phase is a period when people with limited experience, and limited intention of testing and developing the algorithm, are drawn into applying a 'standard' version of the method to many problems. This results in too many poor results, and the method has gained a reputation of being over-hyped. We are now in a position where the Genetic Algorithm method is generally recognised as having advantages and disadvantages, and that it has its place in our repertory of tools for problem solving, something that the early practitioners always knew, but got lost along the way.

2.1. Advantages and Disadvantages

> . . . what sort of problems can a genetic algorithm solve . . . ?

As with any method, the GA has its advantages and disadvantages. These will in large part determine whether the method is appropriate for solving a particular problem (Table 1). If you know something about your problem that can be exploited to solve it, then a GA is probably not what you need. You might need a GA if your parameter space is: large; not perfectly smooth, or if it is noisy; has multiple local optima; or is not well understood. Two other considerations that I find important are: whether I need lots of quite good solutions, rather than one very good solution; and can I make use of the inherent parallelism of the method. I always find it useful to apply a simple local search algorithm after the GA has been terminated. When presenting GAs to industrialists as a method for solving problems, three of its advantages are considered to be very important, even to the point that they out weigh the possibility of using another method. These advantages are: that the algorithm returns multiple solutions for further consideration, this is important when the model does not capture all of the known behaviour; that the algorithm is very robust, this is important if it cannot be guaranteed that the objective function can always be evaluated successfully; that it is possible to easily parallelise the process, this is attractive as many organisations have

TABLE 1

Advantages and disadvantages of Genetic Algorithms

Advantages	Disadvantages
Only uses function evaluations.	Cannot use gradients.
Easily modified for different problems.	Cannot easily incorporate problem specific information.
Handles noisy functions well.	Not good at identifying local optima.
Handles large, poorly understood search spaces easily.	No effective terminator.
Good for multi-modal problems.	Not effective for smooth unimodal functions.
Returns a suite of solutions.	Needs to be coupled with a local search technique.
Very robust to difficulties in the evaluation of the objective function.	
Easily parallelised.	

many computers doing nothing over-night. If you do decide to use a GA, then it is important to make sure that that the version you choose is appropriate to your problem.

2.2. Review of Genetic Algorithms Literature

Genetic Algorithms are one strand of what is generally termed 'Evolutionary Computation'. The other two main strands are 'Evolutionary Strategies' and 'Evolutionary Programming', there are also many minor strands such as 'Genetic Programming'. GAs were invented by John Holland (1975), and his book 'Adaptation in natural and artificial systems' is generally regarded as the seed from which wide spread research into GAs started. The first international conference was held in 1985, the number of research papers published each year has grown dramatically through the last decade. It is now very difficult to stay abreast of all of the developments that are being reported.

Having read this chapter, where should the interested reader go for more information? My first stopping point would be Melanie Mitchell's: An Introduction to Genetic Algorithms (Mitchell, 1998). This is a very readable introduction to many areas of research into Genetic Algorithms, and contains a good selection of references. The best place to start on the internet is the GA-list website at http://www.aic.nrl.navy.mil/galist.

Of the older books that are often referenced, I would avoid initially Holland's book (Holland, 1975) and Goldberg's book (Goldberg, 1989). Both are interesting with many valuable ideas, but in my view are not a good guide to current practice in many areas of GA research. Davies' (1991): Handbook of Genetic Algorithms is worth an early visit.

3. DESIGN OF A GENETIC ALGORITHM

What exactly is a Genetic Algorithm . . . ?

Genetic Algorithms are a group of closely related algorithms that draw upon ideas of Darwinian evolution and genetics. Almost every implementation will be different, and so it is wrong to think of there being a 'standard' GA. In my own work I usually redesign the details of the algorithm for each new problem.

3.1. Terminology

As with any research area, GA has its own terminology, and any discussion of the topic is made much easier if the standard terminology is used and understood. In this section I briefly describe each of the terms that are commonly used.

Search Space: This is a conceptual space that describes all the solutions that are possible to the problem under investigation. The space only contains information about things that can be changed. Depending on the problem this space might be: a continuous Euclidian space (finite or infinite in extent), or a discrete Euclidian space (with finite or infinite numbers of solutions), a combinatoric space with a finite number of individuals. There is often a 'natural' space that describes the possible solutions, and their relationship to one another, in a way that can easily be interpreted.

It is often possible to translate one space into another, but this may result in the relationship between solutions becoming less clear.

Individual: This refers to the resulting model produced by using any one of the possible solutions from the search space. There is often a blurring of the difference between the individual and a point in search space, as it is common to have a one-to-one mapping between the two.

Population: A collection of individuals form a population.

Genome: The information that defines an individual forms the genome. This is composed of two types, the information that comes from the search space, and all the other information that is needed to construct an individual. This second type of information is constant for all the individuals and hence is normally not explicitly considered. The genome is therefore taken to be a representation of the search space. There is a one-to-one mapping between the genome and the search space.

Chromosome: Within the genome information may may be grouped in some way, with each group being considered separately for some operations. Such a group is referred to as a chromosome. In most GAs the genome has only one chromosome, therefore some blurring of the distinction between the two can occur.

Gene: Within a chromosome a group of numbers may jointly code for a specific trait, e.g. eye colour or length of wing, this grouping is known as a gene. In GAs all the information for a single gene is usually collected together in one place, but this need not be so in every case.

Alleles: This is the set of values that an individual gene can take.

Locus: This is the location of the smallest piece of information held in the genome. If we are using a binary code within the genome, then it is the location of just one of those binary bits. In some cases a gene will consist of just one number, which can be stored at a single locus. More often a gene will be spread across many loci.

Alphabet: This is a list of symbols that may appear at a locus. In a binary code genome, the alphabet is the set $\{0, 1\}$. If a real number code is used, then the alphabet is the set of real numbers between the relevant upper and lower bounds.

Genotype: A collection of alleles form a genotype. Depending on how a gene codes into an alleles, then there might be a one-to-one mapping between genome and genotype. In many cases there is no distinction, or difference, between genotype, chromosome and genome which may lead to some confusion.

Phenotype: In nature the phenotype is the physical representation of the genotype. In GAs it is common for there to be a one-to-one mapping between phenotype and genotype, but examples do exist of one-to-many and many-to-one mappings.

Parent and Offspring: A parent is an individual from the current population, the information in their genome will be used along with the information from other parents (normally one) to generate (breed) an offspring (child).

Selection Pressure: How one selects individuals from the current population to become parents is a key element within the GA. It is this process that drives the population towards better solutions. How this pressure is used can be critical to how quickly the method finds good solutions, and just how good those solutions are.

Crossover: This is the process that takes the information, expressed as the genome, from the parents and produces the genome of the offspring. This process depends primarily on the structure of the genome. A good crossover operator will ensure that most of the information in the offspring's genome is directly inherited from one of the parents, in particular one hopes that whole genes are inherited from one parent and that there is a family resemblance at the phenotype level. A poor crossover operator is likely to randomly mix the information from the two genomes, and cause too many non-inherited traits at the phenotype level.

Mutation: This is a random change to an offspring's genome after crossover has been completed. Normally this occurs very infrequently.

Fitness: This is the name given to the objective function. In biological terms the fitter an organism is the more likely it is to produce offspring. So any GA seeks to increase fitness, and this is done by selecting individuals with high fitness to become parents and hopefully over time the general level of fitness in the population increases. In many applications of GAs the fitness is directly related to some objective function. Even to the point that no distinction is made between the two. In other cases fitness simply allows you to compare two individuals and decide which of them is the fitter on the basis of a more complex rule.

Fitness Landscape: This is the term used to describe how the fitness changes over the search space: hills represent either local or global optima, valleys are areas of poor fitness that separate local optima, plains and plateaus are areas where the fitness function has an almost constant value. We then talk about the population exploring the fitness landscape, and migrating from one area to another.

Viability: Ideally any offspring that is produced, via the selection of parents and the use of crossover, will be able to go on to produce children of their own. Sometimes however an offspring produced in this way will violate some additional constraint, such a child is said to be non-viable. Two approach are used when this happens, either abort the offspring, or try to repair the genome. Neither of these approaches works well. It seems to be much better to choose genome and crossover so that these additional constraints are included automatically and that any offspring is viable. The down side of this approach is that you can end up using a genome structure which causes its own problems. Solving this viability problem can be a key element in making GA work successfully.

Diversity: This is a measure of how a population is spread across the fitness landscape. If there is too little genetic diversity then the population will have converged on to a single local optima and little further progress can be expected. If the genetic diversity is large then fit individuals tend to be scattered over many local optima and there is little to be gained from inheriting information from parents. The search is then about as effective as a random search. We aim to have a level of genetic diversity that balances the amount of exploration and exploitation that takes place.

Generation: In most GAs the size of the population from which parents are drawn is fixed. We go through a cycle of selecting parents, producing offspring and then culling some of the individuals. Finishing with the same population size as we started. This cycle is generally known as a generation, although other definitions are sometimes used. In many cases the population of offspring completely replaces the population of parents, this is known as a generational replacement scheme.

Cloning: In a generational replacement scheme, all of the parents are culled, no matter how fit they are, at the end of each generation. In some applications a parent's genome has a chance of being copied, cloned, directly into the succeeding generation.

Elitism: This is a special case of cloning, where only the best individual is cloned directly into the next generation. This is often done for the simple expedient of knowing that the best known solution is kept in the current population.

Niching: Sometimes it is desirable that the population is made of several distinct subpopulations, where each subpopulation occupies a distinct part,or niche, of the fitness landscape.

3.1.1. Example of the various data structures

To try and illustrate the various data structures described above let us consider the following, rather contrived, example. The search space consists of a four dimensional discrete Euclidian space, with four parameters which can take on the following values:

A	B	C	D
0.1	1.1	0	0.005
0.3	1.2	1	0.010
0.7	1.3	2	0.015
1.0	1.4	3	0.020

There are 256 solutions in this search space, and a GA would not be the normal solution method for a problem of this size. For this problem we will use the traditional binary strings as described in detail later, to encode this information into the genome. So each of our variables can be encoded using two digits. We will however use a complex structure for the genome, which will consist of two chromosomes:

Chromosome 1

a_1	a_2
b_1	b_2

Chromosome 2

c_1	c_2	d_1	d_2

The four genes, which correspond to the four parameters, are marked by the lower case letters a, b, c and d. The alleles for genes a, b and d are simply the four values given in the search space. The alleles for gene c are: red, yellow, green and blue. Let us now consider a specific instance instance of the gene

Chromosome 1

0	1
0	0

Chromosome 2

1	1	1	1

The genotype that corresponds to this genome is $(0.3, 1.1, \text{Blue}, 0.020)$. From the genotype we need to construct the phenotype, which means we need to know more about the thing that we are describing. In this case the variables describe a wall: its height at various point along its length, its colour and its thickness. We now consider the different ways of constructing the phenotype.

Phenotype 1: This model produces a one-to-one mapping between genotype and phenotype. Variable 1 is the height at the left hand end of the wall; variable 2 is the height at the right hand end; the height of the wall changes linearly between the two ends; variable 4 is the wall thickness; and variable 3 is the colour the wall is painted.

Phenotype 2: This model produces a one-to-many mapping between genotype and phenotype. Variables 1–4 have the same meaning as in the previous case. However the height of the wall half way between the the ends is now randomly chosen to be between the heights at either end, i.e. the wall is not completely determined by its genome. The wall is constructed so that the height varies linearly between the left hand end and the middle, and again linearly between the middle and the right hand end.

Phenotype 3: This model produces a many-to-one mapping between the genotype and phenotype. Variables 3 and 4 have the same meaning as before. This time the height of the wall is constant along its length and is given by the mean of variables 1 and 2.

How data is represented, manipulated and used within the various parts of the GA is important to the overall performance obtained. The structure and choices related to genome design are discussed in greater detail later.

3.2. Basic Structure

There is an underlying structure that is common to most, if not all, implementations of a GA and this is illustrated in Fig. 1. The first important difference that you will notice between a GA and most other optimisation algorithms is that it uses a population of solutions, rather than a single 'current' solution. The process proceeds by selecting individuals from the population to become parents, from these parents you then breed children, or offspring, using the crossover and mutation operators, these children can then go into the population and the process can be repeated.

To implement a GA it is necessary to make a number of decisions about how to represent solutions, how to manipulate information and how the population is maintained. In my opinion there are eight decisions that need to be considered when designing a GA, these are, in order of importance:

Structure of the genome: How is the information that describes an individual stored. In some problems this may be quite straight forward, but it is possible to encode the information so as to include known properties of the problem. Often the difference between a good GA and a poor GA is determined by the decisions made about the genome.

Crossover operator: Given the structure of the genome and two, or possibly more, parents, how do you combine the information of the parents to produce a child. It is important that the child produced is viable, it may be required to have certain properties if it is to be a valid solution. A good crossover operator will ensure that a child will inherent most of its features from one or other parent, a poor crossover operator will introduce too many random changes. There is usually a balance between the design of the genome and the design of the crossover operator.

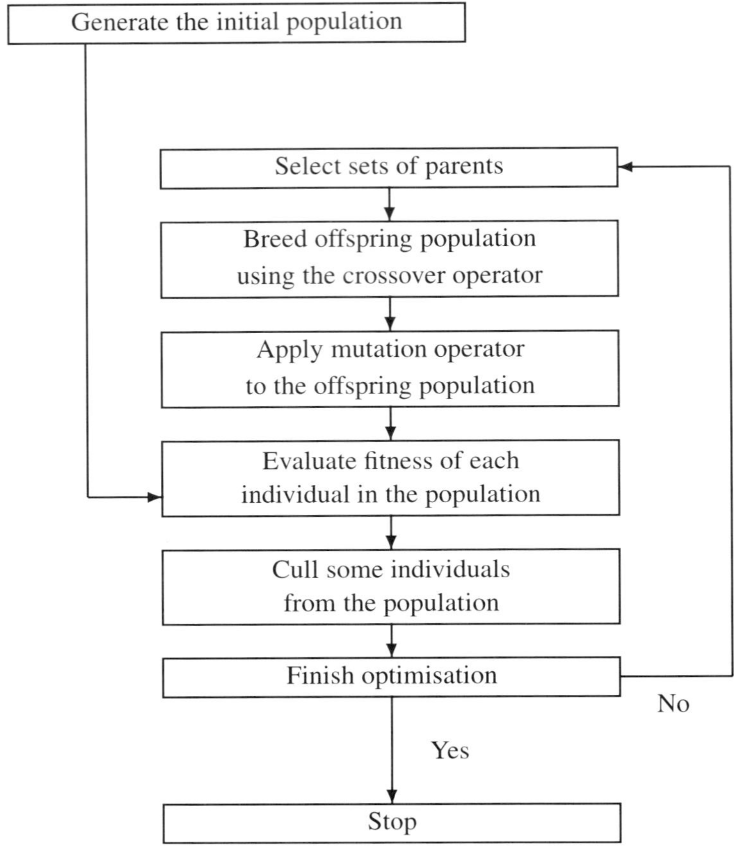

Fig. 1. The general structure of a Genetic Algorithm.

Selection of parents: The progress of a GA is controlled to a large extent by the method of selecting parents from the population that is available. In general, parents are selected on a random basis, with the probability that an individual being selected being a function of its own fitness and the fitness of all other individuals in the population. If no selection pressure is used when selecting parents, then one is not selecting for any particular trait or property, and effectively a random search is carried out. If the selection pressure is to high, and only a few individuals are selected as parents from the population, then the available information is exploited rapidly, but little exploration takes place. This tends to result in rapid convergence to a poor local optima. Depending on the aims and the time available, then selection pressure can be changed to achieve a different final result.

Construction of new populations: As the GA progresses we go around a cycle of having a population of parents, creating a population of offspring, whose size may be greater or less than the parent population, and then combining these two populations to form the parent population for the next cycle. At one end of the options available we have generation replacement, where all the parents are culled and replaced by offspring. At the other end we select just the fittest individuals from both populations. The balance we are seeking is between exploration and exploitation. If we keep only the best individuals, then we maximise our ability to exploit information about the fitness landscape sampled by the population. But we reduce the population's ability to explore the landscape. If we replace all of the parents then we make it harder to exploit information already gathered, but increase the ability for exploration to take place. A balance needs to be found so as to achieve the right amounts of exploration and exploitation. This is why elitism in generational replacement schemes works so well. We maximise the amount of exploration, whilst retaining information about the best place found so far.

Mutation operator: This is a way of introducing information into the population, that does not already exist. It is usually carried out at a very low background level and consists of making a random change to one of the offspring. In studies to test the method it is usually found to have some benefit, but to high a level of mutation can be detrimental to the overall behaviour of the system. It is often thought of as a device for maintaining genetic diversity, although it cannot overcome to use of too much selection pressure.

Population size: A large population allows the search space to be explored, the down side is that many of the individuals will tend not to be very fit and it takes many generations for the population to converge on a good area in the search space. The positive aspect to this is that the population tends to converge on a very good area having had the opportunity to search effectively. A small population tends to converge on a small area of search space quickly, with fewer unfit individuals. However the solution may not be as good as one found with a larger population.

Initial population generation If the time taken to perform an optimisation was not an issue, then a large population well spread through the search space should give the best results. This is not normally the case, so we tend to use a smaller population concentrated in that part of search space that is likely to contain the optimum. We need to find a balance between, sufficient genetic diversity to allow adequate exploration, and sufficient useful information to be available for efficient exploitation. It can be important how the initial population is generated. One method is to use a combination of randomly generated individuals and user specified individuals. The danger with this approach is that the search is biased towards the solution you first thought of by the inclusion of individuals that are significantly fitter than randomly generated individuals. In this case many of the benefits of using a GA is lost.

Parameter settings: There are many parameters to be set within the overall structure of the GA, which can make it seem an unfriendly algorithm for newcomers. It seems that the performance is fairly insensitive to most of these, provided that general guidelines are followed. The efficiency of a GA for a specified problem can be improved by experimenting with the parameters, however the gains are often small and of questionable value given the effort required to achieve them. It usually pays more dividends to concentrate on items at the top of this list.

In this list I have described the main areas that need to be considered when designing a GA and briefly commented on some of the consequences of the choices. In the following sections I consider each area in turn, examine some of the design options available and the consequences of different choices.

3.3. Structure of the genome

The role of the genome is to gather together the information needed to construct the phenotype of an individual. This needs to be done in such a way that when the crossover operator is applied, a viable offspring is produced, with a high level of inheritance at the genotype level. The genome design should also try to preserve as much information about relationships between genes as possible by collecting related genes together in the same chromosome, and even in the same part of the chromosome. The structure of a chromosome can be anything that helps retain important relationships between genes.

If you were to look through the GA literature you could find many different structures used as the genome. The most common are strings of binary digits, there are also strings constructed from other alphabet, the use of real numbers is also fairly common. Some problems that use GAs have quite different structures, an area of research known as ‘Genetic Programming’ (Koza, 1992) uses parse trees, as illustrated in Fig. 2, as the chromosome. Problems from the area of combinatoric optimisation use lists of operations, e.g. (a, g, e, f, d, b, c), where each symbol appears once and only once. A problem from the nuclear industry (Carter, 1997) uses a grid of unique symbols, see Fig. 3. Neural networks have been evolved using the network structure as the chromosome. All of these different chromosomes have one thing in common, for the problem being considered the representation is in some sense a ‘natural’ representation.

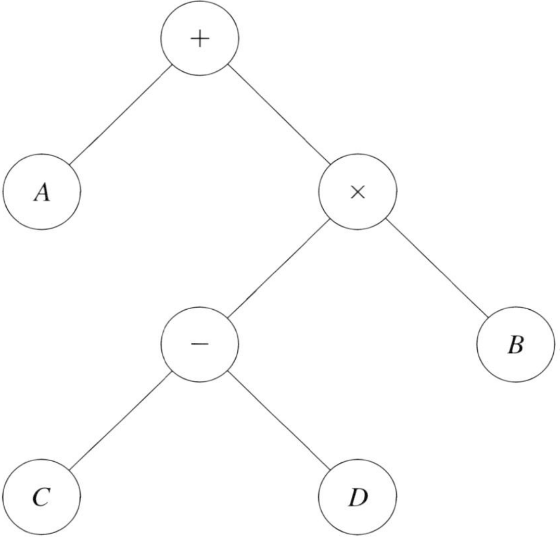

Fig. 2. A parse tree or the function $A + (C - D) \times B$.

A	J	E	K
L	B	F	G
I	H	D	
M	C		

Fig. 3. Genome for a nuclear core reload problem.

However, there is no formal definition of what constitutes a natural representation. The guiding principle that I use is that the representation should be as simple and as meaningfully as possible, and that under crossover and mutation operators, relationships between genes should be preserved. It is likely that the reader of this chapter will be interested in solving problems involving real numbers, so I will now consider the options available for constructing a suitable genome.

Let us assume that all of the variables that we are interested in are real numbers between known upper and lower bounds, and that the search space is a simple hyper-cube in a high dimensional Euclidian space. this is a very common form for a problem to take, we have m variables which take on values

$$x_i(\min) <= x_i <= x_i(\max) \tag{1}$$

There are essentially two options available to us to encode each variable, either we use a real number or we use a binary string which can be decoded to give a number between the known upper and lower bounds. This latter method was the norm in almost all the early work done on GAs, and most of the older texts concentrate on this approach. The reason for this was some early theoretical analysis which suggested that the smaller the alphabet used to encode the problem, the better the results. A binary alphabet is as small as you can get, therefore it should have given the best results. However there have been many examples since which appear to work better with larger alphabets, in particular the use of real numbers. My normal practice is to use real numbers, unless the use of a binary string is going to give me some additional useful control over the optimisation process. If I choose to use binary strings, then there are two options to consider, standard binary encoding or Gray encoding, these are described below.

Let us now consider the following example, as shown in diagrammatic form in Fig. 4, involving the estimation of some variables as part of a seismic experiment. We have a source, S, and three detectors at distances d_1, d_2 and D from the source as shown. We are trying to determine the depth, H, and thickness, h, of a geological layer. The velocities of the seismic waves, (V_p, V_s), also need to be determined. The distance D is known exactly, but all other variables are only known to lie between a lower and upper bound, so the search space is the six dimensional Euclidian space with variables $(d_1, d_2, H, h, V_p, V_s)$. As all six variables are completely independent, there is no advantage in using a complex genome structure since there are no relationships to capture. The most natural way to group the genes is as a linear string. We therefore have three options: real number string, binary string, Gray code string.

Let us assume that the variables have the bounds shown in Table 2.

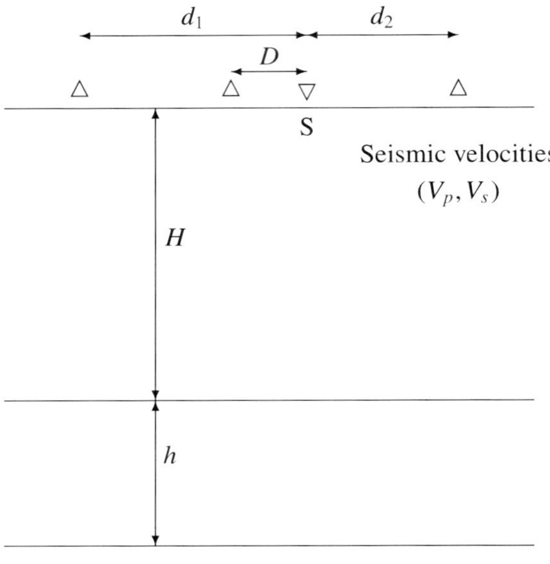

Fig. 4. Simple seismic experiment.

TABLE 2

Lower and upper bounds for the simple seismic variables.

$d_1(\text{min}) = \ \ 120 \le d_1 \le 150 \ \ = d_1(\text{max})$
$d_2(\text{min}) = \ \ 200 \le d_2 \le 250 \ \ = d_2(\text{max})$
$H(\text{min}) = 3000 \le H \le 3500 = H(\text{max})$
$h(\text{min}) = \ \ 200 \le h \le 300 \ \ = h(\text{max})$
$V_p(\text{min}) = 2000 \le V_p \le 3000 = V_p(\text{max})$
$V_s(\text{min}) = 1500 \le V_s \le 2500 = V_s(\text{max})$

Real number string: If we have an individual with the following position in the search space

$$(d_1, d_2, H, h, V_p, V_s) = (111.3, 202.4, 3321.7, 266.4, 1897.5, 2675.0)$$

then the genome is simply

111.3	202.4	3321.7	266.4	1897.5	2675.0

Binary string: When using a binary string we have first to decide how many binary bits we wish to use to represent a variable. If we choose to use N bits, then we have to divide the range that the variable can take such that it is covered by 2^N equally spaced points. If we choose $N = 5$, then we use 32 distinct values for the gene. For

TABLE 3

Binary and Gray codes for the integers 0–31

	binary	Gray		binary	Gray		binary	Gray
0	00000	00000	11	01011	01110	22	10110	11101
1	00001	00001	12	01100	01010	23	10111	11100
2	00010	00011	13	01101	01011	24	11000	10100
3	00011	00010	14	01110	01001	25	11001	10101
4	00100	00110	15	01111	01000	26	11010	10111
5	00101	00111	16	10000	11000	27	11011	10110
6	00110	00101	17	10001	11001	28	11100	10010
7	00111	00100	18	10010	11011	29	11101	10011
8	01000	01100	19	10011	11010	30	11110	10001
9	01001	01101	20	10100	11110	31	11111	10000
10	01010	01111	21	10101	11111			

the first of our variables the 32 values would be given by

$$d_1(i) = d_1(\min) + i \times \frac{(d_1(\max) - d_1(\min))}{2^N - 1} \qquad \forall i \in \{0, 2^N - 1\}$$

Using the binary expansions given in Table 3 and individual in search space with the position

$$(d_1, d_2, H, h, V_p, V_s) = (131.6, 227.4, 3048.4, 277.4, 2032.3, 2435.5)$$

the genome would be given as

0	1	1	0	0	1	0	0	0	1	0	0	0	1	1	1	1	0	0	0	0	0	0	0	1	1	1	1	0	1

Gray code string: As can be seen from Table 3, the Gray coding scheme is an alternative method of using a binary alphabet. The justification for using it are that any two adjacent integers in the Gray code are only one bit different. Therefore there is little difference in terms of the Gray code for two realizations of the variable that are similar. So for variable d_1 the values 134.52 and 135.48 are adjacent alleles, their binary representations are [0|1|1|1|1] and [1|0|0|0|0] which are significantly different. Which may cause problems when one part of the gene is inherited from one parent and the rest from another. Under Gray coding the equivalent representations are: [0|1|0|0|0] and [1|1|0|0|0], which are very similar. Under Gray coding, the example used for the binary coding would become

0	1	0	1	0	1	1	0	0	1	0	0	0	1	0	1	0	1	0	0	0	0	0	0	1	1	0	0	1	1

We have now seen how we might structure the genome for a simple case where all of the variables are completely independent. Let us now consider a variant of the example that has some unknown dependency between some of the variables. The modified example is illustrated in Fig. 5. The simple horizontal surfaces have now been replaced by, a surface that has a variable depth along the x direction, and the layer is also of variable thickness. The two surfaces are defined in terms of four variables each. There

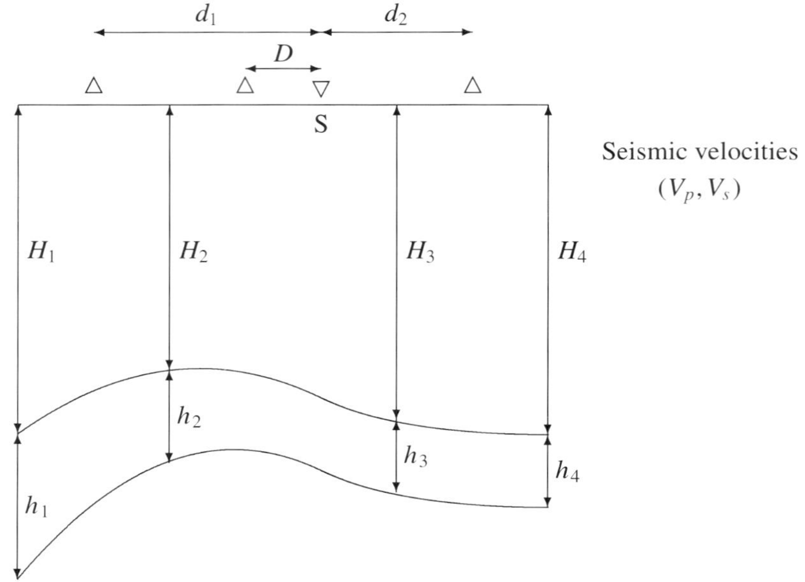

Fig. 5. Less simple seismic experiment.

is therefore a unknown relationship among the H_i's, and also between any H_i and the h_i aligned vertically with it. We could assume that there was no relationship and put all of the variables in a single string. The GA would be able to handle this situation, and would find a solution. However, we can improve the performance by constructing the genome in such a way that the relationships among the variables are preserved. The simplest way this can be achieved is by representing the variables H_i and h_i in a two dimensional array as one chromosome, and putting the other variables into a string as a second chromosome. This is illustrated in Fig. 6, where it is assumed that the genes are represented as real numbers. If we wish to use a binary encoding, instead of real numbers, then the structure can be expanded in exactly the same way as was done for the simple example. Given the change in the structure of the genome, there will need to be a corresponding change in the crossover operator as will be discussed below. The potential advantages of this new structure will become more obvious later.

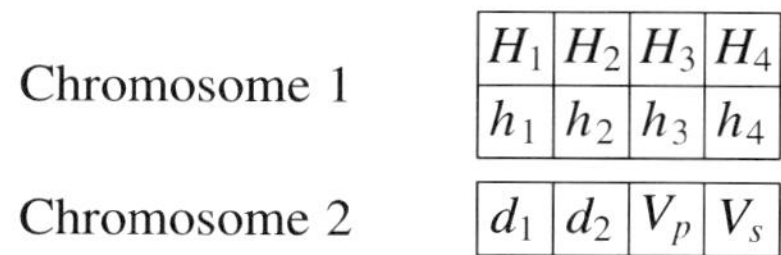

Fig. 6. Genome for less simple seismic problem.

3.4. Crossover operators

Having now selected a structure, and encoding method, for the genome we need to find a way of taking the information from two parents and merging that information to form a offspring. Our aim is that any offspring should inherit most of its genes from one or other of the parents, and that there should be minimal introduction of random variation in the genes.

If we are using a single binary string to encode the genotype, as was the norm in early GA research, then the crossover operator took a fairly simple form. Each of the parents is presented as a string of 0's and 1's, with there being no indication of where one gene ends and the next starts. In nature the machinery that creates the genome does not know about genes, genes being a consequence of creating the phenotype from the genotype. In a similar way crossover operators do not know about genes either. The crossover operator creates the genome of the offspring by taking a few characters from one parent, then the next few from the other parent, and then the next few from the first again, and so on until the new genome is complete. For both of our examples, if coded as single binary strings, then the resulting offspring would be viable. The key question in this process is when to switch the selection from one parent to the other. This is accomplished using the k-point crossover operator.

3.4.1. k-point crossover operator

This operator, also known as parameterised uniform crossover (Spears and de Jong, 1991), is quite flexible in its application. We define a probability, p, that is used to switch the selection from one parent to the other. The creation of the new genome proceeds as follows:

(1) let $k = 1$ (where k is the locus position) and the selected parent be parent 1
(2) generate a random number R
(3) if $R < p$ then swap to other parent
(4) let the locus of the offspring at position k equal that of the selected parent, i.e. offspring(k) = parent(k)
(5) $k = k + 1$
(6) if not finished goto step 2

Because of the use of the random number R, we are never able to predict where the change will occur. However by setting the probability p, we can control the expected length of a continuous section copied from one parent. If $p = 1.0$, then after every copy we will switch to the other parent. This process is very disruptive, with no gene begin inherited from one parent alone. If the probability is reduced to $p = 0.5$ then the expected length of a continuous section from either parent is 2 loci. In most work the probability is set such that the switch, from one parent to the other, is expected to occur once or twice along the whole length of the genome. In the earliest literature this choice was forced and the crossover operator was known as either a one-point crossover operator or a two-point crossover operator. If the chromosome contains many genes then it is reasonable to allow the expected number of crossovers to be slightly higher. A rule of thumb used by some people is that the expected number of crossovers should be equal to the number of genes. The operation of the operator is illustrated in Fig. 7.

Crossover probability = 0.12

Random number	Selected parent	Parent 1		Child		Parent 2
0.93	1	**0**	→	0		1
0.85	1	**1**	→	1		0
0.56	1	**1**	→	1		0
0.44	1	**0**	→	0		1
0.50	1	**0**	→	0		0
0.63	1	**1**	→	1		1
0.81	1	**0**	→	0		1
0.98	1	**0**	→	0		0
0.07	2	0		0	←	**0**
0.42	2	1		0	←	**0**
0.16	2	0		1	←	**1**
0.15	2	0		1	←	**1**
0.35	2	0		1	←	**1**
0.64	2	1		0	←	**0**
0.65	2	1		0	←	**0**
0.42	2	1		0	←	**0**
0.89	2	1		1	←	**1**
0.41	2	0		1	←	**1**
0.00	1	**0**	→	0		0
0.56	1	**0**	→	0		0
0.48	1	**0**	→	0		0
0.31	1	**0**	→	0		1
0.46	1	**0**	→	0		1
0.83	1	**0**	→	0		0
0.69	1	**1**	→	1		0
0.18	1	**1**	→	1		0
0.11	2	1		1	←	**1**
0.63	2	1		1	←	**1**
0.42	2	0		0	←	**0**
0.65	2	1		0	←	**0**

Fig. 7. Example of the k-point crossover operator.

3.5. Crossover operators for real valued genomes

There are many instances in the literature of real valued genomes being used to solve problems. In solving the problems it is often necessary to incorporate additional conditions, this is often done during the crossover operation. It can therefore be difficult to be sure how the crossover is being carried out, and hence it is not simple to reuse it in another application. We also find that the same idea reappears in the literature under a different name and without references to earlier work. This is simply different people reinventing the same algorithm. Having looked through the literature there seem to be four algorithms that get reused, and so these are likely to have some intrinsic value. The algorithms are:

- A variant of the k-point crossover described above. This appears many times under many names, I have not been able to satisfy myself that I have tracked down the first appearance of this in the literature.
- The BLX-α operator of Eshelman and Schaffer (1993).
- The UNDX operator of Ono (Ono and Kobayashi, 1997).
- The SBX operator of Deb (Deb and Agrawal, 1995; Deb and Beyer, 1999).

In the following sections I will describe each of the algorithms, and give a personal opinion of their value.

3.5.1. *k-point crossover for real valued strings*

The algorithm operates in exactly the same way as described above. The only difference being that the child's genome is created by copying a real number from one of the parents. The crossover probability is set so that only a few crossovers occur along the length of the string, this number being well below the number of genes (assuming that each real number represents one gene). The advantage of the approach is that it is very easy to implement and is consistent with my stated aim that offspring should inherit genes from their parents. The disadvantage is that only those genes that are present in the initial population are ever considered, which may make it difficult to converge on to a local optimum. This might not be a problem if the population is large, but any loss of genetic variability may cause significant problems.

3.5.2. *The BLX-α operator*

This operator operates on a variable-by-variable basis, there being an underlying assumption that there is a low level description of every gene that is not explicitly modelled, and that the true crossover happens at this lower level. At the level of the real numbers we only seek to approximate what is happening at the lower level. For each offspring variable y_i we randomly choose a value, using a uniform distribution such that

$$\frac{x_i^{(1)}+x_i^{(2)}}{2} - (\alpha + \tfrac{1}{2})(x_i^{(2)} - x_i^{(1)}) \le y_i \le \frac{x_i^{(1)}+x_i^{(2)}}{2} + (\alpha + \tfrac{1}{2})(x_i^{(2)} - x_i^{(1)}) \qquad (2)$$

where α is a user defined parameter, $x_i^{(1)}$ and $x_i^{(2)}$ are the parental genes. The advantages of the method are that is able to explore the parameter space quite well. The method does seem to have a number of disadvantages though: it seems to have difficulty exploiting information that it has gathered, so as to locate an optimum. This appears to be related to how the fitness landscape is aligned with the search space coordinate system. The other disadvantage is that no genes are directly inherited from either parent and that relationships between genes are totally ignored.

3.5.3. *UNDX operator*

This operator changes very gene simultaneously by making use of a multi-variate normal distribution. This operator is unusual in that it needs three non-identical parents. The first two parents are used to define a dominant axis through the search space of real numbers. The crossover is carried out on a co-ordinate system that is rotated to coincide with this dominant axis. The distance between the two parents is used to define the standard deviation of a normal distribution along this dominant axis. The distance of the third parent from this dominant axis is used to define the standard deviation used for all of the remaining normal distributions. The reader is referred to the original paper (Ono and Kobayashi, 1997) for the details of the implementation.

Its advantages are that it seems to be able to exploit information well and its not effected by the alignment of the fitness landscape with the search space. It's disadvantages are that it more complex to implement, it seems to have a difficulty with exploring the search space, and that it does not allow genes to be directly inherited. I also suspect that it will not perform very well in very high dimensional search spaces.

3.5.4. The SBX crossover operator

This operator works on a gene by gene basis, but all of the genes of an offspring are more related to one of its parents. It has one user defined parameter which controls just how closely related to its parent the offspring is. The values of the offspring variables $y_i^{(1)}$ and $y_i^{(2)}$ are given by

$$y_i^{(1)} = 0.5\left((1+\beta_q)x_i^{(1)} + (1-\beta_q)x_i^{(2)}\right) \tag{3}$$

$$y_i^{(2)} = 0.5\left((1-\beta_q)x_i^{(1)} + (1+\beta_q)x_i^{(2)}\right) \tag{4}$$

where β_q is given by

$$\beta_q = \begin{cases} (2u)^{\frac{1}{\eta+1}} & 0.0 < u \le 0.5 \\ \left(\frac{1}{2(1-u)}\right)^{\frac{1}{\eta+1}} & 0.5 < u < 1.0 \end{cases} \tag{5}$$

and u is a random number with $u \in (0,1)$. η is a non-negative real number set by the user. At one end of the parameter range, $\eta = \infty$, the probability for a gene becomes two delta functions centred on the the gene values of the two parents. At the other extreme, $\eta = 0$, the probability distribution is much less tightly restricted. Figure 8 shows the probability distribution function in one dimension, for three values of the parameter η. The offspring are symmetrically placed around the mean of the parents. In experiments that I have performed, I have found this function to work very successfully.

3.5.5. Comparison of the three crossover operators

It is enlightening to compare the probabilities of where a child's genes will be in search space relative to those of the parents. Figure 9 sketches the probability contour

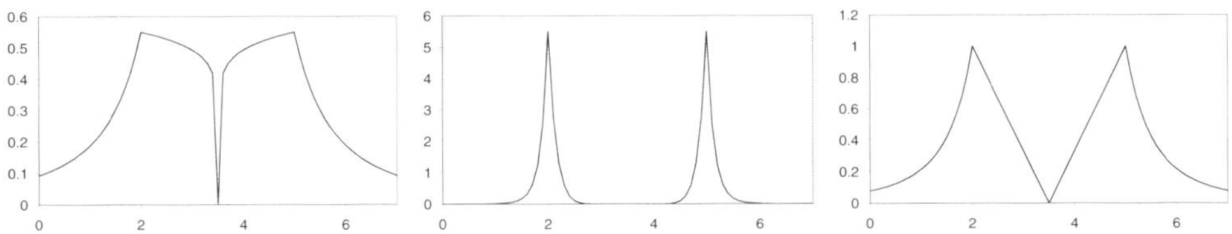

Fig. 8. The pdf of child locations for parents at 2 and 5, for $\eta = (0.1, 1.0, 10)$.

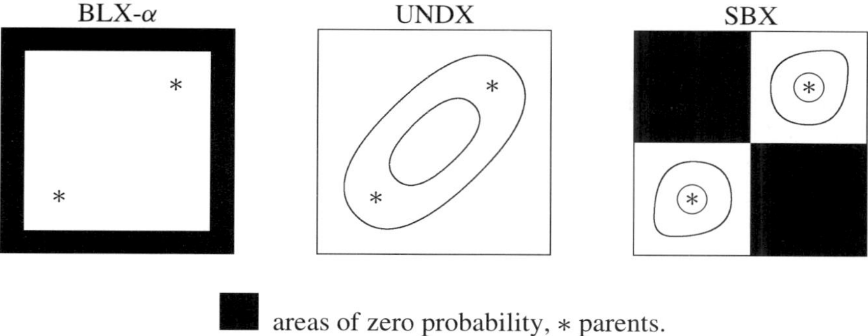

Fig. 9. Lines of constant probability in a two dimension parameter space.

maps for a two variable problem. We can see that the BLX-α operator will put many of its offspring in areas away from either parent. This should be a useful strategy in the early part of a search when exploration is required, but later when exploitation is required the algorithm is not likely to perform as well. The UNBX operator has no areas of zero probability, so exploration is always possible, however most offspring are likely to be concentrated around the mean of the two principle parents, and will not inherit the characteristics of either. This is likely to be acceptable for those occasions when the parents are sufficiently dispersed that small scale local optima are of little concern. But as the population starts to divide into subpopulations on different optima, then this approach is unlikely to do well, unless parents are always chosen so as to maximum the chance that they come from the same subpopulation. The SBX operator has non-zero probabilities only around the two parents. It therefore goes along way towards my requirement that offspring inherit information from a parent, it is less capable of exploring the search space. It also excludes areas close to the mean of the two parents that might well be of interest.

I would like to suggest a modification of the SBX operator that would avoid these zero probability areas, whilst concentrating most of the search close to one or other of the parents. Using the same notation as for the SBX this variant is defined as

$$y_i^{(1)} = \begin{cases} \frac{1}{2}\left((1+\beta_q)x_i^{(1)} + (1-\beta_q)x_i^{(2)}\right) & 0.0 < u \leq 0.5 \\ \frac{1}{2}\left((3-\beta_q)x_i^{(1)} - (1-\beta_q)x_i^{(2)}\right) & 0.5 < u < 1.0 \end{cases} \tag{6}$$

$$y_i^{(2)} = \begin{cases} \frac{1}{2}\left((1-\beta_q)x_i^{(1)} + (1+\beta_q)x_i^{(2)}\right) & 0.0 < u \leq 0.5 \\ \frac{1}{2}\left(-(1-\beta_q)x_i^{(1)} + (3-\beta_q)x_i^{(2)}\right) & 0.5 < u < 1.0 \end{cases} \tag{7}$$

$$\beta_q = \begin{cases} \left(\frac{1}{2u}\right)^{\frac{1}{\eta+1}} & 0.0 < u \leq 0.5 \\ \left(\frac{1}{2(1-u)}\right)^{\frac{1}{\eta+1}} & 0.5 < u < 1.0 \end{cases} \tag{8}$$

The one dimensional pdf and a sketch of the constant probability contours for two dimensions are shown in Fig. 10. This variant form of the SBX operator retains the

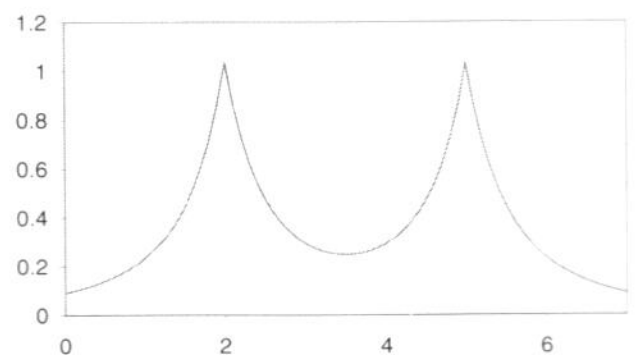

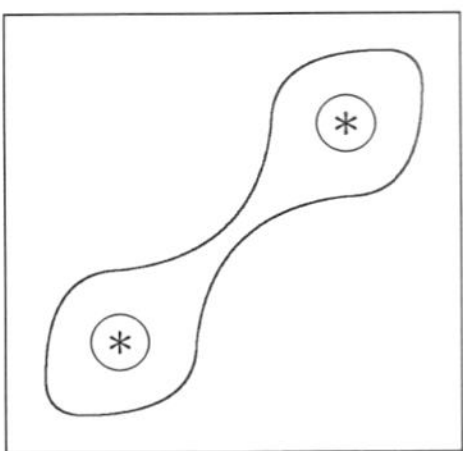

Fig. 10. Pdf distributions for the vSBX crossover operator.

original tendency to place offspring close to one parent. But it avoids having areas of zero probability and hence can explore the search space more easily.

3.6. *Combining k-point and gene-based crossover operators*

The weakness of the k-point operator is that it is unable to introduce new genes into the population. Its strength is its ability to keep useful gene combinations together. The three gene-based operators described have exactly the opposite strength and weakness. By carefully combining the k-point crossover with one of the gene-based operators, we can eliminate the weaknesses and combine the strengths.

This is achieved by using the k-point operator as before. But when the random number indicates that information should be taken from the other parent, the next locus of the offspring is obtained by using a gene-based operator, this is illustrated in Fig. 11.

3.7. *Crossover operator for multi-dimensional chromosomes*

The purpose of having multi-dimensional chromosomes was to allow relationships between genes to be recognised and retained. In one dimension, the k-point operator keeps neighbouring genes together most of the time. in higher dimensions, this generalises to selecting one region of the offspring's chromosome from one parent, and

	Parent 1		Child		Parent 2
	12	→	12		37
	65	→	65		40
Crossover point	84	→	70	←	66
	27		11	←	11
	93		74	←	74
	31		34	←	34

Fig. 11. Combined k-point and gene based crossover.

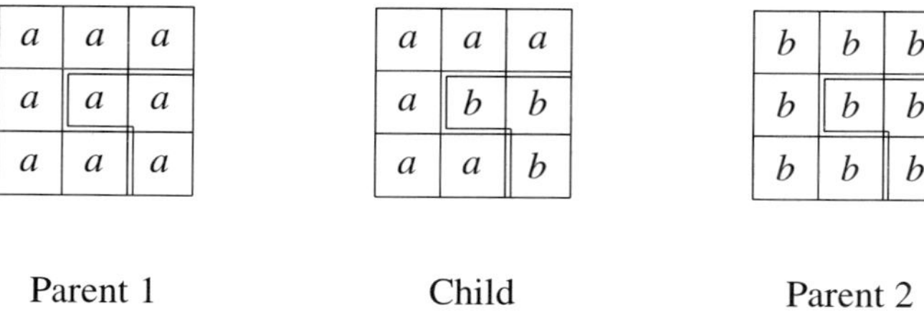

Fig. 12. k-point crossover in two dimensions.

the rest from the other parent. There are many ways that this selection can be made, a simple two-dimensional example is given in Fig. 12. This approach can be extended by using gene-based operators at the interface of the regions.

3.8. Selection of parents

The selection of parents is a key component within any GA. Having designed our genome and crossover operator such that offspring inherit appropriate information from their parents. We now need a method to ensure that the parents have the traits that represent success. In natural evolution success is measured by the number of children that you produce, that are capable of going on to produce children of their own. In our population you can only become a parent if you are selected, therefore an individual needs to be selected as often as possible, hence it needs to satisfy the selection criteria imposed. These criteria need to reflect what we regard as a successful solution.

So our criteria need to select better individuals that are regarded as more successful. If this selection policy is too aggressive, then we will drive the population into an ecological dead-end. This will result in good initial improvement in the function value, but ultimately the solution will probably be quite poor. This is because there will be a rapid loss of genetic diversity, and the population will not be able to explore the fitness landscape.

There are many schemes for selecting the parents based on their fitness values. These include: roulette wheel (Holland, 1975), stochastic universal sampling (Baker, 1987), Boltzmann selection (Goldberg, 1990), rank selection (Baker, 1985) and tournament selection (Goldberg and Deb, 1991). In my opinion, tournament selection is the most useful general purpose selection scheme, and is the only one I shall describe. It has the advantages of being easy to implement, efficient in its operation, and has sufficient flexibility to allow a variety of different selection pressures to be applied.

Tournament selection is implemented as follows:

(1) Select randomly an individual (individual **A**) from the population using a uniform probability distribution function.
(2) Select randomly another individual (individual **B**) from the population using a uniform probability distribution function.
(3) Choose the fitter individual with probability P, choose the less fit individual with probability $(1 - P)$.
(4) The chosen individual becomes individual **A**.

(5) Return to step 2, until a prescribed number N individuals have been compared.
(6) Individual **A** now becomes a parent.

There are three decisions to be made with this scheme: The number of individuals compared to select a parent, the probability P used to choose between individuals, and whether a previously selected individual is replaced into the population at step 2. A commonly used version of this scheme uses only two individuals (2-person tournament), $P = 1$ (i.e. the fittest individual is always chosen), individuals are replaced into the population (this means that it is not necessary to keep track of selected individuals). Other choices can increase or decrease the selection pressure and hence change the balance between exploration and exploitation of the fitness landscape. By choosing not to replace into the population at step 2, one may reduce the rate at which genetic diversity is lost.

3.9. Construction of new populations

Having produced a population of offspring, it is then necessary to combine this population with the parent population in such a way that the new population has the same size as the parent population. The most commonly used approach is to completely replace the parent population with the offspring population. This has the advantage of avoiding difficult questions about how to combine the population. We often combine this approach with elitism, which preserves the best known individual from generation to generation. This approach does not allow any control over how the population develops, and can result in the loss of genetic divergence.

If the whole population is not to be replaced at each generation, then we have to decide which parents are replaced by which offspring. The simple approach of replacing the worst section of the parent population with the offspring population, tends to lead to a rapid loss of genetic diversity and to poor performance. This is because the offspring tend to be related to the best members of the parent population, and so after a few generations all of the individuals are closely related and have similar genotypes.

The alternative is that each offspring has to compete with one, or more, individuals from the parent population for a place in that population. The loser of the tournament is culled from the population. There are a number of variants of this approach described in the literature (Harik, 1995; Mahfoud, 1995). I describe a version that seems to work reasonably well, and produces the desired niching behaviour. It is known as the probabilities crowding algorithm (Mengshoel and Goldberg, 1999).

(1) For each individual, x', in the offspring population
(2) Randomly select w individuals from the parent population.
(3) Choose the individual w' that is closest to to the offspring x' in the search space
(4) On the assumption that we are maximising an objective function f, the the probability of the two individuals winning a two person tournament are:

$$p(w') = \frac{f(w')}{f(w') + f(x')} \tag{9}$$

$$p(x') = \frac{f(x')}{f(w') + f(x')} \tag{10}$$

(5) The winner enters/remains in the parent population, the loser is culled.
The problem with this approach is that f needs to be chosen such that this probabilistic approach is meaningful. To some extent the probability problem can be corrected by keeping track of the worst function evaluation, $f(\text{worst})$, in the current parent and offspring populations, and writing the probabilities as:

$$p(w') = \frac{f(w') - f(\text{worst})}{f(w') + f(x') - 2f(\text{worst})} \tag{11}$$

$$p(x') = \frac{f(x') - f(\text{worst})}{f(w') + f(x') - 2f(\text{worst})} \tag{12}$$

3.10. Mutation operators

The role of mutation in GAs is to maintain genetic diversity by introducing genes that are not present in the population. This process is normally completely random and involves the changing of a locus to another other value of its possible alphabet.

For binary representations the process is particularly simple: randomly choose a locus and switch its value, either $0 \mapsto 1$ or $1 \mapsto 0$. The size of the change that this produces will depend on whether standard binary or Gray coding is used, and where along a particular gene the locus occurs.

For real number loci there is an additional question, "how big a change?". The answer depends in part on what one sees the role of mutation being. If the role of mutation is to sample new areas of the search space, then large jumps are suggested. The problem with this is that most changes will result in individuals with poor fitness and will not improve the process very often. If the role of mutation is to make small changes to the population and help it explore locally then small jumps are preferred. In binary representations this issue is automatically decided.

My personal preference is for large jumps to help the exploration phase. Then to use a simple local search after the GA has been stopped to optimise locally.

There is also the question of how often mutation should occur? There does not seem to be an agreed answer. Although some basic statements have been generally agreed: "mutation must not occur to frequently, nor should it be ignored totally". Both of these extremes cause most GAs to perform less well. In the literature, probabilities of around 0.01 per locus per child are often quoted.

3.11. Population size

The choice of population size is very much a matter of preference and depends to some extent on the problem. The general rule-of-thumb is that the larger the population the more exploration is possible, subject to genome, crossover and parent selection scheme, smaller populations allow fewer function evaluations before selection pressure and exploitation is applied. It often comes down to how many function evaluations can be afforded. As a rule I often use populations of 20–100 individuals, in the literature the common range of reported population sizes is 100–1000.

3.12. Generation of the initial population

To start the whole process going requires an initial population, so how should we generate this population? The answer derives from the fact that we do not know where in the search space the optimum is to be found. The initial population needs to be randomly generated, such that the whole of the search space is covered. Ideally every gene will have each of its alleles represented several times within the population. This will then allow the possibility that offspring can be generated for any point in the search space. This situation is possible for cases where each gene can take on only a small number of values. If the genes are real numbers then this is not possible, and we have to be content with a selection of the possible values.

My normal practice is to generate all of the initial population in a completely random way. The individuals produced in this fashion are almost always not very fit. I sometimes include one individual not generated in this way, which is often a best guess of where an optimum is to be found. There is a danger in inserting into the initial population such an individual. It is likely that this individual will be significantly fitter than the randomly generated individuals, and that all the offspring in later generations will be related to this individual. If this happens then the search space will not have been explored, and we will end up with essentially the solution we first thought of. This may be acceptable, if the main aim of the search is to find better solutions than we have already, rather than optimum solutions. However, if we are really searching for optimum solutions then we are better off exploring the whole search space, and a best guess solution may lead us astray.

3.13. General parameter settings

I have now described all of the items that need to be considered when setting up a GA for a new problem. It can seem quite daunting to make all the right choices for the various schemes available and the choice of parameter values. Fortunately, the algorithm is quite robust to the use of non-optimal schemes and parameters. This has a positive aspect, most forms of a GA work quite well, and a negative aspect, it is difficult to get spectacular performance.

So what would I recommend as a first attempt at a new GA?

Chromosomes: multi-dimensional real numbered chromosome which capture relationships between genes.

Population size: about 100 individuals.

Crossover operator: generalised k-point operator coupled with SBX (or possibly my variant SBX). Two offspring are generated per set of parents.

Mutation operator: random change within known bounds for 1% of genes.

Selection of parents: two person tournament, with fitter individual always winning.

Combining of population: either generational replacement with elitism, or two person probabilistic crowding with 20% of the parent population being tested to find the nearest individual to the offspring.

Initial population: generated completely randomly.

4. CONCLUSIONS

In this chapter I have given a personal view of how to set up a genetic algorithm for a problem that is new to me. The options I have described are not the only ones that are available, and other researchers may well have other preferences. Having constructed a GA for a new problem and tested it on a few short runs, I have no problems with changing the structure of the GA or the component schemes. Due to the stochastic nature of the method it is very unwise to draw a conclusion from a single run of the method. I usually use an absolute minimum of 10 runs of the method to assess mean, best and worst case performance. At the end of a GA run I normally take the best individual found, and use a simple stochastic hill-climber to find the local optimum. Any local search method could be used, however stochastic local search is often easy to implement as it can be carried out using a mutation operator from the GA itself.

REFERENCES

Baker, J.E., 1985. Adaptive selection methods for genetic algorithms. In: Grefenstette, J.J. (Ed.), *Proceedings of the First International Conference on Genetic Algorithms and their Applications*. Erlbaum, 1985.

Baker, J.E., 1987. Reducing bias and inefficiency in the selection algorithm. In: Grefenstette, J.J. (Ed.), *Proceedings of the Second International Conference on Genetic Algorithms and their Applications*. Erlbaum, 1987.

Carter, J.N., 1997. Genetic algorithms for incore fuel management and other recent developments in optimisation. In: Lewins, J. and Becker, M. (Eds.), *Advances in Nuclear Science and Technology*, 25: 113–154.

Davies, L.D., 1991. *Handbook of Genetic Algorithms*. Van Nostrand Reinhold.

Deb, K. and Agrawal, R.B., 1995. Simulated binary crossover for continuous search space. *Complex Systems*, 9: 115–148.

Deb, K. and Beyer, H.-G., 1999. Self adaption in real parameter genetic algorithms with simulated binary crossover. In: Banzhaf, W., Daida, J., Eiben, A.E., Garzon, M.H., Honavar, V., Jakiela, M. and Smith, R.E. (Eds.), *Proceedings of the Genetic and Evolutionary Computation Conference*. Morgan Kauffman, pp. 172–179.

Eshelman, L.J. and Schaffer, J.D., 1993. Real coded genetic algorithms and interval schemata. In: *Foundations of Genetic Algorithms II*, pp. 187–202.

Goldberg, D.E., 1989. *Genetic Algorithms in Search, Optimisation, and Machine Learning*. Addison-Wesley, 1989.

Goldberg, D.E., 1990. A note on Boltzmann tournament selection for genetic algorithms and population oriented simulated annealing. *Complex Systems*, 4: 445–460.

Goldberg, D.E. and Deb, K., 1991. A comparative analysis of selection schemes used in genetic algorithms. In: Rawlins, G. (Ed.), *Foundations of Genetic Algorithms*. Morgan Kaufmann, 1991.

Harik, G.R., 1995. Finding multi-modal solutions using restricted tournament selection. In: Eshelman, L. (Ed.), *Proceedings of the Sixth International Conference on Genetic Algorithms*. Morgan Kaufmann, pp. 24–31.

Holland, J.H., 1975. *Adaption in Natural and Artificial Systems*. University of Michigan Press, 1975.

Koza, J.R., 1992. *Genetic Programming: On the Programming of Computers by Means of Natural Selection*. MIT Press, 1992.

Mahfoud, S.W., 1995. Niching methods for genetic algorithms. *IlliGAL Rep.*, 95001.

Mengshoel, O.J. and Goldberg, D.E., 1999. Probabilistic crowding: deterministic crowding with probabilistic replacement. In : Banzhaf, W., Daida, J., Eiben, A.E., Garzon, M.H., Honavar, V., Jakiela, M. and Smith, R.E. (Eds.), *Proceedings of the Genetic and Evolutionary Computation Conference*. Morgan Kauffman, pp. 409–416.

Mitchell, M., 1998. *An Introduction to Genetic Algorithms*. MIT Press, 1998.

Ono, I. and Kobayashi, S., 1997. A real coded genetic algorithm for function optimisation using unimodal normal distribution crossover. In: Back, T. (Ed.), *Proceedings of the Seventh International Conference on Genetic Algorithms*. Morgan Kaufmann, pp. 246–253.

Spears, W.M. and de Jong, K.A., 1991. On the virtues of parameterized uniform crossover. In: Belew, R.K. and Booker, L.B. (Eds.), *Proceedings of the Fourth International Conference on Genetic Algorithms*. Morgan Kaufmann, 1991.

Developments in Petroleum Science, 51
Editors: M. Nikravesh, F. Aminzadeh and L.A. Zadeh

Chapter 4

HEURISTIC APPROACHES TO COMBINATORIAL OPTIMIZATION

VIRGINIA M. JOHNSON

Lawrence Livermore National Laboratory, Livermore, CA 94551, USA

1. INTRODUCTION

The literature of mathematical optimization is huge and highly confusing to the novice. Because it is a fundamental activity with applications in almost every subject capable of being quantified, there exists a vast array of methods that have evolved to handle both general and specialized characteristics of the problem to be optimized. The discussion which follows attempts to provide an intuitive understanding of the reasoning behind the choice of two heuristic search techniques, the genetic algorithm (GA) and simulated annealing (SA), to optimize the reservoir engineering problem described later in this book. Along the way, mention is made of some of the more commonly known alternative optimization methods; but the primary purpose of the discussion is to explain the aspects of the reservoir engineering problem, namely the nature of the decision variables and the attributes of the objective function, that made the application of the GA and SA desirable.

References on which this discussion is based include general introductions to optimization (Hillier and Lieberman, 1990; Miller, 2000), coverage of specialized methods for combinatorial optimization (Cook et al., 1998), and overviews of the heuristic methods (Reeves, 1993; Pham and Karaboga, 2000). Description and referencing of specific implementation parameters for the GA and SA are covered in the chapter describing the reservoir engineering optimization case study.

2. DECISION VARIABLES

Combinatorial optimization is a term applied to problems where the decision variables, i.e. those aspects of the problem which can be manipulated, are discrete rather than continuous. Take, for example, the diagram of a well field in Fig. 1. Assume that the five points identified as $x_1 \dots x_5$ are potential locations for seawater injection wells for a proposed water flood operation. The existing production wells are identified by black circles. Now suppose that the optimization problem is to identify the subset of the five prospective injection locations which will maximize the productivity of the field, defined perhaps as the field's Net Present Value (NPV). If, for each well, it is possible to select a rate of injection anywhere from 0 to some maximum (such as 20,000 bbl/day), then the decision variables are defined on a continuous scale. If, however, a well can only be injecting or not injecting (a binary situation) or can only inject at selected rates along the continuum (such as 0, 5,000, 10,000, 15,000, or 20,000 bbl/day), then the decision variables are referred to as discrete.

The discretization of the decision variable is not done to simplify the problem. To the contrary, discretization actually complicates the optimization problem by making it difficult to apply the most efficient and powerful search techniques. Rather, the discretization is tolerated when it is the most natural representation of the real-world problem to be solved. In the reservoir engineering application which is described in a later chapter, a large-scale seawater injection plan is described in which each potential location is either pumped to the maximum permitted by the physical characteristics of the reservoir in its vicinity or not at all. This means that, while different locations may inject at different rates, there still only a choice of injecting vs. not injecting at any one location. This feature of the decision variables makes the application a combinatorial optimization problem.

Because combinatorial problems are finite, it is theoretically possible to locate the optimal solution by simply evaluating all possible combinations according to the objective function (e.g. NPV) and selecting the combination showing the most desirable performance. In the binary form of the seawater injection problem in Fig. 1, there are 32 possible combinations of the decision variables as defined by the combination of n things taken k at a time, where n is always equal to 5:

$$\sum_{k=0}^{5} n!/k!(n-k)! \tag{1}$$

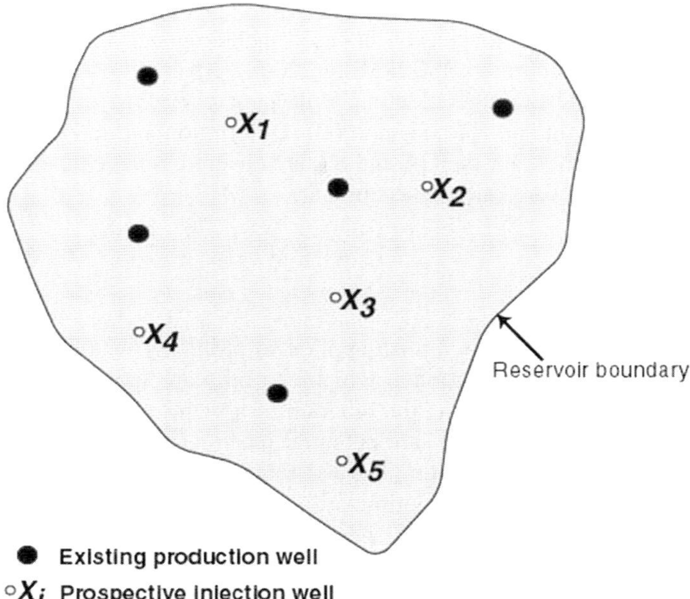

Fig. 1. Five potential injection well locations for the proposed waterflooding of a reservoir with five existing production wells.

While evaluating 32 cases presents no computational problems, the combinatorics of real-world problems can quickly become extreme. For example, the reservoir engineering application described later involves 25 potential locations with $k = \{0 \ldots 4\}$. Such a problem formulation yields 15,276 combinations to evaluate. Had the cap on the number of injection wells that could be drilled been raised to 6, the total number of combinations would have become 245,506. With modern computational resources, this number of cases to evaluate might still be quite manageable depending on the time required to evaluate the objective function for each case. In the reservoir engineering application, where a time-consuming 3D reservoir simulator is being used to predict the performance of each combination, even 15,000 combinations would be prohibitive to evaluate.

Combinatorial problems can be analyzed by a wide variety of optimization techniques, depending on how easily a problem's parameters can be shoe-horned into a particular formulation and on the size of the problem to be solved. Some problems, such as graph and network applications in which the goal of optimization is to obtain the optimal ordering of elements, are best handled by the specialized algorithms with which the term 'combinatorial optimization' has usually been associated. In contrast, the small water injection problem depicted in Fig. 1 could be fit to a linear programming (LP) model (of which the simplex technique is the best-known representative) by simply converting the encoding of the presence/absence of a well from a binary (on/off) to a floating point (0.0/1.0) representation. This will work so long as the objective function, NPV, is a linear function of the decision variables and if obtaining a solution for the five locations such as $x_1 = 0.0$, $x_2 = 1.0$, $x_3 = 0.6$, $x_4 = 0.1$, $x_5 = 0.0$ is reasonable. In some circumstances, it may indeed be feasible to inject into location x_3 at 0.6 of the maximum rate. And in many problems the margin for error is large enough to warrant rounding location x_4 from 0.1 to 0.0.

If solutions such as 0.6 and 0.1 are not acceptable, then integer programming (IP) methods, which permit the decision variables to take on only positive integer values, may be applicable. Although they are not usually categorized as such, integer problems certainly meet the definition of combinatorial optimization, since the decision variables are discrete rather than continuous. However, the main IP method, branch-and-bound, has difficulty accommodating large-scale problems and is still a linear technique. So, while it provides a more realistic representation for the wells in Fig. 1, branch-and-bound still does not solve the second major attribute of the water injection problem.

3. PROPERTIES OF THE OBJECTIVE FUNCTION

Simply put, a linear problem is one in which the whole is the sum of its parts. In the example in Fig. 1, if the NPV of the field is V_1 when x_1 is turned on and V_2 when x_2 is turned on, then NPV should equal $V_1 + V_2$ when both wells are turned on. If the two wells compete with each other, however, so that NPV is less than $V_1 + V_2$ when they are injected simultaneously, the problem becomes nonlinear.

Certain subsurface problems, such as the prediction of groundwater flow as a function of pumping wells, are considered linear and linear optimization techniques can readily be applied. When the objective function to be minimized (or maximized) involves more than one response variable, however, the situation becomes more complex. For example, when prediction of the migration of contaminants is added to the groundwater flow problem, the result is a nonlinear problem, which is why the water resources management literature contains so many applications of nonlinear methods (see the review by Wagner, 1995). In petroleum reservoir simulation, nonlinearity is introduced by the simultaneous prediction of gas/oil/water ratios in the produced fluid. In the modeling of surface facilities coupled to the reservoir simulator, the practice of siphoning off a portion of produced gas to power the compressors needed to withdraw more product introduces another potential nonlinearity. Nonlinearities can also be introduced by constraints imposed on the decision variables. In the large-scale water flood problem described later in this book, portions of the objective function's response surface are undefined because there are engineering constraints which disallow some combinations of wells. As a rule, as the problem formulation grows more complex to better mirror reality, the potential for nonlinearities to creep into the objective function also grows.

In general, the penalty imposed by the existence of these nonlinearities is loss of the guarantee that the optimal solution will be found.

There are many forms of nonlinearity and, therefore, many kinds of nonlinear optimization methods. The more readily a problem can be formulated as a special case, say as a quadratic problem, the more readily can a powerful and efficient method be found to locate an optimal solution. In practice, however, it is often difficult to determine the attributes of the objective function, especially in cases such as the NPV example where there are several steps between the decision variables and the final economic calculations. Consequently, methods whose applicability is not tied to any particular set of assumptions about the function being optimized begin to look attractive. Two such methods are the GA and SA.

4. HEURISTIC TECHNIQUES

The term 'heuristic' has more than one meaning in the computational and mathematical literature. For example, it is often used to refer to the application of domain-dependent knowledge to make a problem easier to solve. For the current discussion, however, a heuristic technique is one that

> *". . . seeks good (i.e. near-optimal) solutions at a reasonable computational cost without being able to guarantee either feasibility or optimality, . . . "* (Reeves, 1993, p. 6).

Given all the things a heuristic technique is unable to guarantee, one might wonder just what the attraction is, after all. However, inspection of Fig. 2 may make the situation somewhat clearer. The surface in Fig. 2 is a rather extreme illustration of some of the difficulties a nonlinear problem may pose. Since the five-well binary problem in Fig. 1 cannot be visualized in a 2D figure, Fig. 2 depicts a two-well problem in which each well's injection rate can take on several discrete values. The effect of selecting different combinations of rates for the two wells on the field's NPV is shown by the height of the

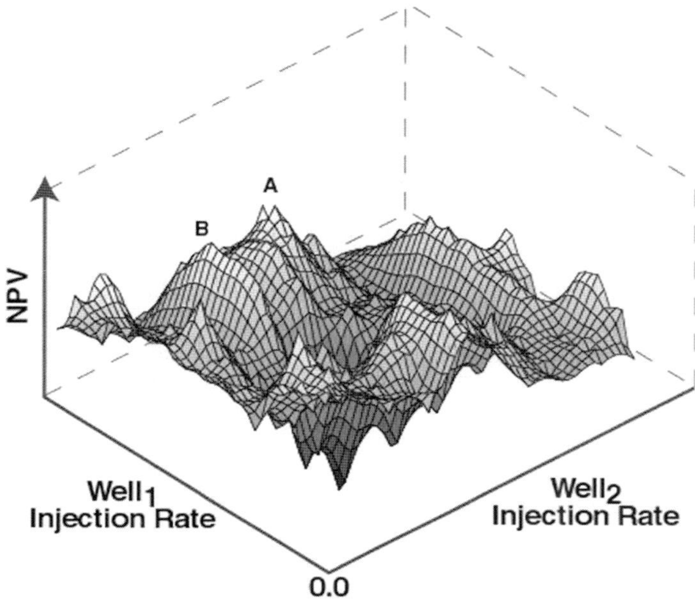

Fig. 2. A response surface showing the Net Present Value (NPV) obtained by various combinations of discrete injection rates at two potential well locations.

surface. This being a combinatorial problem, valid values on the surface only occur at the intersection of the grid lines.

Note, first, that the effect on NPV of increasing injection rates for $Well_1$ and $Well_2$ is highly convoluted. Instead of gradually increasing slopes as injection rates increase, the surface is full of hills and valleys. (Assume that some byzantine economic calculations are producing this phenomenon, in addition to competition between the injection wells and interference with the existing producers in a heterogeneous reservoir.) A convoluted surface such as this one will confound any linear method. The surface is also multi-modal, meaning that it contains several peaks, including two, A and B, which represent the optimal and near-optimal values of NPV that can be achieved. Peak A being the highest point in the entire surface, it represents the globally optimal solution to the 2-well problem. Peak B is a local optimum, since it is the best value in its local neighborhood.

There is no optimization technique which is capable of guaranteeing that the solution at Peak A will be found, short of enumerating all possible combinations of the two wells and evaluating each one. Since that is usually not a practical approach in large-scale problems, some form of sampling must be employed. A random or structured sample over the values that the decision variables can take on is going to be both wasteful and crude. Too much computational time will be spent on unpromising regions of the surface and, if the sampling plan is sparse, the peaks can be missed entirely. Instead, some method is needed to combine random sampling with a sense of direction so that computational resources become focused on combinations in the more promising regions of the surface. *Random, directed search* is the fundamental concept underlying the heuristic techniques. This still does not guarantee that Peak A will be located; but the odds of locating it or something close to it, such as Peak B or some of the lesser solutions on the slopes of Peak A, are improved without expending outrageous amounts of computational time.

Fig. 2 can be used to illustrate how SA algorithms implement random, directed search. Imagine that some initial, randomly-generated combination C_0 of the two wells' injection rates produces an NPV somewhere on the slopes of Peak B. The algorithm decides where next to move by randomly sampling n well-value combinations in the neighborhood of C_0, and moving to the combination C_1 with the best NPV value. C_1 then becomes the new center around which more random sampling is conducted to identify an even better combination C_2. If that were all the algorithm did, however, the inevitable result would be to move up to Peak B and remain trapped there. Unfortunately, the only way to get to Peak A from the slopes of Peak B is to initially move toward combinations that yield lower values of NPV. This is exactly what the algorithm does. According to rules which will be described in detail in the chapter covering the reservoir engineering optimization problem, SA does sometimes move toward less promising combinations to increase the odds of breaking out of local optima.

The GA tries to solve the problem in a different fashion. While the SA charts a path over the surface by moving from point (C_i) to point (C_{i+1}), the GA begins by distributing a population of n points over the surface to be searched, each point effectively searching its own vicinity. This is not, however, just a parallel version of the SA with many different starting points. One can think of the population of combinations in a GA as a bunch of iron shavings initially sprinkled over the surface. The peaks in the surface act like magnets. Over the course of search, individual shavings are drawn to the closest peak of attraction. If there are several peaks with roughly equivalent attractiveness, then at least some of the combinations in the population will have gravitated to each peak. If one peak is substantially more attractive than the others, all of the combinations will migrate to that peak. The specific rules which govern how the population members in a GA move over the surface will, of course, be described later.

The single-point vs multiple-point contrast between the SA and GA has implications for their performance and usage. SA algorithms generally converge on a single optimal combination more quickly while the slower GAs can illuminate more features of the surface. For example, suppose that Peak A contains combinations that will ultimately prove to be impractical to implement for political or regulatory reasons that cannot be encoded into the problem formulation. In an SA implementation, the existence of the near-optimal Peak B combinations may never be discovered. The GA, by dealing in populations rather than single points, is better structured to identify the existence of the Peak B combinations. By tweaking various parameters, however, both search techniques can be made to behave more like the other.

Both the GA and SA can be applied to continuous as well as discrete problem formulations, although the GA's theoretical foundations are rooted in discrete mathematics. Their special appeal for combinatorial problems lies in the flexibility of their problem representations and their robustness with respect to assumptions regarding the nature of the problem to be optimized.

ACKNOWLEDGEMENTS

This work has been performed under the auspices of the U.S. Department of Energy by Lawrence Livermore National Laboratory under contract W-7405-ENG-408.

REFERENCES

Cook, W.J., Cunningham, W.H., Pulleyblank, W.R. and Schrijver, A., 1998. *Combinatorial Optimization*. John Wiley and Sons, New York, NY.

Hillier, F.S. and Lieberman, G.J., 1990. *Introduction to Operations Research, 5th ed.* McGraw-Hill, New York, NY.

Miller, R.E., 2000. *Optimization*. John Wiley and Sons, New York, NY.

Pham, D.T. and Karaboga, D., 2000. *Intelligent Optimisation Techniques: Genetic Algorithms, Tabu Search, Simulated Annealing and Neural Networks*. Springer, New York, NY.

Reeves, C.R., 1993. *Modern Heuristic Techniques for Combinatorial Problems*. Halstead Press, New York, NY.

Wagner, B.J., 1995. Recent advances in simulation-optimization groundwater management modeling. *Rev. Geophys., Suppl., U.S. Natl. Rep. Int. Union Geodesy Geophys.*, 1991–1994, pp. 1021–1028.

Developments in Petroleum Science, 51
Editors: M. Nikravesh, F. Aminzadeh and L.A. Zadeh

Chapter 5

INTRODUCTION TO GEOSTATISTICS

RAJESH J. PAWAR

Los Alamos National Laboratories, Los Alamos, NM, USA

1. INTRODUCTION

Over the past three decades, geostatistics has become an important tool in characterization of heterogeneities in geologic media. Geostatistical methods and tools find their origin in the mining industry. These methods were initially developed by Matheron (1962) to estimate distribution of gold ore in South African mines. Although these methods are entirely based on principles of statistics and random variables, due to their early and wide application in geology they are popularly known as geostatistical techniques. A wide range of applications in diverse fields such as hydrology, petroleum engineering and mining has made geostatistics a central discipline in characterization of spatially heterogeneous media.

Geological media are inherently heterogeneous. The values of various properties are spatially variable. The properties include distribution of ores in mines, seam thickness of coal, as well as properties important for fluid flow such as porosity and permeability, etc. Accurate representation of these properties is important for different applications. For example, if one has to predict the nature of fluid flow in an oil field or a contaminated aquifer, the spatial distribution of permeability should be properly characterized. Usually due to the scale of the domain in question (1000s of feet to 10s of kilometers/miles) and the inherent heterogeneous nature of the problem, the required true representation can only be achieved by measuring the properties at every spatial location in the domain. It will be impractical to perform experimental measurements at this scale. In practice, these properties are usually measured at a limited number of locations, sometimes on the order of a few hundreds. These measurements are subsequently used to estimate the spatial distribution over the entire area of interest. The inherent nature of estimation process results in errors associated with the estimated values. A good estimation process attempts to minimize the errors associated with the estimated values as well as increase the confidence in them by integrating data from multiple sources. Geostatistics is one such suite of tools where the basic principles were developed to minimize the errors associated with estimation. At the heart of all geostatistical tools is a technique called kriging, defined as "a collection of generalized linear regression techniques for minimizing an estimation variance defined from a prior model for a covariance, a measure of spatial variability" (Isaaks and Srivastava, 1989). Early applications of geostatistics were primarily to estimate parameter values based on kriging. Since their early applications, the geostatistical methods have evolved into sophisticated techniques for integrating data from multiple sources to increase confidence in the estimated parameter values. This chapter introduces the basic principles of kriging and geostatistics. Readers are referred to Isaaks and Srivastava (1989) and Deutsch and Journel (1992) for further details.

2. RANDOM VARIABLES

The theory of geostatistics is based on the principles of stochastic theory and random variables. The property whose unknown value is to be estimated is assumed a random variable. A set of random variables defines a random function. The probability of a random variable having a certain value is given by the cumulative distribution function (cdf) of the variable. If the random variable at a location u is denoted by $Z(u)$, then the cdf is defined as:

$$F(u,z) = \text{Probability}\{Z(u) \leq z\} \tag{1}$$

Similar to the cdf for a random variable, the random function is characterized by a set of all the cdfs for the constitutive random variables. The joint cdf is defined for 'k' values at locations $u_1,\ldots,u_k$.

$$F(u_1,\ldots,u_k;\, z_1,\ldots,z_k) = \text{Probability}\{Z(u_1) \leq z_1,\ldots,Z(u_k) \leq z_k\} \tag{2}$$

The joint cdf defines the joint probability that 'k' values $Z(u_1),\ldots,Z(u_k)$ will be observed at 'k' spatial locations. The estimation processes based on stochastic principles attempt to predict the underlying random function as well as cdfs of constitutive random variable. In practice, a number of samples are required to estimate the cdf (at a spatial location). This need of multiple experimental observations can be overcome based on the principles of inference and stationarity. These two are important concepts in the development of theory and principles of geostatistics. Inference says that an unknown value of a variable can be inferred from other known values of the same variable or a different variable. Under the hypothesis of inference, one can determine the cdf at a location using measurements at other locations or by using measurements at the same location but at different times. This becomes possible under the assumption of stationarity of a random function. A stationary random function means that the multivariate cdf given in Eq. (2) remains constant throughout a field of interest. That is to say:

$$F(u_1,\ldots,u_k;\, z_1,\ldots,z_k) = F(u_1+m,\ldots,u_k+m;\, z_1,\ldots,z_k)$$
$$\forall \text{ translation vectors } m \tag{3}$$

If a higher-order cdf is stationary, it follows that all the lower order cdfs and all of their moments are stationary as well. This is a very important assumption as it implies that one can use statistical moments calculated from the available sample data as representative of the statistical moments over the entire field. Thus, an experimental study can be designed where samples are collected at select few locations and the experimental data is used to infer properties of distributions. For example, covariance (one of the moments of the second order cdf) between a random variable and other random variables separated by a separation vector 'h' can be inferred from the covariance value calculated from all the available sample data pairs separated by same separation vector 'h'. Stationarity assumption should be used with caution. Sample data should be carefully examined before using statistical moments based on it to infer statistics of unknown random variables. Relevant available geologic information should be used to justify statistical sampling.

By combining the concepts of stationarity and inference, one can determine the cdf of an unknown variable from the available sample data. The information on spatial variability of a random variable is incorporated in the estimation procedure through second order statistical moments such as covariance. Following section introduces the concepts of covariance and semi-variogram, the measures of spatial variability.

3. COVARIANCE AND SPATIAL VARIABILITY

The basic principles of geostatistics were developed based on two-point statistics. In other words, spatial variability of a random variable is calculated such that at a time relationship between only two random variables is used for calculations. The covariance is one such two-point statistics. It summarizes the joint distribution of pairs of data value as a function of distance. The covariance between two random variables, $Z(u_1)$ and $Z(u_2)$ is defined as,

$$C(u_1,u_2) = E\{Z(u_1)Z(u_2)\} - E\{Z(u_1)\}E\{Z(u_2)\} \tag{4}$$

$E\{Z(u)\}$ is the expected value of the random variable. The above equation can also be written as a function of separation vector 'h', between two random variables,

$$C(h) = E\{Z(u+h)Z(u)\} - [E\{Z(u)\}]^2 \tag{5}$$

Traditionally the semi-variogram has been widely used in geostatistical applications. A semi-variogram is defined as the variance of the difference between two data values separated by a vector 'h'.

$$\gamma(h) = \frac{1}{2}E\{[Z(u) - Z(u+h)]^2\} \tag{6}$$

The semi-variogram measures the average degree of dissimilarity between an unknown value and a nearby data value. The covariance and the semi-variogram are related as follows:

$$\gamma(h) = C(0) - C(h) \tag{7}$$

$C(0)$ is the stationary variance and $C(h)$ is the stationary covariance. Above relationship results due to the assumption of stationarity of the random function (Isaaks and Srivastava, 1989; Deutsch and Journel, 1992). In practice, the sample data values are used to compute a semi-variogram as follows:

$$\gamma(h) = \frac{1}{2N(h)}\sum_{i=1}^{N(h)}(x_i - y_i)^2 \tag{8}$$

$N(h)$ is the number of data pairs x_i and y_i separated by a separation (lag) vector 'h'. The separation vector 'h' is specified by magnitude and direction. Since the expected difference between data values at lag distance of 'zero' (or to say, difference between a data value and itself) is always 'zero', the theoretical value of semi-variogram at zero lag separation is always 'zero'. A semi-variogram is described by three parameters: range, sill and nugget effect.

- *Range:* The range defines the lag distance at which the semi-variogram reaches the maximum value. As separations between the data pairs increase, the dissimilarities between pairs also increase, resulting in higher semi-variogram values. As the separation between two data values increases beyond the 'range', the value of the semi-variogram does not increase any more. This is also saying that the data values separated by lag distance greater than the 'range' do not influence each other. When estimating an unknown value, the sample data values that are at a distance greater than the 'range' are usually not considered in the estimation procedure, unless the number of available data is small.
- *Sill:* The maximum value that the semi-variogram attains at the 'range' is called the 'sill'. The assumption of stationarity implies that the 'sill' is usually equal to the variance of the sample data set.
- *Nugget effect:* Theoretically, a semi-variogram should have value of zero for lag distance of zero. In practice, practical limitations such as sampling errors and short scale variability that can not be measured at extremely low resolution may result in a non-zero value for zero lag separation. The non-zero semi-variogram value at zero lag separation is known as the 'nugget'.

The above-mentioned parameters are shown schematically in Fig. 1. As mentioned before, the semi-variogram measures the degree of dissimilarity between two data values. For a random function, the random variables that are separated by large distances are expected to be dissimilar or independent of one another. For large distance, the covariance, which is a measure of similarity between two data values, reaches zero, while the semi-variogram reaches a maximum value. As the distance increases, the semi-variogram values approach the variance of the random function. The nature of semi-variogram can vary with the data sets. For example, for a sample data set

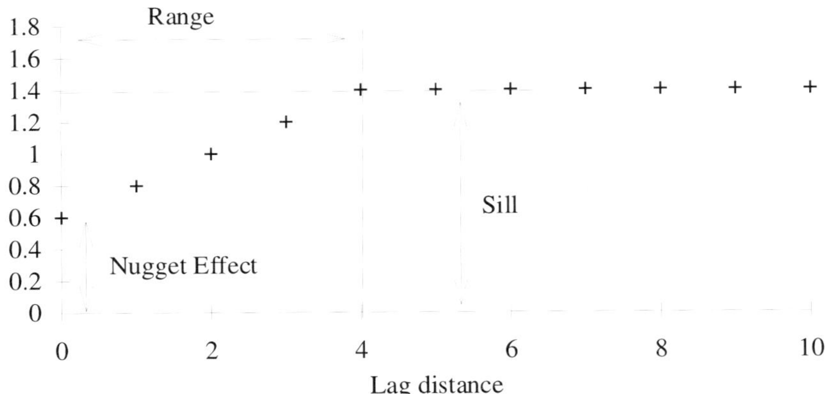

Fig. 1. An example semi-variogram showing variogram parameters.

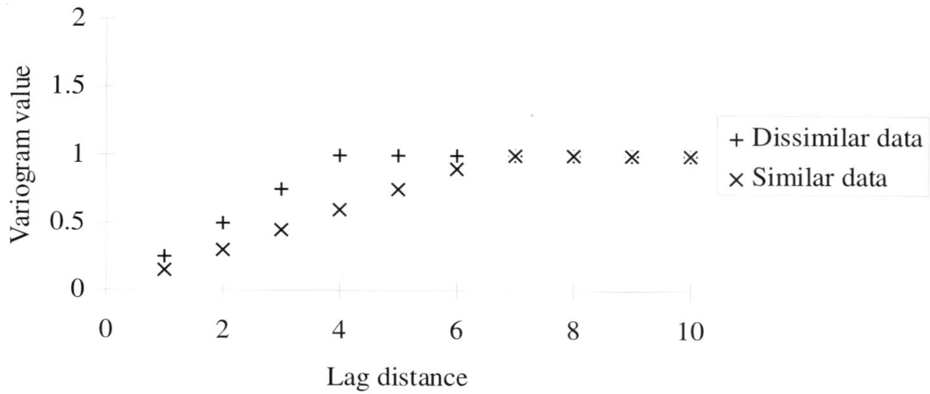

Fig. 2. Comparison of semi-variograms for two types of data sets.

where closely spaced data values are dissimilar, the semi-variogram will approach the maximum or variance rapidly. On the other hand, for a data set with closely spaced similar values, the semi-variogram will reach the maximum value (variance) slowly. This difference is exhibited in Fig. 2. Similar to the 'lag', the semi-variogram is also a vector quantity. Depending on the sample data set, a semi-variogram and the values of associated parameters will change with direction. A directional semi-variogram is calculated by selecting sample data pairs separated by lag vectors that are principally oriented in the direction of interest. The directional dependence of the semi-variogram or spatial variability is a result of spatial distribution of sample data. Directional semi-variograms can be used to find the directions of maximum and minimum spatial continuity of sample variables within the study area. The direction that results in the maximum value for 'range' is the direction of maximum continuity or the principle direction of spatial continuity. The direction in which the semi-variogram has the lowest 'range' is defined as the direction of minimum spatial continuity. The ratio of the minimum to the maximum 'range' is known as the 'anisotropy ratio'. Sometimes, either the semi-variogram values do not show any dependence on direction or the number of available data is limited to determine the principle directions of continuity. For such cases, an omni-directional semi-variogram is calculated. The omni-directional semi-variogram is calculated by using all the available sample data pairs, irrespective of directional orientation of the lag vectors.

The process of choosing a semi-variogram that captures the spatial variability of data is subjective. In practice, a number of different semi-variograms are calculated and the semi-variogram that has an interpretable structure and low variability is chosen as a representative semi-variogram. This semi-variogram may be directional or omni-directional. Once the semi-variogram is chosen, a model describing the functional relationship between the semi-variogram value and the lag distance is selected. The semi-variogram parameters, namely, 'range', 'sill', and 'nugget' are used to define the model. In practice, four standard models are used to describe a semi-variogram.

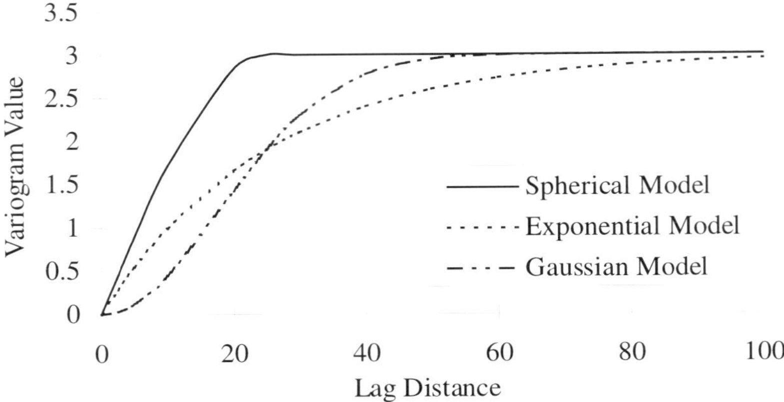

Fig. 3. Three popular semi-variogram models.

- Spherical model

$$\gamma(h) = c \cdot \mathrm{sph}\left(\frac{h}{a}\right)$$
$$\mathrm{sph}\left(\frac{h}{a}\right) = \left[1.5\frac{h}{a} - 0.5\left(\frac{h}{a}\right)^3\right] \quad h \leq a \tag{9}$$
$$\mathrm{sph}\left(\frac{h}{a}\right) = 1 \quad h \geq a$$

- Exponential model

$$\gamma(h) = c \cdot \exp\left(\frac{h}{a}\right) = c \cdot \left[1 - \exp\left(-\frac{h}{a}\right)\right] \tag{10}$$

- Gaussian model

$$\gamma(h) = c \cdot \left[1 - \exp\left(-\frac{h^2}{a^2}\right)\right] \tag{11}$$

- Power model

$$\gamma(h) = c \cdot h^a \tag{12}$$

'c' is the sill and 'a' is the range. A semi-variogram can be fitted with any one of these models or a linear combination of multiple models. The first three are the most common models used in practice. These 3 models are exhibited in Fig. 3. The information provided by the semi-variogram model is incorporated in the estimation procedure based on the principles of 'kriging'.

4. KRIGING

Kriging is a method of estimating an unknown value or a set of unknown values from the available known values. Kriging is a linear estimator, which means that the estimate of the unknown value is a linear combination of known data values. Each sample value is given an appropriate weight, where the weights are chosen such that the estimates are unbiased. The weights are calculated so that the mean residual error (error is defined as the difference between an actual value and the estimate) is equal to zero. In addition, the weights are constrained such that the variance in the error is minimized with respect to the weights. An estimate of an unknown value as the linear combination of 'n' known values is as given below:

$$\hat{z} = \sum_{j=1}^{n} w_j \cdot z_j \tag{13}$$

w_j is the weight assigned to the known value z_j and $\hat{z}$ is the estimated value. The error in the estimate is the difference between the actual value and the estimated value.

$$r_i = \hat{z} - z_i \tag{14}$$

The average error for a set of 'k' estimates will be:

$$m_r = \frac{1}{k} \sum_{i=1}^{k} \hat{z}_i - z_i \tag{15}$$

In practice, the true values of variable will always remain unknown. This means that it will be impossible to calculate the actual error of the estimate or the mean of the actual error of the estimates [as defined in Eqs. (14) and (15)]. Thus, it will not be possible to determine the weights required to calculate the linear estimate defined in Eq. (13). This problem can be overcome by applying the principles of the stochastic theory. It is assumed that the process that generated the random variables is represented by a random function model. Both the unknown and known values can be represented as random variables that result due to the random process. Assumption of stationarity says that each random variable has the same expected value. The joint distribution of a pair of random variables depends only on the separation distance and their covariance is given by $C(h)$. With these assumptions, Eqs. (13) and (14) can be represented in terms of random variables. The estimate for the unknown value, represented by a random variable, is now a weighted linear combination of 'n' random variables.

$$\hat{Z}(x_0) = \sum_{i=1}^{n} w_i \cdot Z(x_i) \tag{16}$$

The error in the estimate is the difference between the true value and the estimated value as follows:

$$R(x_0) = \hat{Z}(x_0) - Z(x_0) = \sum_{i=1}^{n} w_i \cdot Z(x_i) - Z(x_0) \tag{17}$$

The expected value of the error is given by:

$$E\{R(x_0)\} = E\left\{\sum_{i=1}^{n} w_i \cdot Z(x_i)\right\} - E\{Z(x_0)\} = \sum_{i=1}^{n} w_i E\{Z(x_i)\} - E\{Z(x_0)\} \tag{18}$$

The assumption of stationarity says that all the random variables have the same expected value, which results in:

$$E\{R(x_0)\} = \sum_{i=1}^{n} w_i E\{Z\} - E\{Z\} \tag{19}$$

In order to assure unbiasedness, the expected value of the error is set equal to zero. This results in the condition:

$$\sum_{i=1}^{n} w_i = 1 \tag{20}$$

The above equation provides one of the conditions that should be satisfied when choosing the weights. The other condition is obtained by minimizing the variance of the error with respect to the weights. The variance of the error is given by

$$\begin{aligned}\mathrm{Var}\{R(x_0)\} &= \mathrm{Var}\{\hat{Z}(x_0) - Z(x_0)\} \\ &= \mathrm{Cov}\{\hat{Z}(x_0)\hat{Z}(x_0)\} - 2\,\mathrm{Cov}\{Z(x_0)\hat{Z}(x_0)\} + \mathrm{Cov}\{Z(x_0)Z(x_0)\}\end{aligned} \tag{21}$$

Cov{} is the covariance between two random variables. The covariance of a random variable with itself is equal to its variance. After substituting for $\hat{Z}(x_0)$ from Eq. (16) and simplification, above expression reduces to the following equation (Isaaks and Srivastava, 1989).

$$\mathrm{Var}\{R(x_0)\} = \tilde{\sigma}^2 + \sum_{i=1}^{n}\sum_{j=1}^{n} w_i w_j \tilde{C}_{ij} - 2\sum_{i=1}^{n} w_i \tilde{C}_{i0} \tag{22}$$

$\tilde{\sigma}^2$, the variance in the random variables is same for all variables due to the stationarity assumption. $\tilde{C}_{ij}$ denotes the covariance between two sample values at spatial locations x_i and x_j. $\tilde{C}_{i0}$ is the covariance between the value to be estimated and the available sample values. The variance given in Eq. (22) is minimized with respect to the weights w_i. This results in 'n' equations in 'n' unknowns. The additional unbiasedness constraint on the weights [Eq. (20)] results in an over constrained system of equations. The problem is made well posed by introducing a new variable, μ, known as the Lagrange parameter in Eq. (22):

$$\mathrm{Var}\{R(x_0)\} = \tilde{\sigma}^2 + \sum_{i=1}^{n}\sum_{j=1}^{n} w_i w_j \tilde{C}_{ij} - 2\sum_{i=1}^{n} w_i \tilde{C}_{i0} + 2\mu\left[\sum_{i=1}^{n} w_i - 1\right] \tag{23}$$

By adding the right most term in Eq. (22), the equation does not change because of the condition obtained due to the unbiasedness assumption on the weights [Eq. (20)]. Now we have '$n+1$' equations in as many variables. Minimization of the variance with respect to the weights and the Lagrange parameter results in the following set of equations:

$$\begin{aligned}&\sum_{j=1}^{n} w_j \tilde{C}_{ij} + \mu = \tilde{C}_{i0} \quad \forall i = 1,\ldots,n \\ &\sum_{i=1}^{n} w_i = 1\end{aligned} \tag{24}$$

The weights obtained by solving above system of equations are subsequently used to calculate an estimate of the unknown value. In order to solve the above set of equations, different values of covariances are needed. These covariances are calculated from the semi-variogram calculated from the sample data and the relationship between covariance and semi-variogram.

$$\gamma_{ij} = \tilde{\sigma}^2 - C_{ij} \tag{25}$$

The procedure described above is known as Ordinary Kriging. The estimates calculated using this procedure are unbiased towards any particular sample value and they are obtained by minimizing the variance in error. The basic steps required in the estimation procedure using ordinary kriging are as follows:

(1) Calculate the spatial variability through a semi-variogram.
(2) Find a functional relationship between the semi-variogram values and the separation vector.
(3) Calculate the covariance matrix needed in estimation of an unknown value.
(4) For each unknown value form a linear system of equations and obtain the weights required for each sample value.
(5) Estimate the unknown value as the weighted linear average of the sample values.

Kriging forms the basis of all other geostatistical tools. Although kriging provides the best estimate for an unknown value, it is of limited use. The output of ordinary kriging loses its structure and the data is smoothed out (Deutsch and Journel, 1992). Kriging gives precedence to the accuracy of a local estimate. Each estimate is estimated independent of other estimates and without considering the resulting spatial statistics of all the estimates. Another major limitation of kriging is that it provides a single set of deterministic estimates of the unknown values. As mentioned before, the geostatistical techniques are used to model a random process (although the depositional process itself is not random, its representative model is assumed to be random). Use of a single realization to represent the random process may not be wise, as in practice the true representation of the random process will never be achieved. It is important to add a measure of uncertainty to the stochastically generated realizations. One can study the variation in property distributions by generating multiple realizations. Stochastic simulation techniques are specifically structured to overcome the above shortcomings of kriging.

5. STOCHASTIC SIMULATIONS

Unlike kriging, simulations give precedence to the resulting global features and statistics of simulated values. In the simulation approach, an estimate of the unknown value is generated by taking into account all the available data in the neighborhood. The available data includes not only the sample data but also simulated estimates of unknown variables in the neighborhood. Each time a new estimate is generated it is added to the available set of sample data and previously estimated values. Thus, the number of conditioning values goes on increasing as the simulation progresses. It should be noted that in kriging the mean and variance of the global random function are fixed because of stationarity. In simulations, the mean and variance need not necessarily stay stationary as the conditioning data set varies during simulations.

Simulations can also be used to generate multiple alternative models of a random function. The estimates are generated in a random order and the order can be varied to generate multiple realizations. All models are generated with the same procedure and have equal probability of occurrence. A number of different simulation techniques have been developed. Most of the techniques assume a particular model for the random function that is being modeled. For example, due to its simplicity Gaussian random function model has been a popular choice for random function model. The simulation algorithms based on Gaussian random function model include, sequential Gaussian (Journel and Alabert, 1989), Turning Band (Matheron, 1973; Mantoglou and Wilson, 1982) and LU Decomposition (Davis, 1982). Gaussian models have been particularly used for properties whose values are continuous such as porosity or permeability. Other popular random function model is the indicator model. An indicator random variable is defined as a binary variable, where the value of the variable is either 'one' or 'zero'. These variables are popularly used to model properties where one needs to show presence or absence of certain value or property, such as occurrence of gold, or occurrence of sandstone etc. They can also be used to model continuous variables, where the variable can be categorized based on multiple cutoffs (Journel and Alabert, 1990). The simulation algorithms based on this model include, sequential Indicator simulation (Journel and Alabert, 1990), Indicator Principal Components (Suro-Perez and Journel, 1991), etc. Farmer (1989) introduced algorithms based on simulated annealing to generate geologic property distributions. These algorithms are based on the annealing procedure where an objective function is minimized as the simulation progresses (similar to principles of annealing where an object is quenched). The objective function is defined such that the difference between statistics of simulated property distribution and the statistics calculated from sample data is minimized. For example, an objective function where the difference between the observed and model semi-variograms is minimized is as follows

$$Obj = \sum_{h} \frac{\left[\gamma_{\mathrm{observed}}(h) - \gamma_{\mathrm{model}}(h)\right]^2}{\gamma^2_{\mathrm{model}}(h)} \tag{26}$$

The annealing algorithm can also be used to refine simulation results obtained through other simulation algorithms, so that they reproduce certain important statistical properties. Another set of algorithms, known as the Boolean algorithms are used to distribute geometric objects in space (Haldorsen et al., 1988). These algorithms are mainly applied to generate distribution of sand bodies. The geometric objects are distributed based on some predetermined probability laws. All of these techniques offer a suite of tools to characterize spatially heterogeneous data.

One of the important advantages of the geostatistical techniques is their ability to take into account all available information through conditioning. The information can be incorporated in the estimation process through a number of ways. For the grid points where no sample data are available, the estimates are obtained through simulations. For the grid points where sample data are already available, the simulations can be made to honor the observed values. This is called 'hard' conditioning, where the observed sample value is exactly reproduced. Geologic characterization usually results in multiple types of information that are related to a property, either directly or indirectly. For example, the values for porosity at any location can be determined from information from different sources, including, direct experimental measurements on the porous rock core samples or indirect measurements from various logs such as density logs, neutron logs etc. The indirect information is sometimes referred to as 'soft' data. This additional information can be used to increase confidence in the estimated value of a given property. Soft conditioning uses principles of conditional probability or joint probability to generate conditional distributions. The geostatistical techniques are still evolving. New methods that can incorporate and integrate multiple data sources to increase confidence in the estimated values are being continuously developed.

REFERENCES

Davis, M., 1982. Production of conditional simulations via the LU triangular decomposition of the covariance matrix. *J. Math. Geol.*, 19(2): 99–108.

Deutsch, C.V. and Journel, A.G., 1992. *GSLIB: Geostatistical Software Library and User's Guide*. Oxford University Press, New York, NY.

Farmer, C.L., 1989. Numerical rocks. In: *IMA/SPE European Conference on the Mathematics of Oil Recovery*. Cambridge University.

Haldorsen, H.H., Brand, P.J. and MacDonald, C.J., 1988. Review of the stochastic nature of reservoirs. In: Edwards, S. and King, P.R. (Eds.), *Mathematics of Oil Production*. Clarendon Press, Oxford, pp. 109–210.

Isaaks, E.H. and Srivastava, R.M., 1989. *An Introduction to Applied Geostatistics*. Oxford University Press, New York, NY.

Journel, A.G. and Alabert, F., 1989. Non-Gaussian data expansion in the earth science. *Terra Nova*, 1: 123–134.

Journel, A.G. and Alabert, F., 1990. New Method for reservoir mapping. *J. Pet. Technol.*, 42(2): 212–218.

Mantoglou, A. and Wilson, J.L., 1982. The turning bands method for simulation of random fields using line generation by a spectral method. *Water Resour. Res.*, 18(5): 1379–1394.

Matheron, G., 1962. *Traité de geostatistique appliquée, Vol. 1*. Technip, Paris.

Matheron, G., 1973. The intrinsic random functions and their applications. *Adv. Appl. Prob.*, 5: 439–468.

Suro-Perez, V. and Journel, A.G., 1991. Indicator principal component kriging. *Math. Geol.*, 23(5): 759–788.

Developments in Petroleum Science, 51
Editors: M. Nikravesh, F. Aminzadeh and L.A. Zadeh

Chapter 6

GEOSTATISTICS: FROM PATTERN RECOGNITION TO PATTERN REPRODUCTION

JEF CAERS

Department of Petroleum Engineering, Stanford University, Stanford, CA 94305-2220, USA

1. INTRODUCTION

Geostatistics deals with simple but important questions of spatial interpolation and uncertainty assessment. Fig. 1a draws what could be described as one of the common problems posed: how to estimate the variable under study at an un-sampled location given sample observations at nearby locations. This question aims at finding a *single* number, termed estimate, at the un-sampled location. Estimation is possible due to *spatial correlation*, i.e. the underlying geological or physical phenomenon causes observations to be dependent on each other if they are measured close by. If the unknown outcome at the un-sampled location were dependent on the known sample value at another location, then that sample value necessarily carries information about the unknown.

Finding a single estimate is often unsatisfactory in many applications. One is often interested in determining the *uncertainty* about the unknown value at the un-sampled location. In the statistical sciences, uncertainty is quantified through a statistical distribution, describing the frequencies of outcomes one expects at that location. That distribution function will depend on the number of sample data, their spatial configuration, the data values and the specific spatial phenomenon under study. Once that distribution is quantified, probability intervals or other measures of risk can be calculated for certain outcomes.

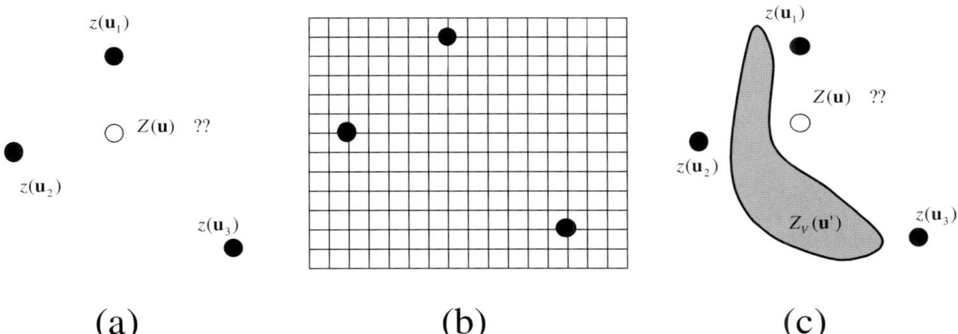

Fig. 1. (a) Estimating unknown Z at $\boldsymbol{u}$ from data (b) estimation on a regular grid (c) estimation/simulation with data of various support.

In practice one is seldom interested in determining an estimate or uncertainty model of a single un-sampled location. One is more interested in the *joint* determination of all the un-sampled values on a specific regular or irregular grid. For example, a 3D permeability model can be used to determine flow response for a given well configuration. A simple way to obtain such grid is to repeat the estimation of the single un-sampled node in Fig. 1a to all the other nodes in Fig. 1b, resulting in a *single* map of estimates. However, this single map should and cannot be used as a representation of the unknown spatial phenomenon. Since the true spatial distribution of all values in the grid is unknown, a single map can only provide a smooth representation of the truth, see Fig. 3a,b. Moreover, to determine that smooth map, estimates at each grid node are determined independent of each other, i.e. the estimate at each node is determined independent of estimates at other nodes. Instead of providing a single map, modern geostatistics aims at providing a set of alternative maps, termed *realizations* that quantify the uncertainty about the joint outcomes of the unknown grid node values. In an era of ever increasing CPU power, *Monte Carlo simulation* has become an increasingly popular geostatistical technique for determining these multiple representations of the unknown truth (Fig. 3c,d).

In this introductory chapter, an overview of the most used geostatistical estimation and simulation methods will be provided. Over the last decade, geostatistics has evolved from an emphasis on estimation to a field of stochastic imaging and data integration through Monte Carlo simulations. Recently, *multiple-point geostatistics* has emerged as a new powerful field for obtaining realistic geostatistical models that can integrate consistently a large variety of different sources of information acting on different scales. This introduction therefore tries to stress the importance of simulation over estimation in current practical applications. Modern geostatistics has essentially become a field of spatial pattern recognition and reproduction. Although most of the discussion and examples relate to reservoir characterization, the theory is presented in a general form. A glossary of important concepts is provided at the end.

2. THE DECISION OF STATIONARITY

Although geostatistics is part of the engineering sciences, it is firmly based on traditional statistical concepts. Statistics of *random variables* is simply based on the fact that when something is unknown or uncertain, one turns it into a random variable Z and describes its possible outcomes z through a probability model, often as a cumulative distribution function (cdf)

$$\Pr(Z \leq z) = F_Z(z) \tag{1}$$

One of statistics most used concepts is that of *stationarity*. In simple words, in order to be able to perform estimation, one requires to have multiple observations of the same variable. For example, in order to estimate the expected value of a population or, to determine the probability of a certain event occurring, or to draw a regression line between two variables, one needs multiple outcomes of the same random variables. This requires that all outcomes are legitimate alternative representations of the same random variable, which precludes a rather strong assumption that all outcomes come from the same population. The stationarity principle is particularly relevant to the study of spatial phenomena. For example, should one pool all porosity measured at all wells of the reservoir into a single histogram, or should one pool information per layer, or maybe per well? In a spatial context only one single outcome z exists at each spatial location $\boldsymbol{u} = (x, y, z)$. The unknown outcome at un-sampled locations $\boldsymbol{u}$ is modeled by a random variable $Z(\boldsymbol{u})$, determined by a distribution function $F_Z(z\boldsymbol{u})$. How can one then estimate the expected value at $Z(\boldsymbol{u})$ or the probability that it exceeds a certain threshold when no alternative outcomes are observed? This calculation requires a *decision of stationarity*. For example: to estimate the mean of Z at $\boldsymbol{u}$, one uses information at nearby locations. These locations are termed the *zone of stationarity*, where statistics about Z are assumed to be similar. Such zone is not only dependent on the particular geological phenomenon under study, but also on the amount of data. Indeed, if one has only a few, say 10 samples, all samples will be pooled together to calculate the mean, otherwise the estimate would be unreliable. It should be stressed that stationarity is not a property of the geological or physical phenomenon, rather it is a model decision that allows us to estimate with spatially varying data. Such decision is based on sound judgment and cannot be statistically tested or objectively rejected/accepted.

3. THE MULTI-GAUSSIAN APPROACH TO SPATIAL ESTIMATION AND SIMULATION

3.1. *Quantifying spatial correlation with the variogram*

Geological genesis of a reservoir causes observations to be correlated in space. The underlying geological continuity causes porosity at location $\boldsymbol{u}$ is to be dependent on porosity at location $\boldsymbol{u}+\boldsymbol{h}$, as long as distance vector $\boldsymbol{h}$ is not too large. In geostatistics, one attempts at quantifying that geological continuity. The simplest way is to consider correlation or degree of association between any two points in space. The *correlogram* is such a measure of spatial continuity and represents the correlation between $Z(\boldsymbol{u})$ and $Z(\boldsymbol{u}+\boldsymbol{h})$

$$\rho\big(Z(\boldsymbol{u}), Z(\boldsymbol{u}+\boldsymbol{h})\big) = \frac{\operatorname{Cov}\big(Z(\boldsymbol{u}), Z(\boldsymbol{u}+\boldsymbol{h})\big)}{\sqrt{\operatorname{Var}\big(Z(\boldsymbol{u})\big)\operatorname{Var}\big(Z(\boldsymbol{u}+\boldsymbol{h})\big)}}$$

with

$$\operatorname{Cov}\big(Z(\boldsymbol{u}), Z(\boldsymbol{u}+\boldsymbol{h})\big) = \mathrm{E}\big[\big(Z(\boldsymbol{u}) - \mathrm{E}[Z(\boldsymbol{u})]\big)\big(Z(\boldsymbol{u}+\boldsymbol{h}) - \mathrm{E}[Z(\boldsymbol{u}+\boldsymbol{h})]\big)\big]$$

To infer this probabilistic measure of association from the data, one requires multiple outcomes of the pair $(Z(\boldsymbol{u}), Z(\boldsymbol{u}+\boldsymbol{h}))$; hence one requires a decision of stationarity to model the experimental or sample correlogram. Indeed, to calculate the *experimental correlogram*, the dataset is searched for various pairs of sample locations that are approximately a *lag distance* $\boldsymbol{h}$ apart. Note that $\boldsymbol{h}$ is a vector, hence pairs of data within a certain distance *and* along a certain direction should be distinguished. In a layered reservoir, one would expect a higher degree of association between observations in any horizontal direction compared to the vertical. The assumption or decision of stationarity precludes pooling all sample information that are $\boldsymbol{h}$ apart into one *scatterplot* from which the correlation coefficient is an estimate of the location independent correlogram

$$\rho\big(Z(\boldsymbol{u}), Z(\boldsymbol{u}+\boldsymbol{h})\big) = \rho(\boldsymbol{h})$$

An estimate of $\rho(\boldsymbol{h})$ provides a quantification of geological continuity along various directions. Note that $\rho(\boldsymbol{h})$ quantifies patterns between any two spatial locations, hence it is termed a *two-point statistic*. In many cases, a reservoir with heterogeneous and complex geology (curvi-linear features or strong connectivities) cannot be described accurately by considering two-point patterns only. Hence, two-point geostatistics will be extended to multiple-point geostatistics in order to describe better the actual geological continuity of the reservoir.

Geostatistics traditionally uses a different, but related measure of spatial continuity than $\rho(\boldsymbol{h})$, which is termed the variogram

$$\gamma(\boldsymbol{h}) = \mathrm{Var}\big(Z(\boldsymbol{u}) - Z(\boldsymbol{u}+\boldsymbol{h})\big) = \mathrm{Var}(Z)\big(1-\rho(\boldsymbol{h})\big)$$

The variogram is a measure of dissimilarity between the property at location $\boldsymbol{u}$ and $\boldsymbol{u}+\boldsymbol{h}$, hence it tends to increase with increasing $|\boldsymbol{h}|$. Fig. 2 shows an example of an experimental variogram. The variogram is small for small $|\boldsymbol{h}|$ (small dissimilarity) and tends to increase with increasing $|\boldsymbol{h}|$ as samples taken at larger distances become uncorrelated (perfectly dissimilar).

The variogram or correlogram are also termed *moments* of the *bi-variate distribution* between properties at any two spatial locations. The bi-variate distribution is a more complete measure of spatial dependency than the variogram:

$$\Pr\big(Z(\boldsymbol{u}) \le z, Z(\boldsymbol{u}+\boldsymbol{h}) \le z'\big) = F_Z(\boldsymbol{h}, z, z')$$

While the variogram considers the overall spatial continuity between any two locations, the bi-variate distribution quantifies spatial continuity between specific classes of the property at any two locations. For example, the bi-variate distribution allows quantifying the spatial correlation between the low property values (when z, z' is low)

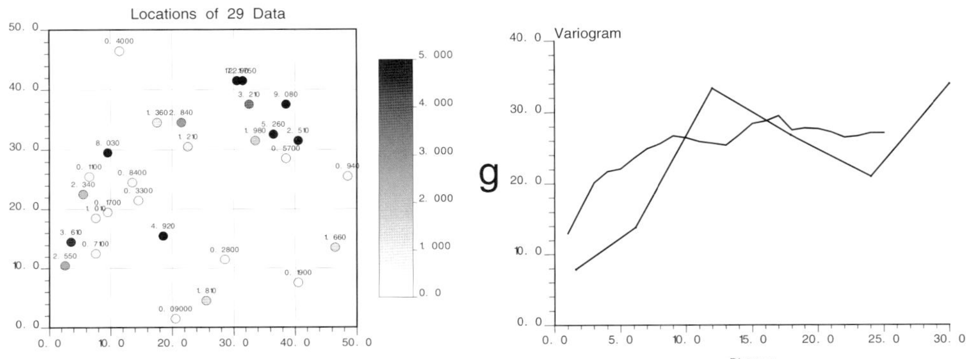

Fig. 2. (a, left) 29 sample data taken from a 'true' field of Figure 3a, (b,right) sample variogram (broken line) of the 29 data versus variogram of the true field of Figure 3a (more continuous line).

separately from the high (when z, z' is high) property values. The overall variogram $\gamma(\boldsymbol{h})$ measures the dissimilarity between all ranges of the data taken together. Note that in general geological continuity of high values need not be the same as geological continuity for low values. To quantify the bi-variate distribution, one employs so-called *indicator random variables*

$$I(\boldsymbol{u}, z) = \begin{cases} 1 & \text{if } Z(\boldsymbol{u}) \leq z \\ 0 & \text{otherwise} \end{cases}$$

One can define multiple of these indicator variables for each threshold z. The indicator correlogram or variogram is related to the bi-variate distribution, namely

$$\Pr\big(Z(\boldsymbol{u}) \leq z, Z(\boldsymbol{u}+\boldsymbol{h}) \leq z\big) = \mathrm{E}\big[I(\boldsymbol{u}, z) \cdot I(\boldsymbol{u}+\boldsymbol{h}, z)\big] = p\big(\rho_I(\boldsymbol{h}, z)(1-p) + p\big)$$

with p the marginal proportion of Z exceeding the cutoff z, hence

$$p = \Pr\big(Z(\boldsymbol{u}) \leq z\big) = F_Z(z)$$

The indicator variograms or correlograms can be estimated from the sample data using the same procedure as outlined for the overall variogram. This would require transforming the z-values (e.g. porosity) measured at wells into multiple indicator data, each describing different ranges of the property value.

4. SPATIAL INTERPOLATION WITH KRIGING

Spatial correlation allows making predictions about the property at un-sampled locations from sample data. If samples were not correlated in space, the best linear estimate of $Z(\boldsymbol{u})$ would be the global mean $m = \mathrm{E}[Z]$. *Kriging*, basically a form of generalized linear regression, is a name for a spatial estimation technique, that uses the variogram as a model of geological continuity and estimates un-sampled locations on that basis. The simplest way of kriging is termed simple kriging and consists of estimating the un-sampled location as a linear combination of the neighboring data values

$$z^*_{\mathrm{SK}}(\boldsymbol{u}) - m = \sum_{\alpha=1}^{n} \lambda_\alpha(\boldsymbol{u})(z(\boldsymbol{u}_\alpha) - m)$$

$\lambda_\alpha(\boldsymbol{u})$ are termed the kriging weights. Kriging provides a single estimate $z^*_{\mathrm{SK}}(\boldsymbol{u})$ of the true unknown value $Z(\boldsymbol{u})$ at location $\boldsymbol{u}$, hence the error of the estimate can be determined in terms of the mean and variance

mean (unbiasedness): $\mathrm{E}\big[Z(\boldsymbol{u}) - Z^*_{\mathrm{SK}}(\boldsymbol{u})\big] = 0$

Error variance: $\mathrm{Var}\big[Z(\boldsymbol{u}) - Z^*_{\mathrm{SK}}(\boldsymbol{u})\big]$

Kriging aims at finding the weights $\lambda_\alpha(\boldsymbol{u})$ that minimize the above error variance. A simple calculation leads to the simple kriging system

$$\sum_{\beta=1}^{n} \lambda_\alpha(\boldsymbol{u}) \mathrm{Cov}\big(Z(\boldsymbol{u}_\alpha), Z(\boldsymbol{u}_\beta)\big) = \mathrm{Cov}\big(Z(\boldsymbol{u}_\alpha), Z(\boldsymbol{u})\big) \qquad \forall \alpha = 1, \ldots, n$$

or under stationarity assumptions

$$\sum_{\beta=1}^{n} \lambda_\alpha(\boldsymbol{u})\mathrm{C}(\boldsymbol{u}_\alpha - \boldsymbol{u}_\beta) = \mathrm{C}(\boldsymbol{u}_\alpha - \boldsymbol{u}) \qquad \forall\alpha = 1,\ldots,n \qquad C(\boldsymbol{h}) = \mathrm{Var}(Z) - \gamma(\boldsymbol{h})$$

which is a system of n linear equations in the n unknown $\lambda_\alpha(\boldsymbol{u})$. The minimum error variance is termed the kriging variance and can be calculated once the linear system is solved

$$\sigma^2_{\mathrm{SK}}(\boldsymbol{u}) = \mathrm{Var}(Z) - \sum_{\alpha=1}^{n} \lambda_\alpha(\boldsymbol{u})C(\boldsymbol{u}_\alpha - \boldsymbol{u})$$

4.1. Stochastic simulation

Kriging essentially leaves the job (of spatial pattern reproduction) unfinished. The kriging map in Fig. 3b is unique and smooth. A simple visual comparison between the true unknown field and the kriging map shows a smooth pattern in the kriging image and an un-smooth pattern in the true image (Fig. 3a).

Since the properties at un-sampled locations are largely unknown, the single kriging estimate should be replaced by a set of alternative models that honor any sample information and that reproduce a geological pattern that is similar to the geological pattern of the true reservoir. Stochastic simulation is a geostatistical technique that allows reproduction of a given geological pattern constrained to various types of data such as '*hard*' sample data and '*soft*' seismic or production data. By 'hard data' it is understood that the variable under study is measured exactly at sample locations. 'Soft data' comprises any type of indirect measurement of the variable to be simulated or estimated. Stochastic simulation generates multiple reservoir models, termed realizations, each reproducing the same patterns, each representing alternative representations of the unknown truth. In the same sense as a set of outcomes drawn from a uni-variate distribution model represent the uncertainty about a random variable, a set of stochastically generated reservoir models represent the uncertainty about the unknown true reservoir. These stochastic realizations are equi-probable, meaning that no single 'best' realization exists.

Geological patterns can be represented by a simple variogram, or as will be explained later, as a *training image* that contains any geological pattern deemed relevant for the reservoir under study.

4.2. Sequential simulation

The first and mathematically most sound stochastic simulation method is termed sequential Gaussian simulation or *sgsim*. sgsim corrects the smoothing effect of kriging

---

Fig. 3. (a) True field and histogram, (b) kriging estimates based on the 29 data of Fig. 2 (smoother than true field); notice that the variance of the kriging estimate is less than the actual variance, (c,d) two sequential Gaussian simulations constrained to the 29 data; note that the histograms of the Gaussian simulations are similar to the true field.

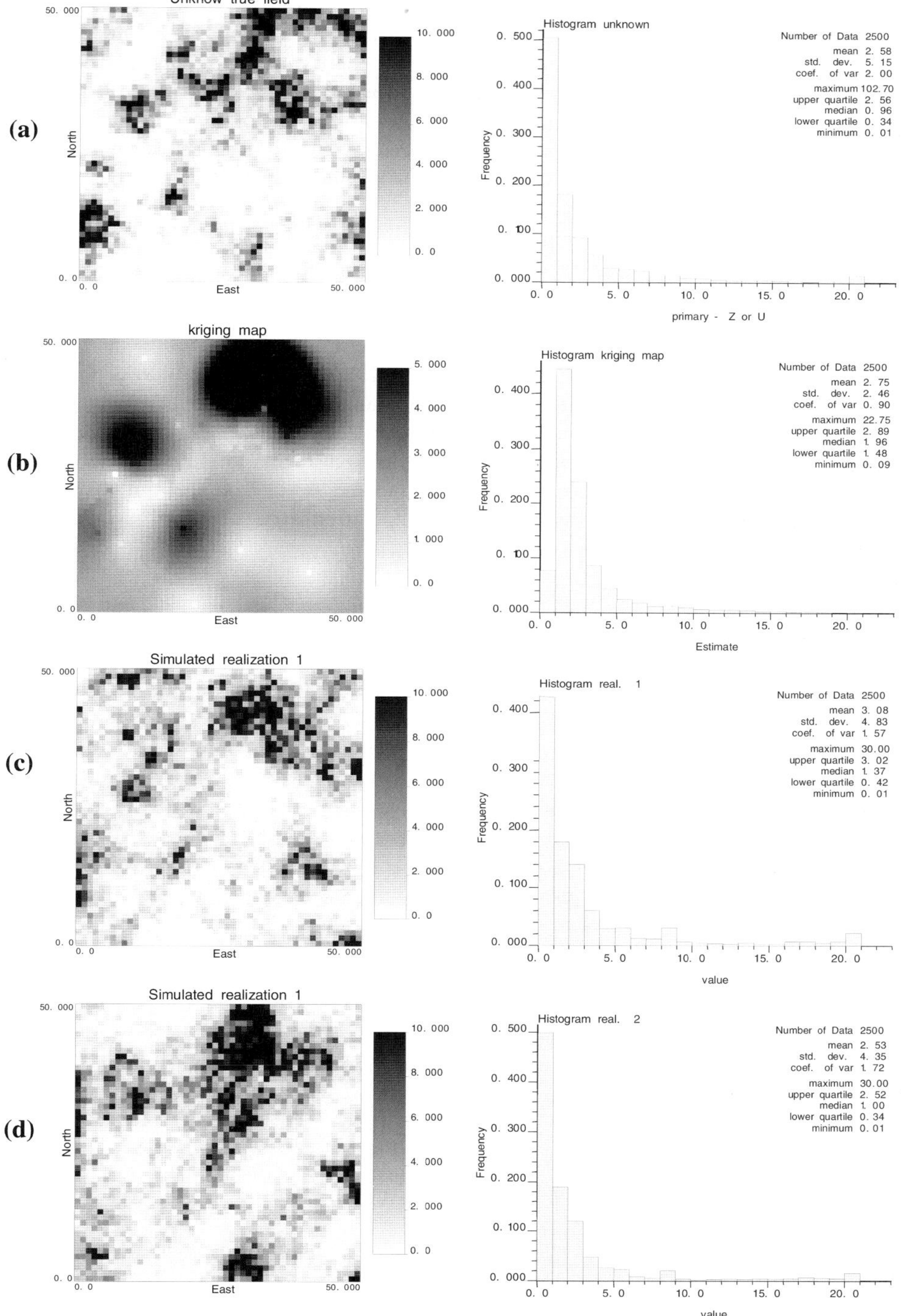
(a)
Unknow true field
North
East
Histogram unknown
Number of Data 2500
mean 2. 58
std. dev. 5. 15
coef. of var 2. 00
maximum 102. 70
upper quartile 2. 56
median 0. 96
lower quartile 0. 34
minimum 0. 01
Frequency
primary - Z or U
(b)
kriging map
North
East
Histogram kriging map
Number of Data 2500
mean 2. 75
std. dev. 2. 46
coef. of var 0. 90
maximum 22. 75
upper quartile 2. 89
median 1. 96
lower quartile 1. 48
minimum 0. 09
Frequency
Estimate
(c)
Simulated realization 1
North
East
Histogram real. 1
Number of Data 2500
mean 3. 08
std. dev. 4. 83
coef. of var 1. 57
maximum 30. 00
upper quartile 3. 02
median 1. 37
lower quartile 0. 42
minimum 0. 01
Frequency
value
(d)
Simulated realization 1
North
East
Histogram real. 2
Number of Data 2500
mean 2. 53
std. dev. 4. 35
coef. of var 1. 72
maximum 30. 00
upper quartile 2. 52
median 1. 00
lower quartile 0. 34
minimum 0. 01
Frequency
value

and generates stochastic realizations that honor a specific geological pattern as quantified by the variogram and histogram. The major default of kriging, which leads to its smoothness, lies in the fact that un-sampled locations are estimated independently from one another. Indeed, in kriging, the value at location $\boldsymbol{u}$ is estimated without taking into account the value previously estimated at $\boldsymbol{u}''$. Evidently, one can therefore never expect kriging to reproduce the correct spatial association, as measured by the variogram. This smoothing effect is apparent is the histogram of kriged estimates which has too low variance (Fig. 3d).

In order to guarantee that the correct geological patterns are reproduced, one needs to define the joint probability model of properties at all grid locations taken together, not one-by-one as done in kriging. A joint distribution is defined as

$$F(z_1, z_2, \ldots, z_N) = \Pr\big(Z(\boldsymbol{u}_1) \le z_1, \ldots, Z(\boldsymbol{u}_N) \le z_N\big)$$

N is the number of grid nodes. Generating a sample from the distribution is equivalent to generating a sample at each individual grid location. However, in specifying this joint distribution, one takes into account the joint dependency between all grid nodes taken together, aiming to reproduce spatial patterns as defined by this joint distribution model. Since it is mathematically too difficult to define and draw from a joint distribution model, one relies on the following general decomposition

$$\Pr\big(Z(\boldsymbol{u}_1) \le z_1, \ldots, Z(\boldsymbol{u}_N) \le z_N\big) = \Pr\big(Z(\boldsymbol{u}_N) \le z_N \mid Z(\boldsymbol{u}_1) \le z_1, \ldots, Z(\boldsymbol{u}_{N-1}) \le z_{N-1}\big) \times \ldots$$
$$\ldots \times \Pr\big(Z(\boldsymbol{u}_2) \le z_2 \mid Z(\boldsymbol{u}_1) \le z_1\big) \times \Pr\big(Z(\boldsymbol{u}_1) \le z_1\big) \qquad (1)$$

which states that any joint distribution can be decomposed in a product of N conditional distribution, hence drawing from a joint distribution is equivalent to drawing from N uni-variate conditional distributions. This decomposition is the basis for sequential simulation, one of the most used geostatistical simulation techniques. It allows generating a single stochastic realization, by drawing sequentially from uni-variate conditional distributions at each grid node, until the entire grid is filled. Note that in sequential simulation, unlike kriging, each node is simulated based on previous simulated nodes. Conditional sequential simulation is referred to the technique whereby one has any amount of hard or soft data, denoted as (n). Each conditional distribution in decomposition (1) is then based on previously simulated nodes and hard/soft data in the grid:

$$\Pr\big(Z(\boldsymbol{u}_1) \le z_1, \ldots, Z(\boldsymbol{u}_N) \le z_N \mid (n)\big) = \Pr\big(Z(\boldsymbol{u}_N) \le z_N \mid (n+N-1)\big) \times \ldots$$
$$\ldots \times \Pr\big(Z(\boldsymbol{u}_2) \le z_2 \mid (n+1)\big) \times \Pr\big(Z(\boldsymbol{u}_1) \le z_1 \mid (n)\big) \qquad (2)$$

The type of the conditional distributions in (2) and the way their parameters are estimated determines the type of sequential simulation algorithm. The general outline of the sequential simulation algorithms for simulating a single realization is therefore as follows

(1) Assign any hard data (n) to the grid
(2) Define a random path visiting all nodes $\boldsymbol{u}$ in the grid
(3) Loop over all nodes $\boldsymbol{u}_i$
    (a) Construct a conditional distribution

$$F_Z\big(\boldsymbol{u}_i, z \mid (n+i-1)\big) = \Pr\big(Z(\boldsymbol{u}_i) \le z \mid (n+i-1)\big)$$

(b) Draw a simulated value $z(\boldsymbol{u}_i)$ from the conditional distribution

$$F_Z\big(\boldsymbol{u}_i, z \mid (n+i-1)\big)$$

(c) Add simulated value to data set $(n+i-1)$

(4) End simulation

A first in a row of sequential simulation techniques is termed sequential Gaussian simulation.

4.3. Sequential Gaussian simulation

Sequential Gaussian simulation is the only sequential simulation where one explicitly assumes a mathematical form for the joint distribution in (1). One assumes that the joint distribution is *multi-variate normal*. This assumption is purely based on mathematical convenience and has no geological motivation. Indeed, any conditional distribution under the multi-Gaussian model is also Gaussian, hence fully determined by two parameters: the mean and variance. This entails that all conditional distributions under the decomposition (1) are also Gaussian. Moreover, it turns out that the mean of each conditional distribution is expressed as a linear combination of previously simulated nodes

$$\mathrm{E}\big[Z(\boldsymbol{u}_i) \mid Z(\boldsymbol{u}_{i-1}) = z_{i-1}, \ldots, Z(\boldsymbol{u}_1) = z_1\big] = \sum_{j=1}^{i-1} \lambda_j(\boldsymbol{u}_i) z(\boldsymbol{u}_j) = z^*_{\mathrm{SK}}(\boldsymbol{u}_i)$$

where the weights $\lambda_j(\boldsymbol{u}_i)$ are determined using simple kriging and where the variance equals the kriging variance

$$\mathrm{Var}\big[Z(\boldsymbol{u}_i) \mid Z(\boldsymbol{u}_{i-1}) = z_{i-1}, \ldots, Z(\boldsymbol{u}_1) = z_1\big] = \sigma^2_{\mathrm{SK}}(\boldsymbol{u}_i)$$

Given these mathematical results, the sequential Gaussian simulation algorithm can be defined as follows:

(1) Transform the sample data to standard normal scores

(2) Assign the data (n) to the grid

(3) Define a random path visiting all nodes $\boldsymbol{u}$

(4) Loop over all nodes $\boldsymbol{u}_i$

(a) Construct a conditional Gaussian distribution

$$G\big(\boldsymbol{u}_i, z \mid (n+i-1)\big) = G\left(\frac{z - z^*_{\mathrm{SK}}(\boldsymbol{u}_i)}{\sigma_{\mathrm{SK}}(\boldsymbol{u}_i)}\right)$$

(b) Draw a simulated value $z(\boldsymbol{u}_i)$ from the conditional distribution

$$G\big(\boldsymbol{u}_i, z \mid (n+i-1)\big)$$

(c) Add simulated value to data set $(n+i-1)$

(5) End simulation

(6) Transform the entire simulation back to the original data histogram

Note the histogram transformation before and after simulation, which is required, since a multi-Gaussian assumption includes also the assumption that the histogram of the data is standard Gaussian.

4.4. Accounting for secondary attributes

In many practical cases various sources of indirect information (*soft or secondary data*) about the *primary variable* to be estimated or simulated. Various seismic attributes indirectly inform porosity within the reservoir. The traditional, non-geostatistical approach to using multiple attributes for prediction, is to determine a, possibly non-linear, regression relation between the primary variable Z_1, and any set of secondary attributes, Z_i, $i = 2, \ldots, N_v$

$$Z_1 = \varphi(Z_2, Z_3, \ldots, Z_{N_v})$$

A neural net function φ can be calibrated for example to model porosity from any set of co-located seismic or well-log attributes. Then, the trained neural net can predict $Z_1(\boldsymbol{u})$ anywhere in the reservoir where $Z_2, Z_3, \ldots, Z_{N_v}$ data is available at $\boldsymbol{u}$. This procedure has various short-comings:

(1) It ignores the spatial variability of the original $Z_1(\boldsymbol{u})$ variable. Indeed in the neural network training and application one does not include any variogram information about $Z_1(\boldsymbol{u})$. Hence the resulting map of $Z_1(\boldsymbol{u})$ estimates using a regression function ϕ will be smooth, essentially like a kriging-type map.
(2) It does not provide a measure of spatial uncertainty about the unknown $Z_1(\boldsymbol{u})$ values.

Geostatistics allows various approaches for integrating secondary variable, but the drawbacks of many of these traditional approaches is that they allow only a single-point relation between $Z_1(\boldsymbol{u})$ and the various $Z_i(\boldsymbol{u})$, $i > 1$. A single point relation is a relation that describes the relation between $Z_1(\boldsymbol{u})$ as a sum/product of single functions of $Z_i(\boldsymbol{u})$, for example $Z_1 = f_2(Z_2) + f_3(Z_3) + \cdots f_{Nv}(Z_{Nv})$ or $Z_1 = f_2(Z_2) f_3(Z_3) \cdots f_{Nv}(Z_{Nv})$ is a single point relation, while $Z_1 = f_2(Z_2, Z_3, Z_4) + f_3(Z_5, Z_6) + \cdots$ expresses a multiple-point relation. Two traditional approaches are presented.

4.5. Secondary data as trend information

A simple but often effective way to introduce secondary data in either kriging or simulation algorithm is to use secondary information as a trend. In *kriging with locally varying mean*, the primary variable is decomposed into a mean and trend component

$$Z_1(\boldsymbol{u}) = m(\boldsymbol{u}) + R(\boldsymbol{u})$$

Since secondary data is often smooth in nature, the mean-component can be used to model the trend

$$\mathrm{E}\big[Z_1(\boldsymbol{u})\big] = m(\boldsymbol{u}) = \varphi(z_2(\boldsymbol{u}))$$

where the function φ needs to be calibrated from data. This requires that secondary data is available everywhere $Z_1(\boldsymbol{u})$ needs to be determined.

4.6. Full co-kriging

In kriging with locally varying mean, the spatial variability of the soft data itself is largely neglected. Often, secondary data has its own particular spatial continuity, or shows distinct patterns of spatial correlation with the primary variable. In such case, one would like to use the full spatial correlation of the secondary data expressed in the correlogram $\rho_{Z_2}(\boldsymbol{h})$ and the spatial correlation between secondary and primary expressed through the cross-covariance

$$\rho_{Z_1 Z_2}(\boldsymbol{h}) = \frac{\mathrm{Cov}\big(Z_1(\boldsymbol{u}), Z_2(\boldsymbol{u}+\boldsymbol{h})\big)}{\sqrt{\mathrm{Var}(Z_1)\,\mathrm{Var}(Z_2)}}$$

This covariance is used to determine the weights $\lambda_\alpha(\boldsymbol{u})$ and $\lambda_\beta(\boldsymbol{u})$ in the full co-kriging estimate $z^*_{\mathrm{coK}}(\boldsymbol{u})$ of $Z_1(\boldsymbol{u})$

$$z^*_{\mathrm{coK}}(\boldsymbol{u}) = \sum_{\alpha=1}^{n} \lambda_\alpha(\boldsymbol{u}) z_1(\boldsymbol{u}_\alpha) + \sum_{\beta=1}^{n} \lambda_\beta(\boldsymbol{u}) z_2(\boldsymbol{u}_\beta)$$

Due to the smoothness, hence spatial redundancy, of secondary data only the co-located secondary is retained to estimate the primary variable at $\boldsymbol{u}$, leading to a kriging termed *co-located co-kriging*. This simplifies considerably the covariance modeling and is therefore the most used form of co-kriging.

4.7. Accounting for scale of data sources

Data from various sources does not come at the same scale. In reservoir characterization, well-log data provides a local, fine-scale (decimeter-scale), usually vertical insight into the reservoir heterogeneity. Seismic provides indirect information on rock properties on a much larger scale and tends to smooth out heterogeneities, particularly any vertical variations. Seismic resolution in the vertical is usually equal to a quarter wavelength, which for current seismic surveying method entails a vertical resolution of 15–20 m for reservoirs at depth of 2 km. Well-test and production data provides information on an even larger scale of the reservoir.

Geostatistics allows integration of data existing at various scales, i.e. with various volume supports. Often, one aims at constructing reservoir models at the very fine scale taking into account coarse scale data, then upscale the fine model into models for flow simulation. Fig. 1c draws a hypothetical case where one wants to estimate or simulate the unknown fine scale value at $\boldsymbol{u}'$ given other fine scale samples $z(\boldsymbol{u}_\alpha)$ and a single coarse scale datum z_V. We assume that the block value (the coarse scale information) is a linear average of all the fine scale values within that block

$$z_V = \frac{1}{V}\int_V z(\boldsymbol{u})\,\mathrm{d}\boldsymbol{u}$$

Block kriging or block sequential simulation allows to estimate, simulate respectively, the unknown fine scale or point value, given the linear average z_V information and the point data using a block kriging system of equations

$$\begin{cases} \sum_{\beta=1}^{n} \lambda_\alpha(\boldsymbol{u}) C(\boldsymbol{u}_\alpha - \boldsymbol{u}_\beta) + \lambda_0 \overline{C}(V, \boldsymbol{u}_\alpha) = C(\boldsymbol{u}_\alpha - \boldsymbol{u}) & \forall \alpha = 1,\ldots,n \\ \sum_{\beta=1}^{n} \lambda_\beta(\boldsymbol{u}) C(\boldsymbol{u}_\beta, V) + \lambda_0 \overline{C}(V, V) = \overline{C}(V, \boldsymbol{u}) & \text{with } C(\boldsymbol{h}) = \mathrm{Var}(Z) - \gamma(\boldsymbol{h}) \end{cases}$$

with $C(\boldsymbol{h})$ the spatial covariance between any two locations at distance $\boldsymbol{h}$ and

$$\overline{C}(V,\boldsymbol{u}) = \frac{1}{V}\int_V C(\boldsymbol{u}-\boldsymbol{u}')\,\mathrm{d}\boldsymbol{u}' \qquad \overline{C}(V,V) = \frac{1}{V^2}\int_V\int_V C(\boldsymbol{u}-\boldsymbol{u}')\,\mathrm{d}\boldsymbol{u}'\mathrm{d}\boldsymbol{u}$$

The block kriging estimate being

$$z^*_{\mathrm{SK}}(\boldsymbol{u}) - m = \sum_{\alpha=1}^{n}\lambda_\alpha(\boldsymbol{u})\big(z(\boldsymbol{u}_\alpha)-m\big) + \lambda_0\big(z_V(\boldsymbol{u}')-m\big)$$

$\overline{C}(\boldsymbol{u},V)$ is the average covariance between the block V and the location to be estimated, it measure the degree of correlation between the known block and unknown point support.

4.8. Beyond multi-variate normality: non-parametric indicator geostatistics

Although the sequential Gaussian simulation technique is a mathematically sound and often used in practical applications it has some severe shortcoming attributed to the multi-Gaussian distribution model. Geologists often criticize reservoir models constructed by sgsim as too disorganized, noisy or simply geologically unrealistic, even though the models honor variogram information extracted from wells or outcrops. The multi-Gaussian model enforces a so-called maximum entropy property on the resulting simulations, meaning that, for a given variogram model, sgsim will construct reservoir models containing maximally disconnected extremes. A closer inspection of sgsim realizations shows that the extreme (both maximum and minimum) values are maximally separated; hence the texture appears disorganized. Often, nature is more organized than assumed by the multi-Gaussian model. Recall that the multi-Gaussian distribution is a model choice, i.e. embedded into the sgsim algorithm and has no connection to reality. To correct this default of sgsim, one needs to model the connectivity of extremes from any sample or outcrop data and enforce that connectivity to exist in the resulting realizations. Note that the connectivity of the lows need not be the same as connectivity of highs, nor need their direction of continuity be the same. In order to quantify connectivity better, one uses so-called indicator random functions as introduced above, i.e. for various thresholds $z_k, k = 1,\ldots,K$ one defines a variable

$$I(\boldsymbol{u},z_k) = \begin{cases} 1 & \text{if } Z(\boldsymbol{u}) \le z_k \\ 0 & \text{otherwise} \end{cases}$$

Each of the K indicator variables allows us then to focus on various ranges of the variable. Taking z_k equal to the median allows quantifying the connectivity of average z-values, while taking z_k equal to some extreme quantile allows quantifying connectivity of extremes. A spatial measure of connectivity between any two points in space separated by a lag-distance $\boldsymbol{h}$ is now the indicator variogram,

$$\gamma_I(\boldsymbol{h},z_k) = \mathrm{Var}\big(I(\boldsymbol{u},z_k) - I(\boldsymbol{u}+\boldsymbol{h},z_k)\big)$$

which can be modeled from the sample data by transforming each sample into a vector of size K of indicator data (zeros and ones). Once the K indicator variograms are determined, one can perform indicator kriging and simulation. However the goal of indicator kriging or indicator simulation is not to estimate or simulate zeros and ones (indicators) but to determine local probability models that do not rely on any Gaussian assumption. Indeed, in indicator methods one relies on the following simple rule that

$$\begin{aligned} \mathrm{E}\big[I(\boldsymbol{u},z_k)\big] &= \Pr\big(I(\boldsymbol{u},z_k)=1\big)\times 1+\Pr\big(I(\boldsymbol{u},z_k)=0\big)\times 0 \\ &= \Pr\big(I(\boldsymbol{u},z_k)=1\big) \\ &= \Pr\big(Z(\boldsymbol{u})\le z\big) \end{aligned}$$

Hence, any estimation of $I(\boldsymbol{u},z_k)$ is also an estimation of $\Pr\big(Z(\boldsymbol{u})\le z\big)$, which means that kriging of an indicator variable is nothing more than determining the local uncertainty about the original Z-variable

$$\mathrm{E}\big[I(\boldsymbol{u},z_k)\mid(n)\big]=\Pr\big(Z(\boldsymbol{u})\le z\mid(n)\big)$$

For example, to determine the probability of porosity at un-sampled location $\boldsymbol{u}$ to be above z_k, given the local well data, one only needs to krige (estimate) the indicator variable using indicator kriging, then that kriged value is a also a probability. The indicator kriging estimate is written as

$$\big[I(\boldsymbol{u},z_k)\big]^*-p=\sum_{\alpha=1}^{n}\lambda_\alpha(\boldsymbol{u})\big(i(\boldsymbol{u}_\alpha,z_k)-p\big)=\big[\Pr\big(Z(\boldsymbol{u})\le z_k\mid(n)\big)\big]^*$$

p is the estimated proportion of data above the cut-off z_k. The indicator kriging weights are determined by solving the indicator kriging equations

$$\sum_{\beta=1}^{n}\lambda_\alpha(\boldsymbol{u})\mathrm{Cov}\big(I(\boldsymbol{u}_\alpha,z_k),I(\boldsymbol{u}_\beta,z_k)\big)=\mathrm{Cov}\big(I(\boldsymbol{u}_\alpha,z_k),I(\boldsymbol{u},z_k)\big)\qquad \forall\alpha=1,\ldots,n$$

By estimating $\Pr\big(Z(\boldsymbol{u})\le z_k\mid(n)\big)$ for various thresholds z_k, indicator kriging provides the uncertainty about the property at an un-sampled location $\boldsymbol{u}$ given any sample data (n). Note that no Gaussian or any parametric distribution model is assumed, the uncertainty is quantified through a series of cumulative probabilities at given thresholds z_k.In the same way simple kriging naturally extends into sequential Gaussian simulation, indicator kriging extends into sequential indicator simulation (sisim). Sisim does not require any distribution assumption, as does sgsim.

(1) Assign the data (n) to the grid
(2) Define a random path visiting all nodes $\boldsymbol{u}$
(3) Loop over all nodes $\boldsymbol{u}_i$
  (a) Construct a conditional distribution $F(\boldsymbol{u}_i,z\mid(n+i-1))$ by estimating $\Pr(Z(\boldsymbol{u})\le z_k\mid(n+i-1))$ for various threshold using indicator kriging
  (b) Draw a simulated value $z(\boldsymbol{u}_i)$ from the conditional distribution $F(\boldsymbol{u}_i,z\mid(n+i-1))$
  (c) Add simulated value to data set $(n+i-1)$
(4) End simulation

Fig. 4 provides an example of an indicator simulation realization.

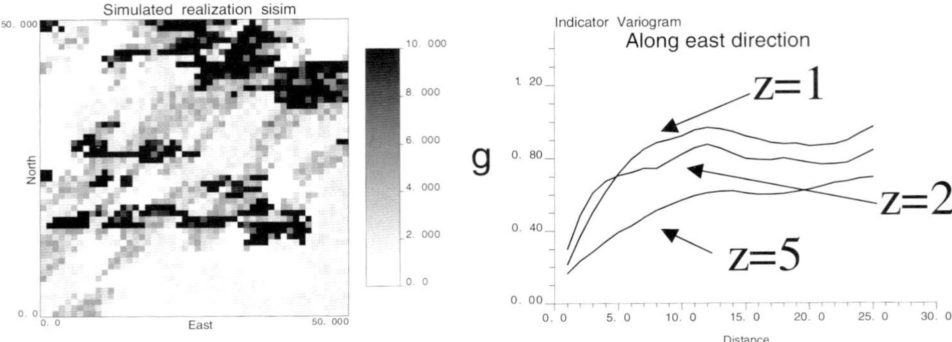

Fig. 4. (a, left) single sisim realization, notice that the high values (black) have a different geological continuity than the low values (white, gray), (b, right) this difference is also observed in the indicator variograms of for various values of z.

5. BEYOND TWO-POINT MODELS: MULTIPLE-POINT GEOSTATISTICS

5.1. *Accounting for geological realism*

In the last decade, geostatistics has evolved from estimation (kriging) to stochastic simulation, essentially driven by applications in the Petroleum geosciences. In its early day, geostatistics was mostly applied to mining problems, where obtaining estimates for blocks to be mined was essential. In reservoir characterization one is less interested in estimating permeability, rather, permeability should be viewed from a fluid flow perspective. This entails that a correct quantification of the permeability connectivity or its spatial pattern is more important than obtaining a locally accurate, but smooth estimate. Unfortunately, the tradition of geostatistics borrowed from the mining days relies on quantifying spatial patterns using the variogram. Variograms are very limited measures of spatial patterns, since they measure the relation between two points in space only. Strong connectivities, or curvi-linear features (channels, cross-bedding), require quantification of patterns, which consist of multiple spatial locations. It has long been a tradition in image analysis to quantify patterns in images by multiple-point templates or windows. The recognition that variograms are essentially too limited quantification of geological continuity has lead to a new field in geostatistics: *multiple-point geostatistics*. Multiple-point geostatistics extends the geostatistical simulation paradigm, namely stochastic imaging consists of two parts: recognize and quantify inherent patterns from the data, then reproduce that pattern set in a set of realization conditional (constrained) to the actual hard and soft data. Hence, in traditional geostatistics, pattern recognition is quantified in the variogram model, while pattern reproduction is obtained through sequential conditional simulation. Multiple-point geostatistics extends this paradigm through the use of a training image or reservoir analog as a mean for quantifying spatial patterns. The training image provides the geologist the opportunity to conceptually draw any patterns deemed important for the field under study. The idea of multiple-point geostatistics is then to capture essential patterns from this training image and reproduce them conditional to data in a set of realizations. While the field of image analysis deals only with the recognition of patterns, the field of multiple-point geostatistics aims at recognition *and* (conditional) reproduction.

5.2. From variogram to training image to multiple stochastic models

In order to go beyond the traditional variogram model or two-point correlation measures one needs to construct a training image or training reservoir model. This training image is an analog of what one would like to observe, in term of geological structures, in the simulated reservoir models. Indeed, as it is already difficult to extract from wells a suitable variogram model it will be even more difficult to determine higher order statistics. Training images provide a dense dataset from which many multiple-point statistics including two-point statistics can be extracted. Training images are often difficult to come by and when available are deemed too specific, not representative for the actual reservoir, hence using them would lead to too subjective modeling of the reservoir. However, in traditional geostatistics, multiple-point information is never quantified and is always implicitly delivered in the reservoir model through the specific algorithm applied. It is not because these higher moments are not explicitly visible, they should be deemed more objective. In fact a training image allows a full quantification of all higher order moments or patterns *before* any geostatistical modeling is attempted. Geological interpretation is therefore best delivered in images, which can be obtained for example using a non-conditional Boolean technique, through dense outcrop sampling, or even simple geological re-constructions.

The sequential simulation methodology can be extended to simulate stochastic models that reflect the geological structure of the training. As outlined above, sequential simulation determines at each grid node to be simulated a conditional probability, written in short as

$$P(A_k \mid \boldsymbol{B})$$

$A_k = \{\text{facies category } k \text{ occurring}\}$
or $A_k = \{\text{petrophysical property within class } [z_k, z_{k+1}] \text{ occurring}\}$
and $\boldsymbol{B} = \{\text{the sample data and previously simulated nodes}\}$.

Recently, a methodology termed *snesim* (single normal equation simulation) has been developed that allows exporting the patterns in the training image to the actual subsurface reservoir model conditioned to local well data. The core idea of the snesim code is to model, from the training image, the conditional distribution $P(A_k \mid \boldsymbol{B})$ for any possible A_k and $\boldsymbol{B}$. Therefore, the snesim program allows, before the geostatistical simulation is started, to store all possible distribution values $P(A_k \mid \boldsymbol{B})$. Sequential simulation then amounts to sequential drawing from these conditional distributions.

Fig. 5 shows a reservoir of sand/shale layering with two vertical wells. The snesim methodology consists of recognizing the elliptical shale patterns in the training image, then simulating multiple realizations constrained by the two facies data from two vertical wells and reproducing patterns similar than the training image.

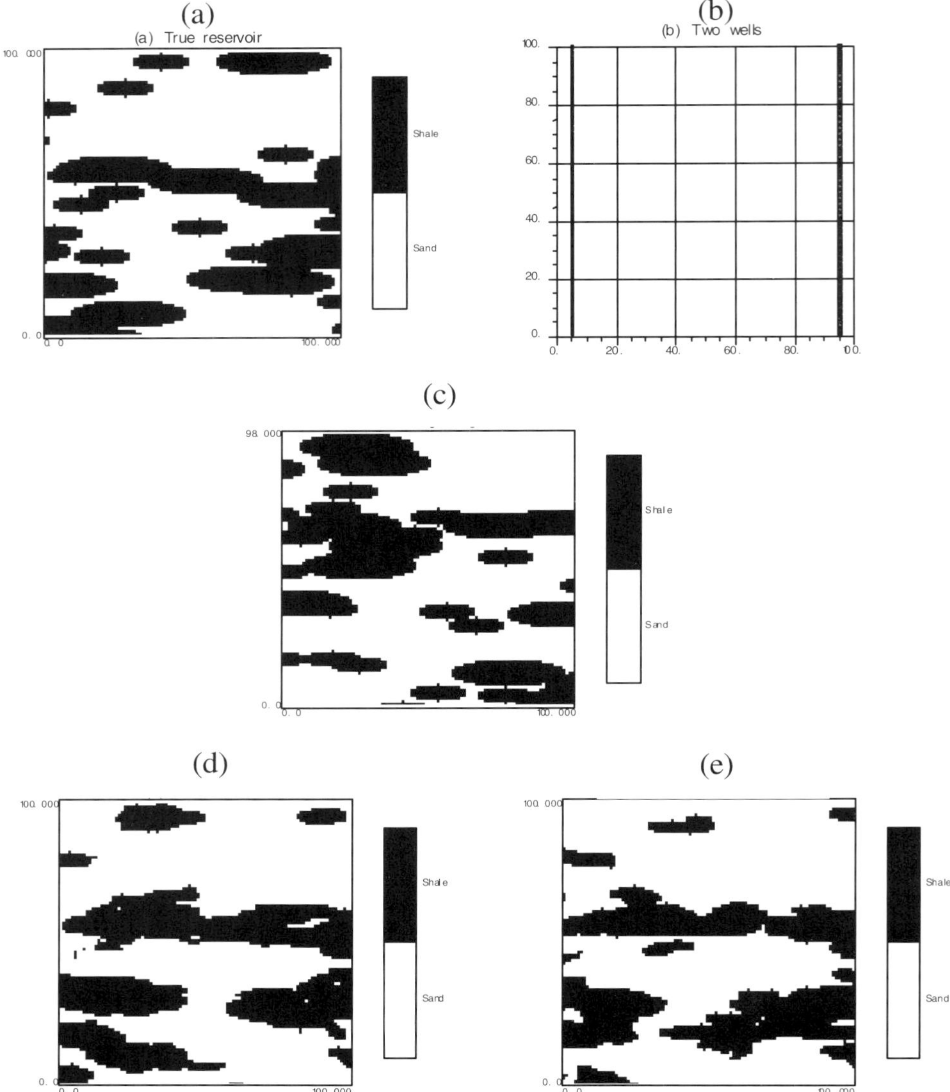

Fig. 5. (a) True ‘unknown’ reservoir with sand and shale, (b) facies sequence observed in two vertical wells, (c) training image, not constrained to any well data, merely conceptual, (d,e) two realizations constrained to the wells, reflecting the geological continuity of the training image.

5.3. Data integration

Another tradition that has emerged in practising geostatistics is the inclusion of so-called soft data (e.g. seismic data) by modeling the two-point cross-correlation between facies/petrophysical properties and the soft data. The cross-variogram or co-variance is used as a structural model to represent the relation between facies and soft data. This procedure largely neglects the spatial pattern of the soft data itself because (1) the cross-variogram is only a limited two-point correlation measure and (2) predictions of facies or petrophysical properties are only based on co-located information. Actual seismic may contain spatial patterns that are related to certain geological events (channeling, shale bodies). A geophysicist interprets such event by observing a pattern change in the seismic, not by observing a single seismic voxel or pixel.

Neural networks have recently emerged as a powerful tool for modeling non-linear and multiple-point relation between the unknown facies/petrophysical property A and multiple seismic attributes $\boldsymbol{C}$. The neural net can provide us with a probability model $P(A \mid \boldsymbol{C})$, essentially quantifying the uncertainty about facies presence at location $\boldsymbol{u}$ from multiple seismic attributes. However, neural nets do not account for any information about the geological continuity of the facies or petrophysical property, as quantified by $P(A \mid \boldsymbol{B})$.

A simple yet effective way to combine the $P(A \mid \boldsymbol{C})$ from the neural net calibration and $P(A \mid \boldsymbol{B})$ from the snesim method to get $P(A \mid \boldsymbol{B}, \boldsymbol{C})$ needs to be determined. Then, $P(A \mid \boldsymbol{B}, \boldsymbol{C})$ is used in snesim to simulate a facies value at nodes, given the well data, seismic data and previously simulated nodes. The following relationship has been proposed as one way to combine various sources of information

$$\frac{x}{b} = \frac{c}{a} \Rightarrow x = \frac{bc}{a} \geq 0$$

where

$$a = \frac{1-P(A)}{P(A)}, \quad b = \frac{1-P(A \mid \boldsymbol{B})}{P(A \mid \boldsymbol{B})}, \quad c = \frac{1-P(A \mid \boldsymbol{C})}{P(A \mid \boldsymbol{C})}, \quad x = \frac{1-P(A \mid \boldsymbol{B}, \boldsymbol{C})}{P(A \mid \boldsymbol{B}, \boldsymbol{C})}$$

Then

$$P(A \mid \boldsymbol{B}, \boldsymbol{C}) = \frac{1}{1+x} = \frac{a}{a+bc}$$

6. CONCLUSIONS

The information we have about a spatially varying phenomenon is usually incomplete. Most often, only few samples of the variable under study are available, next to an abundance of indirect information gathered with remote sensing devices. This implies that one cannot determine with full confidence the exact unknown true outcome of that variable at every location. Geostatistics allows quantifying the uncertainty about the unknown spatial phenomenon in terms of a set of alternative representations, termed realizations, of the unknown truth. The aim of stochastic simulations is to enforce various properties on these realizations

(1) The realizations should have a similar pattern of variability as the unknown truth. These patterns can be quantifies through a variogram model fitted from the data or through a training image modeled from a concept or analog
(2) The realization should be constrained to any local hard sample data or soft data.

Once these realizations are simulated, they can be post-processed using any type of transfer function (flow simulation, mining operation, environmental clean-up) to determine cost/benefit and associated risk.

7. GLOSSARY

Bi-variate distribution: Statistical distribution describing the joint variability of two random variables. In geostatistics, the bi-variate distribution is used to describe the dependence between variables at two spatial locations; hence it also depends on the distance $\boldsymbol{h}$ between those two locations. It is related to the variogram/covariance of the indicators.

Co-kriging (co-located): Refers to a least-square spatial estimation technique that estimates an unknown primary variable at any location $\boldsymbol{u}$ from both primary and secondary data. In co-located co-kriging, one retains only the secondary data at $\boldsymbol{u}$.

Covariance (spatial): is a measure of two-point association between random variables at any location $\boldsymbol{u}$ and $\boldsymbol{u}+\boldsymbol{h}$, for whatever vector $\boldsymbol{h}$.

Correlogram: is the covariance standardized with the variance of that variable.

Indicator random variables: variables that 'indicate' to which class an observation belongs (discrete case) or if or not the variable exceeds a given threshold (continuous case).

Hard data: exact local measurements about the variable one estimates/simulates.

Kriging: a generalized least square technique for spatial estimation.

Monte Carlo simulation: procedure of generating a series of outcomes from a probability model.

Multi-variate normal: specific parametric form of a multi-variate distribution that is completely determined by the means and variances of each of the single variables and all their respective covariances.

Multiple-point geostatistics: area of geostatistics that quantifies spatial continuity by considering multiple locations in space instead of taking them two by two.

Primary variable: variable to be estimated/simulated.

Random variables: variables that model the uncertain outcomes about an event.

Realization: outcome of a Monte Carlo simulation experiment.

Secondary variable: variables bring indirect information about the primary variable.

Soft data: indirect measurements about the variable one estimates/simulates.

Stationarity: a model property stating that statistics of a variable or set of variables are translation invariant in space.

Training image: an conceptual image depicting geological structure to be reproduced in multiple stochastic realizations.

Two-point statistics: any type of statistics between two variables. These variables could for example model the unknown outcomes at two spatial locations.

Variogram: a measure of dissimilarity between random variables at any two spatial locations.

REFERENCES

Chiles, J.-P. and Delfiner, P., 1999. *Geostatistics: Modeling Spatial Uncertainty.* Wiley and Sons, New York, NY.

Deutsch, C.V. and Journel, A.G., 1998. *GSLIB: Geostatistical Software Library and User's Guide.* Oxford University Press.

Goovaerts, P., 1997. *Geostatistics for Natural Resources Estimation.* Oxford University Press.

Isaaks, E., and Srivastava, S., 1989. *An Introduction to Applied Geostatistics.* Oxford Univerity Press.

Journel, A.G. and Huijbregts, C., 1978. *Mining Geostatistics.* Academic Press.

Journel, A.G., 1989. *Fundamentals of Geostatistics in Five Lessons, Volume 8. Short Course in Geology.* American Geophysical Union, Washington, DC.

Journel, A.G., 1999. Conditioning Geostatistical Realization to Non-Linear Volume Averages. *Math. Geol.*, 31, 955–964

Olea, R., 1991. *Geostatistical Glossary and Multilingual Dictionary.* Oxford University Press.

Strebelle, S., 2000. *Sequential Simulation Drawing Structure from Training Images.* Ph.D. Dissertation, Stanford University, Stanford, CA.

PART 2.

GEOPHYSICAL ANALYSIS AND INTERPRETATION

Developments in Petroleum Science, 51
Editors: M. Nikravesh, F. Aminzadeh and L.A. Zadeh

Chapter 7

MINING AND FUSION OF PETROLEUM DATA WITH FUZZY LOGIC AND NEURAL NETWORK AGENTS

MASOUD NIKRAVESH [a,b,1] and FRED AMINZADEH [b,c,2]

[a] *BISC Program, Department of Electrical Engineering and Computer Sciences, University of California, Berkeley, CA 94720, USA*
[b] *Zadeh Institute for Information Technology (ZIFIT)*
[c] *Fact Incorporated, 14019 SW Freeway, Suite 301-225, Sugar Land, TX 77478, USA*

ABSTRACT

Analyzing data from well logs and seismic is often a complex and laborious process because a physical relationship cannot be established to show how the data are correlated. In this study, we will develop the next generation of 'intelligent' software that will identify the nonlinear relationship and mapping between well logs/rock properties and seismic information and extract rock properties, relevant reservoir information, and rules (knowledge) from these databases. The software will use fuzzy logic techniques because the data and our requirements are imperfect. In addition, it will use neural network techniques, since the functional structure of the data is unknown. In particular, the software will be used to group data into important data sets; extract and classify dominant and interesting patterns that exist between these data sets; discover secondary, tertiary and higher-order data patterns; and discover expected and unexpected structural relationships between data sets.

1. INTRODUCTION

In reservoir engineering, it is of great importance to characterize how seismic information (attributes), the lithology and geology of the rocks are related to the well logs such as porosity, density, and gamma ray. However, data from well logs and seismic attributes are often difficult to analyze because of their complexity. A physical relationship cannot usually be established to show how the data are correlated, except by a laborious process, and the human ability to understand and use the information content of these data is limited.

Neural networks provide the potential to establish a model from nonlinear, complex, and multi-dimensional data and find wide application in analyzing experimental, industrial, and field data (Baldwin et al., 1990; Baldwin et al., 1989; Pezeshk et al., 1996; Rogers et al., 1992; Wong et al., 1995a,b; Nikravesh et al., 1996). In recent

[1] E-mail: Nikravesh@cs.berkeley.edu; Fax: +1 (510) 642-5775.
[2] Fax: +1 (281) 265-2512.

years, the neural network literature has stimulated growing interest among reservoir engineers, geologists, and geophysicist (Nikravesh and Aminzadeh, 1998; Boadu, 1997; Nikravesh, 1998; Klimentos and McCann, 1990; Aminzadeh et al., 1994). Boadu (1997) applied artificial neural network successfully to find the relationship between seismic and rock properties for Sandstones Rocks. In our recent studies (Nikravesh and Aminzadeh, 1998), we further analyzed data published by Klimentos and McCann (1990) for Sandstones and recently analyzed by Boadu (1997) using an artificial neural network. It was concluded that neural network model had a better performance than a multiple linear regression model. Neural network, neuro-fuzzy, and knowledge-based models have been successfully used to model rock properties based on well log databases (Nikravesh, 1998). However, using neural networks for identification purposes is more useful when a large number of data are available. In addition, conventional neural network models cannot deal with uncertainty in data due to fuzziness.

In this study, our 'intelligent' software will be built around two technologies: artificial neural network and fuzzy logic (Appendix A). Each of them is expected to remedy one or more short-comings of conventional data mining and knowledge management systems. The software will take advantage of the tolerance for imprecision that fuzzy logic can bring to bear on the process of knowledge acquisition from massive data sets. The software will use fuzzy logic techniques because the data and our requirements are imperfect. Fuzzy logic is considered to be appropriate to deal with the nature of uncertainty on system and human error which are not included in current reliability theories. We will integrate our recently developed algorithms and heuristics to induce fuzzy logic rules from data sets. An artificial neural network will be used to extract and fine-tune the rule base including granulation of variables and the characteristics of the membership functions.

In this study, we begin by mining wireline logs (such as density, gamma ray, travel time, SP, and resistivity). Then, the results will be used to train the neural network models. In addition, we will analyze the data to recognize the most important patterns, structures, relationships, and characteristics based on neural network and neuro-fuzzy models.

2. NEURAL NETWORK AND NONLINEAR MAPPING

In this section, a series of neural network models (Appendix B) will be developed for nonlinear mapping between wireline logs. A series of neural network models will also be developed to analyze actual welllog data and seismic information and the nonlinear mapping between wireline logs and seismic attributes will be recognized.

In this study, wireline logs such as travel time (DT), gamma ray (GR), and density (RHOB) will be predicted based on SP, and resistivity (RILD) logs. In addition, we will predict travel time (DT) based on induction resistivity and vice versa. In this study, all logs are scaled uniformly between -1 and 1 and results are given in scaled domain.

Fig. 1A through E show typical behavior of SP, RILD, DT, GR, and RHOB logs in scaled domain. The design of a neural network to predict DT, GR, and RHOB based on RILD and SP logs starts with filtering, smoothing, and interpolating values (in a

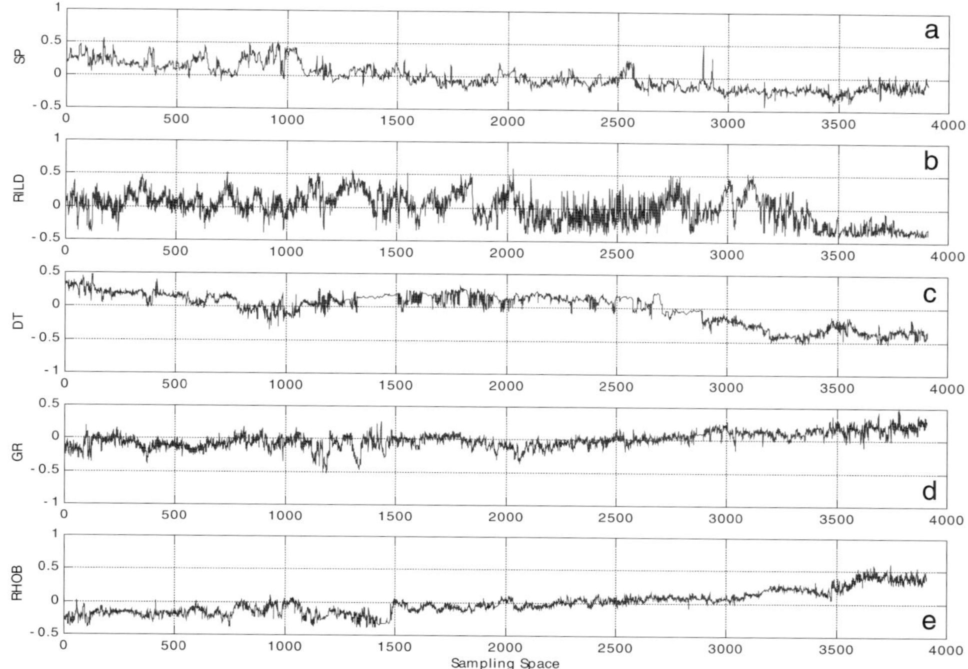

Fig. 1. Typical behavior of SP, RILD, DT, GR, and RHOB logs.

small horizon) for missing information in the data set. A first-order filter and a simple linear recursive parameter estimator for interpolating were used to filter and reconstruct the noisy data. The available data were divided into three data sets: training, testing, and validation. The network was trained based on the training data set and continuously tested using a test data set during the training phase. The network was trained using a backpropagation algorithm and modified Levenberge–Marquardt optimization technique (Appendix C). Training was stopped when prediction deteriorated with step.

2.1. Travel time (DT) prediction based on SP and resistivity (RILD) logs

The neural network model to predict the DT has 14 input nodes (two windows of data each with 7 data points) representing SP (7 points or input nodes) and RILD (7 points or input nodes) logs. The hidden layer has 5 nodes. The output layer has 3 nodes representing the prediction of the DT (a window of 3 data point). Typical performance of the neural network for training, testing, and validation data sets is shown in Fig. 2A–D. The network shows good performance for prediction of DT for training, testing, and validation data sets. However, there is not a perfect match between actual and predicted values for DT is the testing and validation data sets. This is due to changes of the lithology from point to point. In other words, some of the data points in the testing and validation data sets are in a lithological layer which was not presented in

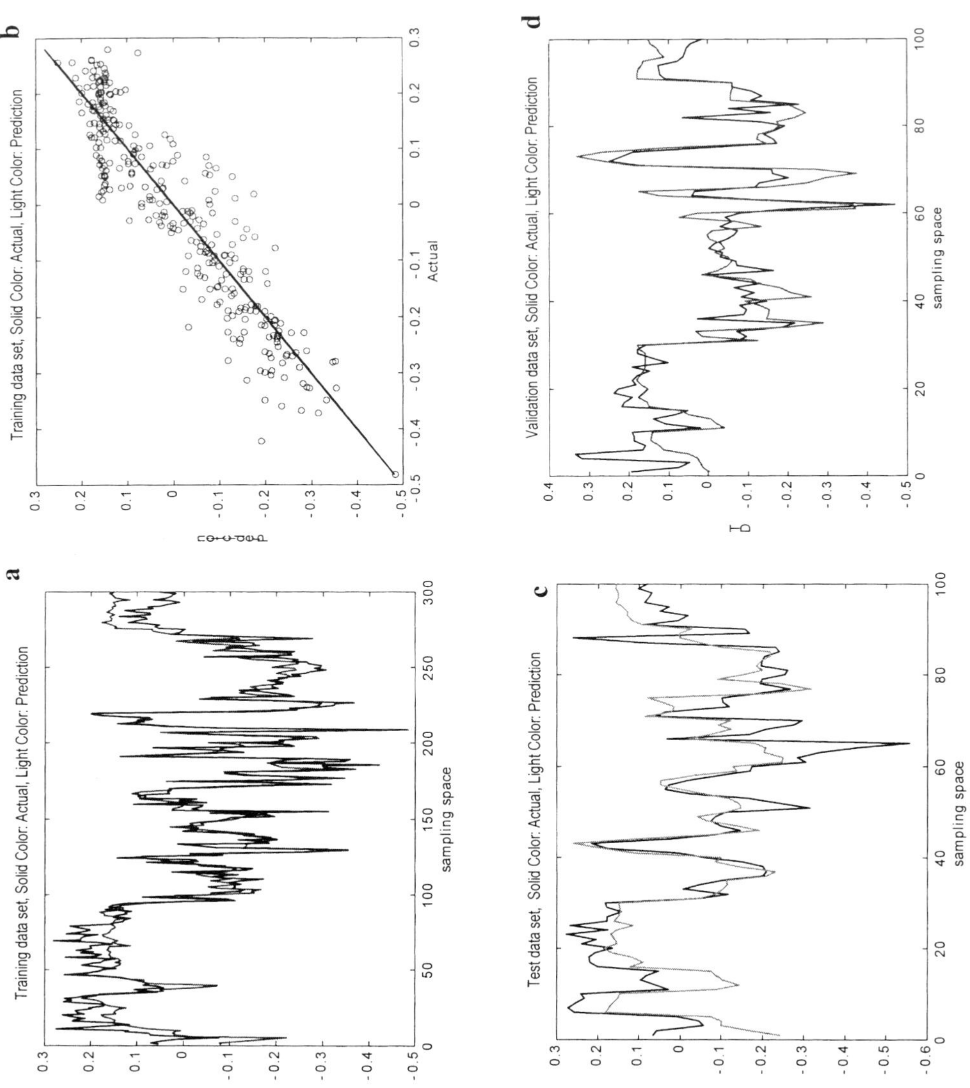

Fig. 2. Typical neural network performance for prediction of DT based on SP and RILD.

the training phase. Therefore, to have perfect mapping, it would be necessary to use the layering information (using other types of logs or linguistic information) as input into the network or use a larger data set for the training data set which represent all the possible behaviors in the data.

2.2. *Gamma ray (GR) prediction based on SP and resistivity (RILD) logs*

In this study, a neural network model is developed to predict GR based on SP and RILD log. The network has 14 input nodes (two windows of data, each with 7 data points) representing SP (7 points or input nodes) and RILD (7 points or input nodes) logs. The hidden layer has 5 nodes. The output layer has 3 nodes representing the prediction of the GR. Fig. 3A through D show the performance of the neural network for training, testing, and validation data. The neural network model shows a good performance for prediction of GR for training, testing, and validation data sets. In comparison with previous studies (DT prediction), this study shows that the GR is not as sensitive as DT to noise in the data. In addition, a better global relationship exits between SP–resistivity–GR rather than SP–resistivity–DT. However, the local relationship is in the same order of complexity. Therefore, two models have the same performance for training (excellent performance). However, the model for prediction of GR has a better generalization property. Since the two models have been trained based on the same criteria, it is unlikely that this lack of mapping for generalization is due to over fitting during the training phase.

2.3. *Density (RHOB) prediction based on sp and resistivity (RILD) logs*

To predict density based on SP and RILD logs, a neural network model with 14 input nodes representing SP (7 points or input nodes) and RILD (7 points or input nodes) logs, 5 nodes in the hidden layer, and 3 nodes in the output layer representing the prediction of the RHOB is developed. Fig. 4A through D show a typical performance of the neural network for the training, testing, and validation data sets. The network shows excellent performance for the training data set as shown in Fig. 4A and B. The model shows a good performance for the testing data set as shown in Fig. 4C. Fig. 4D shows the performance of the neural network for the validation data set. The model has relatively good performance for the validation data set. Therefore, there is not a perfect match between the actual and predicted values for RHOB for the testing and validation data set. Since RHOB is directly related to lithology and layering, and to have perfect mapping, it would be necessary to use the layering information (using other types of logs or linguistic information) as an input into the network or use a larger data set for the training data set which represent all the possible behaviors in the data. In these cases, one can use a knowledge-based approach using knowledge of an expert and select more diverse information which represent all different possible layering as a training data set. Alternatively, one can use an automated clustering technique to recognize the important clusters existing in the data and use this information for selecting the training data set.

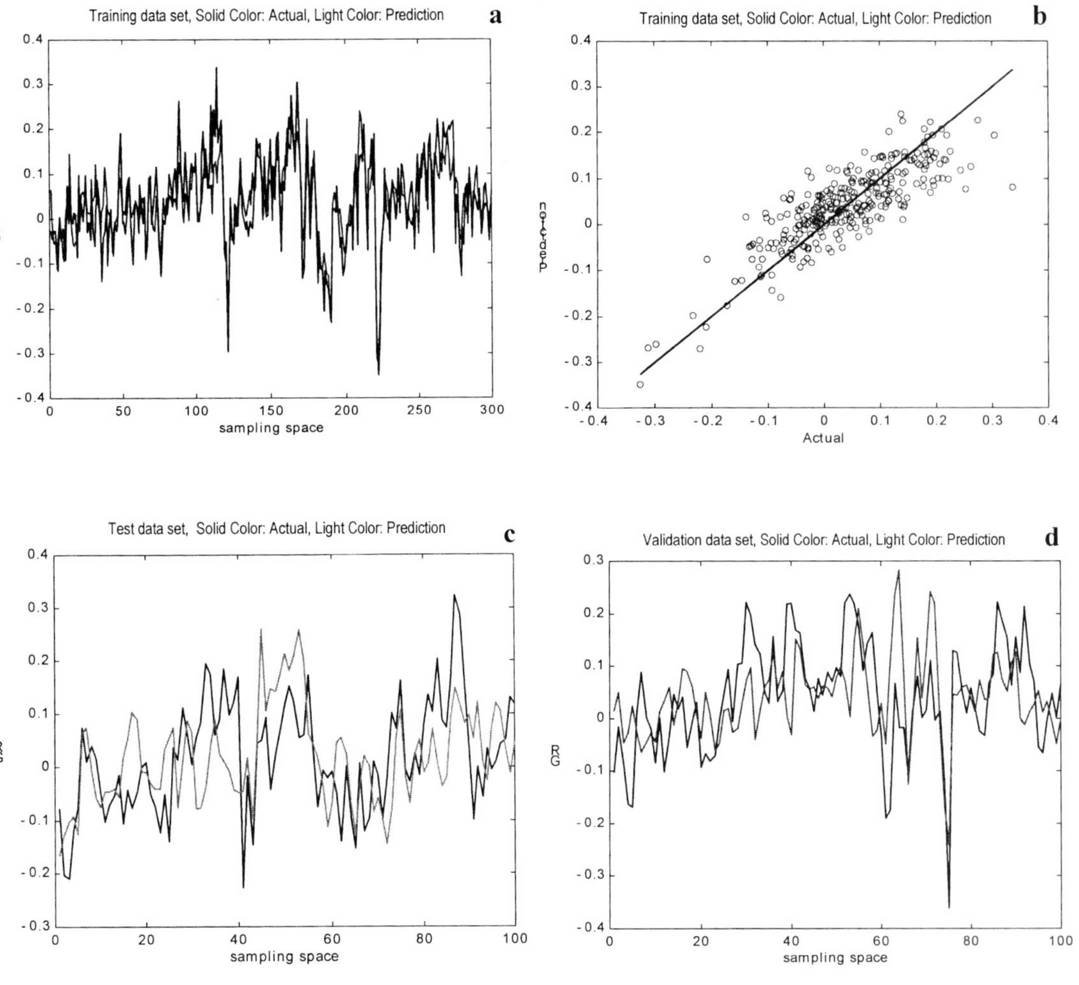

Fig. 3. Typical neural network performance for prediction of GR based on SP and RILD.

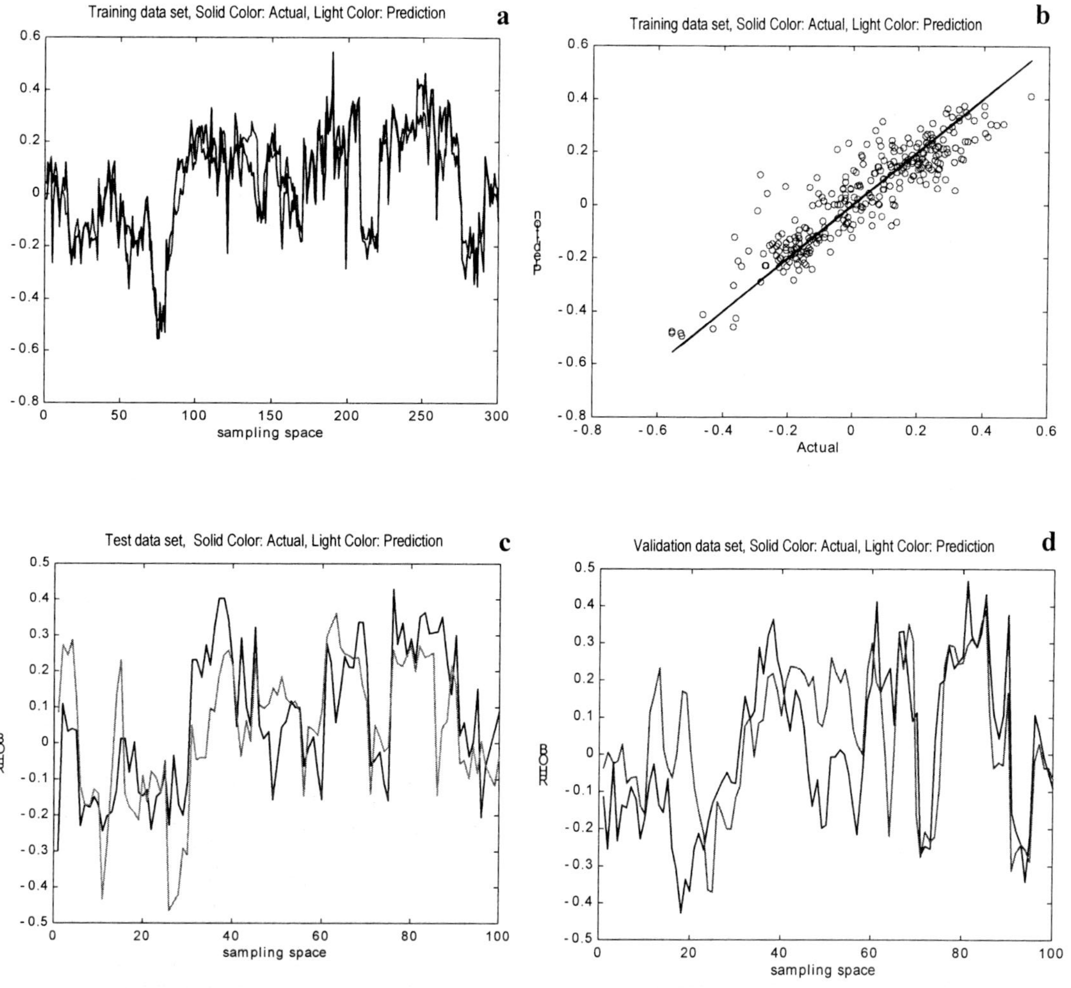

Fig. 4. Typical neural network performance for prediction of RHOB based on SP and RILD.

2.4. *Travel time (DT) prediction based on resistivity (RILD)*

The neural network model to predict the DT has 11 input nodes representing a RILD log. The hidden layer has 7 nodes. The output layer has 3 nodes representing the prediction of the DT. Using engineering knowledge, a training data set is carefully selected so as to represent all the possible layering existing in the data. The typical performance of neural network for the training, testing, and validation data sets is shown in Fig. 5A through D. As expected, the network has excellent performance for prediction of DT. Even though only RILD logs are used for prediction of DT, the network model has better performance than when SP and RILD logs used for prediction of DT (comparing Fig. 2A through D with 5A through D). However, in this study, knowledge of an expert was used as extra information. This knowledge not only reduced the complexity of the model, but also better prediction was achieved.

2.5. *Resistivity (RILD) prediction based on travel time (DT)*

In this section, to show that the technique presented in the previous section is effective, the performance of the inverse model is tested. The network model has 11 input nodes representing DT, 7 nodes in the hidden layer, and 3 nodes in the output layer representing the prediction of the RILD. Fig. 6A through D show the performance of the neural network model for the training, testing, and validation data sets. Fig. 6A and B show that the neural network has excellent performance for the training data set. Fig. 6C and D show the performance of the network for the testing and validation data set. The network shows relatively excellent performance for testing and validation purposes. As was mentioned in the previous section, using engineering knowledge the complexity of the model was reduced and better performance was achieved.

In addition, since the network model (prediction of DT from resistivity) and its inverse (prediction of resistivity based on DT) have relatively excellent performance and generalization properties, a one-to-one mapping was achieved. Therefore, this implies that a good representation of layering was selected based on knowledge of an expert.

3. NEURO-FUZZY MODEL FOR RULE EXTRACTION

In this section, a neuro-fuzzy model (Appendix D) will be developed for model identification and knowledge extraction (rule extraction) purposes. The model is characterized by a set of rules which can be further used for representation of data in the form of linguistic variables. Therefore, in this situation the fuzzy variables become linguistic variables. The neuro-fuzzy technique is used to implicitly cluster the data while finding the nonlinear mapping. The neuro-fuzzy model developed in this study is an approximate fuzzy model with triangular and Gaussian membership functions originally presented by Sugeno and Yasukawa (1993). K-mean technique (Appendix E) is used for clustering and the network is trained using a backpropagation algorithm and modified Levenberge–Marquardt optimization technique (Appendix C).

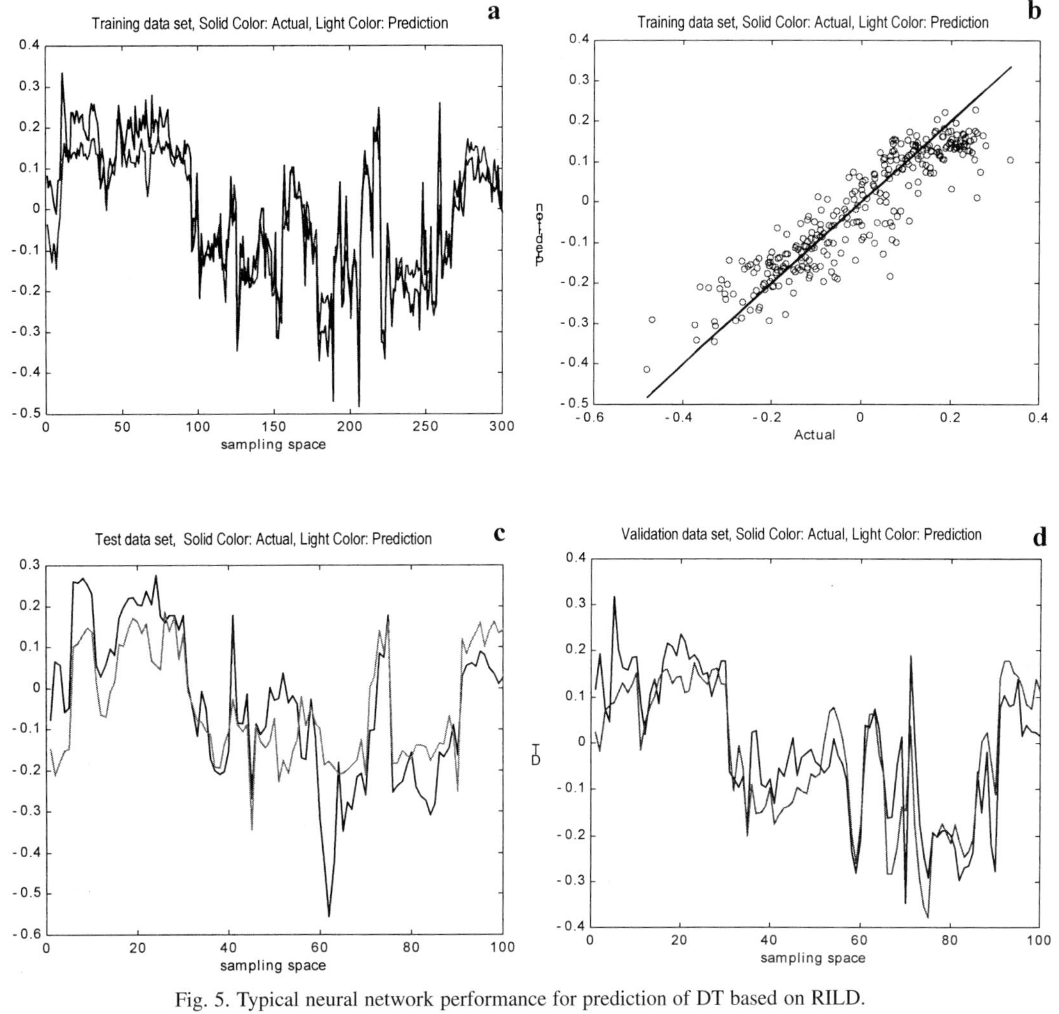

Fig. 5. Typical neural network performance for prediction of DT based on RILD.

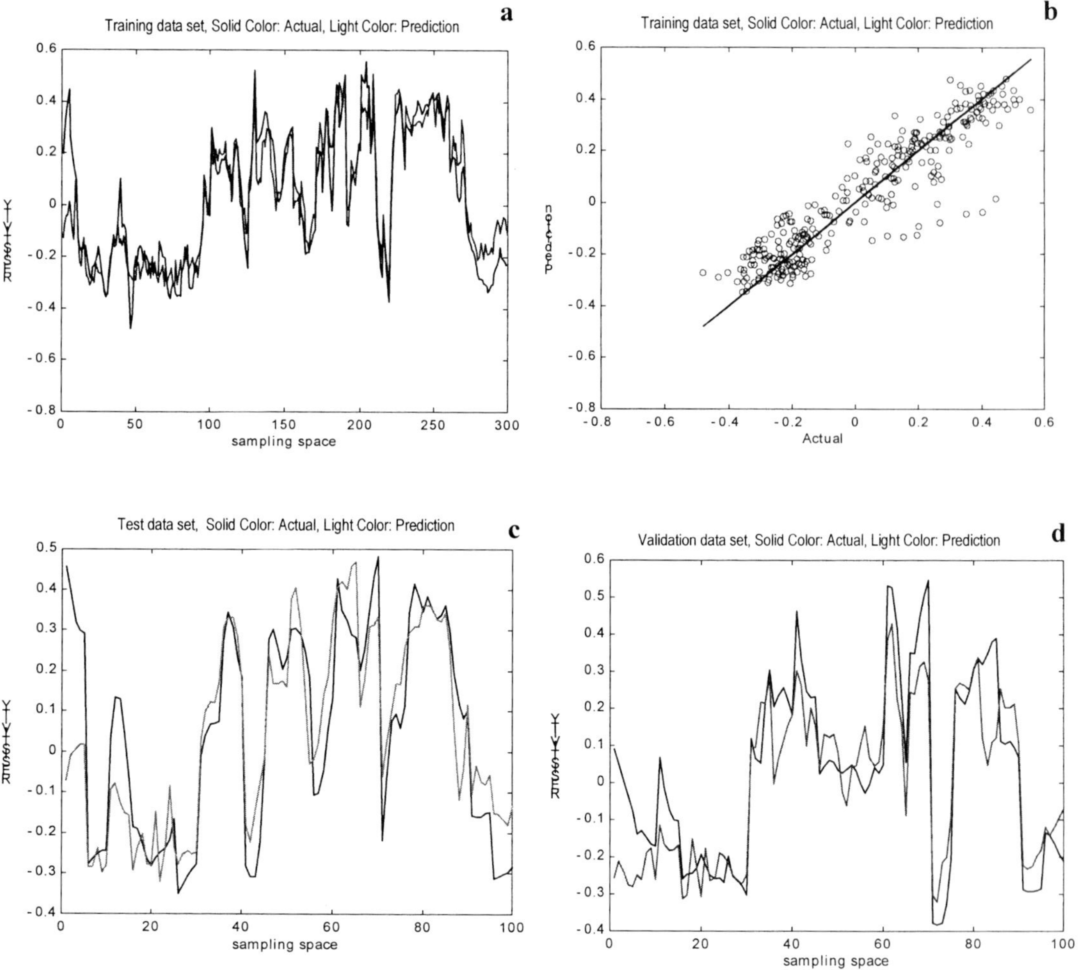

Fig. 6. Typical neural network performance for prediction of RILD based on DT.

In this study, the effect of rock parameters and seismic attenuation on permeability will be analyzed based on soft computing techniques and experimental data. The software will use fuzzy logic techniques because the data and our requirements are imperfect. In addition, it will use neural network techniques, since the functional structure of the data is unknown. In particular, the software will be used to group data into important data sets; extract and classify dominant and interesting patterns that exist between these data sets; and discover secondary, tertiary and higher-order data patterns. The objective of this section is to predict the permeability based on grain size, clay content, porosity, P-wave velocity, and P-wave attenuation.

3.1. Prediction of permeability based on porosity, grain size, clay content, P-wave velocity, and P-wave attenuation.

In this section, a neuro-fuzzy model will be developed for nonlinear mapping and rule extraction (knowledge extraction) between porosity, grain size, clay content, P-wave velocity, P-wave attenuation and permeability. Fig. 7 shows typical data, which has been used in this study. In this study, permeability will be predicted based on the following rules and Eqs. (D.1) through (D.4), see Appendix D:

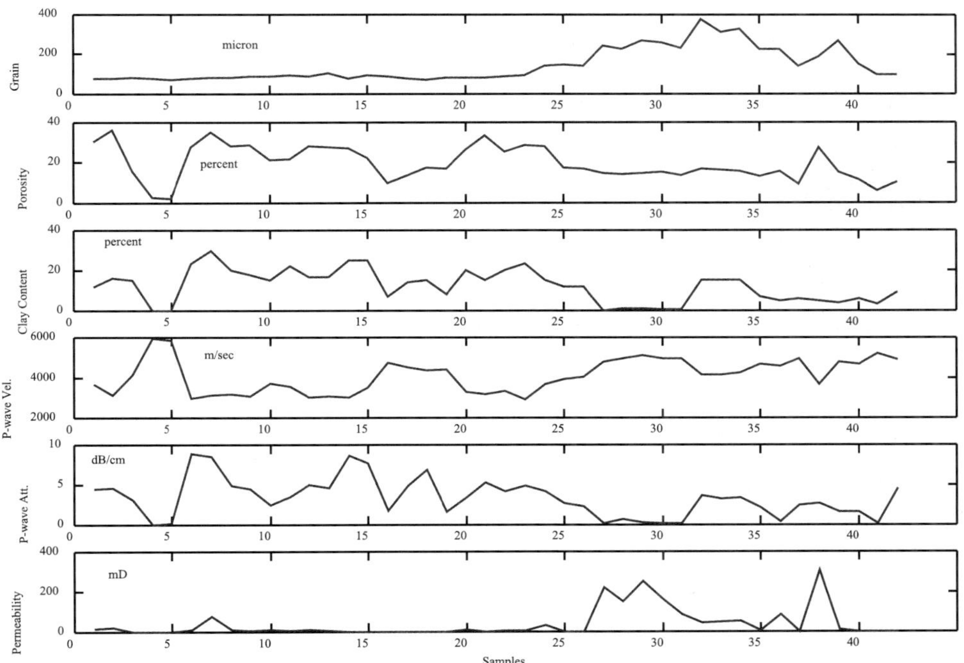

Fig. 7. Actual data (Boadu, 1997).

TABLE 1

Boundary of rules extracted from data

Porosity	Grain size	Clay content	P Wave velocity	P Wave attenuation
[−0.4585, −0.3170]	[−0.6501, −0.3604]	[−0.6198, −0.3605]	[0.0893, 0.2830]	[−0.6460, −0.3480]
[0.4208, 0.5415]	[−0.9351, −0.6673]	[0.2101, 0.3068]	[−0.7981, −0.7094]	[0.0572, 0.2008]
[−0.3610, −0.1599]	[−0.7866, −0.4923]	[−0.3965, −0.1535]	[−0.0850, 0.1302]	[−0.4406, −0.1571]
[−0.2793, −0.0850]	[−0.5670, −0.2908]	[−0.4005, −0.1613]	[−0.1801, 0.0290]	[−0.5113, −0.2439]
[−0.3472, −0.1856]	[−0.1558, 0.1629]	[−0.8093, −0.5850]	[0.1447, 0.3037]	[−0.8610, −0.6173]
[0.2700, 0.4811]	[−0.8077, −0.5538]	[−0.0001, 0.2087]	[−0.6217, −0.3860]	[−0.1003, 0.1316]
[−0.2657, −0.1061]	[0.0274, 0.3488]	[−0.4389, −0.1468]	[−0.1138, 0.1105]	[−0.5570, −0.1945]

IF Rock Type = Sandstones [1]
AND Porosity = [p1,p2]
AND Grain Size = [g1,g2]
AND Clay Content = [c1,c2]
AND P-Wave Vel. = [pwv1,pwv2]
AND P-Wave Att. = [pwa1,pwa2]
THEN Y* = a0 + a1*P + a2*G + a3*C + a4*PWV + a5*PWA.

where P is % porosity, G is grain size, C is clay content, PWV is P-wave velocity, and P-wave attenuation, Y*, is equivalent to f in Eq. (D.1).

Data are scaled uniformly between −1 and 1 and the result is given in the scaled domain. The available data were divided into three data sets: training, testing, and validation. The neuro-fuzzy model was trained based on a training data set and continuously tested using a test data set during the training phase. Training was stopped when it was found that the model's prediction suffered upon continued training. Next, the number of rules was increased by one and training was repeated. Using this technique, an optimal number of rules were selected. Fig. 8A through E and Table 1 show typical rules extracted from the data. In Table 1, column 1 through 5 show the membership functions for porosity, grain size, clay content, P-wave velocity, and P-wave attenuation, respectively. Using the model defined by Eqs. (D.1) through (D.4) and membership functions defined in Fig. 8A through E and Table 1, permeability was predicted as shown in Fig. 9A. In this study, 7 rules were identified for prediction of permeability based on porosity, grain size, clay content, P-wave velocity, and P-wave attenuation (Fig. 8A through E and 9A). In addition, 8 rules were identified for prediction of permeability based on porosity, clay content, P-wave velocity, and P-wave attenuation (Fig. 9B). Ten rules were identified for prediction of permeability based on porosity, P-wave velocity, and P-wave attenuation (Fig. 9C). Finally, 6 rules were identified for prediction of permeability based on grain size, clay content, P-wave velocity, and P-wave attenuation (Fig. 9D). The neural network model shows very good performance

Fig. 8. Typical rules extracted from data, 7 Rules. (A) Porosity; (B) Grain size; (C) Clay content.

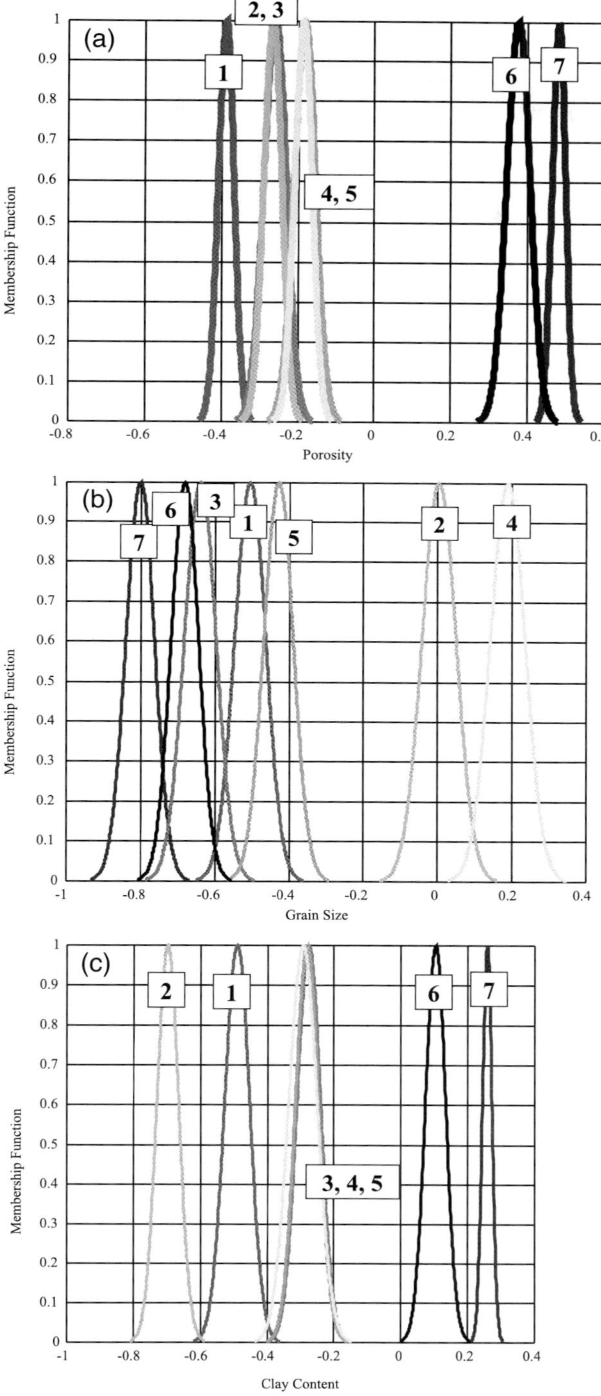
(a)
1
2, 3
4, 5
6
7
Membership Function
Porosity
(b)
7
6
3
1
5
2
4
Membership Function
Grain Size
(c)
2
1
3, 4, 5
6
7
Membership Function
Clay Content

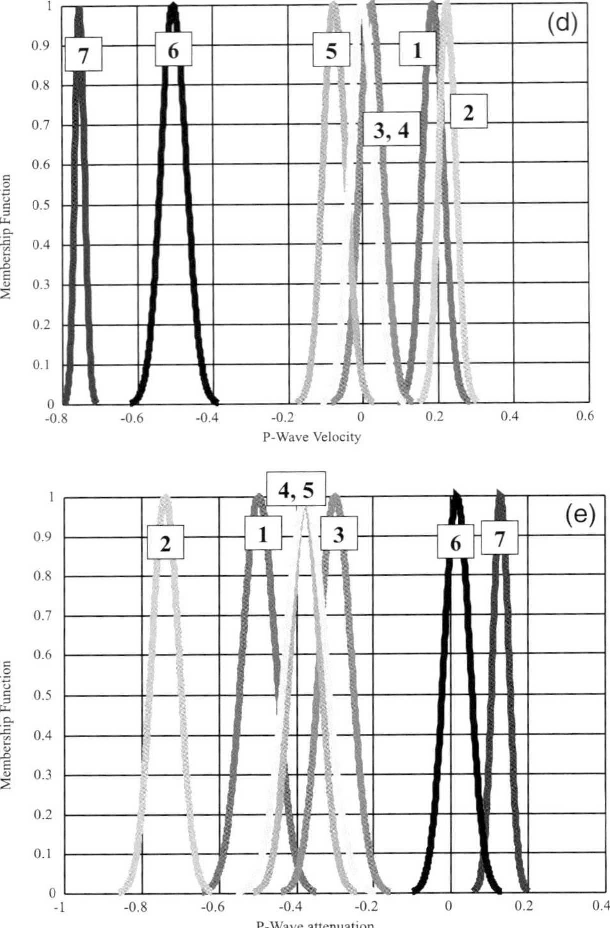

Fig. 8 (continued). (D) P-wave velocity; (E) P-wave attenuation.

for prediction of permeability. In this situation, not only a nonlinear mapping and relationship was identified between porosity, grain size, clay content, P-wave velocity, and P-wave attenuation, and permeability, but the rules existing between data were also identified. For this case study, our software clustered the parameters as grain size,

---

Fig. 9. Performance of Neural-Fuzzy model for prediction of permeability. (A) Performance of Neural-Fuzzy model for prediction of permeability [K = f (P, G, C, PWV, PWA)]; (B) Performance of Neural-Fuzzy model for prediction of permeability [K = f (P, C, PWV, PWA)]; (C) Performance of Neural-Fuzzy model for prediction of permeability [K = f (P, PWV, PWA)]; (D) Performance of Neural-Fuzzy model for prediction of permeability [K = f (G, C, PWV, PWA)].

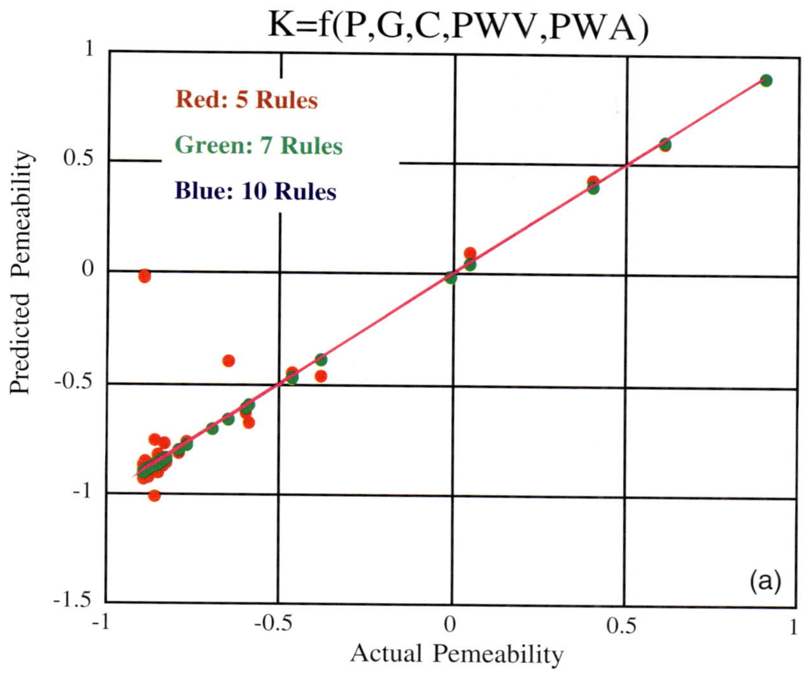
K=f(P,G,C,PWV,PWA)
Red: 5 Rules
Green: 7 Rules
Blue: 10 Rules
Predicted Pemeability
Actual Pemeability
1
0.5
0
-0.5
-1
-1.5
-1
-0.5
0
0.5
1
(a)

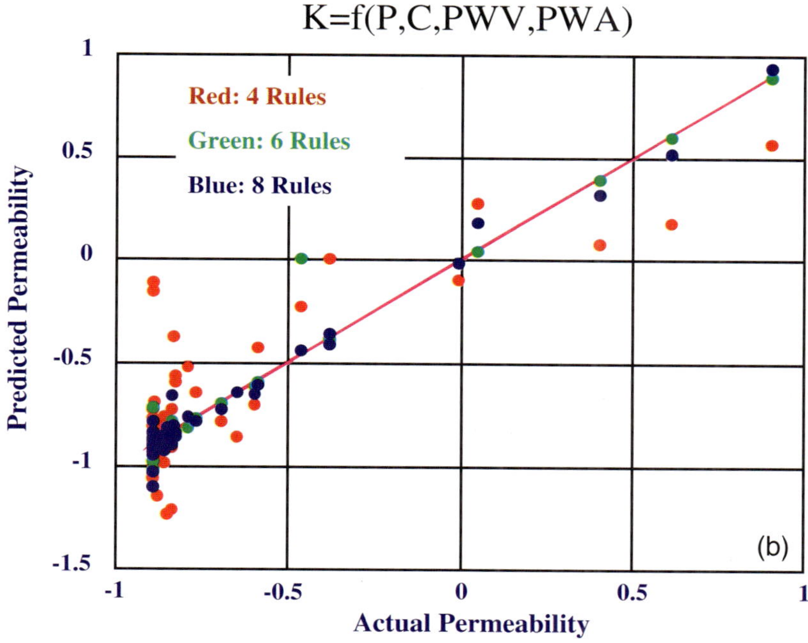
K=f(P,C,PWV,PWA)
Red: 4 Rules
Green: 6 Rules
Blue: 8 Rules
Predicted Permeability
Actual Permeability
1
0.5
0
-0.5
-1
-1.5
-1
-0.5
0
0.5
1
(b)

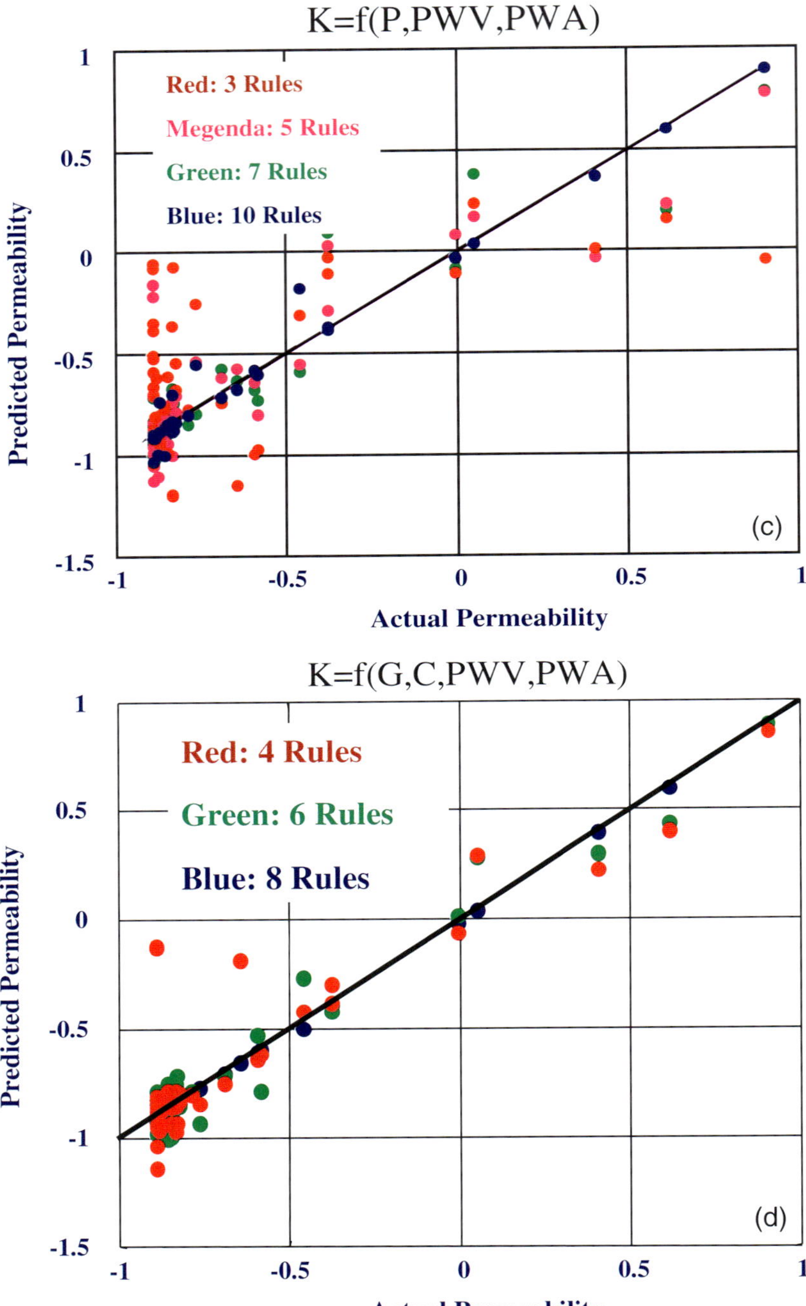
K=f(P,PWV,PWA)
Red: 3 Rules
Megenda: 5 Rules
Green: 7 Rules
Blue: 10 Rules
Predicted Permeability
Actual Permeability
1
0.5
0
-0.5
-1
-1.5
-1
-0.5
0
0.5
1
(c)
K=f(G,C,PWV,PWA)
Red: 4 Rules
Green: 6 Rules
Blue: 8 Rules
Predicted Permeability
Actual Permeability
1
0.5
0
-0.5
-1
-1.5
-1
-0.5
0
0.5
1
(d)

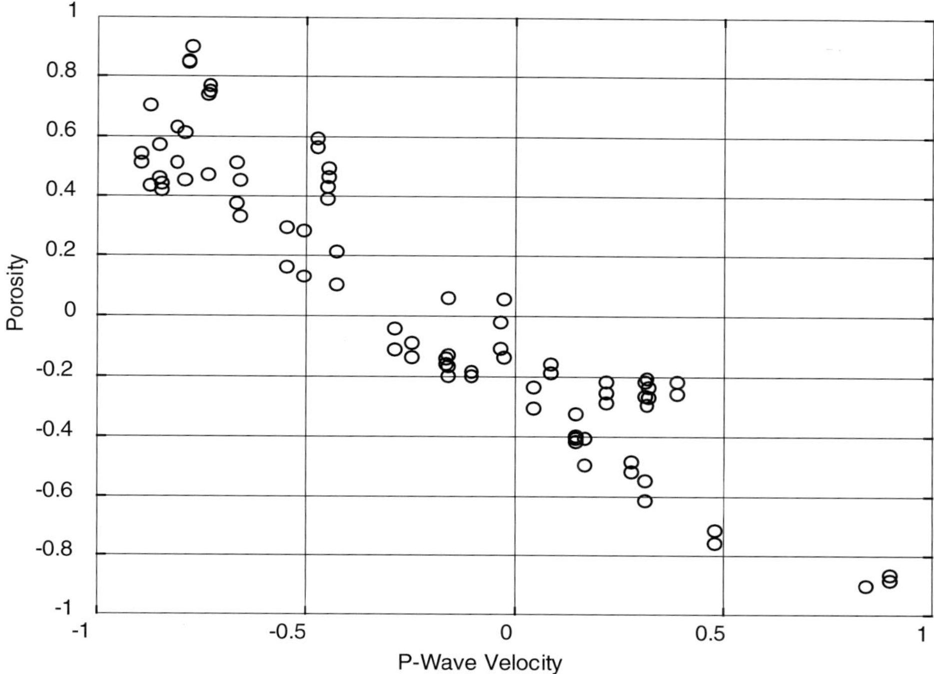

Fig. 10. Relationship between P-Wave velocity and porosity.

P-wave velocity/porosity (as confirmed by Fig. 10 since a clear linear relationship exists between these two variables), and P-wave attenuation/clay content (as it is confirmed by Fig. 11 since an approximate linear relationship exists between these two variables). In addition, using the rules extracted, it was shown that P-wave velocity is closely related to porosity and P-wave attenuation is closely related to clay content. Boadu (1997) also indicated that the most influential rock parameter on the attenuation is the clay content. In addition our software ranked the variables in the order grain size, P-wave velocity, P-wave attenuation and clay content/porosity (since clay content and porosity can be predicted from P-wave velocity and P-wave attenuation).

4. CONCLUSION

In this paper, we developed the next generation of 'intelligent' software that will identify the nonlinear relationship and mapping between well logs and seismic attributes.

Fig. 9 (continued). (C) Performance of Neural-Fuzzy model for prediction of permeability [K = f (P, PWV, PWA)]; (D) Performance of Neural-Fuzzy model for prediction of permeability [K = f (G, C, PWV, PWA)].

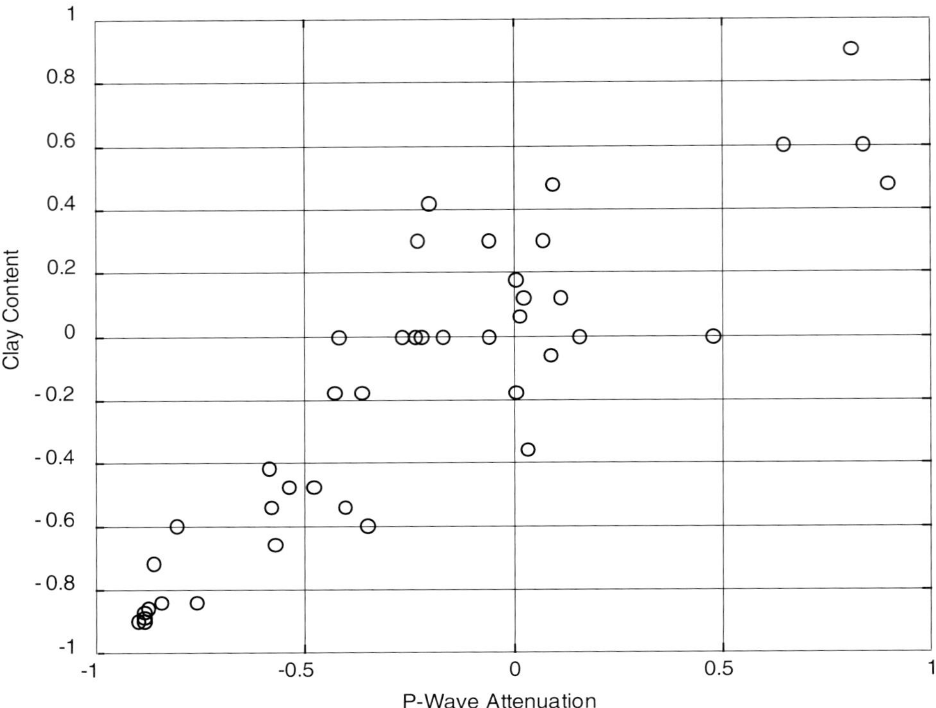

Fig. 11. Relationship between P-wave attenuation and clay content.

We developed a series of neural network models to analyze actual welllogs. Wireline logs such as travel time (DT), gamma ray (GR), and density (RHOB) were predicted based on SP, and resistivity logs. We also predicted travel time (DT) based on induction resistivity and vice versa.

In addition, we developed the next generation of 'intelligent' software that will identify the nonlinear relationship and mapping between rock proprieties and seismic attenuation and extracts rules (knowledge) from these databases. The neuro-fuzzy model was used to implicitly cluster the data while finding the nonlinear mapping. For the example studied here, our software clustered the parameters as grain size, P-wave velocity/porosity, and P-wave attenuation/clay content. In addition our software ranked the variables as grain size, P-wave velocity, P-wave attenuation and clay content/porosity.

ACKNOWLEDGEMENTS

The authors express their thanks to Dr. Roy Adams from EGI-University of Utah for his useful comments and suggestions. Appendix A, in part, is provided by Dr. Roy Adams.

APPENDIX A. BASIC PRIMER ON NEURAL NETWORK AND FUZZY LOGIC TERMINOLOGY

Neural networks. Neural networks are systems that ". . . use a number of simple computational units called 'neurons'. . . " and each neuron ". . . processes the incoming inputs to an output. The output is then linked to other neurons." (Von Altrock, 1995). Neurons are also called 'processing elements'.

Weight. When used in reference to neural networks, 'weight' defines the robustness or importance of the connection (also known as a link or synapse) between any two neurons. Medsker (1994) notes that weights ". . . express the relative strengths (or mathematical value) of the various connections that transfer data from layer to layer".

Backpropagation learning algorithm. In the simplest neural networks, information (inputs and outputs) flows only one way. In more complex neural networks, information can flow in two directions, a 'feedforward' direction and a 'feedback' direction. The feedback process is known as 'backpropagation.' The technique known as a 'backpropagation learning algorithm' is most often used to train a neural network towards a desired outcome by running a 'training set' of data with known patterns through the network. Feedback from the training data is used to adjust weights until the correct patterns appear. Hecht-Nielsen (1990) and Medsker (1994) provide additional information.

Perceptron. There are two definitions of this term (Hecht-Nielsen, 1990). The 'perceptron' is a classical neural network architecture. In addition, processing elements (neurons) have been called 'perceptrons'.

Fuzziness and fuzzy. It is perhaps best to introduce the concept of 'fuzziness' using Zadeh's original definition of fuzzy sets (Zadeh, 1965): "A fuzzy set is a class of objects with a continuum of grades of membership. Such a set is characterized by a membership (characteristic) function which assigns to each object a grade of membership ranging between zero and one." Zadeh (1973) further elaborates that fuzzy sets are ". . . classes of objects in which the transition from membership to non-membership is gradual rather than abrupt." Fuzzy logic is then defined as the ". . . use of fuzzy sets defined by membership functions in logical expressions" (Von Altrock, 1995). Fuzziness and fuzzy can then be defined as having the characteristics of a fuzzy set.

Neuro-fuzzy. This is a noun that looks like an adjective. Unfortunately, 'neuro-fuzzy' is also used as an adjective, e.g. 'neuro-fuzzy logic' or 'neuro-fuzzy systems.' Given this confusing situation, a useful definition to keep in mind is: "The combination of fuzzy logic and neural net technology is called 'NeuroFuzzy' and combines the advantages of the two technologies." (Von Altrock, 1995). In addition, a neuro-fuzzy system is a neural network system that is self-training, but uses fuzzy logic for knowledge representation, the rules for behavior of the system, and for training the system.

Crisp sets and fuzzy sets. "Conventional (or crisp) sets contain objects that satisfy *precise properties* required for membership." (Bezdek and Pal, 1992). Compare this to their definition that 'fuzzy sets' ". . . contain objects that satisfy *imprecise* properties to varying degrees. . . ". Each member of a crisp set is either 'true' or is 'false,' whereas each member of a fuzzy set may have a certain degree of truth or a certain degree of falseness or may have of some degree of each!

APPENDIX B. NEURAL NETWORKS

Details of neural networks are available in the literature (Cybenko, 1989; Hecht-Nielsen, 1989; Widrow and Lehr, 1990; Kohonen, 1987, 1997; and Lin and Lee, 1996) and therefore only the most important characteristics of neural networks will be mentioned. The typical neural network (Fig. B.1) has an input layer, an output layer, and at least one hidden layer. Each layer is in communication with the succeeding layer via a set of connections of various weights, i.e. strengths. In a neural network, nonlinear elements are called various names, including nodes, neurons, or processing elements (Fig. B.2). A biological neuron is a nerve cell that receives, processes, and passes on information. Artificial neurons are simple first-order approximations of biological neurons.

Consider a single artificial neuron (Fig. B.2) with a transfer function ($y1^{(i)} = f(z^{(i)})$), connection weights, w_j, and a node threshold, θ. For each pattern i,

$$z^{(i)} = x_1^{(i)} w_1 + x_2^{(i)} w_2 + \cdots + x_N^{(i)} w_N + \theta \quad \text{for } i = 1, \ldots, \text{P}. \tag{B.1}$$

All patterns may be represented in matrix notation as,

$$\begin{bmatrix} z^{(1)} \\ z^{(2)} \\ \vdots \\ z^{(P)} \end{bmatrix} = \begin{bmatrix} x_1^{(1)} & x_2^{(1)} & \cdots & x_N^{(1)} & 1 \\ x_1^{(2)} & x_2^{(2)} & \cdots & x_N^{(2)} & 1 \\ \vdots & \vdots & \vdots & \vdots & \vdots \\ x_1^{(P)} & x_2^{(P)} & \cdots & x_N^{(P)} & 1 \end{bmatrix} \begin{bmatrix} w_1 \\ w_2 \\ \vdots \\ w_N \\ \theta \end{bmatrix} \tag{B.2}$$

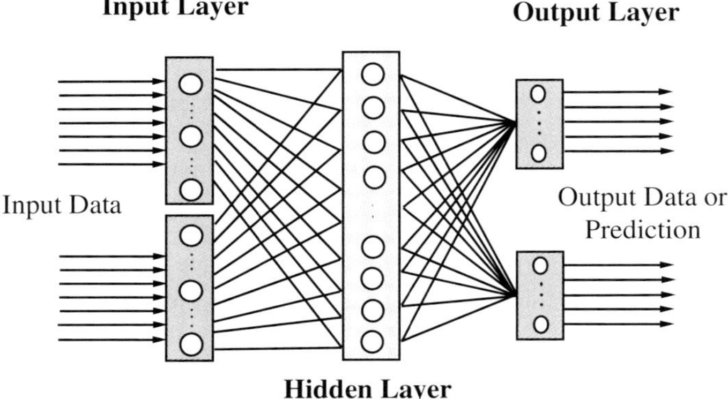

Fig. B.1. Typical neural network model.

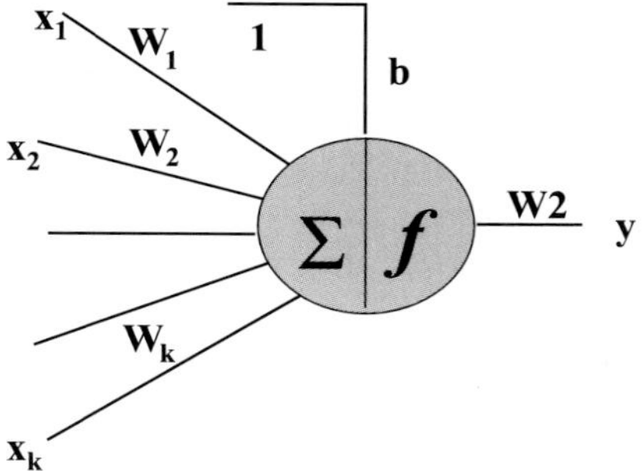

$$\mathbf{y = f\,[\,b + w_1\,x_1 + w_2\,x_2 + \ldots + w_k\,x_k\,]\;.\;w_2}$$

Fig. B.2. Typical neuron.

and

$$\underline{y1} = f(\underline{z}) \tag{B.3}$$

The transfer function, f, is typically defined by a sigmoid function such as the hyperbolic tangent function,

$$f(z) = \frac{e^z - e^{-z}}{e^z + e^{-z}}.$$

In more compact notation,

$$\underline{z} = \underline{\underline{X}}_1 \underline{w}_\theta = \underline{\underline{X}}\,\underline{w} + \underline{\theta} \tag{B.4}$$

where,

$$\underline{w}_\theta = \left[\underline{w}^{\mathrm{T}} \mid \theta\right]^{\mathrm{T}} \tag{B.5}$$

$$\underline{\underline{X}}_1 = \left[\underline{\underline{X}} \mid \underline{1}\right] \tag{B.6}$$

and, $\underline{1}$ = column vector of ones with P rows; $\underline{\underline{X}} = P \times N$ matrix with N input and P pattern; θ = Bias vector, vector with P rows of θ; and w = weights, vector with N rows.

During learning, the information is propagated back through the network and used to update the connection weights (backpropagation algorithm). The objective function for the training algorithm is usually set up as a squared error sum,

$$\mathrm{E} = \frac{1}{2} \sum_{i=1}^{P} \left(y^i_{(\text{observed})} - y^i_{(\text{prediction})}\right)^2 \tag{B.7}$$

This objective function defines the error for the observed value at the output layer, which is propagated back through the network. During training, the weights are adjusted to minimize this sum of squared errors.

APPENDIX C. MODIFIED LEVENBERGE–MARQUARDT TECHNIQUE

Several techniques have been proposed for training the neural network models. The most common technique is the backpropagation approach. The objective of the learning process is to minimize the global error in the output nodes by adjusting the weights. This minimization is usually set up as an optimization problem. Here, we use the Levenberge–Marquardt algorithm, which is faster and more robust than conventional algorithms, but it requires more memory. Using nonlinear statistical techniques, the conventional Levenberge–Marquardt algorithm (optimization algorithm for training the neural network) is modified. In this situation, the final global error in the output at each sampling time is related to the network parameters and a modified version of learning coefficient is defined. The following equations briefly show the difference between the conventional and the modified technique as used in this study.

For the conventional technique:

$$\Delta W = \left(\underline{J}^{\mathrm{T}} J + \mu^2 \underline{I}\right)^{-1} \underline{J}^{\mathrm{T}} e \tag{C.1}$$

whereas in the modified technique

$$\Delta W = \left(\underline{J}^{\mathrm{T}} \underline{\Lambda}^{\mathrm{T}} \underline{\Lambda} J + \Gamma^{\mathrm{T}} \Gamma\right)^{-1} \underline{J}^{\mathrm{T}} \underline{\Lambda}^{\mathrm{T}} \underline{\Lambda} e \tag{C.2}$$

where

$$\underline{\Lambda}^{\mathrm{T}} \underline{\Lambda} = \underline{\hat{V}}^{-1} \tag{C.3}$$

$$\hat{V}_{ij} = \frac{1}{2m+1} \sum_{k=-m}^{m} \hat{e}_{i+k} \hat{e}_{j+k} \tag{C.4}$$

$$\underline{\hat{V}} = \hat{\sigma}^2 \underline{I} \tag{C.5}$$

$$\hat{W} = W \pm k\hat{\sigma} \tag{C.6}$$

APPENDIX D. NEURO-FUZZY MODELS

In recent years, considerable attention has been devoted to the use of hybrid neural network–fuzzy logic approaches (Jang, 1991, 1992) as an alternative for pattern recognition, clustering, and statistical and mathematical modeling. It has been shown that neural network models can be used to construct internal models that capture the presence of fuzzy rules.

Neuro-fuzzy modeling is a technique of describing the behavior of a system using fuzzy inference rules using a neural network structure. The model has a unique feature in which it can express linguistically the characteristics of the complex nonlinear system. In this study, we will use the neuro-fuzzy model originally presented by Sugeno and Yasukawa (1993). The neuro-fuzzy model is characterized by a set of rules. The rules are expressed as follows:

$$\begin{aligned} R^i: \quad & \text{if } x_1 \text{ is } A_1^i \text{ and } x_2 \text{ is } A_2^i \ldots \text{ and } x_n \text{ is } A_n^i \quad \text{(Antecedent)} \\ & \text{then } y^* = f_i(x_1, x_2, \ldots, x_n) \quad \text{(Consequent)} \end{aligned} \tag{D.1}$$

where $f_i(x_1, x_2, \ldots, x_n)$ can be constant, linear, or fuzzy set. For the linear case

$$f_i(x_1, x_2, \ldots, x_n) = a_{i0} + a_{i1}x_1 + a_{i2}x_2 + \cdots + a_{in}x_n \tag{D.2}$$

Therefore, the predicted value for output y is given by:

$$y = \frac{\sum_i \mu_i f_i(x_1, x_2, \ldots, x_n)}{\sum \mu_i} \tag{D.3}$$

with

$$\mu_i = \prod_j A_j^i(x_j) \tag{D.4}$$

where R_i is the ith rule, x_j are input variables, y is output, A_j^i are fuzzy membership functions (fuzzy variables), and a_{ij} are constant values.

In this study, we will use the adaptive neuro-fuzzy inference system (ANFIS) technique (Jang and Gulley, 1995; The Math Works™, 1995). The model uses neuro-adaptive learning techniques. This learning method is similar to that of neural networks. Given an input/output data set, the ANFIS can construct a fuzzy inference system (FIS) whose membership function parameters are adjusted using the backpropagation algorithm or similar optimization techniques. This allows fuzzy systems to learn from the data they are modeling.

APPENDIX E. K-MEANS CLUSTERING

An early paper on k-means clustering was written by MacQueen (1967). K-means is an algorithm to assign a specific number of centers, k, to represent the clustering of N points ($k < N$). These points are iteratively adjusted so that each point is assigned to one cluster, and the centroid of each cluster is the mean of its assigned points. In general, the k-means technique will produce exactly k different clusters of the greatest possible distinction.

The algorithm is summarized in the following:

(1) Consider each cluster consisting of a set of M samples that are similar to each other: $x_1, x_2, x_3, \ldots, x_m$
(2) Choose a set of clusters $\{y_1, y_2, y_3, \ldots, y_k\}$
(3) Assign the M samples to the clusters using the minimum Euclidean distance rule
(4) Compute a new cluster so as to minimize the cost function
(5) If any cluster changes, return to step 3; otherwise stop.
(6) End

REFERENCES

Aminzadeh, F. and S. Chatterjee, S., 1984/85. Applications of clustering in exploration seismology. *Geoexploration*, 23: 147–159.

Aminzadeh, F., Katz, S. and Aki, K., 1994. Adaptive neural network for generation of artificial earthquake precursors. *IEEE Trans. Geosci. Remote Sensing*, 32(6).

Baldwin, J.L., Otte, D.N. and Wheatley, C.L., 1989. Computer emulation of human mental process: application of neural network simulations to problems in well Log interpretation. *Soc. Pet. Eng., SPE Paper* #19619, 481.

Baldwin, J.L., Bateman, A.R.M. and Wheatley, C.L., 1990. Application of neural network to the problem of mineral identification from well logs. *Log Anal.*, 3, 279.

Bezdek, J.C. and Pal, S.K., 1992. *Fuzzy Models for Pattern Recognition*. IEEE Press, p. 539.

Boadu, F.K., 1997. Rock properties and seismic attenuation: neural network analysis. *Pure Appl. Geophys.*, 149: 507–524.

Cybenko, G., 1989. Approximation by superposition of a sigmoidal function. *Math. Control Sig. System*, 2: 303.

Hecht-Nielsen, R., 1989. Theory of backpropagation neural networks. In: *IEEE Proceedings, International Conference on Neural Networks*, Washington DC, I-593.

Hecht-Nielsen, R., 1990. *Neurocomputing*. Addison-Wesley Publishing, p. 433.

Jang, J.S.R., 1991. Fuzzy modeling using generalized neural networks and Kalman filter algorithm. In: *Proceedings of the Ninth National Conference on Artificial Intelligence*, pp. 762–767.

Jang, J.S.R., 1992. Self-learning fuzzy controllers based on temporal backpropagation. *IEEE Trans. Neural Networks*, 3(5).

Jang, J.S.R. and Gulley, N., 1995. *Fuzzy Logic Toolbox*. The Math Works Inc., Natick, MA.

Klimentos, T. and McCann, C., 1990. Relationship among compressional wave attenuation, porosity, clay content and permeability in sandstones. *Geophysics*, 55: 991014.

Kohonen, T., 1987. *Self-Organization and Associate Memory, 2nd Edition*. Springer Verlag, Berlin.

Kohonen, T., 1997. *Self-Organizing Maps, Second Edition*. Springer, Berlin.

Lin, C.T. and Lee, C.S.G., 1996. *Neural Fuzzy Systems*. Prentice Hall, Englewood Cliffs, NJ.

MacQueen, J., 1967. Some methods for classification and analysis of multivariate observation. In: LeCun, L.M. and Neyman, J. (Eds.), *The Proceedings of the 5th Berkeley Symposium on Mathematical Statistics and Probability*. University of California Press, Vol. 1, pp. 281–297.

Medsker, L.R., 1994. *Hybrid Neural Network and Expert Systems*. Kluwer Academic Publishers, Dordrecht, p. 240.

Nikravesh, M., 1998. Neural network knowledge-based modeling of rock properties based on well log databases. 1998 SPE Western Regional Meet., Bakersfield, CA, 10–13 May, SPE Paper #46206.

Nikravesh, M. and Aminzadeh, F., 1998. Data mining and fusion with integrated neuro-fuzzy agents: rock properties and seismic attenuation. JCIS 1998, The Fourth Joint Conference on Information Sciences. Research Triangle Park, NC, 23–28 October.

Nikravesh, M., Farell, A.E and Stanford, T.G., 1996. Model identification of nonlinear time-variant processes via artificial neural network. *Comput. Chem. Eng.*, 20(11): 1277.

Pezeshk, S, Camp, C.C. and Karprapu, S., 1996. Geophysical log interpretation using neural network. J. Comput. Civ. Eng., 10: 136 pp.

Rogers, S.J., Fang, J.H., Karr, C.L. and Stanley, D.A., 1992. Determination of lithology, from well logs using a neural network. *Am. Assoc. Pet. Geol., Bull.*, 76: 731.

Sugeno, M. and Yasukawa, T., 1993. A fuzzy-logic-based approach to qualitative modeling. *IEEE Trans. Fuzzy Systems*, 1(1).

The Math Works™, 1995. The Math Works Inc., Natick, MA.

Von Altrock, C., 1995. *Fuzzy Logic and NeuroFuzzy Applications Explained*. Prentice Hall PTR, 350 pp.

Widrow, B. and Lehr, S.K., 1990. 30 years of adaptive neural networks: perception, madaline, and backpropagation. *Proc. IEEE*, 78(9): 1414.

Wong, P.M.,. Jiang F.X. and Taggart, I.J., 1995a. A critical comparison of neural networks and discrimination analysis in lithofacies, porosity and permeability prediction. J. Pet. Geol., 18: 191.

Wong, P.M., Gedeon, T.D. and Taggart, I.J., 1995b. An improved technique in prediction: a neural network approach. *IEEE Trans. Geosci. Remote Sensing*, 33: 971.

Zadeh, L.A., 1965. Fuzzy sets. *Inf. Control*, 8: 33353.

Zadeh, L.A., 1973. Outline of a new approach to the analysis of complex systems and decision processes. *IEEE Transactions Syst., Man Cybern.*, SMC-3: 244.

Developments in Petroleum Science, 51
Editors: M. Nikravesh, F. Aminzadeh and L.A. Zadeh

Chapter 8

TIME LAPSE SEISMIC AS A COMPLEMENTARY TOOL FOR IN-FILL DRILLING

M. LANDRØ [a,1], L.K. STRØNEN [b], P. DIGRANES [b], O.A. SOLHEIM [a] and E. HILDE [a]

[a] *Statoil Research Centre, Postuttak 7005 Trondheim, Norway*
[b] *Statoil Gullfaks Production, 5021 Bergen, Norway*

ABSTRACT

Time lapse seismic data have improved the drainage understanding of the Gullfaks oil field, and have resulted in several potential new in-fill drilling targets. The baseline survey was acquired in 1985, one year before production start-up. Since then, three more surveys have been acquired, one in 1995, one in 1996, and the last one so far in 1999. Key elements in the process of reducing the uncertainties are cross-disciplinary verifications and detailed geophysical analysis and modeling, where different saturation profiles are input to the model. Waterfront movements predicted by time lapse seismic analysis have been confirmed by well observations. It is estimated that the recovery will most likely be increased by approximately 2% as a result of seismic monitoring, corresponding to a net value contribution of approximately $100 M, assuming an oil price of $15 per barrel.

1. INTRODUCTION

4D seismic is topical and the interest is steadily increasing. There are active 4D seismic studies ongoing at several of the major North Sea oil fields today. Historically, time lapse seismic successes have been reported from small, shallow onshore fields, often monitoring steam injection. Due to the pronounced temperature effects on seismic velocities caused by steam injection these examples have been very encouraging. However, for larger and often more complex North Sea hydrocarbon reservoirs, the 4D results have been less encouraging. Main reasons have been that replacement of water with oil (or gas with oil) generally causes less changes on seismic parameters than for instance a steam injection. During the last couple of years, interesting case histories showing clearly that time lapse seismic can be used to monitor fluid movements have been presented. The Gullfaks 4D case history is perhaps among the best known North Sea examples, but there are several other examples demonstrating that time lapse seismic can also contribute even in a complex North Sea type of reservoir. Another topic that will gain increased focus is the use of well log information and the necessity of acquiring such data simultaneously with the seismic data. Repeated logging as a direct reservoir management tool and as a calibration

[1] Present address: Department of Petroleum Engineering and Applied Geophysics, NTNU, 7491 Trondheim, Norway

tool for the repeated seismic is important. The results from different 4D seismic case studies so far indicate that the areal consistency of various seismic features is robust with respect to noise, and useful information can be obtained even if the baseline survey is old. Some recent examples from the North Sea are the Oseberg study (Johnstad et al., 1995), the Magnus time-lapse study (Watts et al., 1996), the Fulmar study (Johnston et al., 1997), and the Draugen time-lapse study (Gabriels et al., 1999). Earlier presentations of the Gullfaks 4D study can be found in Landrø et al. (1999).

The Gullfaks Field is located in the Norwegian sector of the North Sea, approximately 175 km northwest of Bergen. The reservoirs are sandstones of middle and early Jurassic age, belonging from top to base to the Brent Group, the Cook Formation and the Statfjord Formation. They represent shallow marine to fluvial deposits. Total reservoir thickness is several hundred meters, and the depth to top structure is around 1800 m subsea. Reservoir quality is generally very good, with permeabilities ranging from few tens of millidarcies to several darcies depending on layer and location. Porosities are generally in the range of 30–35%. The structural setting of the field is very complex, representing one of the main factors of uncertainty concerning the drainage of the field. Base recoverable reserves are estimated to be 319 MSm3, of which 80% belong to the Brent Group. Peak production was reached in 1994, and approximately 250 MSm3 have been produced to date. An important goal for the Gullfaks licence is to increase the recoverable reserves by 40 MSm3, of which the Brent Group is expected to yield 30 MSm3 through a recently launched technology programme. Locating the remaining oil is an important part of this programme. Time lapse seismic data is considered to be a key element in this process, and a multidisciplinary team is now in the process of integrating 4D seismic data with all other available reservoir and production information, in order to improve the static as well as the dynamic reservoir description.

2. FEASIBILITY STUDY

A rock physics model calibrated with well log measurements was used to predict the seismic effect of substituting oil with water. The basic equation in the rock physics modeling is the Gassmann equation. Repeated loggings in wells show a change in water saturation from values around 10% in 1989 to values around 70–80% in 1995. According to the model a 60% increase in water saturation should correspond to 8% increase in the P-wave impedance, while the expected change in S-wave impedance is practically zero. In order to understand how pressure changes within (and perhaps above) the reservoir impact the seismic parameters, a set of 29 core plugs from various formations were selected for ultrasonic core measurements. All the measurements were performed on dry core plugs. Typical pore pressure reductions observed in various wells range from 20 to 40 bars. The average increase in P-wave impedance is 4% for a pore pressure drop of 40 bars according to these ultrasonic core measurements. The relevance of such measurements is of course limited, but at least it gives us an indication of how fluid and pressure changes might impact time-lapse seismic data. A summary of the feasibility study showing how various seismic parameters vary with changes in pore pressure and fluid saturation is presented in Fig. 1.

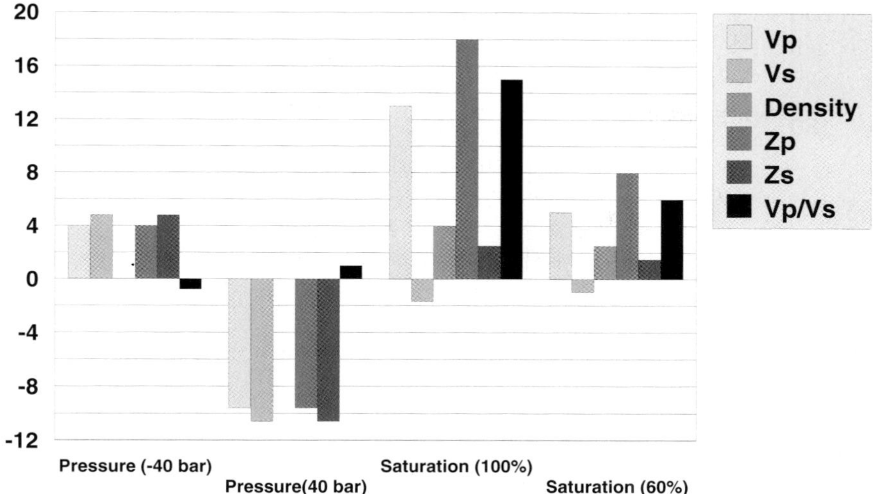

Fig. 1. Summary of the feasibility study showing expected changes in various seismic parameters due to changes in pore pressure and fluid saturation. (Zp and Zs denote P-wave impedance and S-wave impedance, respectively. Figure reprinted with permission from Petroleum Geoscience).

3. 3D SEISMIC DATA SETS

So far four different 3D seismic data sets have been acquired. The baseline survey was acquired in 1985, one year before production start-up. The second one was acquired in 1995, covering approximately 50% of the field. The third survey was acquired in 1996, followed by the fourth survey in 1999, both of them covering the whole field. Similar, but not identical acquisition parameters, have been used. In contrast to the three latest surveys, the baseline survey (1985) was more limited with respect to navigation accuracy. The timing of the four seismic surveys are shown in Fig. 2 together with the production profile. The stacked and 3D-migrated data are considerably more repeatable than the pre-stack data may suggest, and production related changes are generally well above the general background noise level. Apart from a deterministic de-bubble filter applied to the 1985 data (the 1985 source signature had a more pronounced bubble signal), the data sets were processed in the same way. Moreover, the repeatability has been improved by a trace by trace post-stack matching filter. We observe that the repeatability between the 1996 and the 1999 surveys are improved compared to for instance 1996 versus 1985.

4. 4D SEISMIC ANALYSIS APPROACH

It is possible to divide 4D seismic analysis into two complementary techniques: One amplitude oriented approach where the main focus in the interpretation is to study amplitude variations from one survey to the next. The other approach is to study changes in seismic travel times between different surveys, and interpret such timeshifts (often

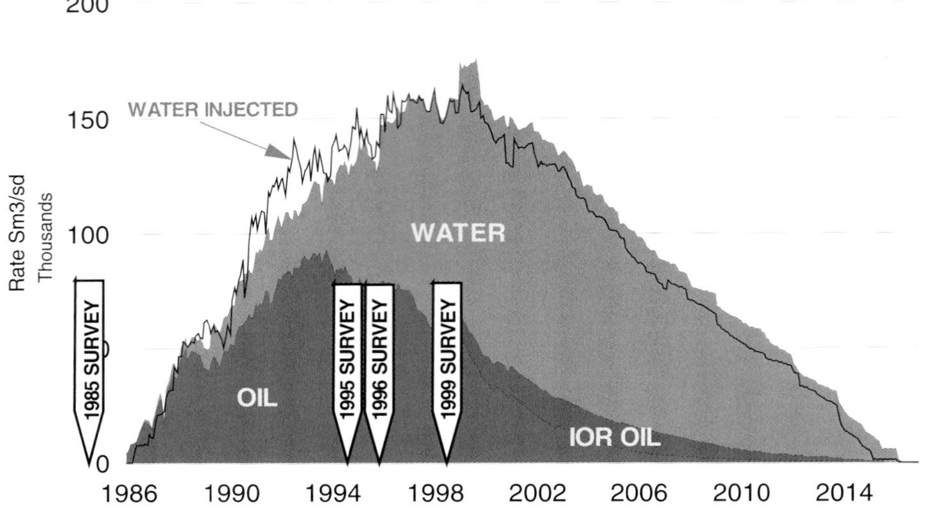

Fig. 2. Production profile together with the timing of the four seismic surveys. Figure reprinted with permission from Petroleum Geoscience.

referred to as pull-up or pull-down effects) in terms of production related effects. The two approaches are schematically illustrated in Fig. 3. For the Gullfaks time lapse study, we found that the timeshift technique was more unstable and even though some observations did fit with observations, the uncertainty of this approach was too high to be used extensively as an interpretation approach. Therefore the main focus in the 4D study at Gullfaks has been on interpretation of amplitude changes only.

The first step in the 4D analysis process has been to qualitatively compare the different 3D seismic data sets on a line-by-line basis, and identify and document areas where the seismic response from the reservoir is unchanged, or partly unchanged. These areas may represent undrained oil and potential targets for in-fill drilling. The next step is a multidisciplinary validation phase, where the observations are cross-checked with other reservoir, production and well data. Afterwards a detailed geophysical analysis is carried out. During this step the areal extent of the observations are mapped, and the results are brought through a new multidisciplinary validation phase. Based on the results from the previous steps, estimates of remaining oil for the different observations are then carried out. Finally, the various observations are associated with different degrees of uncertainty; this is taken care of by presenting the remaining reserves estimates as probability curves. Saturation maps at different reservoir levels are then generated. This is done by a multi-disciplinary approach.

5. SEISMIC MODELING OF VARIOUS FLOW SCENARIOS

One way to reduce the uncertainty in the interpretation of time lapse seismic data is to compute synthetic seismic data sets corresponding to different fluid flow scenarios,

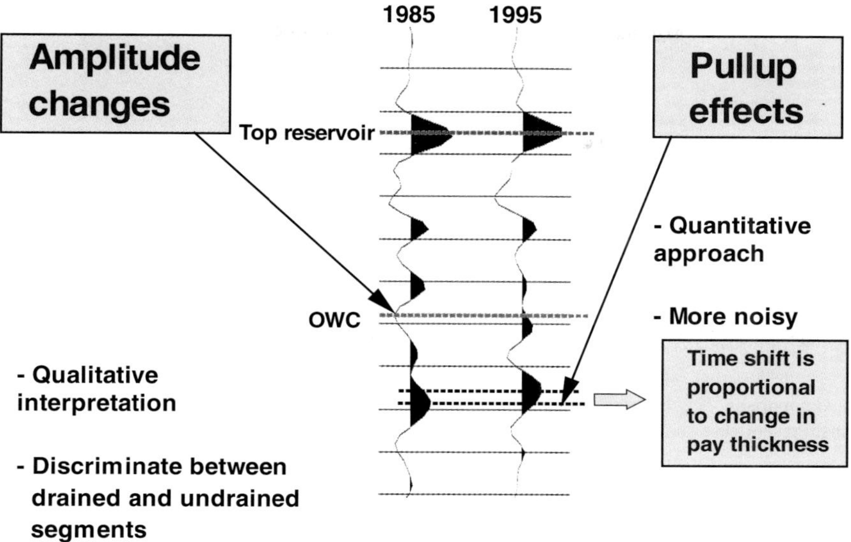

Fig. 3. Two complementary 4D analysis techniques. Schematic illustration of two ways to analyze time lapse seismic data.

followed by a comparison between the various synthetic seismic data sets and real seismic data. An example of such a procedure is shown in Fig. 4, where we have shown the synthetic seismic data that was closest to the real data out of 5 different fluid flow scenarios. The conclusion from this example was that in this segment the Tarbert 2 and 3 formations most likely have been drained, while the Tarbert 1 and Ness formations are still undrained. In this way we can use time lapse seismic data to reduce the uncertainty in the interpreted fluid flow pattern. This approach can also be used to constrain the history matching process: If you have several flow simulations that fit the production data, additional information obtained from modeling of these simulations and comparing with time lapse seismic data can be used to pick the most probable drainage simulation.

6. 4D SEISMIC FOR DETECTING FLUID MOVEMENT

The uppermost frame of Fig. 5 shows a seismic line acquired in 1985 over the northeastern part of the field, the middle frame shows the same line acquired in 1996, and the difference section is presented in the bottom frame. In the 1985 data we observe a well defined oil/water contact as noted on the figure (OWC). The OWC is no longer visible over the central and western part of the line acquired in 1996. This clearly stands out as an anomaly in the difference section (bottom). However, the OWC is

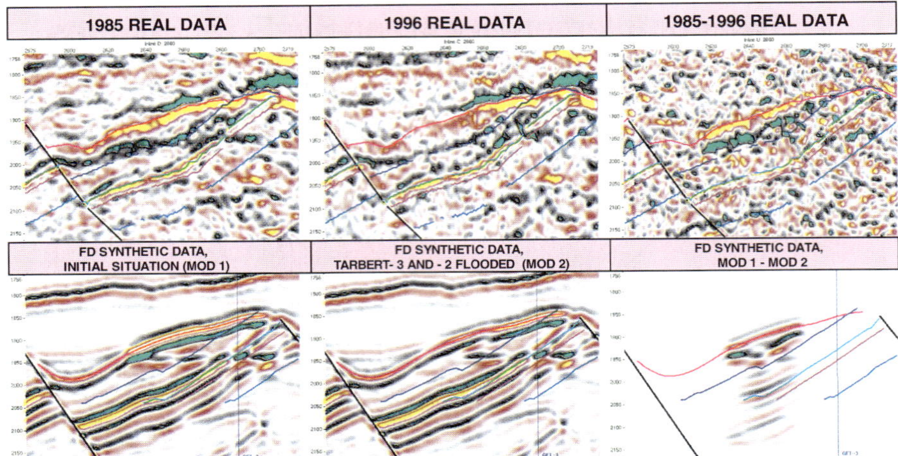

Fig. 4. Comparison between real seismic data and synthetic seismic data based upon various flooding scenarios. Present case shows Tarbert 3 and Tarbert 2 flooded, while Tarbert 1 and Ness are still oil-filled.

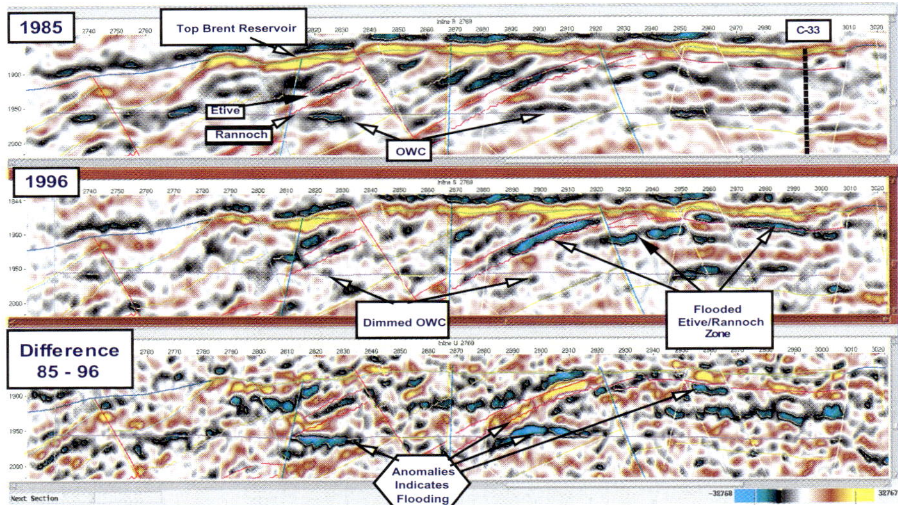

Fig. 5. 4D seismic mapping of waterflooding in the Etive-Rannoch Formations.

still visible in the eastern fault block, indicating remaining oil in this area. Another significant change has occurred in the fault blocks where the OWC has moved. A pronounced seismic response from the Etive Formation within the Brent Group has appeared in 1996, and is clearly expressed on the (bottom) difference section. Well data and seismic modeling indicate that this change is due to water having replaced oil in the homogeneous reservoir unit below the interface, while the more heterogeneous reservoir

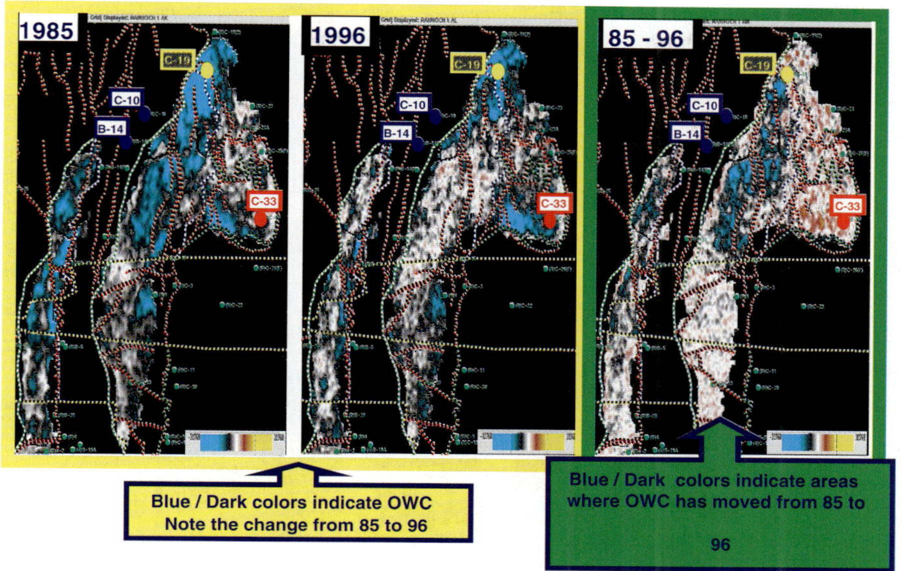

Fig. 6. Mapping oil-water contact movement from 1985 to 1996. The seismic amplitude maps are taken at the original oil-water contact level.

unit above it is still oil-filled. Hence, it is believed that the middle and western fault segments are fully flushed from the original OWC and up to the top Etive Formation.

In the eastern fault segment, where the OWC is still visible in the 1996 data set, there are clear indications of similar changes at the top Etive interface as observed in the two other fault segments. This suggests that we have an 'override' situation in this area, where water has flushed the much more permeable Etive and upper Rannoch formations, leaving oil behind in the lower parts of the Rannoch Formation. This phenomenon has repeatedly been observed on Gullfaks; variously deduced from production data, observed in new and in-fill wells and, now, directly observed on 4D seismic. In the present case, the observation is supported by the well C-33, which was drilled in 1996. The location of the well is marked in Fig. 6, showing the amplitude map along the original OWC in 1985 (left) and in 1996 (middle); the figure to the right shows the amplitude differences between 1985 and 1996. From these maps we clearly observe two well-defined areas where the OWC is also present in 1996. It is uncertain how much of the remaining oil C-33 will be able to produce. In the northern undrained area, C-19 which currently produces from the deeper Statfjord Formation, will shortly be opened in the Rannoch to produce the remaining reserves in the northern end of the segment.

Well C-36 (Fig. 7) was drilled to produce (and inject into) the fault block segments H7 and G6. The amplitude difference maps indicated no clear changes at the top Brent Group level and at the OOWC, supporting the assumption that these segments were undrained. The C-36A well found the OWC in the H7 segment unchanged (in agreement with the 4D seismic data). The second part of the well (C-36AT3) was drilled

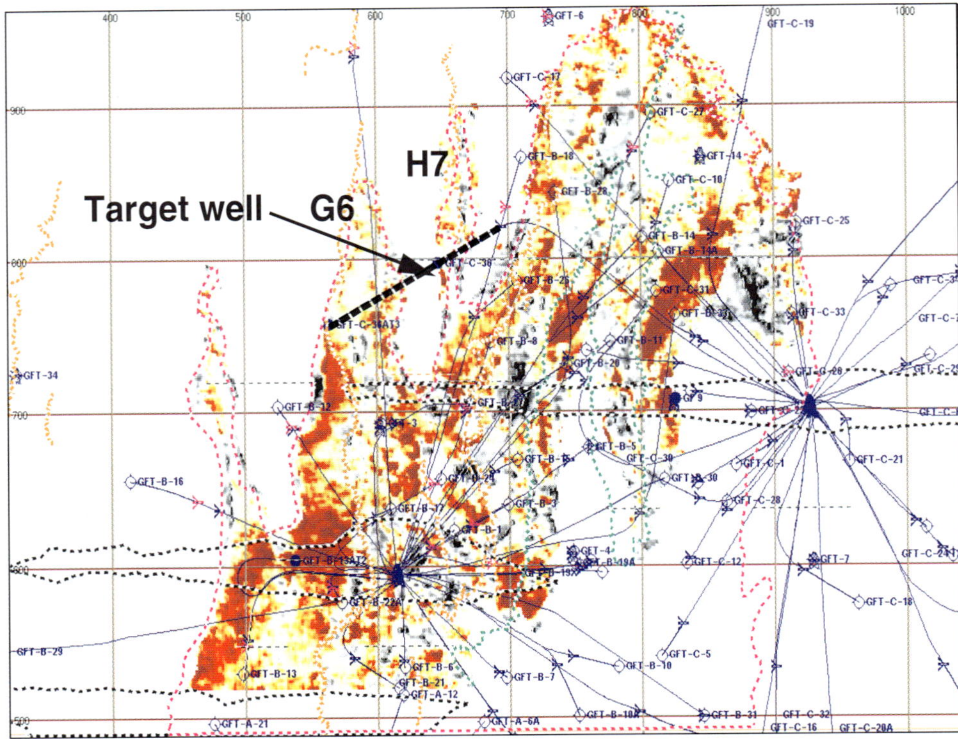

Fig. 7. Difference map taken at the original oil water contact from the northern part of the Gullfaks Field. Red colors indicate significant seismic changes, and are interpreted as areas with significant water flooding. The target well shown on the map encountered oil in both the H7 and the G6 segment.

horizontally through the reservoir section in the G6 segment approximately 30 m above the initial OWC. It proved that the Ness and Tarbert Formations were undrained at this level. Fig. 8 shows the evolution of seismic differences within one year, focusing at the G6 segment (the G6 segment is marked in Fig. 7). By comparing the 1995–1985 difference map with the 1996–1985 difference map, we see that the waterfront has moved approximately 100 m from 1995 to 1996. In February 1998, the water front reached the southern part of well C-36, an observation that fits well with the difference maps shown in Fig. 8.

7. 4D SEISMIC FOR DETECTING PORE PRESSURE CHANGES

From the feasibility study (Fig. 1) we found that a pore pressure increase causes a larger change in seismic parameters than a corresponding decrease in pore pressure.

This indicates that it should be easier to detect and identify an area or compartment of a reservoir that has experienced a pore pressure increase than an area that has gone

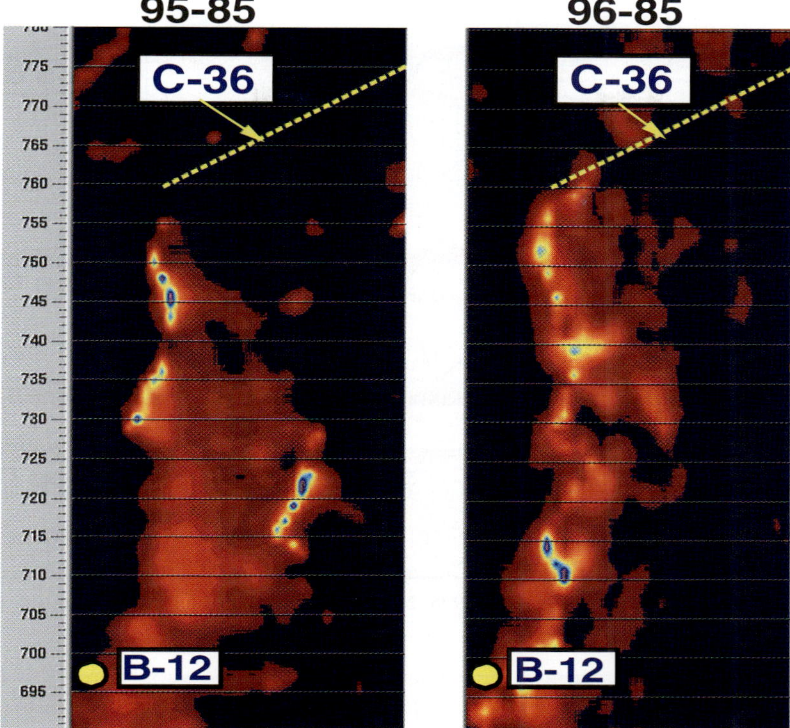

Fig. 8. Map view showing seismic differences taken at the original oil-water contact level in the G6-segment between 1995 and 1985 (left) and 1996 and 1985 (right). The extension of the difference anomaly observed between the two images is interpreted as a movement of the waterfront by approximately 100 m from 1995 to 1996. The B-12 water injection started in 1991. Total distance from B-12 to waterfront in 1996 is roughly 700 m, which corresponds to 140 m movement of waterfront per year (Figure reprinted with permission from Petroleum Geoscience).

through a pore pressure decrease. This is exactly what has been observed at the Gullfaks Field. Typical pore pressure decreases are of the order of 20–30 bars, and it has been hard to detect and map such areas directly by using repeated seismic data. On the other hand, in areas where pore pressure increases up to 50–60 bars have been measured, clear and mappable anomalies have been observed on the time lapse seismic data. An example from the Cook Formation is shown in Fig. 9, where a clear increase in seismic amplitude level is observed when comparing the 1995 section with the 1996 section. The same amplitude anomaly is shown as an amplitude difference (map view) in Fig. 10. Within this segment we know that there have been both pore pressure changes and fluid saturation changes in the time interval between the seismic surveys. Time lapse AVO (Amplitude Versus Offset) variations might contain information enabling us to distinguish between two effects. Several algorithms have been suggested to discriminate between pressure and fluid effects. Brevik (1999), Tura and Lumley (1999) and Landrø

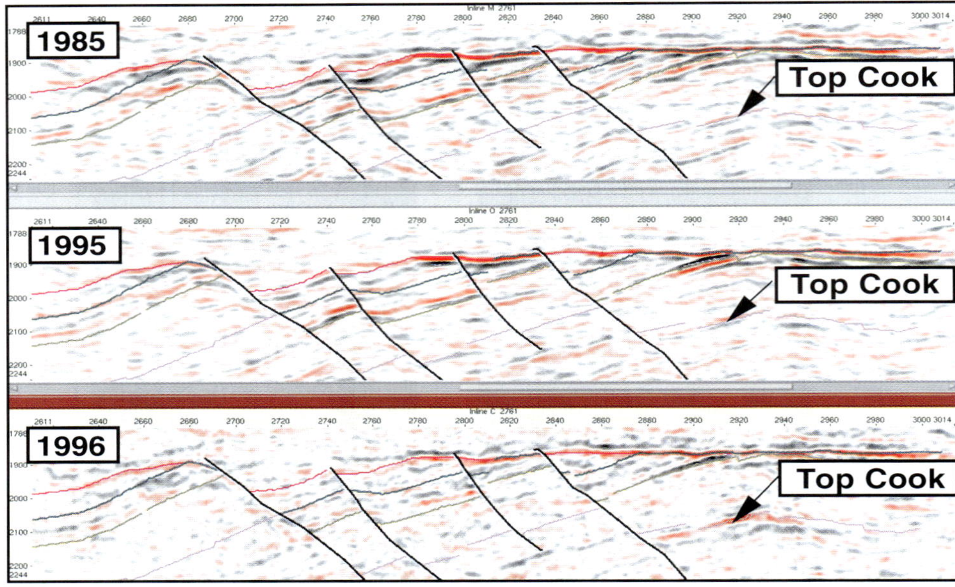

Fig. 9. Seismic data from 1985, 1995 and 1996. Notice the strong amplitude increase at the top Cook interface between 1995 and 1996.

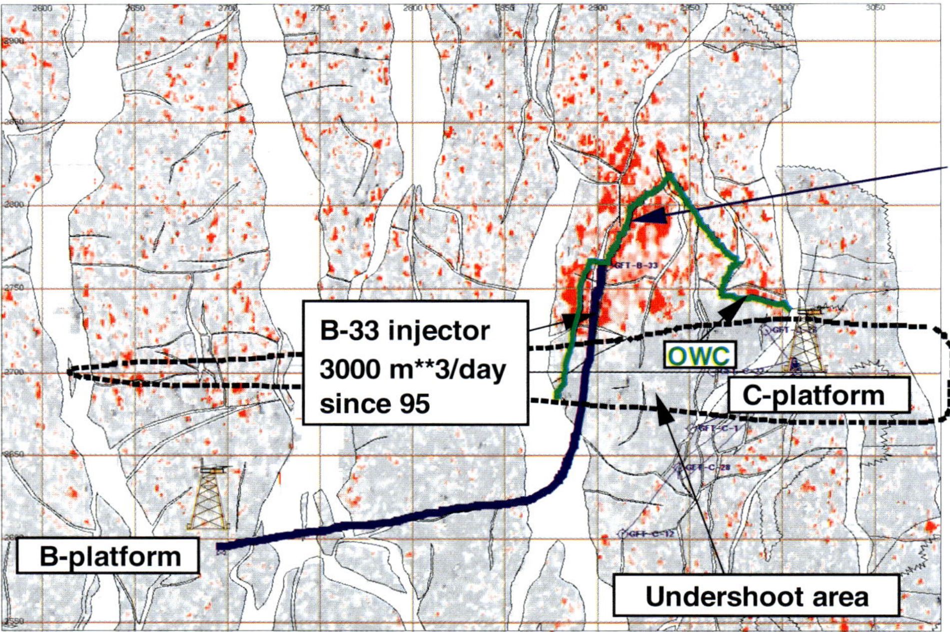

Fig. 10. Map view of seismic differences between 1996 and 1985 for the top Cook interface. The original oil-water contact is shown in green solid line.

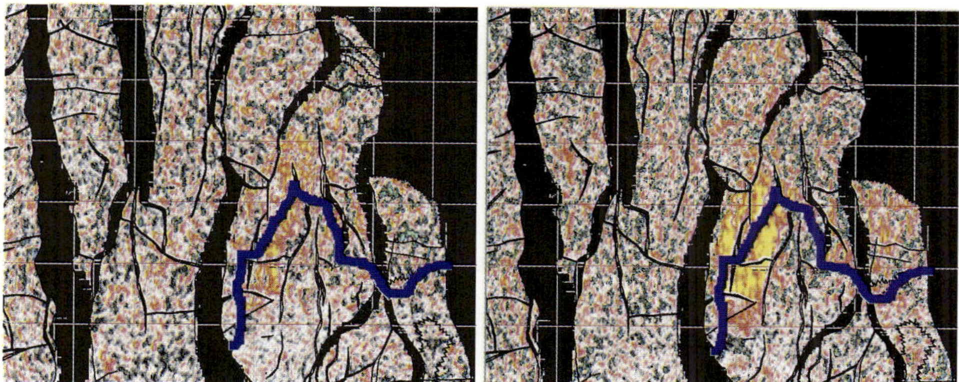

Fig. 11. Estimated saturation changes (left) and pore pressure changes (right) for the Cook Formation at Gullfaks. The solid blue line shows the original oil-water contact within this segment. Notice that the saturation changes follow this line in the southern part of the segment, but not in the northern part. Also note that the pressure attribute (right) crosses the oil-water contact, and terminates close to the major faults.

(1999) are some of the recent contributors within this area. Fig. 11 shows the result of the algorithm presented in Landrø (1999) for this segment. We observe that the extension of the pressure anomaly shown in Fig. 11 coincides with the major fault pattern, a result that seems realistic. 27% of the estimated recoverable reserves were produced in 1996 for this segment. The overpressure in this area has been caused by water injection (from well B-33 as shown in Fig. 10). We know that the pore pressure in the areas south of the major fault are considerably lower than in the mapped areas, again a fact that supports the observed anomaly.

8. 4D SEISMIC AND INTERACTION WITH THE DRILLING PROGRAM

Two major benefits have been achieved by extensive use of time lapse seismic data at Gullfaks:

First of all in-fill drilling into hydrocarbon pockets that have not been properly drained. Secondly, to prevent new drilling into locations that have been sufficiently drained. So far, one well has been removed from the drilling program, and two to three new wells have been proposed based on interpretation of repeated seismic data, and included in the well drilling program. One such example is shown in Fig. 12, where the seismic signal between 1985 and 1996 remain unchanged at the top reservoir reflector, and this structural high was therefore interpreted to be undrained. This was confirmed by in-fill drilling in 1999; the well encountered an oil column of 50 m at this location.

Interpretation of the 4D seismic data shot late in 1996 commenced early in 1998. By the time indications of remaining reserves were identified and ranked, and action could be taken to implement solutions, nearly three years had passed. The validity of some of the observations from 1996 were therefore uncertain, and whether these still warrant action or not. This would particularly be the case for areas of the field

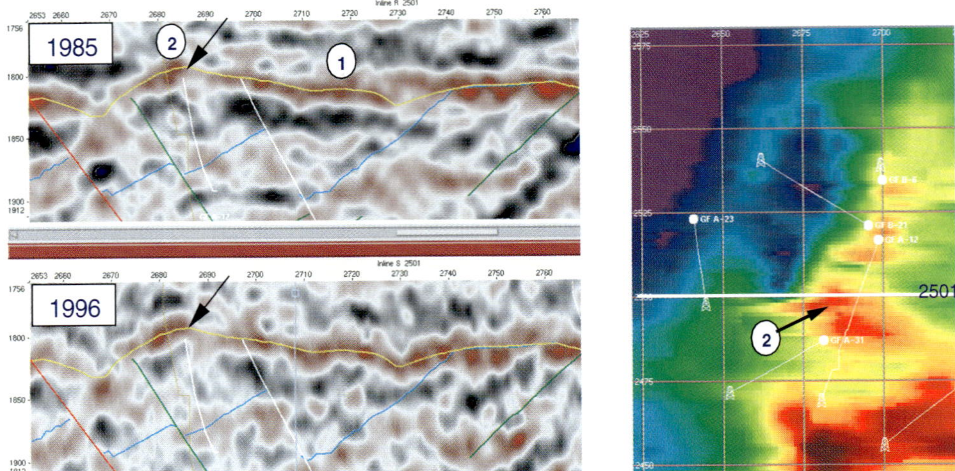

Fig. 12. Seismic profile showing a structural high (2) that was interpreted as undrained. The well drilled into location (2) on the map to the right confirmed this interpretation.

that have experienced changes in the drainage pattern since 1996. Faster acquisition and processing is therefore clearly needed. A new 3D seismic survey covering the whole field was acquired during summer 1999 in order to answer these questions. The interpretation of this new data set started in November 1999, which means that the turnaround time for time lapse studies has been reduced considerably, compared to the first seismic time lapse analysis performed at Gullfaks.

9. CONCLUSIONS

Time lapse seismic data have improved the drainage understanding of the Gullfaks Field, and have resulted in both new and accelerated in-fill drilling targets. The process of identifying new drilling targets is in progress and several potential targets have already been selected for multidisciplinary verification. Future work will focus on quantifying 4D effects within the different reservoir formations, both laterally and vertically. The main conclusion is that interpretations of undrained areas based on time lapse seismic data have been verified in several wells, and that the Gullfaks 4D study has contributed considerably in the reservoir management of the field. A recent acquisition of new seismic data over the field, demonstrated that the turnaround time can be reduced considerably, since the data was ready for interpretation only 3–4 months after the survey. The estimated net value contribution of 4D seismic at Gullfaks is approximately \$100 M, assuming an oil price of \$15 per barrel.

ACKNOWLEDGEMENTS

Statoil and the Gullfaks licence partners, Norsk Hydro and Saga Petroleum, are gratefully acknowledged for permitting this paper to be presented.

REFERENCES

Ånes, H.M., Haga, O., Instefjord, R. and Jakobsen, K.G., 1991. The Gullfaks Lower Brent Waterflood Performance. Paper presented at the 6th European Symposium on Improved Oil Recovery, 21–23 May 1991 in Stavanger, Norway.

Brevik, I., 1999. Rock model based inversion of saturation and pressure changes from time lapse seismic data. *69th Annu. Int. Meet., Soc. Explor. Geophys.*, Expanded Abstracts, pp. 1044–1047.

Fossen, H. and Hesthammer, J., 1998. Structural Geology of the Gullfaks Field, northern North Sea. In: Coward, M.P., Daltaban, T.S. and Johnson, H. (Eds.), *Structural Geology in Reservoir Characterization. Geol. Soc. London, Spec. Publ.*, 127: 231–261.

Gabriels, P.W., Horvei, N.A., Koster, J.K., Onstein, A. and Staples, R., 1999. Time lapse seismic monitoring of the Draugen Field. *69th Annu. Int. Meet., Soc. Explor. Geophys.*, Expanded Abstracts, pp. 2035–2037.

Johnstad, S.E., Seymour, R.H. and Smith, P.J., 1995. Seismic reservoir monitoring over the Oseberg field during the period 1989–1992. *First Break*, 13(5): 169–183.

Johnston, D.H., McKenny, R.S. and Burkhart, T.D., 1997. Time-lapse seismic analysis of the North Sea Fulmar Field. *67th SEG Meet.*, Abstracts, pp. 890–893.

Landrø, M., 1999. Discrimination between pressure and fluid saturation changes from time lapse seismic data. *69th Annu. Int. Meet., Soc. Explor. Geophys.*, Expanded Abstracts, pp. 1651–1654.

Landrø, M., Solheim, O.A., Hilde, E., Ekren, B.O. and Strønen, L.K., 1999. The Gullfaks 4D seismic study. *Pet. Geosci.*, 5(3).

Tura, A. and Lumley, D.E., 1999. Estimating pressure and saturation changes from time-lapse AVO data. *69th Annu. Int. Meet., Soc. Explor. Geophys.*, Expanded Abstracts, pp. 1655–1658.

Watts, G.F.T., Jizba, D., Gawith, D.E. and Gutteridge, P., 1996. Reservoir monitoring of the Magnus Field through 4D time-lapse seismic analysis. *Pet. Geosci.*, 2: 361–372.

Developments in Petroleum Science, 51
Editors: M. Nikravesh, F. Aminzadeh and L.A. Zadeh

Chapter 9

IMPROVING SEISMIC CHIMNEY DETECTION USING DIRECTIONAL ATTRIBUTES

KRISTOFER M. TINGDAHL

Department of Earth Sciences – Marine Geology; Göteborg University, Box 460, SE-405 30 Göteborg, Sweden

1. INTRODUCTION

In seismic object detection variations in the seismic response are enhanced to improve the geological interpretation of geometrical objects. Target objects of interest are, for example, flat spots, faults or chimneys, the subject of this chapter. A key problem in seismic object detection is to find attributes originating from the seismic signal that correlate with the object of interest. Many multi-trace attributes are sensitive to dipping strata (Fig. 1). In this chapter we will present a technique to improve the quality of multi-trace attributes by taking local dip information into account. This chapter has three parts:

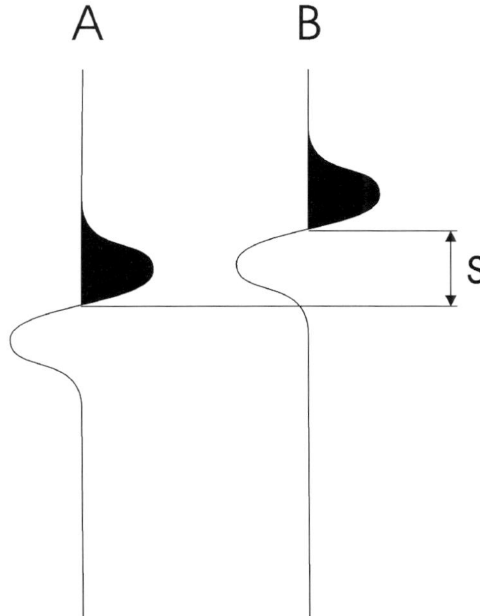

Fig. 1. The effect of dip on trace similarity. Trace segments A and B are quite different when compared horizontally; A has high values when B has low. If account is taken to the dip, trace B is shifted downwards s milliseconds before the comparison. The two trace segments will then become quite similar.

First, a method to estimate dip from 3D seismic data is presented. Secondly, a method is presented to use the dip to locally follow a seismic event through the cube. Finally, these methods are applied to improve seismic attributes used in chimney detection, and the result of this application is presented. Understanding of the dip estimation algorithm is not required to understand the other parts of the chapter. The dip estimation algorithm is presented in order to give an example of such an algorithm. My method is relatively new and has advantages over other algorithms. Any algorithm that estimates dip from 3D seismic data can be used in the algorithms presented in the latter parts of this chapter.

1.1. Introduction to seismic chimney detection

Seismic chimneys (Fig. 2) are the spatial link between source rock, reservoir trap, spill-point, and shallow-gas anomalies. Detailed interpretation helps to unravel a basin's hydrocarbon history, distinguish between charged and non-charged prospects, and to detect geo-hazards (Heggland, 1998). Chimneys are caused by saturated fluids and/or free gas migrating through porous rocks. As the fluids move up the pressure drops

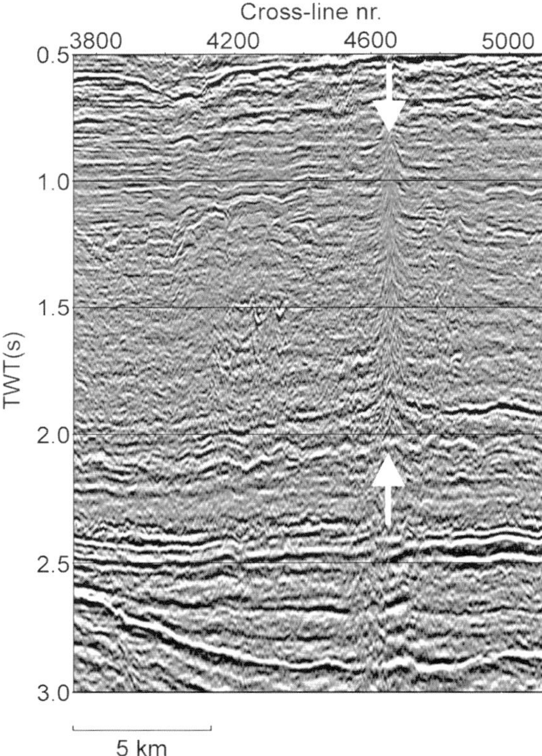

Fig. 2. An arbitrary in-line showing a seismic chimney pointed out by the two white arrows. Note the low similarity and chaotic appearance of the seismic response inside the chimney. The trace-to-trace distance is 25 meters.

and solution gas is released. Some gas stays in the pores, thus changing the acoustic properties of the rock. This connate gas affects especially the P-wave velocity. Two other reasons for fluid migration paths in the seismic record are diagenetic processes caused by the migrating fluids that may have changed the rock itself and over-pressured fluids that may have cracked the rocks causing scattering of the seismic waves.

Scanning through seismic sections and time/horizon slices and looking for vertical disturbances with associated seepage-related features as pock-marks and mud-volcanoes is the traditional way of interpreting chimneys (for instance, Heggland, 1997, 1998; Hovland and Judd, 1988). This is a time-consuming process that only a few specialists are able to perform. The chimney cube (Meldahl et al., 1998, 1999; Heggland et al., 1999) is a semi-automated chimney detection method that has greatly reduced the chimney interpretation time. Moreover, hydrocarbon migration path details are seen in the cube that could not be interpreted before.

The chimney detection is done by a backpropagation artificial neural network (ANN) with one hidden layer, which is given a set of attributes, calculated from the seismic cube. An experienced interpreter selects locations in the seismic cube with recognized chimneys as well as counter examples outside chimney bodies. Attributes at all example locations are calculated and given to the neural network to classify the data in the two classes: chimney or non-chimney. The desired output vectors are $1,0$ and $0,1$ for the respective classes. The training is done on 80% of the picked locations. The network is then applied to the remaining locations in order to test whether the network predicts the desired output for these locations.

The trained neural network is subsequently applied to the entire data volume to yield the chimney cube. At each input sample location all attributes are calculated and given to the network. The network produces two output values that mirror each other. It is therefore sufficient to pass only the output value for the chimney node. A high value thus indicates a high probability for a chimney at that location.

The original neural network used to generate a chimney cube is displayed in Fig. 3. The gray levels in the input layer indicate that much of the neural network's prediction is based on similarity between trace segments. These are computed without comparison for eventual structural dip. This implies that the network's accuracy is expected to deteriorate in the presence of strong dips. In this chapter I discuss how the accuracy of the chimney prediction is improved by taking structural dip information into account.

1.2. Introduction to dip calculations

If dip should be used to enhance the neural network's ability to detect seismic objects using this setup, there is a need for an algorithm that estimates the dip accurately at every position in the data at high speed. It is also desired that it should be possible to specify the lateral validity of the algorithm, that is the 'localness' of the calculated dip.

In seismic processing, dip is defined as the difference in time between two points, divided by the distance between the points:

$$p = \frac{\delta t}{d} \tag{1}$$

where δt is the time difference and d is the distance between the points. In 3D seismic

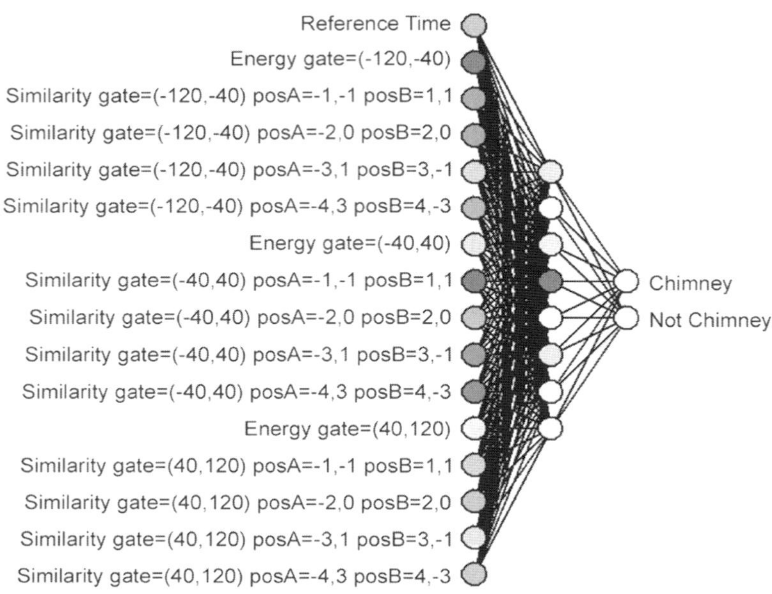

Fig. 3. The neural network used to produce the original chimney cube. The grayscale indicates the relative contribution of each input node to the classification result. A description of the input attributes (to the left) is given in the text. The input attributes' time-gates are specified in milliseconds and their positions are specified as trace positions in in-line and cross-line respectively. The outputs (to the right) predict the presence and absence of seismic chimneys respectively.

data, the dip is defined with two parameters, namely:

$$p_x = \frac{\delta t}{\delta x} \tag{2}$$

and

$$p_y = \frac{\delta t}{\delta y}, \tag{3}$$

where δx and δy are the distance between the points in the x and y dimensions respectively.

The first calculation of dip was performed by computing it from an auto-tracked horizon (Dalley et al., 1989). The dip is then the time difference between two nodes on the horizon, divided by the distance between the nodes. This method only calculates the dip at these horizons, which disqualifies it for volume-based seismic processing.

Without a horizon it is still possible to calculate the dip. One method is to compare the similarity of two adjacent trace segments versus time offset. The dip is then the time difference at which the trace segments are most similar to each other, divided by the lateral distance between the trace segments. The similarity between the trace segments can be quantified by any expression that measures similarity between vectors, for instance cross correlation (Bahorich and Farmer, 1995) or distance in hyperspace. Both the power and the disadvantage of the similarity method are that it, in its original form, only evaluates the similarity between two or three traces, which makes the resulting dip quite local. If a more regional dip were desired, the searching would become rather complex and time consuming.

One approach is to take the similarity along an arbitrary plane (or a disc) through the signal. A set of dips is tested, and the dip for the plane with the highest similarity among the samples on the disc is assumed to be the dip of that location (Marfurt et al., 1998). The problem with this approach is that the dip can only obtain values from the set of dips. If a high resolution is desired, the number of tested dips must be high, leading to a slow algorithm.

Dip can also be estimated by various forms of covariance analysis (Marfurt and Kirlin, 2000) but these methods are often computer time consuming and are therefore not considered further in this context.

2. DIP CALCULATIONS

Here I will describe a method that fulfills all the stated requirements and calculates the dip in frequency–wavenumber domain (f–k domain). The f–k domain is a multidimensional representation of a signal's frequencies. The aim is to calculate a dip in every position in the 3D signal, and the frequency representation must therefore be local at every position. Consequently, we want to focus on how the signal's local representation in f–k domain varies with space and time.

A sampled 3D seismic signal can be written:

$$u(x,y,t) \qquad x,y,t \in Z \tag{4}$$

Seismic signal are by default real-valued. The Fourier transform of a real-valued signal have a symmetry where each frequency will be represented by two peaks in k–f domain. It is therefore beneficial to use analytic (complex-valued) signals that lack this symmetry. The signal's imagery part is calculated by:

$$\mathrm{Im}\big(u(x,y,t)\big) = \mathcal{H}\big(\mathrm{Re}\big(u(x,y,t)\big)\big) \qquad x,y,t \in Z, \tag{5}$$

where $\mathcal{H}$ denotes the Hilbert transform.

By taking the Fourier transform on the entire signal, the obtained frequency representation will represent the entire signal's frequencies. In order to get a local spectrum, the signal must be windowed, that is all samples outside the area of interest must be zeroed. Multiplying the signal with a window function can do this:

$$u_w(x,y,t) = u(x,y,t)w(x-x_0, y-y_0, t-t_0) \qquad x,x_0,y,y_0,t,t_0 \in Z, \tag{6}$$

where x_0, y_0 and t_0 define the center of the volume of investigation. The window function should be designed to give values close to one when the function's variables are close to zero and values approaching zero when the function's variables divert from zero. The window function's role is twofold. First, it makes it possible to calculate a localized frequency representation of the signal. Secondly, by choosing an adequate window function, spurious high-frequency peaks from the window's edges can be avoided. There are numerous windows that are suitable for this, for instance the 3D Hanning window:

$$\begin{cases} w(x,y,t) = \frac{1}{2} + \frac{1}{2}\cos(\pi r) & r \leq 1 \\ w(x,y,t) = 0 & r > 1, \\ r = \sqrt{\frac{x^2}{X^2} + \frac{y^2}{Y^2} + \frac{t^2}{T^2}} \end{cases} \tag{7}$$

where X, Y and T are the length of the desired window in the x, y and t dimensions respectively.

The signal should then be transformed into f–k domain by the discrete Fourier transform:

$$\hat{u}(u,v,f) = \sum_{x=0}^{X}\sum_{y=0}^{Y}\sum_{t=0}^{T} u_w(x,y,t)\exp\left[-2\pi i\left(\frac{xu}{X} + \frac{yv}{Y} + \frac{tf}{T}\right)\right], \tag{8}$$

where X, Y and T are the length of the windowed signal in the x, y and t dimensions respectively and u, v and f are the frequencies in the x, y and t dimensions respectively. In order to be able to compare the different frequencies, the signal's spectrogram must be calculated. This is performed by:

$$P(u,v,f) = \left|\hat{u}(u,v,f)\right|^2. \tag{9}$$

The standard method to calculate the dip is to perform the Radon transform in either frequency or time domain. Since dip can be estimated from the f–k domain as:

$$p_x = \frac{u}{f} \tag{10}$$

and

$$p_y = \frac{v}{f} \tag{11}$$

the Radon transform in f–k domain becomes:

$$\check{P}(p_x, p_y, f) = P(fp_x, fp_y, f), \tag{12}$$

where p_x and p_y are the dips in the signal. The frequency variable can be removed by integrating over all frequencies:

$$\check{P}(p_x, p_y) = \int \check{P}(p_x, p_y, f)\,\mathrm{d}f. \tag{13}$$

The Radon transform does thus gives a power for each dip. The dip can then be calculated by using the Radon spectrum as a weight function for the dips. Calculating the weighted average dips performs this:

$$\overline{p}_x = \frac{\iint p_x \check{P}(p_x, p_y)\,\mathrm{d}p_y\mathrm{d}p_x}{\iint \check{P}(p_x, p_y)\,\mathrm{d}p_y\mathrm{d}p_x} \tag{14}$$

and

$$\overline{p}_y = \frac{\iint p_y \check{P}(p_x, p_y)\, \mathrm{d}p_x \mathrm{d}p_y}{\iint \check{P}(p_x, p_y)\, \mathrm{d}p_y \mathrm{d}p_x}. \tag{15}$$

The problem is that if there are two or more peaks in the Radon spectrum, indicating two possible dips in the signal, the calculated dip will not be one of the dips, but somewhere between them. This can be avoided if the dips with the maximum power were selected.

It is probable that the highest peak in the f–k spectrogram is located along the same dip as the highest peak in the Radon spectrum. Given that this is true, we can localize the maximum dip already in the f–k spectrogram and save much computation time. The problem is that the resolution of the f–k spectrogram is limited, and by just picking the position (u, v, f) in the spectrogram with the maximum value will only give a few possible values that the dip could take. This can be solved by an interpolation in the f–k spectrum. The value at a position in the spectrogram can be seen as the probability for the dip that that position represents [by Eqs. (10 en (11)] is the best estimation of the dip. Given the sampled, one-dimensional spectrum in Fig. 4 and that the spectrum is assumed do be smooth, one could assume that the 'true' maximum would be somewhere between value five and six. If a smooth function with a continuous first derivative is estimated from the maximum value and its two adjacent values in the spectrum, the function can be derived and searched for its maximum position. It has been found (Tingdahl, 1999) that the Gaussian clock function is the best function for this application on both synthetic and real seismic data. In one dimension, the Gaussian clock function is written:

$$f(x) = c\exp\left(-\tfrac{1}{2}a^2(b-x)^2\right). \tag{16}$$

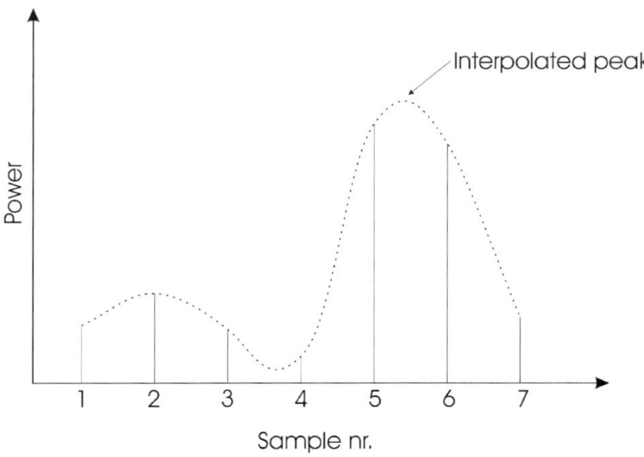

Fig. 4. A one-dimensional powerspectrum. A smooth function with a continuous first derivative can be fit to the maximum sample and its two neighbors. The maximum position can then be derived from that function. The 'true' maximum can in this case be assumed to be between samples five and six.

The equation is fit to a one-dimensional spectrogram's peak by:

$$\begin{cases} f(q-1) = c\exp\left(-\frac{1}{2}a^2(b+1)^2\right) \\ f(q) \quad\;\; = c\exp\left(-\frac{1}{2}a^2b^2\right) \\ f(q+1) = c\exp\left(-\frac{1}{2}a^2(b-1)^2\right) \end{cases}, \tag{17}$$

where $f(q)$ is the spectrograms maximum value and $f(q-1)$ and $f(q+1)$ are the values adjacent to the maximum, respectively. The system of equations is solved for b:

$$b = \frac{1}{2} - \frac{\ln\left(\dfrac{f(q)}{f(q+1)}\right)}{\ln\left(\dfrac{f(q)^2}{f(q-1)f(q+1)}\right)}. \tag{18}$$

The estimated maximum position is then:

$$x_{\max} = q + b. \tag{19}$$

This interpolation schedule can easily be extended to 3D signals in order to estimate all three temporal frequencies and the dips can then be calculated with Eqs. (10) and (11).

The sketched algorithm was applied to a 3D signal with linearly increasing dip in the y direction, and constant dip in the x direction (Fig. 5). The spectrogram was calculated in small sub-cubes with a side size of seven samples. The dip in the y-direction is displayed in Fig. 6. It can be clearly seen that the dip decreases linearly from left to

Fig. 5. A section through a 3-D signal with linearly increasing dip in y direction. The dip goes from 220 μs/m (left) to −220 μs/m (right).

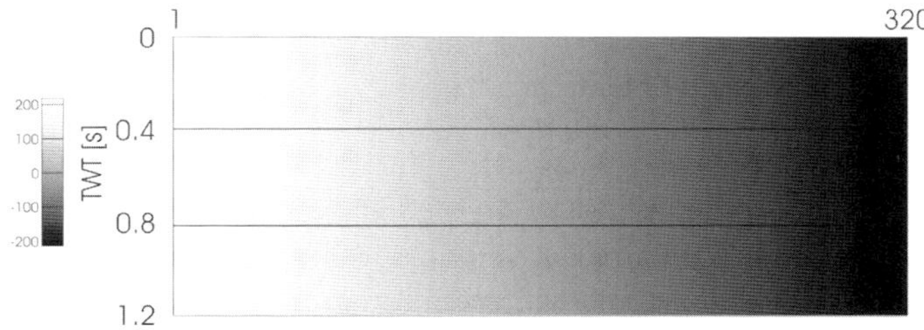

Fig. 6. The estimated dip in the y-direction from Figure 5. The dip was calculated in sub-cubes of $7 \times 7 \times 7$ points. The dip is given in microseconds per meter.

right, just as the input signal does. The algorithm performs quite well on non-synthetic data as well as on the displayed synthetic data.

3. DIP STEERING

Once the dip is known at a position (x,y,t), the dip plane can be followed until it intersects with another trace, at the position $(x+a,y+b,t_t)$, where:

$$t_t = t + p_x(x,y,t)\cdot a + p_y(x,y,t)\cdot b \tag{20}$$

and $p_x(x,y,t)$ and $p_y(x,y,t)$ are the apparent dips in x and y directions, respectively.

If we assume that the dip is accurate, the waveforms at (x,y,t) and at intersected position $(x+a,y+b,t_t)$ should be similar and should thus have the same phase. However, regardless of how the dip is calculated, it will only be correct up to a certain precision. I introduce the term phase-locking, as the process of fine-tuning the intersect-time by searching a trace for the time where the trace's phase equals the phase at the original position. To avoid loop-skips the search is limited to a time aperture centered at t_t. Let the phase at the original position be defined as $\varphi(x,y,t)$ and the target trace be positioned at $(x+a,y+b)$. Then we find the time of equal phase, t_{ph}, by solving:

$$\begin{cases} \varphi(x,y,t) = \varphi(x+a,y+b,t_{\mathrm{ph}}) \\ t_t - \tau \le t_{\mathrm{ph}} \le t_t + \tau \end{cases}, \tag{21}$$

where 2τ is the width of the aperture. If the same phase occurs more than once within the aperture, the solution closest to t_t is chosen. If no solutions are found within the aperture, we assume t_t to be the optimum estimation of the intercept time. The size of the aperture is a critical parameter. Too small values will restrict the effectiveness. Too large values may produce loop-skips.

I introduce the term dip-steering (Fig. 7) as the process of following the dip from any position on a trace to a position on another trace. This can be done with or without phase-locking to any trace distance away from the starting trace. Starting at a given position, the dip is followed (optionally using phase-locking) to an adjacent trace. At this trace the dip is calculated and followed to the next trace, and so on. Dip steering works like a local autotracker that localizes the same seismic event on neighboring traces as observed at the original position. This trace to trace process allows a higher precision than if the intercept time was calculated directly without the intermediate steps.

4. CHIMNEY DETECTION

The neural network is fed with a number of attributes. The original chimney cube used Similarity, Energy, and Spatial Position (Fig. 3). Spatial Position is defined by two-way reference time and optionally in-line and cross-line location. Spatial position enables the network to detect spatial trends in the prediction. Similarity and Energy are described in the following sections.

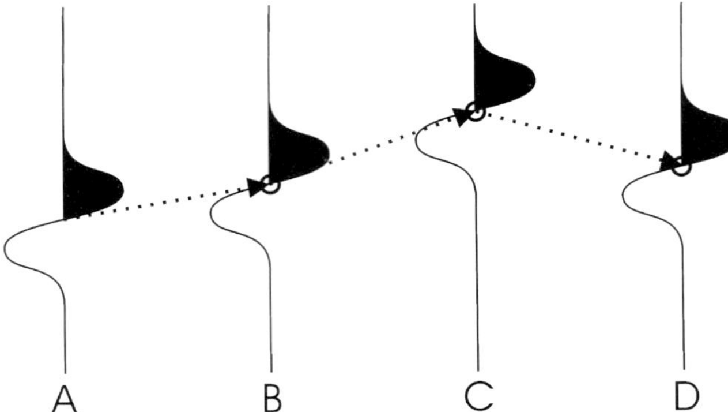

Fig. 7. The principle of dip-steering. The dip is calculated at trace A and followed to trace B, optionally with phase-locking. At the intercepttime with trace B, the dip is calculated and followed, optionally with phase-locking, to trace C. In the same manner, the dip is calculated at the intercepttime with trace C and followed, optionally with phase-locking, to trace D.

4.1. Similarity

Similarity quantifies how similar two segments of seismic traces $u(x,y,t)$ are. The similarity between two trace segments at (x_A,y_A) and (x_B,y_B), centered at the time t, is defined as the distance in hyperspace between the segments' vectors, normalized to the sum or the vectors' length:

$$S = 1 - \frac{|a-b|}{|a|+|b|}, \tag{22}$$

where:

$$a = \begin{bmatrix} u(x_A,y_A,t+t_1) \\ u(x_A,y_A,t+t_1+\mathrm{d}t) \\ \vdots \\ u(x_A,y_A,t+t_2-\mathrm{d}t) \\ u(x_A,y_A,t+t_2) \end{bmatrix}, \tag{23}$$

$$b = \begin{bmatrix} u(x_B,y_B,t+t_1) \\ u(x_B,y_B,t+t_1+\mathrm{d}t) \\ \vdots \\ u(x_B,y_B,t+t_2-\mathrm{d}t) \\ u(x_B,y_B,t+t_2) \end{bmatrix}, \tag{24}$$

$\mathrm{d}t$ is the sampling interval, t_1 is the relative start-time of the comparison window, and t_2 is the relative stop-time of the comparison window. The similarity measure gives results ranging from zero to one, where values close to zero should be interpreted as low similarities and values close to one should be interpreted as nearly identical trace segments. Compared to cross correlation, this measure gives a result that is sensitive not only to waveform but also to differences in amplitude.

Since seismic chimneys are vertical disturbances of the seismic response, the similarity is lower inside chimneys as compared to the surroundings.

4.2. *Energy*

The energy inside chimneys is lower than its surroundings. The energy at (x, y, t) is calculated as the sum of the squared amplitudes inside a window:

$$E = \sum_{\tau=t_1}^{t_2} u(x, y, t+\tau)^2, \tag{25}$$

where, t_1 is the relative start-time of the energy window and t_2 is the relative stop-time of the energy window.

5. DIP-RELATED ATTRIBUTES

The only multi-trace attribute in the original chimney cube is similarity. Consequently, similarity is the only attribute in the original chimney cube that is sensitive to structural dips. The similarity calculated by Eq. (22) decreases with increasing dip and increasing distances between the traces.

5.1. *Dip-steered similarity*

The dip-steered intercept time (calculated above) yields the time of the seismic event at the target trace that was tracked from the starting position. Trace segments can be extracted around these calculated intercept times to compute the dip-steered similarity given by:

$$S_{\text{dip}} = 1 - \frac{|a_{\text{dip}} - b_{\text{dip}}|}{|a_{\text{dip}}| + |b_{\text{dip}}|}, \tag{26}$$

where:

$$a_{\text{dip}} = \begin{bmatrix} u(x_A, y_A, t_A + t_1) \\ u(x_A, y_A, t_A + t_1 + \mathrm{d}t) \\ \vdots \\ u(x_A, y_A, t_A + t_2 - \mathrm{d}t) \\ u(x_A, y_A, t_A + t_2) \end{bmatrix}, \tag{27}$$

$$b_{\text{dip}} = \begin{bmatrix} u(x_B, y_B, t_B + t_1) \\ u(x_B, y_B, t_B + t_1 + \mathrm{d}t) \\ \vdots \\ u(x_B, y_B, t_B + t_2 - \mathrm{d}t) \\ u(x_B, y_B, t_B + t_2) \end{bmatrix}, \tag{28}$$

where t_A and t_B are the dip-steered times going from the position (x, y, t) to the traces at (x_A, y_A) and (x_B, y_B) respectively.

Fig. 8 shows two time slices with normal similarity and dip-independent similarity from an area with dipping reflectors. It is clear that dip-steering reduces low similarities caused by dipping layers. Classification of chimneys and non-chimneys will benefit from dip-independent similarities, because a major reason for low similarity other than being inside a chimney is reduced.

5.2. *Dip variance*

Similarity is the only attribute in the original chimney classification network that can be improved by taking directionality into account. Directionality by itself or via derived statistical properties can also be used to improve the classification result.

The inner parts of chimneys are characterized by a rather chaotic texture, without coherent reflectors. Hence, it is probable that the calculated dips inside a chimney are high, that they vary from sample to sample and are not very accurate. This behavior is quantified by calculating the statistical variance of the dips inside a small subcube:

$$\mathrm{var}(p_x) = \frac{1}{n-1} \sum_{\beta=-x_s}^{x_s} \sum_{\alpha=-y_s}^{y_s} \sum_{\tau=t_1}^{t_2} \left(p_x(x+\alpha, y+\beta, t+\tau) - \bar{p}_x\right)^2, \tag{29}$$

where:

$$\overline{p}_x = \frac{\sum_{\beta=-x_s}^{x_s} \sum_{\alpha=-y_s}^{y_s} \sum_{\tau=t_1}^{t_2} p_x(x+\alpha, y+\beta, t+\tau)}{n}, \tag{30}$$

t_1 is the relative start-time for the sub-cube, t_2 is the relative stop-time, n the total number of samples in the sub-cube, and x_s and y_s are the sub-cube's lateral stepout in x and y directions respectively. In the same manner the dip-variance in the y direction, $\mathrm{var}(p_y)$, can be calculated for a sub-cube. It can be advantageous to combine $\mathrm{var}(p_x)$ and $\mathrm{var}(p_y)$ into one variable for the overall dip-variance by taking the average:

$$\overline{p}_{\text{var}} = \frac{\mathrm{var}(p_x) + \mathrm{var}(p_y)}{2}. \tag{31}$$

Hereafter, the term dip-variance relates to this average. Fig. 9 shows that the dip-variance is much greater inside chimneys as compared to outside.

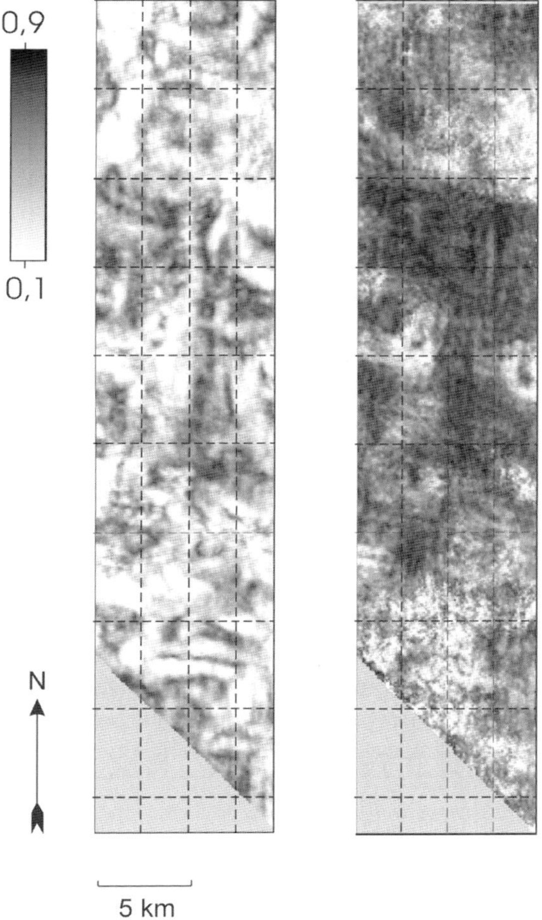

Fig. 8. A timeslice of the original similarity (left) and corresponding dip-steered similarity (right). Values close to zero should be interpreted as low similarities and values close to one should be interpreted as nearly identical trace segments. The figure shows that the dip-steered similarity is much higher than the original one in this dipping area.

6. PROCESSING AND RESULTS

Statistics based on misclassifications is one unbiased way of comparing neural network performances. However, the final judgement on how good a classification neural network is compared to another is not only a matter of statistics. It is also based on a subjective visual comparison between original seismic data and neural network predicted results. I compare resolution and the shape of predicted chimney bodies in this assessment.

In this study, various attribute and neural network configurations were tested on North Sea data and the results were compared; see Tingdahl, 1999 for details. In all tests the same example points were used to train and test the neural networks.

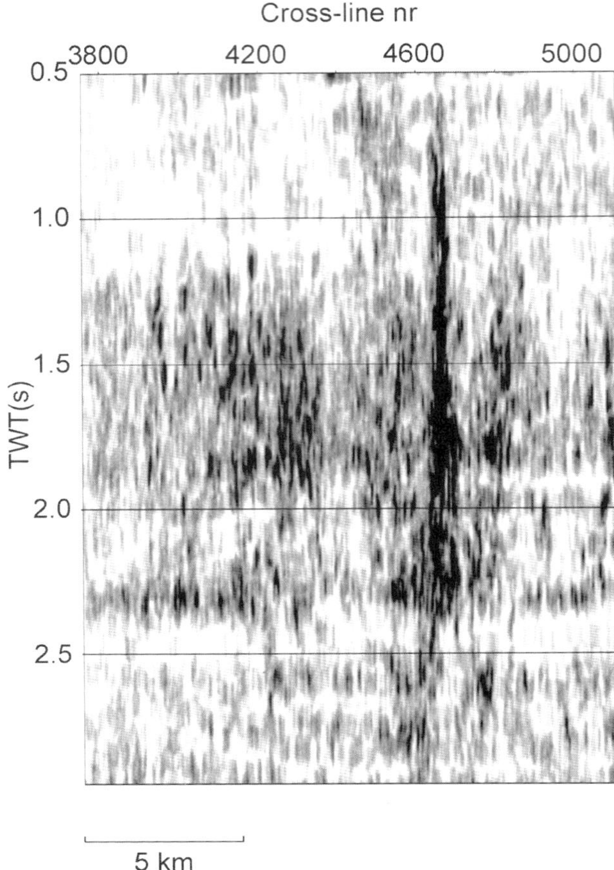

Fig. 9. The dip variance calculated from the in-line in Figure 2. Black indicates high variance while white indicates low variance.

The networks used to generate the original chimney cube and the one that gave the optimal improvement are displayed in Fig. 3 and Fig. 10, respectively. The best result was achieved when the original attributes were amended with one dip-steered similarity per time-gate, one dip-variance per time-gate, and the in-line and cross-line position. The dip-steered similarity performed better without phase-locking disabled than with phase-locking enabled. In our opinion different areas and surveys may have their own optimum set of attributes. Therefore, the presented set is not necessarily the ultimate solution for all cases.

The original and the improved chimney cube are displayed in Fig. 11. It is evident that the improved chimney cube exhibits a higher resolution and contrast as compared to the original one. The improved network also has a lower misclassification rate for the test locations: 14.5% for the improved network compared to 17.4% for the original network.

7. DISCUSSION AND CONCLUSIONS

In this chapter I have described ways to improve a seismic classification method aimed at detecting seismic chimneys. The improvements were achieved by including local dip information. I introduced two new terms: dip-steering and phase-locking. Dip-steering is the process of following a seismic event from any starting position to a position on an adjacent trace and beyond. Phase-locking is the process of finding the intercept time at the target trace by searching along the dip for the same phase as the phase at the starting position. The failure of the phase-locking is believed to be due to that the phase is quite sensitive to noise, and does therefore have a lower accuracy than the dip itself.

By using dip-steering I was able to compute dip-independent similarities, which are better than 'horizontal' similarities in areas with high structural dip (Fig. 8). The classification in terms of chimney and non-chimney benefits from this improvement. The main improvement however, is contributed to the new dip-variance attribute as can be judged from the node gray level in the neural network input layer (Fig. 10).

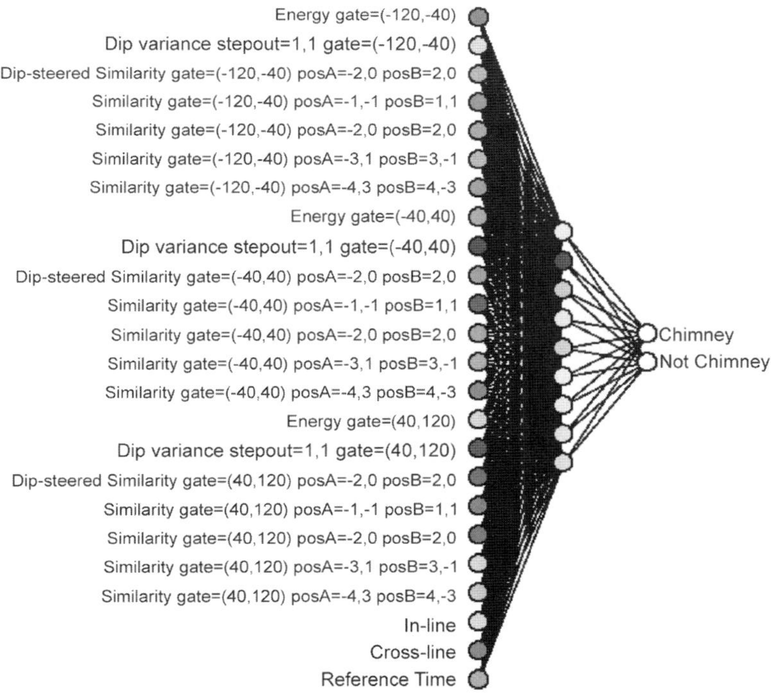

Fig. 10. The neural network used to predict the improved chimney cube. The grayscale indicates the relative contribution of each input node to the classification result. A description of the input attributes (to the left) is given in the text. The input attributes' time-gates are specified in milliseconds and their positions are specified as trace positions in in-line and cross-line respectively. The outputs (to the right) predict the presence and absence of seismic chimneys respectively.

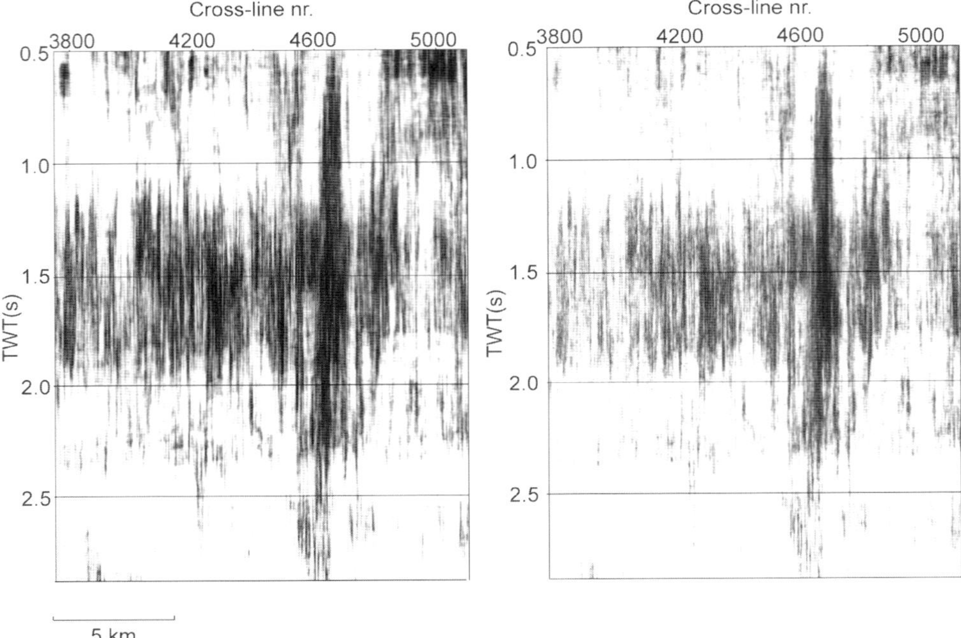

Fig. 11. The original chimney cube (left) and the improved chimney cube (right), calculated from the in-line in Figure 2. The two sections have the same grayscale, where black indicates high chimney probabilities and white indicates low chimney probabilities. It can clearly be observed that the improved prediction exhibits a higher resolution and a higher contrast than the original one.

ACKNOWLEDGEMENTS

Den Norske Stats Oljeselskap A/S (Statoil) and de Groot-Bril Earth Sciences (dGB) are thanked for the permission to publish this chapter and the use of their data. A patent has been applied for (Meldahl et al., 1998).

REFERENCES

Bahorich, M. and Farmer, S., 1995. 3-D seismic discontinuity for faults and stratigraphic features; the coherence cube. *Leading Edge* (Tulsa, OK), 14(10): 1053–1058.

Dalley, R.M. et al., 1989. Dip and azimuth displays for 3D seismic interpretation. *First Break*, 7(3): 86–95.

Heggland, R., 1997. Detection of gas migration from a deep source by the use of exploration 3D seismic data. *Mar. Geol.*, 137(1–2): 41–47.

Heggland, R., 1998. Gas seepage as an indicator of deeper prospective reservoirs. A study based on exploration 3D seismic data. *Mar. Pet. Geol.*, 15(1): 1–9.

Heggland, R., Meldahl, P., Bril, A.H. and de Groot, P.F.M., 1999. The chimney cube, an example of semi-automated detection of seismic objects by directive attributes and neural networks. Part II: Interpretation. *Soc. Explor. Geophys., 69th Annu. Meet.*, Tulsa, OK.

Hovland, M. and Judd, A.G., 1988. *Seabed Pocketmarks and Seepages – Impact on Geology, Biology and the Marine Environment*. Graham and Trotman, London.

Marfurt, K.J. and Kirlin, R.L., 2000. 3-D broad-band estimates of reflector dip and amplitude. *Geophysics*, 65(1): 304–320.

Marfurt, K.J., Kirlin, R.L., Farmer, S.L. and Bahorich, M.S., 1998. 3-D seismic attributes using a semblance-based coherency algorithm. *Geophysics*, 63(4): 1150–1165.

Meldahl, P., Heggland, R., Bril, A.H. and de Groot, P.F.M., 1999. The chimney cube, an example of semi-automated detection of seismic objects by directive attributes and neural networks. Part I: Methodology. *Soc. Explor. Geophys., 69th Annu. Meet.*, Tulsa, OK.

Meldahl, P., Heggland, R., de Groot, P.F.M. and Bril, A.H., 1998. Method of seismic body recognition. Patent GB 9819910.02.

Tingdahl, K.M., 1999. *Improving Seismic Detectebility Using Intrinsic Directionality*. Earth Sciences Centre, Göteborg University, Göteborg, Sweden.

Developments in Petroleum Science, 51
Editors: M. Nikravesh, F. Aminzadeh and L.A. Zadeh

Chapter 10

MODELING A FLUVIAL RESERVOIR WITH MULTIPOINT STATISTICS AND PRINCIPAL COMPONENTS

P.M. WONG [a,1] and S.A.R. SHIBLI [b,2]

[a] *School of Petroleum Engineering, University of New South Wales, Sydney NSW 2052, Australia*
[b] *Landmark Graphics (M) Sdn. Bhd., Menara Tan and Tan, 55100 Kuala Lumpur, Malaysia*

ABSTRACT

Traditional reservoir modeling techniques use oversimplified two-point statistics to represent geological phenomena which are typically curvilinear and have other complex geometrical configurations. Use of multipoint statistics has shown some improvement in recent years to reduce such limitations. This paper compares the performance of the use of conventional and multipoint data for estimating porosity from seismic attributes in a fluvial reservoir using neural networks. According to the results of the study, the neural network trained on multipoint data gave smaller error and higher correlation coefficient of porosity in a blind test. Further improvement is also obtained by reducing the dimensionality of the input space using principal components. This study shows a successful integration of neural networks and principal components for modeling multipoint data in practical reservoir studies.

1. INTRODUCTION

Realistic description of 2D/3D rock properties is an important issue for fluid-flow simulation. One of the current challenges in reservoir modeling is to maximise the use of information. For instance, generating a 2D porosity map from only isolated well data is too simplistic and many uncertainties exist between wells. If suitable and correlatable seismic attributes (e.g. interval velocity and amplitude) are available, it is best to incorporate such information as soft data in order to guide the interpolation procedure. In many cases, this approach can help to reduce the uncertainty of reservoir predictions between wells. Geostatistical techniques (e.g. co-kriging) are popular and are able to provide an uncertainty indicator (e.g. kriging variance) for each prediction to aid in risk analysis.

Despite the success of geostatistical applications, three major theoretical and practical limitations are noted: (1) the conventional techniques assume the existence of a linear relationship between the well and soft data; (2) only one or two soft data types can be incorporated in practice because of the difficulties in covariance and cross-covariance

---

[1] Present address: Veritas DGC Inc., 10300 Town Park Drive, Houston, TX 77072, USA.
[2] Present address: Maersk Oil and Gas, Esplanaden 50, DK-1263 Copenhagen K, Denmark.

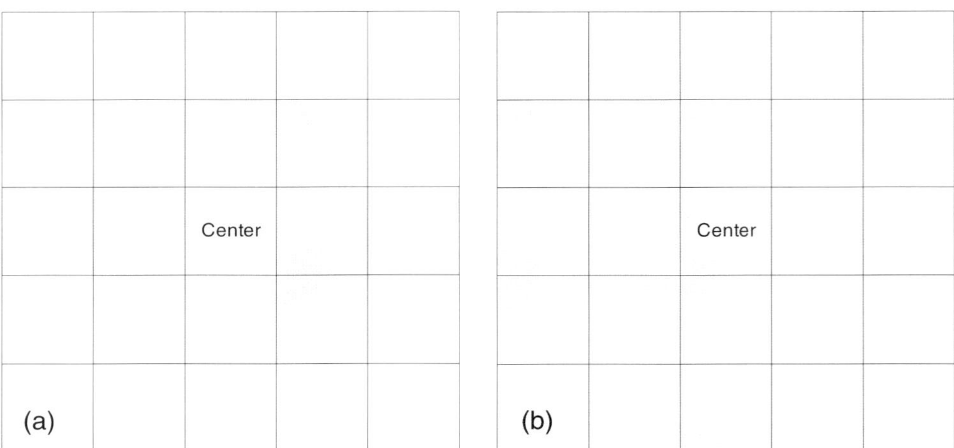

Fig. 1. Example templates for neighborhood searching. (a) A 5 × 5 star-shape template. (b) A 5 × 5 square template.

modeling; and (3) the conventional techniques use only two-point statistics (e.g. variogram and covariance). In response to points (1) and (2), artificial neural networks have provided a more practical alternative. These models can learn non-linear spatial relationships between hard and multivariate collocated soft data, without tedious cross-covariance modeling (Wang et al., 1999). Point (3), however, has not been addressed extensively and it certainly has great potential for future research and development.

Two-point statistics represent gross oversimplifications of how neighboring values (two, three, or even more points) relate in a spatial sense. The practical limitations of two-point statistics in reservoir modeling are several. One inherent limitation is that it does not represent complex curvilinear shapes (e.g. curvilinear fluvial channels, sand dune structures, and cross-bedded reservoir units) in the underlying geology accurately. Results derived from conventional two-point techniques look typically unrealistic and artificial. The constraints introduced by dense soft or secondary information do offer some improvement, but even the cross-covariance is still a two-point statistic.

In Caers and Journel (1998), the authors presented a new approach for capturing multipoint statistics from a training image using neural networks. It uses a (2D) search window ('template') to extract neighboring data around a center location. Fig. 1a shows an example star-shape template with 12 neighbors. The training set was constructed using the soft data (e.g. seismic measurements) at the neighboring locations as inputs and the hard data (e.g. rock properties) at the center as target output. A neural network was set up to learn the multipoint relationships. After training, a conditional distribution of the rock properties can be estimated and can be used for subsequent simulation (Caers, 2000, 2001). The choice of the template geometry is important, and the user must conduct sensitivity analyses to obtain the best results. A similar problem is also observed in well logging when petrophysicists try to incorporate log signals above and below the depth of interest for estimating rock properties from core analysis (Wong, 1999).

The objective of this paper is to address a number of technical issues in the use of neural networks for extracting multipoint statistics. We will first revisit some basic concepts of neural networks. Based on the use of a real data set (Wong and Shibli, 2000), we will condition the porosity distribution of a fluvial channel to two seismic attributes (velocity and attributes). We will compare the performance of neural networks trained by the 'conventional training set' and the 'multipoint training set'. Since the larger the template, the more the inputs, we will also attempt to reduce the input space using principal components. A number of numerical experiments will be performed and the predictions will be compared with the actual porosity values in a blind test.

2. NEURAL NETWORKS REVISITED

Artificial neural networks (ANN) are computer models that attempt to use simple functions of human nervous system and solve real life complex problems. These models are adaptive, multivariate and highly non-linear which are able to develop associations, transformations and mappings between objects or data. They have been used for solving many reservoir-related problems (Tamhane et al., 2000).

Supervised learning is common for solving reservoir modeling problems as it allows the mapping of multivariate data on the known rock properties. In simple words, supervised neural networks use a set of known input–output patterns (training patterns) to optimise the coefficients (weights) of a highly non-linear equation, which has been proven to be a universal approximator. Iterative learning techniques such as the popular 'backpropagation' algorithm (Rumelhart et al., 1986) can be used to optimise the weights. This paper will use the so-called 'backpropagation neural network' (BPNN) due to its popularity, although more advanced algorithms can be applied. After the weights are derived, the trained network can be applied to situations where actual measurements are not available.

In practice, the convergence of the error minimisation algorithm depends on the number of training patterns and unknowns (weights). When the number of inputs increases, the number of weights also increases. A small number of training patterns may not give satisfactory performance. When one attempts to model multipoint statistics, it is also important to note that the larger the template, the smaller number of training patterns will be extracted (due to missing neighbors at the edges). The problem is worse if multiple soft data types are used. The number of weights (including biases), Nw, for a multiple-input and single-output neural network can be defined as:

$$Nw = (Ni + 1) \cdot Nh + (Nh + 1)$$

$$Ni = Nv \cdot (Nt)^2$$

where Nv is the number of soft data types, Nt is the size of a square template (Fig. 1b), Ni is the number of inputs and Nh is the number of hidden neurons.

For example, a neural network with 2 soft data types, a template size of 5 pixels and 10 hidden neurons will have a total of 521 weights. Therefore, it is necessary to maintain a balance for effective training. The ratio, Np/Nw (number of training patterns required for each unknown to be estimated), gives a measure of degree of freedom in the network. The higher the ratio (say, $Np/Nw > 5$), the better the model; the smaller the ratio (say, $Np/Nw < 1$), the worse the performance. For instance, if we have 521 weights and set $Np/Nw = 5$, we will need to have a total of more than 2500 patterns available for training the network.

If it is necessary to have a large template, a practical solution is to apply principal component analysis (PCA) in order to reduce the number of inputs. PCA aims to reduce high dimensional data to fewer orthogonal dimensions for viewing and analysis with a minimum loss of total variance observed in the original data set (Davis, 1986). It works by extracting eigenvalues and eigenvectors of the variance–covariance matrix of the data. Very often, the PCs with the first few largest eigenvalues almost invariably account for a large proportion of the total variance of the original data set. Some previous studies using principal components as inputs to neural networks have shown good prediction performance and/or good model stability (Aminzadeh et al., 2000; Wong and Cho,

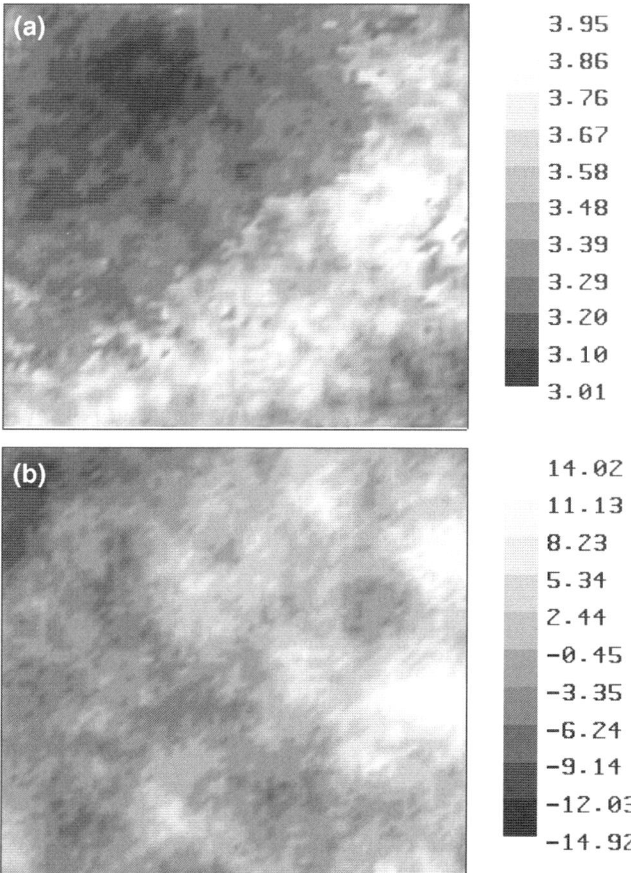

Fig. 2. Seismic attribute maps. (a) Velocity map (in km/sec). (b) Amplitude map (in metres).

2001). In this paper, we will investigate if the use of a 'reduced training set' using principal components will offer any improvement in this study.

3. CASE STUDY

This section shows an application of BPNN in a reservoir with 294 wells (Wong and Shibli, 2000). The formation consists of a series of fluvial channel sediments that accumulated in an environment much like the modern day Mississipi. Fine-grained sediments settled along the inside of meander belts and eventually graded into shale barriers. For each of the 294 wells, one zone average was derived for shale content, porosity and permeability from the existing log and core data. Maps of 2D seismic velocity and amplitude on 70 × 70 pixels are available and are shown in Fig. 2. This paper will focus on porosity estimation on the same pixels.

By gathering the data, we obtained 294 patterns with known seismic velocity and amplitude as inputs and porosity as output. To start the analysis, we will first randomly divide the original data set into three subsets: training (200), test (50) and validation (44). The training set is used to develop the neural network model; the test set is used to determine the optimal model configuration (e.g. no. of hidden neurons, termination criteria, etc.); and the validation set is used for blind testing.

A conventional training set was built using well coordinates (easting, northing), seismic velocity and amplitude as inputs (4 inputs) and porosity as output (1 output). After several trials, it was concluded that the optimum number of hidden neurons was four (4), and using this configuration the root mean square error (RMSE) and the correlation coefficient (R2) of the blind test results were found to be 1.00 and 0.84, respectively. The corresponding porosity map is displayed in Fig. 3.

Several points are worth mentioning upon inspection of the porosity map in Fig. 3.

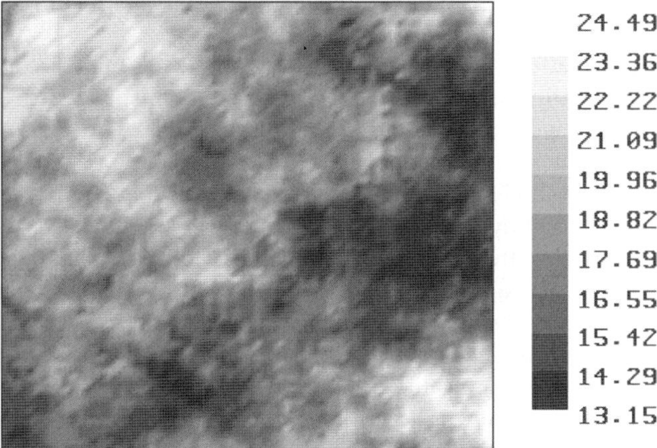

Fig. 3. Porosity map from the base case.

The expected deterioration in reservoir porosities along the NW/SE plane, though expected to be gradual, appears to be patchy in a lot of areas. Although the velocity trends have somewhat been captured, the gradual transition between the higher porosity at the NW and the SE of the reservoir is not modelled accurately, i.e. the curvilinear porosity transition has not been resolved to a great degree of accuracy, as evident from seismic. These results are typical also of cokriging and kriging with an external drift, all of which rely on two-point statistics to infer the shape of the underlying property distribution. Such differences are subtle, but can nonetheless be a cause of anguish for those trying to capture such subtleties in the resulting prediction maps. Multipoint statistics attempts to reduce such limitations.

3.1. Procedures

The essential steps for training with multipoint data are listed below:

(1) Define the template or window size. In this paper, we used $Nt = 5$.
(2) Based on the center (well) locations, extract the neighboring seismic patterns ($Nv = 2$) and construct the corresponding multipoint data sets.
(3) Train a BPNN, conduct the blind test and make a porosity map.
(4) Extract the principal components from the multipoint data sets and construct the corresponding 'reduced' data sets.
(5) Redo the analysis as in step (3).

3.2. Results

In this study, we added the multipoint data to the conventional training set. The input dimension for the multipoint data set (including the spatial coordinates of the center location) became $Ni + 2$. This was done for two reasons: (1) we can directly compare the results of the network derived from the conventional training set; and (2) we found that the network without the spatial coordinates as inputs did not converge to a satisfactory level. Hence, the inclusion of spatial coordinates provides additional information on the spatial relationships of all the templates.

Due to the edging effect from the template ($Nt = 5$), the number of patterns in the training, test and validation sets reduced to 185, 44 and 41 respectively. The corresponding input dimension was 52 (25 velocity, 25 amplitude and 2 coordinates). The number of pixels in the resulting porosity map became $66 \times 66 = 4356$. Based on the minimum error on the test set, seven (7) hidden neurons was the optimum. This configuration gave a total of 379 weights, and the Np/Nw ratio became 0.5, which was very small.

The blind test results are tabulated in Table 1. Compared to the base case, the RMSE was smaller (0.86) and the R2 was higher (0.86). This suggested that the use of multipoint data gave more accurate results and hence better generalisation. A simple sensitivity study was conducted for examining the relative significance of each input. This was achieved by calculating the change of RMSE when we remove one input at a time from the trained network. The larger the change, the more significant that input generally is.

Fig. 4 shows two bubble plots displaying the relative significance of each 25 velocity

TABLE 1

Neural network configurations and blind test results

Case	Training					Blind test		
	Ni	*Nh*	*Nw*	*Np*	*Np/Nw*	*Np*	RMSE	R2
Base	4	4	25	200	8.0	44	1.00	0.84
Multipoint	52	7	379	185	0.5	41	0.86	0.86
PC	4	6	37	185	5.0	41	0.75	0.89

and amplitude data around and including the center location (3,3). The size of the circle is proportional to the relevance of the input. As shown, each seismic attribute contributes differently at different directions (and lags). Note that some neighbors are more relevant to the collocated data at the center (3,3). In essence, not all of these complex representations can be captured by any two-point statistics. The corresponding porosity map is shown in Fig. 5. The major features are similar to those presented in Fig. 3, but with less patchiness and a more faithful representation of the velocity information.

Despite the success of the use of multipoint data, the Np/Nw ratio was too small and deemed unfavourable. We subsequently applied PCA to the input vectors (only the 50 neighbors of the seismic attributes). The analyses showed that the first two PCs accounted for more than 88% of the total variance. We then used the first two PCs together with the spatial coordinates as inputs to train another neural network. The optimal number of hidden neurons was 6. This gave a total of only 37 weights, and the Np/Nw ratio became 5. Fig. 6 shows the maps of the first two PCs. The results are also shown in Table 1.

Surprisingly, the RMSE (0.75) and R2 (0.89) were even more favourable than the multipoint case. This was due to the ability of PCA for simplifying the input space that eventually gave a higher probability of searching for a better solution. Fig. 7 shows the porosity map based on PCs. Again, the map contains all the major features as in the previous ones. We can also observe some smoothing in the map. This was due to the use of only two PCs, and hence there was a small lost of data variance.

3.3. Discussion

This study shows great potential use of search template for extracting multipoint statistics. The shape and size of the template has been an issue as discussed in Wang (1996) and Caers (2000). While Wang et al. (1999) showed that the isotropic templates perform well in reproducing even complex anisotropic structures, Caers (2000) claimed that the star-shape is the best overall. This study shows that a simple analysis on the trained network gives relative significances of all the inputs. This may help us to define the optimal shape and size of the template.

When both the template and the number of soft data types are large, it may become difficult to derive the optimal network configuration. The performance of any neural networks is often improved with a large Np/Nw ratio. When Np/Nw is small, PCs reduce

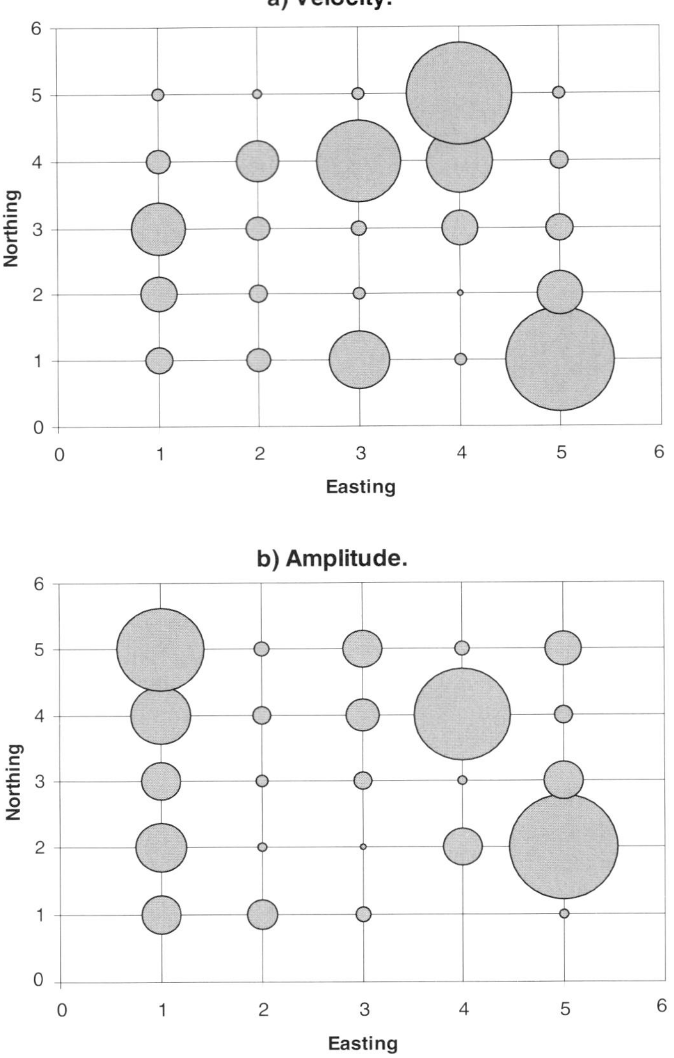

Fig. 4. Relevance of each neighbor for each seismic attribute around the center (3,3).

the dimensionality of the input space and effectively increases the Np/Nw ratio. This also dramatically reduces the computational time and improves the chance of finding the optimal solution. The proposed technology integration is general and is applicable to many areas of stochastic simulation (e.g. sequential simulation, facies simulation, etc).

Although PCs offer many practical advantages for training neural networks, there will be some loss of variance. The significance may vary from applications to applications. Moreover there may be some hidden geological information in the PCs (see Fig. 6),

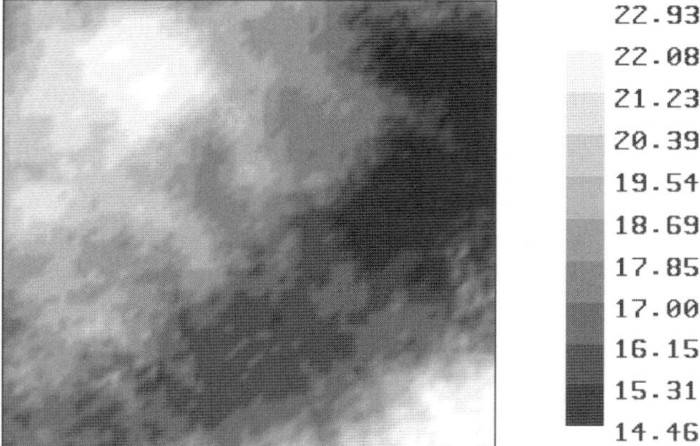

Fig. 5. Porosity map from the multipoint case.

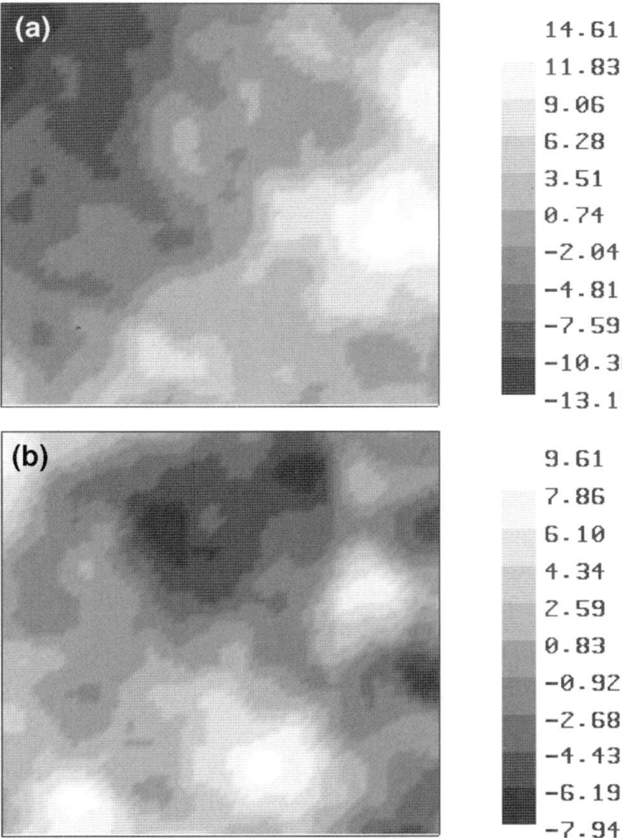

Fig. 6. Maps of principal components. (a) PC1 map. (b) PC2 map.

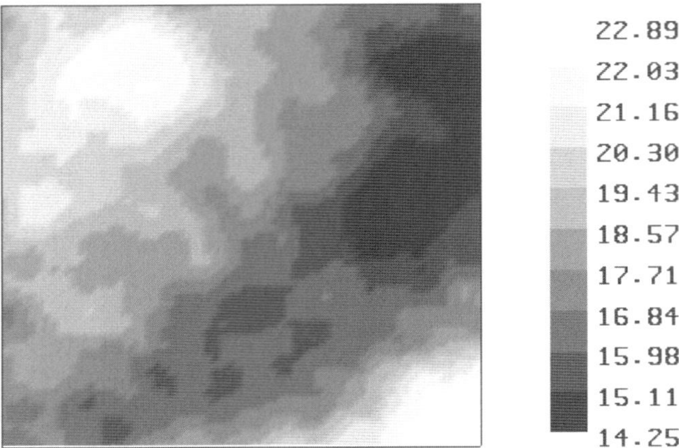

Fig. 7. Porosity map from the PC case.

which are outside the scope of the present work. The PCA used in this paper was the simplest model based on linear reduction. Although the PCs are orthogonal (uncorrelated) vectors, they may not be totally independent. It is mainly because the standard PCA works only with second-order covariance matrix. Future studies will investigate the potential of 'independent component analysis' or ICA (Lee, 1998), an extension of PCA, taking into account higher-order dependencies and independencies.

4. CONCLUSIONS

This paper compares the performance of the use of conventional and multipoint data for estimating porosity from seismic attributes using neural networks. According to the case study presented in a fluvial reservoir, the results show that the neural network trained by multipoint data gave smaller error and higher correlation coefficient in a blind test. Due to the unfavourable Np/Nw ratio, we apply principal component analysis to the high-dimensional input vectors containing the multipoint statistics. The use of only the first two components provides further improvement in the blind test. This study successfully shows that neural network training with principal components offer many practical advantages in reservoir modeling. It provides a more faithful representation of the underlying property distribution, based on typically dense training information to infer information about the shape of such a distribution. The next step is to apply such techniques to even more complex curvilinear shapes based on data such as outcrop measurements for various geological configurations.

REFERENCES

Aminzadeh, F., Barhen, J., Glover, C.W. and Toomarian, N.B., 2000. Reservoir parameter estimation using hybrid neural network. *Comput. Geosci.*, 26: 869–875.

Caers, J., 2000. The A, B, C of a modern geostatistics. *Proc. of the 20th gOcad Annu. Meet.*, Nancy, 4 pp.

Caers, J., 2001. Geostatistical reservoir modeling using statistical pattern recognition. *J. Pet. Sci. Eng.*, in press.

Caers, J. and Journel, A.G., 1998. Stochastic reservoir simulation using neural networks. *SPE Annu. Tech. Conf. and Exhibition*, New Orleans, SPE #49026, pp. 321–336.

Davis, J.C., 1986. *Statistics and Data Analysis in Geology*. John Wiley and Sons, New York, NY.

Lee, T.-W., 1998. *Independent Component Analysis – Theory and Applications*. Kluwer Academic Publishers, Boston, MA.

Rumelhart, D.E., Hinton, G.E. and Williams, R.J., 1986. Learning representations by back-propagation errors. *Nature*, 323: 533–536.

Tamhane, D., Wong, P.M., Aminzadeh, F. and Nikravesh, M., 2000. Soft computing for intelligent reservoir characterization. *SPE Asia Pacific Conf. on Integrated Modelling for Asset Management*, Yokohama, SPE #59397, 11 pp.

Wang, L., 1996. Modeling complex reservoir geometries with multipoint statistics. *Mathematical Geol.*, 28: 895–908.

Wang, L., Wong, P.M., Kanevski, M. and Gedeon, T.D., 1999. Combining neural networks with kriging for stochastic reservoir modelling. *In Situ*, 23: 151–169.

Wong, P.M., 1999. Prediction of permeability and its reliability from well logs using a windowing technique. *J. Petroleum Geol.*, 22: 215–226.

Wong, P.M. and Cho, S., 2001. Permeability prediction from well logs and principal components. EAGE/SEG Research Workshop on Reservoir Rocks, Pau, PAU27, 4 pp.

Wong, P.M. and Shibli, S.A.R., 2000. Combining multiple seismic attributes with linguistic reservoir qualities for scenario-based reservoir modelling. SPE Asia Pacific Oil and Gas Conference and Exhibition, SPE 64421, 5 pp.

PART 3.
COMPUTATIONAL GEOLOGY

Developments in Petroleum Science, 51
Editors: M. Nikravesh, F. Aminzadeh and L.A. Zadeh

Chapter 11

THE ROLE OF FUZZY LOGIC IN SEDIMENTOLOGY AND STRATIGRAPHIC MODELS

ROBERT V. DEMICCO [a], GEORGE J. KLIR [b] and RADIM BELOHLAVEK [c]

[a] *Department of Geological Sciences and Environmental Studies, Binghamton University, Binghamton, NY 13902-6000, USA*
[b] *Center for Intelligent Systems, Watson School of Engineering and Applied Science, Binghamton University, Binghamton, NY 13902-6000, USA*
[c] *Institute for Research and Applications of Fuzzy Modeling, University of Ostrava, Brafova 7, 70103, Czech Republic*

ABSTRACT

There has been a recent explosive growth in the theory and application of fuzzy logic and other related 'soft' computing techniques, opening new ways of modeling based on knowledge expressed in natural language. Fuzzy logic systems (based on fuzzy set theory), produce realistic sedimentation dispersal patterns in sedimentologic simulations in general and stratigraphic models in particular. The purposes of this paper are: (1) to present the basic concepts of fuzzy sets and fuzzy logic; and (2) to employ those concepts in an increasingly complex set of sedimentation models. The sedimentation models vary in temporal and spatial scales and employ fuzzy logic systems to model sediment dispersal systems. Models described here include: (1) a two-dimensional model of reef development over the last 80,000 y of variable sea level; (2) a three-dimensional hypothetical flood-plain delta simulation with either variable or constant sea level; (3) a two-dimensional model of carbonate sediment production on the Great Bahama Bank west and northwest of Andros Island; and (4) a model reproducing facies found in a deep core taken from the mixed chemical and siliciclastic sediments in the central basin of Death Valley. The final model of Death Valley makes use of the 'learning ability' of fuzzy logic systems coupled with an adaptive neural network. Stratigraphic models wherein fuzzy logic models the sedimentary portions of the model have the potential to accurately model subsurface distribution of sedimentary facies (not just water depths of deposition) in terms of the natural variables of geology. This method offers an alternative to the statistical modeling of subsurface geology. It is more computationally efficient and more intuitive for geologists than complicated models that solve coupled sets of differential equations.

1. INTRODUCTION

In recent years, two-dimensional and three-dimensional computer-based models of sedimentary basin-filling have become increasingly important tools for research in geological science, both applied and theoretical (Burton et al., 1987; Tetzlaff and Harbaugh, 1989; Angevine et al., 1990; Bosence and Waltham, 1990; Franseen et al., 1991; Bosscher and Schlager, 1992; Flint and Bryant, 1993; Bosence et al., 1994; Slingerland et al., 1994; Mackey and Bridge, 1995; Forster and Merriam, 1996; Leeder et al., 1996; Nittrourer and Kravitz, 1996; Nordlund, 1996; Wendebourg and Harbaugh, 1996; Whitaker et al., 1997; Harbaugh et al., 1999; Harff et al., 1999). These models produce synthetic stratigraphic cross-sections that are of great value for two reasons. First they give us a predictive picture of the subsurface distribution of rocks (sedimentary facies) whose petrophysical properties are useful in oil exploration, gas exploration, groundwater exploitation, groundwater remediation, and even naval warfare. Second synthetic stratigraphic models increase our theoretical understanding of how sediment accumulation varies in time and space in response to external driving factors (such as eustasy and tectonics) and internal driving factors (such as compaction, isostatic adjustments, and crustal flexural adjustments) made in response to tectonic loading and sedimentary accumulation (cf. Angevine et al., 1990).

The thorniest problem faced by stratigraphic modelers is simulating sediment erosion, sediment transportation, and sediment accumulation within a forward model (what Wendebourg and Harbaugh, 1996, refer to as 'sedimentary process simulators'). For example, in coastal and shallow marine systems, waves, wave-induced currents, tidal currents and storm-induced (i.e. 'event') waves and currents lead to ever-changing patterns of sediment erosion, transportation, and accumulation. Modeling such events entails handling physical laws and empirically derived relationships (cf. Slingerland et al., 1994). These physical laws and empirical relationships are generally described by nonlinear, complex sets of partial differential equations (Slingerland, 1986; Li and Amos, 1995; Wendebourg and Harbaugh, 1996; collected papers in Acinas and Brebbia, 1997; Harff et al., 1999). Moreover, these equations must be coupled during solution. Furthermore, some parameters that cannot be easily formalized, such as antecedent topography and changing boundary conditions, and incorporation of 'rare' events need to be taken into account. When we consider carbonate depositional systems, we are also confronted by the in situ formation of the sediments themselves both as reefs (cf. Smith and Kinsey, 1976; Buddemeier and Smith, 1988, 1992), and bank-interior sediments (cf. Broecker and Takahashi, 1966; Morse et al., 1984).

Coastal oceanographic modelers have made great strides in dealing with the complexities of coupled solutions as well as wave dynamics, current dynamics and sediment transport. However, finite difference and finite element numerical simulations such as those in Acinas and Brebbia (1997) and Harff et al. (1999) have two drawbacks when applied to stratigraphic models. First, they are site specific and depend on rigorous application of boundary conditions, initial conditions, and wave and tidal forcing functions over a discrete domain. Secondly, these process-response models operate at tens to hundreds of year time scales, which are very short in comparison to basin-filling models. As a result, the effects of large, complex storm events, which are suspected of being important agents in ancient depositional systems, are only rarely included in coastal models. Indeed, such complexities lead Pilkey and Thieler (1996) to question the applicability of even short-term coastal models built around dynamic sedimentary process simulators.

Early siliciclastic sedimentary process simulators employed either the diffusion equation to represent sediment dispersal (see discussion in Wendebourg and Harbaugh, 1996, p. 4) or used linear approximations of more complicated sediment dispersal. The two-dimensional code of Bosence and Waltham (1990), Bosence et al. (1994), the 'Dr. Sediment' code of Dunn (1991), the 2-dimensional alluvial architecture code of Bridge and Leeder (1979), the 3-dimensional update of that code by Mackey and Bridge (1995) and the 'CYCOPATH 2D' code of Demicco (1998) all use such an approach. Finally, there exist a number of sophisticated, sedimentary process simulators that employ numerical solutions of the fundamental, dynamical, physical equations coupled with empirical and semi-empirical equations. Such integrated flow and sediment transport models involve calculations of bed shear stress along the bottom of a circulation model. The bed shear stress from that model would then be used as input to solve the temporal and spatial terms in bedload and suspended load sediment transport equations. Examples of such models are the STRATAFORM family of models (Syvitski and Alcott, 1995; Nittrourer and Kravitz, 1996), the SEDSIM models of Wendebourg and Harbaugh, 1996 (see page 11; see also Tetzlaff and Harbaugh, 1989), and the river avulsion model of Slingerland and Smith (1998). Although these models have been successful, they can be computationally quite complex.

We have been developing *fuzzy logic* models of sediment production, erosion, transportation and deposition based on qualitatively and quantitatively defined observational rules. Nordlund (1996) and Fang (1997) suggested that fuzzy logic could be used to overcome some of the difficulties inherent in modeling sediment dispersion. There is a wealth of observational data on flow and sediment transport in the coastal zone, in river systems, on carbonate platforms, and in closed basin settings. Nordlund (1996) refers to this as 'soft' or qualitative information on sedimentary dynamics. However, we also have a fair amount of quantitative information on some sedimentary processes (e.g. the volumetric production of lime sediment per year on different areas on carbonate platforms – see Broecker and Takahashi, 1966; Morse et al., 1984). Examples of qualitative information would be "beach sands tend to be well sorted and are coarser than offshore sands", or "carbonate sediment is produced in an offshore carbonate 'factory' and is transported and deposited in tidal flats". Such statements carry information, but are not easily quantified. Indeed, these types of qualitative statements are commonly the exact kind of information that is obtained by studies of ancient sedimentary sequences. Moreover, with the development of 'seismic stratigraphy' and 'sequence stratigraphy', applied and academic geologists have both moved into an arena where there is commonly a complex blend of 'hard' and 'soft' information. Hard data might include seismic (or outcrop-scale) geometric patterns of reflectors or bedding geometries whereas soft information would include description of rock types, interpretations of depositional settings, and their positions within 'system tracts' (cf. Vail et al., 1977; Wilgus et al., 1989; Schlager, 1992, 1999; Loucks and Sarg, 1993; Emery and Myers, 1996).

Fuzzy logic allows us to formalize and treat such information in a rigorous, mathematical way. It also allows quantitative information to be treated in a more natural, continuous fashion. The purpose of this paper is to present a number of simulations of increasing complexity, where we have used fuzzy logic to model sediment dispersal in three-dimensional stratigraphic models wherein sea level changes, subsidence, isostasy, and crustal flexure are modeled using conventional mathematical representations (Turcotte and Schubert, 1982; Angevine et al., 1990; Slingerland et al., 1994). Our preliminary results along with the model FLUVSIM (Edington et al., 1998) and the modeling of the Smackover Formation described by Parcell et al. (1998) suggest that fuzzy logic may be a powerful and computationally efficient alternative technique to numerical modeling for the basis of a sedimentary process simulator. It has the distinct advantage in that models based on fuzzy logic are robust, easily adaptable, computationally efficient, and can be easily altered internally allowing many different combinations of input parameters to be run in a sensitivity analysis in a quick and efficient way.

2. BASIC PRINCIPLES OF FUZZY LOGIC

2.1. Fuzzy sets

Fuzzy logic is based on the concept of *fuzzy sets* (Zadeh, 1965; Klir and Yuan, 1995). In a conventional *crisp set*, an individual is either included in a given set or not included in it. This distinction is often described by a *characteristic function*. The value of either 1 or 0 is assigned by this function to each individual of concern, thereby discriminating between individuals that either are members of the set (the assigned value is 1) or are not members of the set (the assigned value is 0). Fig. 1A is an example the crisp set

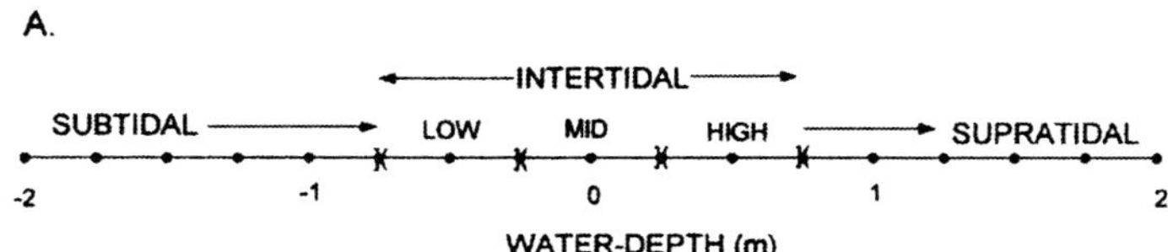

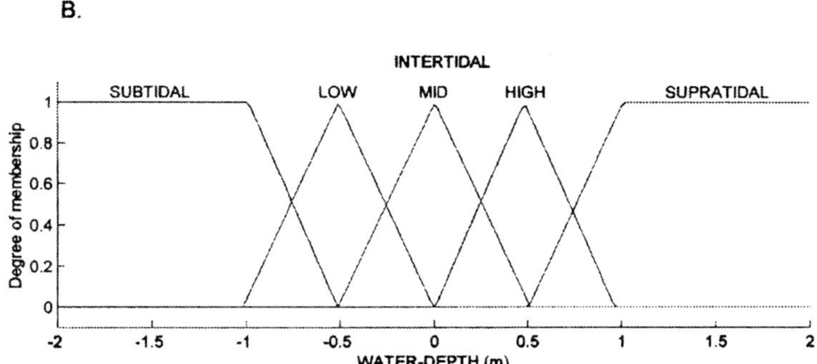

Fig. 1. Comparison of a crisp set description of the variable 'tidal range' (above) with a fuzzy set description (below). Mean low water = −1.25 m, mean sea level = 0 m, and mean high water = 0.75 m. The fuzzy set representation better captures natural variations (implied by the adjective 'mean') due to periodic tidal curve changes resulting from the ebb–neap–ebb cycle, and non-periodic, random variations such as storm flooding, etc.

concept of 'water depth' applied to an intertidal setting. The domain of this variable ranges from 2 m below mean sea level to 2 m above mean sea level. This continuum is generally divided into a number of crisp sets: subtidal, intertidal and supratidal with the intertidal being further subdivided into high-intertidal, mid-intertidal, and low-intertidal areas (Reading and Collinson, 1996, p. 213). In the example shown in Fig. 1A, the characteristic function $A(x)$ of the crisp set 'mid-intertidal' is for example:

$$A(x) = \begin{cases} 1 & \text{when } -0.25 \text{ m} < x \leq 0.25 \text{ m} \\ 0 & \text{otherwise} \end{cases} \tag{1}$$

However, on modern tidal flats, these boundaries are constantly changing due to periodic variations in over a dozen principle tidal harmonic components (cf. table 11.1 in Knauss, 1978). More importantly, it is commonly flooding due to anomalous 'wind tides' and 'barometric tides' (Knauss, 1978) that is important for erosion and deposition in beaches, tidal flats, etc.

A *standard fuzzy set* conveys the inherent imprecision of arbitrary 'pigeon hole' boundaries. In a standard fuzzy set the characteristic function is generalized by allowing us to assign not only 0 or 1 to each individual of concern, but also any value between 0 and 1. This generalized characteristic function is called a *membership function* (Fig. 1B). The value assigned to an individual by the membership function of a fuzzy set is interpreted as the degree of membership of the individual in the standard fuzzy set. The membership function $B(x)$ of the standard fuzzy set 'mid-intertidal' represented in Fig. 1B is:

$$B(x) = \begin{cases} 0 & \text{when } x \leq -0.5 \text{ m} \\ \dfrac{x+0.5}{0.5} & \text{when } -0.5 \leq x \leq 0 \text{ m} \\ \dfrac{0.5-x}{0.5} & \text{when } 0 \leq x \leq 0.5 \text{ m} \\ 0 & \text{when } 0.5 \text{ m} \leq x \end{cases} \tag{2}$$

The fuzzy set description of tidal range given in Fig. 1B better captures the essence of the gradations between locations on beaches, tidal flats, etc. Similarly, 1–2 m below sea level is certainly shallow, but where does a carbonate platform or siliciclastic shelf become 'deep' or 'open' (cf. Nordlund, 1996)? Using fuzzy sets, there can be a complete gradation between all these depth ranges. Each membership function is represented by a curve that indicates the assignment of a membership degree in a fuzzy set to each variable within the domain of the variable involved (e.g. the variable 'water depth'). The membership degree may also be interpreted as the degree of compatibility of each value of the variable with the concept represented by the fuzzy set (e.g. subtidal, low-intertidal, etc.). Curves of the membership functions can be simple triangles, trapezoids, bell-shaped curves, or have more complicated shape.

Contrary to the symbolic role of numbers 1 and 0 in characteristic functions of crisp sets, numbers assigned to individuals by membership functions of standard fuzzy sets have clearly a numerical significance. This significance is preserved when crisp sets are viewed (from the standpoint of fuzzy set theory) as special fuzzy sets. Other, nonstandard types of fuzzy sets have been introduced in the literature (Klir and Yuan, 1995). In this paper, however, we consider only standard fuzzy sets in which degrees of membership are characterized by numbers between 0 and 1. Therefore the adjective 'standard' is omitted.

Another example of the difference between crisp and fuzzy sets is provided by the concept of 'grain size'. The domain of this variable ranges over at least 6 orders of magnitude from particles that are micron-size to particles that are meter-size. Because of this spread in the domain of the variable, grain size is usually represented over a base 2 logarithmic domain. This continuum is generally divided into four crisp sets; clay, silt, sand and gravel. The characteristic function $A(x)$ of sand is for example:

$$A(x) = \begin{cases} 1 & \text{when } \frac{1}{16}\text{ mm} \leq x \leq 2\text{ mm} \\ 0 & \text{otherwise} \end{cases} \tag{3}$$

In this crisp set representation of grain size a grain with diameter of 1.9999 mm would be classified as sand, whereas a grain with diameter of 2.0001 mm would be classified as gravel. If fuzzy sets are used instead of crisp sets, than the artificial classification boundaries are replaced by gradational boundaries and the two grains described would *share* membership in both sets, described by the linguistic terms 'coarse sand' and 'gravel'. With increasing diameter of grains, the membership in 'gravel' will increase and the membership in 'sand' will decrease in some way that depends on the application context. The basic idea is that the membership in a fuzzy set in not a matter of affirmation or denial, as it is in a classical set, but a matter of degree. The membership functions in this case have complicated formulas because the domain is represented by a logarithmic function whereas the range is on an arithmetic scale. Other examples of fuzzy sets relevant to sedimentary systems might include sediment type, and the thickness of sediments eroded and deposited. In this context, terms such as 'produce some', 'erode a little', or 'deposit a lot' have meaning.

2.2. *Fuzzy logic systems*

The purpose of fuzzy logic is to formalize reasoning in natural language. This requires that propositions expressed in natural language be properly formalized. In fuzzy logic, the various components of natural-language propositions (predicates, logical connectives, truth qualifiers, quantifiers, linguistic hedges, etc.) are represented by appropriate fuzzy sets and operations on fuzzy sets (Zadeh, 1975, 1976). Each of these fuzzy sets and operations is strongly content dependent and, consequently, must be determined in the context of each application (Klir and Yuan, 1995). The most common *fuzzy logic systems* are sets of fuzzy inference '*if–then*' rules. These are conditional and unqualified fuzzy propositions that describe dependence of one or more output-variable fuzzy sets to one or more input-variable fuzzy sets. A simple fuzzy if–then rule assumes the canonical form:

If x is A then y is B

where A and B are linguistic values defined by fuzzy sets on the universal sets X and Y, respectively. The 'if' part of the rule 'x is A' is referred to as the *antecedent* or premise whereas the 'then' part of the rule 'y is B' is referred to as the consequent or conclusion.

2.3. Application of 'if–then' rules to coral reef growth

Coral animals are capable of rapid fixation of $CaCO_3$ from seawater because of symbiotic photosynthetic algae within their tissues. Thus, carbonate production of these animals is, in some way, related to light penetration into the shallow ocean. Fig. 2 shows data on growth rates of the main Caribbean reef-building coral *Montastrea annularis* (from Bosscher and Schlager, 1992, their fig. 1, p. 503). Bosscher and Schlager (1992), following Chalker (1981), fit the following equation to this data,

$$G = G_{\mathrm{m}} \tanh(I_0\, \mathrm{e}^{-kz}/I_k) \tag{4}$$

Here z is water depth, G is growth rate at a given depth (z), G_{m} is maximum growth rate (G at $z = 0$), I_0 is surface light intensity, I_k is saturation light intensity, and k is the extinction coefficient given in the Beer–Lambert law,

$$I_z = I_0\, \mathrm{e}^{-kz} \tag{5}$$

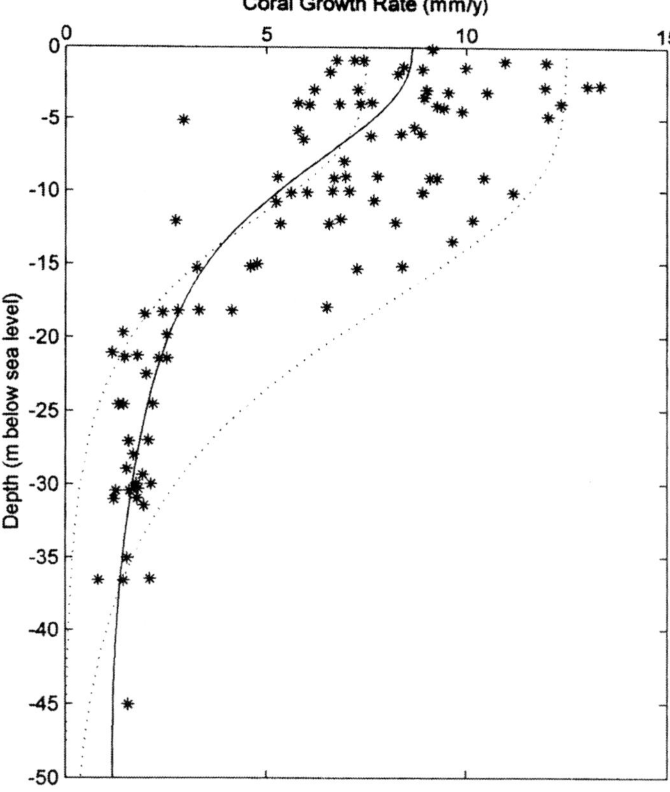

Fig. 2. Measured growth rates of the main Caribbean reef-building coral *Montastrea* annularis (from Bosscher and Schlager, 1992, their fig. 1). The 2 dotted lines are solutions of Eq. (4). The solid curve is the result of the fuzzy logic system described in the text.

In Fig. 2, the two dotted curves are fit to the data with Eq. (4) using different values of the parameters G_{m}, I_0, I_k, and k (see Bosscher and Schlager, 1992, for details).

A fuzzy logic system approach to this data set would use natural language to capture the essence of the data: "If the water is shallow, then coral growth rate is fast. If the water is deep, then the coral growth rate is slow". The input or antecedent parameter here is water depth whereas the output or consequent variable is coral growth rate. Both of these variables can be represented by fuzzy sets. The upper plot in Fig. 3 shows 2 possible membership functions for the fuzzy sets 'shallow' and 'deep' for the input variable depth over the domain 0 to 50 m. The lower plot in Fig. 3 shows 2 possible membership functions for the fuzzy sets 'fast' and 'slow' for the variable growth rate over the domain 0 to 15 mm y^{-1}.

The standard (so-called 'Mamdani') interpretation (Mamdani and Assilian, 1975) of the if–then rules ("if the water is shallow, then coral growth rate is fast; if the water is deep, then the coral growth rate is slow") is shown in Fig. 4 for a water depth of 10 m. The left hand column represents the input variable water depth whereas the right column represents the output variable growth rate. The upper row is the rule: "if depth is shallow, then growth rate is fast". The second row is the rule: "if depth is deep then growth rate is slow". The input variable (10 m) is evaluated simultaneously for each water depth and a *truth value* = degree of membership of the input variable in each of the potential input sets ('shallow' and 'deep') is calculated. These truth values truncate

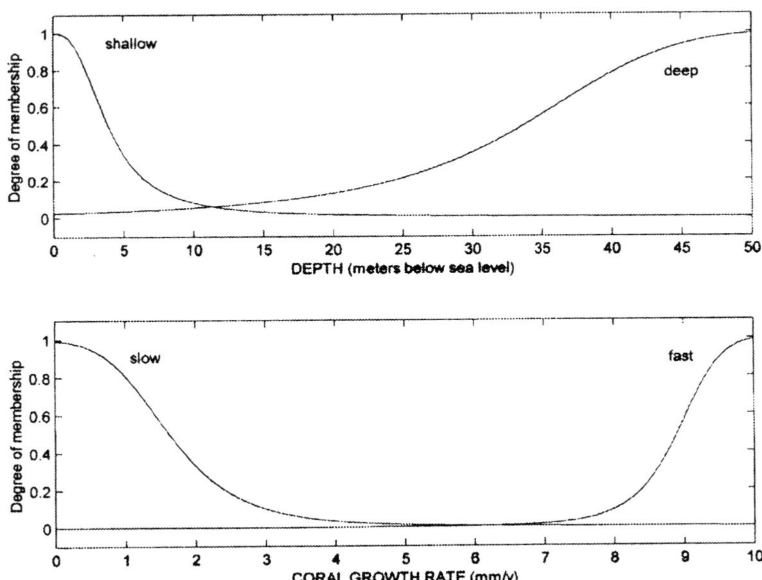

Fig. 3. Upper plot shows 2 membership functions for the fuzzy sets 'shallow' and 'deep' for the input variable depth over the domain range 0 to 50 m. The lower plot shows 2 membership functions for the fuzzy sets 'fast' and 'slow' for the variable growth rate over the domain 0 to 15 mm y^{-1}. Membership functions were adjusted by hand to produce the visual 'best fit' curve in Fig. 2.

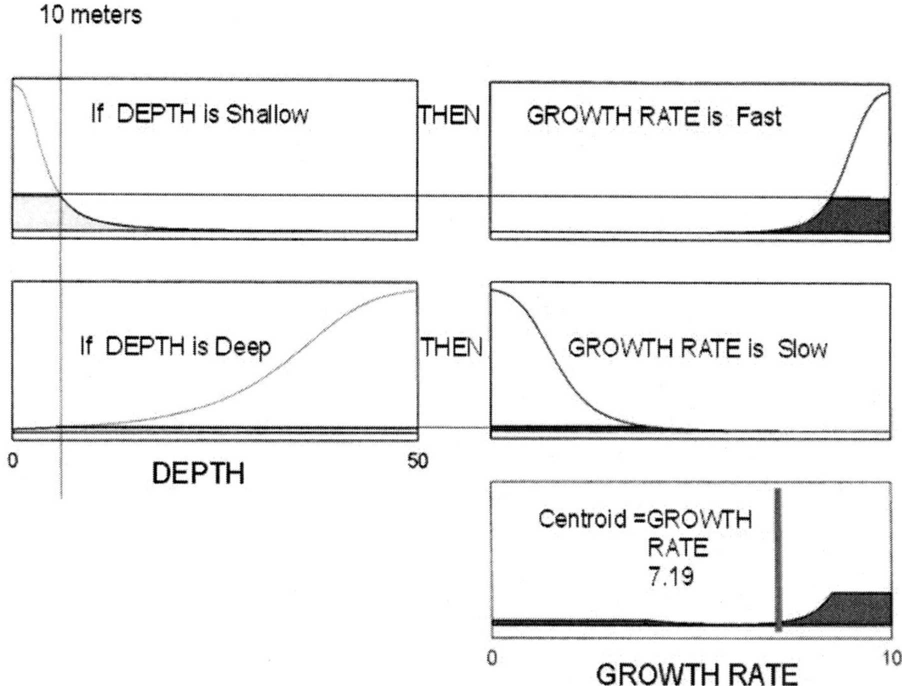

Fig. 4. A standard ('Mamdani') interpretation of the if then rules ("if the water is shallow, then coral growth rate is fast; if the water is deep, then the coral growth rate is slow") is shown for a water depth of 10 m. The input variable is evaluated for each water depth and a *truth value* = degree of membership of the input variable in each of the potential input sets ('shallow' and 'deep') is calculated. These truth values truncate the membership functions of the appropriate output variable. For each water depth, the truncated membership functions of the output variable are summed, and the centroid of the appropriate curve is taken as the 'defuzzified' output value.

the membership functions of the appropriate output variable. For each water depth, the maximum of the two truncated membership functions of the output variable is taken, and the centroid of the appropriate curve is taken as the 'defuzzified' output value. The solid curve in Fig. 2 shows this fuzzy inference system evaluated over the depth 0 to 50 m.

Fig. 5 compares a two-dimensional forward model of reef development based on Eq. (4) (Fig. 5A) with a model based on the fuzzy inference system of Fig. 4 (Fig. 5B). Bosscher and Schlager (1992) developed a numerical model of the geologic history of coral reefs growing on the Atlantic shelf–slope break of Belize by step-wise solution of the following differential equation.

$$\mathrm{d}h(t)/\mathrm{d}t = G_{\mathrm{m}} \tanh\big(I_0 \exp\{-k[h_0 + h(t)] - [s_0 + s(t)]\}/I_k\big) \tag{6}$$

Here $\mathrm{d}h(t)/\mathrm{d}(t)$ is the change in the height of the coral surface with time, h_0 is the initial height of the surface at the start of a time step, $h(t)$ is the growth increment in that time step, s_0 is the initial sea level position for a time step and $s(t)$ is the variation in sea level for that time step. The forward model simulation for coral reef growth based on

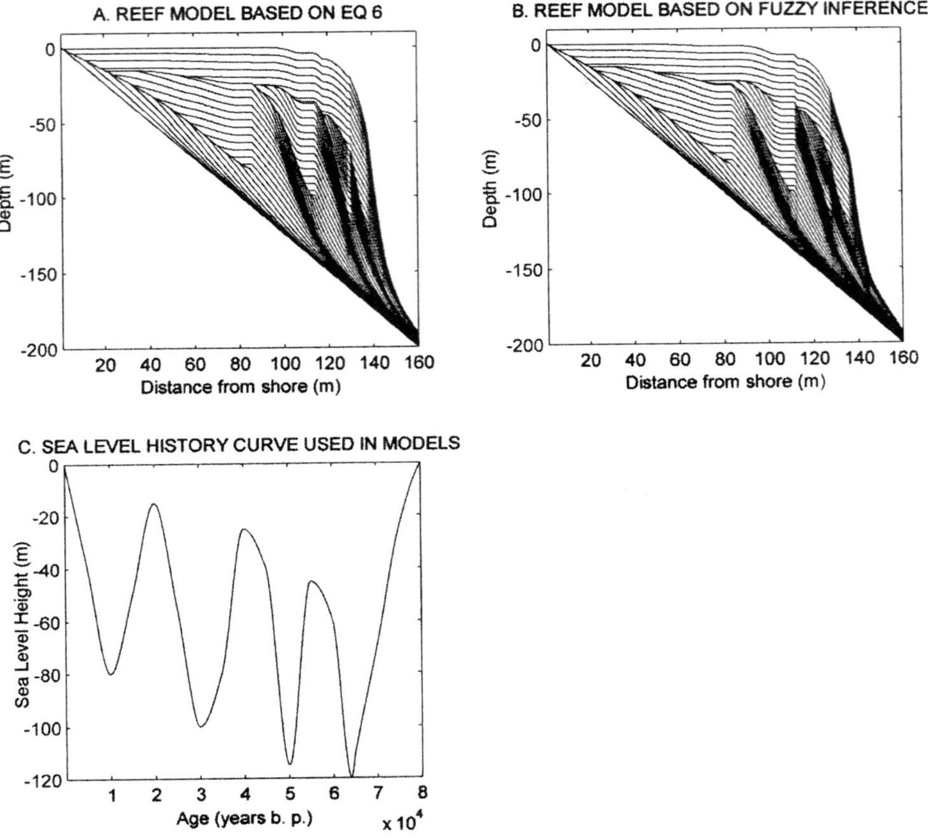

Fig. 5. Comparison of 2-dimensional models of the geologic history of coral reefs growing on the Atlantic shelf-slope break of Belize. (A) Stepwise solution of differential equation (6) (Bosscher and Schlager, 1992). The forward model solution for coral reef growth assumes an initial starting slope, initial values of G_m (maximum growth rate), I_0 (initial surface light intensity), k (extinction coefficient). (B) Variable sea level curve of the past 80,000 years input into the two models. (C) Model of reef growth based on the same sea level curve, same starting slope, and same initial value of G_m, but with the fuzzy inference system described in the text replacing the differential equation for coral growth production.

this equation is shown in Fig. 5A. The solution assumes an initial starting slope, initial values of G_m (maximum growth rate), I_0 (initial surface light intensity), k (extinction coefficient), and the variable sea level curve over the last 80,000 years (Fig. 5C). Fig. 5B compares a simulation wherein the historical sea level curve, starting slope, and initial value of G_m are the same as in the Bosscher and Schlager (1992) model, but the fuzzy inference system described in Fig. 4 replaces the differential equation expression for coral growth.

It is important to note that fuzzy logic systems are very versatile and, indeed, can be more versatile then deterministic equations. So far we have been using *ordinary fuzzy sets* wherein for a given input value there is one output value. In general, although we will not use them in this paper, we can generalize ordinary fuzzy sets into *second-order* fuzzy sets (Mendel, 2001), where the membership function does not assign to each element of the universal set one real number, but a fuzzy number (a fuzzy set defined on the real numbers in the unit interval) or by a closed interval of real numbers between the identified upper and lower bound. Clearly, this approach would be warranted by the spread in the initial data on coral growth rates versus depth in Fig. 2.

2.4. Application of multi-part 'if–then' rules to a hypothetical delta model

A hypothetical, greatly simplified river flood plain and deltaic system based on the modeling of Nordlund (1996) begins with a simple geometry and a grid of cells 125 by 125 each 1 km^2 (Fig. 6A). The colors in Fig. 6 reflect the caliber of the sediments that

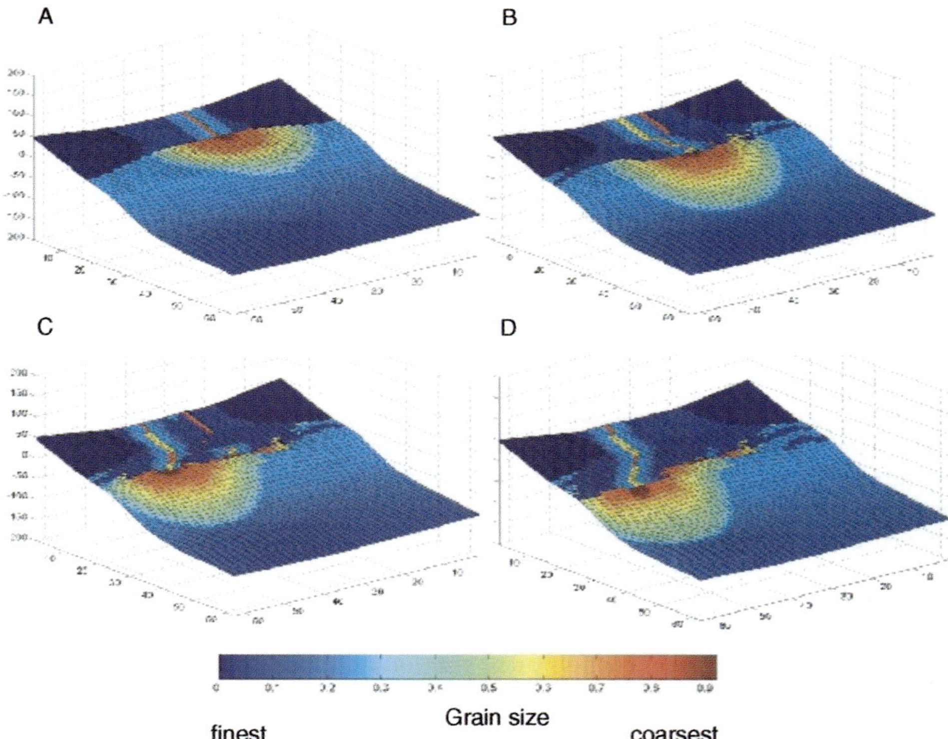

Fig. 6. Isometric view of hypothetical delta simulation at different time steps. Dark blues represent finest floodplain muds (0 on colorbar) and deepest marine muds whereas reds denote clean sands in the river and at the river mouth (0.9 on colorbar). Deltaic dispersion cone varies from sands (0.7) through muddy sands (0.5) to sandy muds (0.3) of the deeper shelf. (A) Sediment surface at start of simulation. (B) Sediment surface at end of 20,000 years after one sea level rise and fall. (C) Sediment surface at the end of 28,000 years during sea level fall. (D) Sediment surface at the end of 30,000 years at the end of sea level fall. Note how the river deposits of abandoned channels sink into the floodplain surface. Also note how the river 'meanders' across the floodplain as it seeks the lowest path to the shoreline. See text for further details.

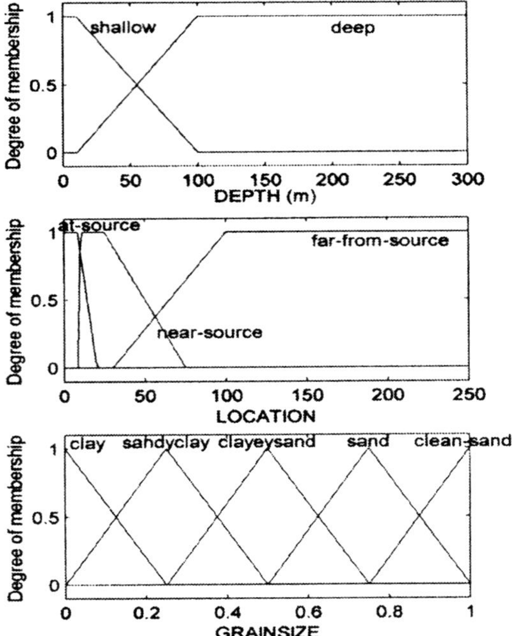

Fig. 7. Upper plot shows 2 membership functions for the fuzzy sets 'shallow' and 'deep' for the input variable depth over the domain range 0 to −300 m. Middle plot shows 3 membership functions for the fuzzy sets 'at-source', 'near-source' and 'far-from-source' for the input variable location over the domain 0 to 250 km. (C) Lower plot shows 5 membership functions and their designations for the output variable grain size over a normalized domain.

are deposited in each cell at each time step. The dark blue represents the finest-grained, flood-plain mud. The red through light blue hues indicates a simple deltaic sediment cone, the river that feeds it, and the river's levee–crevasse-splay system. Red represents the best-sorted and coarsest sands, yellows and greens sands through silts, and lighter blues crevasse-splay and offshore mud.

Four fuzzy inference systems, 2 for the delta and 2 for the river, control sediment deposition in the model. In each case, one fuzzy logic system controls sediment grain size and one fuzzy logic system controls thickness of sediment deposited in each cell. The fuzzy logic system that controls grain size for the subaqueous, deltaic deposition is further described in Figs. 7 and 8. There are two antecedent variables in this fuzzy logic system, water depth and location (distance from the mouth of the river), and one consequent or dependent variable, grain size (Fig. 7). The depth domain is 0 to 300 m below sea level. Depth is described by two trapezoidal membership functions: shallow and deep. The membership function 'deep' is here the standard *fuzzy complement* of the membership function 'shallow', i.e. for any depth: deep = shallow − 1. Thus the membership function 'deep' could be just as easily replaced with 'not shallow'. In fuzzy logic, the adjective 'not' implies the fuzzy complement of the fuzzy set it modifies. The domain of the variable location varies from 0 to 250 km from the river mouth. Location

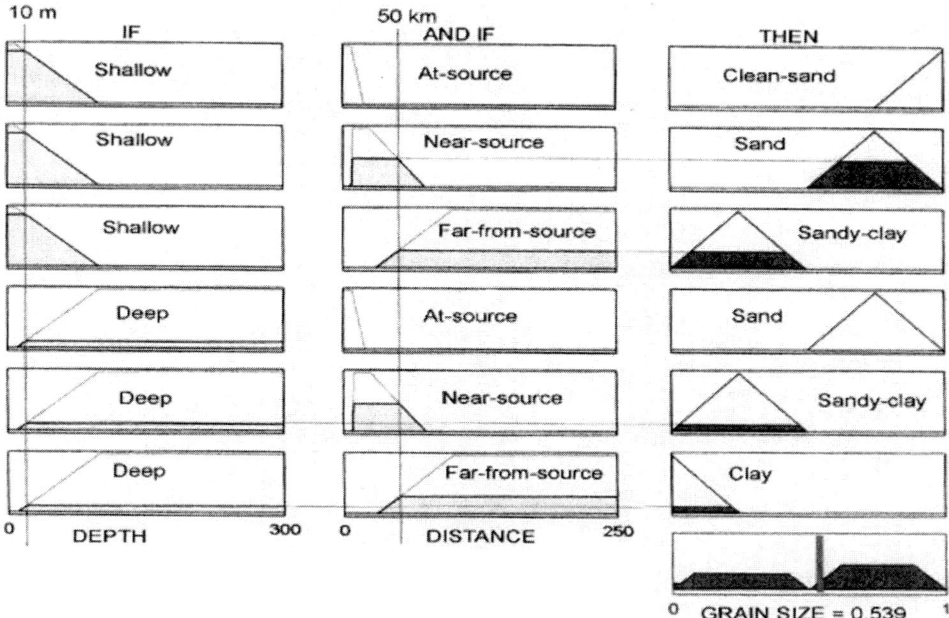

Fig. 8. A standard ('Mamdani') interpretation of the 'if–then' rules controlling subaqueous grain size in the deltaic model. The two left columns are the 2 antecedent variables and the right column is the consequent variable. Each row represents one fuzzy inference rules. See text for further details.

is characterized by 3 trapezoidal membership functions: at-source, near-source, and far-from-source. Grain size here is normalized over the interval 0 to 1 and characterized by five 5 triangular membership functions: clay, sandy-clay, clayey-sand, sand, and clean sand. The six rules of the fuzzy logic system and how they would be evaluated for a depth of 10 m and a point 50 km from the river mouth are shown in Fig. 8. The columns are the 2 input and one output variables: depth, distance, and grain size and the 6 rows represent the 6 rules. For example the rule in the first row is: "if the depth is shallow AND if the distance is at-source, then the grain size is clean-sand". The 'and' connector in this rule implies the intersection of the two sets, which, in this standard interpretation, means truncating the appropriate output function in each rule by the lowest membership value of the input functions. Thus, in the second row, a depth of 10 m truncates the shallow membership function at a value of about 0.9. A distance of 50 km from the river mouth truncates the membership function near-source to a degree of about 0.5. The 'and' implies truncating the output membership function by the lowest of the input degrees. Again, all of the rules are evaluated simultaneously, the maximum of the truncated output functions is taken, and the centroid of that outcome (last box in the right hand column) yields a 'defuzzified' grain size of 0.539 for 10 m at a distance of 50 km from the river mouth.

Two model runs each simulating 50,000 years of sedimentation at 200 year time steps are presented here: (1) with a superimposed, simple sinusoidal oscillation of sea level with amplitude of 10 m and duration of 20,000 years; and (2) with no sea level change. The starting point of the models is the same geometry (Fig. 6A). In both models, the subsidence is a maximum in the center of the model and falls off to the edges of the model. Subsidence remains constant through the simulation time. At each time step, the sediment surface subsides. Next, the values of the input variables for each fuzzy logic system are calculated numerically (e.g. water depth, distance from river mouth, distance from river channel) and those values are fed into the appropriate fuzzy logic system to calculate the thickness and type of sediment deposited in each cell. The sediment surface at 3 arbitrary times step from the variable sea level model are shown in Fig. 6B–D. Finally at each time step, the thickness and type of sediment deposited in each cell is stored for later visualization of the resulting deposits.

In addition, there are random (in time) avulsions of the river upstream of the model.

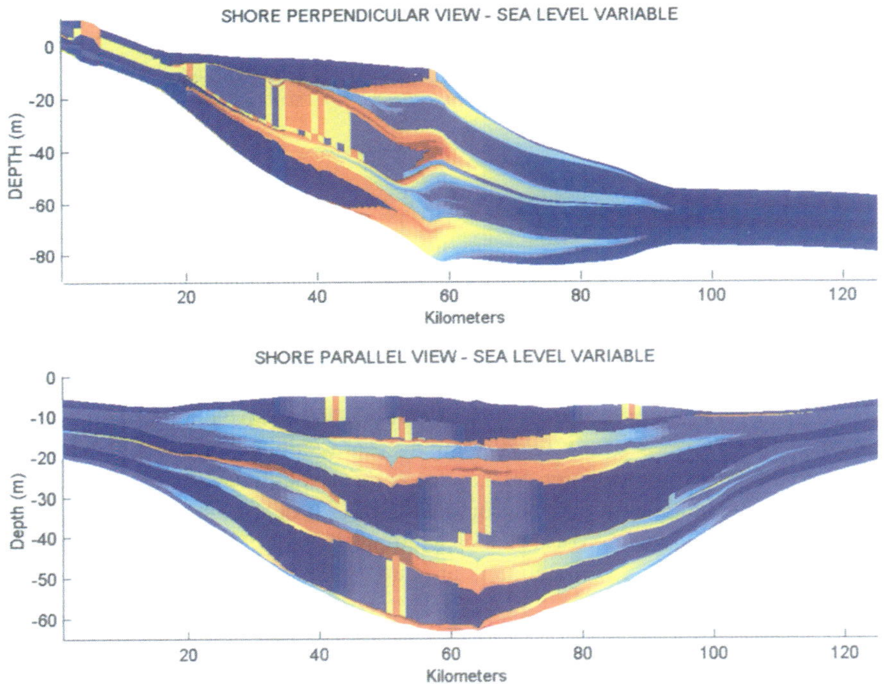

Fig. 9. Synthetic stratigraphic cross sections generated by the fuzzy logic model with sinusoidal sea level oscillation of 10 ky and 10 height. View perpendicular to shore (above) and parallel to shore out in the basin along the strike axis of the basin (below) produced at the end of the model. In the parallel section, marine deltaic deposits (red sands through yellow silts to light blue marine mud) alternate with the dark blue fluvial mud and the aggrading channel deposits (vertical red bars flanked by yellow levee deposits). In particular, note the four avulsion events preserved in the top-most fluvial section in the shore-parallel. The fluvial to marine cycles ten meters or so thick are dictated by the external sea level driver and are thus known as 'allocycles'. The other, smaller scales of cycles are driven by avulsion and kilometer-scale shift of the mouth of the delta between time steps. These kinds of cycles, driven by internal variability in sedimentation are called autocycles. See text for details.

After an avulsion, the river enters the upstream end of the model at the lowest point. The interplay of subsidence with floodplain aggradation history and location of the prior channel belts will determine this low point. From the lowest point at the upstream end of the model the river finds the lowest set of adjacent cells to reach the shoreline. When sea level is rising, the location of the river mouth simply backtracks up the river course, whereas when sea level is at a still stand or falling, the river seeks the lowest adjacent cells in front of it until it reaches 0 elevation, i.e. sea level. There are a number of interesting things to note on sediment surfaces reproduced in Fig. 6B–D. Note that abandoned channels sink into the floodplain and are buried by later flood basin deposits. Also note how the buildup of an off-shore delta platform influences the position of the river after and avulsion.

Figs. 9 and 10 show synthetic cross-sections through the deltaic deposit at the end of the model run. The upper panel in each diagram is a shore perpendicular view whereas the lower panel is a shore parallel view out in the basin: colors have the same meanings as in Fig. 6. Fig. 9 shows the case with variable sea level whereas Fig. 10 shows

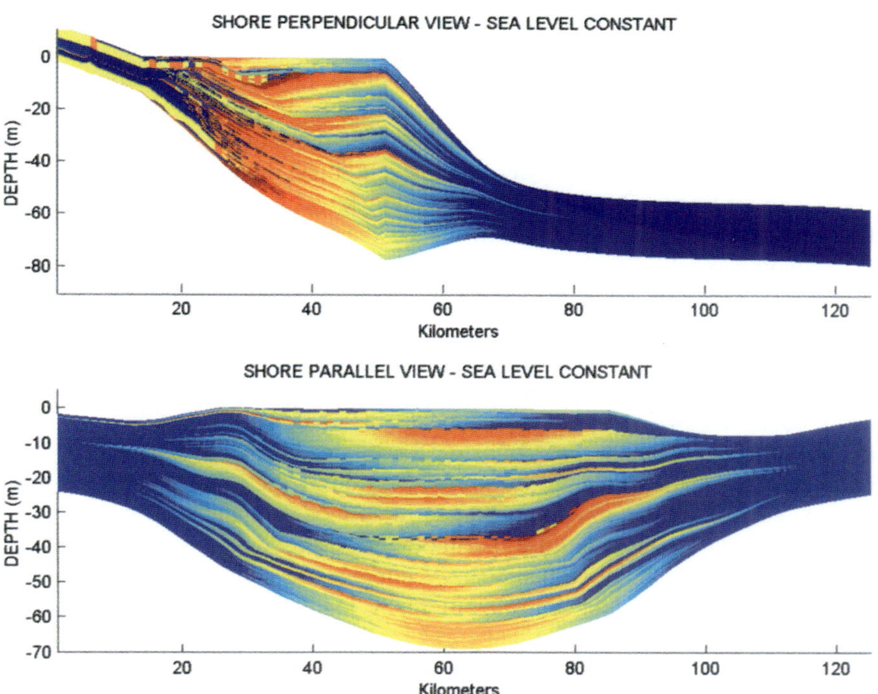

Fig. 10. Synthetic stratigraphic cross-sections generated by the fuzzy logic model with a constant sea level. View perpendicular to shore (above) and parallel to shore out in the basin along the strike axis of the basin (below) produced at the end of the model. In the parallel section, marine deltaic deposits (red sands through yellow silts to light blue marine mud) alternate with the dark blue fluvial mud and the aggrading channel deposits (vertical red bars flanked by yellow levee deposits). All scale of cycles here are generated by sedimentary processes within the model (a.k.a. autocycles).

the constant sea level scenario. There are a number of scales of both autocycles and allocycles in Fig. 9. The largest scale of cycles (tens of meters) is due to the sea level driver that interleaves fluvial deposits with marine deltaic sediments. The next smaller cycles are autocycles due to the avulsion of the river. Finally, there are meter-scale cycles due to the river seeking the lowest path to the sea between time steps. This alters the position of the delta by a few kilometers in each time step and produces the small-scale cycles. In contrast, all of the cycles shown in Fig. 10, the deposits of the delta are autocycles. The third order cycles here are due to river avulsions and the smaller cycles are due to shifts in the river mouth due to the interplay of subsidence and sedimentation at each time step.

3. FUZZY INFERENCE SYSTEMS AND STRATIGRAPHIC MODELING

There are a number of distinct advantages to employing fuzzy inference systems to model sediment dispersal in stratigraphic models. First, fuzzy sets describe systems in 'natural language' and provide the tools to rigorously quantify 'soft' information. Second, fuzzy inferences systems are more computationally efficient than finite-element or finite-difference models, and can even run faster than even a simple linear interpolation scheme. Last, and most importantly, the shapes of the membership functions can easily be changed by small increments, thereby allowing rapid 'sensitivity analysis' of the effects of changing the boundaries of the fuzzy sets. In robot control algorithms, where fuzzy logic was first developed, systems could self adjust the shapes of the membership functions and set boundaries, until the required task was flawlessly performed. This aspect of fuzzy systems, commonly facilitated via the learning capabilities of appropriate neural networks (Kosko, 1992; Klir and Yuan, 1995; Nauck and Klawonn, 1997) or by genetic algorithms (Sanchez et al., 1998), is one of their great advantages to numerical solution approaches.

The two models discussed above have been fixed with somewhat arbitrary membership functions predetermined by our interpretation of the characteristic of the variables in the models. Below we present 2 models of 'real world' sedimentary systems, sediment production on the Great Bahama Bank, and Pleistocene/Holocene chemical sedimentation in Death Valley, California. In Section 3.1 below we present an example where we 'hand tune' a fuzzy logic system for a model of lime sediment production on the Great Bahama Bank and use that model to discuss computational efficiency. In Section 3.2 we use an adaptive neuro-fuzzy inference system to determine the shapes of membership functions and the rules of fuzzy logic systems involved in modeling sedimentation in Death Valley.

3.1. Production of carbonate sediment on the Great Bahama Bank

Where the fuzzy logic system is composed of 2 antecedent variables and one consequent variable and standard fuzzy sets are used, the two input variables 'map' to a singular value of the output variable that comprises a 3-dimension surface. This is a particularly powerful way to envision a fuzzy logic system. For example, Fig. 11 is

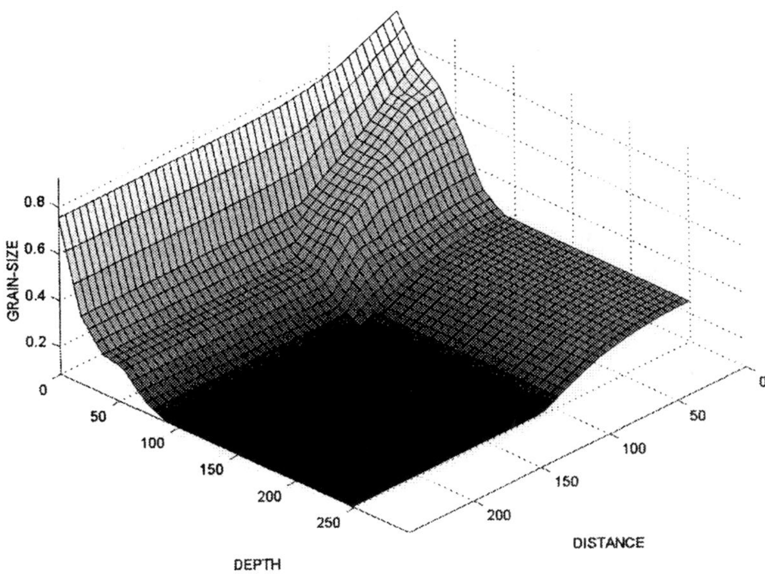

Fig. 11. Three-dimensional representation of the fuzzy logic system described in Fig. 8. The $[x\ y]$ axes are the input variables water depth and distance, the z scale is the output variable grain size determined by the fuzzy logic system.

the three-dimensional representation of the fuzzy logic system described in Fig. 8. The $[x\ y]$ axes are water depth and distance, and grain size is $[z]$. Notice that this rather simple fuzzy logic system generates a rather complicated non-planar relationship among depth, distance and grain size. The shape of the output surface can be quickly changed by many operations the two most common of which are: (1) varying the shapes and boundaries of the membership functions; and (2), changing the rule connectors between fuzzy intersection ('and') or fuzzy union ('or') rules. If the connector 'or' is used instead of 'and', fuzzy union is implied. In this case the appropriate output membership functions are truncated at the highest degree of membership of the input functions. We can 'warp' the output surface to any arbitrary shape we want by varying the shapes of the membership functions and the connectors in the rule system.

Carbonate sediment production is commonly modeled as being linearly depth-dependent as in the example of coral growth discussed above. However, the pioneering work of Broecker and Takahashi (1966) showed that sediment production on shallow carbonate platforms is not only dependent on shallow depths, but also on the residence time of the water, which, practically speaking, translates into distance away from the nearest shelf margin. Fig. 12 shows the shallow bathymetry of the Great Bahama Bank northwest of Andros Island. The dotted contours on Fig. 12 are carbonate production contours in kg m^{-2} y^{-1} taken from Broecker and Takahashi (1966, their fig. 10, p. 1585). Broecker and Takahashi (1966) did not make any measurements that allowed them to compute carbonate production at the steep edge of the platform. However, where such measurements have been made they cluster in the range of 1 to 4 kg m^{-2} y^{-1}. We

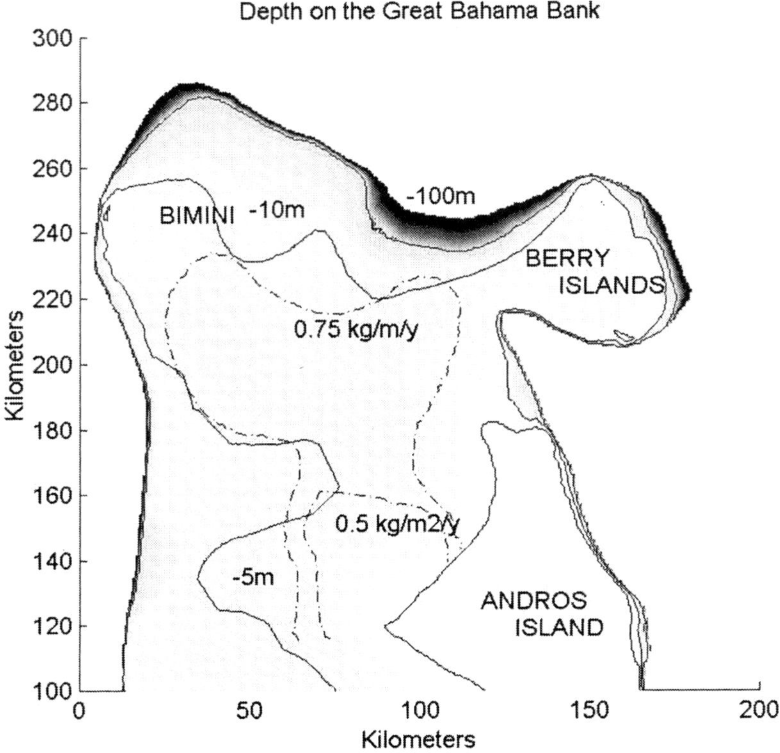

Fig. 12. Bathymetry of the Great Bahama Bank northwest of Andros Island. Edge of the platform is taken as the −100 m sea level contour. The dotted lines are contours of carbonate sediment production in kg m^{-2} y^{-1} taken from Broecker and Takahashi (1966, their fig. 10, p. 1585).

have chosen 1 kg m^{-2} y^{-1} as a target figure for carbonate sediment production at the margin of the Great Bahama Bank.

Our models of sediment production use a 1 km^2 grid of bathymetric data (Fig. 12), compute sediment production in each cell, and graph the results. The first model (Fig. 13A) uses a simple linear interpolation (Fig. 14A) of sediment production with normalized distance from the bank margin:

$$\text{production} = -0.5 \times \text{distance} + 1 \tag{7}$$

The second model (Fig. 13B) uses a simple linear function of sediment production with depth shown in Fig. 14B. This function has a maximum at 10 m and more or less resembles the data set of coral growth rates in Fig. 2.

$$\begin{array}{ll} \text{if depth} < -10\text{ m} & \text{production} = 0.01111 \times \text{depth} + 1.11111 \\ \text{else} & \text{production} = -0.07 \times \text{depth} + 0.3 \end{array} \tag{8}$$

Neither of these models does a particularly good job in predicting carbonate sedimentation production patterns of the Great Bahamas Bank. The third model combines

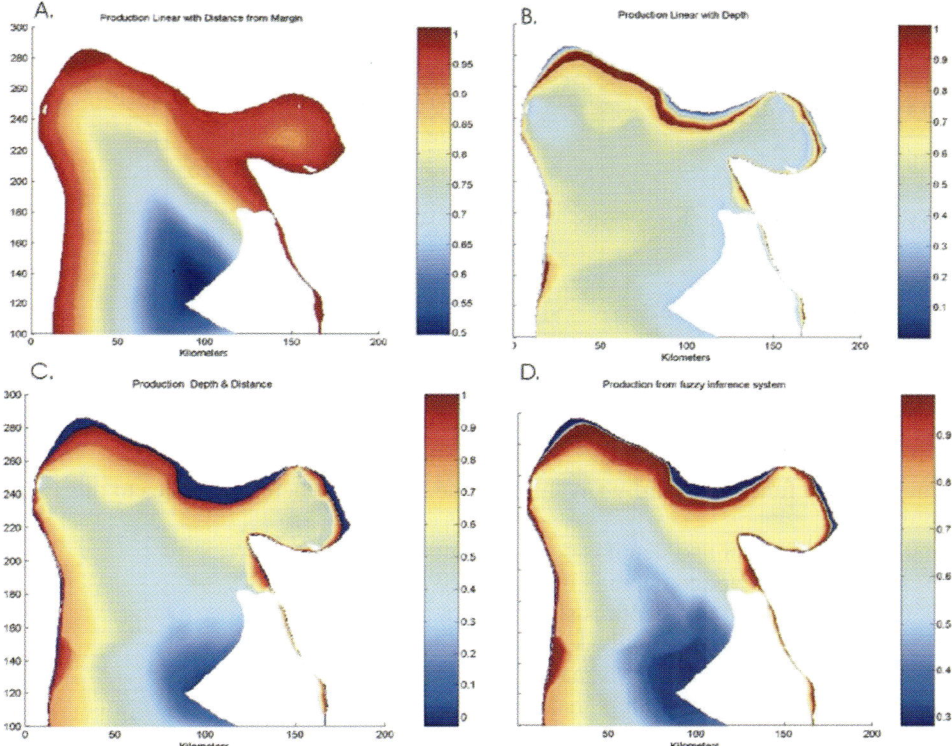

Fig. 13. Models of carbonate sediment production on the Great Bahama Bank. (A) Sediment production modeled as a linear decrease with (normalized) distance from bank margin according to Eq. (7). (Also see Fig. 14A.) (B) Sediment production modeled using piecewise linear relationship of Eq. (8) relating sediment production to water depth. (Also see Fig. 14B.) (C) Sediment production modeled using piecewise planar relationships of Eq. (9) relating sediment production to both water depth and distance from bank margin. (Also see Fig. 15A.) (D) Sediment production modeled using fuzzy logic system described in text. (Also see Fig. 15B and Fig. 16.)

the two linear models:

$$\begin{aligned} &\text{if depth} < -10\text{ m} \quad \text{production} = (0.011111 \times \text{depth} + 1.11111) - (0.5 \times \text{distance}) \\ &\text{else} \qquad\qquad\quad\;\; \text{production} = (-0.05 \times \text{depth}) + 0.3) - (0.5 \times \text{distance}) \end{aligned} \tag{9}$$

Fig. 15A is a graph of this function. This function is not easily altered to fit observations and is not that transparent to someone unfamiliar with the problem.

Contrast this piecewise approach to a fuzzy logic system of the same problem. Fig. 16 shows the two input variables (normalized distance from shelf edge and depth) and the output variable production. Distance is characterized by two gaussian membership functions near and far whereas depth is characterized by two trapezoidal memberships functions (deep and shallow) and one triangular membership function (maximum production depth, abbreviated max on the figure). The output variable production comprises

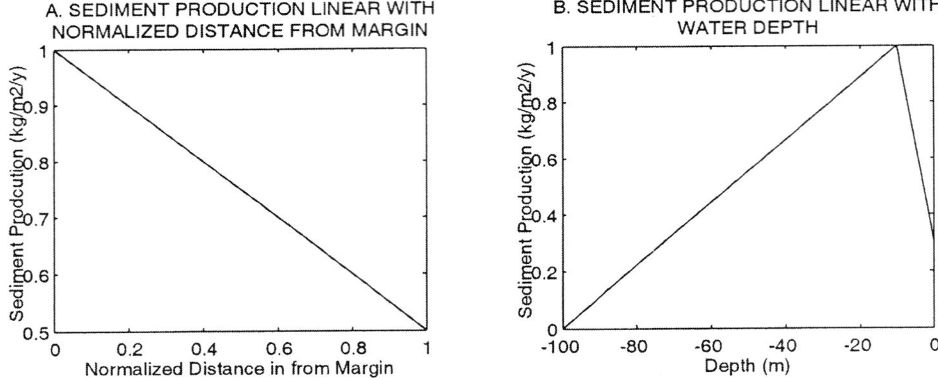

Fig. 14. Graphs of linear sediment production functions. (A) Graph of Eq. (7). (B) Graph of Eq. (8).

4 membership functions, hardly any, little, some, and lots. There are 6 rules to this fuzzy logic system:

(1) if distance is near and depth is deep produce hardly any;
(2) if distance is near and depth is max produce lots;
(3) if distance is near and depth is shallow then produce some;
(4) if distance is far and depth is deep produce hardly any;
(5) if distance is far and depth is max produce little;
(6) if distance is far and depth is shallow produce hardly any

Fig. 15B is a graph of the production versus depth and distance determined by this fuzzy logic system next to the piecewise planar approximation. Both of these models do a fairly good job in reproducing the carbonate sediment production pattern on the Great Bahama Banks northwest of Andros Island. We have adjusted the boundaries and shapes of the depth function to tune this model. It is important to note that tuning the model

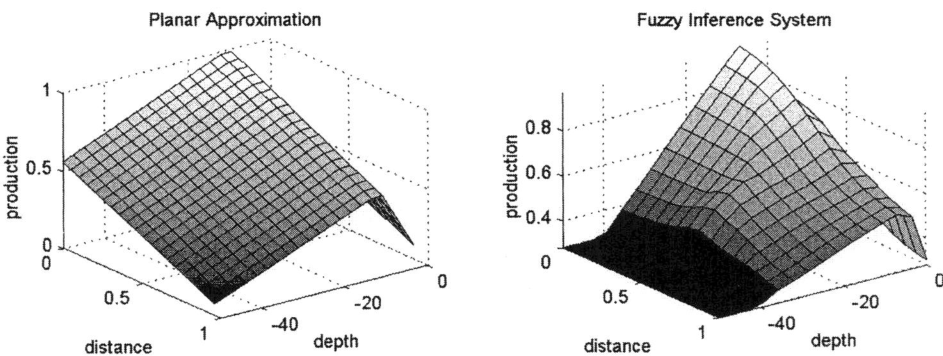

Fig. 15. Models of sediment production based on distance and water depth. (A) Graph of Eq. (9). Sediment production modeled as piecewise planes. (B) Graph of sediment production according to fuzzy logic system described in text and Fig. 16.

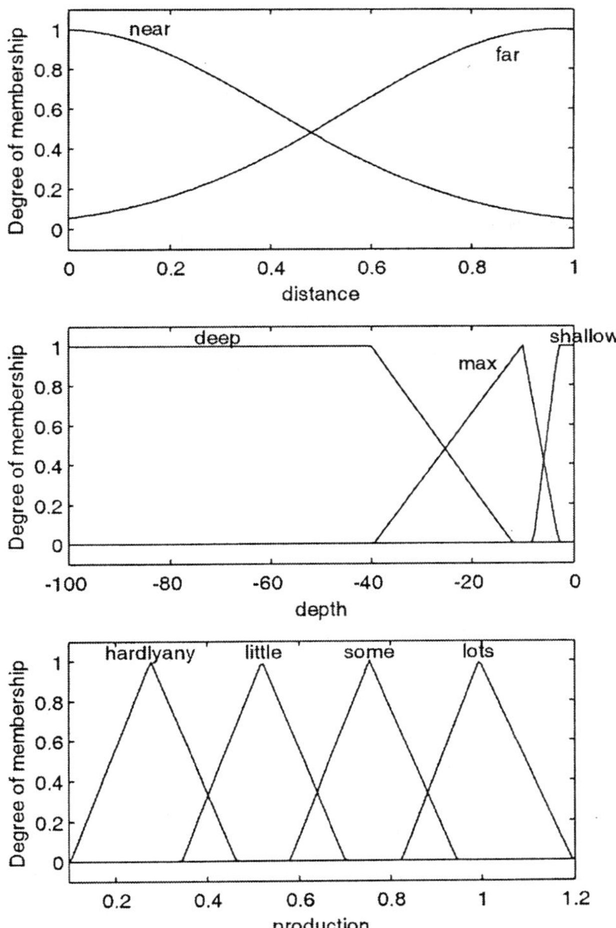

Fig. 16. Membership functions comprising part of the fuzzy logic system that generated the solution in Fig. 13D. The input variables are (normalized) distance (modeled with gaussian membership functions) and water depth (modeled by both trapezoidal or triangular membership functions). Rules relating these variables are given in text and the solution is shown in Fig. 15B.

by adjusting the membership functions is relatively easy versus trying to recalculate the piecewise approximating equations.

3.2. Death Valley, California

Death Valley is currently an arid closed basin located in the southwestern United States. The basin is a half graben approximately 15 km across and 65 km long. The center of the basin is currently a nearly flat complex of saline pans and playa mudflats approximately 100 m below sea level. Gravel alluvial fans radiate from streams along

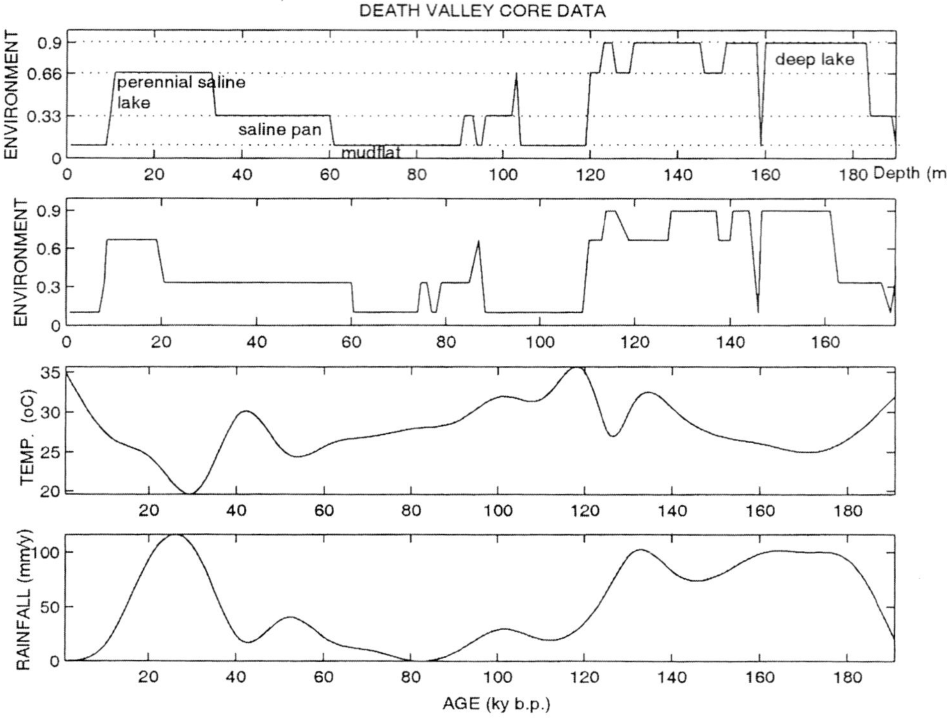

Fig. 17. Data from the sediment core taken from near the center of Death Valley.

steep mountain fronts on the east side of the basin where the active fault is inferred. These fans are steep and grade out to the floor of the basin over a few kilometers. The mountain front on the west side of the basin is gentler. Alluvial fans issuing from streams on this side of the basin have a lower gradient than those on the east side and extend nearly halfway across the basin floor. A 175 meter long, well dated core taken through the central portion of the basin sampled the basin floor sediments deposited during the last 191 thousand years (Roberts and Spencer, 1995; Li et al., 1996; Lowenstein et al., 1999; and references cited therein). Fig. 17 summarizes the data from the core. The top panel is a plot of thickness of the different deposits found with depth in the core. Thick sections of chemical sediments (principally halite) deposited in saline lakes and mudflats are interbedded with muds deposited in desiccated playa mudflats or fossil-rich muds deposited in a deep lacustrine setting. Interpolated age of the deposits based on U series chronology of samples of the various halites comprises the second panel down. The third panel down is a smoothed curve through paleotemperatures measured from brine inclusions preserved in primary halite deposits, principally from saline lakes and saline pan deposits. The last panel is an interpreted record of paleorainfall based on a number of proxy measurements from the core and surrounding areas (e.g. dated lacustrine tufas, shorelines, and etc. found in the basin, Lowenstein, personal communication).

Our modeling here starts with the premise that the deposits in the floor of the basin are directly related in some way to a combination of temperature and rainfall. This is not an unreasonable interpretation for closed basin deposits (see Smoot and Lowenstein, 1991). Indeed it is a prerequisite for using sedimentary records of lakes and other continental environments for research into paleoclimates.

In robot control algorithms, where fuzzy logic was first developed, systems could self adjust the shapes of the membership functions and set boundaries, until the required task was flawlessly performed. This aspect of fuzzy systems, commonly facilitated via the learning capabilities of appropriate neural networks (Kosko, 1992; Klir and Yuan, 1995; Lin and Lee, 1996; Nauck and Klawonn, 1997) or by genetic algorithms (Sanchez et al., 1998), is one of their great advantages to numerical solution approaches. To generate a fuzzy logic system for the Death Valley Core data, we employed the adaptive neuro-fuzzy system that is included in the Fuzzy Logic Toolbox of the commercial high-level language MATLAB.

The MATLAB adaptive neuro-fuzzy system is a program that utilizes learning capabilities of neural networks for tuning parameters of fuzzy inference systems on the basis of given data. However, as explained below, the type of fuzzy inference systems dealt with by this program are not the classical Mamdani type. The program implements a training algorithm employing the common backpropagation method based on the least square error criterion (see Klir and Yuan, 1995, appendix A).

All of the fuzzy logic systems we used to this point in the paper have been so-called ‘Mamdani’ fuzzy inference systems wherein the output variable is divided into standard fuzzy sets. In final step of a Mamdani-type fuzzy logic system, a two-dimensional fuzzy set that has been generated by aggregating the appropriate truncated membership functions of the output variable has to be ‘defuzzified’ by some averaging process (e.g. finding the ‘centroid’). Contrary to a Mamdani fuzzy inference system, an alternative approach to formalizing fuzzy inference systems, developed by Takagi and Sugeno (1985) employs a single ‘spike’ as the output membership functions. Thus, rather than integrating across the domain of the final output fuzzy set, a Takagi–Sugeno type fuzzy inference system employs only the weighted average of a few data points. There is a clear computational advantage to employing a Takagi–Seguno fuzzy logic system. Moreover, the adaptive neuro-fuzzy inference engine of MATLAB only supports Takagi–Seguno type output membership functions. Fig. 18 shows the two antecedent membership functions used to ‘adjust’ the output linear functions. The training algorithm generates 9 linear output functions. Fig. 19 is the surface generated by this fuzzy logic system. Fig. 20 is a direct comparison between our modeling results and the facies data both plotted against age.

Demicco and Klir (2001) show an earlier three-dimensional model of Death Valley deposition over the past 191 ky. The model was a grid 14 km across and 65 km long represented by approximately 1900 active cells each 0.5×1.0 km in size. The modern topography was the starting point for elevation at each cell in the model. This model used 4 hand-tuned mamdani-type fuzzy inference systems. Two fuzzy logic systems produced basin floor sediments. One of these generated the sediment type and the other generated the sediment thickness. The input variable in both of these fuzzy logic systems were the temperature and rainfall signal determined from the core by

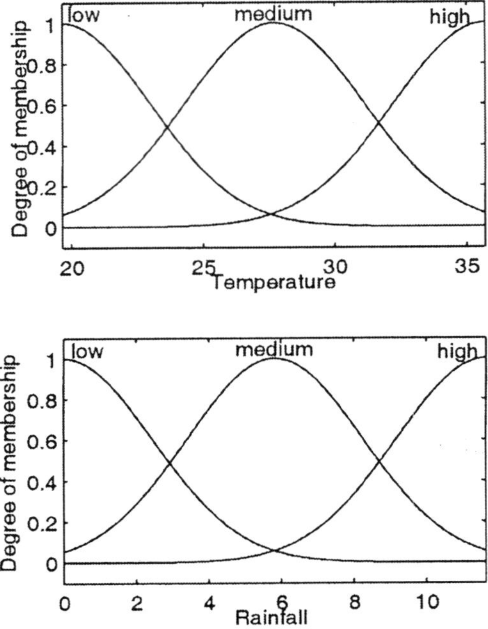

Fig. 18. Antecedent membership functions used as input to 'tune' an adaptive neuro-fuzzy model relating temperature and rainfall to sedimentary environments.

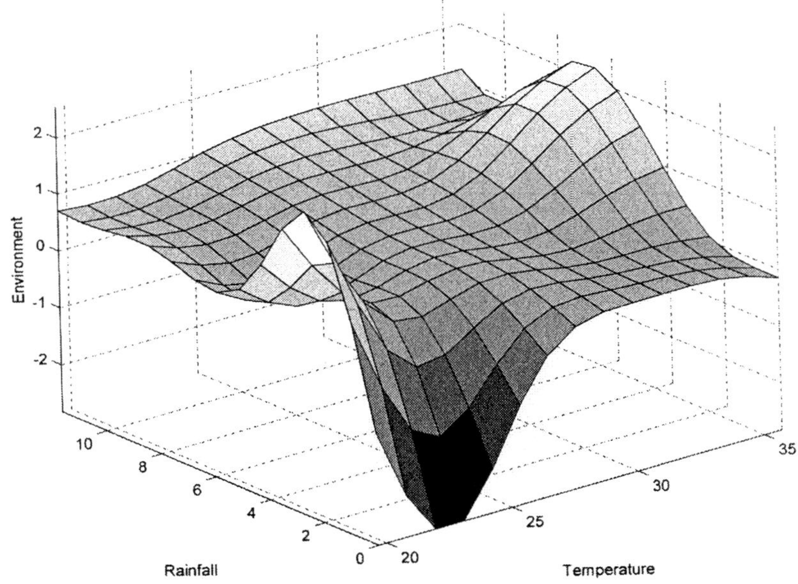

Fig. 19. Solution surface of the adaptive, neuro-fuzzy logic model generated from MATLAB.

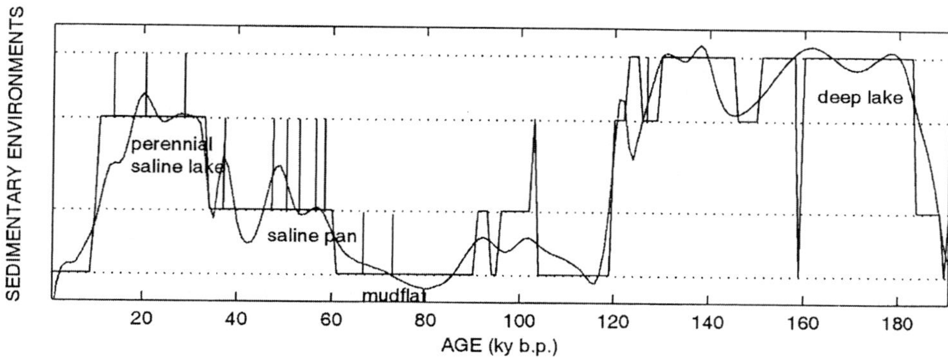

Fig. 20. Direct comparison between sedimentary record of Death Valley salt core with synthetic stratigraphy generated from the adaptive neuro-fuzzy logic model described in the text.

Lowenstein (personal communication). In the original model, alluvial input from the sides arose from the canyon locations at the heads of the main modern alluvial fans around the basin margin. Deposition on the alluvial fan to playa mudflat drainage ways was modeled by 2 mamdani fuzzy logic systems. The inputs variables to both models were distance from canyon mouth and slope of the sediment surface in each cell. These input variables controlled the particle size of the deposit and thickness of the alluvial deposits in each cell. Subsidence was -0.2 m ky^{-1} along the edges of the model and increased to 1 m ky^{-1} along the steep (eastern) margin of the basin halfway down the axis of the basin.

Fig. 21 shows two synthetic stratigraphic cross-sections of this model rerun with all conditions being the same, except the fuzzy logic systems that control deposition in the center of the basin. Fig. 21A is a cross-section across the basin in a west–east orientation, whereas Fig. 21B is a cross-section along the north–south long-axis of the basin. The sedimentary environment in the basin center is controlled by the machine-developed fuzzy logic system described above. The thickness of sediment and the water depth in the basin center were controlled by two additional machine-developed fuzzy logic systems. In the synthetic cross-valley section (Fig. 21A) short steep fans on the eastern side of the basin comprise coarser gravels (red-orange) and contrast to the long, lower gradient fans on the western side of the basin that are generally comprised of finer sediment (yellow and green). The alluvial input into the basin ultimately leads to the deposition of playa muds in the floor of the basin. The basin floor sediment is color-coded: deep freshwater lake and playa mud flats are blue, saline pan is red and saline lake is shades of yellow and orange. Playa mud flats develop in the floor of the basin when the chemical or lacustrine sediments are minimal.

4. SUMMARY AND CONCLUSIONS

It is clear that fuzzy logic systems have the potential to produce very realistic sedimentation dispersal patterns when used as ‘expert’ systems. By expert systems we

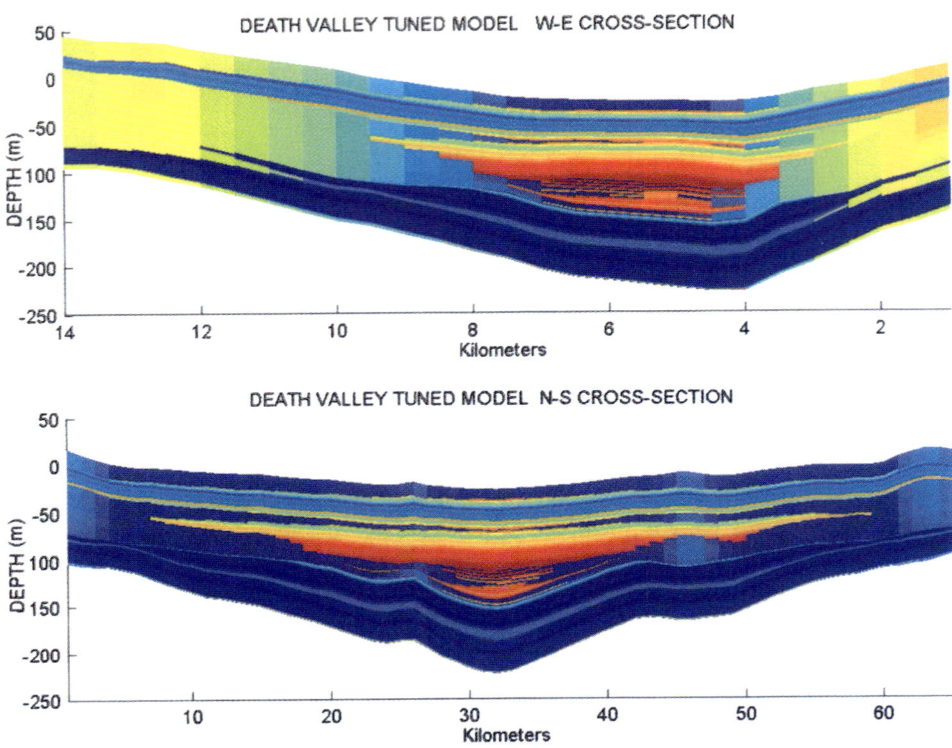

Fig. 21. Synthetic stratigraphic models of Death Valley, see text for details.

imply a knowledge 'engineer' (i.e. someone familiar with the techniques described in this paper) and a sedimentary geologist intuitively or (quantitatively) familiar with a modern sedimentary environment or an ancient stratigraphic succession. If the sedimentary geologist can distill the key points of their depositional model into the types of "if A and if B then C" propositions described above then the knowledge engineer can translate them into mathematically rigorous fuzzy logic systems. It is important to note that there is only a practical limit to the number of antecedent propositions in a fuzzy logic statement. A statement such as "if A and if B and if C then D" would map three input variable into a region of space. Moreover, there is currently an explosive growth in the theory and application of fuzzy logic and other related 'soft' computing techniques, opening new ways of modeling based on knowledge expressed in natural language.

Fuzzy logic models hold the potential to accurately model subsurface distribution of sedimentary facies (not just water depths of deposition) in terms of the natural variables of geology. As exploration moves further into use of three-dimensional seismic data gathering, the utility of easy to use, flexible three-dimensional forward models is obvious. Such models could be used to produce synthetic seismic sections. Moreover, the 'learning ability' of fuzzy logic systems coupled with neural networks offers the long-term possibility of self tuning sedimentary models that can match three-dimensional seismic subsurface information in a 'nonhuman' expert system. This method offers an alternative to the statistical modeling of subsurface geology. It is more computationally efficient and more intuitive for geologists than complicated models that solve coupled sets of differential equations.

ACKNOWLEDGEMENTS

N.S.F. Grant EAR9909336 supported this research. We would like to thank Dr. Masoud Nikravesh for inviting us to submit a paper to this volume.

REFERENCES

Acinas, J.R. and Brebbia, C.A., 1997. *Computer Modelling of Seas and Coastal Regions*. Computational Mechanics, Inc., Billerica, MA.

Angevine, C.L., Heller, P.L. and Paola, C., 1990. *Quantitative Sedimentary Basin Modeling. Continuing Education Course Notes Series, 32*. American Association of Petroleum Geologists, Tulsa, OK.

Bosence, D. and Waltham, D., 1990. Computer modeling the internal architecture of carbonate platforms. *Geology*, 18: 26–30.

Bosence, D.W.J., Polmar, L., Waltham, D.A. and Lankester, H.G., 1994. Computer modeling a Miocene carbonate platform, Mallorca, Spain. *Am. Assoc. Pet. Geol. Bull.*, 78: 247–266.

Bosscher, H. and Schlager, W., 1992. Computer simulation of reef growth. *Sedimentology*, 39: 503–512.

Bridge, J.S. and Leeder, M.R., 1979. A simulation model of alluvial stratigraphy. *Sedimentology*, 26: 617–644.

Broecker, W.A. and Takahashi, T., 1966. Calcium carbonate precipitation on the Bahama Banks. *J. Geophys. Res.*, 71: 1575–1602.

Buddemeier, R.W. and Smith, S.V., 1988. Coral reef growth in an era of rapidly rising sea level: predictions and suggestions for long-term research. *Coral Reefs*, 7: 51–56.

Burton, R., Kendall, C.G. St. C. and Lerche, I., 1987. Out of our depth: on the impossibility of fathoming eustasy from the stratigraphic record. *Earth-Sci. Rev.*, 24: 237–277.

Chalker, B.E., 1981. Simulating light-saturation curves for photosynthesis and calcification by reef-building corals. *Mar. Biol.*, 63: 135–141.

Demicco, R.V., 1998. CYCOPATH 2-D, a two-dimensional, forward-model of cyclic sedimentation on carbonate platforms. *Comput. Geosci.*, 24: 405–423.

Demicco, R.V. and Klir, G., 2001. Stratigraphic simulation using fuzzy logic to model sediment dispersal. *J. Pet. Sci. Eng.*, in press.

Dunn, P.A., 1991. *Diagenesis and Cyclostratigraphy – An Example from the Middle Triassic Latemar Platform, Dolomite Mountains, Northern Italy*. Ph.D. Thesis, The Johns Hopkins University, Baltimore, MD.

Edington, D.H., Poeter, E.P. and Cross, T.A., 1998. FLUVSIM; a fuzzy-logic forward model of fluvial systems. *Abstr. with Programs, Geol. Soc. Am. Annu. Meet.*, 30: A105.

Emery, D. and Myers, K.J., 1996. *Sequence Stratigraphy*. Blackwell Science Ltd, Oxford.

Fang, J.H., 1997. Fuzzy logic and geology. *Geotimes*, 42: 23–26.

Flint, S.S. and Bryant, I.D., 1993. *The Geological Modelling of Hydrocarbon Reservoirs and Outcrop Analogues. Int. Assoc. Sedimentol., Spec. Publ. 15*. Blackwell Scientific Publications, Oxford.

Forster, A. and Merriam, D.F., 1996. *Geological Modeling and Mapping*. Plenum Press, New York, NY.

Franseen, E.K., Watney, W.L., Kendall, C.G. St. C. and Ross, W., 1991. Sedimentary modeling: computer simulations and methods for improved parameter definition. *Kansas Geol. Surv. Bull.*, 233.

Harbaugh, J.W., Watney, L.W., Rankey, E.C., Slingerland, R., Goldstein, R.H. and Franseen, E.K., 1999.

Numerical Experiments in Stratigraphy: Recent Advances in Stratigraphic and Sedimentologic Computer Simulations. Soc. Econ. Paleontol. Mineral., Spec. Publ., 62. SEPM, Tulsa, OK.

Harff, J., Lemke, W. and Stattegger, K., 1999. *Computerized Modeling of Sedimentary Systems.* Springer, New York, NY.

Klir, G.J. and Yuan, B., 1995. *Fuzzy Sets and Fuzzy Logic – Theory and Applications.* Prentice Hall, Englewood Cliffs, NJ.

Knauss, J.A., 1978. *Introduction to Physical Oceanography.* Prentice Hall, Englewood Cliffs, NJ.

Kosko, B., 1992. *Neural Networks and Fuzzy Systems.* Prentice Hall, Englewood Cliffs, NJ.

Leeder, M.P., Mack, G.H. and Peakall, J., 1996. First quantitative test of alluvial stratigraphic models: Southern Rio Grande rift, New Mexico. *Geology*, 24: 87–90.

Li, J., Lowenstein, T.K., Brown, C.B., Ku, T.-L. and Luo, S., 1996. A 100 ka record of water tables and paleoclimates from salt core, Death Valley, California. *Palaeogeogr., Palaeoclimatol., Palaeoecol.*, 123: 179–203.

Li, M.Z. and Amos, C.L., 1995. SEDTRANS92: a sediment transport model of continental shelves. *Comput. Geosci.*, 21: 553–554.

Lin, C.-T. and Lee, G., 1996. *Neural Fuzzy Systems.* Prentice Hall, Englewood Cliffs, NJ.

Loucks, R.G. and Sarg, J.F., 1993. *Carbonate Sequence Stratigraphy. Am. Assoc. Pet. Geol. Mem., 57.* American Association of Petroleum Geologists, Tulsa, OK.

Lowenstein, T.K., Li., J., Brown, C., Roberts, S.M., Ku, T.-L., Luo, S. and Yang, W., 1999. 200 k.y. paleoclimate record from Death Valley salt core. *Geology*, 27: 3–6.

Mackey, S.D. and Bridge, J.S., 1995. Three-dimensional model of alluvial stratigraphy: theory and application. *J. Sediment. Res.*, B65: 7–31.

Mamdani, E.H. and Assilian, S., 1975. An experiment in linguistic synthesis with fuzzy logic controller. *Int. J. Man-Machine Studies*, 7: 1–13.

Mendel, J.M., 2001. Uncertain *Rule-Based Fuzzy Logic Systems: Introduction and New Directions.* Prentice Hall, Englewood Cliffs, NJ.

Morse, J.W., Millero, F.J., Thurmond, V., Brown, E. and Ostlund, H.G., 1984. The carbonate chemistry of Grand Bahama Bank waters: after 18 years another look. *J. Geophys. Res.*, 89(C3): 3604–3614.

Nauck, D. and Klawonn, F., 1997. *Foundations of Neuro-Fuzzy Systems.* John Wiley and Sons, New York, NY.

Nittrourer, C.A. and Kravitz, J.H., 1996. STRATAFORM: A program to study the creation and interpretation of sedimentary strata on continental margins. *Oceanography*, 9: 146–152.

Nordlund, U., 1996. Formalizing geological knowledge – with and example of modeling stratigraphy using fuzzy logic. *J. Sediment. Res.*, 66: 689–712.

Parcell, W.C., Mancini, E.A., Benson, D.J., Chen, H. and Yang, W., 1998. Geological and computer modeling of 2-D and 3-D carbonate lithofacies trends in the Upper Jurassic (Oxfordian), Smackover Formation, Northeastern Gulf Coast. *Abstr. with Programs, Geol. Soc. Am. Annu. Meet.*, 30: A338.

Pilkey, O.H. and Thieler, E.R., 1996. Mathematical modeling in coastal geology. *Geotimes*, 41: 5.

Reading, H.G. and Collinson, J.D., 1996. Clastic coasts. In: Reading, H.G. (Ed.), *Sedimentary Environments: Processes, Facies and Stratigraphy.* Blackwell Scientific, Oxford, pp. 154–231.

Roberts, S.M. and Spencer, R.J., 1995. Paloetemperatures preserved in fluid inclusions in halite. *Geochim. Cosmochim. Acta*, 59: 3929–3942.

Sanchez, E., Shibata, T. and Zadeh, L.A., 1998. *Genetic Algorithms and Fuzzy logic systems.* World Scientific, Singapore.

Schlager, W., 1992. *Sedimentary and Sequence Stratigraphy of Reefs and Carbonate Platforms. Continuing Education Course Note Series, 34.* American Association of Petroleum Geologists, Tulsa, OK.

Schlager, W., 1999. Sequence stratigraphy of carbonate rocks. *Leading Edge*, 18: 901–907.

Slingerland, R., 1986. Numerical computation of co-oscillating paleotides in the Catskill epeiric Sea of eastern North America. *Sedimentology*, 33: 487–497.

Slingerland, R. and Smith, N.D., 1998. Necessary conditions for a meandering-river avulsion. *Geology*, 26: 435–438.

Slingerland, R., Harbaugh, J.W. and Furlong, K., 1994. *Simulating Clastic Sedimentary Basins. Physical Fundamentals and Computer Programs for Creating Dynamic Systems.* Prentice Hall, Englewood Cliffs, NJ.

Smith, S.V. and Kinsey, D.W., 1976. Calcium carbonate production, coral reef growth, and sea level change. *Science*, 194: 937–939.

Smoot, J.P. and Lowenstein, T.K., 1991. Depositional environments of non-marine evaporates. In: Melvin, J.L. (Ed.), *Evaporites, Petroleum and Mineral Resources: Developments in Sedimentology.* Elsevier, Amsterdam, pp. 189–347.

Syvitski, J.P.M. and Alcott, J.M., 1995. RIVER3: simulation of river discharge and sediment transport. *Comput. Geosci.*, 21: 89–115.

Takagi, T. and Sugeno, H., 1985. Fuzzy identification of systems and its application for modeling and control. *IEEE Trans. Syst., Man Cybern.*, 15: 116–132.

Tetzlaff, D.L. and Harbaugh, J.W., 1989. *Simulating Clastic Sedimentation.* VanNostrand Reinhold, New York, NY.

Turcotte, D.L. and Schubert, G., 1982. *Geodynamics – Applications of Continuum Physics to Geological Problems.* John Wiley and Sons, New York, NY.

Vail, P.R., Mitchum Jr., R.M., Todd, R.G., Widmier, J.M., Thompson III, S., Sangree, J.B., Bubb, J.N. and Hatleid, W.G., 1977. Seismic stratigraphy and global changes in sea level. In: Payton, C.E. (Ed.), *Seismic Stratigraphy – Application to Hydrocarbon Exploration. Am. Assoc. Pet. Geol. Mem.*, 26: 49–62.

Wendebourg, J. and Harbaugh, J.W., 1996. Sedimentary process simulation: a new approach for describing petrophysical properties in three dimensions for subsurface flow simulations. In: Forster, A. and Merriam, D.F. (Eds.), *Geological Modeling and Mapping.* Plenum Press, New York, NY, pp. 1–25.

Whitaker, F., Smart, P., Hague, Y., Waltham, D. and Bosence, D., 1997. Coupled two-dimensional diagenetic and sedimentologic modeling of carbonate platform evolution. *Geology*, 25: 175–178.

Wilgus, C.K., Hastings, B.S., Kendall, C.G. St. C., Posamentier, H.W., Ross, C.A. and van Wagoner, J.C., 1989. *Sea Level Changes: an Integrated Approach. Soc. Econ. Paleontol. Mineral., Spec. Publ., 42.* SEPM, Tulsa, OK.

Zadeh, L.A., 1965. Fuzzy sets. *Inf. Control*, 8: 94–102.

Zadeh, L.A., 1975. The concept of a linguistic variable and its application to approximate reasoning. *Inf. Sci.*, 8: 199–249; 8: 301–357; 9: 43–80.

Zadeh, L.A., 1976. The concept of a linguistic variable and its application to approximate reasoning. *Inf. Sci.*, 8: 199–249; 8: 301–357; 9: 43–80.

Developments in Petroleum Science, 51
Editors: M. Nikravesh, F. Aminzadeh and L.A. Zadeh

Chapter 12

SPATIAL CONTIGUITY ANALYSIS. A METHOD FOR DESCRIBING SPATIAL STRUCTURES OF SEISMIC DATA

A. FARAJ [a,1] and F. CAILLY [b,2]

[a] *Institut Français du Pétrole, 1–4 Avenue de Bois-Préau, 92500 Rueil Malmaison, France*
[b] *Beicip Franlab, 232 Avenue Napoléon Bonaparte, 92500 Rueil Malmaison, France*

1. INTRODUCTION

The seismic data are acoustic images of the subsurface geology. We try to identify and to analyze spatial structures contained in these images in order to evaluate and interpret subjacent geologic structures. Such data are naturally spatial; observations being described by their co-ordinates on a geologic location map. It is then essential, in order to account for all of the structures inherent in these data, to take their statistical characteristics as well as spatial contiguity relations between observations into consideration. We use spatial contiguity analysis (SCA) both for spatial multidimensional description and filtering of seismic images. Compared to classical methods, such as principal component analysis (PCA), SCA is more efficient for multivariate description and spatial filtering of this kind of images. We compare PCA and SCA results. This data set gives us the opportunity to show the interest of preliminary spatial analysis of initial variables, and the effects of spatial direction and distance on the data decomposition in elementary structures.

In the present article, we first present the SCA using the geostatistical formalism developed by Matheron (1963). A preliminary spatial analysis of initial variables is required. Made with the help of variogram curves, this permits to underline spatial properties of these variables and defines contiguity distance and direction to apply SCA. We show the importance of variogram curves in the description of spatial properties of the variables analyzed. SCA, as we apply it, consists in calculating linear combinations f of these variables which optimize the $[\gamma_f(h)]/\sigma_f^2$ ratio (where $\gamma_f(h)$ is the value of the directional variogram at point $h > 0$ and σ_f^2 the variance of f). The directional variogram curves are a great help for determining the suitable direction and distance h (i.e. those giving the best decomposition of data into elementary spatial structures).

A series of mathematical tools is defined. They allow to quantify the information held by initial variables and factorial components in terms of variance and spatial variability and exhibit data spatial structures on different scales. They are used for interpreting the spatial components obtained in terms of variance and spatial variability. This allows to select, among all the spatial components obtained, those corresponding to structures that we call local (small spatial scale structure) or regional (large spatial scale structures) in order to perform filtering(s) of the initial variables.

---

[1] E-mail: abdelaziz.faraj@ifp.fr
[2] E-mail: frederic.cailly@ifp.fr

SCA is presented, in the present article, both from a geostatistical and factorial angle. This procedure allows to combine tools specific to geostatistics with the multivariable descriptive approach of factorial techniques, which endows it with a dual, both multidimensional and spatial, data and result interpretation specificity.

After a brief review of the state of the art, we present, in Section 3, the geostatistical as well as statistical notations and definitions of the data. Section 4 describes the principle of SCA and presents properties of the factorial components resulting therefrom. These properties thereafter allow to define, in Section 5, the mathematical tools by means of which the results of a SCA can be interpreted. We present in Section 6 an example of application of SCA to a seismic data set. This example allows to illustrate the use of the technique by presenting analysis stages. By means of this application, we show how to use the variogram curves of the analyzed variables in order to define the suitable direction and distance for optimum use of SCA. The results are then analyzed with the help of the tools defined. Finally, Section 7 relates to conclusions and remarks, in a general context, concerning both exploratory analysis of the data and their spatial and multidimensional description.

2. STATE-OF-THE-ART

SCA was defined by Lebart (1969; Burtschy and Lebart, 1991) to analyze geographic structures inducing a graph relationship between observations. It is a multidimensional factorial method characterized, like most of these methods, by its rather descriptive aspect. It applies to spatial data or to data related by a contiguity graph and it is based on calculation of a linear combination of initial variables that optimizes Geary's (1954) contiguity relation.

A SCA variant is presented by Switzer (Switzer and Green, 1984) under the name of 'min/max autocorrelation factors' as a spatial filtering method applied within the scope of image processing. This method is then generalized and presented under the name of MNF (for maximum noise fraction) as a spectral component extraction technique (Green et al., 1988). The authors present it as a method maximizing the signal-to-noise ratio, thus allowing better separation between noise and 'useful' information. The noise to be extracted is identified and designated by a specific model.

Under the name of spatial proximity analysis, SCA has been applied for filtering of geochemical and geologic data (Royer, 1984) as well as seismic data (Faraj and Fournier, 1993; Faraj, 1994b; Wloszczczowski et al., 1998). As a spatial filtering method, SCA can be compared to kriging factorial analysis (Matheron, 1982; Wackernagel, 1989; Goovaerts, 1992). Both tend to isolate the structures on different spatial scales. They, however, differ from one another in that implementation of kriging factorial analysis requires a spatial model (based on a theoretical model of the variogram) that takes globally account of the data, whereas spatial proximity analysis uses the experimental variogram and applies to selected contiguity levels.

As it takes neighborhood relations between observations into consideration, SCA is close to geostatistical techniques. Geary's contiguity coefficient corresponds to the point of the omnidirectional (or isotropic) variogram curve associated with the value of the spatial distance equal to 1. The variogram curve is commonly used in geostatistics.

It exists a direct extension of SCA to a discriminant analysis generalization (Faraj, 1994a). This method – called generalized discriminant analysis (GDA) – can be viewed as an extension of classical factorial discriminant analysis to more than one qualitative variable. The notion of spatial contiguity between two observations is replaced by their simultaneous belonging to the same category of a qualitative variable. The intersection of the classes defined by the categories of the qualitative variables is not strictly empty. These overlapping classes can therefore be compared to the spatial neighborhoods defined in the contiguity analysis. By extension, notions of 'spatial contiguity' and 'simultaneous belonging to the same set' have the same meaning.

A large overview on SCA extensions is given in the book of Lebart (Lebart et al., 1995, pp. 327–335).

3. LOCAL VARIANCE AND COVARIANCE BETWEEN STATISTICS AND GEOSTATISTICS

3.1. Variogram–crossed covariogram

Consider n objects located at points x_i $(i = 1,\ldots,n)$ on a location map and described by J variables $\boldsymbol{Z}^j$ $(j = 1,\ldots,J)$ so that Z_i^j is the measurement of variable $\boldsymbol{Z}^j$ on observation i. $Z = \left[Z_i^j\right]$ is the matrix of the data.

In conventional statistical data analysis, such data are considered as variables defined on a set of observations. The n measurements Z_i^j are realizations of random variable $\boldsymbol{Z}^j$. This representation does not take account of the spatial nature of the data.

It is customary, in geostatistics, to represent such data by a set of *regionalized variables* considered to be realizations of a family $\boldsymbol{Z}(x) = \left\{\boldsymbol{Z}^j(x);\ j = 1,\ldots,J\right\}$ of random functions of point x. (i.e. $\boldsymbol{Z}(x)$ is a vector random function having values in $\boldsymbol{R}^J$). Thus, $Z^j(x_i)$ – that is denoted by Z_i^j – is a realization of random function $\boldsymbol{Z}^j(x)$ of point x. The n multidimensional measurements $(Z^1(x_i), Z^2(x_i), \ldots, Z^J(x_i))$ are thus realizations of n different random vectors of $\boldsymbol{R}^J$ associated each with x_i in the location map.

The term 'regionalized' was proposed by Matheron (1963) to qualify a phenomenon that spreads in space and exhibits a certain structure therein. Matheron defines a regionalized variable as an irregular function of point x that shows two contradictory (or complementary) aspects:

- one, structured, associated with a spatial organization (on a more or less large scale),
- the other, random, linked with unpredictable irregularities and variations from one point to the next.

These structures, when they exist, describe spatial organizations of the data on different spatial scales. It is the whole of these structures that we propose to extract in order to describe and to filter the initial variables. These structures, given the multidimensional aspect of the data, can be redundant insofar as they would be common to the J initial variables. Once established, they can be used with a view to a typology of these variables.

A direction is set on the data location map and a distance is considered in this direction. h denotes both this distance (which is a scalar) and the vector defined by the direction and the length h.

$$\Delta \boldsymbol{Z}(h) = \boldsymbol{Z}(x+h) - \boldsymbol{Z}(x) \tag{1}$$

denotes the multivariable increment of $\boldsymbol{Z}$ between points x and $x+h$ separated by distance h in the previously set direction.

$\boldsymbol{Z}$ is assumed to be stationary of order 2, which means that increment $\Delta\boldsymbol{Z}(h)$ has a zero average ($E(\Delta\boldsymbol{Z}(h)) = 0$) and a variance $E(\Delta\boldsymbol{Z}(h)^2)$ that depends only on h. Under these conditions, the *variograms–crossed covariograms* matrix (intrinsic codispersion matrix) $\Gamma_{ZZ}(h)$ is defined as the variance–covariance matrix of increments $\Delta\boldsymbol{Z}(h)$:

$$\Gamma_{ZZ}(h) = \tfrac{1}{2}E\big[\Delta\boldsymbol{Z}(h)^t \cdot \Delta\boldsymbol{Z}(h)\big] = \tfrac{1}{2}C_{\Delta Z \Delta Z}(h) \tag{2}$$

It is a matrix of dimension $J \times J$ whose general term is:

$$\gamma_{jj'}(h) = \tfrac{1}{2}E\big[\big(\boldsymbol{Z}^j(x+h) - \boldsymbol{Z}^j(x)\big) \cdot \big(\boldsymbol{Z}^{j'}(x+h) - \boldsymbol{Z}^{j'}(x)\big)\big] \tag{3}$$

The direction on the data location map being set, the diagonal term of $\Gamma_{ZZ}(h)$ then designates the *semivariogram* of $\boldsymbol{Z}^j$ whose value measures the *spatial variability* of $\boldsymbol{Z}^j$ at distance h. The experimental variogram is used as a curve depending on distance h. In the finite or discrete case, it could also be written as follows:

$$\gamma_j(h) = \gamma_{jj}(h) = \frac{1}{2m(h)} \sum_{d(i,i')=h} (Z_i^j - Z_{i'}^j)^2 \tag{4}$$

where $m(h)$ is the number of pairs made up of observations h away from one another.

$\gamma_j(h)$ represents the variability of variable $\boldsymbol{Z}^j$ linked with the spatial fluctuations of the point pairs h away from one another. Considered to be a curve depending on h, it is conventionally used in geostatistics. Its shape gives information about the spatial behavior of variable $\boldsymbol{Z}^j$ (Isaaks and Srivastava, 1989; Journel and Huijbregts, 1991; Wackernagel, 1995).

3.2. Local variance and covariance

Distance h induces all of the pairs of points connected by the contiguity graph $G(h) = [g_{ii'}(h)]$ of dimension $n \times n$ whose general term is defined by:

$$g_{ii'}(h) = \begin{cases} 1 & \text{if } d(i,i') = h \\ 0 & \text{otherwise} \end{cases} \tag{5}$$

The expression of the variogram is therefore:

$$\gamma_j(h) = \frac{1}{2m(h)} \sum_{i,i'} g_{ii'}(h)(Z_i^j - Z_{i'}^j)^2 \tag{6}$$

where

$$m(h) = \sum_{i,i'} g_{ii'}(h) \tag{7}$$

designates all the pairs $\{i,i'\}$ of observations connected by graph $G(h)$.

The crossed covariogram of two variables $\mathbf{Z}^j$ and $\mathbf{Z}^{j'}$ is defined by the expression:

$$\gamma_{jj'}(h) = \frac{1}{2m(h)} \sum_{i,i'} g_{ii'}(h)(Z_i^j - Z_{i'}^j)(Z_i^{j'} - Z_{i'}^{j'}) \tag{8}$$

If we consider that the shortest distance between two observations on the location map is equal to 1, the local variance of a variable $\mathbf{Z}^j$, in the isotropic case, is defined by:

$$\gamma_j(1) = \frac{1}{2m(1)} \sum_{i,i'} g_{ii'}(1)(Z_i^j - Z_{i'}^j)^2 \tag{9}$$

The local covariance of two variables $\mathbf{Z}^j$ and $\mathbf{Z}^{j'}$ is defined by:

$$\gamma_{jj'}(1) = \frac{1}{2m(1)} \sum_{i,i'} g_{ii'}(1)(Z_i^j - Z_{i'}^j)(Z_i^{j'} - Z_{i'}^{j'}) \tag{10}$$

The local variance–covariance matrix of the data is denoted by $\Gamma_{ZZ}(1)$. It is the matrix of dimension $J \times J$ whose general term is $\gamma_{jj'}(1)$ (with $\gamma_{jj}(1) = \gamma_j(1)$)

If σ_j^2 designates the empirical global variance of variable $\mathbf{Z}^j$, the contiguity coefficient (Geary, 1954) is defined by:

$$c_j(1) = \frac{\gamma_j(1)}{\sigma_j^2} \tag{11}$$

It measures the part of the variance explained by graph $G(1)$. This ratio is all the lower as the values of variable $\mathbf{Z}^j$ are homogeneously distributed on graph $G(1)$. This means that variable $\mathbf{Z}^j$ exhibits a low spatial variability between contiguous observations.

Whereas $c_j(h)$ defined by:

$$c_j(h) = \frac{\gamma_j(h)}{\sigma_j^2} \tag{12}$$

designates the contiguity coefficient for $h \geq 1$. Similar to $c_j(1)$ allowing to test the hypothesis of a significant influence between neighboring observations, the values of $c_j(h)$ – for $h \geq 1$ – will allow to test to what extent this influence is significant.

We use $c_j(h)$ outside the anisotropic case, i.e. we shall use it for distances h taken in any direction set on the data location map.

4. SPATIAL PROXIMITY ANALYSIS: A PARTICULAR SCA

SCA consists in seeking linear combinations of initial variables of minimum contiguity. It thus consists in calculating vector u that minimizes quotient $(u^t\Gamma_{ZZ}(1)u)/(u^tC_{ZZ}u)$ where C_{ZZ} is the variance matrix of the initial variables. Spatial proximity analysis (Royer, 1984) generalizes SCA for a contiguity level $h \geq 1$. It consists in calculating a variable f (called spatial component), linear combination of the initial variables:

$$f = Zu = \sum_j u_j \mathbf{Z}^j \tag{13}$$

depending on h – which realizes the minimum of expression

$$\frac{\gamma_f(h)}{\sigma_f^2} = \frac{u^t \Gamma_{ZZ}(h) u}{u^t C_{ZZ} u} \tag{14}$$

i.e. which has both a minimum spatial variability and a maximum variance.

There are J variables $f^1, \ldots, f^\alpha, \ldots, f^J$ solutions to this problem corresponding to eigenvectors $u^1, \ldots, u^\alpha, \ldots, u^J$ of matrix $C_{ZZ}^{-1}\Gamma_{ZZ}(h)$ – where $u^\alpha = (u_1^\alpha, u_2^\alpha, \ldots, u_J^\alpha)^t$ $\boldsymbol{R}^J$ – arranged in ascending eigenvalues:

$$\lambda_1 \leqslant \ldots \leqslant \lambda_\alpha \leqslant \ldots \leqslant \lambda_J \tag{15}$$

Eigenvalue λ_α is equal to $\gamma_{f^\alpha}(h)/\sigma_{f^\alpha}^2$ contiguity coefficient at level h associated with f^α, αth spatial component [1].

Royer (1984) refers to the components associated with the highest eigenvalues (i.e. the last components) as local components. They describe purely random structures. He refers to the components associated with low eigenvalues (i.e. the first components) as regional components. They describe regionalized structures, i.e. slowly variable in space.

Experience shows that this is not always the case, as we shall see for the application presented at the end of the present article.

In actual fact, although calculated from matrices C_{ZZ} and $\Gamma_{ZZ}(h)$ that include both statistical and spatial interdependencies of the data, the λ_α represent each isolated information specific to the αth factorial component. Unlike principal component analysis, for example, where the eigenvalues (those of C_{ZZ}) represent the part of the total data variance explained by the αth component, the sum of the λ_α, eigenvalues of SCA, is equal to $\text{tr}[C_{ZZ}^{-1}\Gamma_{ZZ}(h)]$ which, as for discriminant factorial analysis (Rencher, 1995; Romeder, 1973 pp. 46 and 47), would be meaningless.

It is rather $\text{tr}[\Gamma_{ZZ}(h)]$ – sum of the variogram values for distance h of the initial variables – which is of interest, and not $\text{tr}[C_{ZZ}^{-1}\Gamma_{ZZ}(h)]$. It therefore seemed necessary to us to define new criteria for quantifying the statistical and spatial information contained in the SCA components. This is possible with the help of result interpretation aid tools that we have been able to define and to which a paragraph is devoted. These tools have been established by means of the statistical and spatial properties of the SCA components.

4.1. Statistical and spatial properties of SCA components

SCA factorial components form a statistically and spatially orthogonal base. It seems that this second property has not been exploited by the various authors who have worked on this subject. It is precisely this property that implies the original aspect of SCA. It allows to build result interpretation tools.

[1] SCA results are closely linked with the neighborhood h set. We have however deliberately omitted to indicate it in the writings so as not to overload them by using notations f^α, λ_α, ... instead of $f^\alpha(h)$, $\lambda_\alpha(h)$, ...

Let $F = \left[f_i^\alpha\right]_{i=1,\dots,n}^{\alpha=1,\dots,J}$ the rectangular table of the measurements of the J SCA factorial components (in columns) on the n observations (in rows).

We show that the variance–covariance matrix C_{FF} of F, if the variables are centered, is written as follows:

$$C_{FF} = \tfrac{1}{n} F^t F = I_J \tag{16}$$

where I_J is the identity matrix $J \times J$. The variograms–crossed covariograms matrix associated with the factorial components is written as follows:

$$\Gamma_{FF}(h) = \Lambda \tag{17}$$

where Λ is the diagonal matrix of eigenvalues λ_α.

In other words, the correlation coefficient of two components f^α and f^β is written as follows:

$$\operatorname{cor}(f^\alpha, f^\beta) = \begin{cases} 1 & \text{if } \alpha = \beta \\ 0 & \text{otherwise} \end{cases} \tag{18}$$

and their crossed covariogram at point h:

$$\gamma_{f^\alpha f^\beta}(h) = \begin{cases} \lambda_\alpha & \text{if } \alpha = \beta \\ 0 & \text{otherwise} \end{cases} \tag{19}$$

The factorial components are therefore orthonormal. Furthermore, the value of their variogram for distance h is equal to the eigenvalue. The value of the crossed covariogram for distance h of two distinct variables is zero. They are therefore spatially independent. We deduce therefrom interesting relations concerning decomposition of the variograms–crossed covariograms matrix of the initial variables. We show [2] that its general term is written is the following form:

$$\gamma_{jj'}(h) = \sum_\alpha \operatorname{cor}(\mathbf{Z}^j, f^\alpha)\operatorname{cor}(\mathbf{Z}^{j'}, f^\alpha)\lambda_\alpha \tag{20}$$

and more particularly the variogram of variable $\mathbf{Z}^j$:

$$\gamma_j(h) = \sum_\alpha \operatorname{cor}^2(\mathbf{Z}^j, f^\alpha)\lambda_\alpha \tag{21}$$

And since $\sum_\alpha \operatorname{cor}^2(\mathbf{Z}^j, f^\alpha) = 1$, the variogram of a regionalized variable $\mathbf{Z}^j$ is written as a weighted mean of the variograms of the SCA factorial components.Generally speaking, the variograms–crossed covariograms matrix, considering (21), is written in the form:

$$\Gamma_{ZZ}(h) = C_{FZ}^t \Lambda C_{FZ} \tag{22}$$

where Λ is the diagonal matrix of the eigenvalues of $C_{ZZ}^{-1}\Gamma_{ZZ}(h)$ and C_{FZ} that of the covariances between factorial components f^α (in rows) and initial variables $\mathbf{Z}^j$ (in columns).

---

[2] By writing $\mathbf{Z}^j = \sum_{\alpha=1}^{J} \operatorname{cor}(\mathbf{Z}^j, f^\alpha) \cdot f^\alpha$ in the expression of $\gamma_{jj'}(h)$ while taking account of equality (19).

5. SCA RESULT INTERPRETATION AID TOOLS

In the remainder of the paper, we assume that SCA has been applied for a set h on the data location map. All the SCA results (eigenvalues λ_α, factorial components $f^\alpha,\ldots$) thus depend on h.

The writing of $\gamma_j(h)$ in (21) shows that the expression that we define as follows:

$$\text{cnt}(\mathbf{Z}^j, f^\alpha) = \text{cor}(\mathbf{Z}^j, f^\alpha)\left|\text{cor}(\mathbf{Z}^j, f^\alpha)\right| \frac{\lambda_\alpha}{\gamma_j(h)} \tag{23}$$

ranges between -1 and 1. It measures the *signed contribution of variable* $\mathbf{Z}^j$ *to the spatial variability of the* α*th factorial component*. Such a definition allows to have negative as well as positive values of the contribution, thus conferring a role similar to that of the correlation thereon. It will therefore be profitable to represent initial variables on circles that we shall call *contribution circles* (circle of radius 1) whose use is identical to that of the correlation circles used in principal component analysis. It can furthermore be noted that, for any variable $\mathbf{Z}^j$

$$\sum_{\alpha=1}^{J} \left|\text{cnt}(\mathbf{Z}^j, f^\alpha)\right| = 1 \tag{24}$$

In fact, if the value of $|\text{cnt}(\mathbf{Z}^j, f^\alpha)|$ is close to 1, we can say that component f^α greatly contributes to the *spatial variability* of variable $\mathbf{Z}^j$.

Besides, we define:

$$\text{cnt}(f^\alpha) = \frac{\sum_{j=1}^{J} \text{cor}^2(\mathbf{Z}^j, f^\alpha)\lambda_\alpha}{\sum_{j=1}^{J} \gamma_j(h)} = \frac{\sum_j \text{cor}^2(\mathbf{Z}^j, f^\alpha)\lambda_\alpha}{\sum_\beta \left(\sum_j \text{cor}^2(\mathbf{Z}^j, f^\beta)\lambda_\beta\right)} \tag{25}$$

This expression – ranging between 0 and 1 – measures the *absolute contribution of the* α^{th} *factorial component to the spatial variability of the data*[3]. The value of $\text{cnt}(f^\alpha)$ is all the higher (i.e. close to 1) as component f^α shows spatial structures common to a great number of initial variables.

The first index $\text{cnt}(\mathbf{Z}^j, f^\alpha)$ is useful, on the one hand, for interpretation of factorial components as a function of the initial variables and, on the other hand, for typology of the latter. A variable $\mathbf{Z}^j$ which, from a spatial point of view, is the most similar to component f^α, is a variable for which the value of $|\text{cnt}(\mathbf{Z}^j, f^\alpha)|$ is close to 1. Such a result is visualized by means of the contribution circle by representing the initial variables on a factorial map (f^α, f^β) by their co-ordinates $\text{cnt}(\mathbf{Z}^j, f^\alpha)$ and $\text{cnt}(\mathbf{Z}^j, f^\beta)$.

The second index $\text{cnt}(f^\alpha)$ is a global measurement of the spatial variability reproduced by component f^α. It is a descriptor of the spatial information contained in the structure revealed by this component. This criterion will help us select the most

[3] Taking (21) into account, we have $\sum_{\alpha=1}^{J} \text{cnt}(f^\alpha) = 1$.

significant components. It is on the basis of this criterion, and not of the eigenvalues, that we arrange the factorial components in descending spatial contributions.

These indices are however not sufficient to describe the data globally. They do not take account of the data variance, which is essential for measuring the part of the 'statistical information' reproduced by the components.

We therefore keep the following index:

$$\mathrm{cntV}(\mathbf{Z}^j, f^\alpha) = \mathrm{cor}(\mathbf{Z}^j, f^\alpha)\left|\mathrm{cor}(\mathbf{Z}^j, f^\alpha)\right| \tag{26}$$

ranging between -1 and 1 – which measures *the signed contribution of variable* $\mathbf{Z}^j$ *to the variance of the αth factorial component*. Definition of such an index follows from the fact that $\sum_{\alpha=1}^{J} \mathrm{cor}^2(\mathbf{Z}^j, f^\alpha) = 1$.

And, more generally, index

$$\mathrm{cntV}(f^\alpha) = \frac{\sum\limits_j \mathrm{cor}^2(\mathbf{Z}^j, f^\alpha)\sigma_j^2}{\sum\limits_j \sigma_j^2} \tag{27}$$

ranging between 0 and 1 – measures *the contribution of the* α^{th} *factorial component to the variance of the data*[4]. We thus have, for each component, the measurement of the variance part it contains.

We propose to arrange components f^α of the SCA, on the one hand, in descending order of the $\mathrm{cntV}(f^\alpha)$, i.e.:

$$\mathrm{cntV}(f^{\alpha(1)}) \geq \mathrm{cntV}(f^{\alpha(2)}) \geq \ldots \geq \mathrm{cntV}(f^{\alpha(J)}) \tag{28}$$

and, on the other hand, in descending order of the $\mathrm{cnt}(f^\alpha)$, i.e.:

$$\mathrm{cnt}(f^{\beta(1)}) \geq \mathrm{cnt}(f^{\beta(2)}) \geq \ldots \geq \mathrm{cnt}(f^{\beta(J)}) \tag{29}$$

The first p components kept in (28) will be called *regional components* because they express the structure of the data on a large scale.

The first q components kept in (29) will be called *local components* because they account for the structure of the data on a small scale.

Some components can be both regional and local. Components f^α at the end of (28) and (29) (i.e. for which we both have $\mathrm{cntV}(f^\alpha) \approx 0$ and $\mathrm{cnt}(f^\alpha) \approx 0$) are considered as random noise.

The part of the spatial variability of data explained by the first q local components is:

$$S_{\mathrm{local}} = \sum_{k=1}^{q} \mathrm{cnt}(f^{\beta(k)}) \tag{30}$$

The part of the data variance explained by the first p regional components is:

$$\vartheta_{\mathrm{regional}} = \sum_{k=1}^{p} \mathrm{cntV}(f^{\alpha(k)}) \tag{31}$$

[4] We have $\sum_{\alpha=1}^{J} \mathrm{cnt}V(f^\alpha) = 1$.

NB: The prerequisite condition for calculating contributions $\text{cntV}(f^\alpha)$ for a family of factorial components is that this family must be orthogonal. It is more particularly possible to define $\text{cntV}(f^\alpha)$ for each factorial component f^α of PCA; the latter being orthogonal. We show, in this case, that $\text{cntV}(f^\alpha) = \lambda_\alpha / \sum_\beta \lambda_\beta$ inertia percentage measuring the representation quality of f^α. In this expression, λ_α corresponds to the αth eigenvalue of the variance–covariance matrix of the initial variables (diagonalized in PCA). $\text{cntV}(f^\alpha)$ is therefore the criterion of the variance directly available in PCA. In the case of SCA, its value for all the spatial components is lower than the inertia percentage of the first factorial component of PCA.

6. APPLICATION TO SEISMIC IMAGE DESCRIPTION AND FILTERING

6.1. Seismic images

A seismic image is an acoustic representation of the subsurface obtained by means of waves produced at the surface. Caused by shots, these waves spread into the subsurface where they are reflected by the various geologic structures met (Mari et al., 1999). The signals thus reflected are picked up at the ground surface. Processed and analyzed, these images allow to understand the subsurface geology. A 3D seismic image consists of a set of vertical signals (referred to as *seismic traces*) distributed horizontally at the nodes of a regular grid. Measurements characterizing these signals are called seismic amplitudes. Such images have a better horizontal resolution but, unlike geologic information (example of cores from boreholes), they lack vertical precision. The more or less high amplitude values of seismic traces account for the geologic structures of the subsurface. From this point of view, a seismic image can be considered to be a remote echography showing the subsurface geologic structures.

A 3D seismic image – referred to as seismic cube or volume – comes in the form of a three-dimensional volume. From a statistical point of view, we consider this cube as a pile of horizontal images representing seismic horizons in cases where the geologic structures are horizontal tabular. If the geologic structures are not of the horizontal tabular type, they can be made as such by means of suitable processing (flattening, . . .).

Each horizon induces a variable whose measurements – which are the seismic amplitudes – are located at the nodes of the regular grid defined at the surface during seismic measuring surveys. There are then as many variables as there are horizons. Observations correspond to the nodes of the grid.

Various analyzes – *seismic processing* – are applied to these data in order to improve the signal-to-noise ratio. This aims to improve the ‘vision’ of subjacent geometric structures making such images geologically interpretable. Most of these techniques come within the field of signal processing (Lavergne, 1989; Glangeaud and Mari, 1994).

Once processed, the seismic data are used for (structural or stratigraphic) geologic interpretation performed upon discovery or characterization of a reservoir. More recently, lithologic interpretation of seismic data (Dumay and Fournier, 1988; Fournier and Derain, 1994; Déquirez et al., 1995) consists in analyzing, on the reservoir scale, the volume portions limited to the reservoir. A certain number of attributes – obtained by various transformations of seismic amplitudes – (Justice et al., 1985) are used to perform classifications (supervised or not) of the seismic traces considered to be observations whose attributes define the variables. These classifications – based on statistical or neural pattern recognition methods – allow to group the seismic traces into classes, thus defining what is referred to as *seismic facies*. The map thus obtained, thanks to the very good horizontal distribution of the seismic data and to the geologic and seismic information well tie, allows to predict geology between the wells so as to optimize drilling.

Factorial methods (such as PCA) are often used in a preliminary stage of seismic image description and filtering. However, they do not take account of spatial properties, which is the reason why we use SCA. The factorial components obtained with this method, by means of their statistical and spatial properties, allow to account for the basic spatial structures intrinsic to the data. Once established, these structures are used to describe and to filter the initial images.

6.2. Analyzed data

The seismic cube that we analyze consists of 6 horizons that will define the 6 variables of the analysis. These variables are represented by the six images Z1 to Z6 in Fig. 1. The gradations of colors show the intensity of the seismic amplitudes. These data have been simulated to illustrate the method stages and implementation. The distance between seismic traces in the case of real data is some 10 m (between 10 and 50 m). In our simulations, we select this distance arbitrarily equal to 1 m with a view to simplification.

Three spatial structures have been generated on these images and fitted into each other: two large-scale structures and a third, small-scale structure. The first large-scale structure consists of lenticular shapes of east–west axis offset in the north–south orientation. This structure is common to images Z1, Z2 and Z3. The second large-scale structure consists of lenticular shapes of north–south axis offset in the east–west orientation. It is common to images Z4, Z5 and Z6. Random noise (inducing local spatial heterogeneities) is generated and mixed with the two north–south and east–west structures. Local spatial heterogeneities are more extensive in images Z2, Z4 and Z6.

The north–south and east–west structures are large-scale spatial organizations whereas random noise rather appears on a smaller scale. This is what Matheron defines by the expression 'random, unpredictable from one point to the next'.

Furthermore, although structurally similar to one another, the images in group Z1, Z2 and Z3 (respectively Z4, Z5 and Z6) are weakly correlated with one another – within the same group – because of the apparent offset of the lenticular shapes forming each image [see correlation table (Table 1) below]. As a result of this weak correlation, separation of these structures by means of a conventional multivariable method (such as PCA) based on correlations and that does not take spatial distances between observations into consideration is difficult – or even impossible.

By applying PCA to the 6 variables, we obtain principal components CP1 to CP6 (Fig. 2). The variance percentages corresponding thereto are given in Fig. 3.

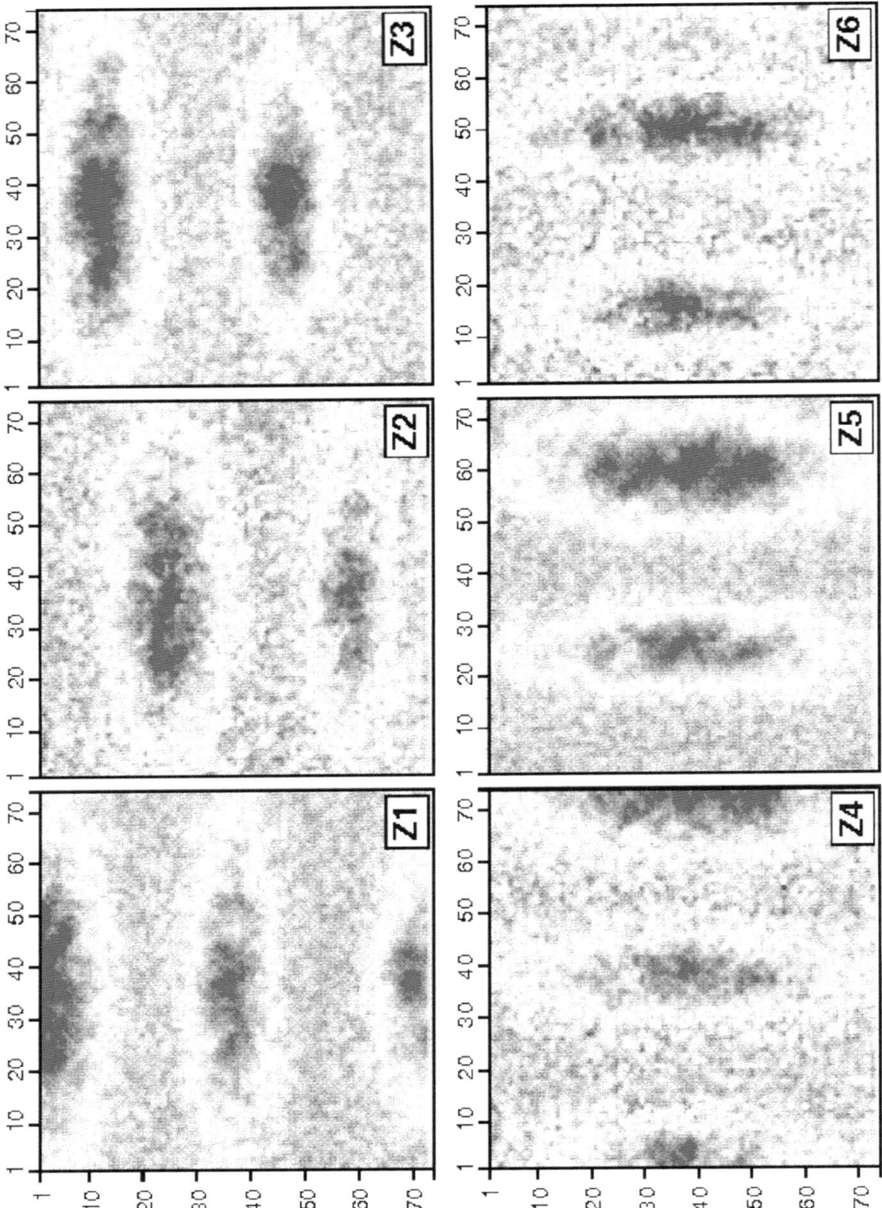

Fig. 1. 6 images corresponding to the 6 horizons of the seismic cube analyzed.

Paradoxically, the first three principal components are the most locally heterogeneous among the 6 components. This is due to the fact that the random noise, as it is the case for seismic data, is generally not correlated laterally (i.e. on the same horizon) but it is

TABLE 1

Correlations of the 6 images

	Z1	Z2	Z3	Z4	Z5	Z6
Z1	1.00	−0.26	0.11	−0.11	0.05	−0.02
Z2	−0.26	1.00	0.14	0.27	0.07	**0.41**
Z3	0.11	0.14	1.00	−0.05	0.13	0.00
Z4	−0.11	0.27	−0.05	1.00	−0.15	0.17
Z5	0.05	0.07	0.13	−0.15	1.00	0.02
Z6	−0.02	0.41	0.00	0.17	0.02	1.00

vertically (i.e. from one horizon to another), so that it is 'captured' by the first principal components. It is the first limitation of PCA. Another limitation of this method comes from the fact that it does not take account of spatial relations between observations. All this leads to poor separation of the spatial structures of the data on different scales. Besides, the random noise is mixed with the east–west and north–south structures on these first three principal components. The following components, although comprising less noise, cannot separate the two large-scale spatial structures.

6.3. *Descriptive preliminary geostatistical analysis of initial variables*

The aim is to determine the direction and distance h for which SCA provides adequate results. By 'adequate', we mean results that satisfy the user insofar as he tries to extract, by means of SCA, spatial structures he considers to be relevant.

This is done by means of the variogram curves during a preliminary stage of geostatistical analysis of the initial variables. Such an analysis is called variographic analysis. It consists in studying the variogram curves in various direction of the data location map in order to establish certain spatial properties of the variables.

The value of $\gamma_j(h)/\sigma_j^2$ defined by expression (12) measures the part of the variance of variable $\mathbf{Z}^j$ explained by spatial fluctuations of pairs of observations h apart. It corresponds to the value of the variogram curve, for distance h, of the reduced variable $\mathbf{Z}^j$. This curve is conventionally used by geostatisticians as a global descriptor of the spatial behaviour of a regionalized variable.

The jump at the origin – referred to as *nugget effect* – of this curve shows, if it exists, the presence of local measurement heterogeneities. It corresponds to value $\gamma_j(1)/\sigma_j^2$ equal to Geary's contiguity coefficient, in the isotropic (or omnidirectional) case.

It measures the part of the variance explained by the pairs of neighboring observations at level '1'. This jump at the origin of the curve is all the higher as the values of variable $\mathbf{Z}^j$ are distributed heterogeneously in the local spatial neighborhoods (i.e. for small values of h), which means that variable $\mathbf{Z}^j$ locally has a high spatial variability.

The use we make of the variogram curves of initial variables is not within the scope of conventional geostatistics. Our approach remains essentially descriptive whereas, in a conventional geostatistical study, these curves are used to develop spatial models of regionalized variables. These models, that may be considered to be an interpolation

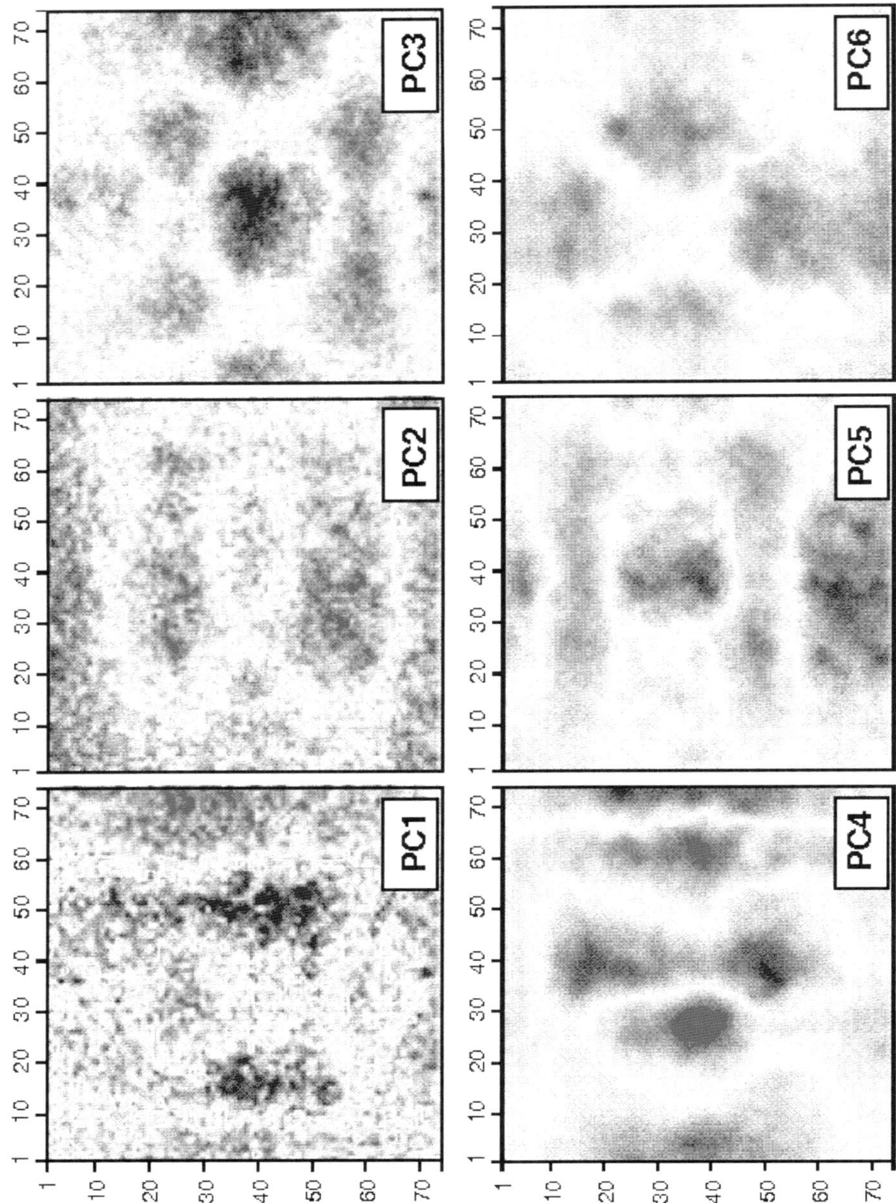

Fig. 2. Principal components obtained by PCA of the 6 images Z1 to Z6.

of the experimental variogram by a deterministic function, are thereafter used to make geostatistical estimations (kriging or co-kriging) or simulations (Isaaks and Srivastava, 1989; Wackernagel, 1995).

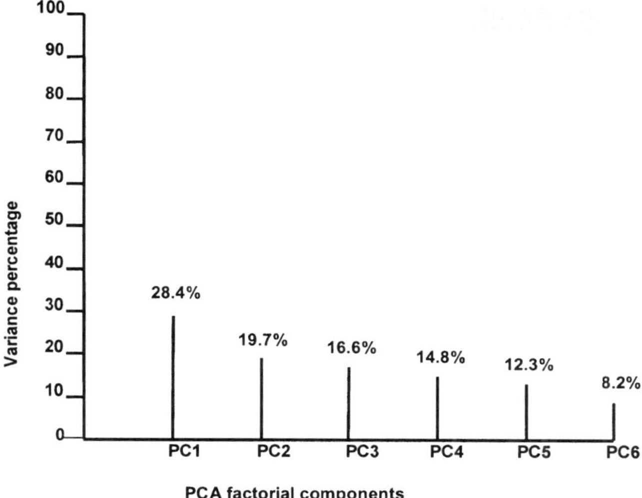

Fig. 3. Variance percentage of the PCA components.

In our data description and exploration approach, we try to bring out spatial behaviour groups by comparing the directional variograms of the J variables with one another in order to bring out spatial structure families. This furthermore allows us to determine the suitable distance(s) for implementation of SCA.

Fig. 4 show the variogram curves of the 6 images in 5 directions (N–S, N22°E, N45°E, N67°E and E–W). Each figure (Figs. 4a–e) corresponds to a direction in which the 6 curves of the variograms corresponding to variables Z1 to Z6 are represented.

Direction N45°W does not allow to distinguish the 6 variogram curves from one another. Whereas the N–S and E–W directions, which discriminate these curves best, naturally show two homogeneous and well-separate spatial behaviour families, notably for values of h ranging between 5 and 25 m. We distinguish the periodicities of the horizontal (images Z1, Z2 and Z3) and vertical events (Z4, Z5 and Z6) on the corresponding variograms. The latter even allow to give this periodicity an order of magnitude ($\approx$35 m) which actually corresponds to the mean thickness of the lenticular shapes.

It can also be noted that the jump in the neighborhood of the origin (all variograms taken into account) is different from one image to the next. The latter are arranged in ascending order of the value at the origin as follows: Z1, Z5, Z3, Z2, Z4 and Z6. It is precisely in this order that the images are arranged according to ascending local heterogeneity.

Selection of the direction and of contiguity distance h for applying PCA depends on the expected results. The user's choice is directed towards the spatial structure he wishes to extract from his data. These structures, as described above, are generally spotlighted by the preliminary variographic study of the initial variables.

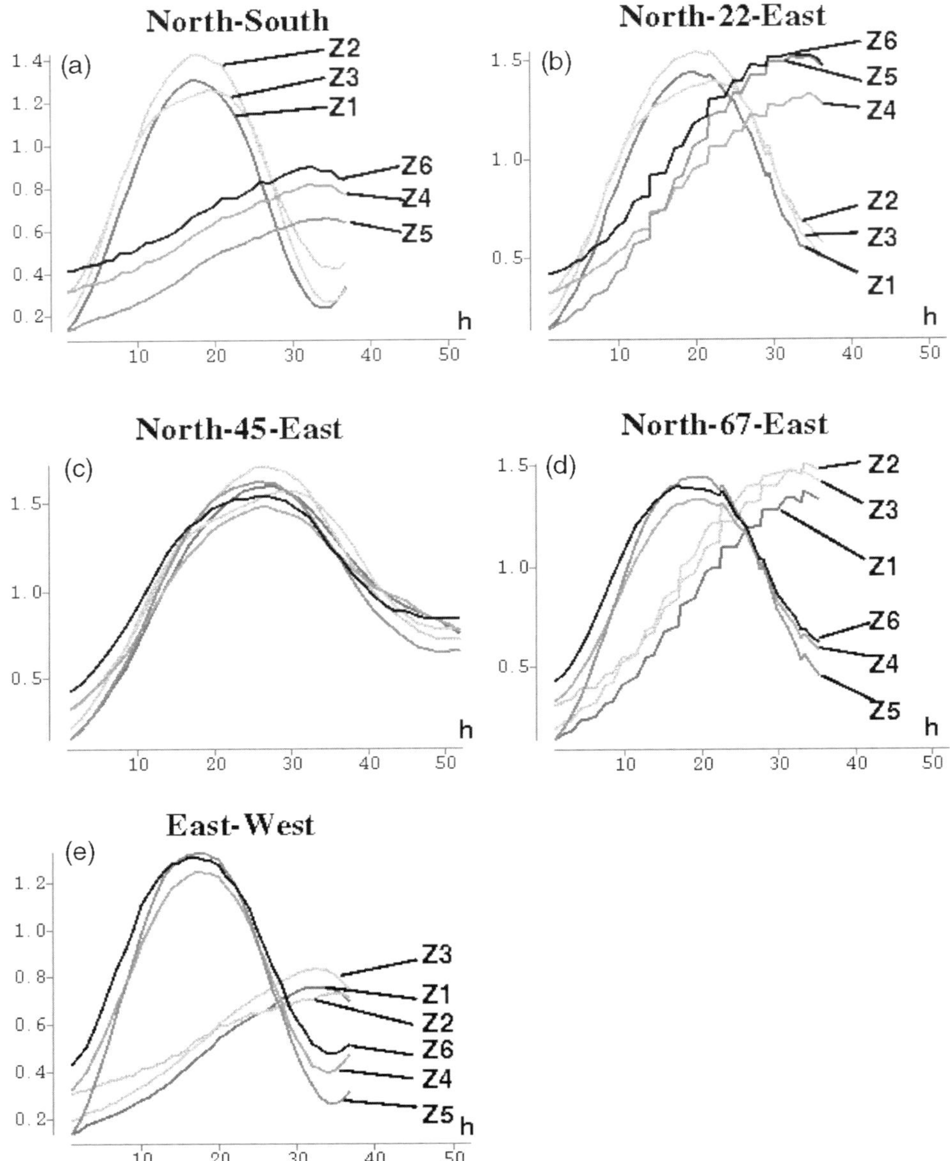

Fig. 4. Directional variogram curves corresponding to variables Z1 to Z6 in 5 directions of the map.

We are hereafter going to apply SCA for two values of h – first for 1 m in the isotropic (or omnidirectional) case, then for 15 m in the E–W direction – to show that the results obtained are very different according to the value of h. The spatial structures extracted can be quite distinct.

6.4. *SCA results in the anisotropic case for $h = 1$ m*

Applied for low values of h, SCA allows to separate variables of high local irregularity from locally homogeneous variables.

Applying SCA in the anisotropic case for $h = 1$ meter coincides with the method as developed by Lebart (taking account of the pairs of neighboring observations at contiguity level '1' in all the directions of the location map). For the data that we analyze, this contiguity value corresponds to the value, on the variogram curves, for which the 6 initial variables are arranged according to ascending local heterogeneity (in order Z1, Z5, Z3, Z2, Z4 and Z6).

Although it is well-suited for local heterogeneity filtering, this value of h does not give suitable results concerning better extraction of large-scale spatial structures of the variables. This clearly appears on spatial components CS3 to CS6 shown in Fig. 5. The first two components CS1 and CS2 whose contributions to the spatial variability of the data are maximum (respectively 66.03 and 30.82% – see Fig. 6) 'capture' mainly the random noise (data heterogeneity). The following components allow more or less to capture large-scale structures without separating them quite distinctly. These two large-scale structures are particularly mixed in component CS4 (see Fig. 5).

The 6 spatial components are however not very differentiated in terms of variance. The highest contribution to the data variance is reached by CS1 with cntV(CS1) = 23.76% and the lowest contribution is reached by CS5 with cntV(CS5) = 9.93% (see Fig. 6b).

The circles relative to the contributions of the SCA components to spatial variability of the variables (Fig. 7) allow to position the initial variables in relation to the spatial components. This positioning plays a part of interpretation of the (initial and factorial components) variables in relation to one another, identical to that of correlation circles.

Spatial variability contribution circles only allow to bring out two homogeneous and well-separated classes among the initial variables. The first class is made up of horizons Z2, Z4 and Z6 that greatly contribute (negatively) to the spatial variability of component CS1 whereas the second class is made up of horizons Z1, Z3 and Z5 that greatly contribute (negatively) to the spatial variability of component CS2 (see factorial map CS1–CS2). None of the other maps, apart from CS1–CS2, allows good initial variable discrimination. The high values of signed contribution of variables Z2, Z4 and Z6 to the spatial variability of component CS1 (considered as local component of the analysis) underline the high local heterogeneity specific to these three variables.

6.5. *SCA results in the E–W direction for $h = 15$ m*

We present hereafter results of the SCA applied for distance $h = 15$ m in the E–W direction. This distance and this direction correspond to the best discrimination between the two structures; the variogram curves of the two families being still better discriminated in the E–W direction than in the N–S direction. The spatial components are given in Fig. 8.

The two spatial components CS1 and CS2 have relatively high values as regards contribution to the spatial variability of the data (respectively 32.22 and 30.65% – see

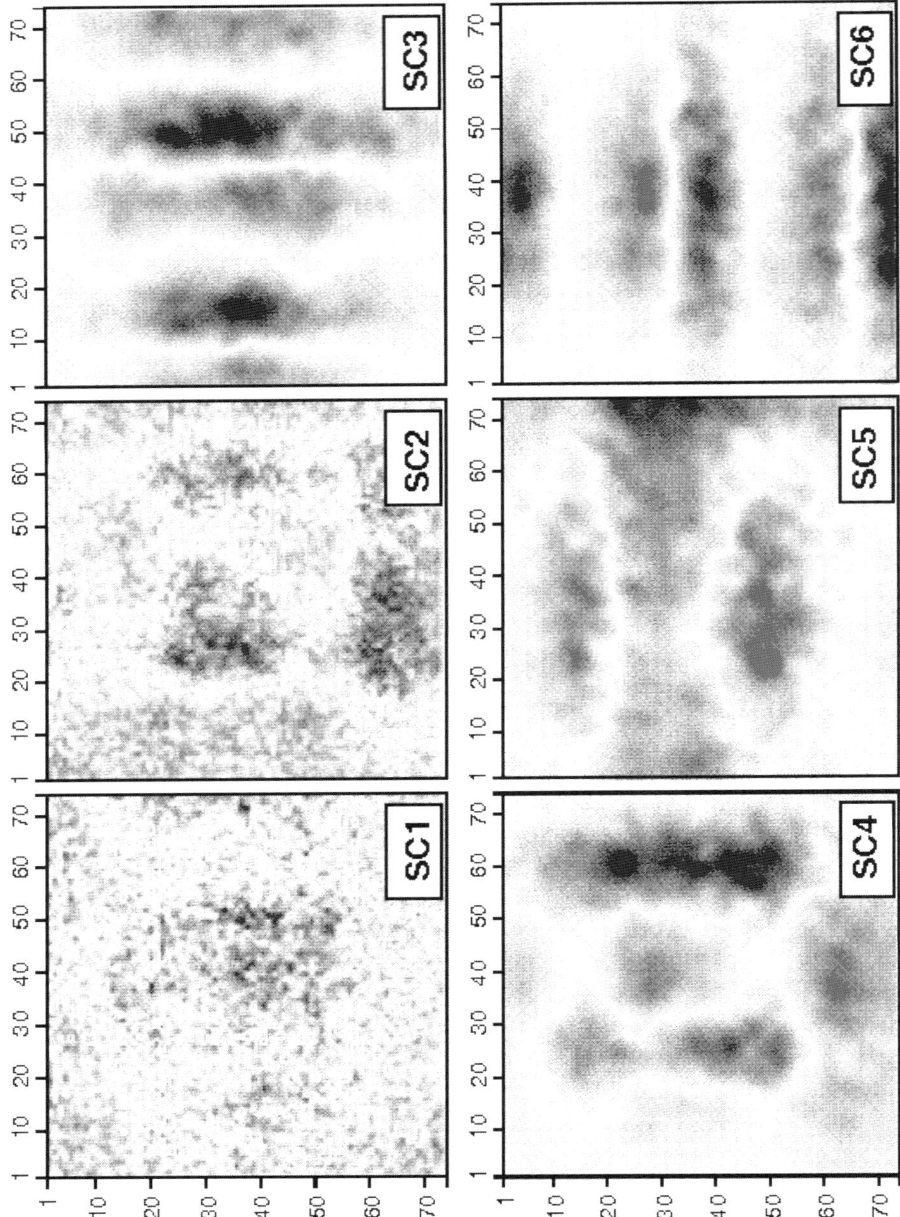

Fig. 5. Spatial components of SCA (omnidirectional – $h = 1$ m) applied to the 6 seismic horizons.

Fig. 9), which allows to qualify them as local components. Components CS5 and CS6, whose contributions to spatial variability are low (respectively 4.35 and 4.27%), are regional components. The former (CS1 and CS2) exhibit more spatial variability in the

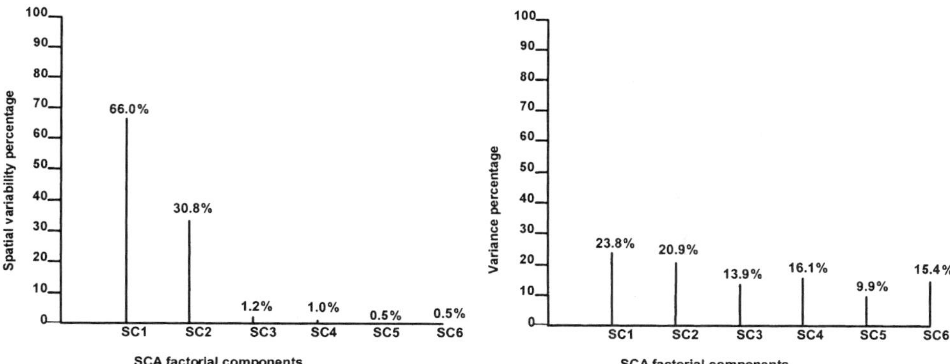

Fig. 6. Bar graph of the contributions of components to spatial variability and variance (omnidirectional SCA – $h = 1$ m)

E–W direction and for distance $h = 15$ m. The latter (CS5 and CS6) in fact have a higher regularity (spatial continuity) in this direction.

The contribution circle associated with factorial map CS1–CS2 (Fig. 10a) allows to relate the three variables Z4, Z5 and Z6 to the two components CS1 and CS2. Similarly, the contribution circle corresponding to factorial map CS5–CS6 (Fig. 10e) allows to relate the three variables Z1, Z2 and Z3 to the two components CS5 and CS6. CS1 and CS2 thus correspond to the vertical structures whereas CS5 and CS6 correspond to the horizontal structures. Intermediate components do not capture the local heterogeneity of the initial variables as well.

These results show that analysis in the E–W direction for $h = 15$ m allows better discrimination of the two large-scale structures; however, it does not allow to extract the random noise.

In the present case, on account of the two large-scale structures having orthogonal orientations, when one appears as a regional structure, the other appears as a local structure. This is of course not always the case. In most cases, the suitable direction and distance would have to be found for ‘capture’ of a suitable structure.

6.6. Optimum extraction of large-scale structures and random noise from the spatial components obtained from the two analyzes

Tables 2 and 3 give, for the two analyzes (respectively omnidirectional SCA for $h = 1$ m and SCA in the E–W direction for $h = 15$ m), the spatial variability S_{local}, defined by (30), and variance $\varphi_{\text{regional}}$, defined by (31), percentages of the initial variables restored by the most significant spatial components obtained during each analysis.

During the first analysis (omnidirectional SCA for $h = 1$ m), only the two components CS1 and CS2 allow good discrimination of the initial variables in two classes. They appear as local components. As we have seen, these two components ‘contain’ the local spatial heterogeneities of the initial variables that they divide into two classes; the first one, made up of Z2, Z4 and Z6, related to component CS1 and the second, made up of

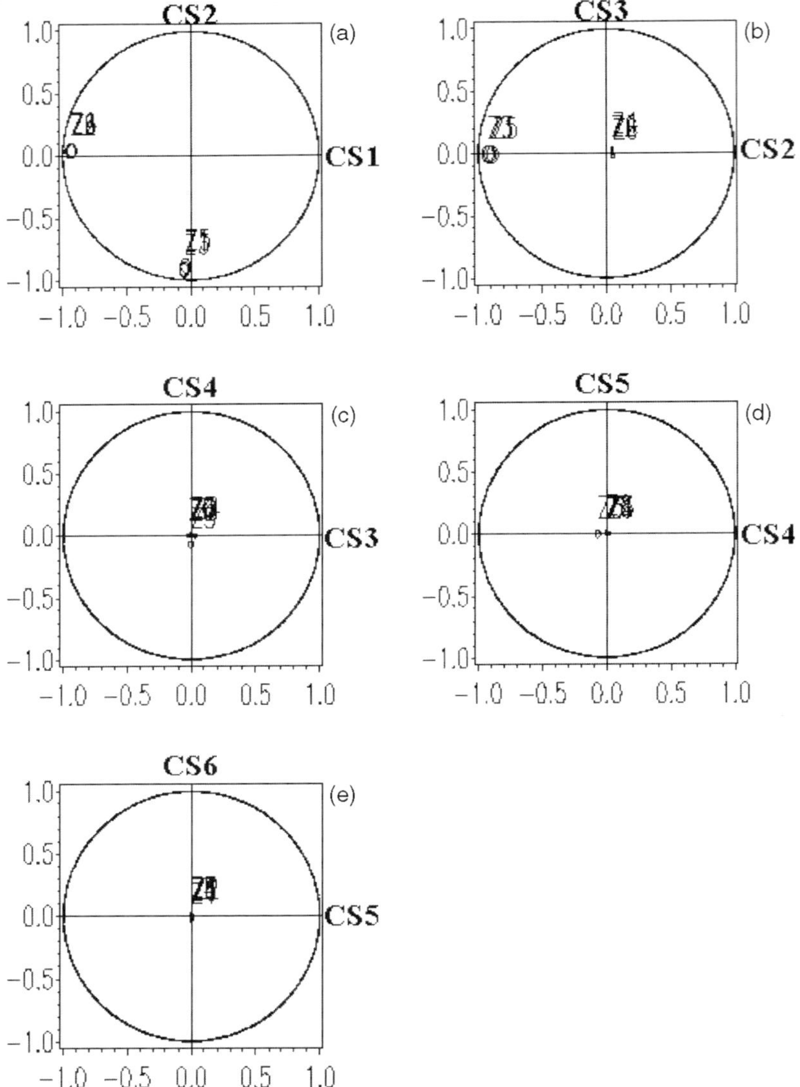

Fig. 7. Circles of initial variable contributions to spatial variability of SCA components (omnidirectional – $h = 1$ m).

Z1, Z3 and Z5, related to component CS2. The random noise can therefore be extracted by means of these two components (Fig. 11).

During the second analysis (SCA in the east–west direction for $h = 15$ m), components CS1 and CS2, on the one hand, and CS5 and CS6, on the other hand, are the most relevant for large-scale structures. The two former (corresponding to the lenticular shapes of N–S axis contained in CS1–CS2) reveal a first class of variables (Z4, Z5 and

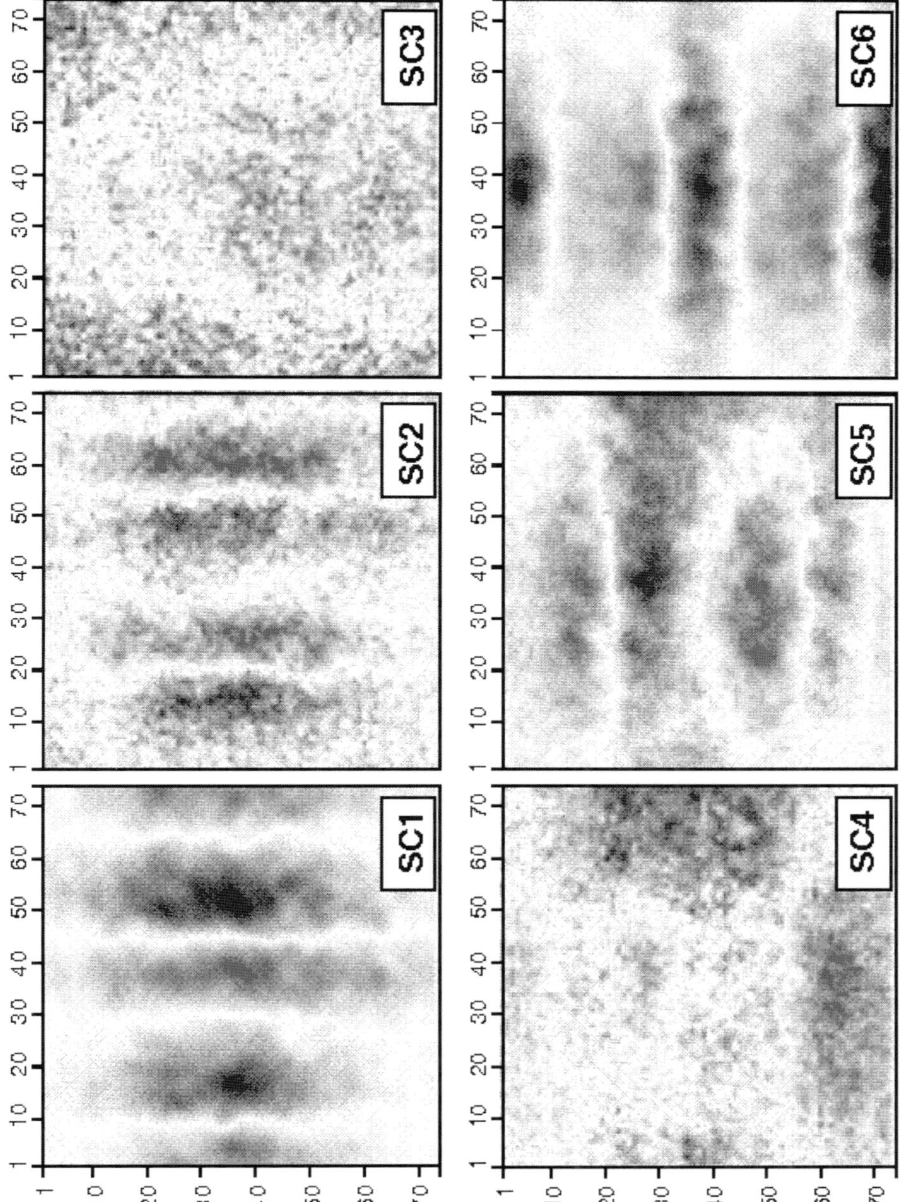

Fig. 8. Spatial components of SCA (East–West direction – $h = 15$ m) applied to the 6 horizons Z1 to Z6.

Z6), whereas the two latter (corresponding to the lenticular shapes of E–W axis contained in CS5–CS6) reveal a second class of variables (Z1, Z2 and Z3). This therefore allows to extract these two large-scale structures alone from the initial variables (Fig. 12).

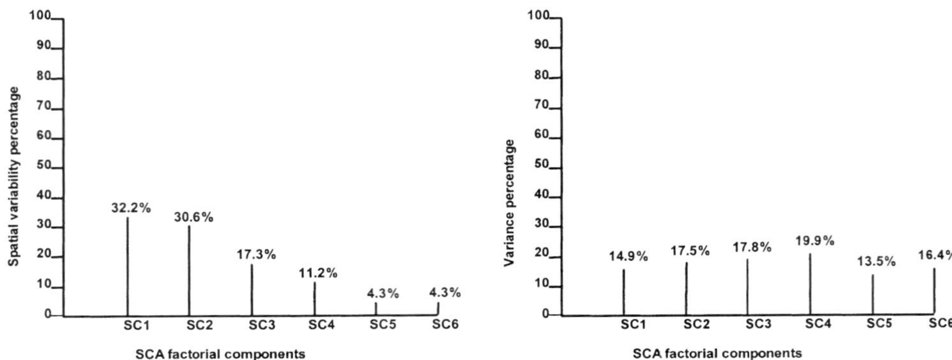

Fig. 9. Bar graph of the contributions of spatial components to the spatial variability and variance of the data (SCA East–West direction – $h = 15$ m).

TABLE 2

Anisotropic (or omnidirectional) SCA for $h = 1$ meter: spatial variability (S_{local}) and variance ($\vartheta_{\text{regional}}$) of the initial images restored by the spatial components

	SCA nisotropic – $h = 1$ m.			
	Initial variables restoration by CS1		Initial variables restoration by CS2	
	S_{local}	$\vartheta_{\text{regional}}$	S_{local}	$\vartheta_{\text{regional}}$
Z1	5%	1%	**90%**	**32%**
Z2	**93%**	**41%**	4%	4%
Z3	5%	1%	**92%**	**47%**
Z4	**92%**	**43%**	5%	4%
Z5	4%	1%	**88%**	**33%**
Z6	**93%**	**56%**	5%	6%

TABLE 3

SCA in the east-west direction for $h = 15$ meter: spatial variability (S_{local}) and variance ($\vartheta_{\text{regional}}$) of the initial images restored by the spatial components

	SCA east–west – $h = 15$ m.			
	Initial variables restoration by CS1 and CS2		Initial variables restoration by CS5 and CS6	
	S_{local}	$\vartheta_{\text{regional}}$	S_{local}	$\vartheta_{\text{regional}}$
Z1	2%	1%	**61%**	**34%**
Z2	3%	1%	**68%**	**46%**
Z3	2%	0%	**71%**	**51%**
Z4	**81%**	**55%**	18%	40%
Z5	**89%**	**75%**	9%	20%
Z6	**82%**	**64%**	19%	36%

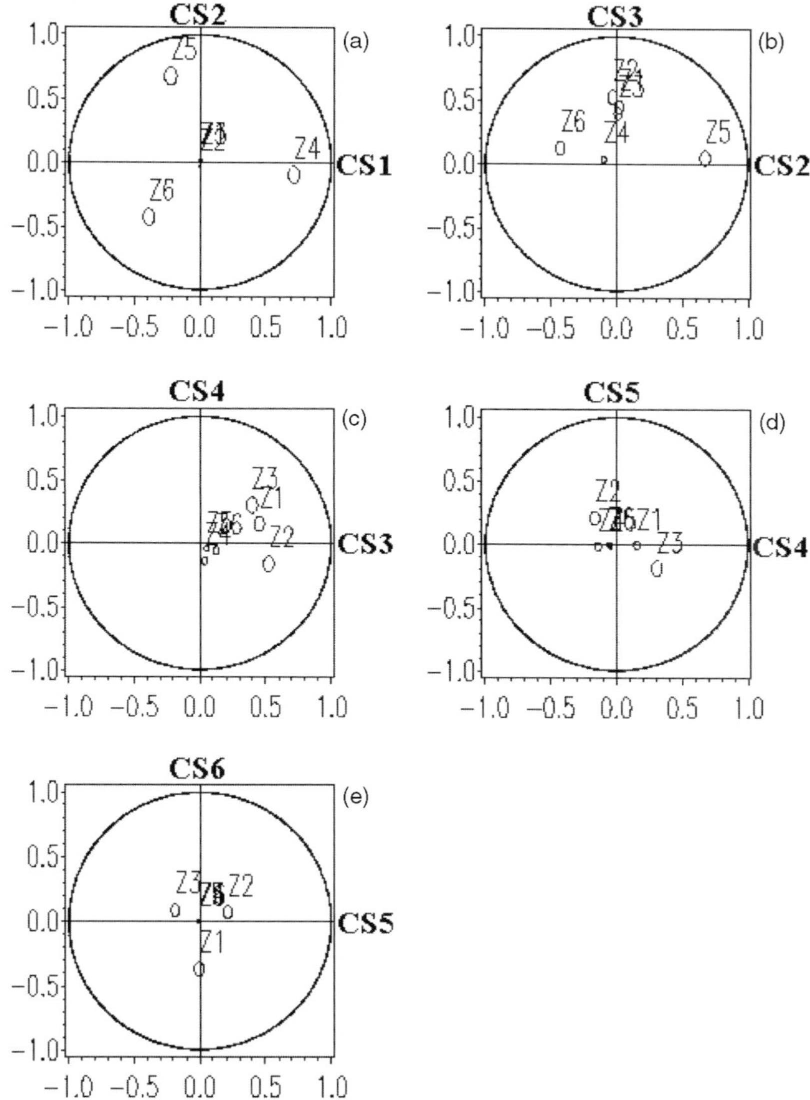

Fig. 10. Circles of initial variable contributions to the spatial variability of SCA components (East–West direction – $h = 15$ m).

7. CONCLUSION

Spatial contiguity analysis, thanks to its dual geostatistical and multifactorial aspect, allows to combine tools specific to geostatistics (variograms, crossed covariograms) with interpretation aid tools. The components obtained by SCA show elementary structures

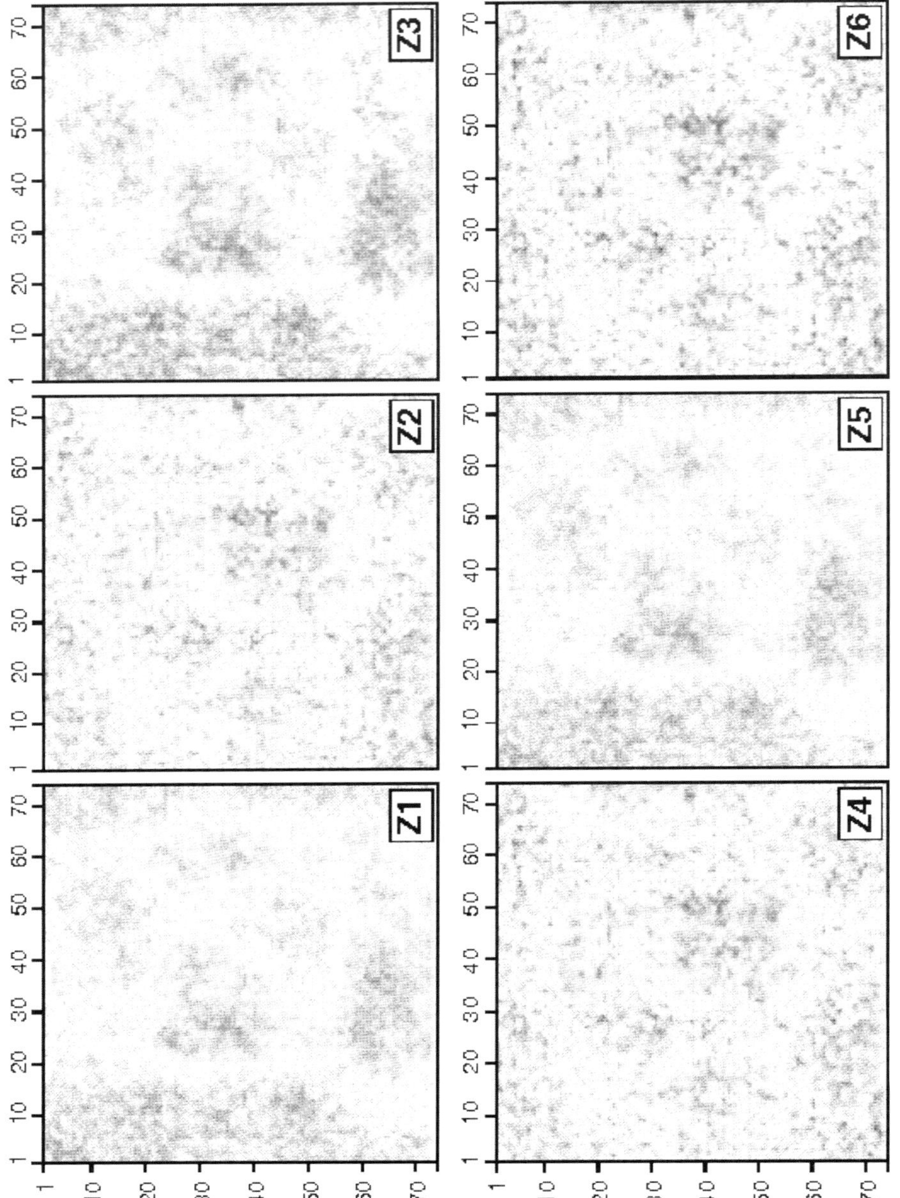

Fig. 11. Restoration of the noise in the initial variables by means of spatial components CS1 and CS2 (omnidirectional SCA – $h = 1$ m).

on different spatial scales. Concerning seismic data, these structures can be related to subsurface geologic structures or interpreted as such.

The tools that we present allow to establish connections between factorial com-

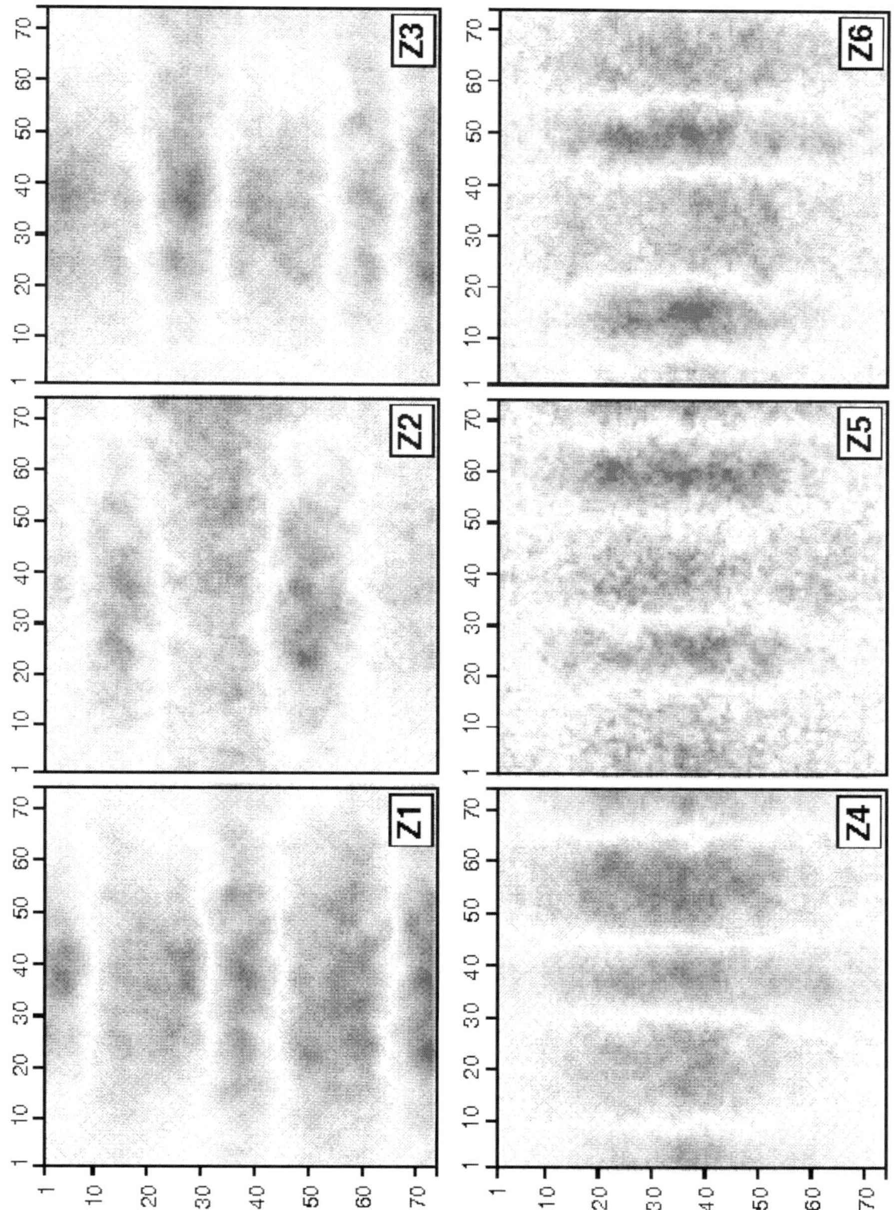

Fig. 12. Restoration of the large-scale structures of variables Z1, Z2, Z3 by means of components CS5 and CS6, and of variables Z4, Z5, Z6 by means of components CS1 and CS2 (SCA – east–west direction – $h = 15$ m).

ponents and subjacent geology. They allow to characterize the components obtained both in terms of variance and of spatial variability. These two notions complement one another and underline large-scale (regional structures) and small-scale (local structures) data structures.

SCA thus provides a coherent analysis framework allowing to relate initial data to results obtained by means of conventional tools such as correlation circles, but also new tools specific to the data we analyze (i.e. related by spatial contiguity constraints). The signed contributions of initial variables to the spatial variability of SCA factorial components, presented on contribution circles, thus allow to bring out the spatial typology of initial variables differently from correlation circles.

Selection of a spatial contiguity direction and distance influences the structures shown by the factorial components obtained. This provides a wide analysis selection range, an analysis being associated with a contiguity distance and direction pair. Each analysis gives different results. The user of the method is directed by criteria falling within his sphere of expertise and in connection with the very nature of his data. The tools used allow him to concentrate on essentials and not to get lost in the maze of a method which, without these tools, would quickly become tedious because of the great number of results to which it may lead.

REFERENCES

Burtschy, B. and Lebart, L., 1991. Contiguity analysis and projection pursuit. In: Gutieretz, R. and Valderrama, J. (Eds.), *Applied Stochastic Models and Data Analysis*. World Scientific, Singapore, pp. 117–128.

Déquirez, P.Y., Fournier, F., Blanchet, C., Feuchtwanger, T. and Torriero, D., 1995. Integrated stratigraphic and lithologic interpretation of the East Senlac heavy oil pool. *65th Annu. Int. Meet., Soc. Explor. Geophys.*, Expanded Abstracts, pp. 104–107.

Dumay, J. and Fournier, F., 1988. Multivariate statistical analyses applied to seismic facies recognition. *Geophysics*, 53(9): 1151–1159.

Faraj, A., 1994a. Interpretation tools for generalized discriminant analysis. In: *New Approaches in Classification and Data Analysis, IFCS Meeting, August 31–September 4, Paris, 1993.* Springer-Verlag, Heidelberg, pp. 285–291,

Faraj, A., 1994b. Application of spatial contiguity analysis to seismic data filtering. *64th Annu. Int. Soc. Explor. Geophys. Meet., Los Angeles, October 23–28 1994*, Expanded Abstracts, Vol. 1, pp. 1584–1587, Paper SP5.7.

Faraj, A., 1999. Statistics and data analysis. In: Mari, J.-L. et al. (Eds.), *Geophysics of Reservoir and Civil Engineering*. Éditions Technip, Paris.

Faraj, A. and Fournier, F., 1993. Proximity analysis and principal components analysis: two filtering techniques. In: Fabri, A.G. and Royer, J.J. (Eds.), *3rd CODATA Conference on Geomathematics and Geostatistics. Science de la Terre, Sér. Inf., Nancy*, 32, pp. 153–166.

Fournier, F. and Derain, J.F., 1994. A statistical methodology for deriving reservoir properties from seismic data. Rapport interne Institut Français du Pétrole, No. 41 133.

Friedman, J.H. and Tuckey, J.W., 1974. A Projection Pursuit Algorithm for Exploratory Data Analysis. *IEEE Trans. Comput., Ser. C*, 23: 881–889.

Geary, R.C., 1954. The contiguity ratio and statistical mapping. *Incorporated Statistician*, 5: 115–145.

Glangeaud, F. and Mari, J.L., 1994. *Wave Separation*. Éditions Technip, Paris.

Goovaerts, P., 1992. Factorial kriging analysis: a useful tool for exploring the structure of multivariate spatial information. *J. Soil Sci.*, 43: 597–619.

Green, A., Berman, M., Switzer, P. and Craig, M., 1988. A transformation for ordering multispectral data in terms of image quality with implications for noise removal. *IEEE Trans. Geosci. Remote Sensing*, 26(1).

Isaaks, E.H. and Srivastava, R.M., 1989. *Applied geostatistics*. Oxford University Press, Oxford.

Journel, A.G. and Huijbregts, Ch. J., 1991. *Mining Geostatistics*. Academic Press.

Justice, J.H., Hawkins, D.J. and Wong, G., 1985. Multidimensional attribute analysis and pattern recognition for seismic interpretation. *63rd Annu. Meet. and Int. Exp., Soc. Expl. Geophys.*, Expanded Abstracts, pp. 285–288.

Lavergne, M., 1989. *Seismic Methods*. Éditions Technip, Paris.

Lebart, L., 1969. Analyse statistique de la contiguite. *Publ. Inst. Stat., Paris*, VIII: 81–112.

Lebart, L., Morineau, A. and Piron, M., 1995. *Statistique exploratoire multidimensionnelle*. Dunod, Paris.

Mari, J.-L., Arens, G., Chapellier, D. and Gaudiani, P., 1999. *Geophysics of Reservoir and Civil Engineering*. Éditions Technip, Paris.

Matheron, G., 1963. *Principles of Geostatistics*. Éditions Technip, Paris.

Matheron, G., 1982. Pour une analyse krigeante des données régionalisées. Centre de Géostatistique, Fontainebleau, Publ. CGMM N-732, 22 pp.

Rencher, A.C., 1995. *Methods of Multivariate Analysis*. Wiley, New Yrok, NY.

Romeder, J.-M., 1973. *Méthodes et programmes d'analyse discriminante*. Dunod, Paris.

Royer, J.-J., 1984. Proximity analysis: a method for geodata processing. Proc. of the Int. Coll. Computers in Earth Sciences for Natural Resources Characterization, April 9–13, Nancy, France. *Sciences de la Terre, No. 20.*

Switzer, P. and Green, A., 1984. Min/Max autocorrelation factors for multivariate spatial imagery. Dept. of Statistics, Standford University, Tech. Report, No. 6.

Wackernagel, H., 1989. Geostatistical techniques for interpreting multivariate spatial information. In: Chung, C.F. et al. (Eds.), *Quantitative Analysis of Mineral and Energy Resources*. pp. 394–409.

Wackernagel, H., 1995. *Multivariate Geostatistics*. Springer-Verlag, Berlin.

Wloszczczowski, D., Gou, Y. and Faraj, A., 1998. 3D acquisition parameters: a cost-saving study. *Soc. Explor. Geophys., 68th Annu. Int. SEG Meet., New Orleans, September 13–18, 1998*, Expanded Abstracts, Vol. 1, 70–73, Paper ACQ.1.5.

Developments in Petroleum Science, 51
Editors: M. Nikravesh, F. Aminzadeh and L.A. Zadeh

Chapter 13

LITHO-SEISMIC DATA HANDLING FOR HYDROCARBON RESERVOIR ESTIMATE: FUZZY SYSTEM MODELING APPROACH

E.A. SHYLLON [1]

Department of Geomatics, University of Melbourne, Parkville, Victoria 3010, Australia

ABSTRACT

Fuzzy system modeling provides a strict mathematical environment in which vague conceptual phenomena can be rigorously studied. For hydrocarbon reservoir, its data model consists of parameters such as location identifier, time as well as attributes such as porosity, saturation, hydrocarbon and formation volume factor. These are usually developed from the litho-seismic data of an oilfield. In particular, Fuzzy descriptions are obtained for the main parameters that define the structure and model of the hydrocarbon formation. From these, the membership grade functions of the fuzzy subsets are determined using an interval [0, 1]. *To simplify the model formulation of the ill-defined problem of oilfield services, the results of core analysis are expressed with linguistic quantifiers such as minimum, maximum or most likely porosity, saturation hydrocarbon, etc. Our model provides a new approach for tackling the sustainable development and management of hydrocarbon reservoir on stream and enhanced recovery. In this example, the potential acreage is mapped and the reservoir estimate is obtained easily using de-fuzzifier such as mean of maxima.*

1. INTRODUCTION

Several factors are considered in determining the economical potential of a new reservoir. However, in any such exercise a major consideration involves an estimate of possible hydrocarbon reserves. Then the development or production geophysicist must answer questions such as:

- *"What is the extent and size of the new field?"*
- *"What is the optimal estimate of the field on stream?"*

For sustainable development and exploitation of hydrocarbon reservoir on stream and enhanced recovery, a fast and economical modeling approach for estimating the field characteristic is desirable, especially one that can take into account the ill-defined problem of formation estimate.

[1] E-mail: eashyllon@hotmail.com

Gathering the necessary data is a major and important step in oil reservoir characterization studies. The well log information is generally reliable for modeling because the measurements are made in the formation with a variety of tools that relate directly to the reservoir properties of interest. For obvious reasons the well to seismic calibration is full of assumptions.

This chapter presents a fuzzy system modeling approach for solving ill-defined or vague problems of formation estimate as found in the oilfield services. This process explains the data structure required for fuzzy system modeling. To begin with, this chapter examines several economical ways of analyzing non-linear dynamical nature of hydrocarbon reservoir estimate.

Section 2 discusses uncertainty in hydrocarbon reservoir estimate, which are obtained during the litho-seismic data acquisition, processing and interpretation. Section 3 treats the issue of litho-seismic data handling and restructuring by introducing the use of linguistic quantifiers. In Section 4, the chapter gives the fuzzy system modeling approach and shows how the input parameters are fuzzified. It also treats multiplication operation on the fuzzy subsets of litho-seismic data sets and explains the defuzzification of results using mean of maxima.

2. UNCERTAINTIES IN HYDROCARBON RESERVOIR ESTIMATE

Hydrocarbon reservoir is a real world system that has locations and specific attributes. There is rich interactions among the attributes which are complex and nonlinear and dynamic. A systematic analysis is presented for the understanding of uncertainties associated with hydrocarbon reservoir estimate for sustainable management.

2.1. Types of uncertainties in hydrocarbon reservoir estimate

2.1.1. Uncertainty in data acquisition

It is obvious that there is no measurement that is absolutely free of errors. This means ultimately that all measurements, no matter how precise, admit the possibility of errors. Even the most precise measurements are uncertain. In geostatical mapping for reservoir characterization, data uncertainty may arise as a result of:

- *Mistake:* This is often referred to as gross error in the interpretation of data sets
- *Systematic errors:* are often referred to as errors of known sources such as operator, instrument, weather conditions, etc. However, several techniques are supposedly being used to eliminate or minimize them.
- *Random errors:* For each measurement there is an error, which is considered as event.

In case a large number of observations, the conventional method use the theory of probability and statistics. Random sampling of reservoir parameters where there is susceptibility of oil deposits is another source of uncertainty.

2.1.2. Uncertainty in model formulation

Various modeling tools are used for the analysis of reservoir data; (for parameter estimation) some are deterministic, probabilities, etc. Parameter estimation is a measure that expresses the uncertainty regarding the actual value of variable under investigation. It is presented in terms of following 'vague' prepositions:

. . . ϕ is 15%, or
ϕ is about 15%; or
ϕ is around 15%; or
ϕ is approximately 15%; or
ϕ is mostly likely to be 15% . . .

Some other models that are uncertain include: root mean square (RMS) error, statistical mean, etc.

2.1.3. Uncertainty due to linguistic imprecision

Another source of uncertainty is in the expression of some reservoir parameters by linguistic variables, which are not taken care of in the data analysis. This leads to systematic errors. The magnitude of which cannot be estimated, but play a significant role in interpretation of results. Such variables are often used to quantity the entire process in well lithology.

For example:

- Effective porosity; 'open' and 'closed' porosity
- Total porosity, etc.
- Vertical/horizontal porosity

2.1.4. Uncertainty due to resolution limit of the equipment

Usually, it is given as part-per-million (ppm). It is user defined. This leads to systematic errors in reservoir characterization. It includes the uncertainty of the graduated scales of the unit of measurement. They are reduced to random errors by application of standardization (calibration) corrections. Such equipment includes wireline technology, sensors, etc. (Fig. 1). There are various types of well logging instruments but they are based on the same principle. They have various resolution limits. As a consequence, accurate measurements are also difficult to obtain.

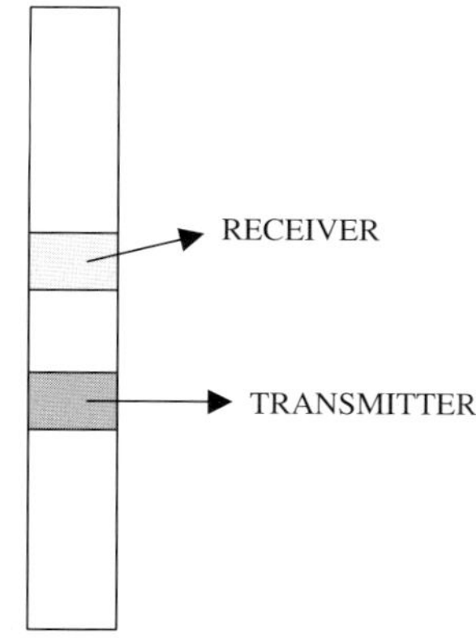

Fig. 1. Example of wireline logging instrument.

The interpretation of well logs for reservoir characterization is usually qualitative. Often pay zones are mapped and correlated from well logs. For this task, many well logging techniques are employed and enormous amounts of data are combined. Such combination includes mapping or of overlay conventional electric logs. (SP, lateral, micro-log, normal, etc.) to locate, correlate and identify formations of interest. Further combination of other logs (caliper, acoustic, etc.) provide quantitative estimate of porosity and hydrocarbon/water ratio. This is a case of decision making under uncertainty.

2.1.5. Uncertainty due to incomplete information

Many mapping applications (regional or local) that use geometrical or physical models are more or less approximation of reality. In case of hydrocarbon reservoir on stream, the production geoscientist must give the estimate of the reservoir for enhanced recovery (secondary or ultimate). The traditional modeling tools presume certainty of Litho-seismic data set used in deriving the parameters of hydrocarbon estimate.

The conclusion drawn from such analysis has limited bearing on reality, because these parameters are necessarily uncertain or vague in a number of ways, viz.: (1) They cannot describe the estimate precisely. The complete description of the estimate often would require more detailed data set than can possibly be measured, processed and interpreted accurately.

In most cases, the petrophysical properties are averaged to provide a single statistic for values of parameters such as for example porosity or other parameters for the reservoir zone. Many mapping applications assume that the mean and variance of the reservoir properties derived from the well location are representative of the entire field. Hirsche et al. (1998) discuss some of the issues of uncertainty and assumptions implicit in hydrocarbon reservoir characterization and geostatistical mapping.

Another issue is the fact that the main parameters that define the structure and model the hydrocarbon formation have rich interaction among one another. In essence, the estimate is fairly complex, nonlinear and dynamic. Empirically, the hydrocarbon pore volume or pre-drill estimate is obtained as (Archer and Wall, 1986), (Shyllon, 1993) and (Jahn et al., 1998):

$$V_{\text{in-place}} = f(\phi, \beta, S, L, B, H). \tag{1}$$

and that of the material balance is obtained as:

$$V_{\text{balance}} = V_{\text{in-place}} + V_{\text{mn}} - V_{\text{mp}} + V_{\text{mi}} \tag{2}$$

where the parameters of the estimate are: ϕ – porosity; β – inverse of formation volume factor; S – saturation hydrocarbon; H – net thickness; L – length of the oilfield; B – breadth of the oilfield; $V_{\text{in-place}}$ – volume of oil in place; V_{balance} – volume of balanced material; V_{mn} – volume of material (natural influx); V_{mp} – volume of material withdraw (on stream); V_{mi} – volume of material injected (on enhanced recovery).

The sense of uncertainty represented by fuzziness, however is the uncertainty resulting from the imprecision of the meaning of a concept expressed by a linguistic term in a natural language, such as '*about*', '*approximately*' and the like. It is obvious that the concept captured by 'approximate porosity' and 'average saturation hydrocarbon', is uncertain and fuzzy, because some numbers or values on either side of the central value are included. Usually such quantifiers represent an interval of values (Klir et al., 1997).

2.2. Magnitude of errors and uncertainty

In conventional modeling tools where probability theory and statistics are used for data analysis, the magnitude of error or uncertainty is also user defined. This is expressed as:

- Confidence level or limit: the choice of this level is optional (67–95%);
- Error ellipses (ellipsoids): these are used to express the spread of errors.

These quantifiers do not convey the reality, since the exact value of estimate is not known.

3. LITHO-SEISMIC DATA HANDLING

3.1. Seismic data

In a geophysical exploration, geological features are defined by using exploration seismology. It is a "science concerned with observing and recording the generation and propagation of elastic waves in the earth". As a result of seismic survey, maps of subsurface geological structures can be produced. Two main techniques are used (Telford et al., 1990):

- *Refraction seismology:* a seismic survey method which records the seismic waves that are refracted from bed boundaries after a seismic disturbance has been produced on the surface. It is used especially for shallow beds.
- *Reflection seismology:* is a seismic survey method which records the seismic waves that are reflected from bed boundaries after a seismic disturbance has been produced on the surface. It is used especially for deep beds.

Usually, a three-dimensional survey technique is employed for oil exploration and sustainable reservoir management. A two-way travel time from the source through the formations and back to the receivers is recorded. This is followed by data processing and interpretation. Finally, seismic section map data are produced (Fig. 2).

3.2. Well lithology

There are various types of wells. These include exploratory and development wells. An exploration well is drilled to permit more information about subsurface structure. A development well is drilled in a proven territory for the production of oil and gas.

A well logging for reservoir characterization involves measuring the physical properties of surrounding rocks with a sensor located in a borehole (Telford et al., 1990). The principal objectives of these tasks are:

- Identification of geological formations
- Identification of fluid formation in the pores
- Evaluation of the production capabilities of a reservoir formation.

Finally, map data are produced. Usually, about five well logged are combined for an

Fig. 2. Part of seismic section of the area of interest.

investigation such as: formation thickness (lithology, Fig. 3); porosity and permeability; and saturation water and hydrocarbon.

Some types of well logging that are listed in Telford et al. (1990) include:

- *Electrical resistivity logging:* determines the type of fluid, which occupies a rock's pore space. The relative saturation of oil and water can be determined using spontaneous potential (SP) logging.
- *Radioactivity logging:* involves gamma ray and density loggings. This measures the natural radiation of the rocks that have been penetrated. It can be combined simultaneously with neutron and sonic loggings to define lithology and correlate other wells.
- *Auxiliary logging* (includes sonic logging): which uses sound pulse, which is sent into the formation. The time required for the pulse to travel through the lock is measured and used to determine the porosity.

An analysis of litho-seismic data involves the determination of patterns of data associated with location and manipulation of location-related data to derive new information from existing data. Moreover, it is concerned with geo-spatial patterns defining the locations relationships (vertically or horizontally) among points, lines and surfaces; and spatial processes that define the dynamical nature of these features. Then analysis of clusters of reservoir rock types (sandstone, shale, etc.) and reservoir properties such as (porosity, saturation, hydrocarbon, etc.) involve the determination of patterns that include non-random occurrences for the preparation of facies maps. After data acquisition, processing and interpretation the results are presented as *approximately or about*. This is because such derived information cannot be characterized in terms of absolutely precise numbers. For example: "The porosity is about 0.15 or approximately 15% ", etc., or "The saturated hydrocarbon is about 75%".

3.3. Litho-seismic data restructuring

The basic data sets that contribute to the main parameters of the estimate of the hydrocarbon reservoir include:

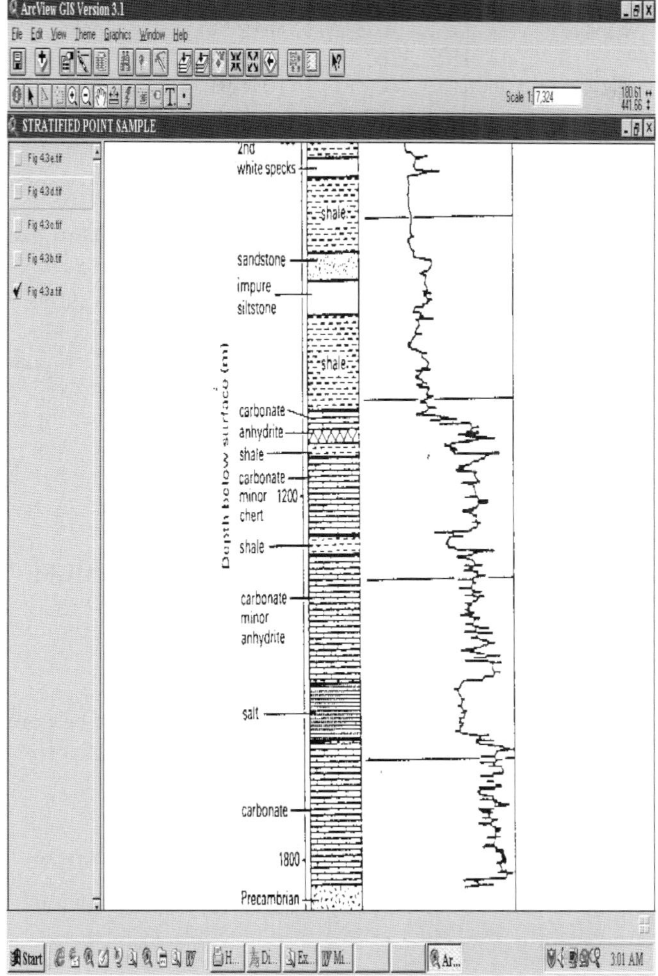

Fig. 3. Well lithology displayed in Arcview GIS.

- Core analysis at intervals of 20 m
- Seismic records with bin size of 20–25 m (Fig. 2)
- Well lithology (Fig. 3).

For the purpose of the estimate the litho-seismic data are restructured as follows:

3.3.1. Acreage

The approximate acreage is illustrated in Fig. 4. A quick browse of the seismic section gives an indication of approximate length and breadth of the oil field. The minimum, maximum and most likely values of the acreage are evaluated.

The interpreted seismic section revealed the extent of the oilfield. The length and

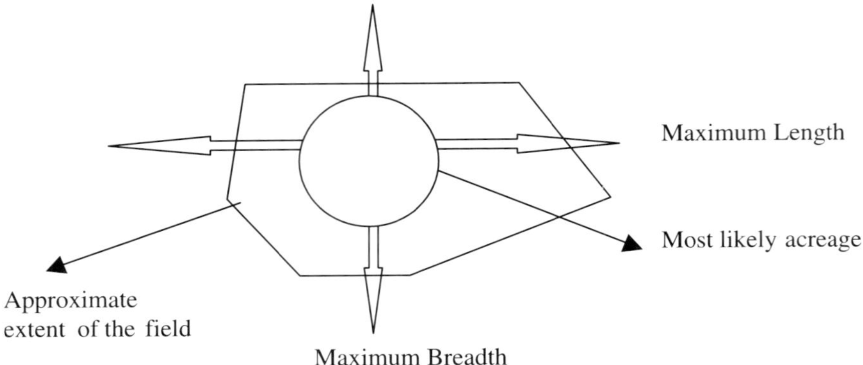

Fig. 4. Approximate acreage of the reservoir of interest.

breadth data sets are generated:

$$L = \{L_1, \ldots, L_n\}$$

then the data set is ordered by finding the MINIMUM and MAXIMUM:

$$\begin{aligned} L_{\min} &= \mathrm{Min}\{L_1, \ldots, L_n\} \\ L_{\max} &= \mathrm{Max}\{L_1, \ldots, L_n\} \\ L &= \{L_{\min}, \ldots, L_{\max}\} \end{aligned}$$

For the breadth:

$$B = \{B_1, B_2, B_3, \ldots, B_n\}$$

then the data set is ordered by finding the MINIMUM and MAXIMUM:

$$\begin{aligned} B_{\min} &= \mathrm{Min}\{B_1, B_2, B_3, \ldots, B_n\} \\ B_{\max} &= \mathrm{Max}\{B_1, B_2, B_3, \ldots, B_n\} \\ B &= \{B_{\min}, \ldots, B_{\max}\}; \end{aligned}$$

3.3.2. Most likely porosity

The porosity is obtained from the interpreted well logs, a domain of porosity data set is generated:

$$\phi = \{\phi_1, \phi_2, \phi_3, \ldots, \phi_n\}$$

then the data set is ordered by finding the MINIMUM and MAXIMUM:

$$\begin{aligned} \phi_{\min} &= \mathrm{Min}\{\phi_1, \phi_2, \phi_3, \ldots, \phi_n\} \\ \phi_{\max} &= \mathrm{Max}\{\phi_1, \phi_2, \phi_3, \ldots, \phi_n\} \\ \phi &= \{\phi_{\min}, \ldots, \phi_{\max}\} \end{aligned}$$

3.3.3. Saturation hydrocarbon

The saturation is obtained from the interpreted well logs, a domain of saturation data set is generated:

$$S = \{S_1, \ldots, S_n\}$$

then the data set is ordered by finding the MINIMUM and MAXIMUM:

$$\begin{aligned} S_{\min} &= \text{Min}\{S_1, S_2, S_3, \ldots, S_n\} \\ S_{\max} &= \text{Max}\{S_1, S_2, S_3, \ldots, S_n\} \\ S &= \{S_{\min}, \ldots, S_{\max}\} \end{aligned}$$

3.3.4. Formation volume factor

From the Laboratory test of core analysis, the formation volume factor data set is generated:

$$\beta = \{\beta_1, \beta_2, \ldots, \beta_n\}$$

then the data set is ordered by finding the MINIMUM and MAXIMUM:

$$\begin{aligned} \beta_{\min} &= \text{Min}\{\beta_1, \beta_2, \beta_3, \ldots, \beta_n\} \\ \beta_{\max} &= \text{Max}\{\beta_1, \beta_2, \beta_3, \ldots, \beta_n\} \\ \beta &= \{\beta_{\min}, \ldots, \beta_{\max}\} \end{aligned}$$

3.3.5. Net thickness

From the set of all well lithology the net thickness is given as:

$$H = \{h_1, h_2 \ldots h_n\}$$

then the data set is ordered by finding the MINIMUM and MAXIMUM:

$$\begin{aligned} H_{\min} &= \text{Min}\{H_1, H_2, H_3, \ldots, H_n\} \\ H_{\max} &= \text{Max}\{H_1, H_2, H_3, \ldots, H_n\} \\ H &= \{H_{\min}, \ldots, H_{\max}\} \end{aligned}$$

The MINIMUM and MAXIMUM values of the data sets are obtained easily using appropriate computer algorithm (see Plates 1–6).

3.4. Training data set

The main parameters that define the structure and model of the hydrocarbon formation in case of an *ideal oil sand* are considered as training data sets. These values are then the 'most likely' such as porosity = 15%, formation volume factor = 1.3 and saturation hydrocarbon $S = 70\%$. They are required for the optimal estimation, simulation and monitoring of the reservoir.

4. FUZZY SYSTEM MODELING APPROACH

Fuzzy sets are data sets with imprecise boundary. The membership in a fuzzy set is a matter of degree (Zadeh, 1997). Fuzzy system provides:
(1) meaningful and powerful representation of measurement uncertainty;
(2) meaningful modeling technique of ill-defined problems or vague concepts such as hydrocarbon reservoir estimate.

4.1. Fuzzy system

A fuzzy system is any system whose variable(s) range over states that are approximate. The fuzzy set is usually an interval of real number and the associated variables are linguistic variable such as most likely, about, etc. Due to the finite resolution of any measuring instrument, appropriate quantization, whose coarseness reflects the limited measurement resolution, is inevitable whenever a variable represents a real-world attribute.

Fuzzy system approach exploits the tolerance for uncertainty, imprecision and partial truth of various types to achieve tractability, low solution cost and robustness, and better rapport with reality (Esogbue and Kacprzk, 1996). It is concerned with the effect of applying approximate methods to imprecisely formulated problems of computational complexity. The primary aim of this approach is to develop computational methods that produce acceptable approximate solutions at low cost. A fuzzy system modeling is distinguished into the following stages (see Fig. 5).

4.1.1. Fuzzification of hydrocarbon reservoir parameters

There are several methods of constructing membership functions in the literature. Such methods include piecewise, bell-shaped, triangular-shaped function and neural-networks. To simplify the model formulation of the ill-defined problem, triangular membership grade functions are used for all the input parameters. The restructured litho-seismic data (Sections 3.1.1–3.1.5) are used to derive the 'support' of the parameters in the fuzzy system.

The α-cut is the property of a fuzzy set, which allows the representation based on specific assignments of numbers in $\alpha \in [0, 1]$ to obtain a crisp set. It is a restriction of membership degrees that are greater than or equal to some chosen value of α. When this restriction is applied to a fuzzy set, a subset is obtained (Klir et al., 1997). It follows that the α-cut of any fuzzy sets of the hydrocarbon reservoir is a range of crisp data set as follows:

4.1.1.1. Porosity. The report of analysis of a combination of well logs information from one or multiple experts, which describes the concept of porosity that is presented as "the average porosity is between 10 and 20% and varies linearly top down the formation thickness". The fuzzy membership function is generated using triangular function. The α-cut of $\phi(x)$ is derived from the triangular membership function for $\alpha \in [0, 1]$ as (see Plate 1):

$$^{\alpha}\phi = [0.14\alpha + 0.01, 0.20 - 0.05\alpha] \tag{3}$$

This α-cut ($^{\alpha}\phi$) is a closed interval and it is referred to as the *support* when $\alpha = 0$; and *core* when $\alpha = 1$.

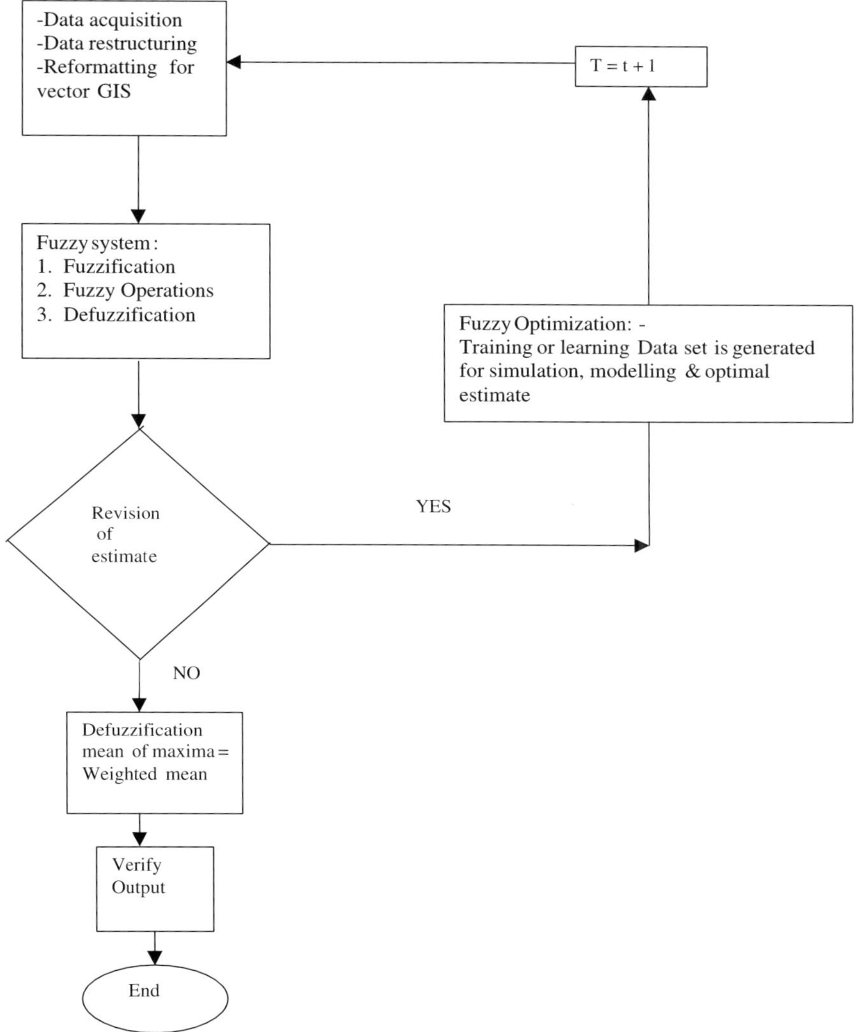

Fig. 5. Fuzzy system modeling approach for hydrocarbon reservoir estimate.

4.1.1.2. Formation volume factor (see Plate 2). The report of analysis of a core data from the wells that is presented by one (or multiple) expert(s) describes the concept of formation volume factor as a scalar. This is the ratio of the volume of hydrocarbon at reservoir condition to the same at standard condition of temperature and pressure. The fuzzy membership function is generated using triangular function. The α-cut of $\beta(x)$ is derived from the triangular membership function for $\alpha \in [0, 1]$ as (see Plate 2):

$$^{\alpha}\beta = [0.15\alpha + 0.60, 0.95 - 0.15\alpha] \tag{4}$$

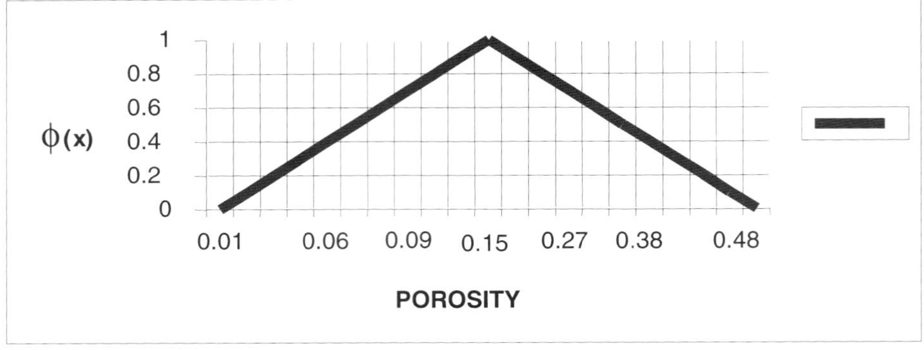

$\phi = \{\phi_1 \text{-----}, \phi_n\}$

The Ordered Set becomes:

$\phi = \{\phi_{min} \text{-----}, \phi_{max}\}$

$\phi = \{1\% \text{-----}, 48\%\}$

while effective porosity is 15%

Construct the membership function

$$\phi(x) = \begin{cases} (x-0.01)/0.14 & \text{for } 0.01 < x < 0.15 \\ (0.48-x)/0.23 & \text{for } 0.15 < x < 0.48 \\ 0 & \text{otherwise} \end{cases}$$

α -Cut of φ (x)

$^{\alpha}\phi = [0.05\alpha + 0.10, 0.20 - 0.05\alpha]$ for $\alpha \in [0,1]$

Plate 1. Creating fuzzy subset for porosity.

4.1.1.3. Saturation hydrocarbon (see Plate 3). The report of laboratory tests and log analysis from one (or multiple) expert(s) shows that saturation varies linearly throughout the net thickness. The fuzzy membership function is generated using triangular function. The α-cut of $S(x)$ is derived from the triangular membership function for $\alpha \in [0, 1]$ as (see Plate 3):

$$^{\alpha}S = [0.15\alpha + 0.40, 0.99 - 0.15\alpha] \tag{5}$$

4.1.1.4. Length of oilfield (see Plate 4). In today's 3D seismic surveying, coverage is usually 12.5 meters by 25 meters cells called stack bins. The inline spacing is sufficiently dense to allow the locations of the reflection points (emerging from between lines of subsurface structure) to be measured (Fig. 6). The average length is about 15 km. The fuzzy membership function is generated using triangular function. The α-cut of $L(x)$ is derived from the triangular membership function for $\alpha \in [0, 1]$ as (see

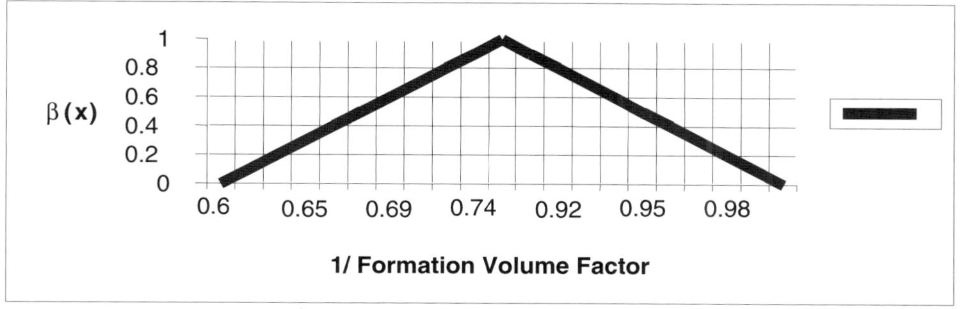

$\beta = \{ \beta_{1,\ldots\ldots}, \beta_n \}$

The Ordered Set becomes:

$\beta = \{ \beta_{min,\ldots\ldots}, \beta_{max} \}$

$\beta = \{60\% \ldots\ldots, 95\% \}$

The average formation volume factor is about 1.33, while $\beta = 1/\ 1.33 = 0.75$.

Creating of membership function

$$\beta(x) = \begin{cases} (x-0.6)/0.15 & \text{for } 0.60 < x < 0.75 \\ (0.95-x)/0.20 & \text{for } 0.75 < x < 0.95 \\ 0 & \text{otherwise} \end{cases}$$

α -Cut of $\beta(x)$

${}^{\alpha}\beta = [0.15\alpha + 0.60,\ 0.95 - 0.15\alpha]$ for $\alpha \in [0,1]$

Plate 2. Creating fuzzy subsets for formation volume factor.

Plate 4):

$$^{\alpha}L = [5\alpha + 10, 25 - 10\alpha] \tag{6}$$

4.1.1.5. Breadth of the oilfield (see Plate 5). Also, the cross-line spacing is sufficiently dense to allow the locations of the reflection points (emerging from between lines of subsurface structure) to be measured (Fig. 6). The average breadth is about 12 km. The fuzzy membership function is generated using triangular function. The α-cut of $B(x)$ is derived from the triangular membership function for $\alpha \in [0, 1]$ as (see Plate 5):

$$^{\alpha}B = [4\alpha + 8, 15 - 3\alpha] \tag{7}$$

4.1.1.6. Net thickness (see Plate 6). The report of analysis of a combination of well logs and seismic sections information from one or multiple experts is used to describe the formation thickness. The fuzzy membership function is generated using triangular function. The α-cut of $H(x)$ is derived from the triangular membership function for $\alpha \in [0, 1]$ as (see Plate 6):

$$^{\alpha}H = [200\alpha + 300, 750 - 250\alpha] \tag{8}$$

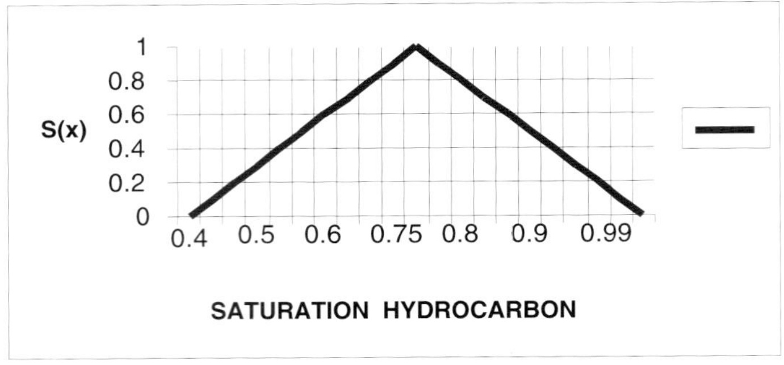

$S = \{S_1, \text{---------}, S_n\}$

The Ordered Set becomes:

$S = \{S_{min}, \text{---------}, S_{max}\}$

$S = \{40\% , \text{---------}, 99\%\}$

S is approximately 75% or more

Creating of membership function

$$S(x) = \begin{cases} (x - 0.4)/0.35 & \text{for } 0.40 < x < 0.75 \\ (0.99 - x)/0.25 & \text{for } 0.75 < x < 0.99 \\ 0 & \text{otherwise} \end{cases}$$

α -Cut of S(x)

$${}^{\alpha}S = [0.15\alpha + 0.40, 0.99 - 0.15\,\alpha] \qquad \text{for } \alpha \in [0,1]$$

Plate 3. Creating fuzzy subsets for saturation hydrocarbon.

4.1.2. Operation on fuzzy subsets

4.1.2.1. Fuzzy numbers. A fuzzy number is a fuzzy set that is defined on the set 'R' of real numbers. The membership functions of these sets have the form (Klir et al., 1997):

$$A(x) : \mathrm{R} \rightarrow [0, 1]$$

The parameters that constitute the hydrocarbon reservoir estimates are restructured. Hence, they are fuzzy numbers defined on the set 'R' of real numbers. Membership functions of these sets have the above form. They capture the concepts of approximate numbers or intervals, such as numbers that are close to a given real number. The fuzzy subsets satisfy the properties of fuzzy numbers, which are discussed below:

(1) They are normal fuzzy sets – this implies that the concept of a set of real number close 'r' is fully satisfied by 'r' itself, then the membership grade of 'r' in any fuzzy set that attempt to capture this concept is 1.

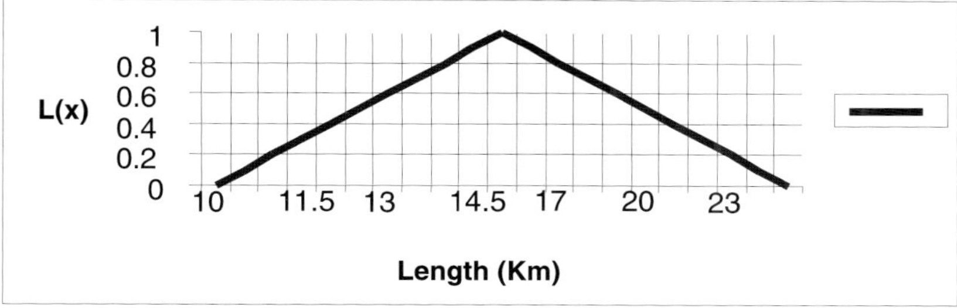

The Ordered Set becomes:

$$L = \{ L_{min}, \text{---------}, L_{max} \}$$

$$L = \{10\text{km}, \text{---------}, 25\text{km} \}$$

The average Length is a about 15 km

Creating of membership function

$$L(x) = \begin{cases} (x - 10)/5 & \text{for } 10 < x < 15 \\ (25 - x)/10 & \text{for } 15 < x < 25 \\ 0 & \text{otherwise} \end{cases}$$

α -Cut of L(x)

$${}^{\alpha}L = [5\alpha + 10, 25 - 10\alpha] \qquad \text{for } \alpha \in [0,1]$$

Plate 4. Creating fuzzy subsets for length of the oil field.

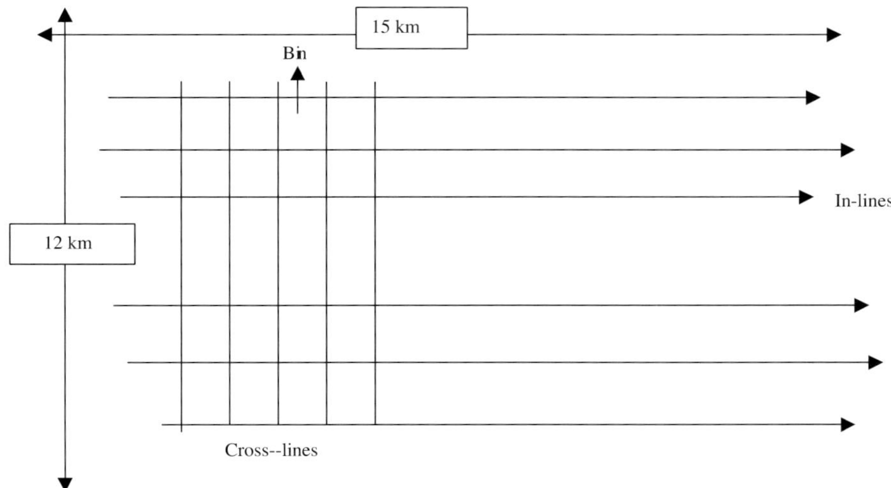

Fig. 6. Length and breadth of oilfield.

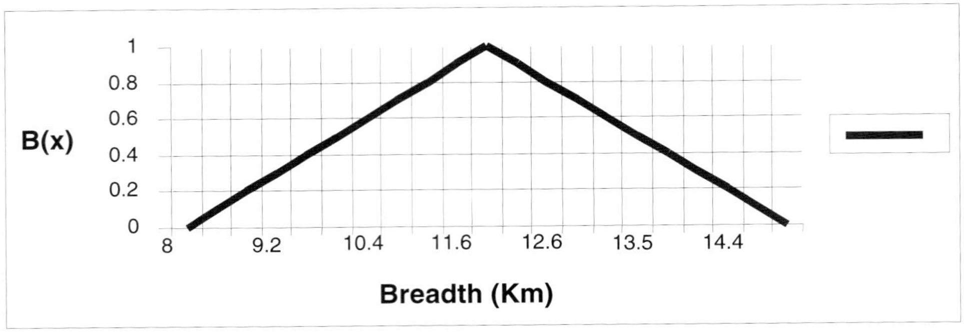

The Ordered Set becomes:

B= { B_{min} ,---------, B_{max} }

B= {8km ,---------, 15km }

The average Breadth is a about 12km

Creating of membership function

$$B(x) = \begin{cases} (x-8)/4 & \text{for } 8 \le x \le 12 \\ (15-x)/3 & \text{for } 12 \le x \le 15 \\ 0 & \text{otherwise} \end{cases}$$

α -Cut of L(x)

$${}^{\alpha}B = [4\alpha + 8,\ 15 - 3\alpha] \qquad \text{for } \alpha \in [0,1]$$

Plate 5. Creating fuzzy subsets for breadth.

(2) The α-cuts of these parameters have closed interval for every $\alpha \in [0, 1]$. This implies that they are convex fuzzy sets.

(3) The support of these parameters are bounded i.e. the bounded support of these fuzzy numbers and all their α-cuts for $\alpha \neq 0$ are closed intervals. This property allows meaningful arithmetic operations on these fuzzy numbers in terms of standard arithmetic operations on closed intervals, which is well established in classical interval mathematics.

4.1.2.2. Reservoir estimate – fuzzy approach. The parameters are expressed as fuzzy numbers. They are fully and uniquely represented by their α-cuts. Also, the α-cuts are closed intervals of real numbers for all $\alpha \in [0, 1]$. These properties enable arithmetic operations on their α-cuts (i.e. interval analysis on these closed intervals).

The reservoir estimate involves multiplication operation on the closed interval of the fuzzified parameters of hydrocarbon reservoir. The closed intervals are defined already

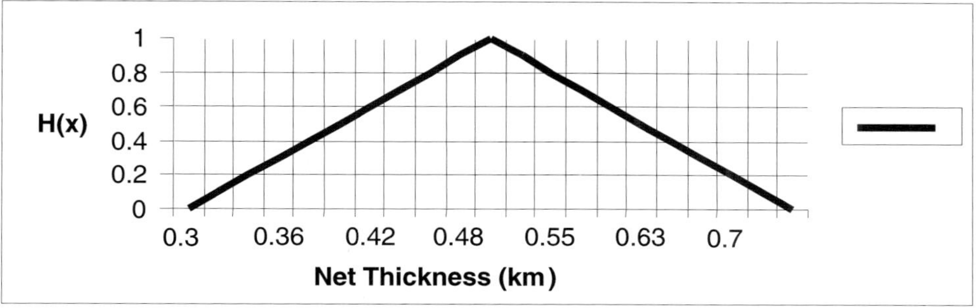

$$H = \{H_{1}, \text{---------}, H_n\}$$

The Ordered Set becomes:

$$H = \{H_{min}, \text{---------}, H_{max}\}$$

$$H = \{200m, \text{---------}, 750m\}$$

The average net thickness is a about 500m

Creating of membership function

$$H(x) = \begin{cases} (x-300)/200 & \text{for } 300 < x < 500 \\ (750-x)/250 & \text{for } 500 < x < 750 \\ 0 & \text{otherwise} \end{cases}$$

α-Cut of H(x)

$${}^{\alpha}H = [200\alpha + 300, 750 - 250\alpha] \qquad \text{for } \alpha \in [0,1]$$

Plate 6. Creating fuzzy subsets for net thickness.

by the α-cut equations (3)–(8).

Let ${}^{\alpha}A = [a,b]$ and ${}^{\alpha}B = [d,e]$

For multiplication operation:

$$\text{Then } {}^{\alpha}A * {}^{\alpha}B = [a,b] * [d,e]$$
$$= [\min(ad,ae,bd,be), \max(ad,ae,bd,be)]$$

Klir et al. (1997) states that the multiplication operations satisfy the following properties:

Let $A = [a1,a2]$, $B = [b1,b2]$ and $C = [c1,c2]$

1. $A * B = B * A$ – commutative
2. $(A * B) * C = A * (B * A)$ – associative

These properties are used to find the fuzzy subset of the estimate (Table 1):

$${}^{\alpha}V_{\text{in-place}} = {}^{\alpha}\phi * {}^{\alpha}\beta * {}^{\alpha}S * {}^{\alpha}L * {}^{\alpha}B * {}^{\alpha}H \tag{9}$$

4.1.3. Defuzzification of the result

In order to obtain the most likely value of the estimate, the fuzzy subset of the volume (in-place) is defuzzified using mean of maxima. This is a weighted mean of the

TABLE 1

Fuzzy subset of the estimate

α	$V_{\text{in-place}}$	α	$V_{\text{in-place}}$	α	$V_{\text{in-place}}$	α	$V_{\text{in-place}}$
0.000000	4.528080	0.410000	13.858428	0.820000	35.295881	0.780000	82.974045
0.010000	4.666382	0.420000	14.205181	0.830000	36.045166	0.770000	84.746046
0.020000	4.808140	0.430000	14.559061	0.840000	36.807552	0.760000	86.548596
0.030000	4.953421	0.440000	14.920182	0.850000	37.583220	0.750000	88.382104
0.040000	5.102293	0.450000	15.288661	0.860000	38.372353	0.740000	90.246982
0.050000	5.254827	0.460000	15.664617	0.870000	39.175135	0.730000	92.143647
0.060000	5.411092	0.470000	16.048169	0.880000	39.991754	0.720000	94.072519
0.070000	5.571161	0.480000	16.439438	0.890000	40.822397	0.710000	96.034022
0.080000	5.735105	0.490000	16.838546	0.900000	41.667255	0.700000	98.028585
0.090000	5.903000	0.500000	17.245617	0.910000	42.526520	0.690000	100.056641
0.100000	6.074918	0.510000	17.660776	0.920000	43.400386	0.680000	102.118627
0.110000	6.250937	0.520000	18.084151	0.930000	44.289050	0.670000	104.214983
0.120000	6.431132	0.530000	18.515868	0.940000	45.192708	0.660000	106.346155
0.130000	6.615582	0.540000	18.956057	0.950000	46.111562	0.650000	108.512592
0.140000	6.804366	0.550000	19.404850	0.960000	47.045814	0.640000	110.714747
0.150000	6.997562	0.560000	19.862379	0.970000	47.995666	0.630000	112.953079
0.160000	7.195253	0.570000	20.328779	0.980000	48.961326	0.620000	115.228049
0.170000	7.397520	0.580000	20.804184	0.990000	49.943000	0.610000	117.540124
0.180000	7.604447	0.590000	21.288731	1.000000	50.940900	0.600000	119.889775
0.190000	7.816118	0.600000	21.782561	1.000000	50.940900	0.590000	122.277476
0.200000	8.032618	0.610000	22.285812	0.990000	52.136427	0.580000	124.703707
0.210000	8.254034	0.620000	22.798626	0.980000	53.354473	0.570000	127.168952
0.220000	8.480454	0.630000	23.321147	0.970000	54.595365	0.560000	129.673700
0.230000	8.711966	0.640000	23.853520	0.960000	55.859433	0.550000	132.218442
0.240000	8.948661	0.650000	24.395891	0.950000	57.147012	0.540000	134.803677
0.250000	9.190629	0.660000	24.948409	0.940000	58.458440	0.530000	137.429906
0.260000	9.437963	0.670000	25.511222	0.930000	59.794057	0.520000	140.097635
0.270000	9.690757	0.680000	26.084483	0.920000	61.154208	0.510000	142.807375
0.280000	9.949105	0.690000	26.668345	0.910000	62.539242	0.500000	145.559642
0.290000	10.213103	0.700000	27.262961	0.900000	63.949510	0.490000	148.354956
0.300000	10.482849	0.710000	27.868490	0.890000	65.385367	0.480000	151.193841
0.310000	10.758441	0.720000	28.485087	0.880000	66.847173	0.470000	154.076827
0.320000	11.039978	0.730000	29.112914	0.870000	68.335290	0.460000	157.004448
0.330000	11.327561	0.740000	29.752132	0.860000	69.850086	0.450000	159.977243
0.340000	11.621293	0.750000	30.402904	0.850000	71.391929	0.440000	162.995756
0.350000	11.921277	0.760000	31.065394	0.840000	72.961193	0.430000	166.060535
0.360000	12.227618	0.770000	31.739770	0.830000	74.558257	0.420000	169.172132
0.370000	12.540422	0.780000	32.426199	0.820000	76.183501	0.410000	172.331107
0.380000	12.859795	0.790000	33.124852	0.810000	77.837311	0.400000	175.538022
0.390000	13.185847	0.800000	33.835900	0.800000	79.520076	0.390000	178.793444
0.400000	13.518688	0.810000	34.559518	0.790000	81.232188	0.380000	182.097947

volumes and the weighs are taken as the $\alpha \in [0,1]$:

$$V_{\text{in-place}} = \sum \left({}^{\alpha}V_{\text{in-place}} * \alpha \right) / \sum \alpha \tag{10}$$

(Table 1 expresses the fuzzy subset of the volumes):

$$V_{\text{in-place}} = 7220.439237/50.500000$$

$$V_{\text{in-place}} = 142.978995 \text{ bbl}$$

5. INTERPRETATION OF RESULT

5.1. Most likely estimate

A crisp value is obtained by defuzzifier (mean of maxima). There are several advantages of this technique over the traditional ones. The main advantage is that every data element contributing to a fuzzy set is considered as an object, and its main attribute is the degree of membership in that fuzzy set. The most likely estimate is 142.978 bbl.

5.2. Optimal estimate – good estimate

The parameters of the hydrocarbon formation in case of an ideal oil sand are used to train the model (Section 3.4). Specific values of the estimate are obtained by assigning various values of α, as $\alpha \to 1$. The strong α-cut is generated with $\alpha = 0.7$. These values are evaluated from Plates 1–3:

Strong α-cut of porosity $\phi(x)$:

$${}^{0.7+}\phi = [0.05\alpha + 0.10, 0.20 - 0.05\alpha] \qquad \text{for } \alpha \in (0.7, 1]$$

Strong α-cut of formation volume factor $\beta(x)$:

$${}^{0.7+}\beta = [0.15\alpha + 0.60, 0.95 - 0.15\alpha] \qquad \text{for } \alpha \in (0.7, 1]$$

Strong α-cut of Saturation hydrocarbon $S(x)$:

$${}^{0.7+}S = [0.15\alpha + 0.40, 0.99 - 0.15\alpha] \qquad \text{for } \alpha \in (0.7, 1]$$

$${}^{\text{optimal}}V_{\text{in-place}} = {}^{0.7+}\phi * {}^{0.7+}\beta * {}^{0.7+}S * {}^{\alpha}L * {}^{\alpha}B * {}^{\alpha}H$$

A multiplication operation is performed on the closed interval as in Section 4.1.2.2.

5.3. Very good estimate

Linguistic hedge 'very' is used to give a '*concentration*' of the semantics of the reservoir estimate. This is applied on the parameters of the model of hydrocarbon formation. Let the optimal estimate be a good estimate of the reservoir, where $\alpha = 0.7$; since the parameters of ideal oil sand are used to train the model. Then the hedge '*very*' good estimate is evaluated as follows:

For porosity $\phi(x)$, the hedge is evaluated as: $(\phi(x))^2$

For formation volume factor $\beta(x)$, the hedge is evaluated as: $(\beta(x))^2$

For saturation hydrocarbon $S(x)$, the hedge is evaluated as: $(S(x))^2$

These are implemented simply as $\alpha^2 = (0.7)^2 = 0.49$. This means that a very good estimate = 148.354956 bbl (Table 1).

5.4. Slightly good estimate

Linguistic hedge 'slightly' is used to give a '*dilution*' of the semantics of the reservoir estimate. This is applied on the parameters of the model of hydrocarbon formation. Let the optimal estimate be a good estimate of the reservoir, where $\alpha = 0.7$; since the parameters of ideal oil sand are used to train the model. Now, the hedge 'slightly' good estimate is evaluated as follows:

$$\sqrt{\alpha} = \sqrt{0.7}$$

This means that the membership grade = 0.84.

For porosity $\phi(x)$, the hedge is evaluated as: $(\phi(x)) = 0.84$

For formation volume factor $\beta(x)$, the hedge is evaluated as: $(\beta(x)) = 0.84$

For saturation hydrocarbon $S(x)$, the hedge is evaluated as: $(S(x)) = 0.84$

Then a slightly good estimate = 72.961193 bbl (Table 1)

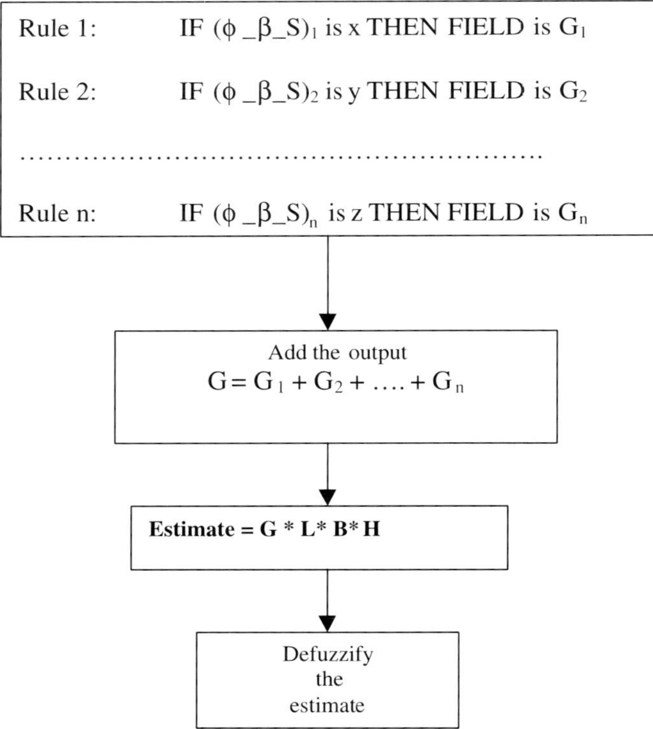

Fig. 7. Fuzzy rules based estimate – fuzzy logic.

5.5. *Rule-based estimate*

In the field of artificial intelligence, there are many ways of representing knowledge. One common way of representing knowledge is to form it into natural language expression: 'IF premise (antecedent) THEN conclusion'. This is used for the simulation and monitoring of the reservoir estimate. The knowledge is developed from the training data set. Afterwards, the rules are developed from the knowledge as in Fig. 7.

6. CONCLUSION

This approach provides meaningful and powerful representation of measurement uncertainties of litho-seismic data sets. It gives faster numerical method of estimating the material balance. It has established an economical way of analyzing hydrocarbon reservoir estimate on stream and enhanced recovery. The computational cost is low. It is implemented using computer algorithms. The potential acreage is mapped as a range of values with the weighted mean as the most likely.

Linguistic hedges are important components of the fuzzy system. They are used to model closely the semantics of the knowledge of the estimates. However, the outcome of the results depends on the type of hedge that is used (very, slightly good, etc.). It is obvious that the estimate that is obtained by using the linguistic hedge '*very good*' is close to reality. The result obtained from the fuzzy system technique is satisfactory.

7. C CODES TO COMPUTE THE ESTIMATES

```
#include  <stdio.h>
#include  <stdlib.h>
#include  <conio.h>
#include  <ctype.h>
#include  <string.h>
#include  <math.h>

#define  MAX  200
#define  TRUE   1
#define  FALSE  0

double minimum (double a,  double b,  double c,  double d) ;
double maxmum (double a,  double b,  double c,  double d) ;
void DoAlphaCut (void);

double  atab [MAX] ;
double  btab [MAX] ;
double  alphas [MAX] ;
double  step=0.01,   low=0.0,   high=1.0 ;

void main (void)
{
     char choice;

     while (TRUE)
```

```
    {
       do
        {

                       clrscr() ;
                       printf(" [S]  et Parameters\n") ;
                        printf (" [V] iew current Parameter Settings\n") ;
               printf(" [D] o Alpha Cut\n") ;
                    printf(" [E] xit\n") ;
               printf("?  :  ") ;
               choice = toupper(getch() ) ;
       }
       while( !strchr("SVDE",  choice)) ;
                    switch(choice)
       {
           case  'S' :
                       clrscr() ;
                       printf("Present Step Size = %f\n", step) ;
                  printf("Enter New Value  :  ") ;
                  scanf("%1f",  &step) ;
                          printf("Present Lower Bound = %f\n",  low) ;
                  printf("Enter New Value  :  ") ;
                  scanf("%1f",  &low) ;
                          printf("Present Lower Bound = %f\n",  high) ;
                   printf("Enter New Value  :  ") ;
                  scanf("%1f",  &high) ;
                  break;
               case  'V'  :
                       clrscr() ;
                  printf ("current Parameter Settings\n") ;
                   printf ("====================\n") ;
                  printf("Step Size           :  %f\n",  step) ;
                  printf("Lower Bound  :  %f\n",  low) ;
                  printf("Upper Bound  :  %f\n",  high) ;
                  break;
               case  'D' :
                  DoAlphaCut();
                  break;
                   case  'E' :
                       clrscr() ;
                  exit(0) ;
       }

      }

  }

void DoAlphaCut(void)
{

        double alpha,  a,  b,  d,  e;
        double  a1,  a2,  a3,  a4,  a5;
        double  b1,  b2,  b3,  b4,  b5;
        double  ad,  ae,  bd,  be;
  double sAlpha, sumfz, defz;
  FILE *fp;
  char filename[80] ;
          int  i,  j,  n;
```

```
  sAlpha  =  sumfz  =  0.0;
  printf(computing...\n") ;
        for(alpha=low, j=0;  alpha<high+step;  alpha+=step, j++)
        {
                a  =  5.0*alpha  +  10.0;
                b  =  25.0  -  10.0*alpha;
                d  =  4.0*alpha  +  8.0;
                e  =  15.0  -  3.0*alpha;

                a2  =  200.0*alpha  +  300.0;
                b2  =  750.0  -  250.0*alpha;
                a3  =  0.9*alpha  +  0.01;
                b3  =  0.55  -  0.35*alpha;
                a4  =  0.2*alpha  +  0.4;
                b4  =  1.0  -  0.1*alpha;
a5  =  0.3*alpha  +  0.5;
b5  =  1.1  -  0.3*alpha;

                for(i=1;  i<=4;  i++)
                {

                        ad  =  a*d;
                        ae  =  a*e;
                        bd  =  b*d;
                        be  =  b*e;
                        a1  =  minimum(ad,  ae,  bd,  be);
                        b1  =  maxmum(ad,  ae,  bd,  be);

                        a  =  a1;
                        b  =  b1;
                        switch(i)
                        {

                                case 1:
                                        d  =  a2;
                                        e  =  b2;
                                        break;
                                case 2:
                                        d  =  a3;
                                        e  =  b3;
                                        break;
                                case 3:
                                        d  =  a4;
                                        e  =  b4;
                                        break;
                                case 4:
                                        d  =  a5;
                                        e  =  b5;
                                        break;
                        }
                        ad  =  a*d;
                        ae  =  a*e;
                        bd  =  b*d;
                        be  =  b*e;
                        a1  =  minimum(ad,  ae,  bd,  be);
                        b1  =  maxmum(ad,  ae,  bd,  be);

        sAlpha  +=  alpha;
        sumfz   +=  alpha*(a1+b1) ;
```

```
        atab[j]  =  a1;  btab[j]  =  b1;  alphas[j]  alpha;
            }
printf("done...\n") ;
defz  =  sumfz/sAlpha;
printf("Alpha,  Volumes\n") ;
for(alpha=low, i=0;  alpha<high+step;  alpha+=step, i++)
{
        printf("%f,  %f\n",  alphas[i],  atab[i] );
}
getch();
n=i;
        for(alpha=high, i=n-1;  alpha>0.0;  alpha-=step, i--)
{
        printf("%f,  %f\n",  alphas[i],  btab[i] );
}
printf("sumfz = %f,  sAlpha = %f,  defz = %f,  defz *6.289 = %f \n",
       sumfz, sAlpha, defz, defz*6.289);
printf("Save To File  (Y/N)  ?\n") ;
if(toupper(getch() )=='Y')
{
        printf("filename  :  ") ;
       scanf("%s",  filename) ;
                fp = fopen(filename,  "wt") ;
                if(ferror(fp) )
                {
                perror(filename) ;  getch() ;
             exit(1) ;
             }
        fprintf(fp,  "Alpha,  Volumes\n") ;
        for(alpha=low, i=0;  alpha<high+step;  alpha+=step, i++)
        {
                fprintf(fp,  "%f,  %f\n",  alphas[i],  atab[i] );
             }
       for(alpha=high, i=n-1;  alpha>0.0;  alpha-=step, i--)
      {
                fprintf(fp,  "%f,  %f\n",  alphas[i],  btab[i] );
        }
      fprintf(fp,  "sumfz = %f,  sAlpha = %f,  defz = %f,  defz *6.289 = %f \n",
              sumfz,
      sAlpha,  defz,  defz*6.289) ;
                        fclose(fp) ;
       }
}

double minimum(double a,  double b,  double c,  double d)
{
                double minVal;

                minVal  =  a;
        if(b<minVal)   minVal  =  b;
        if(c<minVal)   minVal  =  c;
        if(d<minVal)   minVal  =  d;

                return minVal;
}

double maximum(double a,  double b,  double c,  double d)
{
                double maxVal;
```

```
          maxVal   =  a;
    if(b<maxVal)    maxVal  =  b;
    if(c<maxVal)    maxVal  =  c;
    if(d<maxVal)    maxVal  =  d;

          return maxVal;
}
```

REFERENCES

Archer, J.S. and Wall, C.G., 1986. *Petroleum Engineering Principles and Practice.* Graham and Totman, pp. 360–362.

Esogbue, A.O. and Kacprzk, J., 1996. Fuzzy dynamic programming: a survey of main development and applications. *Arch. Control Sci.*, Vol. 5(XLI, 1–2): 39–59.

Hirsche, K., Boerner, S., Kalkomey, C. and Gastaldi, C., 1998. Avoiding Pitfalls in Geostatistical Reservoir Characterization. *The Leading Edge*, 17(4): 493–504.

Jahn, F., Cook, M. and Graham, M., 1998. *Hydrocarbon Exploration and Production.* Elsevier Science, Amsterdam, pp. 153–171.

Klir, G.J., Ute St. Clair and Bo Yuan, 1997. *Fuzzy Set Theory: Foundations and Applications*, Prentice Hall.

Shyllon, E.A., 1993. *Mapping Information System: Spatial Data Management for Offshore Oilfields Services.* M.Phil. Thesis (unpublished), pp. 107–120.

Telford, W.M., Geldart, L.P. and Sheriff, R.E., 1990. *Applied Geophysics, 2nd edition.* Cambridge University Press.

Zadeh, L.A., 1997. Toward a theory of fuzzy information granulation and its centrality in human reasoning and fuzzy logic. *Fuzzy Sets and Systems, Int. J. Soft Comput. Intelligence*, 90(2): 111–127.

Developments in Petroleum Science, 51
Editors: M. Nikravesh, F. Aminzadeh and L.A. Zadeh

Chapter 14

NEURAL VECTOR QUANTIZATION FOR GEOBODY DETECTION AND STATIC MULTIVARIATE UPSCALING [1]

A. CHAWATHÉ [a,2] and M. YE [b]

[a] *New Mexico Petroleum Recovery Research Center*
[b] *Equator Technologies Inv.*

ABSTRACT

Stochastic reservoir models generate large grids, sometimes in excess of millions of cells. These models need to be upscaled to reduce the number of grid blocks for practical flow simulations. As a consequence, the volumes of the upscaled simulation grids are represented by corresponding coarser values. This may result in a detrimental correlation structure and variance which may be different from that obtained from the fine grid. Various upscaling algorithms have been proposed, ranging from simple averaging to conservation of fluid flux, and more recently, wavelet transforms. These algorithms are principally focused towards upscaling tensorial reservoir properties such as permeability. Scalar properties, such as porosity, are generally volume-averaged. This approach of alienating porosity from permeability for upscaling purposes questions the geological underpinnings of the depositional/diagenetic model being upscaled. In this paper, we propose a completely new perspective to address the issue of upscaling: neural vector quantization. This new approach offers opportunities such as multivariate upscaling, where many reservoir properties can be upscaled simultaneously.

Neural vector quantization (NVQ) is used to achieve data compression in vector spaces. In NVQ upscaling, the components of the input vector are the different reservoir properties (permeability, porosity, water saturation) to be upscaled simultaneously. The purpose of vector quantization is to categorize a given set, or a distribution of input vectors, into several clusters called Voronoi tessellations. Input vectors in the same Voronoi tessellation are considered to be similar and fall into the same cluster. The vector corresponding to the centroid of the tessellation is the globally averaged value of all the input vectors in that Voronoi tessellation. This vector is called the reference vector. The reference vector yields the upscaled values of the physical properties of interest. In this study, we achieve vector quantization using a new self-organizing, competitive-clustering, adaptive-structure neural network algorithm. Since we have specifically designed this algorithm for petroleum engineering applications, it performs clustering on highly-disjointed data sets. We call it the Heterogeneous Space Classifier (HSC) algorithm.

---

[1] Originally presented as SPE 37989 at the 1997 SPE Reservoir Simulation Symposium, held June 8–11, in Dallas, TX.
[2] Now at ChevronTexaco Exploration and Production Technology Co., San Ramon, CA.

We tested HSC on a simple $25 \times 25 \times 4$ *three-phase, black-oil model. The model contained one injector and one producer. We compared the reservoir performance for the upscaled and fine grids by evaluating the average reservoir pressure, cumulative production, and fluid flow rates for 10 years. We upscaled the model to a* $10 \times 10 \times 4$ *grid without compromising the flow behavior of the system.*

1. INTRODUCTION

Constrained models have dominated recent efforts in reservoir modeling (Deutsch et al., 1996; Doyen et al., 1996; Behrens et al., 1996. Some of the initial studies concentrated on automatic history-matching algorithms, which result in pseudoization of reservoir properties, especially the relative permeabilities. Such methods were severely criticized by petrophysicists and geologists since the final 'pseudo' property did not honor the experimentally measured value. In the past decade, the emphasis shifted to building spatially descriptive models by honoring the point-data as well as the inherent correlation structures observed between the point-located data. These geostatistical techniques, constrained to 3D seismic or well-test data, can yield geologically consistent models (Johann et al., 1996; Reynolds et al., 1997). Some of these models have fine geological descriptions built into the model through iterative interaction between the geologist, the geophysicist, and the reservoir engineer. Nowadays, models containing geological entities, such as braided channels and alluvial fans, are commonly encountered in reservoir simulation literature (Knox and Barton, 1997; Schatzinger and Tomutsa, 1997). These beautifully detailed models capture the depositional and diagenetic nuances of the reservoir, but from a simulation standpoint, they demand a prohibitive number of simulation cells. Constrained, stochastic models can frequently exceed a million cells.

Although impressive, current computing technology limits us from simulating multi-million-cell models. In fact, the routine demands simulation of grids that are one order lower in magnitude (Christie, 1996). This requires a translation of the detailed grids to a coarser, albeit a more manageable, level without sacrificing the anticipated reservoir performance. This translation is commonly known as upscaling.

While it may not be obvious, geostatistics (or statistics) and upscaling are fundamentally similar. In statistics, 'upscaling' is performed when a set of data is described by its 'representative' value – the mean or variance. In geostatistics, the semi variogram takes over the role of the 'representative' value by describing the spatial correlation structure.

Two approaches are associated with geostatistical modeling of reservoirs. The common approach is cell-based, where reservoir property realizations are created from the probability density functions of the observed data. Properly implemented, this approach generates stochastically stationary property fields. A drastic change of property, e.g, transition between channel and non-channel sands, during sampling may result in violation of the stationarity principle. In such cases, a realization is created that honors the local statistical parameters associated with the disjoint field. This realization is then treated independently from the rest of the field, only to be recombined with the whole in the final analysis (Seifert et al., 1997).

Most current techniques strive to upscale fields generated by the cell-based geostatistical modeling approach. Scalar properties such as porosities are arithmetically averaged using volume or pore volume-weighted averages. A significant upscaling effort is directed towards the coarsening of tensorial properties, specifically permeability. The simplest tensorial upscaling technique – the pressure-solver – has been applied to upscale single-phase flow systems (Begg et al., 1989). This direct approach requires specific boundary conditions for which effective permeabilities are calculated that honor the line-scale grid performance. These effective permeabilities result in a diagonal tensor that can be directly incorporated into a reservoir simulator. Full-tensor representations have been tried but their application in conventional simulators is limited (King, 1993; Durlofsky et al., 1994). Other techniques such as renormalization (King, 1989) offer faster solutions at the expense of accuracy. Renormalization is a two-step process that initially involves the calculation of effective permeabilities of a small group of finely gridded cells. The single-valued effective permeability is then replaced as a representative value for those finely gridded cells. This process is repeated over the entire line grid. Peaceman recently relaxed the restrictive renormalization prerequisites that the grids be isotropic and rectangular (Peaceman, 1996).

Durlofsky et al., 1995 emphasizes the need for flow simulations to be an intrinsic part of upscaling. In his studies, he uses the homogenization theory, which provides a rigorous mathematical basis for upscaling. In the homogenization theory, the property to be scaled must have a correlation length much less than the typical reservoir length scale (well-spacing). The fine-scale description is coarsened by solving the partial differential equations, provided by the homogenization theory, that replace the full-permeability tensor by its effective equivalent. In case the correlation length assumption is violated by a specific lithology, e.g., a high permeability streak, the fine-scale description is maintained. The flow-dominating lithologies are identified by fast single-phase simulations.

The wavelet transform (WT) has very recently pervaded the field ofupscaling (Panda et al., 1996; Chu et al., 1996). The properties of WTs that make them so attractive for heterogeneous space upscaling is their multiresolution framework and compact support. The multiresolution framework allows us to upscale properties varying at various scales, whereas the compact support property localizes the effect of the transform. The wavelet transform may be looked upon as the Fourier transform in heterogeneous space. The Fourier transform is characterized by its orthogonal basis functions – the sine and cosine. The equivalent basis functions for WT are the scaling functions and the wavelets. Initial studies related to upscaling, using WT, indicate promising results from an overall flow behavior perspective, as well as the preservation of localized heterogeneity.

The French investigators, Guérillot and Verdière (1997) have proposed an interesting alternative to conventional *a priori* upscaling. They contend that upscaling should be a dynamic process, which is dictated by the mechanics of the fluid flow in the simulation model. Guérillot and Verdière elegantly exploit the characteristics of the IMPES (IMplicit Pressure and Explicit Saturation) solver used in some reservoir simulators. The IMPES solver is computationally intensive when solving for the non linear pressure solution. The saturation determination is fast since its solution is explicit. Additionally, for most reservoir simulation problems, the pressure-solution surface is generally smooth. The saturation front, on the other hand, may have discontinuities, and hence requires fine-resolution grids for stable approximations. Guërillot and Verdière, thus, propose solving the simulation problem with a multi-grid approach. The pressure solution is obtained from a coarse, upscaled grid, which now results in a computationally efficient solution, The saturation profiles through the reservoir are solved on a finer grid.

All the upscaling procedures discussed above are well-suited for the cell-based gridding model. Cell-based modeling can effectively describe reasonably homogenous or gradually varying lithologies such as alluvial fans. Deutsch has recently addressed the object-based modeling strategy for discrete lithologies (Deutsch and Meehan, 1996; Deutsch and Wang, 1996). Rahon et al. have also pin-pointed the necessity of honoring lithological entities for effective simulation (Rahon et al., 1996). In addition to providing a new perspective to reservoir modeling, the object-based modeling approach may gracefully circumvent the stationarity issue.

The new HSC algorithm is designed for object-based upscaling. Since HSC is a NVQ-type clustering algorithm, it delineates various lithological entities (geobodies) from each other and assigns them upscaled values, which may be directly substituted in a reservoir simulator. Permeabilities and porosities are not separated but treated as geobody identifiers. The algorithm works well in highly heterogenous vector spaces, e.g., braided or meandering channels. The level of grid resolution is theoretically dependent on the smallest distinct geobody encountered in the geostatistical grid.

2. CONCEPTS IN NEURAL VECTOR QUANTIZATION

Data compression, for a given fidelity requirement, may be looked upon as upscaling of reservoir properties for a certain anticipated reservoir simulation performance. Henceforth, we will use data compression to mean upscaling.

Realistic problems in various fields such as petroleum engineering, geophysics, satellite imaging, CT/NMR imaging, contain large amounts of data that constitute a N-dimensional real-valued vector space. Often, so that researchers can synthesize and understand the information contained in the data, the vector space needs to be partitioned into its corresponding feature space. This partitioning is achieved through several representative subsets or clusters. When the transformation from data space to feature space occurs automatically, it is called self organization (SO) (Chawathé, 1994). One classic, naturally occurring example of SO is human vision. The visual input can contain various details of an image, which are filtered by the cerebral cortex so that only the most important features are stored. These features usually form a good representation of the original visual input. Some times, the visual input may contain highly disjointed and fractured subsets, which result in a heterogenous vector space.

The purpose of vector quantization is to categorize a distribution of vectors (fractured subsets) into several clusters called Voronoi tessellations. Input vectors in the same Voronoi tessellation are considered to represent similar properties. The centroid of the tessellation is the globally averaged value of all the input vectors in that tessellation. The vector representing this centroid is called the reference vector. The reference vector forms the condensed representation of the cluster because it characterizes the minimum

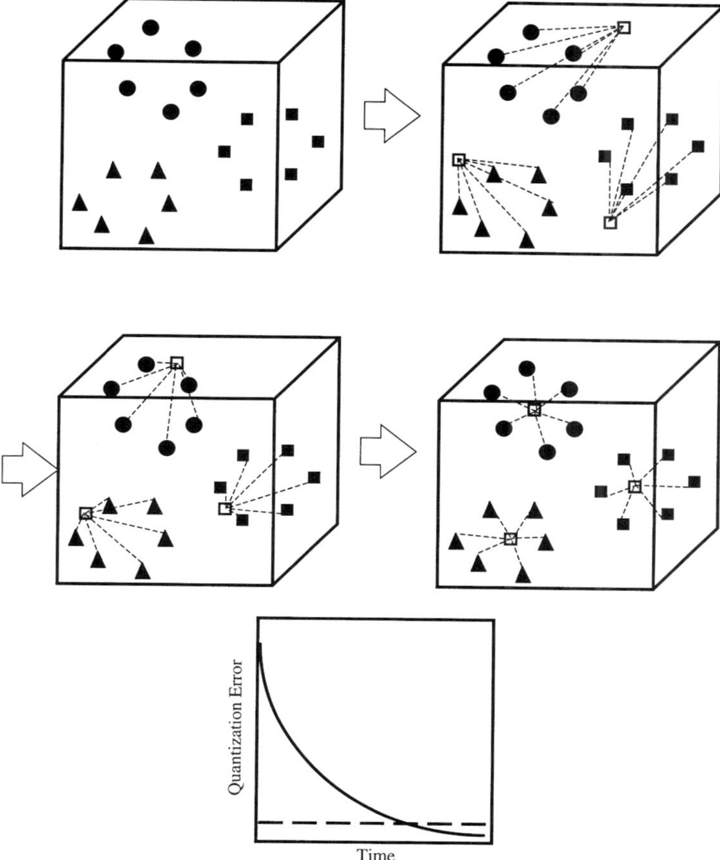

Fig. 1. The process of vector quantization. The figure shows heterogeneous data in the input space. The randomly initialized vectors (squares) converge asymtotically to form reference vectors. The reference vectors characterize the minimum quantization error and may be used as characteristic values for static upscaling.

error of quantization. From an upscaling perspective, the reference vector yields the upscaled values to be used in the simulation. Fig. 1 depicts quantization in a 3D input space.

The traditional method for vector quantization was proposed by Teuvo Kohonen in 1982 and is called the Kohonen feature map (KFM) (Kohonen, 1988). To clarify some of the notation used in this paper, we will discuss the basic structure and operation of the conceptually simple KFM. A typical KFM architecture is shown Fig. 2. There are n continuous-valued inputs, x_i to x_n defining a point X in an N-dimensional space. The output neurons represented by circles are arranged in a rectangular lattice, and are fully connected via connection weights, w_{if} to the inputs. The connection weights for a neuron form vector, W_i, called the synaptic vector. Inputs are presented in a random

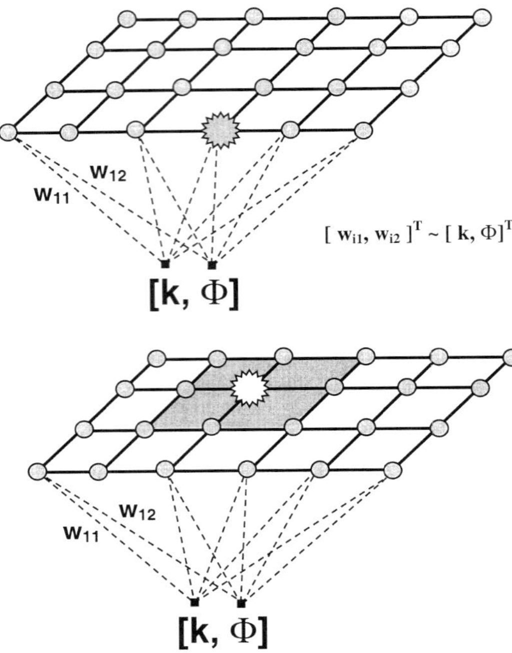

Fig. 2. The Kohonen Feature Map Architecture. Neurons are laid out on a fixed lattice and are fixed in number throughout the training phase. A neuron wins when the input vector is similar to the neuron's weight vector in the Euclidean distance sense. Here, the winning neuron is shown larger than the other neurons. The lower figure shows the development of a cluster that represents a geobody in the input space.

order to ensure correct winning probabilities of all neurons in the network. A neuron wins when its corresponding weight vector is the most similar to the input vector. The similarity association is based upon the Euclidean norm estimation between all the synaptic vectors and the input vector presented at that instant. Presenting all the inputs to the network constitute a complete unsupervised training cycle. As the training progresses, various clusters (Voronoi tessellations) are formed, which transform the data space in the feature space (self organization). The synaptic vectors evolve and converge to the centroid of the Voronoi tessellations and become reference vectors, which comprise the asymptotical values of the synaptic vectors. For a uniformly distributed input space, the KFM yields good results. However, for heterogenous input space which is typically composed of disjunct and fractured data items, KFM performs poorly due to its fixed-structure constraint.

Recently, adaptive-structure neural network models have been proposed by several researchers. Fritzke proposed the 'growing cell structures' (GCS) network and Lee and Peterson suggested the 'space partition network' (SPAN) (Fritzke, 1991; Lee and Peterson, 1990). Their work demonstrated that the adaptive-structure networks are more efficient in transforming the data space into feature space. The adaptive-structure networks can dynamically insert or delete neurons and hence are not constrained by fixed-structure limitations. These structures can also change the neighborhood relations between the neurons during the training process according to the detected complexity of the input space and some prescribed fidelity requirement. The GCS and SPAN are, however, limited in certain aspects. GCS has some flaws in its neuron insertion/deletion criteria, and SPAN operates inefficiently in heterogeneous vector spaces. Both the GCS and SPAN were primarily designed for image analysis.

Data compression (upscaling) always leads to a loss of fidelity (simulation performance). For image analysis, the fundamental objective of data compression is to maximize compression for a given fidelity. Image analysis generally does not involve gaining information about the relative positions of the Voronoi tessellations. Some tessellations may be grouped together and form a region, which is separated from the regions formed by other tessellations in the feature space. The information about the relative locations of each tessellation helps us recognize significant changes in physical properties represented by the input vectors, and is important for reservoir simulation applications.

In this study, we have developed a self-organizing, competitive-clustering, adaptive-structure network to perform vector quantization for heterogeneous vector spaces. The model is tailored to satisfy specific requirements in petroleum engineering and is called the 'heterogenous space classifier' (HSC).

2.1. The HSC algorithm

In this section, we discuss the development of the HSC algorithm. The network is designed for competitive clustering in heterogenous vector spaces. The network is essentially a hybrid of the GCS and SPAN networks.

2.2. Cluster delineation

The network initially consists of one neuron and grows as self-organization proceeds. The inputs are randomly presented to ensure a correct approximation of the winning probability, P_i, of each neuron i. Every time a neuron wins, the Euclidean distance between the winner and the current input is checked. If this distance is greater than some predefined threshold, a new network cluster is created. The cluster contains the new neuron whose synaptic vector is initialized to the current input. The predefined threshold is the smallest distance between the conceptual data clusters, (data clusters that are sufficiently far from each other in the input space). This implies that the data items belonging to different clusters represent significantly different physical properties.

2.3. Neuron accumulation/merging

If the Euclidean distance between the winner and the input is less than the predefined threshold then a new neuron will be created in the same cluster if

$$d_i P_i < \frac{\varepsilon_d}{N}$$

where, d_i is the expectation of the quantization error in the Voronoi tessellation represented by neuron i, ε_d is the allowable system quantization error, and N is the number

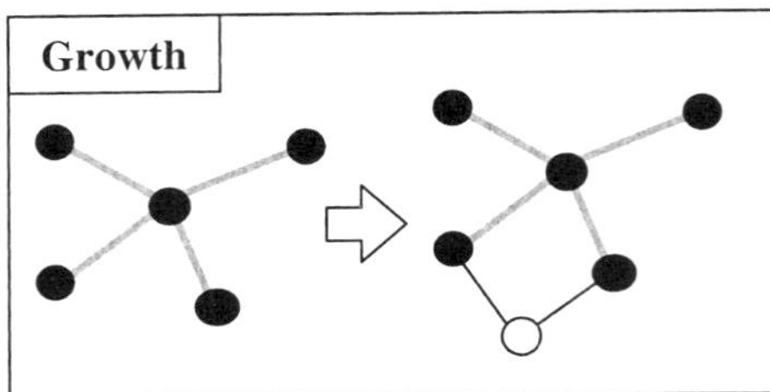

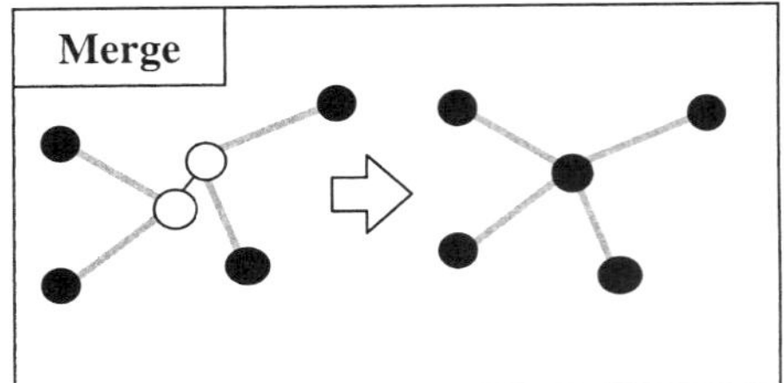

Fig. 3. Neuron growing and merging process. In the growth phase, a neuron is added to honor the probability distribution and to reduce the quantization error. If two neurons are created in immediate proximity, they are merged to form new neurons. New clusters may result when very long connections are deleted.

of neurons in the network. The new neuron will be placed in the same cluster as the winner's and its synaptic vector will also be initialized to be the current input. The new neuron will be a neighbor of the winner and will inherit those neighbors of the winner that are in the proximity of the new neuron. The new neuron will share some of the burden of its parent neuron by sharing some of its quantization errors. A complete presentation of all the inputs constitutes a training cycle. After each training cycle, the neurons are checked individually. The neurons that did not win during the training cycle are considered to be inactive and will be deleted. After several training cycles needed for the full development of the neighborhood adaptation, the neighbor connections are checked. Long neighbor connections are deleted and neurons having sufficiently short neighbor connections are merged to be one neuron whose synaptic vector, for simplicity, is initialized to be the arithmetic average of the merged neurons' synaptic vector. The updating of the neighborhood connections helps spread the neurons evenly inside each cluster.

Fig. 3 illustrates the neuron-growing process and the neuron-merging process. The new neuron (open circles) becomes a neighbor of the winner and inherits those neighbors of the winner that are in the proximity of the new neuron. Neurons having sufficiently short neighborhood connections are merged into one neuron. Fig. 4 displays snapshots of the cluster identification process. The network is initialized to contain only one neuron and new neurons are inserted during the training process. Similar neurons fall into the same cluster. Each network cluster falls into one of the data regions represented by the geometrical shapes. The geometrical shapes exemplify discrete lithological entities.

Compared to GCS, which uses connected cells of triangles and SPAN, which uses lattice network structure, the new model has no specific requirements on the initial shape of the network structure. Shaped networks structures are not necessary for topology preservation of the input space and the preassigned shape will be destroyed whenever neurons are inserted and/or deleted. There is no efficient method to keep the shaped structure consistent and at the same time preserve reasonable neighborhood relations.

In the new model, it is necessary to initialize the synaptic vector of the neuron in a new cluster to be exactly the same as the current input vector. This initialization method ensures that the new neuron will be located at the correct position in the feature space and will never fall into an area of zero-probability distribution. This is especially important for reservoir simulation. The vector space formed by the reservoir property

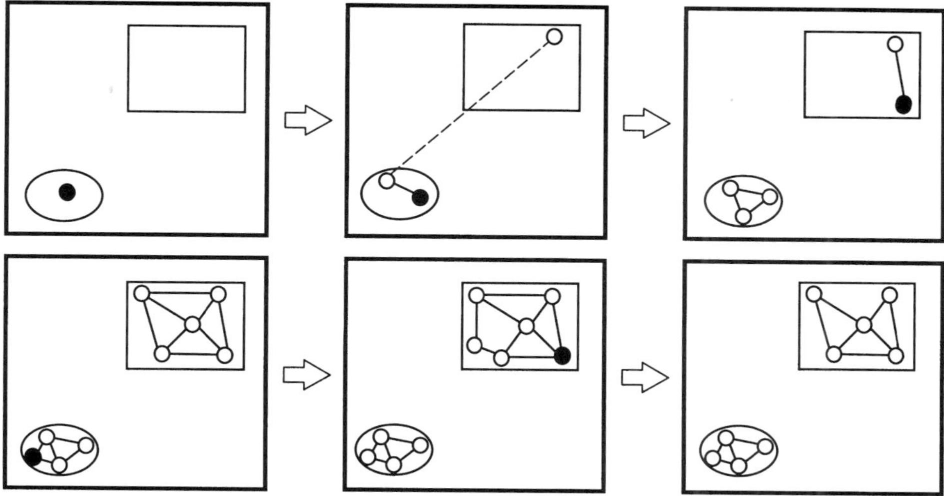

Fig. 4. Snapshots of the neighborhood adaptation and cluster identification. Neurons are created, long connections broken to create clusters, and short connections are merged reducing the overall number of neurons. This process, known as neighborhood adaptation also ensures an even distribution of the neurons.

data usually contains a fractured cluster that may be distinct from the other clusters and has a low probability density, e.g., a fault. If the new neuron is not initialized to fall into the data cluster of low probability density, very few neurons, if any, will eventually fall into that data cluster. This is because of very low probability of synaptic vector updating towards that data cluster.

HSC ensures that data items falling into the same cluster are similar and data items falling into distinct clusters are dissimilar. This is especially true for vector spaces consisting of disjunct and far-separated data clusters. Although HSC is primarily designed for heterogeneous vector spaces, initial studies indicate that it also excels in homogeneous data spaces as long as the inputs are presented randomly.

3. HSC PERFORMANCE

We will give two examples to demonstrate the performance of the HSC. In the first simple example, we will prove the superiority of the HSC for vector quantization over the standard KFM. In the second example we will demonstrate the application of HSC to upscaling.

3.1. Application 1

Fig. 5 shows the distribution of artificially generated data items in input space. There are a total of 108 data items, depicted as dark circles, distributed among four distinct data clusters. The probability distribution of the input vectors (data items) is

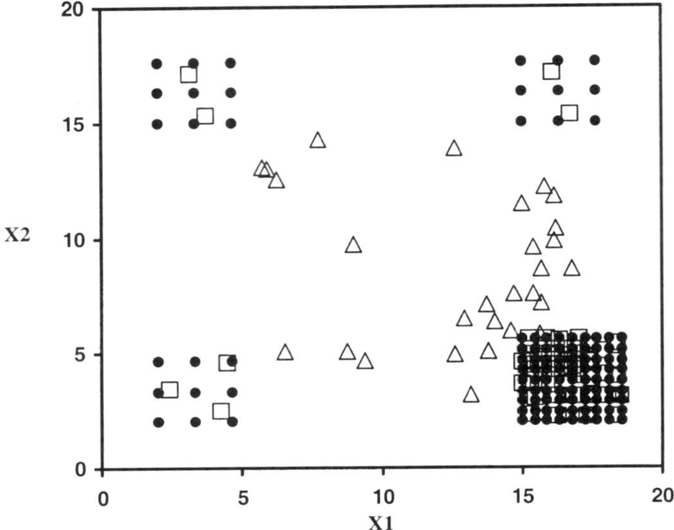

Fig. 5. Data distribution in a heterogeneous input space (solid circles). This figure also shows the results of KFM after 10 training cycles. Dead neurons (open triangles) are in the zero-probability region and active neurons (open squares) are engendered in the input space. The dead neurons result in poor convergence characteristics of the KFM.

uniform within each data cluster and zero elsewhere. Three of the data clusters have the same probability density, which is one-ninth of the remaining high probability density data cluster. Since the KFM performs well only in homogeneous vector spaces, its performance is expected to deteriorate in such a data distribution.

Fig. 5 also shows the neuron positions in the input space after 10 training cycles for KFM using an 8×8 neuronal lattice. Fig. 6 shows neuron positions in the input space for the HSC model after 10 training cycles. Note that KFM clusters poorly with 64 neurons, whereas the HSC model does better cluster delineation with fewer (57) neurons. In Fig. 5, the dark circles represent the data items and the squares represent the synaptic vectors of the active neurons. The triangles represent dead neurons engendered by KFM during training. It is impossible to create dead neurons with the HSC algorithm. The averaged quantization errors at the end of training are 0.578 and 0.2965 for the KFM model and HSC, respectively.

The synaptic vectors of the active neurons are initialized to be randomly selected input vectors. In KFM, some neurons fall into the zero-probability area after very few training cycles and thus become dead neurons. As shown in Fig. 5, 29 out of 64 neurons are dead. These dead neurons are dislocated from the input data distribution and thus will never win. As long as the number of neurons and the network frame are fixed, i.e., a static structure, the quantization error cannot reduce no matter the how many training cycles are added. This is a significant drawback of fixed-structure models.

As seen in Fig. 5, neurons are dragged towards the area of highest probability density. Because of this many dead neurons result and very few neurons are left in the areas

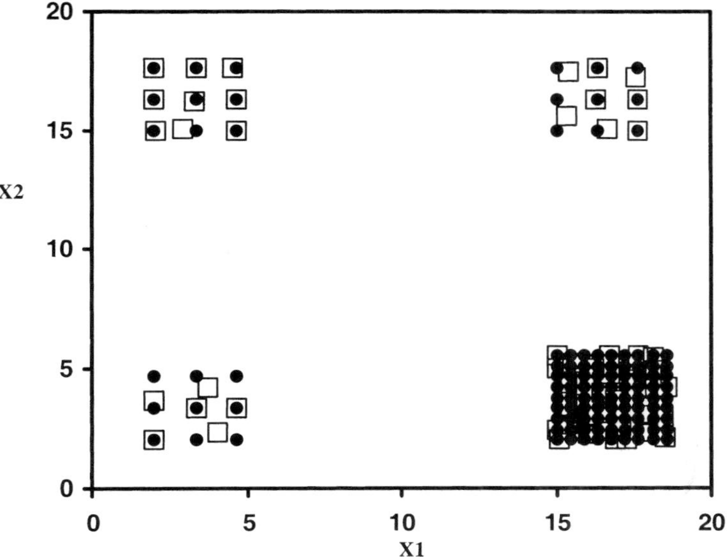

Fig. 6. Cluster delineation using the HSC algorithm. Dead neurons are not engendered in the HSC algorithm, resulting in excellent convergence characteristics and improved probability modeling.

of low probability distributions. Comparing Fig. 5 with Fig. 6, it is clear that dead neurons do not result in the HSC model. A reasonable amount of neurons reside in the areas of low probability distributions. The local density of synaptic vectors in the input vector space approaches the probability density of the input vector distribution. The HSC model makes efficient use of neurons by eliminating the possibility of engendering dead neurons in the system.

Figs. 7 and 8 show the training progress of the KFM and the HSC model, respectively. We can see from these graphs that the KFM model is unable to converge. We tested the KFM model using larger lattice structures, but the KFM failed to converge in all cases (Fig. 7). Table 1 lists some of the KFM error reports using different number of output neurons at the end of 40 cycles. The quantization error remains high for KFM due to its inability to converge. On the other hand, the system quantization error was 32.03 after

TABLE 1

Reports for KFM using different numbers of out put neurons after 40 cycles

Grid structure	Number of neurons	System quantization error	CPU time (s)
8 × 8	64	55.788254	13.48
9 × 9	91	49.845253	17.88
10 × 10	100	42.450687	21.65
11 × 11	121	38.546196	25.72

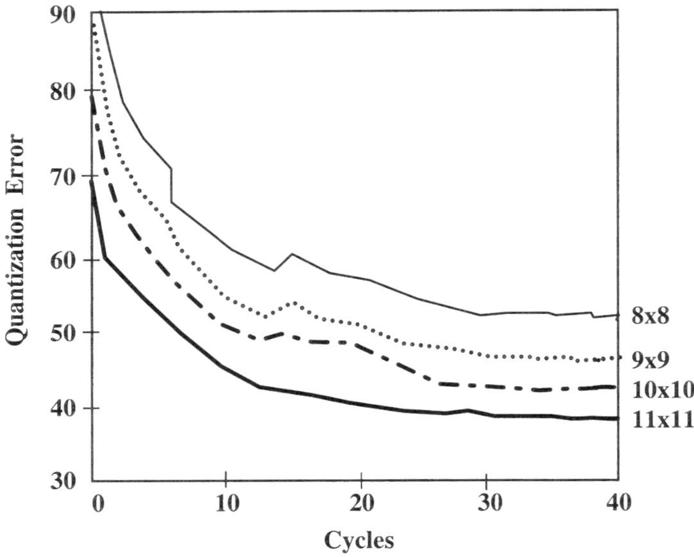

Fig. 7. Training reports for KFM. The KFM network fails to converge in heterogeneous vector spaces

10 training cycles for the HSC model (Fig. 8). This low error value is obtained by using fewer neurons than required by the KFM model and the corresponding CPU time was just 1.61 s.

3.2. Application 2

Due to limited access to a reservoir simulator, we will demonstrate the application of static upscaling using HSC on a very simple problem. Static upscaling could be used as a precursor for flow-based upscaling. We created an artificial simulation data set for a grid comprising 25x25x4 blocks. Porosity and permeability were spatially distributed over the grid by randomly sampling four gaussian distributions with specified mean and variance. Each gaussian distribution represented a geologic layer in the four-layer simulation model. The simulation model had one producer and one injector placed diagonally opposite to each other. We ran the simulation in a three-phase mode with the initial pressure at 4842 psia and the bubble point at 4014 psi.

After simulating the $25 \times 25 \times 4$ system for 10 years, the HSC algorithm was implemented to obtain the upscaled reservoir property values. The two-dimensional input vector components used in the HSC were porosity and permeability. The HSC algorithm was successful in identifying the four different layers (lithologies). It also identified two additional clusters, which were ignored because of their extremely sparse distributions. Once the HSC had identified the clusters successfully, the reference vectors corresponding to these lithologies were extracted. As mentioned earlier, these reference vectors represented the upscaled values. These upscaled values were substituted throughout a new $10 \times 10 \times 4$ upscaled grid, which represented the coarser description

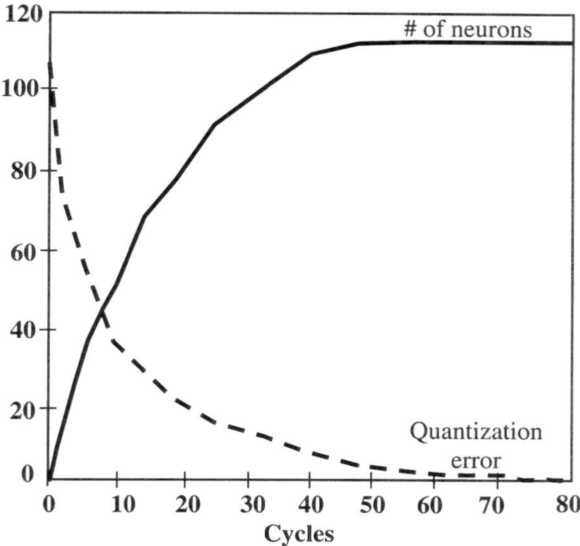

Fig. 8. Training reports for HSC. HSC shows excellent convergence characteristics with smaller number of neurons than the 11×11 KFM network

of the reservoir. The choice of the $10 \times 10 \times 4$ grid was arbitrary, since all the gaussian-distributed values were now represented by property values provided by the reference vectors. The property values were identical within a layer but differed between layers. In other words, the grid coarsening choice is now entirely up to the user. Generally, the smallest grid block is determined by the size of the smallest geobody present in the simulation grid. This makes the HSC upscaled values compatible with local grid refinement, which preserves the local heterogeneities as suggested by Durlofsky et al. (1995).

The locations of the injector and the producer needed some adjustment due to the consequent increase in the simulation block size due to upscaling. All other parameters used in the fine-scale simulation were kept constant. Fig. 9 compares the results from the fine-scale and the coarse-scale models. From this figure, it is evident that the performance of both the systems was very similar with respect to reservoir pressure, fluid flow rates, and cumulative production. The saturation and pressure profiles (not shown) were also consistent.

4. CONCLUSIONS

We have developed a self-organizing, competitive-clustering, adaptive-structure neural network that may be used for efficient geobody detection and static multivariate upscaling of reservoir properties. The neural network, called Heterogeneous Space Classifier (HSC), is well-suited for petroleum engineering applications because of its

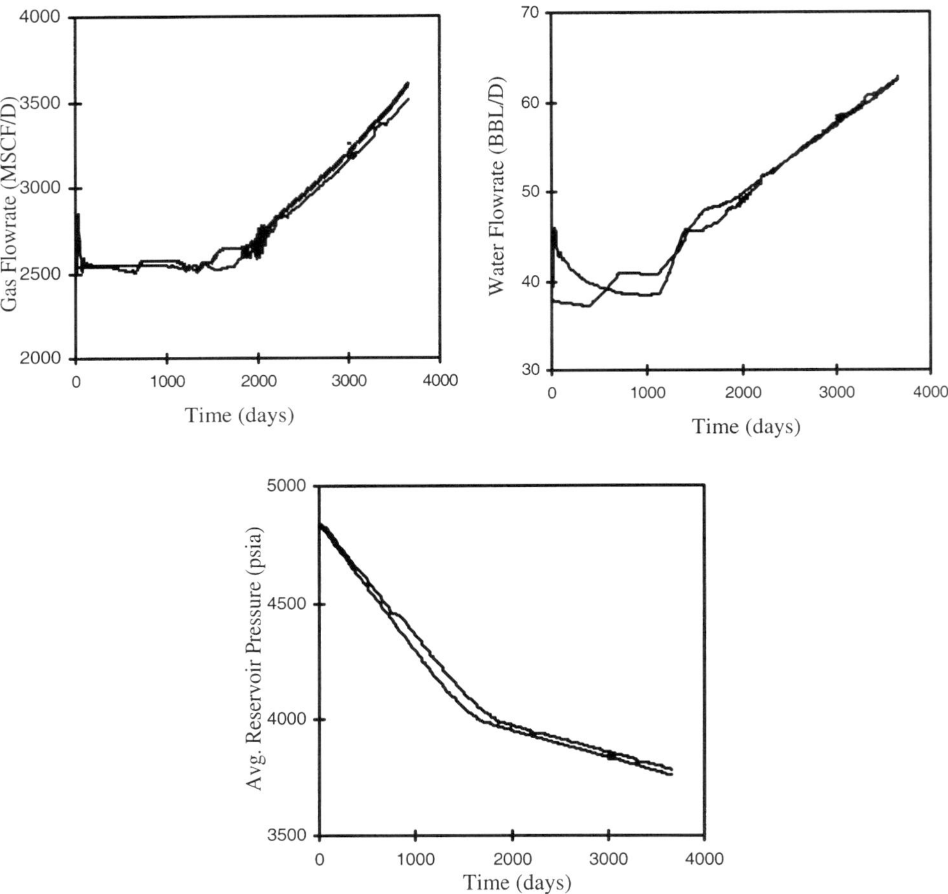

Fig. 9. Comparison of fluid flow rates and reservoir pressure for the upscaled and the original models.

excellent classification characteristics in heterogeneous spaces. The HSC algorithm is designed from an object-based modeling perspective and is valuable in identifying geobodies distributed throughout the reservoir. In addition to the geobody identification, the reference vectors that represent the geobodies may be used to provide upscaled reservoir properties. The reference vectors form the best representations of the geobodies from a minimized quantization error standpoint. The HSC algorithm is superior to the standard KFM clustering algorithm because it is neuron-efficient, fast, and has better convergence characteristics.

The HSC algorithm was tested on a simple static upscaling example comprising a reservoir simulation data set of $25 \times 25 \times 4$ grid blocks. The upscaling reduced the system size to $10 \times 10 \times 4$ grid blocks without compromising the overall fluid flow characteristics.

REFERENCES

Begg, S.H., Carter, R.R. and Dranfield, P., 1989. Assigning effective values to simulator gridblock parameters for heterogeneous reservoirs. *SPERE* (Nov. 1989): 455.

Behrens, R.A., MacLeod, M.K. and Tran, T.T., 1996. Incorporating seismic attribute maps in 3D reservoir models. *1996 SPE Annual Techechnical Conference and Exhibition*, Denver, CO, Oct. 6–9, Paper SPE 36499.

Chawathé, A., 1994. The application of Kohonen type self-organizing algorithm to formation evaluation. *1994 Eastern Regional Conference and Exhibition of the Society of Petroleum Engineers*, Charleston, WV, Nov. 8–10, Paper SPE 29179.

Christie, M.A., 1996. Upscaling for reservoir simulation. *J. Pet. Technol.*, (Nov. 1996): 48.

Chu, L., Schatzinger, R.A. and Tham, M.K., 1996. Application of wavelet analysis to upscaling of rock properties. *1996 SPE Annual Technical Conference and Exhibition*, Denver, CO, Oct. 6–9, Paper SPE 36517.

Deutsch, C. and Meehan, N., 1996. Geostatistical techniques improve reservoir management. *Pet. Eng. Int.*, Mar. 1996: 21.

Deutsch, C. and Wang, L., 1996. Hierarchical object-based geostatistical modeling of fluvial reservoirs. *1996 SPE Annual Technical Conference and Exhibition*, Denver, CO, Oct. 6–9.

Deutsch, C.V., Srinivasan, S. and Mo, Y., 1996. Geostatistical reservoir modeling accounting for precision and scale of seismic data. *1996 SPE Annual Technical Conference and Exhibition*, Denver, CO, Oct. 6–9, Paper SPE 36497.

Doyen, P.M., den Boer, L.D. and Piltet, W.R., 1996. Seismic porosity mapping in the Ekofisk Field using a new form of collocated cokriging. *1996 SPE Annual Technical Conference and Exhibition*, Denver, CO, Oct. 6–9, Paper SPE 36498.

Durlofsky, L.J. et al., 1994. A new method for the scale up of displacement processes in heterogeneous reservoirs. *Proc. of the Fourth European Conference on the Mathematics of Oil Recovery*, Roros, Norway, June 1994.

Durlofsky, L.J., Behrens, R.A., Jones, R.C. and Bernath, A., 1995. Scale up of heterogeneous three dimensional reservoir descriptions. *1995 SPE Annual Technical Conference and Exhibition*, Denver, CO, Oct. 22–25, Paper SPE 30709.

Fritzke, B., 1991. Unsupervised clustering with growing cell structures. *Proc. of the International Joint Conference on Neural Networks*, p. 2.

Guérillot, D. and Verdière, S., 1997. Adaptive upscaling using the dual mesh method. *Fourth International Reservoir Characterization Technical Conference*, Houston, TX, Mar. 1.

Johann, P., Fournier, F., Souza, O., Eschard. R. and Beucher, H., 1996. 3-D stochastic reservoir modeling constrained by well and seismic data on a turbidite field. *1996 SPE Annual Technical Conference and Exhibition*, Denver, CO, Oct. 6–9, Paper SPE 36500.

King, M.J., 1993. Application and analysis of tensor permeability to cross-bedded reservoirs. *Proc. of the Seventh European IOR Symposium*, Moscow, CIS.

King, P.R., 1989. The use of renormalization for calculating effective permeability. *Transp. Porous Media*, 4: 37.

Knox, P.R. and Barton, M.D., 1997. Predicting interwell heterogeneity in fluvial-deltaic reservoirs: outcrop observations and applications of progressive facies variation through a depositional cycle. *Fourth International Reservoir Characterization Technical Conference*, Houston, TX, Mar. 1997.

Kohonen, T., 1898. *Self-Organization and Associative Memory*. Springer-Verlag, New York, NY.

Lee, T.C. and Peterson, A.M., 1990. Adaptive vector quantization using a self-development neural network. *IEEE J. Select. Areas Commun.*, 8(8).

Panda, M.N., Mosher, C. and Chopra, A.K., 1996. Application of wavelet transforms to reservoir data analysis and scaling. *1996 SPE Annual Technical Conference and Exhibition*, Denver, CO, Oct. 6–9, Paper SPE 36516. .

Peaceman, D.W., 1996. Effective transmissibilities of a gridblock by upscaling – why use renormalization? *1996 SPE Annual Technical Conference and Exhibition*, Denver, CO, Oct. 6–9, Paper SPE 36722. .

Rahon, Blanc, G. and Guërillot, D., 1996. Gradients method con strained by geological bodies for history matching. *1996 SPE Annual Technical Conference and Exhibition*, Denver, CO, Oct. 6–9, Paper SPE 36568.

Reynolds, A.C., He, N. and Oliver, D., 1997. Reducing uncertainty in geostatistical description with well testing pressure data. *Fourth International Reservoir Characterization Technical Conference*, Houston, TX, Mar. 1997.

Schatzinger, R.A. and Tomutsa, L., 1997. Multiscale heterogeneity characterization of tidal channel, tidal delta, and foreshore facies, Almond Formation outcrops, Rock Springs Uplift, Wyoming. *Fourth International Reservoir Characterization Technical Conference*, Houston, TX, Mar. 1997.

Seifert, D., Newberry, J.D.H., Ramsey, C. and Lewis, J.J.M., 1997. Evaluation of field development plans using 3-D reservoir modeling. *Fourth International Reservoir Characterization Technical Conference*, Houston, TX, Mar. 1997.

Developments in Petroleum Science, 51
Editors: M. Nikravesh, F. Aminzadeh and L.A. Zadeh

Chapter 15

HIGH RESOLUTION RESERVOIR HETEROGENEITY CHARACTERIZATION USING RECOGNITION TECHNOLOGY

MAHNAZ HASSIBI [a], IRAJ ERSHAGHI [b] and FRED AMINZADEH [a]

[a] *FACT, Inc., 14019 SW FWY, Suite 301-225 Sugar Land, TX 77478, USA*
[b] *University of Southern California, Los Angeles, CA 90007, USA*

ABSTRACT

In reservoirs producing from laminated systems, such as turbidite sequences, rock fabrics exhibit great variations both vertically and in lateral direction. In such formations, patterns observed on lithological logs can show considerable differences. Moreover, the nature of shale stringers and their lateral continuity control cross-flow and vertical communication. Constructing rational and meaningful correlation models consistent with all the lithological signatures can be an enormous task before conceptualization of data on a 3D geologic model. The proposed method can serve as an important pre-processing step to minimize the requirements of the human expert input in the sub-marker detection for substantial number of well logs. Incorporation of micro lamination markers in a cohesive way to map continuity or discontinuity of shale breakers requires detail pattern studies. The automated techniques of pattern classification enhance the characterization and identification of stratigraphic features of laminated type reservoirs. The approach proposed in this paper works on basic lithological logs and marker information. It comprises noise filtering and pattern recognition that lead to identification of reservoir compartments. This helps in delineation of the lateral continuity and discontinuity of reservoir sand and shale laminations.

1. INTRODUCTION

Lithological profiles or ordinary well log responses are efficiently used to compose the major framework of the input to the system for geological modeling. These data sets are huge masses of useful and valuable information that require efficient processing techniques. Cross-correlation and similarity characterization studies can be effectively used to scrutinize the lithological patterns.

Understanding the lithological structure of a formation is a first step for initiation of deterministic or stochastic geologic modeling. The complexities observed on these signals are directly related to the degree of heterogeneity of the formation. Fig. 1a shows a schematic of a cross-section where the correlation among individual wells is clear and deterministic. However, correlation studies can become quite complicated in formations consisting of sand intervals separated by thin shale layers. Fig. 1b is an example of a very complex sedimentary structure, in which considerable variation of

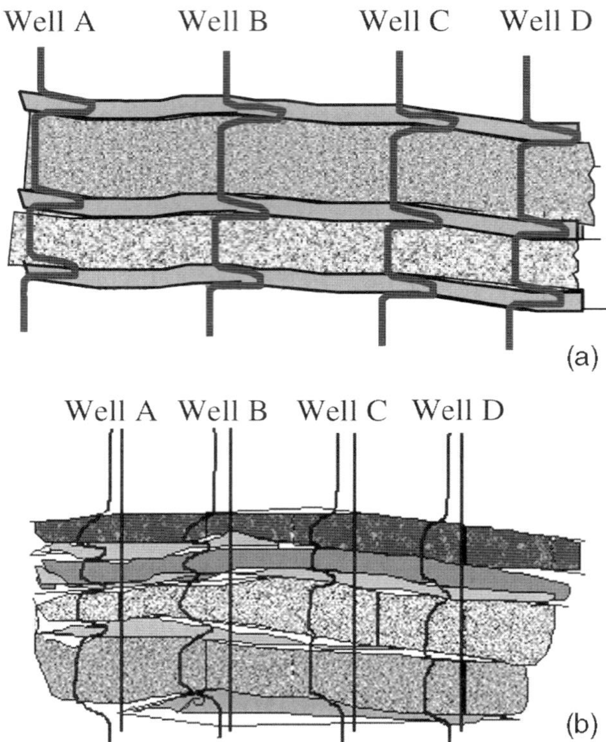

Fig. 1. (a) An example of a formation with clean intervals. (b) An example of a formation with small and dirty intervals.

sedimentation process is observed. Both a gravitational sequencing and the influence of distance from the source material cause these unique characteristics. In cases that small laminations are involved, lithologic log responses as a function of sedimentation processes also exhibit cyclical variation. In this type of formation, identifying different zone boundaries can become very cumbersome. Each zone represents a segment of reservoir, where similarity of log responses can be expected.

Fig. 2 illustrates a reservoir where the process of geological description for reservoir modeling is much more complicated. Without considering the existence of different compartments, estimation errors for prediction purposes can be substantial. Further more, in 3D mapping of thinly bedded strata the identification of major and minor geologic markers is very essential. Because of discontinuity, changes in lithological composition, and various tectonically related processes, exact correlation of the minor markers becomes a difficult task.

Fig. 3 is an example of a 3D geological model that describes the Fault Block II of the Wilmington field in Long. Beach, California. This model merely is an illustration of major markers, however the interbedded shale intervals are not incorporated in its construction. In Fig. 4, four hypothetical markers are depicted, in which two of them are

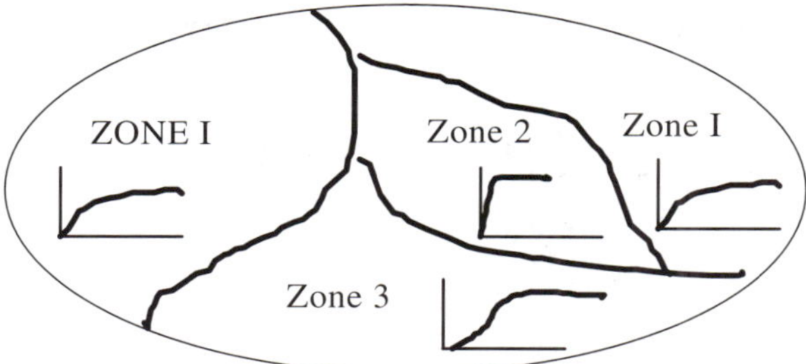

Fig. 2. A case with zonation problem.

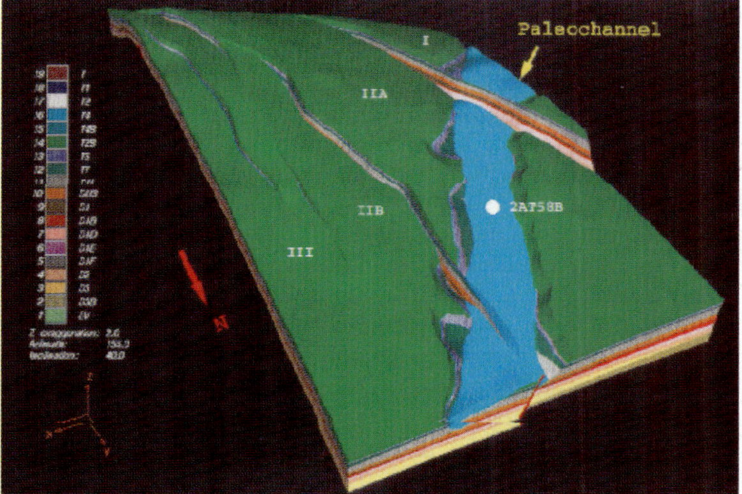

Fig. 3. 3D geological model from Wilmington Field.

analyzed for small lamination characteristics. Apparently, the middle selected marker 'C–D' can be considered as a consistent marker, because of the lateral uniformity of lamination pattern. However, marker 'A–B' does not exhibit similar characteristic, and in fact it is representing an inconsistent marker. Similarity analysis allows the investigation of inconsistency and discontinuity in terms of lateral connectivity.

Commonly, lithological pattern classification is a task handled by human experts. However, in complex situations the result is not always satisfactory. In fact, manual manipulation of quantitatively large data sets is a very complicated task subject to misinterpretation. A more practical alternative is to get assistance from a computer aided technique to automate the pattern classification and similarity analysis processes.

The goal is to develop an automated processing system that receives the log signals from one specified major marker, analyzes the patterns, and as output provides the identified reservoir compartments and extended and terminated layers. The final result

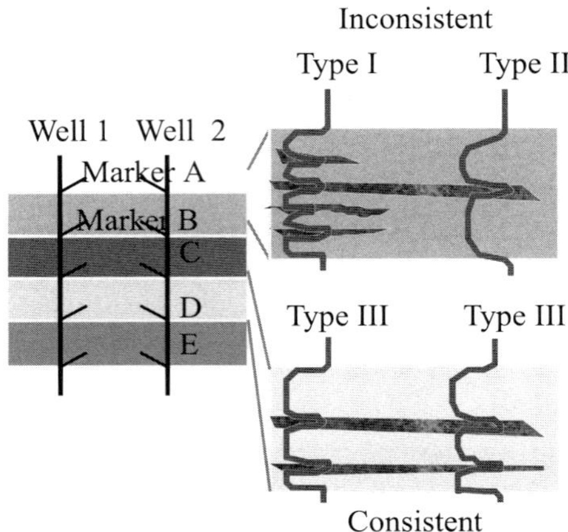

Fig. 4. Examples of consistent and inconsistent markers.

and output from the system will be the identified sub-markers. Fig. 5 describes the general concept of the proposed solution.

Application of automated pattern classification in reservoir compartmentalization and cross-correlation is a new approach to provide fast and reliable preprocess for geological modeling. This approach can help in building detailed geological images, which are essential for accurate illustration of the continuous or discontinuous sedimentary deposits. Important tools for pattern recognition dealing with images are the technology of artificial intelligence, neural network, and statistical pattern classification such as K-means and vector quantization.

Prior studies by other researchers in application of neural networks focused on well logs can be divided into two major areas: (1) classification of different lithofacies, and (2) prediction of reservoir properties such as porosity and permeability.

For lithofacies classification, the input patterns are composed of characteristic log values from typical log responses corresponding to various facies. Also the same input

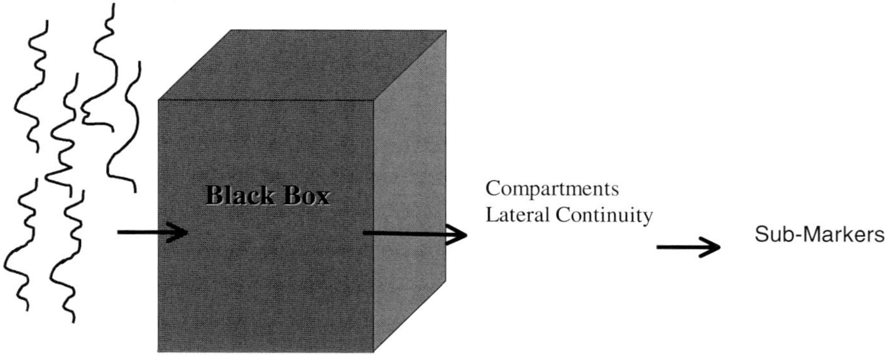

Fig. 5. The proposed processing system.

patterns can be used for reservoir characterization; however, in some studies, the combination of log data and lithofacies information is used to predict reservoir properties. Baldwin (1991) referred to the use of bidirectional associative Neural Network to identify facies from logging data. He used gamma ray log, SP, shallow/medium/deep resistivities, neutron porosity, and other logs as the key for facies classification. Chawathe (1994) proposed another technique to classify the lithofacies. In his approach, input vectors consist of six different log data. Each input log in a certain interval was split into maximum and minimum log values. The neural network worked on a log value range as opposed to single log values. By using self-organize neural network, Chawathe was able to come up with 8 different rock sequences. Baldwin (1989), in another study, discussed the prediction of total porosity and apparent grain density using log data and lithology input. Wong and Taggart (1995) proposed two approaches of non-genetic and genetic reservoir characterization. The non-genetic approach treats an entire reservoir as a whole in the prediction of petrophysical properties, while the genetic approach seeks to identify and treat different lithofacies of a reservoir separately. First he classified different facies by using log data. Then he used both log data and lithofacies information to predict the porosity. In the next step, porosity was added to the previous information in order to predict permeability. In summary, above studies focused on rock mineralogy. The emphasis was to distinguish among different facies and to predict physical properties in a column of rock.

To set the basis for this work, one needs to ponder whether the definition of a particular geologic column is sufficient to establish correlation and similarities across a reservoir. The main objective of this work is to use an automated pattern classification technique as a tool to distinguish among different lithological patterns, and also to classify similar patterns in one group. In previous work by Hassibi and Ershaghi (1996) the application of feed forward neural network with error back propagation was used to classify a set of log signals extracted from a particular interval in Wilmington Field. That technique is a supervised learning approach (Pandya and Macy (1996) and Zurada (1992)), which requires a preliminary knowledge about existing models. However, this knowledge is not generally available for complex systems. For this particular study, the self-organize vector quantization technique (unsupervised pattern classification) is used for pattern recognition purpose (Pandya and Macy, 1996).

Raw and unprocessed data are not considered as sufficiently qualified input for similarity analysis. Before any pattern classification process, another processing system is needed to extract the required information for constructing suitable input patterns. This preprocessing phase of work consists of marker selection, noise filtering (smoothing process), rescaling, and signal simplification processes. A smoothing algorithm is developed to eliminate insignificant fluctuations. Smoothed patterns are used to extract maximum and minimum points, and based on these points shale–sand indicator signals are generated to identity segregated sand and shale intervals. Shale–sand indicator signals are very useful to built input patterns with smaller sizes.

Once the preprocessing phase is completed, system is ready to initiate the proposed pattern classification process to distinguish similar lithological patterns. Pattern recognition allows the identification of reservoir compartments and their boundaries.

In next step lateral correlation process identifies continuous and discontinuous laminations among compartments. This method itself includes another similarity analysis process.

2. COMPLEX SEDIMENTARY ENVIRONMENTS

Sedimentary environments can be divided into various classes such as alluvial fans, rivers and their flood plains, marginal-marine (deltas, alongshore sand bodies), and marine (shelf, submarine fans, turbidite sequences). In this work the main objective is to study the deep-marine environments, which are mostly formed by turbidity flows. Deep marine facies, in particular those comprising fan channels and lobes, constitute some important hydrocarbon reservoirs worldwide. Many factors contribute to the observed heterogeneities in turbidite reservoirs. Among those the proximity of the lobes to the source material and the nature of the sand lobes, determine the vertical and horizontal variations in the sequence. Because of the complex nature of this type of formations, sand continuity and lateral correlation become important issues. In turbidite sediments, rapid vertical and lateral variations in rock type disturb the uniformity and consistency of sand intervals. Fig. 6 illustrates a 3D model, which shows the development of a submarine fan and also depicts the variation of sedimentation in different locations. These changes and variations can be projected on the well log responses. This fact implies that well log signals are function of the sedimentary patterns. For example in Fig. 7 bigger and cleaner intervals for proximal type of sediments and thinner and

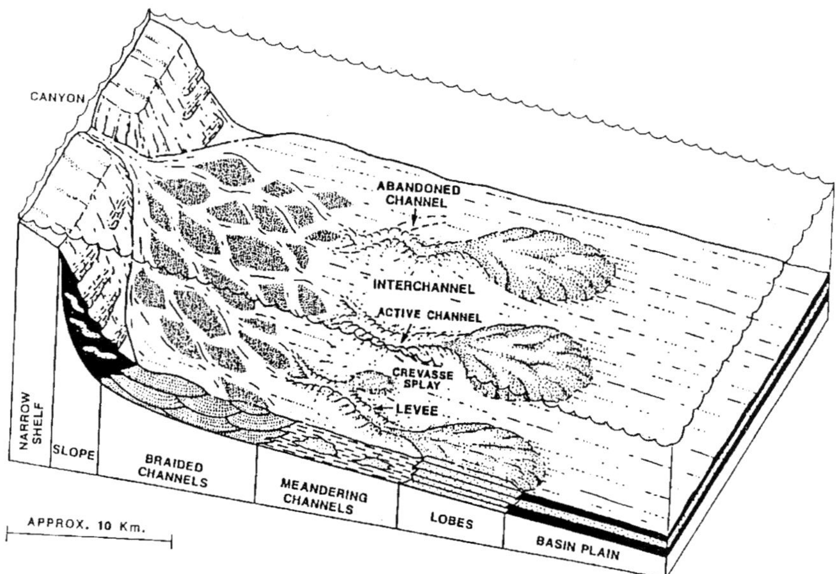

Fig. 6. 3D model of a submarine fan. (From Shanmugam et al., 1988).

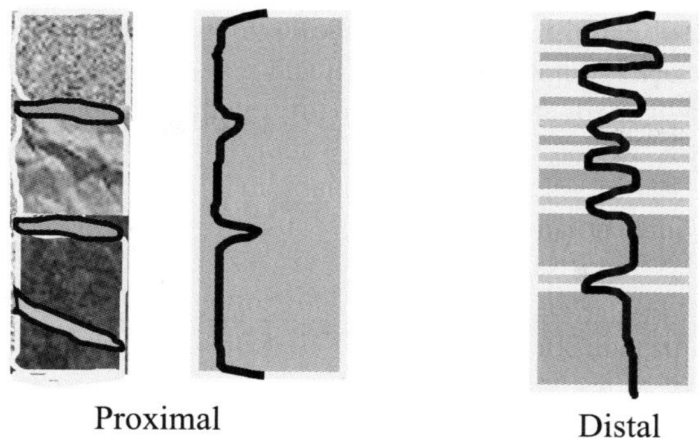

Fig. 7. Relationship between log response and sedimentation pattern.

dirtier intervals for distal deposits are commonly observed on the corresponding well log signals. The main objective of the proposed approach is to automatically inspect the reservoir compartments and lateral correlation for complex sedimentary systems based on geophysical log signals.

3. PATTERN CLASSIFICATION TECHNIQUES

A pattern is an arrangement that can resemble some existing structure. A set of patterns that share some common characteristics are regarded as a pattern class. Automated pattern recognition deals with techniques in which patterns are assigned to their respective classes. In a broad sense, pattern recognition helps to partition input patterns into meaningful categories. In pattern classification procedure different neural and non-neural techniques can be employed. Learning process can be divided into two categories of supervised and unsupervised. Supervised learning has significant usage in systems, which the clusters or classes and the general characteristics of them are clearly defined, prior to any classification process. Conversely, in unsupervised learning, available clustering information is either deficient or completely missing. As to the significant complexities involved in turbidite sediments and sequences, the identification of all the existing models and preparing adequate training set are oftentimes impossible. Thus, any supervised technique such as feed forward with error back propagation neural network might encounter difficulties in the process of pattern recognition. On the other hand, unsupervised techniques such as self-organize vector quantization can categorize lithological profiles in a self-organizing manner. In this paper vector quantization approach is employed for pattern recognition process.

3.1. Vector quantization

This approach is an unsupervised clustering technique based on distance functions within Euclidean space. Vector quantization starts with no allocated cluster center. First input pattern generates the very first cluster center. Thereafter, new clusters will be generated if any new pattern is not classified into any of the pre-existing clusters. Euclidean distance between new input pattern and any allocated clusters is the measure of similarity. Following equation describes the mathematical meaning of Euclidean distance measurement.

$$\|X - Z\| = \left[\sum_{i=1}^{n}(x_i - z_i)^2\right]^{1/2}$$

where X and Z vectors are respectively new input pattern and pre-existing cluster center.

4. ESSENTIAL PRE-PROCESSES

Selecting data with higher resolution, assigning markers, choosing proper smoothing criteria, normalizing patterns, and generating simplified input patterns are essential steps before conducting any classification process.

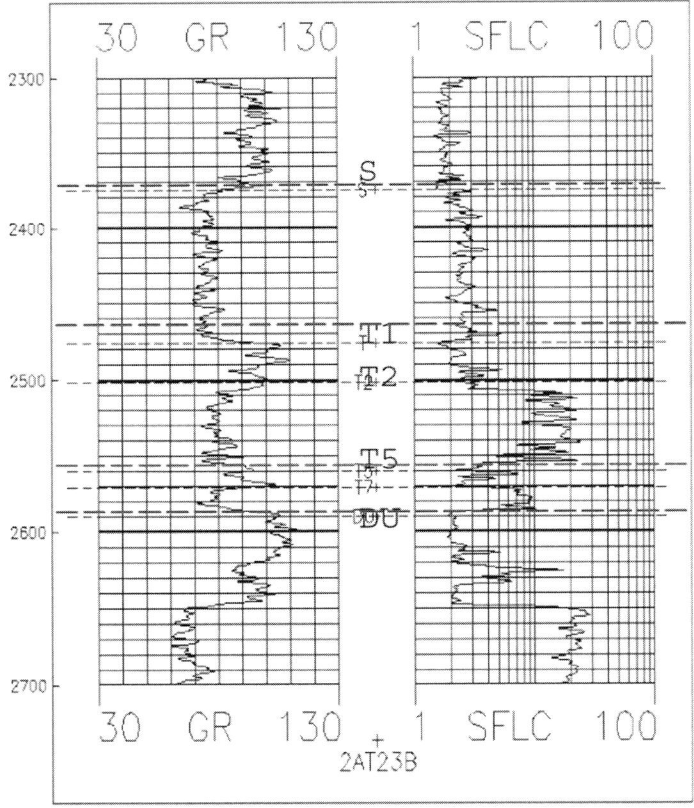

Fig. 8. Typical well log responses in the TAR zone of the Wilmington Field, California.

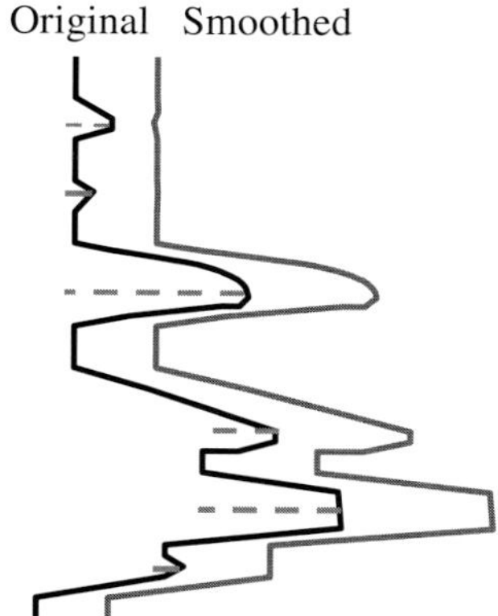

Fig. 9. Typical Results of the Smoothing process.

(1) The resolution of well log signals is one of the most important aspects to discriminate different lithological patterns. To have sharply defined bed boundaries, logging devices must be sensitive to small changes in the sedimentary structure of a formation. Lack of accuracy will result in deficiency of log interpretations.
(2) Marker assignment ensures that all the processing data belong to the same interval in the reservoir. Fig. 8 is a typical well log response, identifying markers.
(3) Noisy log signals are not considered as appropriate input for data analysis. Noise will give misleading messages about the stratigraphic features of the formation. A smoothing criterion should be used to eliminate insignificant variations in each signal. Typical results from the smoothing methodology used in this paper are depicted in Fig. 9.
(4) Fig. 10 illustrates an example of smoothing process along with the simplified form of the max–min pattern, in which the location and magnitude of major picks are displayed.
(5) Normalization can provide a more appropriate range of data for the classification process. The following is a simple equation for normalization

$$X_n = \frac{(X_o - X_{min})}{(X_{max} - X_{min})} \times X_{range} + X_{start}$$

where X_{min}, X_{max} = minimum and maximum of the old range; X_n = new value, X_o = old value; X_{range} = new range of data; X_{start} = start point of new range.
(6) Shale–sand indicator patterns are the modified version of the smoothed patterns, signifying the sand and shale boundaries and their associated log magnitude. Fig. 11 depicts unique characteristic of these signals based on which, input patterns are developed for pattern recognition process. Since thickening and thinning of laminations do not imply any changes in sedimentary structure or continuity issue, therefore only magnitude of shale and sand intervals are good enough to construct input patterns. Thicknesses of layers are used for final 3D geologic modeling.

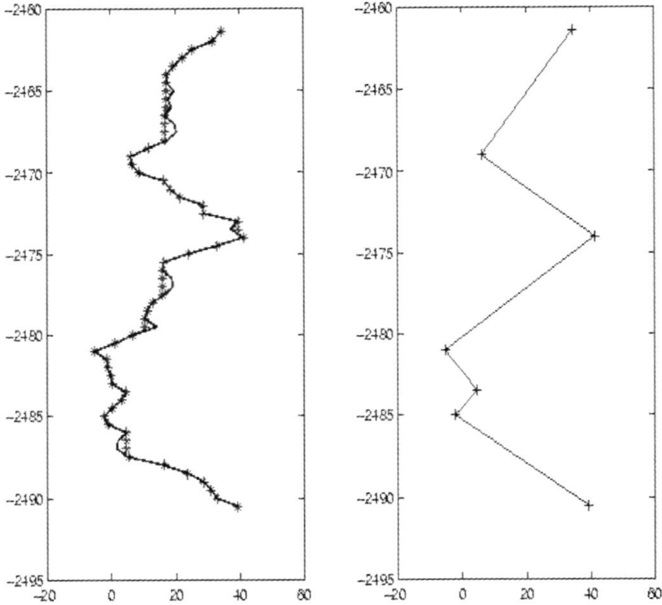

Fig. 10. Original ('——'), Smoothed ('∗∗∗'), and maximum and minimum (right plot) signals for interval T5 of well 2AT37B0 of the Tar zone.

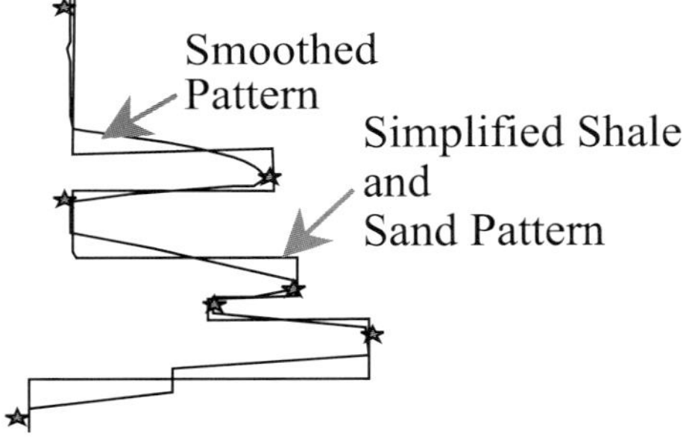

Fig. 11. Shale–Sand indicator pattern.

5. RESERVOIR COMPARTMENTALIZATION AND CONTINUITY CORRELATION

Similarity characterization process using pattern recognition techniques delineates dominant reservoir compartments (Fig. 12). Reservoir compartmentalization helps in better understanding of sedimentary structure of a reservoir. Continuity correlation within each compartment is already defined, because each compartment represents a region with specific sedimentary feature that is consistently observed over the region. However, from one compartment to another there might exist some uncertainties about the different ways that layers are extended.

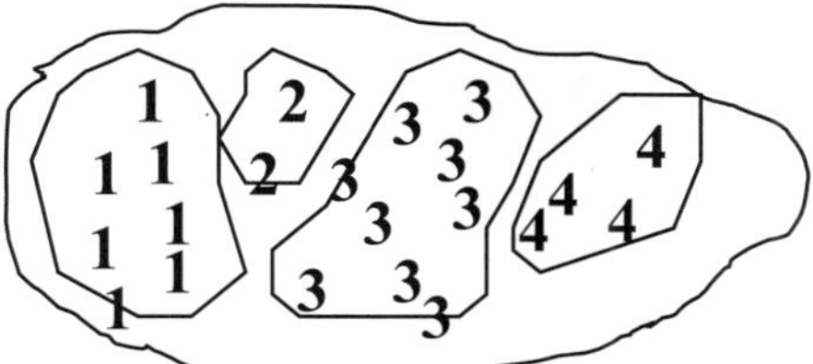

Fig. 12. Reservoir compartmentalization.

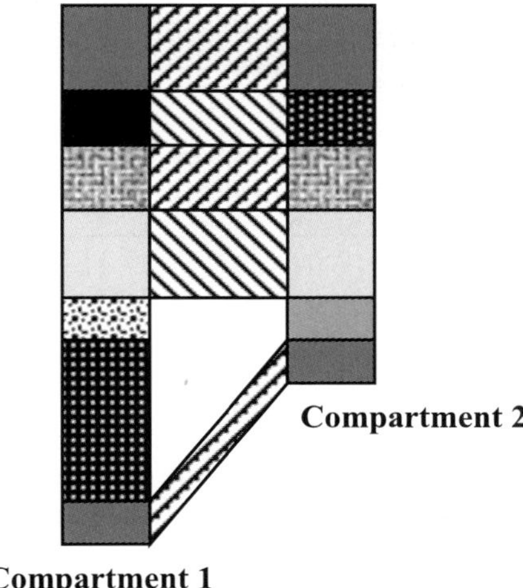

Fig. 13. Lateral correlation

Fig. 13 illustrates the main framework of lateral correlation process. In this figure, each column represents lithological structure of a specified compartment. The fundamental basis, that guides this process, stems from the geological fact that the oldest rock body (in terms of geological time) always appears at the bottom of the sedimentary structure. This means that layers cannot intersect each other. Based on this fact, the proposed algorithm seeks possible ways of lateral continuity among compartments. This approach delineates the extension of sedimentary layers.

6. SYNTHETIC AND REAL FIELD DATA

6.1. Synthetic data

Working with synthetic fields allows assessing the accuracy of the obtained results. If the program successfully handles the compartmentalization and cross-correlation processes for synthetic fields, it can be also applied to a set of real field data. In this paper a synthetic field and the result of reservoir compartmentalization and lateral correlation are illustrated.

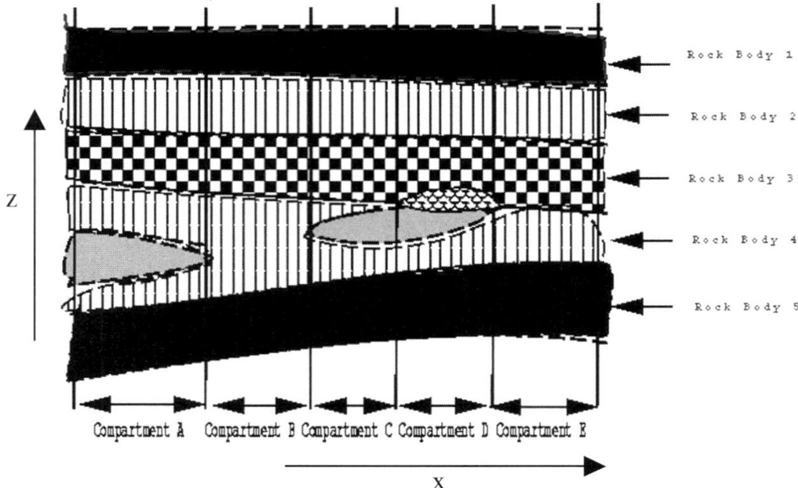

Fig. 14. The synthetic field with five compartments

Fig. 14 represents a heterogeneous reservoir. This field is a combination of uniform and nonuniform laminations. Three major bodies of rock from the top and one from the bottom are uniformly extended throughout the formation, while nonuniform lamination is observed in fourth major interval. Presence of anomalies in fourth interval results in reservoir compartmentalization. According to the figure, five compartments with different characteristics should be recognized for this field. Preprocessing procedure is not required for synthetic field data, because data are not real and noise elimination and all other modifications are already included in construction of the data set. Therefore first step in the procedure would be pattern recognition process that identifies existing lithological models. Mapping the areal distribution of these models can help to verify the compartment boundaries.

Fig. 15 illustrates how this reservoir is partitioned into five compartments. In next step the lateral correlation program searches for possible continuity patterns for each pair of

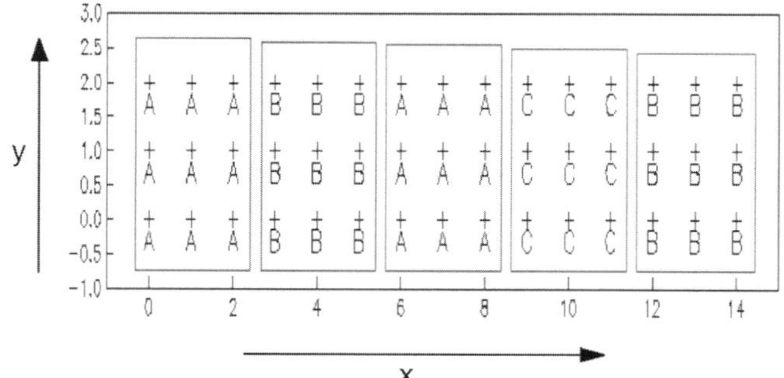

Fig. 15. Areal map showing five distinct compartments.

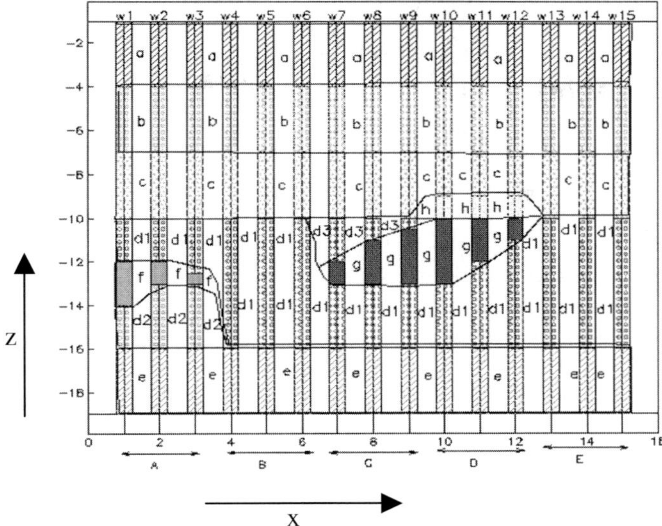

Fig. 16. A cross-section defining the synthetic geologic sequence.

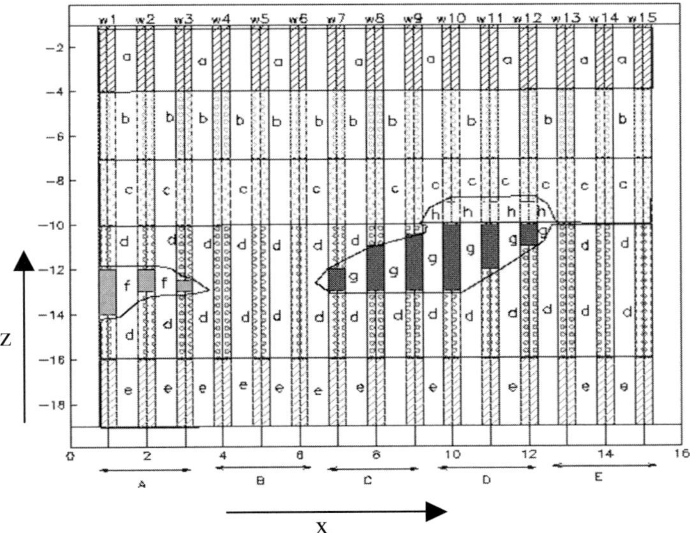

Fig. 17. A cross-section defining the synthetic field.

compartments. Because of the heterogeneous structure of this field the lateral correlation algorithm gives more than one solution. Figs. 16 and 17 are two cross-sections that are built based on the lateral correlation results. Comparison shows that first cross-section does not completely match the original sedimentary structure of the presented field. On the other hand, second cross-section exhibits a good match. Based on the results, program was able to successfully classify the lithological columns and also to identify the lateral continuity patterns for this particular field. The chances of obtaining several viable solutions increase with the complexity of the problem. Under these circumstances, the output from the proposed approach can be cross-examined with other data sources.

6.2. Real field data and results

Considering the important markers, conventional geological models can be generated. However, micro laminations are not incorporated to illustrate the sand and shale continuity and discontinuity within major markers. If small intervals and micro laminations are important, main markers should be scrutinized carefully for consistency matter. To avoid erroneous reservoir classification, micro scale information about shale and sand discontinuity and distribution are required. The proposed approach intends to proceed with automatic lithological pattern classification process for identification of reservoir compartments, and followed by that lateral correlation process to illuminate the extension and delimitation of sedimentary layers within major markers.

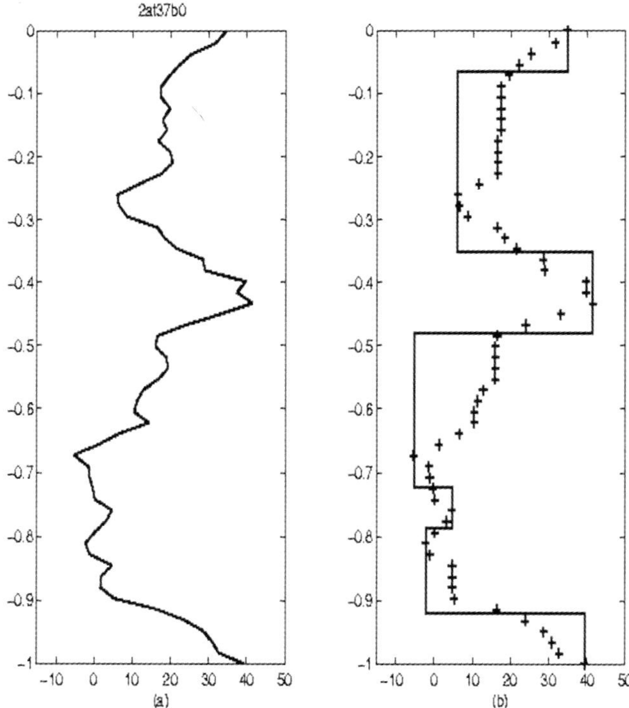

Fig. 18. Original (left signal), smoothed (star sign), and shale–sand indicator (rectangular shape) signals, representing model 'A' in intervalT5 of well 2at37.

Real Field data set used for this study is obtained from the Wilmington Field in Long Beach, California. This field consists of about 6000 feet of interbeded turbidite sand and shale at the top of fractured shale and basement rock. The main focus in our study is the top layer of this formation called the Tar zone. This zone is divided into smaller intervals. As it was mentioned earlier, Fig. 8 illustrates GR and SFLC log responses of well '2AT23' in fault block II of Wilmington field, and identifies major markers such as S, T1, T2, T5, and etc. In this study the interval between markers T5 and DU is selected for pattern classification and reservoir compartmentalization. Log responses from this interval are processed with smoothing program. The smoothed patterns are subjected to the shale–sand indicator process. Finally, these shale–sand indicator patterns are used as input to the pattern classification algorithm. Figs. 18 and 19 consist of original (solid line on the left), smoothed (star), and shale–sand indicator (rectangular shape signal) patterns belonging to two different wells (2AT37B0 and UP908B0). The lithological patterns from these two wells are recognized as similar patterns by the program.

The proposed pattern recognition technique will categorize the similar patterns and makes it possible to anticipate the compartment boundaries, see Fig. 20. Four major

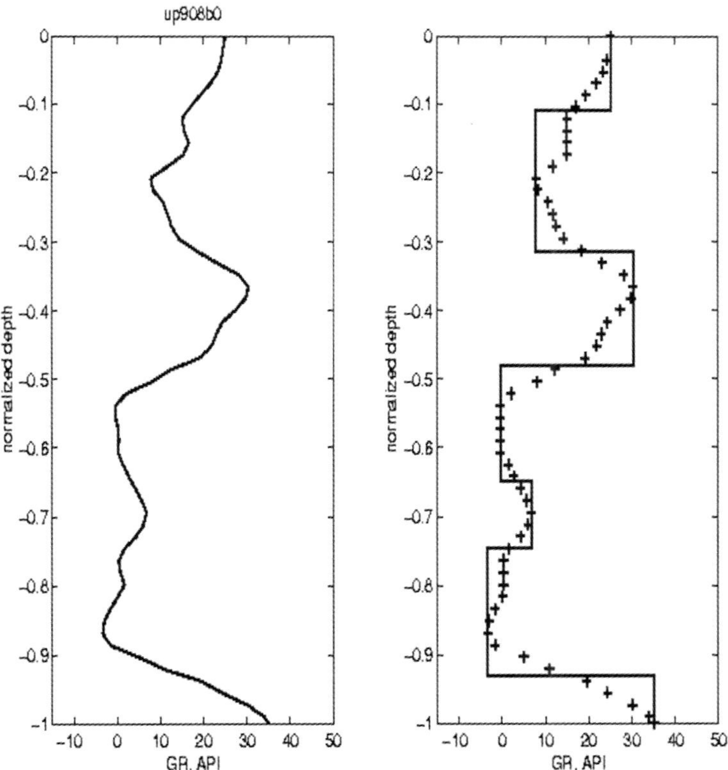

Fig. 19. Original (left signal), smoothed (star sign), and shale–sand indicator (rectangular shape) signals, representing model 'A' in intervalT5 of well up908.

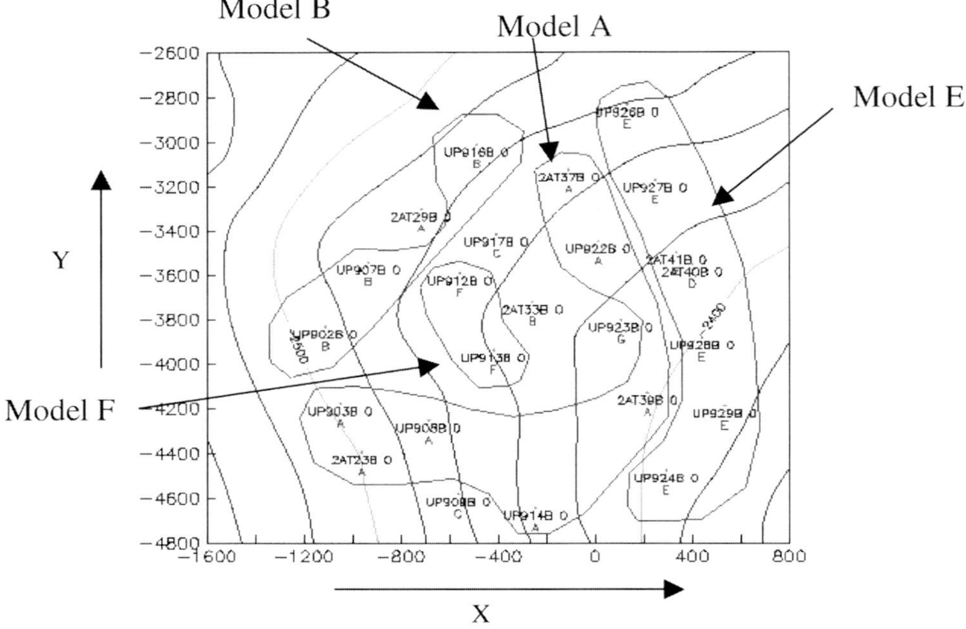

Fig. 20. Areal distribution of lithological models.

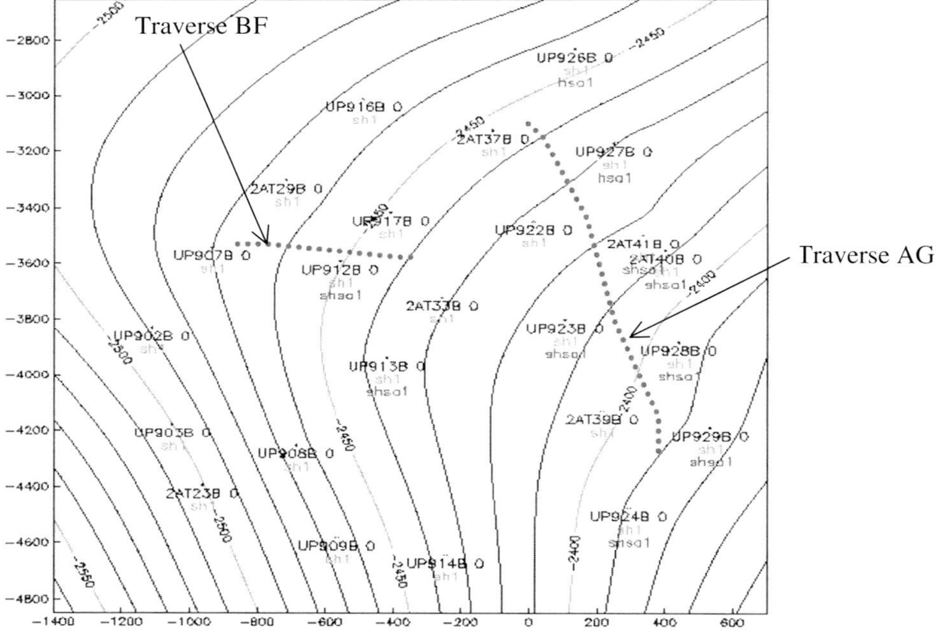

Fig. 21. An areal map showing the appearance of 'sh1' and 'shsa1' layers at different locations.

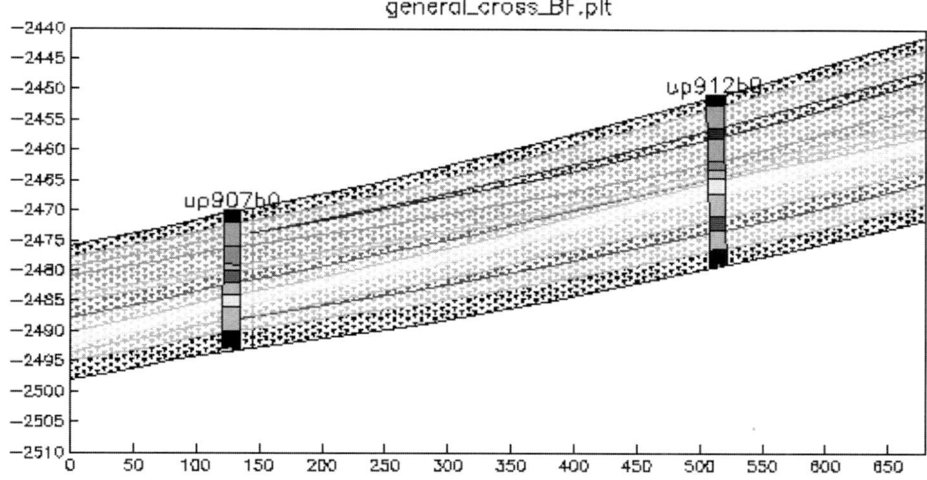

Fig. 22. Cross-section BF of a major marker with associated sub-markers.

distinguished compartments are identified in this figure. Because each compartment represents one particular lithological pattern, thus the lateral correlation is already completed within compartments. However, the lateral correlation from one compartment to another should be analyzed to obtain a complete set of distinguished sub-markers.

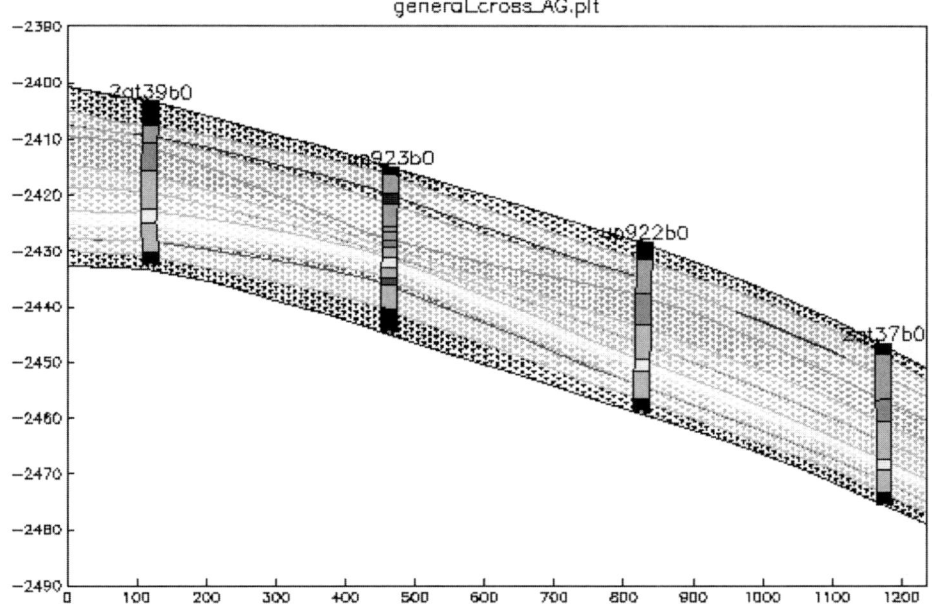

Fig. 23. Cross-section AG of a major marker with associated sub-markers.

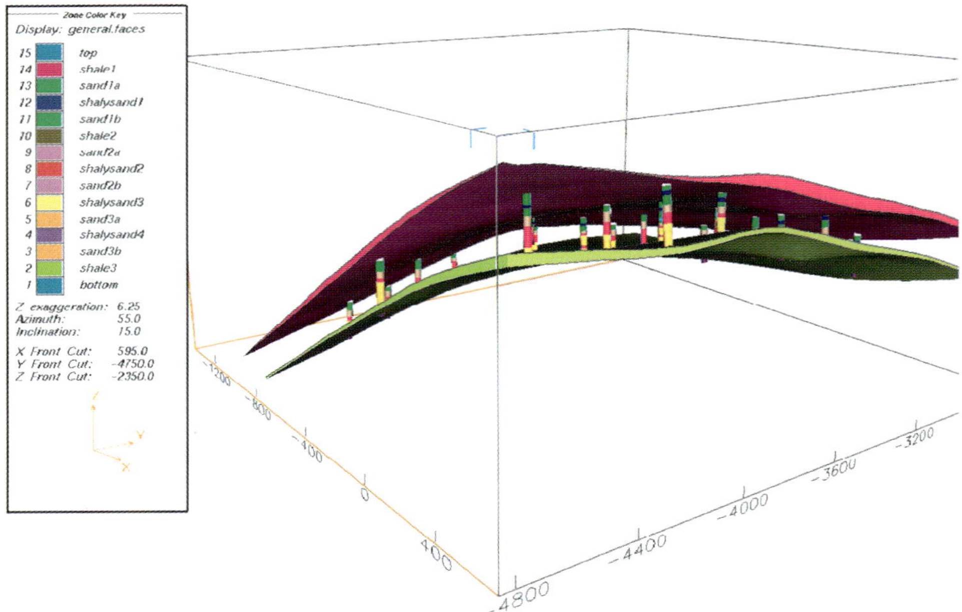

Fig. 24. 3D Model illustrating top and bottom of a major marker and the corresponding sub-markers.

This task is achieved using an automated algorithm that compares sub-patterns from compartments, and identifies all the possible ways that micro-laminations could be extended. This technique allows identifying the sub-markers. In Fig. 21, the results from lateral correlation illustrate the areal extension of sub-marker 'sh1' and 'shsa1' in the reservoir.

Automated cross-correlation algorithm helps to construct the reservoir cross-sections, showing lateral continuity of the interval. As it is expected, more than one solution is detected matching the existing data. In Figs. 22 and 23, two different representing cross-sections are depicted corresponding to BF and AG traverses in Fig. 21. They specifically illustrate the way that different laminations are communicating across designated cross-sections. These detailed cross-sections can help to distinguish micro lamination distribution in a major interval.

Using this information more comprehensive and accurate 2D or 3D geological models can be developed. These models are very important in observing extension and termination of sub-markers in different directions. Fig. 24 is a comprehensive 3D illustration of a major interval and the identified sub-markers.

7. CONCLUSIONS

In this paper an automated approach for lithological pattern classification is developed for complex sedimentary structures. The reservoir compartmentalization process emphasizes on smallest depositional units, and classifies different lithological patterns in each identified marker and characterizes different reservoir compartment boundaries. Moreover, lateral correlation process delineates extension and elongation of continuous layers among compartments. This multiphase procedure offers a methodology that receives well log signals as input patterns and conducts similarity characterization and cross-correlation process. The output of this algorithm can be used to condition geological modeling by incorporating micro-lamination information.

ACKNOWLEDGEMENTS

The authors would like to thank Tidelands Oil Production Company and DOE for their support during the course of this study.

REFERENCES

Baldwin, J.L., 1989. Computer emulation of human mental process: application of neural network simulators to problems in well log interpretation. *64th Annual Technical Conference and Exhibition of SPE*, Oct. 8–11, 1989, SPE 19619. .

Baldwin, J.L., 1991. Using a simulated bidirectional associative neural network memory with incomplete prototype memories to identify facies from intermittent logging data acquired in a siliciclastic depositional sequence: a case study. *66th Annual Technical Conference and Exhibition of SPE*, Oct 6–9, 1991, SPE 22843.

Chawathe, A., 1994. The application of Kohonen type self organization algorithm to formation evaluation. *Eastern Regional Conference of SPE*, SPE 29179.

Hassibi, M. and Ershaghi, I., 1996. A neural network approach for correlation studies in a complex turbidite sequence. *1996 SPE Annual Technical Conference and Exhibition*, Oct 6–9, 1996, SPE 36720.

Mutti, E., Normark, W.R. 1990. Comparing examples of modern and ancient turbidite systems: problems and concepts. *Soc. Econ. Paleontol. Mineral., Pac. Sect., Bakersfield, Short Course Vol.*, 66: 199–246.

Pandya, A.S. and Macy, R.B., 1996. *Pattern Recognition with Neural Networks in C++*. CRC Press, Baton Rouge, FL.

Shanmugam, G., Moiola, R.J., McPherson, J.G. and O'Connell, S., 1988. Comparison of modern Mississippi fan with selected ancient fans. *Trans. Gulf Coast Assoc. Geol. Soc.*, 38: 157–165.

Walker, R.G., 1979. *Facies Models*. Geological Association of Canada.

Wong, P.M., 1995. A critical comparison of neural networks and discriminant analysis in lithofacies, porosity and permeability predictions. *J. Pet. Geol.*, 18(2): 191–206.

Wong, P.M. and Taggart, I.J., 1995. Use of neural network methods to predict porosity and permeability of a petroleum reservoir. *J. AI Appl.*, 9(2): 27–37.

Zurada, J.M., 1992. *Introduction to Artificial Systems*. West Publishing Company.

Developments in Petroleum Science, 51
Editors: M. Nikravesh, F. Aminzadeh and L.A. Zadeh

Chapter 16

EXTENDING THE USE OF LINGUISTIC PETROGRAPHICAL DESCRIPTIONS TO CHARACTERISE CORE POROSITY

TOM D. GEDEON [a,1], PATRICK M. WONG [b,2], DILIP TAMHANE [b] and TAO LIN [c]

[a] *School of Information Technology, Murdoch University, Perth, Australia*
[b] *School of Petroleum Engineering, University of New South Wales, Sydney, Australia*
[c] *Mathematical and Information Sciences, CSIRO, Canberra, Australia*

ABSTRACT

There are many classification problems in petroleum reservoir characterisation, an example being the recognition of lithofacies from well log data. Data classification is not an easy task when the data are not of numerical origin. This paper compares three approaches to classify porosity into groups (very poor, poor, fair, good) using petrographical characteristics described in linguistic terms. The three techniques used are an expert system approach, a supervised clustering approach, and a neural network approach. From the results applied to a core data set in Australia, we found that the techniques performed best in decreasing order of their requirement for significant user effort, for a low degree of benefit achieved.

1. INTRODUCTION

Many forms of heterogeneity in sedimentary rock properties, such as porosity, are present in clastic reservoirs. Understanding the form and spatial distribution of these heterogeneities is fundamental to the successful characterisation of petroleum reservoirs. From a geological viewpoint, the anatomy of reservoir heterogeneity requires two major pieces of information: component lithofacies (and their hydraulic properties) and their internal architecture. Poor understanding of lithofacies distribution results in inaccurate definitions of reserves and improper management schemes. Mapping the continuity of major lithofacies is, therefore, of great importance in reservoir characterisation studies. It is, however, impossible to start this mapping exercise until the major types of lithofacies have been recognised and identified.

Lithofacies recognition is often done in drilled wells where suitable well logs and core samples are available. Pattern recognition techniques, such as k-means cluster analysis (Wolff and Pelissier-Combescure, 1982), discriminant analysis (Jian et al., 1994; Wong et al., 1995), artificial neural networks (Rogers et al., 1992), and fuzzy logic methods (Wong et al., 1997) can be used for classifying well log data into discrete classes. Some hybrid

[1] E-mail: t.gedeon@murdoch.edu.au
[2] Present address: Veritas DGC Inc., 10300 Town Park Drive, Houston, TX 77072, USA.

techniques (Chang et al., 2000; Wong et al., 2000) are also available. These methods, however, cannot be applied without a prior understanding of the lithological descriptions of the core samples typically available in routine core analysis.

The recognition of major lithofacies is not an easy task in heterogeneous reservoirs. Rock characteristics such as petrophysical, depositional (or sedimentary), and diagenetic (or textural) features are common parameters that are used to define lithofacies. However, geologists with different field experiences often create different lithofacies sets based on the same observational information. These diverse definition occur because no quantitative measurements, but only a series of qualitative or linguistic statements, are provided in lithological descriptions. Thus a subjective decision must be made about how many dominant lithofacies are present and what these lithofacies are.

The objective of this paper is to introduce a systematic approach for the handling of linguistic descriptions of core samples, by contrasting a number of approaches to classify porosity into groups using petrographical characteristics. The three techniques used are an expert system approach, a supervised clustering approach, and a neural network approach. We will briefly describe each technique and provide results. We first review the basics of lithological descriptions and describe each technique. We then demonstrate the value of these techniques using a data set available for an oil well in a reservoir located in the North West Shelf, offshore Australia. We then apply the methods to porosity classification based on core descriptions, and validate the model using unseen cases with known porosity classes.

2. LITHOLOGICAL DESCRIPTIONS

Classifying geological data is a complicated process because linguistic descriptions dominate the results of core analysis studies. The problem is worse for lithological descriptions. Each core sample is usually described by a number of petrographic characters (e.g. grain size, sorting and roundness) in linguistic terms. A typical statement for a core sample could be:

"Sst: med dk gry f-med gr sbrndd mod srt arg Mat abd Tr Pyr Cl Lam + bioturb abd"

which means, 'Sandstone: medium, dark gray, fine-medium grain, sub-rounded, moderate sorting, abundant argillaceous matrix, trace of pyrite, calcareous laminae, and abundant bioturbation.'

Although these statements are subjective, they do provide important indications about the relative magnitudes of various lithohydraulic properties such as porosity and permeability. It is, however, difficult to establish an objective relationship between, say, porosity levels (e.g. poor, fair or high) and the petrographic characters.

3. DATA DESCRIPTIONS

An oil well located in the North West Shelf, offshore Australia, provided a routine core analysis report for this field study. There were 226 core plug samples taken from a total of 54 metres of cores obtained from three intervals. The reservoir is composed of sandstones, mudstones, and carbonate cemented facies. The porosity and permeability values ranged from 2 to 22 percent and from 0.01 millidarcy to 5.9 darcies, respectively.

The report includes porosity measurements from helium injection as well as detailed lithological descriptions on each core sample. The lithological descriptions were summarised into six porosity-related characters: grain size, sorting, matrix, roundness, bioturbation, and laminae. Each character was described by a number of attributes. A total of 56 attributes were used. Table 1 tabulates the character-attributes relationships used in this study.

The objective of this study is to demonstrate how intelligent techniques can be applied in classifying linguistic descriptions of core samples into various porosity classes. We will first develop the knowledge base, implemented for the three methods as expert system, clustering diagram and neural networks, respectively. The knowledge base is

TABLE 1

Character and attributes used for porosity classification

Character (No. of attributes)	Descriptions	Attributes
Grain size (12)	The general dimensions (e.g. average diameter or volume) of the particles in a sediment or rock, or of the grains of a particular mineral that made up a sediment or rock.	Very Fine, Very-Fine to Fine, Fine, Fine to Medium, Medium, Fine to Coarse, Medium to Fine, Medium to Coarse, Fine to Very Coarse, Coarse to Very Coarse, Very Fine with Coarse Quartz, Fine with Coarse Quartz.
Sorting (6)	The dynamic process by which sedimentary particles having some particular characteristic (e.g. similarity of size, shape, or specific gravity).	Well, Moderate to Well, Moderate to Poor, Moderate, Poor to Moderate, Poor.
Matrix (14)	The smaller or finer-grained, continuous material enclosing, or filling the interstices between, the larger grains or particles of a sediment or sedimentary rock.	Argillaceous (Arg), Sideritic (Sid), Siliceous (Sil), Sid with Arg, Sid with Sil, Arg with Sil, Sil with Arg, Carbonaceous, Calcareous, Pyritic with Arg, etc.
Roundness (8)	The degree of abrasion of a clastic particle as shown by the sharpness of its edges and corners as the ratio of the average radius of curvature of the maximum inscribed sphere.	Sub-angular (subang), Angular (Ang) to Subang, Subang to Sub-rounded (subrndd), Subrndd to Ang, Subang, Subrndd, etc.
Bioturbation (6)	The churning and stirring of a sediment by organisms.	Abundant bioturbation (bioturb), Increase bioturb, Bioturb, Decrease bioturb, Minor bioturb, Trace of bioturb.
Lamina (10)	The thinnest or smallest recognisable unit layer of original deposition in a sediment or sedimentary rock	Irregular argular, Irregular Calcareous, Trace of Calcareous, Less Traces, Argillaceous, Calcareous, Irregular Silt, Thick, Irregular.

developed using a number of known porosity cases (training data). The knowledge base will then be tested using an unseen set of core descriptions (test data). The performance can be evaluated by comparing the predicted porosity classes with the actual classes using the correct-recognition rate (i.e. number of correct classifications divided by total number of samples).

4. EXPERIMENTS

In the first phase of the experiment, the porosity values were discretised into four classes: 'Very Poor' (<5%); 'Poor' (5–10%); 'Fair' (10–15%); and 'Good' (>15%). Each sample was characterised by the six characters (with the corresponding attributes) and paired with a porosity class.

For the expert system, we chose a total of 140 samples out of the original 226 samples as the training and test data. This was done because the remaining samples lacked descriptions of some of the characters and were not able to be processed by the initial setup of the expert system, and could perhaps be considered as unrepresentative cases. The 140 cases were randomly divided into two data sets: Set #1 and Set #2. Each data set contained 70 cases. We first used the Set #1 data as the training data to develop the knowledge base. The training stage established new rules and updated old rules until the system gave a 100% correct-recognition for the Set #1 data. Then, the Set #2 data was used as the unseen test data, and the corresponding correct-recognition rate was calculated. Note that the existing rules were not updated at the testing stage and hence some results were 'no conclusion.'

We also swapped the usage of both data sets, that is, the Set #2 data were used for training and the Set #1 data for testing, and the whole process was repeated. The objective of the swapping experiment was to determine if there was a simulation bias associated with the random data-splitting procedure.

For the clustering algorithm and neural networks, we will perform the same experiments with the same data arrangement.

In the following sections the three techniques are briefly described, followed by the results sections for each experiment, followed by our conclusions, and suggestions for future work.

5. EXPERT SYSTEM

We have used an expert system knowledge acquisition and maintenance technique, to establish new rules (acquire knowledge) and to update existing rules (maintain knowledge) when suitable observations are obtained. Rules are formulated in the conventional form: IF [conditions] THEN [conclusion]. Knowledge is added to the system only in response to a case where there is an inadequate (i.e. none) or incorrect classification. This technique of 'ripple down rules' has been used in ion chromatography (Mulholland et al., 1993). The notion of basing classification on keystone cases has previously been used in petrography (Griffith, 1987). In cases of an incorrect classification, a human expert needs to provide a justification, in terms of the difference(s) associated with the case that shows the error or prompts the new rules, that explains why his/her interpretation is better than the interpretation given for such cases. Hence, the approach is able to adapt new rules or knowledge without violating previously established rules, and hence, all rules are consistent within the system.

The basic logic is simple and interpretable. There is only one requirement to develop the rule bases: all the cases must be described with a fixed set of descriptive characters. The rules can be viewed as binary decision trees. Each node in the tree is a rule with any desired conjunctive conditions. Each rule makes a classification, the classification is passed down the tree, and the final classification is determined by the last rule that is satisfied. The technique is very simple and has no further complications beyond the description given here. Its benefits derive from its simplicity, and its applicability without the need for an expert system specialist to build the knowledge base. There are some deficiencies, which we describe in the context of our results.

6. SUPERVISED CLUSTERING

A supervised clustering technique was also used. Clustering techniques are generally unsupervised. The benefit of the supervised approach is that the expert can label as acceptable clusters which make suitable distinctions in the data classification. Clusters which are not suitable can be labelled for further clustering. A portion of the data is held out (as for all the three techniques used) from the technique so that the success rate can be validated using this unseen data.

Visual Clustering Classifier (VC+) is a visual system through which users can conduct clustering operations to generate classification models. Clustering as an unsupervised learning mechanism has been widely used for clustering analysis (Jain and Dubes, 1988). Clustering operations divide data entities into homogeneous groups or clusters according to their similarities. As a clustering algorithm, k-means algorithm measures the similarities between data entities according to the distances between them. Lin and Fu (1983) applied a k-means based clustering algorithm for the classification of numerical data entities. To apply clustering algorithm to data mining applications, two important issues need to be resolved: large data set and categorical attribute. Extended from k-means algorithm, k-prototype algorithm (Huang, 1998) has resolved these two issues.

This k-prototype algorithm is based on an assumption that the similar data entities should be located closer than other data entities. Those similar data entity groups are normally called 'clusters'. A classification divides a data set into a few groups that are normally called 'classes'. The classes are determined either by human experts or a few data fields of the data entities, such as the application discussed in this paper. Therefore clusters and classes are not equivalent. To apply k-prototype algorithm for classification, the class distribution of the data entities in the generated clusters must be considered.

Two steps are required for the development of a classification model using VC+: cluster hierarchy construction; and classification model generation. Once the training data set has been loaded into VC+, a root cluster node for the cluster hierarchy is generated. The root contains the entire training data set. The user can apply the

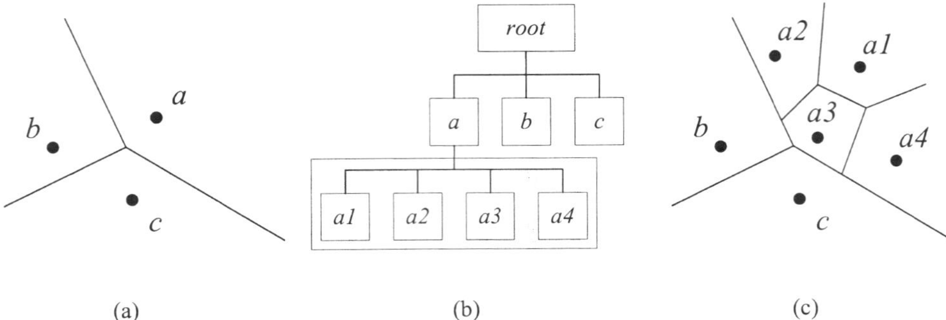

(a) (b) (c)

Fig. 1. Cluster hierarchy construction: (a) clustering result on root; (b) cluster hierarchy; and (c) result of the clustering on node *a*.

clustering operation on the data set to generated clusters that will be the children nodes of the root node. A leave cluster node in the cluster hierarchy will be further partitioned if the shape of distribution is not good or there is no a dominant class in the data entities in this cluster. Fig. 1 illustrates the procedure for generating a classification model. Firstly three clusters that have centers: *a*, *b* and *c* are generated by a clustering operation on root node. The cluster hierarchy will be generated. This cluster hierarchy will be expended after node *a* is further partitioned.

If there is a dominant class in the data entities in a leave cluster node, the center of this cluster will be marked as this class. The classification model generated by VC+ consists of all the leave nodes that have been marked. The class of the cluster in the classification model which has the shortest distance to a given data entity will determine the class of this data entity. If there is no dominant class for the data entities in a leave node and this leave node cannot be further partitioned due to the number of data entities contained, this leave node will be left unmarked and will not be included in the classification model.

To apply *k*-prototype clustering for classification, there are many non-deterministic criteria that directly affect the classification result, such as the number of clusters, the start cluster centers, and the chosen features. However it is out of computational power if all of the combination of these criteria were taken considered. VC+ provides various visualisation tools to display data entities, statistical results and also allows users to compare the results of different clustering operations. It also adopts visualisation techniques to incorporate users' expertise in the procedure for the generation of classification models. This approach increases the exploration space of the mining system. This approach has advantages on handling noise and outliers.

7. NEURAL NETWORKS

Neural networks can perform supervised classification. In this study, a standard 12 input $\times$ 7 hidden $\times$ 4 output backpropagation neural network was used. The input data was encoded by means of a linguistic encoding technique into 12 numeric inputs.

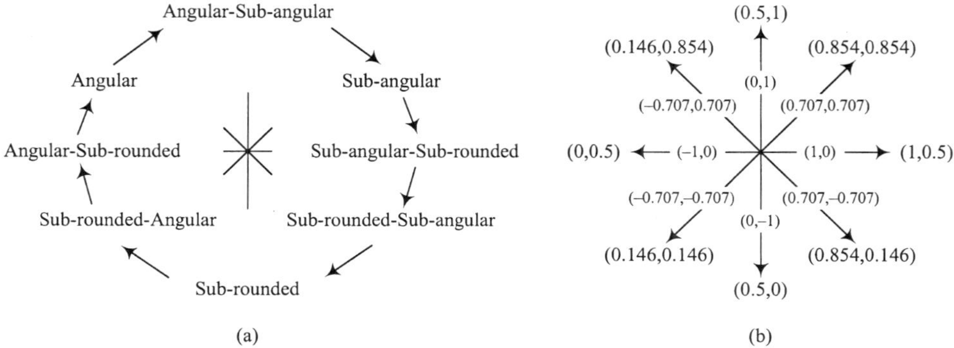

Fig. 2. (a) Circular encoding of roundness (sphericity); (b) normalisation to sine and cosine values.

The simplest case is for 'Sorting', where the characters of 'Poor – Poor-moderate – Moderate-poor – Moderate – Moderate-well – Well-moderate – Well' are easy to place in a sequence, and allocated values evenly distributed from 0 to 1.

For some of the fields more complicated encoding was necessary. For example, in the case of a circular linguistic term ordering, two variables are required to be able to encode the values. The values of the sine and cosine for an even distribution around a circle are required. This is illustrated for 'Sphericity' and 'Roundness' in Fig. 2.

As there are eight values, the familiar points of 0°, 45°, 90° and so on are used. The (sin, cos) tuples are shown in Fig. 2. The values are in the range from -1 to 1, which are then normalised to the range 0 to 1. The property of this circular encoding is that for all adjacent points the sum of the absolute values of the changes to the values is the same.

8. RESULTS

The data set contains 140 data records. We randomly divided the data set into two sets with equal size (70 each): Set #1 and Set #2. The classification matrixes generated for these two data sets are shown respectively in Tables 2–4. For each table, the first sub-table shows the blind test results using Set #1 for training and Set #2 for testing, and the second sub-table shows the blind test with the sets swapped. Note that for the supervised clustering we could not do this, since the human experimenter was making the supervision choices, it was not possible to do a second proper blind test. For the neural network we can again perform this swapped blind training and testing cycle.

As can be seen from the tables, the three techniques performed fairly similarly, with the supervised clustering algorithm performing the best, followed by the expert system technique and then the neural network.

Note that the results presented here are those achieved after some preliminary experiments, particularly with the neural network model to discover a reasonably successful architecture and an appropriate input encoding, with the expert system and supervised clustering models to discover the degree of cognitive effort required to

TABLE 2

Porosity classification results using expert system

Actual class	Predicted class					Total	% Correct
	VP	PR	FR	GD	NC		
(a) Blind test results on Set #1							
VP	10	0	0	0	1	11	90.9
PR	1	10	0	0	7	18	55.6
FR	1	0	7	5	7	20	35.0
GD	0	0	1	17	3	21	81.0
Overall % Correct = 62.9%							
(b) Blind test results on Set #2							
VP	9	6	1	0	1	17	52.9
PR	0	10	1	1	6	18	55.6
FR	0	0	8	1	1	10	80.0
GD	0	0	4	19	2	25	76.0
Overall % Correct = 65.7%							

VP = Very Poor; PR = Poor; FR = Fair; GD = Good; NC = No Conclusion; and % Correct is the correct-recognition rate.

TABLE 3

Porosity classification results on Set #1 using supervised clustering

Actual class	Predicted class				Total	% Correct
	VP	PR	FR	GD		
VP	8	2	1	0	11	72.7
PR	4	11	2	1	18	61.1
FR	1	4	14	1	20	70.0
GD	1	4	1	15	21	71.4
Overall % Correct = 68.6%						

VP = Very Poor; PR = Poor; FR = Fair; GD = Good; and % Correct is the correct-recognition rate.

achieve good results. While these are hard to quantify, it was clear that the supervised clustering required the most attention, followed by the expert system technique. The neural network preliminary experiments needed to be done initially, subsequently there was no intellectual effort required. This was not the case for the other techniques.

9. EXTENSIONS

In the second phase of the experiment, the expert system setup was modified to allow missing data, so the full data set could be used. Some modifications to the data encoding were tried, and the porosity values were discretised into 5 classes, adding a ‘Very Good Porosity’ class. This has the effect of reducing the random guess success rate from 25% to 20%.

TABLE 4

Porosity classification results using neural networks

Actual class	Predicted class				Total	% Correct
	VP	PR	FR	GD		
(a) Blind test results on Set #1						
VP	8	1	2	0	11	72.7
PR	4	11	3	0	18	61.1
FR	2	1	10	7	20	50.0
GD	2	3	3	13	21	61.9
Overall % Correct = 60.0%						
(b) Blind test results on Set #2						
VP	8	8	0	1	17	47.1
PR	0	11	5	2	18	61.1
FR	1	1	6	2	10	60.0
GD	0	2	4	19	25	76.0
Overall % Correct = 62.8%						

VP = Very Poor; PR = Poor; FR = Fair; GD = Good; and % Correct is the correct-recognition rate.

The experiments were re-run using the full date set, split 2/3 training, 1/3 testing, using all three techniques. The overall results were very similar. The supervised clustering algorithm produced 64.2% accuracy, the neural network result on the test set was 60%, and the expert system result was 59.7%.

Note that the expert system required some user effort in manual pre-processing to discover plausible rules and sequencing the data appropriately, to compensate for missing parameters. This is due to the system relying on cornerstone cases, which is prone to bias from the sequence of presentation of examples. Qualitatively, this appeared to be a greater cognitive burden than the equivalent task of encoding the inputs for the neural network, as that had to be done once only and did not require perusal of the entire training set and extraction of significant patterns.

Some extra experiments were performed using the expert system technique to discover the significance of such user pre-processing.

In the first of these extra experiments, very specific rules were created for each pattern, choosing all of the available non-null characters. This produced a result of 51.6% on the test set. This indicates that the previous effort in manual pre-processing had some significant effect, and the difficulty of doing this. The next experiment was to include the null fields for each pattern in each rule. Thus, if for a pattern no 'Sorting' character was reported, the rule specified that the value for this field be 'None'. This produced a result of 38.7%, verifying our belief that the system was providing some generalisation, and demonstrating the importance of making sensible rules. At the same time, we discovered the minimum possible error on the test set (with this data split) of 15% as there are some patterns with identical characters and different category.

Some extra experiments were also performed with the neural network, to modify the output encoding. The initial encoding was a standard one, with an output neuron active for each category. This binary encoding is known to be less than ideal where there is some continuity or ordering between output classes. This is clearly the case for our experiment, 'Good porosity' > 'Fair Porosity' and so on. When the values of the outputs close to class boundaries were modified so that the difference from adjacent values of the pairs of adjacent categories was minimised, an average of 7.1% improvement was found. That is, for example two patterns on either side of the 'Good' and 'Fair' porosity which initially would have been 1,0 and 0,1 now become 0.51,0.49 and 0.49,0.51. This reduces the artificiality of our decision as to the location of the class boundaries and hence better classification ensues. Note that the fuzzy encoding means that it is no longer a simple calculation to determine number of correct classifications. We have chosen to defuzzify the network target and actual outputs to category numbers in the appropriate range, and used a category number difference of <0.5 to indicate a category match. If we allow a catgory match <1.0 (we accept the adjacent category as suitable) then the results for the test sets on partitions 1 and 2 rise to 92.2% and 93.9% respectively. While accepting the entire net category as suitable is excessive, use of this technique allows a determination of the degree to which the incorrect results are plausible, particularly for value between 0.5 and 1.0.

The next step is therefore to modify the unsatisfactory expert system technique to allow such 'fuzzy' output encodings. This will require a major rework of the system, and is a limited form of the integration we propose below.

10. CONCLUSIONS

We have contrasted three techniques for using linguistic information from core analysis reports for classification. We have found that the use of pre-processing and clustering, and output encodings improve the results of the neural network. This kind of effort is required to satisfactory results from the expert system and supervised clustering techniques, which both require a major cognitive effort on the part of the user.

We can conclude that at this stage it is clear that the neural network technique is the best choice. The results are marginally worse, but the results are reproducible without a significant burden on the user.

To be fair, the expert system produced results using symbolic inputs essentially the same as the neural network on the numerically encoded inputs. This suggests that with the use of this encoding further improvements may be achieved. The benefit of expert system technique is that a rule trace is possible for every decision, so failures can be accounted for and successes understood by users. This tends to be an issue in the wider use of neural networks, where the 'black box' nature of predictions is unacceptable, mistrusted or merely not preferred.

The next stage in our work will be to properly integrate the three techniques. Thus, a neural network will be used to learn the significant properties of the data, which can then be examined and verified by the use of the clustering technique, and the training file constructed for the expert system technique. Even further down the track, we can envisage an on-line interactive use of the three techniques. Thus, when a new rule is required in the expert system, the neural network can be run on the as yet uncategorised patterns remaining to suggest some rules, and the clusters of patterns correctly or incorrectly classified be visualised on screen.

The use of these techniques systematically will allow the incorporation of such linguistic information with numeric well logs for improved results.

REFERENCES

Chang, H.-C., Kopaska-Merkel, D.C., Chen, H.-C. and Durrans, S.R., 2000. Lithofacies identification using multiple adaptive resonance theory neural networks and group decision expert system. *Comput. Geosci.*, 26: 591–601.

Griffith, C.M., 1987. Pigeonholes and Petrography. In: Aminzadeh, F. (Ed.), *Pattern Recognition and Image Processing*. Geophysical Press, pp. 539–557.

Huang, Z., 1998. Extension to the *k*-means algorithm for clustering data sets with categorical values. *Data Mining Knowl. Discovery*, 2: 283–304.

Jain, A.K. and Dubes, R.C., 1988. *Algorithms for Clustering Data*. Prentice Hall, Englewood Cliffs, NJ.

Jian, F.X., Chork, C.Y., Taggart, I.J., McKay, D.M. and Barlett, R.M., 1994. A genetic approach to the prediction of petrophysical properties. *J. Pet. Geol.*, 17(1): 71–88.

Lin, Y.K. and Fu, K.S., 1983. Automatic classification of cervical cells using a binary tree classifier. *Pattern Recognition*, 16(1): 68–80.

Mulholland, M., Delzoppo, G., Preston, P., Hibbert, B. and Compton, P., 1993. An expert system for ion chromatography implemented in ripple down rules. *Proc. of the 12th Australian Symposium on Analytical Chemistry*, Perth.

Rogers, S.J., Fang, J.H., Karr, C.L. and Stanley, D.A., 1992. Determination of lithology from well logs using neural networks. *Am. Assoc. Pet. Geol. Bull.*, 76(5): 731–739.

Wolff, M. and Pelissier-Combescure, J., 1982. FACIOLOG: Automatic electrofacies determination. *Annual Logging Symposium of the Society of Professional Well Log Analysts*, Paper FF.

Wong, P.M., Taggart, I.J. and Jian, F.X., 1995. A critical comparison of neural networks and discriminant analysis in lithofacies, porosity and permeability predictions. *J. Pet. Geol.*, 18(2): 191–206.

Wong, P.M., Gedeon, T.D. and Taggart, I.J., 1997. Fuzzy ARTMAP: A new tool for lithofacies recognition. *AI Applications*, 10(3): 29–39.

Wong, P.M., Tamhane, D. and Aminzadeh, F., 2000. A soft computing approach to integrate well logs and geological clusters for petrophysical prediction. *Third Conference and Exposition on Petroleum Geophysics*, New Delhi, 4 pp.

PART 4.

RESERVOIR AND PRODUCTION ENGINEERING

Developments in Petroleum Science, 51
Editors: M. Nikravesh, F. Aminzadeh and L.A. Zadeh

Chapter 17

USING GENETIC ALGORITHMS FOR RESERVOIR CHARACTERISATION

C. ROMERO [a] and J.N. CARTER [b]

[a] *PDVSA Intevep, P.O. Box 76343, Caracas 1070-A, Venezuela*
[b] *Department of Earth Science and Engineering, Imperial College of Science Technology and Medicine, South Kensington, London, SW7 2BP UK*

1. INTRODUCTION

Reservoir characterisation is the process of describing a hydrocarbon reservoir, in terms of the parameters of a numerical model, so that its performance can be predicted. One carries out this process as an aid to reservoir management, which is about making decisions on future production plans and investment so as to maximise the net present value of the hydrocarbon asset. A well executed reservoir characterisation will provide a good description of the reservoir, a prediction of production under a given production plan and an understanding of the uncertainty in both the description and the prediction. To provide an adequate characterisation of a reservoir is a formidable task, and is an active area of research. The work described in this chapter is just one piece of a large interconnecting puzzle. Our work depends on methodologies developed by others for parts of the puzzle, and in turn our methodology may be used to help complete the process.

In the following sections we will describe the methodologies we use to construct a realisation of the numerical model and how they relate to our GA methodology. Our aim is to give sufficient insight into the interactions between the methodologies such that reader could implement our GA methodology, but using different methods for the reservoir description The second section describes the reservoir modelling background to the work, this is followed by sections that describe: the details of our GA methodology; a description of the reservoir model; the results we have obtained and some concluding remarks.

2. RESERVOIR MODELLING

To make a prediction of the likely hydrocarbon production that will result from a given production plan, requires the use of a numerical model of some sort. The numerical models that are most commonly used to predict the reservoir performance, divide the reservoir into many small grid blocks. The mass flow from one grid block to another is calculated by solving the relevant partial differential equations (pde's). This is done using finite difference approximations to the partial differential equations, which results in the solving of large linear matrix systems. The results of the calculations are: the pressure distribution throughout the reservoir; the distribution of the different fluids present throughout the reservoir; and the flow into, and out of, the reservoir through the wells. This information is generated for many times as the pde's are solved at a sequence of times.

To perform a calculation of this sort we need to specify the initial conditions in the reservoir and the boundary conditions at the wells through time. Reservoir characterisation is the process of specifying those initial conditions and well boundary conditions. We are seeking sets of conditions which are consistent with: all of the measurements that have been made on the reservoir; a range of physical principles; and with the geological processes that created the hydrocarbon reservoir.

In reservoir engineering at the current time the numerical models used will typically divide the reservoir into 20,000–100,000 grid blocks. Although examples as small as 1000 grid blocks, or as large as 1,000,000 grid blocks are not uncommon. The size of the model is often determined by three factors: the perceived geological complexity of the reservoir, the computational resources and the time available for a study to be completed.

For each grid block we need to specify the following initial conditions: its location in space; its size; the fraction of the total volume that is occupied by fluids (porosity); the fraction of the pore volume occupied by each type of fluid present (saturation); the initial pressure; the ease with which fluids can flow horizontally and vertically (permeability and relative permeability); the fraction of non-reservoir rock (V-shale). All of these initial conditions must be consistent with some physical properties: e.g. the grid blocks must form a continuous, non-overlapping, space; oil/water mixtures are found above water only regions; pressures must be consistent with principles of fluid hydrostatics. The state-of-the-art at the moment usually assumes that reservoir characterisation is a process that determines four variables for each grid block, these being: horizontal permeability; vertical permeability, porosity and the volume of shale fraction (V-shale or net-to-gross ratio). A number of other boundary conditions may also be subject to the reservoir characterisation process: properties of geological faults, connate water saturation, irreducible oil saturations, well skin factors. It is further assumed that all other initial/boundary conditions can be determined sufficiently accurately from measurements such that they need not be included within the reservoir characterisation process. It is widely accepted that this final assumption is incorrect, and that on some occasions may lead to erroneous predictions. Individual reservoir studies may make slightly different choices about what to change, and what to keep constant, but the general approach taken is the same.

Our task is therefore to determine the values of approximately 100,000 data values, which determine the initial conditions, based on a very small number of data measurements. The reservoir used in this study was 11 km $\times$ 3 km $\times$ 200 m, and contained 11 production wells and six injection wells. We therefore have an areal coverage of one measurement of each property for every 2 km^2. Each measurement is valid for a volume of about 1 m^2 and 20 m in height, we have a total of 510 measurements. We also have measurements of water/oil flow rates through each well every month for four years (a total of not more than 1344 well measurements). In short the problem we are trying to solve is severely under-determined. It follows that many non-unique solutions exist, each of which equally well matches the available measurements. Added to this are problems with rather inaccurate measurements, which all adds to the non-uniqueness problem.

On the positive side we also have "soft" geological information that the reservoir description must be consistent with. That is to say the spatial distribution of properties must be consistent with the statistical properties of the geological processes that create the reservoirs. Many of the distributions of reservoir properties (e.g. permeability and porosity) that are consistent with the measurements of oil/water flow rates are either inconsistent, or very improbable, when viewed from a geological perspective.

We therefore need to determine which distributions of initial conditions are consistent with the expected geological statistics and which lead the numerical model (simulator) to predict measurements that are sufficiently close to the measured data. We have moved from trying to find descriptions that match the measurements, to finding descriptions that have a high probability of being consistent with the available knowledge. In many cases we satisfy ourselves with trying to identify the description with the highest probability.

3. SURVEY OF PREVIOUS WORK

The standard approach, within the petroleum industry, to creating a model that matches the available data remains for an experienced reservoir engineer to hand-craft a model. Computer assisted history matching and automated history history matching are making only slow inroads into traditional practice. It is widely accepted that eventually the reservoir characterisation (history matching) process will be highly automated, but that point has not been reached yet. Despite this slow take up by the industry, research into automatic history matching has had a long track record.

Many previous attempts have been made to automate the reservoir characterisation process, posing the inverse problem as an optimisation problem and using some optimisation technique to match the numerical results to the measurements. Early work focused on nonlinear optimisation and optimal-control theory. Jacquard and Jain (1965) introduced a method for the calculation of influence coefficients (sensitivity coefficients) of a linear system relating the differences between measured and simulated pressures to changes in the reservoir properties of a two-dimensional single-phase, synthetic model. They applied the modified steepest descent method to minimise those differences. The modelling was made under the mathematical theory based on the analogy of the electric analyser, once an important tool in reservoir engineering, where permeability is represented as resistance, porosity as capacity, and production as intensity of current. They reported the successful use of the method although they found that results were highly dependent on the zonation of the permeability zones. Jahns (1966) proposed the use of the Gauss–Newton method for history matching based on the approach defined by Jacquard and Jain, and the calculation of statistical measures of the reliability of the calculated reservoir properties. He tested the method in a five-spot pilot area and on an oil field with an assumed barrier at one side and a natural water drive on the other side. He reported that the method produced average reservoir properties in agreement with other available data, although it displayed a tendency for extreme values as well as poor estimates of the aquifer properties. Coats et al. (1970) further developed the method introduced by Jacquard and Jain by proposing a new approach for the estimation of a linear relationship between the error and reservoir property vectors from a number of simulation runs, using randomly-generated reservoir properties. However, the strong assumption that the pressure errors are linearly related to the values of the reservoir properties seems to limit the applicability of the method.

Generally, these nonlinear regression methods require the calculation of the partial derivatives (sensitivity coefficients) of the reservoir control variables (e.g. pressure) with respect to the reservoir parameters (e.g. porosity) which are usually obtained by numerical differentiation. This is achieved at each iteration by independently perturbing each parameter and running a full flow simulation to calculate the sensitivity of the reservoir control variables to the perturbed reservoir parameter. This implies the run of $(n+1)$ flow simulations if n reservoir parameters are to be estimated. As there are usually many reservoir parameters to be estimated, these requirements lead to significant computing time. Slater and Durrer (1971) presented a modification of the method proposed by Jacquard and Jain that reduced the time required to obtain a solution. This was achieved by using the changes on the residual to determine whether the perturbation was made in the correct direction. However, the computing requirements still remained high. Thomas et al. (1971) presented a technique based on the Gauss–Newton which minimises the differences between simulated and measured pressures by determining a vector for the changes in the reservoir parameter values and scaling this vector at each iteration. A totally different approach was proposed by Veatch and Thomas (1971). They presented a direct method that considered the finite difference equations within the reservoir simulator as a linear system of equations in the unknown reservoir parameters. In this way, if the pressure and saturation profiles are known at n different times, the method can be used to solve directly for n values of porosity and permeability. Its main disadvantage is that it requires pressure and saturation functions at every grid location at several times. The method is also very sensitive to errors in the measurements, because the solution is required to fit all the data exactly.

Carter et al. (1974) used two linear programming methods to minimise the error function based on the influence coefficients defined by Jacquard and Jain, but using predetermined constraint intervals for the values of the reservoir parameters. They found the procedures to be about equally effective in reducing the error function, while guaranteeing the finding of reasonable parameter values, although their main drawback is that they become less efficient near the solution.

However, not all methods are necessarily fully automatic. Hirasaki (1975) proposed a semi-automatic method for the estimation of reservoir properties using relationships between sensitivity coefficients (derivatives of cumulative oil with respect to reservoir properties) and dimensionless fluid production. The method can be useful for simple problems such as the matching of oil production only, but becomes intractable and tedious for complex reservoirs.

Chen et al. (1974) and Chavent et al. (1975) used optimal-control theory for automatic history matching of single phase models, taking the control variable (such as pressure) as the state variable and the reservoir properties (such as permeability) as the forcing variable. The method requires the equivalent of two simulation runs at each iteration and therefore is cheaper than second-derivative methods. It can also be cheaper than first-derivative methods, but the method has only linear convergence properties and a very large number of iterations are required for highly non-linear problems. Watson et al. (1997) reported a successful application of this approach on a two-dimensional two-phase model. Yang and Watson (1987) proposed the use of two variable metric (or quasi-Newton) methods for improving the rate of convergence when using the optimal control method, and incorporating the effectiveness of the variable metric methods in handling parameter inequality constraints. They successfully applied the methods on two synthetic, two-phase, one and two-dimensional test models.

More advanced gradient methods came in the late 1980's. Anterion et al. (1989) proposed a rigorous and computationally efficient method for gradient calculations. The method makes use of the derivatives already computed for the construction of the linear matrix system used to solve the flow equations, to obtain approximate values of the sensitivity coefficients. This technique requires a few additional calculations carried out within the reservoir simulator but eliminates the need for additional simulation runs (accounting for most of the computing time) to calculate the derivatives of the reservoir control variables with respect to the matching reservoir properties. They used the methodology in two simple, synthetic cases to help in a minimisation procedure based on an interactive graphical system and several extrapolation laws. A fully automatic procedure for history matching using the sensitivity coefficients calculated during the flow simulation run was introduced by Tan and Kalogerakis (1991). They successfully implemented a modified Gauss–Newton method for the history match of a three-phase, three-dimensional, synthetic model. They reported the algorithm able to retrieve the original values of permeability and porosity from a wide range of initial guesses, and that the computing requirements were reduced by at least an order of magnitude in single phase problems (1991). Bissell et al. (1994) applied the method on two real, three-phase reservoir models in a stepwise history matching of pore volume and transmissibility multipliers in one case, and fault transmissibilities in the other. They defined grad-zones (grouped grid cells) on which multipliers of the reservoir properties were applied. In general, the number of these grad-zones or matching parameters were less than 20. They reported results obtained within 20–30 simulations and comparable with those previously obtained by hand. Further research on the gradient optimisers using the approach by Anterion et al. (1989) was carried out by Deschamps et al. (1998). They tested six different gradient optimisers in two, relatively simple, two-dimensional models, one purely synthetic and another from a real field case. The gradient optimisers tested were: a Gauss–Newton/steepest descent method (ConReg), quasi-Newton (QN), a ConReg/QN hybrid, the Dennis–Gay–Welsch (DGW) method, the Al Baali–Fletcher (ABF) method, and the Double Dog Leg Strategy (DDLS). They found the hybrid Gauss–Newton methods (ConReg and DDLS) to be the best optimisers. However, they warned that for large history matching problems (e.g. 10,000 grid-cells and 100 parameters) the quasi-Newton method would be the only practical method, although it would nevertheless be unable to determine weakly dependent parameters.

Bayesian history matching was introduced by Gavalas et al. (1976) as an alternative of the zonation (grad-zones) approach. In the method, the inverse problem becomes statistically better determined because a priori statistical information on the unknown parameters is defined, and a penalty term incorporating prior geological information is added to the objective function. They applied the method to a problem of porosity and permeability estimation in a one-dimensional, one-phase, synthetic reservoir model, and found the accuracy of the Bayesian estimates dependent on the accuracy of the prior statistics used. Craig et al. (1996) applied a Bayesian linear strategy for pressure history matching, using data from fast approximations to the reservoir flow simulator. They provide a detailed account of all the assumptions used in the methodology. Bayesian approaches for history matching, however, require the specification of means, variances and covariances to formally combine the reservoir engineer's beliefs.

Incorporation of geological features into the automatic history matching process was the next big step. Bissell (1996) proposed a methodology for the history matching of reservoir models by the positioning of geological objects such as channels. He defined a modified objective function which incorporates a geological component. The method uses a Gauss–Newton optimising algorithm to find successive positions of the channel that progressively reduce the modified objective function using the derivatives of the well control variables with respect to the position or angle of rotation of the channel. He reported its successful application on a synthetic model.

Bissell et al. (1997) proposed and successfully applied the pilot point method as defined in de Marsily et al. (1984) to history-match the porosity values in a synthetic reservoir. In the method, the influence of the adjusted pilot points (geostatistical control points or pseudo-wells) points is communicated through the variogram to neighbouring grid-cells during the geostatistical simulation. The pilot point values are adjusted accordingly to the objective function using an optimisation algorithm. This method is described in detail later.

Wu et al. (1998) applied the method proposed by Tan and Kalogerakis for the conditioning of rock properties fields to a prior geostatistical model (variogram) and production data. They reported excellent matches when the method was applied to the inversion of permeability and porosity in a 25×25 grid block model with five control points, conditioned to flowing bottom hole pressure and water-oil ratio data. A more comprehensive procedure was proposed by Landa and Horne (1997) for the integration of dynamic and geostatistical data. The procedure uses the method proposed by Anterion et al. and a Gauss–Newton optimiser to determine the distribution of permeability and porosity in a synthetic model by integrating information from different sources: pressure measurements (including well tests), production history, interpreted 4-D seismic maps, permeability-porosity correlations, variogram models and geological object modelling. For a simple case with 100 parameters, they concluded that the method can be used as a tool for reservoir characterisation. A gradual deformation method was proposed by Hu and Blanc (1998) for constraining a reservoir facies model to dynamic data. The method gradually deforms a Gaussian-related stochastic model while preserving its spatial variability, and producing a generally smooth objective function that can be minimised with an optimisation algorithm. The iterative procedure is progressively used until a satisfactory calibration is reached.

Similar approaches have been used in ground-water modelling for the determination of transmissivity (related to permeability) and piezometric (related to pressure) data. Gomez-Hernandez et al. (1997) provide an excellent account of the level of development of parameter identification methods in ground-water modelling. Further developments in history matching of hydrocarbon reservoirs using gradient-based methods have been made using streamline-based inversion, as in Wang and Kovscek (2000).

However, the main drawback of most of these techniques is that they perform local searches only, in other words, they can be trapped in local minima. Global optimisation methods, and in particular Simulated Annealing (SA), were introduced in reservoir characterisation by Farmer (1989). Further work by Ouenes (1992) confirmed the robustness of Simulated Annealing for the simultaneous generation and conditioning of petrophysical property fields. He reported good results when applying SA to a set of inverse problems (including real cases) while gradient-based methods failed to approach the global optimum. Sen et al. (1995) applied two different versions of SA to a set of outcrop and tracer flow data: the commonly used Metropolis algorithm, and the heat-bath algorithm. They found that for relatively small problems, the version of SA using the Metropolis algorithm was the fastest. For larger problems, however, the heat-bath algorithm equaled and often outperformed the Metropolis algorithm. Portella and Frais (1999) combined SA and the pilot point method as proposed by de Marsily et al. to reduce the number of adjustable parameters and to introduce geological features into the matching procedure. They used a total of 210 geostatistical control points for the estimation of permeability at 5000 grid blocks constrained by oil and water production rates, on a synthetic, three-dimensional reservoir model. They reported a satisfactory minimisation of the objective function although they observed a bias in the results of the constrained images. They attributed this bias to the wrong permeability average provided by the static data (wells), and suggested adding the average and standard deviation of the permeability to the set of matching parameters. In addition, they found the procedure to be reliable but slow to converge. The tunneling method was implemented by Gomez et al. (1999) using the quasi-Newton method as the local optimiser. They tested the method in both a synthetic and real field cases. They carried out the history matching of production data (bottom hole pressure, water-oil and gas-oil ratios) with a small number (less than 30) of matching parameters (porosity and transmissibility). They reported that the method was able to successively find new minima and the computational cost of the tunnelling phase was comparable to that of the local minimisation.

GAs have also been applied to study inverse problems in many fields including those encountered in reservoir engineering. Stoffa and Sen (1991) applied a GA for the inversion of plane-wave seismograms and reported very encouraging results. Cartwright and Harris (1993) reported significant advances with a GA approach to the source apportionment problem in airborne pollution. In electro-magnetics, Tanaka et al. (1993) applied a modified GA for the estimation of static two-dimensional current distributions in materials using the observed external magnetic field. GAs were also used in crop yield studies by Pachepsky and Acock (1998) for the problem of generating random fields with an accurate reproduction of the input statistics of soil parameters. In ground-water modelling, Mayer and Huang (1999) applied a GA to search for a global or near-global solution in a coupled flow-mass transport inverse problem. They reported that the genetic algorithm used, although provided apparently robust solutions, was considerably less computationally efficient than the quasi-Newton algorithm.

Several attempts have been made within reservoir engineering to address the reservoir characterisation problem. Sen et al. (1995) applied GAs in a fairly standard way to the generation of stochastic permeability fields from outcrop and tracer flow data. The permeability values at grid points (matching parameters) were coded in a string of binary numbers (chromosomes). They worked with population sizes of typically more than 200 and using a random procedure to assign the initial values. Other details of their GA formulation are: standard fitness-proportionate selection, based on a fitness function stretched using an equation analogous to that used in SA for the acceptance of points with lower quality; k-point crossover; and bit-flip mutation. They also defined an update probability to control the influence of the models from the previous generations. They found the performance of the GA to be highly dependent on the choice of population size and the probabilities of crossover, update and mutation. They reported that the global optimum was closely reached when a moderate value of crossover probability (0.60), a small value of mutation probability (0.01), and a large value of update probability (0.90) were used.

Bush and Carter (1996) developed a modified GA that included a non-standard binary encoding and modified niching and breeding strategies. The approach was tested on a synthetic, vertical cross section with only three parameters: sand and shale permeability, and fault throw. These parameters were encoded in a binary string of variable length. They started with a (3×4) bit chromosome to identify only the large structures, and then switched to (3×8) bit chromosome after ten generations, to allow for the identification of smaller structures. They used an inter-leaving rather than a concatenation procedure for the chromosome construction because they initially encountered difficulties due to ghost clusters. They also used a steady-state GA rather than a generational one, halving the size of the parent population at each iteration, and using a modified rank selection with elitism. A random k-point crossover was used to produce two offspring (with k strongly biased towards low values) and mutation was applied. They reported significant gains compared to the standard, simple GA.

Guerreiro et al. (1998) successfully applied a GA to the identification of properties of heterogeneous reservoirs through the matching of tracer breakthrough profiles using six parameters. These parameters were the geometry (position and sides dimensions) of a rectangular insertion and the porosities inside and outside the insertion, of a synthetic model constructed based on a heterogeneous quarter of five-spot. Similarly to the previous works, the parameters were encoded in a binary chromosome that was 45 bits long. Rank-based selection with elitism was carried out using fitness values which are the summation of weighted absolute differences between the target and candidate solutions. Three crossover operators were applied simultaneously: one-point, two-point, and uniform crossovers with probabilities 0.08, 0.48, and 0.24 respectively, leading to a combined crossover probability of 0.8. The bit-flip mutation was set to 0.02, roughly $1/L$, where L is the chromosome length. A population of 200 individuals was evolved for 45 generations. They reported that the GA successfully identified the solution plus another caused by symmetry.

Very brief descriptions of the works by Bush and Holden on the PUNQ-S3 case are given in Bos et al. (1999). Bush used a GA to find the optimal values for the normal score transforms of porosity at pilot points locations defined in the model, whereas Holden applied the GA for the direct estimation of porosity, and horizontal and vertical permeability at each grid block of the simulation grid. Both briefly reported that the methods lead to good matches of the production history.

However, none of the previously described implementations of the GAs to the reservoir characterisation has exploited the relationship between the variables. Particularly, the works by Holden and by Bush, which were applied to a relatively complex reservoir, used one-dimensional arrays for representing all the variables. This approach implies that unrelated information is recombined by the crossover operators, which is not by any means wrong but makes the problem more difficult for the GA. It is expected that the GA will ultimately succeed in minimising the objective function although it might require a very large number of function evaluations.

4. METHODOLOGIES FOR RESERVOIR MODELLING

In this section we describe the methodologies that we have used to construct our reservoir models. It would be wrong to claim that the methodologies we have chosen are the only ones that should ever be used. Different geologies will probably need different methodologies to describe them. However, our choices are expected to work well for the reservoir that we have been studying. They should also work for a wide range of other geologies. The reservoir used in this work is described in more detail in section 5.

We assume that the only variables that the reservoir characterisation process has to select are:

- The variables that describe the distribution of three field properties (porosity, permeability and V-shale) throughout the reservoir.
- Some variables that describe the flow of fluids across faults.
- Variables for the mechanical skin factors in each of the wells.
- Relative permeability end-points.

Fig. 1 shows in diagrammatic form the complete reservoir characterisation process. In this section we will attempt to define everything between the GA and the flow simulation.

The approach we are using to build our numerical simulation model is to define the

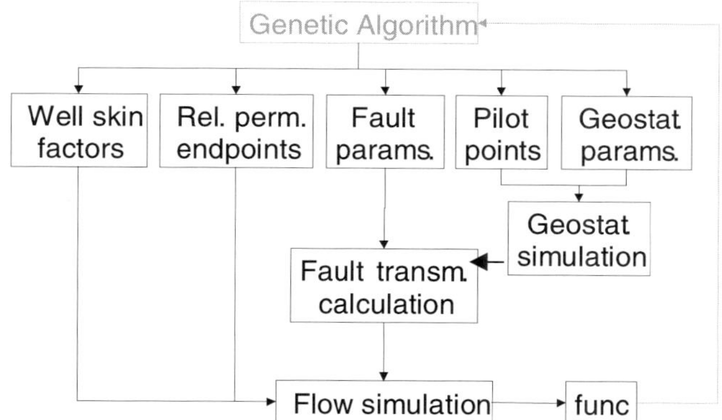

Fig. 1. Schematic of the reservoir characterisation process.

geology directly at the scale of the simulation model. An alternative approach would be to define the geology at a much finer scale, and then to use up-scaling techniques to convert from one scale to another. For this reservoir the fine scale model, known as the geological model, has dimensions of $60 \times 220 \times 197$ grid blocks which gives a total of 2.6×10^6 grid blocks. The simulation model has dimensions of $12 \times 44 \times 10$, that is 5280 grid blocks. The methodology we use to describe the geology works equally well at both scales. We work exclusively at the coarse scale, partly due to considerations of computer time, but mainly because we are interested more in the design of the GA.

To specify each of the three field wide parameter distributions for every grid block would require 15,840 variables, such a large number of variables is clearly impractical given the amount of constraining data. It is also unlikely that the resulting distribution of properties would be regarded as geologically acceptable. We therefore reduce the dimensionality of the problem by using a network of pilot points and geostatistical interpolation methods. The basic idea is that if we specify the value of a property field at a number of points, then we can fill in the missing information using some interpolation scheme. By careful choice of the interpolation scheme, it becomes possible to honour the necessary geological statistics.

In our model we have 57 pilot, or control, points on each of the nine active layers of the simulation model. The values of the variables are specified at these points by the characterisation method, in our work these values will be chosen by a GA optimiser. The remaining 471 grid blocks on each layer have their values determined by the interpolation process. We have chosen to use sequential Gaussian simulation as the interpolation process. The 57 pilot points in each layer are of two types, 17 are related to the positions of wells, the remaining 40 are spread around so that the pilot points cover the whole of the reservoir. The difference between the two types is that at the wells we have measurements taken directly from the reservoir, at the other points there are no direct controls over the values that it is possible for the variables to take.

From Fig. 1 you will see that two sets of variables are needed to carry out the geostatistical simulation. The first is the values at the pilot points, the second are a set of geostatistical parameters. The geostatistical parameters (variogram constants) are: maximum correlation range, anisotropy constant, maximum correlation direction angle and nugget effect constant. These four numbers describe the statistical distributions that the field properties have to conform to. To estimate the field properties for a whole layer in the simulation model involves a two step process: first, the parameter values are specified at the pilot points and values are given for the four geostatistical parameters; second, sequential Gaussian simulation is used to estimate values for the remaining grid blocks.

4.1. Geostatistical simulation

It would be inappropriate to fully describe the geostatistical method used, however it is useful to present a brief explanation of the method.

The combination of statistical methods and geological knowledge to create statistically plausible realisations of variable distributions in the earth sciences is addressed within geostatistics. Geostatistical methods for the spatial description of properties were introduced for mining applications, and the theoretical aspects are described by Journel and Huijbregts (1978). Stochastic conditional simulation is a geostatistical method that uses the available information to generate equiprobable descriptions of the reservoir properties. The method is stochastic because the reservoir properties are generated by a hybrid method that is part deterministic (the same input produces the same output) and part random (the same input yields different, independent output). It is also conditional, because the reservoir properties simulated honour the available data at the sampled locations.

The stochastic conditional simulation generates a random field consisting of values in $j = 1, \ldots, J$ cells based on $i = 1, \ldots, I$ sampled data points. There are therefore $(J - I)$ points to be estimated or simulated initially based on I conditioning points. There are various procedures to perform a conditional simulation. In this work, a widely used method is implemented: the Sequential Gaussian Simulation.

Sequential Gaussian Simulation (SGS) is a procedure that uses the Kriged mean variance to generate a Gaussian field. Slightly different versions of the procedure can be found in the literature (Jensen et al., 1997; Deutsch, 1998) and a rather arbitrary, but reasonable, choice has been made for the version used in this work. The entire procedure is outlined below.

The sampled data is transformed to be Gaussian, usually through the normal score technique, which uses a nonlinear transform to transform a continuous non-Gaussian histogram into a Gaussian cdf. The sample data consist of the petrophysical property values at the wells and the pilot points, which are assumed to be representative of the grid blocks that contain them. The values in the conditioned cells are kept constant throughout the procedure.

A random path through the field is defined so that each unconditioned cell is visited once and only once. At each visited cell, a pre-specified number of surrounding conditioning data are identified. This local neighbourhood is selected within the ellipse range on the semivariogram model, and may contain data from previously simulated cells. The semivariogram model gives a measure of the variance of sample increments measured at given distances units apart, and in this work is defined by the equation:

$$\gamma(h) = \begin{cases} (\sigma^2 - \sigma_o^2)\left(1.5\dfrac{h}{r} - 0.5\left(\dfrac{h}{r}\right)^3\right) + \sigma_o^2 & h \leq r \\ \sigma^2 & h \geq r \end{cases} \tag{1}$$

which is a spherical model defined by a correlation range r, positive variance contribution or sill value σ^2, and nugget effect σ_o^2. Anisotropy is accounted for by including nested structures of the semivariogram model that require the specification of the direction of maximum correlation and an anisotropy ratio, which is the ratio of the range in the maximum correlation direction to the range in the minimum correlation direction. Further details can be found in Deutsch (1998).

Kriging is then performed at each location used in combination with the semivariogram model to determine the Gaussian distribution (1998). Kriging is an estimation technique that provides a minimum error-variance estimate of any unsampled value, and it tends to smooth out the details and extreme values of the original data set. In Kriging, an estimate of a property Z^* is sought at an unmeasured location based on the Z_i,

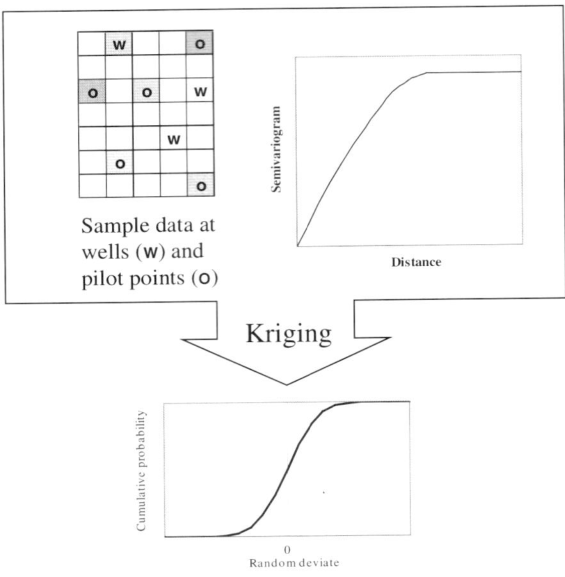

Fig. 2. Determination of the Gaussian distribution.

$i = 1, \ldots, I$ known values:

$$Z^* = \sum_{i=1}^{I} \lambda_i Z_i \tag{2}$$

There are several methods to calculate the Kriging weights (λ_i). In this work, Simple Kriging (without constraints) is used, which minimises the simple Kriging variance σ^2_{SK}:

$$\sigma^2_{\mathrm{SK}} = \mathrm{E}\left[(Z - Z^*)^2\right] \tag{3}$$

where Z^* is the estimate of Z at a single point. The process is illustrated in Fig. 2.

Then, a random number in the interval $[0, 1]$ from a uniform distribution is drawn and used to sample the local cumulative Gaussian distribution. The corresponding transformed value is the simulated value at that cell. The simulated value is incorporated to the data set, and the same procedure is repeated until all cells have been visited. Finally, all the values are back-transformed to their original distribution. This process can be repeated for as many reservoir descriptions as required.

4.2. Fault properties

From Fig. 1 you can see that the field properties are used in two ways, directly in the flow simulation and as an input to the calculation of the fault properties. Again we give a brief description of the methodology we use, the reader should consult the appropriate texts for a full explanation.

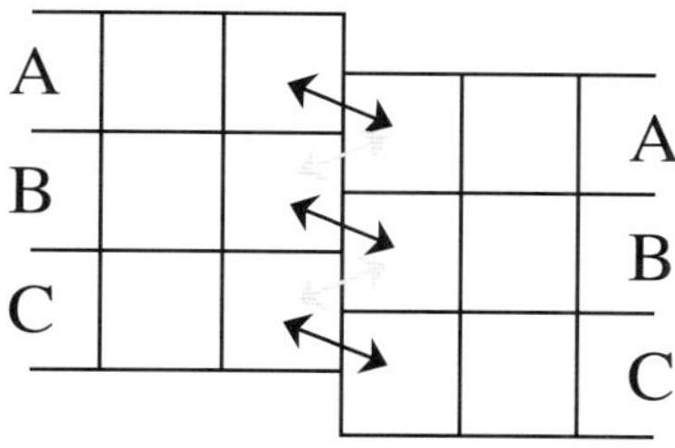

Fig. 3. Schematic of the reservoir characterisation process.

Geological faults are common features in hydrocarbon reservoirs, and numerical models therefore must include geological faults. Sealing capacity of major faults is one of the most influential parameters contributing to the overall model uncertainty (1997).

Faults influence flow in a reservoir simulation model in two ways. They modify the connectivity of sedimentological flow units, because displacements across faults cause partial or total juxtaposition of different flow units. They also affect flow because of the petrophysical properties of the fault-rock. Both thickness and fault permeability are physically observable properties of fault zones.

This section summaries the methodology developed by Manzocchi et al. (1999) that is used in this work to calculate the fault transmissibility multiplier. It is based on a new, geologically-driven method for determining fault transmissibility multipliers as a function of known properties of the reservoir model. The method aims to predict fault zone properties and to capture the influence of unresolved fault zone structure in sandstone/shale sequences using a simple algorithm. The fault zone properties are represented in the form of transmissibility multipliers between pairs of grid blocks. These transmissibility multipliers control the flow between grid blocks in the reservoir model and are calculated in the methodology for two types of connections: normal transmissibility connections and non-neighbour transmissibility connections, Fig. 3 illustrates the two types of connections.

In doing so, the method makes several assumptions and approximations, and require some quantitative data to condition the resultant model. The method makes use of an inverse relationship found between the Shale Gouge Ratio (G_R), and the log of the permeability of the fault-rock, based on various reservoir and outcrop fault-rock samples. It assumes that G_R is equivalent to the shale content of the fault gouge (1997), and also takes into account the fault displacement. The empirical relationship between the fault zone permeability as a function of shale content for a given displacement can be written as

$$\log k_f = -4G_R - \tfrac{1}{4}\log(D)(1 - G_R)^5 \tag{4}$$

where k_f is the fault permeability (in mD) and D is the fault displacement (in metres).

From fault outcrop data, an approximately linear relationship was found between fault zone displacement and fault rock thickness (t_f) with thickness values distributed over about two orders of magnitude for a particular displacement. The thickness of the fault zone is defined as the separation between the outermost slip surfaces (where more than one are present) minus the thickness of undeformed lenses. In the method, the relationship

$$t_f = \frac{D}{66} \tag{5}$$

was used to define the median thickness value (in metres) of a lognormal thickness distribution, with standard deviation of 0.9.

Walsh et al. (1998) assumed that both thickness and permeability vary over the area of a grid block according to a lognormal distribution. For any particular G_R and displacement, it is assumed that the median permeability and thickness values are given by Eqs. (4) and (5), in each case with $\sigma = 0.9$.

For permeability, the log-normal distribution with a log-variable mean

$$\mu = -4G_R - 0.25\log(D)(1 - G_R)^5 \tag{6}$$

and standard deviation $\sigma = 0.9$ gives the arithmetic average permeability as a function of G_R and displacement:

$$\log k_{fa} = 0.4 - 4G_R - \tfrac{1}{4}\log(D)(1 - G_R)^5 \tag{7}$$

whereas for thickness, $\mu = \log(D/66)$, $\sigma = 0.9$ give the harmonic average thickness as a function of displacement:

$$t_{fh} = \frac{D}{170} \tag{8}$$

Equations 7 and 8 give the fault zone permeability and thickness averages to be incorporated in the reservoir flow simulator. However, flow in reservoir models is calculated as a function of transmissibilities between pairs of grid-blocks. By combining the equations for the transmissibility, Trans_{ij}, between two grid blocks i and j separated by: a transmissibility multiplier, and a discrete thickness of fault-rock, the equation for the transmissibility multiplier (T_{ij}) can be written as a function of the dimensions and permeability of the grid-blocks and the thickness and permeability of the fault:

$$T_{ij} = \left(1 + t_f \frac{\dfrac{2}{k_f} - \dfrac{1}{k_i} - \dfrac{1}{k_j}}{\dfrac{L_i}{k_i} + \dfrac{L_j}{k_j}}\right)^{-1} \tag{9}$$

where L_i and L_j are the length of grid blocks i and j respectively.

The geological association between sedimentological and fault properties is maintained by calculating the transmissibility multipliers based on the underlying geological property field (shale volume content) and the global relationships linking G_R and fault displacement (which is assumed to be known) to permeability and thickness.

We calculate the fault thickness and permeability, for each connection, as a function of the shale-volume content and permeability maps, using the relationships $t_f = D/a$ and $\log k_f = b - 4G_R - c\log(D)(1 - G_R)^5$ respectively, where a (fault throw/thickness

TABLE 1

Prior definition of the global fault parameters

	Distribution type	Mean	Standard deviation
Constant a	log normal	170	0.55
Constant b	normal	0.4	0.625
Constant c	log normal	0.25	0.24

sealing parameter), b (fault permeability scaling constant) and c (fault permeability scaling multiplier) are constants. Table 1 gives the distributions used in the PUNQ-Complex model (Bos et al., 1999) and in this work. Finally, we calculate the transmissibility multipliers for the flow simulation model.

4.3. Well skin factors

The final group of parameters that appear in Fig. 1 are the well skin parameters. In any well there is a region immediately around the well-bore that has its properties changed by the process of drilling and completing the well. This region can vary in size from a few millimeters to a few metres, but it is alway small in comparison to the reservoir size. The changes that occur in reality are complex and non-uniform, however they are modelled by a single parameter. The effect of this parameter is to change slightly the pressure in the well-bore compared to what would have been the case if no change had occurred. A positive skin factor will cause a lowering of the well-bore pressure, assuming a fixed flow rate, or a reduction in flow rate for a fixed well-bore pressure.

4.4. Summary of reservoir description

In this section we have briefly described how a numerical model of a reservoir is constructed. It has not been our intention to provide a definitive definition of this process, but simply to indict which methodologies we have used and how they have been implemented. We describe a reservoir using five groups of variables:

- Three sets of reservoir property fields (porosity, permeability and V-shale), each of which are defined only at a set of 513 pilot points.
- Four variables (maximum correlation range, maximum correlation direction angle nugget effect and anisotropy).
- Three variables (fault throw/thickness, sealing parameter, permeability scaling constant, permeability scaling multiplier) defining the properties of the faults.
- One well skin factor for each well, a total of 17 variables.
- Four relative permeability end-points on each of nine active layers, a total of 36 variables.

Our ten-layer numerical model has one inactive layer, which reduces the number of layers to be modelled to nine. We also choose to use a single geostatistical description for the whole reservoir, rather than one for each active layer. Our complete reservoir description is therefore controlled by a total of 1599 variables. Our object in the reservoir characterisation process is therefore to find values for these variables that cause the output from the flow simulator to be a good approximation to the measured historical data from the real reservoir.

5. RESERVOIR DESCRIPTION

In this work we have used a synthetic reservoir model for two main reasons. A synthetic model was already available in the PUNQ Project framework (PUNQ, 1999) reducing the time required to set up a test model. The second reason is that the PUNQ model does not have certain problems typical of real field cases such as scarcity of data, structural uncertainty, etc., that could render intractable the evaluation of the algorithm.

5.1. PUNQ complex model

The first stage in the construction of the model of the original PUNQ-Complex model (1999) involved specifying a geologically reasonable deposition. Porosity (ϕ), permeability (k), and volumetric shale fraction (Vshale) were specified for each grid block of a very fine $60 \times 220 \times 197$ grid (2.6 million grid blocks) with each grid block being of size 50 m $\times$ 50 m $\times$ 1 m. In stage two the model was subjected to realistic changes in structure, due to seismic and tectonic activity, to produce a faulted/domed trap which was allowed to fill with oil. The two last stages were to upscale the model onto a $20 \times 60 \times 20$ grid blocks model, and to design a realistic production plan and simulate the expected production using a fluid flow simulator (Geoquest, 1998) on that scale.

In this work, we have modified the PUNQ model for our particular needs. We have used a $12 \times 44 \times 10$ version of the structural grid. We have also used upscaled data from the very fine scale model for a substitute for measurements at the wells. The number of wells has been reduced from 23 to 17 (11 producers and 6 injectors), but the complex production plan developed for the PUNQ model has been retained. The permeability, porosity and V-shale have been recalculated for this work using a single set of geostatistical parameters.

The generation of the production history and the history match were done on the same scale, thus avoiding the problems of production allocation and the inaccuracy between the grid and well completions.

5.2. Reservoir model

The structure of the PUNQ-Complex model is fairly typical of North Sea fault-bounded trap reservoirs. The reservoir is bounded by a large reservoir normal fault. Fig. 4 shows the large scale structure of the PUNQ-Complex model. Intra-reservoir normal faults have down-throw directions which are both synthetic and antithetic to those of the main fault. Individual faults show displacement variations along their length. The faults are vertical and therefore the top reservoir, fault and structure, contour map

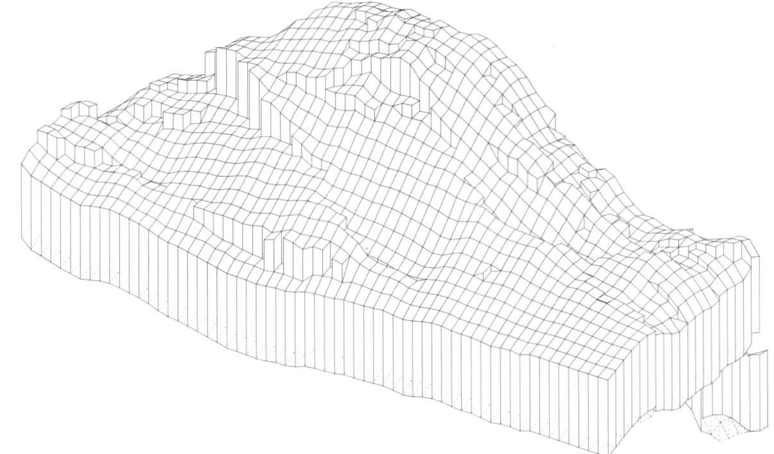

Fig. 4. Reservoir Structure

were taken as the uppermost surface from which the entire reservoir sequence hangs. Fault maps are therefore the same for all horizons, which is a reasonable approximation given the lengths of the fault traces (usually greater than a few hundred meters) and the thickness of the reservoir (ca 200 m). The resulting structure is one in which structural closure provides an oil column, with an oil-water contact which closes at both ends of the field.

The structural part of the $12 \times 44 \times 10$ grid blocks model is as accurate as can be achieved with the coarseness of the grid used; all the large faults were included.

To generate the 'truth' case of our reservoir model, we use exactly the methodology used to generate models within the reservoir characterisation process. The list of variables that we have to specify is:

- well skin factors for 17 wells,
- relative permeability end-points on each of nine layers,
- fault parameters,
- geostatistical parameters,
- porosity, permeability and V-shale on each of nine active layers at each of the 17 wells.

For the variables in the first four points on this list, we have available to us the range of values used as a priori data for the PUNQ complex model. We have simply chosen a value from each range to be the 'truth' value. The final group of variables was obtained from the upscaled data from the PUNQ complex model. Exact details of the choices made can be found in the full report of the work (Romero, 2000).

We then combine the $12 \times 44 \times 10$ structural model with the calculation of the petrophysical properties, and the production plan. This is then processed through the flow simulator to obtain 4 years of production data. The 'exact' measurements obtained from the reservoir simulator then have noise add to them so as to obtain the actual measurements that would be used in the reservoir characterisation.

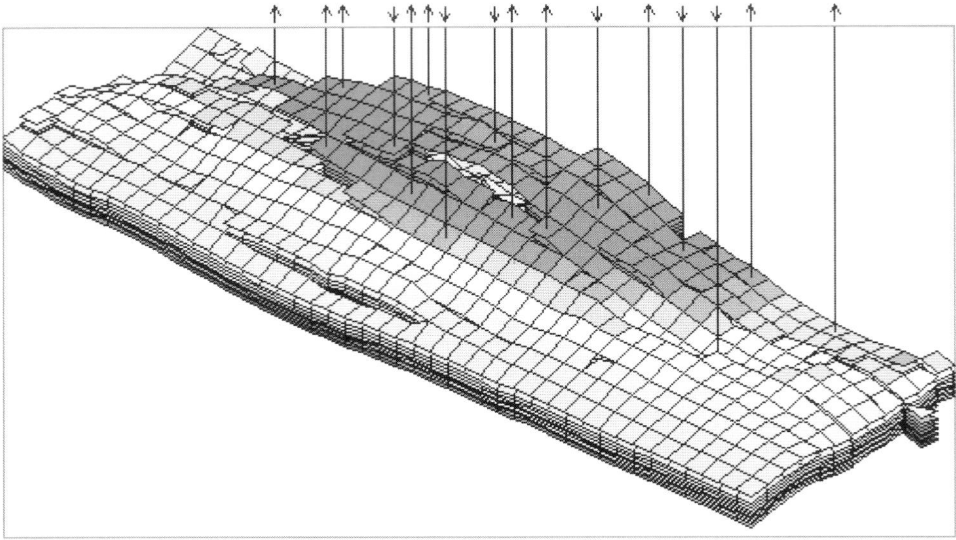

Fig. 5. Location of wells and major faults.

5.3. *Production plan and well measurements*

The production plan was designed with two objectives in mind: firstly, it should be realistic; secondly, it should provide adequate data for history matching. To achieve these objectives a number of simplifying assumptions were made:

- all wells would be purely vertical,
- there would only be one drilling phase,
- operational constraints would be identical each year.

The production plan is defined by the following items:

- the number and location of production and injection wells,
- well and field wide production and injection rates,
- pressure limits for all wells,
- logging and work-over policy for wells.

As with any field development a number of options were proposed and their economics examined. The final plan being chosen on the grounds that it would probably have been implemented had this been a real field.

As the reservoir is highly faulted, it would be very difficult to drain all of the producible oil with just a few production wells. As a consequence the plan requires 11 producer wells and 6 injector wells, the locations of these are shown in Fig. 5. Due to the compartmentalised nature of the reservoir, it was found to be prudent to maintain the reservoir pressure from the start of production. Four producers and two injectors were pre-drilled, into the crest of the reservoir, prior to the start of production. Thereafter a new well is brought on-line every two months.

The field wide operating conditions were quite simple, with a maximum liquid handling rate of 15000 m^3/day for production, and a voidage replacement policy for

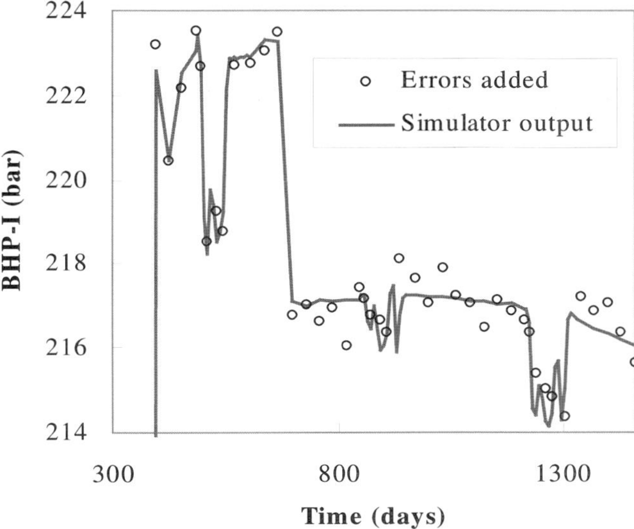

Fig. 6. Typical profile for BHP-I, with and without errors added.

water injection. Individual injection wells are subject to maximum injection rates and maximum bottom hole pressure (BHP) limits, while the production wells are subject to maximum liquid production rates and minimum bottom hole pressures, so as to avoid free gas being liberated in the reservoir. Production wells are also subject to a minimum economic production rate of 80 m^3/day. To allow data to be collected each well is shut-in for two weeks in the middle of each year, during which it is worked over to shut-off any completions producing with water cuts of over 80%.

A large amount of data was collected to allow a variety of data sources to be used for history matching studies. At the field averaged level we have recorded the oil and water production totals and the water injection total. At the well level we have recorded BHP and instantaneous oil/water production rates and water injection rates.

Once the production profiles were generated using the true models outlined in the previous subsection, Gaussian errors were added to the measurements. For the field oil rates, the measurement error was assumed to be random Gaussian, with a standard deviation of 0.25% of the measurement value. A standard deviation of 0.5 bar was assumed for the bottom hole pressure measurements. The standard deviation for each of the three produced fluids is 7.5% of the measured value at the wells, and 2% of the measured value for the injection rate. Fig. 6 shows an example of a typical profile for the bottom hole pressure at a injector (BHP-I).

6. DESIGN OF THE GENETIC ALGORITHM

Fig. 1 shows us using a genetic algorithm to control the inputs to the reservoir description stage. Almost all of the previous research in this area have used more traditional numerical optimisation algorithms to complete this part of the process. One might be tempted to ask why the genetic algorithm is as good as, or possibly a better, choice than the algorithms used elsewhere?

To answer this question we need to look at some of the details of what we are trying to do. The problem, as described, in the preceding sections has the following properties:

- The size of the search space is large, with of the order of 1,600 variables.
- The characterisation process will not be sensitive to every variable, but it is difficult to decide in advance which ones can be safely ignored.
- There is not a one-to-one mapping between the variables chosen and the description obtained. One set of variable values can be mapped to many different reservoir descriptions. Although every reservoir description can be generated from one, and only, one set of variable values.
- The problem is non-linear, highly underdetermined and may well be mutli-modal.
- Gradient information is not easily obtainable for every variable.
- The cost of obtaining a function evaluation is relatively high, and in a real world study one would want to minimise the number of function evaluations used.
- The opportunity exists to parallelise the process cheaply, by simply running multiple simulations on separate machines.
- There exists structural properties within the modelling framework that could be exploited.

From these observations we can draw some conclusions: due to the one-to-many mapping, the lack of gradient information for some parameters, and the uncertainty of parameter sensitivities, gradient based algorithms, or algorithms that approximate gradients, are not appropriate; we therefore need to consider algorithms that use function evaluations only. The structural properties of the modelling framework can easily be handled within a complex data structure, which can then be exploited by a genetic algorithm. The possibilities for coarse grained parallelism would recommend the use of the Genetic Algorithm. None of the issues raised above would present a problem to a Genetic Algorithm based method.

6.1. General parameters

Through out our study we have used a fixed design for the Genetic Algorithm. The details of that design are:

- We use a generational replacement scheme, with the whole population being replaced by their offspring. The population size is generally 20 individuals.
- We use an elitism strategy, where the best individual is cloned, copied into the next generation, and re-evaluated. Due to the one-to-many mapping from genome to reservoir description, this re-evaluation of the individual will cause it to have a different function value.
- Parents are chosen using two person tournaments, with the fitter individual always winning. Two parents, selected independently with replacement, are needed to produce offspring. Each set of parents produces just one child, therefore at each generation 19 offspring are produced in this way.
- A child is constructed by using the crossover operators described below, it may then be subjected to mutation.

- Our approach uses specially designed genome, crossover and mutation operators and has a particular method for generating the initial population.

6.2. Design of the genome

The normal approach to genome design is to use a single chromosome with all of the variables expressed as either real numbers, or binary codes. The chromosome takes the form of a single one dimensional array. You can then apply a single crossover operator and/or a single mutation operator, since every part of the chromosome has the same form. The operators do not differentiate over the different properties of the variables within the single chromosome.

In our work we take a quite different approach. We have seven sets of parameters, which have relationships that can be exploited during the crossover operation to achieve better convergence. The seven sets are: permeability, porosity, V-shale, fault parameters, geostatistical parameters, well skin factors, and relative permeability end-points. We allocate each set to a different chromosome, each chromosome can be design to allow any structure that exists to be exploit, each chromosome is then dealt with separately during crossover and mutation using operators that have been designed specifically for that chromosome.

The first three chromosomes share the same complex three dimensional structure which includes many extrons. This will be described in detail below. The fault parameters are encoded as three real numbers, with known upper and lower bounds. The geostatistical parameters are encoded as four real numbers, again with known upper and lower bounds. The skin factors are encoded as 17 real numbers, finally the relative permeability end-points are encoded as 36 real numbers, with all of the end-points for a particular layer grouped together in one part of the chromosome. Our non-standard structure for the genome is summarised in Fig. 7.

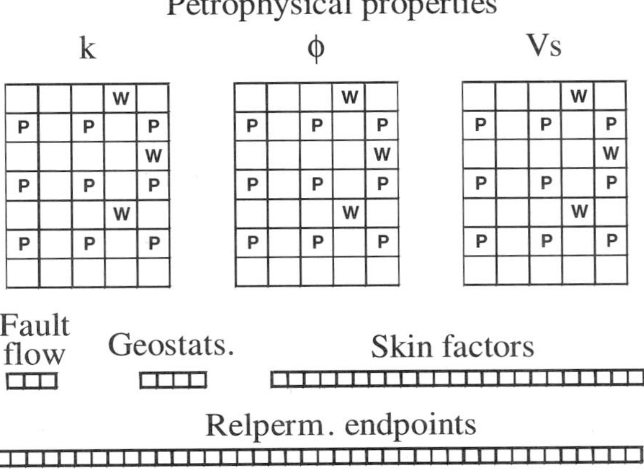

Fig. 7. Non-standard genome structure.

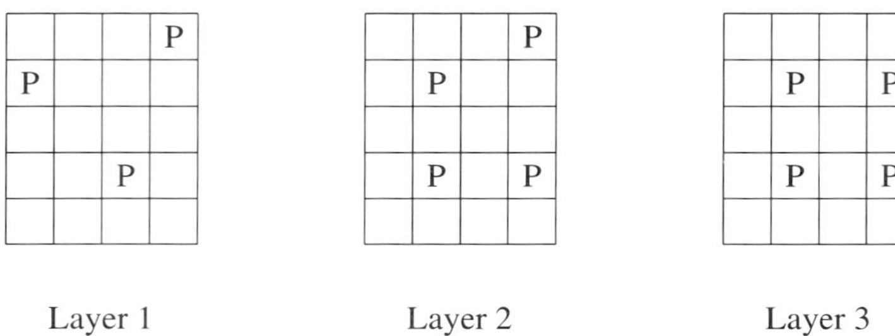

Fig. 8. Example of a field property chromosome.

6.2.1. *Chromosome for reservoir property fields*

From a reservoir engineering point of view, each of the reservoir property fields (permeability, porosity and V-shale) have a very similar structure. The complete fields, i.e. after application of the geostatistical interpolation methods, fill space completely. In the case of our numerical model this means that every grid block has been allocated a value. The value at any point is related to the values in the surrounding three dimensional grid. Some of the points in this three dimensional space are the pilot points that we have described in section 4, and which form the variables on which the GA is expected to work. The influence of any particular pilot point will depend on where it is within the the three dimensional structure, and how far it is from other pilot points. In previous work (Carter, 1997) we have found that it is very advantageous to design a chromosome to take advantage of any spatial relationships that exits. We therefore believe that better results will be obtained by designing the chromosome so that account is taken of the relative locations of all of the pilot points. The chromosome that we use has the same dimensions as the simulation grid on which we are working. The only places within this large structure that have meaningful information are the locations of the pilot points, everywhere else contains extrons (Levenick, 1985), Fig. 8 illustrates what a chromosome would look like for a simple reservoir model.

6.3. *Crossover operators*

We have used two crossover operators in this work. For the one dimensional chromosomes we have used simple k-point crossover, where the probability of switching to extracting information from the other parent is at some constant value for each chromosome. This is a quite standard approach used within GAs. We have designed a special crossover operator for our three dimensional chromosomes.

6.3.1. *Crossover for three dimensional chromosomes*

The purpose of the any crossover operator is to decide from which parent one should select a particular variable. For a one dimensional operator one can produce a map of the form

$$(+,+,+,-,-,-,+,+,+,+,-,+,+)$$

+	+	−	−
+	−	−	+
+	−	+	−
−	+	−	−
+	−	+	+

Bit-flip crossover

−	−	−	+
−	−	−	+
−	−	−	+
+	+	+	+
+	+	+	+

Generalised k-point crossover

Fig. 9. Examples of two dimensional crossover maps.

where the + symbol indicates that you select that variable from parent 1, and the − symbol indicates that you should select the variable from parent 2. Our aim is to produce the equivalent for our three dimensional chromosomes. Fig. 9 shows two possible crossover maps: the first has the symbols allocated in an unbiased random fashion with equal probabilities for each symbol, we call this a bit-flip crossover; the second has all of the symbols set to one of the two possible, except for a rectangular region within which the other symbol is used. We have tried many different ways of constructing these maps, but only report results of one here.

6.4. Mutation operators

In this work we have used three mutation operators, two of them can be applied to any of the chromosomes, the third is only applicable to the well skin factors. The first two operators are defined as operations on real numbers and are comparable to traditional mutation operators, as they cause a random change in part of a chromosome. The third is a deterministic operation that is applied to every parameter within the well skin chromosome, it is also applied after the flow simulation has been completed, but before the function evaluation is made.

6.4.1. Jump mutation

Binary chromosomes are mutated, i.e. an individual bit is flipped ($0 \mapsto 1$, or $1 \mapsto 0$) accordingly to a given probability (Spears, 1993). For real number chromosomes, jump mutation is applied by randomly resetting the value of the gene to a value determined by assuming a uniform pdf with the appropriate upper and lower bounds. It follows that the new parameter θ' is calculated using the equation

$$\theta' = \theta_{\min} + R\Delta\theta \tag{10}$$

where $\theta_{\min}$ is the lower bound, R is a random number between 0 and 1, and $\Delta\theta$ is the range of the relevant variable.

6.4.2. Creep mutation

For binary chromosomes, creep mutation was defined as the mutation causing the smallest possible change in the value of a gene. This is achieved by decoding the relevant gene, moving the real number value one discrete position away, and encoding it back to binary. In the case of real number chromosomes, the value of the gene is randomly changed by a small random quantity assuming a quadratic pdf centred on the current value. It follows that the new parameter θ' is calculated using the equation

$$\theta' = \theta + R^2 \Delta\theta C \tag{11}$$

where θ is the original value of the variable, R is a random number in the range $(0, 1)$, $\Delta\theta$ is the variability range of the relevant parameter, C is $+1$ if a second random number is less than 0.5, and -1 otherwise.

6.4.3. Shift mutation

It is arguable whether a mutation that modifies a gene in a deterministic way can still be considered a mutation. In this work, a mutation operator has been defined that modifies skin factor parameters so that a better match in bottom hole pressures is achieved.

This can be done because, in a producing well the bottom hole pressure is inversely proportional to the well skin factor (S), assuming steady-state flow and that the well is producing at maximum potential, i.e., there are no restrictions in the production facilities. However, the bottom hole pressure is a function of well and block parameters across all layers connected to the well. If a given bottom hole pressure (a point in the production history) must be attained, an iterative method can be used to find a value S that minimises the errors. By equating the fluid rates at each well zone, the equation

$$\sum_{i=1}^{n_c} \frac{\text{const}_i}{\ln\left(\dfrac{r_{oi}}{r_w}\right) + S^{(1)}} \Delta P_i^{(1)} = \sum_{i=1}^{n_c} \frac{\text{const}_i}{\ln\left(\dfrac{r_{oi}}{r_w}\right) + S^{(2)}} \Delta P_i^{(2)} \tag{12}$$

can be obtained, where const_i is proportional to the transmissibility, r_{oi} is the 'pressure equivalent radius', r_w is the well-bore radius, and $\Delta P_i^{(1)}$ and $\Delta P_i^{(2)}$ are the difference between the nodal pressure at the grid block and the bottom hole pressure, under $S^{(1)}$ and $S^{(2)}$, respectively, and n_c is the number of well connections in that particular well. The same equation holds for the injecting wells, where the bottom hole pressure is directly proportional to the skin factor, under the same assumptions.

For a time interval, an average error between the historical and the simulated bottom hole pressures can be estimated, and the aforementioned calculation can be done for a given point within the interval. Essentially, this minimises the average error in the interval by 'shifting' the bottom hole pressure curve by a certain amount (the average error). This approach is recommended over the former, because it avoids the possible use of outliers in the history that can produce a change in the opposite direction. An illustrative example is shown in Fig. 10.

This shift mutation or skin factor recalculation can occur at a certain rate controlled by a given probability. A switch for extreme values avoidance was added for the cases where the new S lies beyond the upper or lower bounds of the parameter. In those cases, S is set to the relevant bound if the extremes avoidance switch is active, or remains at its original value otherwise.

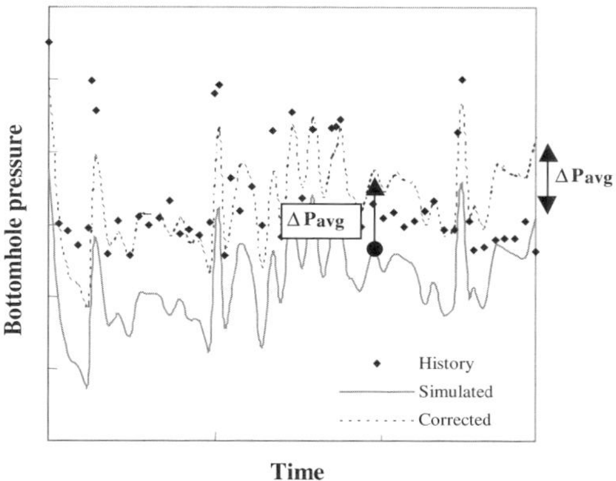

Fig. 10. Illustration of the shift mutation operator.

6.5. Function evaluation

The previous section described the synthetic model on which an algorithm for reservoir characterisation is to be tested. The reservoir characterisation procedure makes use of measurements made on the field to restrict the range of values that the modelling parameters might take. The measurements used can be wide-ranging and include seismic data, data from geological analogues, core and log data from wells, well test data, and production data. However, in this work only measurements from well logs and production profiles are used. Every measurement has associated with it a degree of uncertainty. An stochastic formalism to assess the uncertainty in the production characteristics in hydrocarbon reservoirs can be found in Omre et al. (1993).

In the present work, the inverse problem is also addressed within a Bayesian statistic framework (Sivia, 1996). Under this approach, a probability P_i is defined to measure the likelihood that a particular reservoir description gave rise to the measurement observed. If errors are assumed to be Gaussian distributed, the relationship

$$P_i = \frac{1}{\sigma_i\sqrt{2\pi}} \exp\left(-\frac{1}{2}\left(\frac{o_i^{(\mathrm{m})} - o_i^{(\mathrm{s})}}{\sigma_i}\right)^2\right) \tag{13}$$

can be defined, where $o_i^{(\mathrm{m})}$ is a particular measurement, $o_i^{(\mathrm{s})}$ is the equivalent value from the model representation (the reservoir simulator in this case), and σ_i is an indication of the uncertainty associated with the measurement. Making reasonable assumptions about the independence of the measurement errors, and assuming uniform priors, the reservoir characterisation aims at finding those reservoir descriptions that maximise the likelihood function given by

$$L = \prod_{i=1}^{n} P_i \tag{14}$$

If only the production profiles are included in the calculation, the likelihood function can be written as

$$L = c_1 \exp\left(-\frac{1}{2}\sum_{i}^{n_p}\sum_{j}^{n_w}\sum_{k}^{n_t}\left(\frac{o_{ijk}^{(\mathrm{m})} - o_{ijk}^{(\mathrm{s})}}{\sigma_{ijk}}\right)^2\right) \tag{15}$$

where c_1 is a constant, i runs over the production data types, j over the wells, and k over the report times, with n_p, n_w, n_t being the respective number of samples.

This likelihood function measures the likelihood that the historical data can be produced by a particular reservoir description. High values indicate that the response of the reservoir model resembles the historical data, and therefore the historical data is more likely to come from the given reservoir model than it would be if the likelihood function values is low.

Traditionally, the objective function is defined as

$$F = c_2 - \frac{1}{2}\sum_{i}^{n_p}\sum_{j}^{n_w}\sum_{k}^{n_t}\left(\frac{o_{ijk}^{(\mathrm{m})} - o_{ijk}^{(\mathrm{s})}}{\sigma_{ijk}}\right)^2 \tag{16}$$

where c_2 is another constant, and which is analogous to taking $-\ln(P)$ (Sivia, 1996). In this work, the contributions to the objective function have been grouped according to the production data type, and some averaging has been performed for convenience of the analysis, leaving

$$F = \sum_{j}^{n_p}\frac{1}{n_{oj}}\sum_{i}^{n_{oj}}\left(\frac{o_{ij}^{(\mathrm{m})} - o_{ij}^{(\mathrm{s})}}{\sigma_i}\right)^2 \tag{17}$$

where $n_p = 4$, because in this work only the the bottom hole pressure at the producers and the injectors (BHP-P and BHP-I respectively), water production rates (WPR), and gas production rates (GPR) are included in the objective function, n_{oj} is the number of observations available for each individual control variable. The field oil rate was not included in the objective function. It follows that F provides the sum of the average squared normalised errors for each control variable. The two other production quantities measured at the field, i.e. oil production rate (OPR) and water injection rate (WIR) do not contribute to the objective function because they are specified as boundary conditions in the reservoir simulation and therefore the observed and simulated values are the same.

Assumed standard deviations were as follows: 0.5 bars for BHP-P and BHP-I; 7.5% of the measured values of OPR, WPR and GPR; and 2% of the measured value of WIR. These standard deviations values correspond to those used to add measurement errors to the 'truth' history. Water and gas production rates are preferred over gas-oil ratio (GOR) and water cut (WCT) indicators because the variances of the formers are easier to define. All the reporting times (history points) are taken into account along the production history.

6.6. Generation of the initial population

An optimisation algorithm requires the specification (initialisation) of at least an initial point from which the iterative process can start. This initialisation process can be carried out in different ways depending on how much knowledge is available. If no information is available, the most usual approach is to randomly choose a starting point. At the other end of the spectrum, a starting point that conveys all available information can be carefully specified, in other words, starting from the best estimate. However, if some general and/or specific knowledge is available, then a quasi-random initialisation can be used.

For each of the 1599 variables that we are using, we have available a probability distribution function (pdf) that uses a priori information to estimate the value that each variable might take. Each individual in the initial population is generated randomly, but uses the appropriate pdf. All of the variables are generated directly, except the reservoir property fields which are generated using a less direct approach.

The pilot-point variables can be split into two groups: those at the wells; and those between the wells. The variables at the wells will be constrained by their measured values. That is to say that the uncertainty of these variables is much lower than for equivalent variables at non-well pilot-points. The variables at the pilot-points are generated in a two stage process. First, the a priori pdfs are used to generate the variable values at the wells. This is followed by a geostatistical simulation to generate values for all of the remaining grid blocks. We then extract the variable values at the remaining non-well pilot-points. For full details of the process and the values used the reader is referred to the full report of this study (Romero, 2000).

7. RESULTS

In this section we present four groups of results: a discussion of the general progress of the optimisation; the variability inherent in the methodology used to construct the numerical models; an examination of the quality of the match obtained well-by-well; and a comparison of using the GA and other simple optimisation schemes.

7.1. *Progression of the optimisation*

Fig. 11 shows the typical behaviour of the objective function through a GA run of population size 20 and 30 generations. In general we have seen in our experiments a steep decline of the objective function followed by a less steep decline. We also see quite alot of noise, particularly at the beginning of the optimisation, although there exist sporadic peaks throughout the optimisation. The comparison of curves is very difficult because of this noise, which has two sources. One is the general behaviour of GAs, where there is always a spread of objective function values due to the genetic diversity in the population. The second is due to the one-to-many mapping from genome to reservoir description. We have attempted to assess the contribution to the noise from the creation of the reservoir description by generating multiple realisations of the reservoir from a single set of parameter values. The results of this are shown in Table 2. The parameter set we choose was the generator for the truth case, which is the first one in the table, the other 20 are generated from the same parameters as the truth case but using a different geostatistical seed. For each realisation, four numbers are quoted. The first is

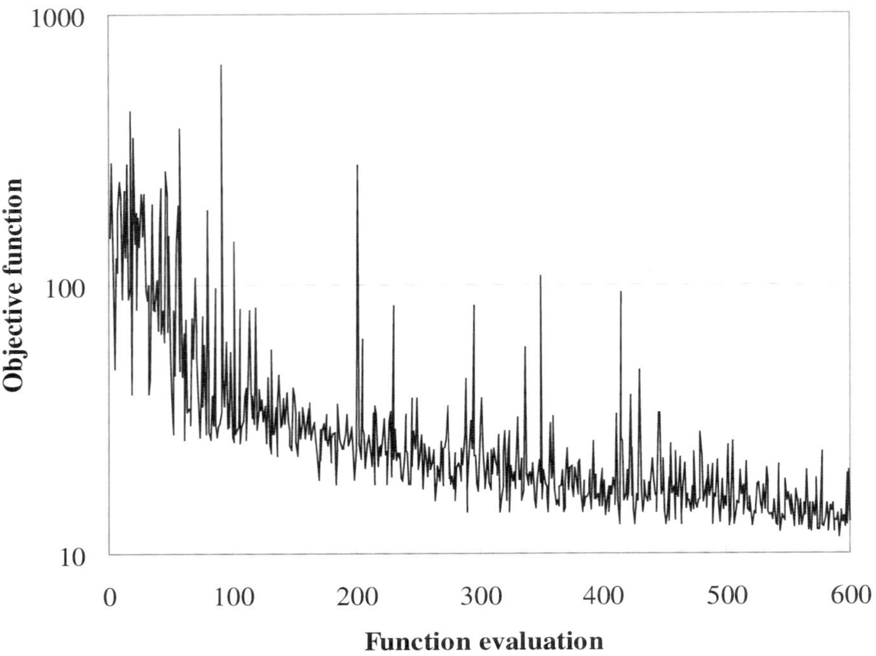

Fig. 11. Typical progress of a GA optimisation.

the objective function value, the other three numbers are the contribution from: the BHP of the producers; the water production; and the BHP of the injectors. The table clearly shows that there is a very wide range of function evaluations that are possible from a single individual. We also see that in many cases the error from the water production part of the function gives the biggest contribution to the function. This is particularly true for those cases with high function values. It seems likely that much of the noise seen in the later parts of the optimisation will be down to variations in the generation process. The best function value that we might expect is about 3.9.

7.2. Analysis of results for each well

Figs. 12–16 show how successful our approach has been in matching the measurements at individual wells. Figs. 12–15 show the results for the 11 production wells, whilst Fig. 16 has the results for the injection wells. For each of the production wells we present the BHP and the water cut, although the calculation of the objective function used water production rate. For the injection wells we show only the BHP. Each graph has the same basic structure: the x-axis displays the 1460 days of the production history from first oil production; open circles are the true measurements obtained from the base case with added noise; the line marked "initial" came from the best model in the initial population of 20 individuals; Model 1 is the result obtained from the best model in generation 20; model 2 is the best model overall and which occurred in generation 18.

Some of the most obvious changes are the improvements in the BHP values. Wells P-1, P-2, P-3, I-1 and I-5 are particularly note worthy. Wells I-4 and P-5 show less good matches on the BHP, whilst the rest of the wells show limited improvements. Matching the water production is an inherently harder task, and the results in this area are some what mixed. In wells P-3, P-9 and P-10, models 1 and 2 are are significantly better than the initial model. Whilst wells P-4 and P-5 show a significant worsening in the water cut obtained, the other wells show either no change or limited differences. Overall we are of the opinion that these results are comparable with what might be achieved by a reservoir engineer. We could improve the quality of our match to the BHP, whilst having minimal effect on water production, by using the shift mutation operator. If we were to do this then the BHP measurements for wells P-3, P-4, P-5, P-9, P-11, I-3, I-4 and I-6, would become very good. If this process produced skin factors that were not believable, then a similar effect can be obtained with very local changes of permeability around the wells.

7.3. Comparison with other optimisation schemes

For comparative purposes, three other search techniques have been implemented: Simulated Annealing, a global optimisation method; hill-climbing, a deterministic, local search technique; and random search. Without intending to be comprehensive, these commonly used methods are applied using fairly standard implementations. This fact must be kept in mind if comparisons are to be made against the modified GA. Additionally, some tests were carried out using hill-climbing to perform local optimisation after a GA run was conducted. In the runs reported in this section, a fixed geostatistical seed was used for the geostatistical simulations, corresponding to the one used to generate the reference petrophysical property fields. The effect of which is to have a one-to-one mapping which reduces the variability seen in the optimisation.

TABLE 2

Influence of the geostatistical seed on objective function

F	BHP-P	WPR	BHP-I	F	BHP-P	WPR	BHP-I
3.8848	0.7201	1.6082	1.1845				
420.7280	40.4580	361.9459	17.9516	31.0346	4.9214	20.3455	5.3958
156.2662	15.9509	136.1200	3.8236	443.0309	93.4493	272.9644	76.2282
72.0467	9.0359	34.0781	28.5609	10.6641	4.1243	2.3045	3.8634
35.1925	6.7948	2.8049	25.2208	61.8122	11.1648	4.1648	46.1107
72.5432	22.1975	25.3801	24.5938	27.0826	15.1759	2.7288	8.8060
722.3554	29.5240	685.6838	6.7752	200.4091	8.9012	184.3918	6.7441
116.3195	19.2038	77.2499	19.4944	42.5094	5.9918	25.6794	10.4663
20.7681	6.0202	3.6501	10.7259	91.3432	34.6119	39.6636	16.6958
56.2835	6.7873	39.7746	9.3496	48.7402	5.8459	35.8186	6.7037
39.8263	22.1039	2.9818	14.3687	66.7608	24.192	4.6406	37.5564

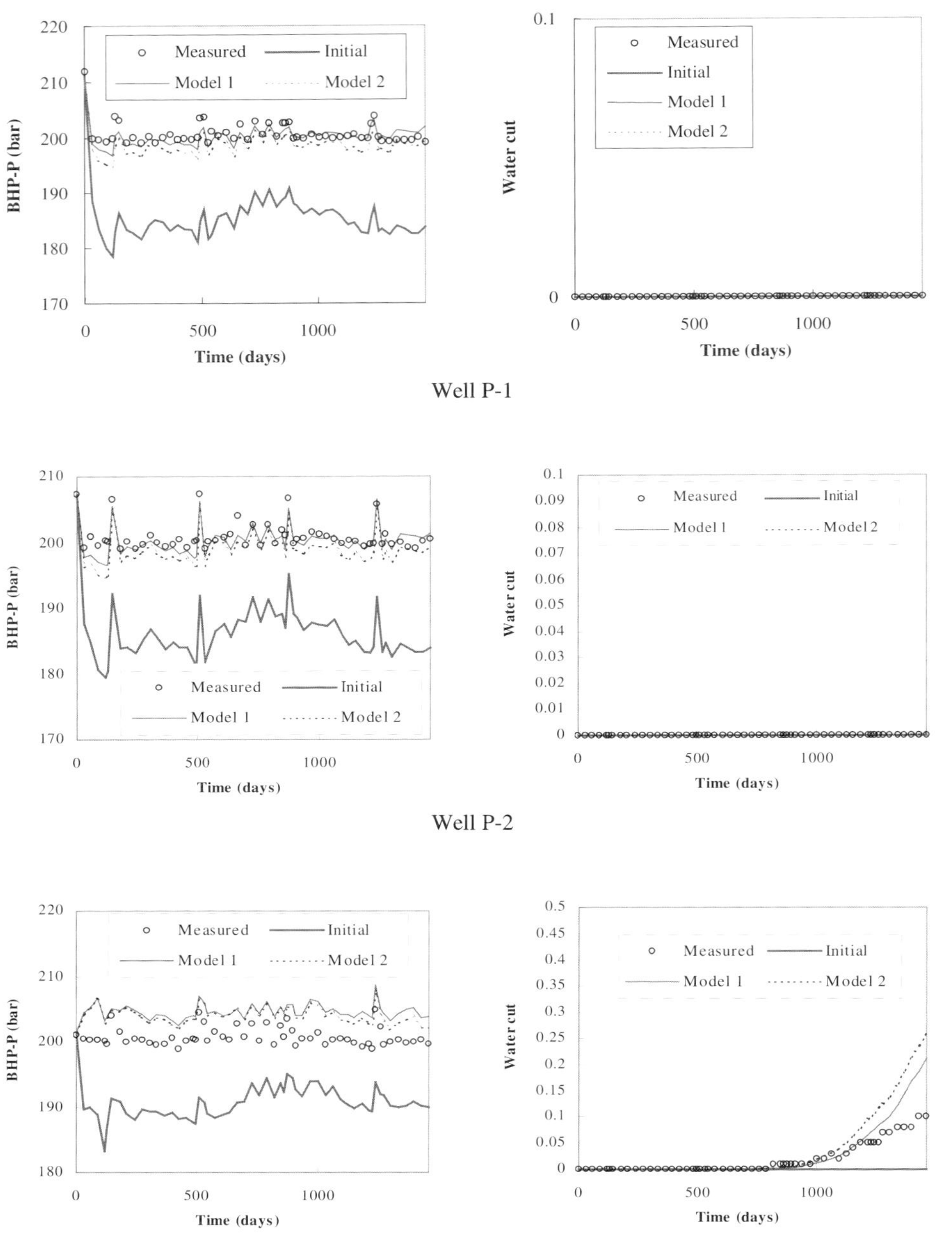

Fig. 12. History match for production wells P-1, P-2 and P-3

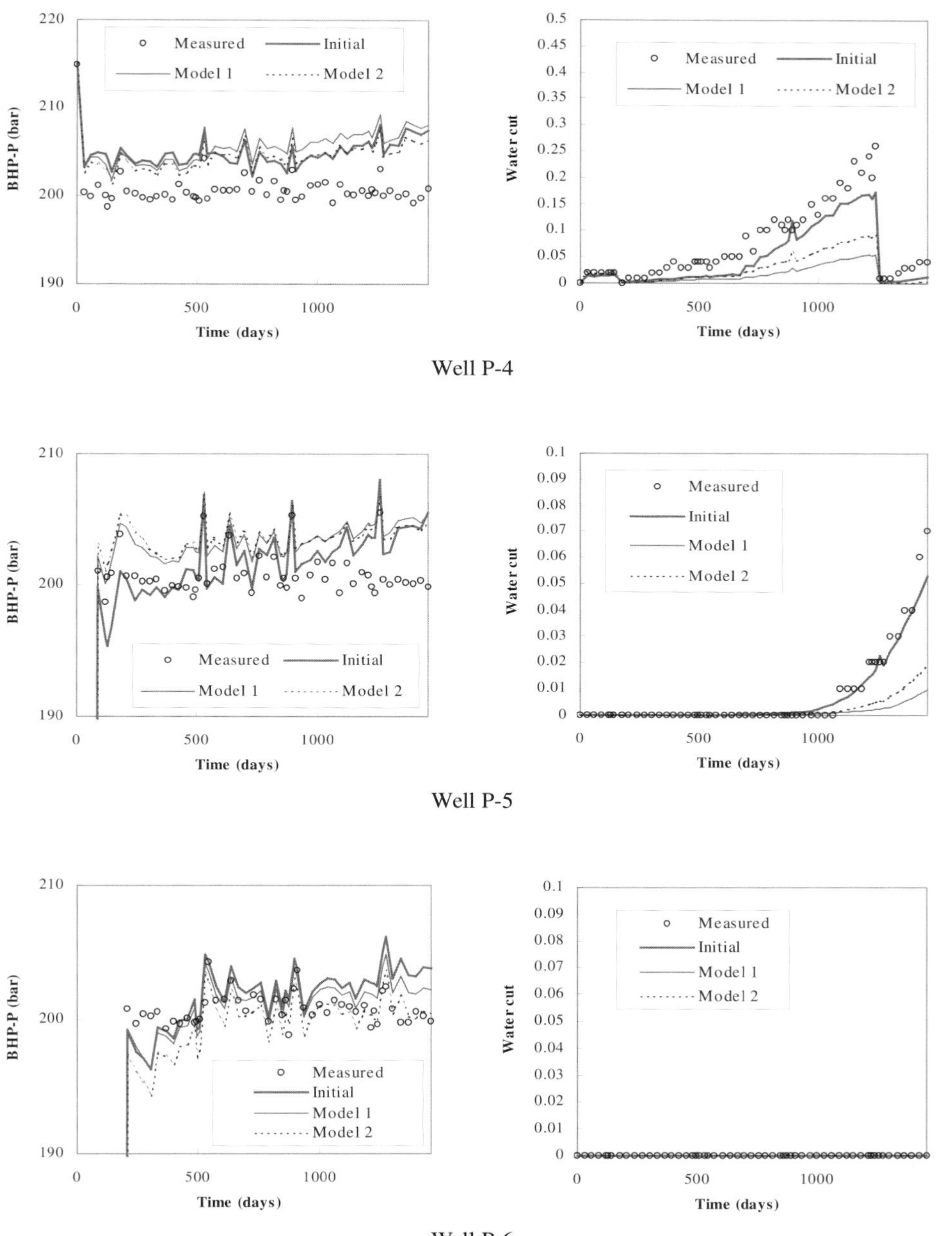

Fig. 13. History match for production wells P-4, P-5 and P-5.

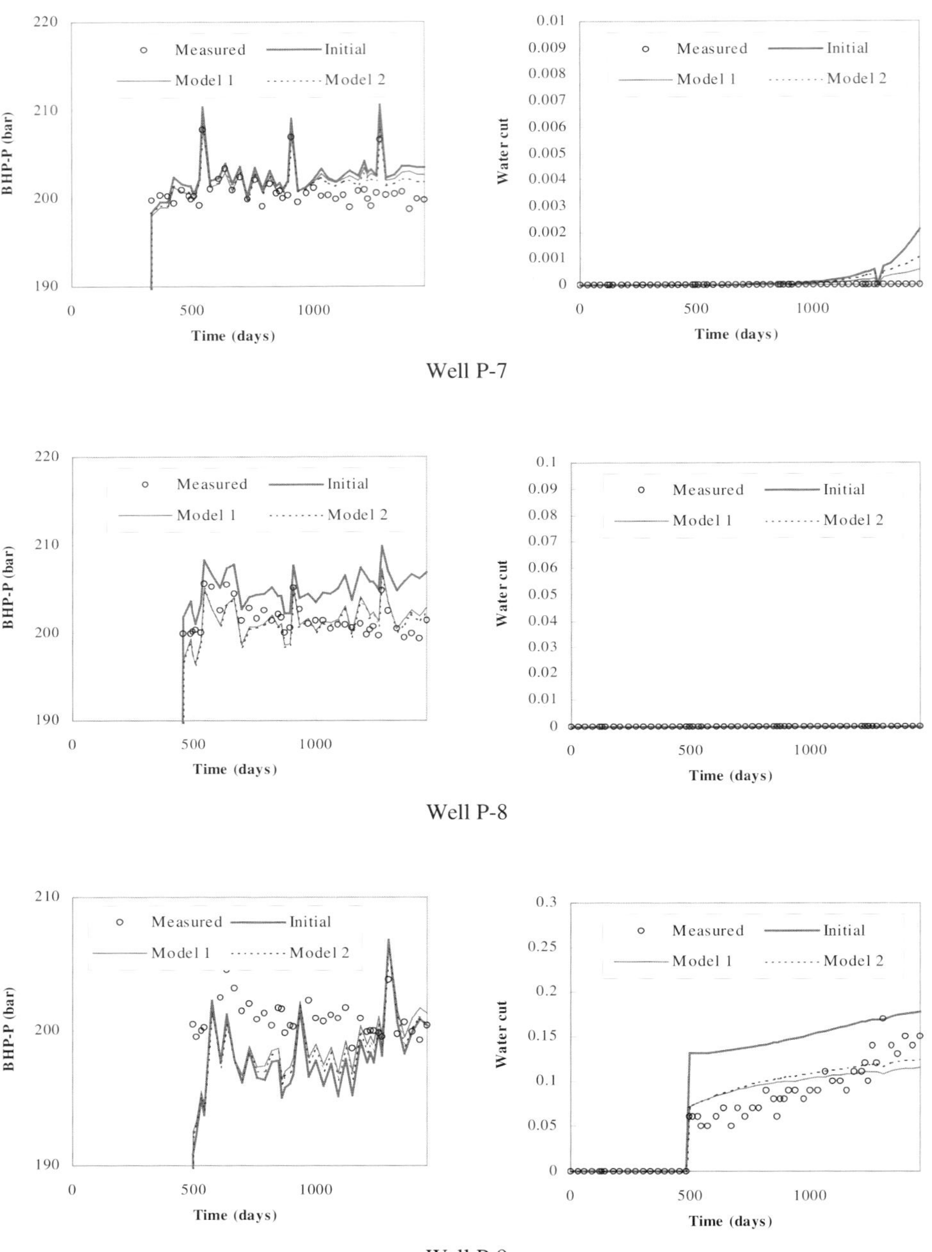

Fig. 14. History match for production wells P-7, P-8 and P-9.

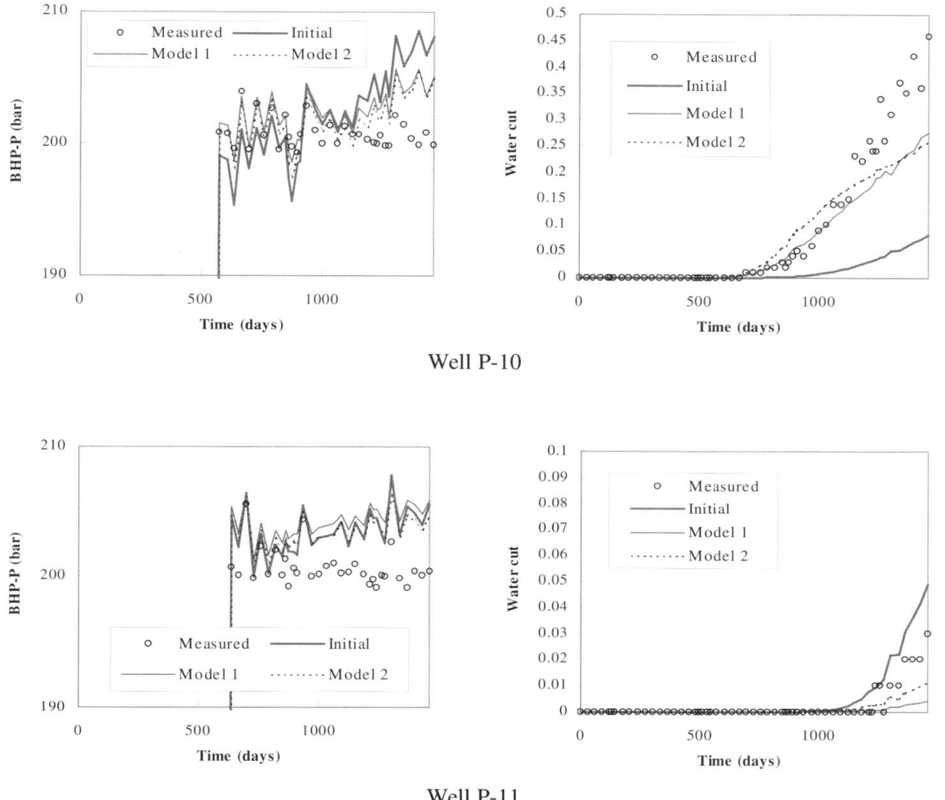

Fig. 15. History match for production wells P-10, and P-11.

7.3.1. *Simulated Annealing and random search*

Simulated Annealing (SA) is an optimisation method inspired by the physical process that reduces the temperature of a system to its minimum energy (annealing), making the analogy of this minimum energy to the global optimum of an objective function. In this work, an SA with an automatic cooling schedule developed by Huang et al. (1986) has been implemented using the creep mutation operator, with a probability of a gene being mutated of 0.02, and using the best model in the first GA generation as the starting point, a rather advantageous start.

The random search generated 400 reservoir realisations by repeating the initialisation process.

The results of using SA and random search are compared to a GA in Fig. 17. The GA shows slightly reduced levels of noise compared with Fig. 11, this is due to the use of the fixed geostatistical seed. Otherwise its behaviour is similar to that seen in the previous figure. The SA was started from the best solution in the initial population from the GA,

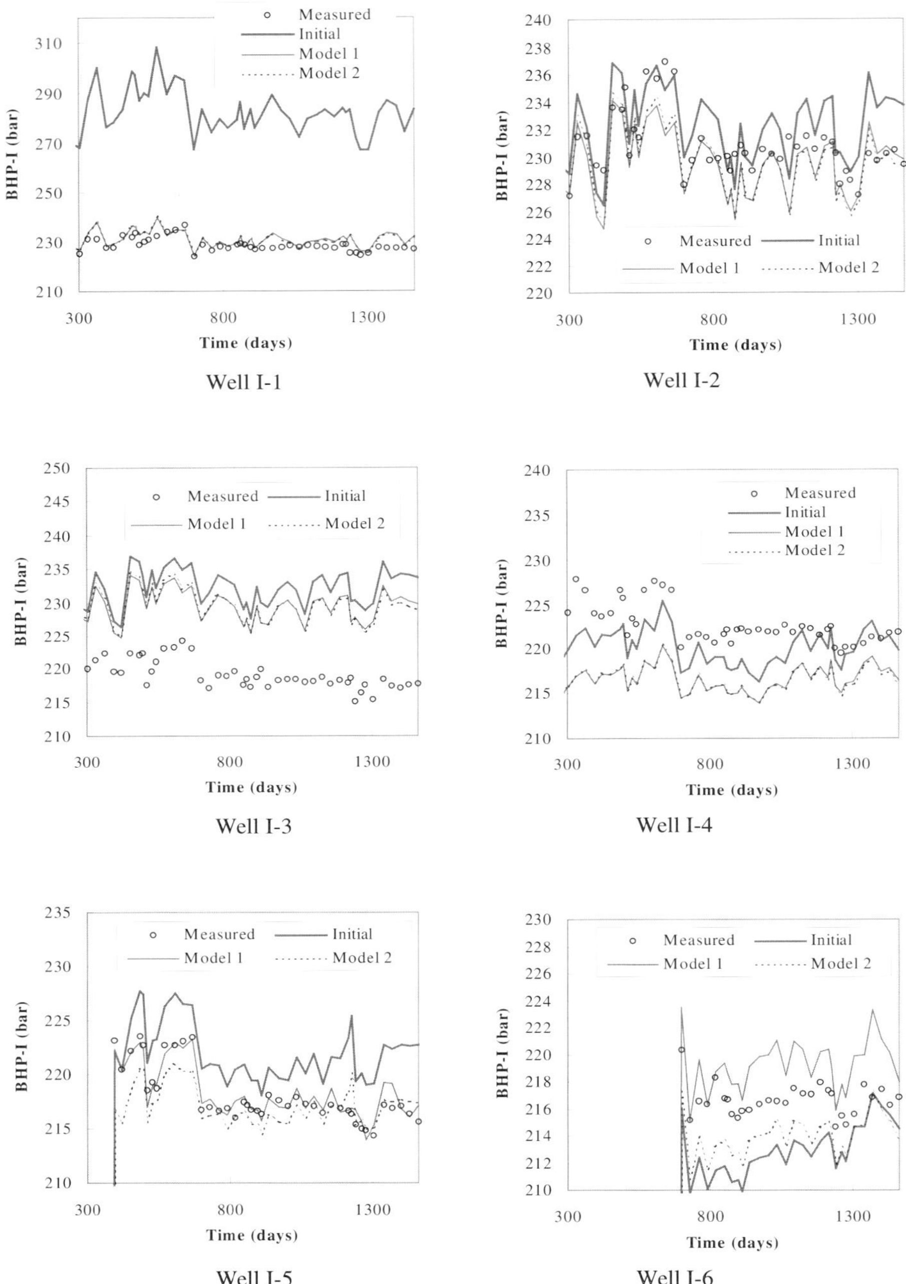

Fig. 16. History match for injection wells.

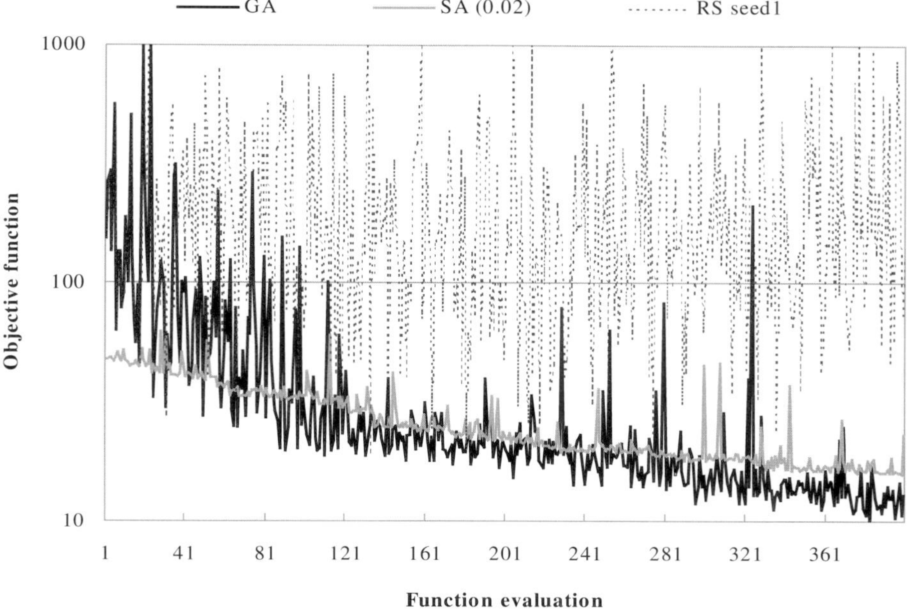

Fig. 17. Comparative performance of the GA with simulated annealing (SA) and random search (RS).

it therefore gains a slight advantage. It shows a steady decline, with significantly lower levels of noise compared to the GA. The random search shows a very high degree of variability with some low function values and some very high function values. None of the solutions is as good as those obtained by either of the two other methods.

We conclude from these comparisons that both the GA and the SA are working successfully. The GA is achieving better overall results and is searching more widely than the SA. The results are summarised in Table 3.

7.3.2. *Hill-climber*

A hill-climber is an optimisation algorithm in which changes made to the parameters are accepted only if they lead to improvements in the objective function. In this work,

TABLE 3

Results for GA, SA and random search

	Best in first 20	Best in first 400
RS	46.491	19.653
SA	46.491	15.953
GA	46.491	13.720

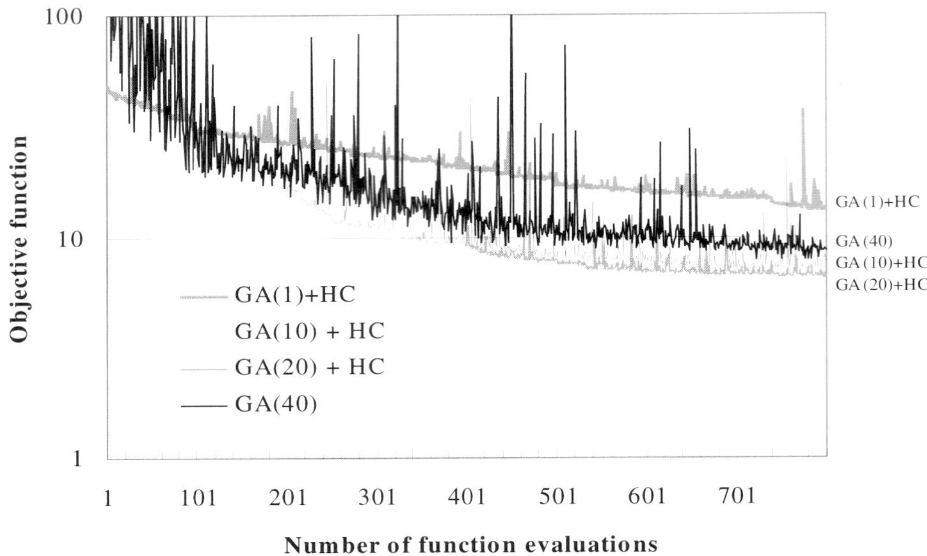

Fig. 18. Hill climber combined with a GA.

a hill climber procedure was tested using the creep mutation operator with each gene being mutated with probability 0.01. In order to test the hill-climbing procedure as a local optimiser after the GA has been used to perform a global optimisation, three strategies using 800 function evaluations are reported here, starting from the best models at three stages of the GA run: from the first, tenth and twentieth generations; the results are compared against an extended version of the GA with 20 additional generations.

The first tests starts from the best model at the initial GA generation, a rather advantageous position, and takes 820 function evaluations. Fig. 18 shows the progress of the optimisation for the four strategies. The best performance is achieved by hill-climbing from the best model at generation 20, slightly outperforming the case starting from the tenth generation. However, from the point of view of optimisation only marginal improvements in the objective function are achieved by extending the GA run. For instance, the run using the combination GA(10)+HC reaches objective function values below 10.0 before the 320th function evaluations, which is only achieved by the GA alone after approximately 400 function evaluations. The results are summarised in Table 4.

8. CONCLUSIONS

We have shown that it is possible to use a Genetic Algorithm to automate the process of reservoir characterisation. We have obtained results that are comparable to what might be obtained by a reservoir engineer using traditional techniques. This result is almost surprising given that we have been dealing with a geologically complex reservoir, needing about 1600 variables to characterise the reservoir, and we have only used 400 function evaluations (simulations). The result obtained could be very easily improved, by a competent reservoir engineer, by the modifications close to the wells so as to correct some of the bottom hole pressures. Discussions of the results of a wide range of comparisons carried out as part of this work can be found in the full report (Romero, 2000).

Whilst the results so far are very encouraging from the reservoir engineering point of view, there are some major short comings from the optimisation point of view. The worse problem is that we present results for only a single run of the GA, and only on one reservoir. Our experience leads us to believe that our result is representative of the general behaviour to be expected. However, this should be tested as part of a wider study. We have only used small populations of 20 individuals, which has resulted in very little exploration of the search space. The population in the final generation has retained quite a lot of genetic diversity, and variability in the function value, although over all the functions values have reduced significantly between the first and last generations. It seems likely that the GA has yet to converge to a limited volume of variable space, which would represent a single optimum. Tests involving larger populations and more generations are needed to allow the potential of the method to be properly assessed.

The method has several advantages:

- The algorithms have the potential for returning a suite of possible solutions (corresponding to different realisations of the reservoir model) from which the reservoir team can select a representative group for further analysis. Although niching was not implemented in this work because priority was given to the most common GA strategies, the use of niching strategies could enhance further the diversity in the reservoir descriptions. Therefore, GAs can be used to obtain multiple possible solutions without repeating the history matching process.
- The method is relatively easy to implement for computer-aided history matching, and robust with respect to lost or corrupted solutions as opposed to the approaches using gradient-based optimisation.
- It can be easily parallelised because it is inherently suited to parallelisation. Additionally, the method is cheap in terms of computational costs, and efficient, in that it requires only a modest number of forward simulations to obtain relatively good solutions.

TABLE 4

Summary of results for hill-climber and GA.

Strategy	Best initial	Best in 800
GA(1)+HC	46.491	13.287
GA(10)+HC	46.491	7.690
GA(20)+HC	46.491	6.532
GA(40)	46.491	7.928

- Furthermore, the method appears to be reasonably insensitive to the parameter settings used to control the GA, which makes it suitable as a general automatic reservoir characterisation algorithm. Minor fine-tuning of parameters does not cause significant improvements, which are only achieved when a dramatically different approach is applied.

8.1. Suggestions for further work

We have identified three large areas for future research:

- To test the method in a field case. This might require the use of more sophisticated modelling techniques such as geological object modelling, stochastic imaging, etc., at a much finer scale, as well as the use of upscaling steps. Nevertheless, the applicability of the method holds.
- The impact on reservoir management of using the ensemble of solutions (e.g. the final population) returned by the GA.
- Another difficult task that could yield important benefits is the use of more complex representation techniques such as probabilistic chromosomes to account for the uncertainty in the parameters, and the use of advanced embryogenies to exploit repetition, hierarchy, etc., that are normally seen in geological data.

Additional lines of general research are:

- The study in detail of the behaviour of the GA applied to reservoir characterisation. A more simple case study comprising only a few layers and wells, or a model taken from previous studies reported in the literature (to allow for comparisons), might be used.
- The evaluation of the impact of the incorporation of prior information in the reservoir characterisation process, including the finding of initial matches by conventional means such as material balance and aquifer analysis. A redefinition of the objective function might be required. In addition, and depending on the particularities of the history matching problems, alternative well controls might be implemented (e.g. specifying bottom hole pressure in the simulator).
- The use of hybrid methods, i.e. the combination of Genetic Algorithms with local optimisers such as hill climbers or gradient methods to further improve the models.
- The incorporation of additional parameters such as capillary pressure, PVT properties, aquifer characteristics, etc. Some of the aspects of the GA formulation that could be studied are mentioned in the following list. The list does not intend to be exhaustive, and aspects are not necessarily listed in order of importance. Nevertheless, efforts must also be directed to evaluate the representativity of the results obtained.
  - Chromosome specification, such as binary bit length, alphabet, etc.
  - The use of concepts such as dominance, diploid, sexual differentiation, deletion, inversion, etc.
  - Alternative selection strategies as well as different versions of those tested in this work, such as sigma-scaling, and Boltzmann selection, and including selection procedures customised for the particular case of reservoir characterisation (e.g., based on the contributions of the control variables).
  - Other population strategies, such as niching, that might lead to clusters of highly fit individuals (solutions).

- Injection Island GAs (Eby et al., 1997)
- Implementation of a steady-state GA.
- Re-starts from the best individual from several shorter GA runs, or from runs using limited production history (to shorten the flow simulation run).
- Determination of the optimal mutation operators as well as the optimum mutation rates for each individual chromosomes.
- Determination of the optimal crossover operators and their rates for both the one-dimensional and three-dimensional chromosomes, as well as the evaluation of the implications of the crossover operators for the three-dimensional chromosomes on the stationarity of the mean and variance of the Gaussian fields.
- Adaptation of the GA code to parallelisation.

REFERENCES

Anterion, F., Eymard, R. and Karcher, B., 1989. Use of parameter gradients for reservoir history matching, SPE 18433, SPE Symp. on Reservoir Simulation, Houston, Texas, February 6–8.

Bissell, R.C., Sharma, Y. and Killough, J.E., 1994. History matching using the methods of gradients: two case studies, SPE 28590, SPE 69th Annu. Tech. Conf. and Exhibition, New Orleans, Louisiana, Sept. 25–28.

Bissell, R., 1996. History matching a reservoir model by the positioning of geological objects, Proc. 5th Eur. Conf. on the Mathematics of Oil Recovery, Leoben, Austria, Sept. 3–6.

Bissell, R.C., Dubrule, O., Lamy, P. Swaby, P. and Lepine, O., 1997. Combining geostatistical modelling with gradient information for history matching: the pilot point method, SPE 38730, 1997 SPE Annu. Tech. Conf. and Exhibition, San Antonio, Texas, Oct. 5–8.

Bos, C., Floris, F., Nepveu, M., Roggero, F., Omre, H., Holden, L., Syversveen, A.R., Zimmerman,,R.W., Carter, J.N., Frandsen, P.E., Bech, N., Geel, ?, Walsh, J., Manzocchi, T., Barker, J., Cuypers, M. and Bush, M., 1999. Final Report on the production forecasting with uncertainty quantification (PUNQ-2) project, Netherlands Organisation for Applied Scientific Research (TNO), Delft, The Netherlands, NITG-99-255-A.

Bush, M.D. and Carter, J.N., 1996. Application of a modified genetic algorithm to parameter estimation in the petroleum industry, in: Dagli et al. (Eds.), Intelligent Engineering Systems through Artificial Neural Networks 6, ASME Press, New York, 397–402.

Carter, R.D., Kemp, L.F., Pierce, A.C. and Williams, D.L., 1974. Performance matching with constraints, *SPE J.*, April, pp. 187–196.

Carter, J.N., 1997. Genetic algorithms for incore fuel management and other recent developments in optimisation, in: Lewins, J. and Beckerand M. (Eds.), Advances in Nuclear Science and Technology 25, Plenum Press, New York, pp. 113–154.

Cartwright, H.M. and Harris, S.P., 1993. Analysis of the distribution of airborne pollution using genetic algorithms, *Atmospheric Environ.*, 27AN (12): 1783–1791.

Chavent, G., Dupuy, M. and Lemonier, P., 1975. History matching by use of optimal theory, *SPE J.*, Feb., pp. 74–86.

Chen, W.H., Gavalas, G.R., Seinfeld, J.H. and Wasserman, M.L., 1974. A new algorithm for automatic history matching, *SPE J.* pp. 593–608.

Coats, K.H., Dempsey, J.R. and Henderson, J.H., 1970. A new technique for determining reservoir description from field performance data, *SPE J.*, March, pp. 66–74.

Craig, P.S. Goldstein, M. and Seheult, A.H., 1996. Bayes linear strategies for matching hydrocarbon reservoir history, in: Berger et al. (Eds.) Bayesian Statistics 5, Oxford University Press.

Deutsch, C.V. and Journel, A.C. (Eds.), 1998. GSLIB: Geostatistical Software Library and User's Guide, 2nd edn., Oxford University Press, New York.

Deschamps, T., Grussaute, T., Mayers, D. and Bissell, R., 1998. The results of testing six different gradient optimisers on two history matching problems, Proc. 6th Eur. Conf. on the Mathematics of Oil Recovery, B24.

Eby, D. Averill, R., Gelfand, B., Punch, W., Mathews, O. and Goodman, E., 1997. An injection island GA for flywheel design optimization, in: 5th Eur. Congr. on Intelligent Techniques and Soft Computing EUFIT '97, Vol. 1, Verlag Mainz, Aachen, pp. 687–691.

Farmer, C.L., 1989. The mathematical generation of reservoir geology, Joint IMA/SPE Eur. Conf. on the Mathematics of Oil Recovery, Robinson College, Cambridge University, UK, July 25–27.

Gavalas, G.R. Shah, P.C. and Seinfield, J.H., 1976. Reservoir history matching by Bayesian estimation, *SPE J.*, Dec., pp. 337–350.

Gómez-Hernández, J.J., Sahuquillo, A. and Capilla, J.E., 1997. Stochastic simulation of transmissivity fields conditional to both transmissivity and piezometric data – I: Theory; II: Demonstration on a synthetic aquifer, *J. Hydrol.*, 203: 162–188.

Gomez, S., Gosselin, O. and Barker, J.W., 1999. Gradient-based history-matching with a global optimization method, SPE 56756, 1999 SPE Annu. Tech. Conf. and Exhibition, Houston, Texas, October 3–6.

Guerreiro, J.N.C., Barbosa, H.J.C., Garcia, A.F.D, Loula, E.L.M. and Malta, S.M.C., 1998. Identification of reservoir heterogeneities using tracer breakthrough profiles and genetic algorithm, *SPE Reservoir Evaluation & Engineering*, June, pp. 218–223.

Hirasaki, G.J., 1975. Sensitivity coefficients for history matching oil displacement processes, *SPE J.*, Feb., pp. 39–49.

Hu, L-Y. and Blanc, G., 1998. Proc. 6th Eur. Conf. on the Mathematics of Oil Recovery, Peebles, Scotland, 8–11 September, B-01.

Huang, M.D., Romeo, F. and Sangiovanni-Vincentelli, A., 1986. An efficient general cooling schedule for simulated annealing, Proc. IEEE Int. Conf. in Computer Aided Design.

Jacquard, P. and Jain, C., 1965. Permeability distribution from field pressure data, *SPE J.*, Dec., pp. 281–294.

Jahns, H.O., 1966. A rapid method for obtaining a two-dimensional reservoir description from Well Pressure Response Data, *SPE J.*, Dec., pp. 315–327.

Jensen, J., Lake, L., Corbett, P. and Goggin, D., 1997. *Statistics for Petroleum Engineers and Geoscientists*, Prentice Hall Petroleum Engineering Series, Englewood Cliffs, New Jersey.

Journel, A.G. and Huijbregts, C.J., 1978. *Mining Geostatistics*, Academic Press, New York.

Landa, J.L. and Horne, R.N., 1997. A procedure to integrate well test data, reservoir performance history and 4-D seismic information into a reservoir description, SPE 38653, 1997 SPE Annu. Tech. Conf. and Exhibition, San Antonio, Texas, Oct. 5–8.

Levenick, J.R., 1985. Inserting introns improves genetic algorithm success rate: Taking a cue from biology, in: Grefenstette, J.J. (Ed.), Proc. 1st Int. Conf. on Genetic Algorithms and Their Applications (Erlbaum), p. 123.

Lia, O., Omre, H., Tjelmeland, H., Holden, L. and Egeland, T., 1997. Uncertainties in reservoir production forecasts, *AAPG Bulletin*, 81(5): 775–802.

Manzocchi, T., Walsh, J.J., Nell, P. and Yielding, G., 1999. Fault transmissibility multipliers for flow simulation models, *Petr. Geosci.*, 5.

de Marsily, G., Lavedan, G., Boucher, M. and Fasanino, G., 1984. Interpretation of interference tests in a well field using geostatistical techniques to fit the permeability distribution in a reservoir model, In: Verly, G. et al. (Eds.), Geostatistics for Natural Resources Characterization, Reidel, pp. 831–849.

Mayer, A.S. and Huang, C., 1999. Development and application of a coupled process parameter inversion model based on the maximum likelihood estimate method, *Adv. in Water Resourc.* 22: 841–853.

Omre, H., Tjelmeland, H., Qi, Y. and Hinderaker, L., 1993. Assessment of uncertainty in the production characteristics of sand stone reservoir, reservoir characterization III – Proc. 3rd Int. Reservoir Characterization Tech. Conf. 1991, Tulsa, Oklahoma, Pennwell Books, Tulsa, Oklahoma.

Ouenes, A., 1992. Application of simulated annealing to reservoir characterization and petrophysics inverse problems, PhD Thesis, New Mexico Institute of Mining and Technology.

Pachepsky, Y. and Acock, B., 1998. Stochastic imaging of soil parameters to assess variability and uncertainty of crop yield estimates, *Geoderma* 85: 213–229.

Portella, R.C.M. and Frais, F., 1999. Use of automatic history matching and geostatistical simulation to improve production forecast, SPE 53976, 1999 SPE Latin American and Caribbean Petroleum Eng. Conf., Caracas, Venezuela, April 21–23.

PUNQ: Production Forecasting with Uncertainty Quantification, A research project funded in part by The European Commission under the Non-Nuclear Energy Programme (JOULE III), contract F3-CT95-0006, http://www.nitg.tno.nl/punq.

Romero, C., A genetic algorithm for reservoir characterisation using production data, Ph.D. Thesis, November 2000, University of London.

Sen, M.K., Datta-Gupta, A., Stoffa, P.L., Lake, L.W., and Pope, G.A., 1995. Stochastic reservoir modeling using simulated annealing and genetic algorithms, *SPE Formation Evaluation*, (March, pp. 49–55.

Sivia, D.S., 1996. *Data Analysis: A Bayesian tutorial*, Oxford University Press.

Slater, G.E. and Durrer, E.J., 1971. Adjustment of reservoir simulation models to match field performance, *SPE J.*, Sep., pp. 295–305.

Spears, W.M., 1993. Crossover or mutation?, In: Whitley, L.D. (Ed.), Foundations of Genetic Algorithms 2, Morgan Kaufmann, San Mateo, California.

Stoffa, P.L. and Sen, M.K., 1991. Nonlinear multiparameter optimization using genetic algorithms: Inversion of plane-wave seismograms, *Geophys.*, 56: 1794–1810.

Tan, T.B.S., 1991. Parameter estimation in reservoir simulation, PhD Thesis, Dept. of Chemical and Petroleum Engineering, Univ. of Calgary.

Tan T.B. and Kalogerakis, N., 1991. A fully implicit, three-dimensional, three-phase simulator with automatic history-matching capability, SPE 21205, 11th SPE Symp. on Reservoir Simulation, Anaheim, California, Feb. 17–20.

Tanaka, Y., Ishiguro, A. and Uchikawa, Y., 1993. A genetic algorithms application to inverse problems in electromagnetics, Proc. 5th Int. Conf. on Genetic Algorithms, Morgan Kaufmann, San Mateo, California, p. 656.

Thomas, L.K., Hellums, L.J. and Reheis, G.M., 1971. A nonlinear automatic history matching technique for reservoir simulation models, SPE 3475, SPE 46th Annu. Fall Meet., New Orleans, Louisiana, Oct. 3–6.

Veatch, R.W. and Thomas, G.W., 1971. A direct approach for history matching, SPE 3515, SPE 46th Annu. Fall Meeting, New Orleans, Louisiana, Oct. 3–6.

Wang, Y. and Kovscek, A.R., 2000. A streamline approach for history-matching production data, SPE 59370, 2000 SPE/DOE Improved Oil Recovery Symp., Tulsa, Oklahoma, April 3–5.

Walsh, J.J., Watterson, J., Heath, A.E. and Childs, C., 1998. Representation and scaling of faults in fluid flow models, *Pet. Geosci.*, 4.

Watson, A.T., Seinfield, J.H. and Gavalas, G.R., 1979. History matching in two-phase petroleum reservoirs, SPE 8250, SPE 54th Annu. Fall Tech. Conf. and Exhibition, Las Vegas, Nevada, Sept. 23–26.

Wu, Z., Reynolds, A.C. and Oliver, D.S., 1999. Conditioning geostatistical models to two-phase production data, *SPE J.*, 4(2).

Yang, P-H. and Watson, A.T., 1987. Automatic history matching with variable-metric methods, SPE 16977, SPE 62nd Annu. Tech. Conf. and Exhibition, Dallas, Texas, Sept. 27–30.

Yielding, G., Freeman, B. and Needham, D.T., 1997. Quantitative fault seal prediction, *Am. Assoc. Pet. Geol. Bull.*, 81.

Developments in Petroleum Science, 51
Editors: M. Nikravesh, F. Aminzadeh and L.A. Zadeh

Chapter 18

APPLYING SOFT COMPUTING METHODS TO IMPROVE THE COMPUTATIONAL TRACTABILITY OF A SUBSURFACE SIMULATION–OPTIMIZATION PROBLEM

VIRGINIA M. JOHNSON and LEAH L. ROGERS

Lawrence Livermore National Laboratory, Livermore, CA 94551

ABSTRACT

Formal optimization strategies normally evaluate hundreds or even thousands of scenarios in the course of searching for the optimal solution to a given management question. This process is extremely time-consuming when numeric simulators of the subsurface are used to predict the efficacy of a scenario. One solution is to train artificial neural networks (ANNs) to stand in for the simulator during the course of searches directed by some optimization technique such as the genetic algorithm (GA) or simulated annealing (SA). The networks are trained from a representative sample of simulations, which forms a re-useable knowledge base of information for addressing many different management questions.

These concepts were applied to a water flood project at BP's Pompano Field. The management problem was to locate the combination of 1–4 injection locations which would maximize Pompano's simple net profit over the next seven years. Using a standard industry reservoir simulator, a knowledge base of 550 simulations sampling different combinations of 25 potential injection locations was created. The knowledge base was first queried to answer questions concerning optimal scenarios for maximizing simple net profit over three and seven years. The answers indicated that a considerable increase in profits might be achieved by expanding from an approach to injection depending solely on converting existing producers to one involving the drilling of three to four new injectors, despite the increased capital expenses.

Improved answers were obtained when the knowledge base was used as a source of examples for training and testing ANNs. ANNs were trained to predict peak injection volumes and volumes of produced oil and gas at three and seven years after the commencement of injection. The rapid estimates of these quantities provided by the ANNs were fed into net profit calculations, which in turn were used by a GA to evaluate the effectiveness of different well-field scenarios. The expanded space of solutions explored by the GA contained new scenarios which exceeded the net profits of the best scenarios found by simply querying the knowledge base.

To evaluate the impact of prediction errors on the quality of solutions, the best scenarios obtained in searches where ANNs predicted oil and gas production were compared with the best scenarios found when the reservoir simulator itself generated those predictions during the course of search. Despite the several thousand CPU hours required to complete the simulator-based searches, the resulting best scenarios failed to match the best scenarios uncovered by the ANN-based searches. Lastly, results obtained from ANN-based searches directed by the GA were compared with ANN-based searches employing an SA algorithm. The best scenarios generated by both search techniques were virtually identical.

1. INTRODUCTION

1.1. Statement of the problem

Reservoir simulation is a well-established component of reservoir management throughout much of the petroleum industry. Black oil simulators and more complex compositional, thermal, and chemical models are used as forecasting tools in both the day-to-day operational management of production facilities and longer-term field development planning. As yet, however, little use has been made of reservoir simulation coupled with systematic optimization techniques. The main advantage of applying these mathematical tools to decision-making problems is that they are less restricted by human imagination than conventional case-by-case comparisons. As the number of competing engineering, economic, and environmental planning objectives and constraints increases, it becomes difficult for human planners to track complex interactions and select a manageable set of promising scenarios for examination. Using optimization techniques, the search can range over all possible combinations of variables, locating strategies whose effectiveness is not always obvious to planners. Optimization can also generate *sets* of promising scenarios from which planners can choose.

The single biggest obstacle to the application of optimization techniques using a reservoir simulator as the forecasting tool is the computational time required to complete a single simulation. Even the examination of 10 variations on a well-field design becomes cumbersome when a single run requires hours or days to complete. Coupling these simulators to optimization methods requiring hundreds or thousands of simulations poses a computational problem bigger than most organizations are willing or able to tackle. The ANN–GA/SA solution to this problem is to train artificial neural networks (ANNs) to predict selected information that the simulator would normally predict. A heuristic search engine, either the genetic algorithm (GA) or simulated annealing (SA), searches for increasingly better strategies (such as the most productive in-fill drilling pattern or the best distribution of steam injection wells), using the trained networks to evaluate the effectiveness of each strategy in place of the original simulator. This substitution has been shown to reduce the time needed to evaluate pump-and-treat groundwater remediation strategies by a factor of nearly a million, enabling the evaluation of millions of strategies in a matter of days on conventional workstations. After analysis of the results of the search, the best-performing strategies are submitted to the original simulator to confirm their performance.

This paper is a detailed case study of the application of the ANN–GA/SA approach to the proposed water flood of a deep water reservoir that had been in production for a little less than 3 years. The reservoir management goal was to identify the best set of injection well locations to maximize some economic measure of performance over a seven-year planning horizon. The ANN–GA/SA methodology was originally developed for use on 2D groundwater remediation problems. The primary purpose of the application described here was to determine how well these methods would translate to 3D simulation of multiphase flow.

1.2. ANN–GA/SA approach to optimization

Simulation–optimization, a term which refers to the coupling of models to optimization drivers, has received extensive attention in the groundwater remediation literature. The goal of optimization for this type of problem is usually to find one or more combinations of extraction and injection well locations that will at least contain and preferrably clean up the contamination at minimum cost or time. Although the number of well combinations is potentially infinite, it has been customary in groundwater optimization work to pre-specify a grid of potentially good well locations and then formulate the search to locate the most time- or cost-effective subset of those locations which meets remediation goals.

Early optimization work at the Superfund site at which the ANN–GA/SA methodology was developed used 20 pre-selected extraction locations with fixed pumping rates and searched for the subset producing the smallest volume of treated water which contained the contamination over a 40-year planning period (Rogers and Dowla, 1994). Later work focused on 28 fixed-rate extraction and injection locations in a multiple-objective search which balanced cost-efficiency with mass-extraction performance, while meeting a containment constraint over a 50-year planning period (Rogers et al., 1995).

Regardless of the problem formulation or the type of search technique employed, key components of the cost function for a particular well pattern are evaluated by a contaminant transport model which assesses the impact of the well pattern on the distribution of the contamination over some period of time. Even in 2D, numerical models of this sort can take hours to evaluate a single pattern on a conventional workstation. As the resolution, dimensionality, and heterogeneity of the models increase, the time required for this evaluation can extend to days. Since even the most efficient, deterministic optimization techniques usually need to evaluate hundreds of patterns, the modeling step becomes a major computational bottleneck in the optimization of realistic environmental engineering problems. Much of the work in this area has accepted the modeling bottleneck as a given, sometimes simplifying the situation by analyzing smaller-scale problems or using simpler models (as illustrated by examples in Wagner, 1995) or seeking to reduce the number of times the model must be called by increasing the efficiency of the search itself (see, for example, Karatzas and Pinder, 1993).

Work intended to confront the modeling bottleneck directly falls into one of two camps. The first approach involves reducing the execution time required by the model through parallel algorithms and computer architectures (Dougherty, 1991; Tompson et al., 1994). This represents a ‘rich man’s’ approach because of the costs normally associated with gaining access to computer resources of this kind. The ANN–GA/SA approach, in contrast, confronts the problem by training neural networks to predict selected model results. The trained networks, rather than the original model, supply the predictions needed to calculate the objective function in fractions of a second.

The network architecture used for this prediction task is a multilayer perceptron, trained by the standard back-propagation learning algorithm (Rumelhart et al., 1986). Training and testing examples are obtained by associating well pattern variations with selected outcomes such as the amount of contamination that has been removed, the highest remaining concentrations after treatment is complete, and whether or not contamination has spread beyond certain boundaries. The examples are drawn from a knowledge base initially created by running the contaminant transport model on a *representative sample* of well patterns. Since there are no dependencies among the model runs, they can be distributed over a network of processors using only the basic remote file system and execution facilities that are now a standard part of most network environments.

Although the trained nets can be coupled with a variety of search techniques, heuristic search techniques (namely, the genetic algorithm as described in Goldberg, 1989, and simulated annealing as detailed in Kirkpatrick et al., 1983) have been the methods of choice for three reasons. First, heuristic methods seek to find reasonably good solutions in a reasonable amount of time, which can be an advantage when complex, real-world problems with significant nonlinearities are being evaluated. Second, since they employ direct function evaluation rather than derivatives of functions, they allow more complex integration of different components of the objective function, which again is more reflective of real-world problems. Finally, the heuristic methods represent a philosophy of search that is especially well suited to design optimization problems (Reeves, 1993). The contaminant transport models used to evaluate the effectiveness of each well pattern are crude approximations of reality. Their utility lies more in outlining broad hydrological design principles applicable to a given site than in predicting precise outcomes. Furthermore, there are many practical engineering, managerial, and political constraints that cannot easily be quantified in a cost function. Consequently, employing a search strategy oriented toward producing one or a handful of best solutions is not likely to be well-received by engineers and planners. Instead, a search technique generating a wide range of potentially effective solutions, which are subsequently analyzed for their common properties, is generally more useful. Designers can then select especially interesting solutions to incorporate into their detailed designs or simply follow the general principles suggested by the analyses.

The components of the ANN–GA/SA methodology are shown in Fig. 1. It is important to note that, since the set of optimal solutions generated by the search engines is obtained by an ANN estimation process that introduces a certain degree of error, the final step in the methodology is to submit that optimal set to the original simulator for verification. The updated performance measures supplied by the simulator on this manageable set of scenarios are the ones which are used in subsequent decision-making.

1.3. Design optimization in petroleum engineering

Reservoir simulation is now a well-established component of reservoir management, as indicated by the role it is given in both general discussions (Breitenbach, 1991;

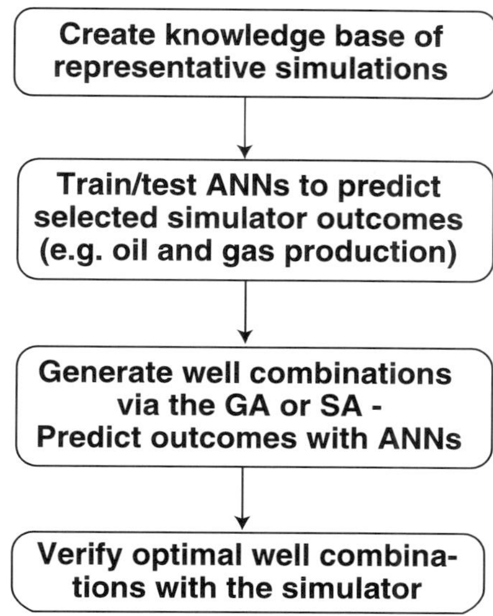

Fig. 1. Components of ANN–GA/SA methodology.

Behrenbruch, 1993) and case studies (York et al., 1992; Aadland et al., 1994) of reservoir management practices. But the use of these simulators in a structured, formal search for more effective recovery strategies is still in its infancy. Typically, the simulator is used to evaluate a small set of development scenarios that has been selected to test specific hypotheses (see, for example, Kumar and Ziegler, 1993; Coskuner and Lutes, 1996; Kikani and Smith, 1996). A few researchers, however, have examined more structured approaches.

Aanonsen et al. (1995) applied concepts from experimental design and response surfaces to optimize a reservoir response variable (e.g. oil production rate) according to reservoir management parameters (e.g. well location and flow rates). Their largest example involved 240 one-hour runs of a 5500 grid block 3D model of a fluvial reservoir. The goal was to build a response surface of discounted oil production from sample inputs consisting of the x and y coordinates of a single producer and the x coordinate of a single injector. To account for uncertainties in the flow field, these three inputs were crossed, as in an experimental design, with eight different realizations of the deposition of channel sands. The response surface was examined for distinct maxima, which became the optimal solutions to the problem. This work is similar to the ANN–GA/SA methodology in that the results of a sample of simulations are used to build surfaces which are then searched for solutions. In the ANN–GA/SA approach, however, the sampling is performed to create a re-usable knowledge base, providing the examples from which many different networks figuring in many different searches are drawn.

Wackowski et al. (1992) employed decision analysis techniques to examine over 2500 expansion, investment, operational, and CO_2 purchase/recompression scenarios to maximize net present value of a project at the Rangely Weber Sand Unit. This ambitious, long-range project pulled together information from many sources (including expert opinion, economic spreadsheet models and reservoir models) into decision trees, from which the highest probability paths were selected. The reservoir model combined the vertical response of a single detailed cross-section with the areal response of a full-field streamtube model to obtain full-field forecasts of injected and produced fluids. Since several techniques were used to reduce the number of paths in the decision tree which required full examination, it is unclear how many scenarios the simulator actually evaluated. This approach to optimization is similar to the ANN–GA/SA methodology in that they both examine very large numbers of alternatives. The techniques, however, are quite dissimilar in their identification of optimal solutions. Unless it is exhaustive of all possibilities, which is unlikely in a real-world problem, a decision tree can only select solutions from paths that have been anticipated by its designers. Optimization techniques, in contrast, can uncover combinations of inputs which produce results which were not anticipated.

A classic application of optimization techniques to facility design is given by Fujii and Horne (1994). They compared three different search techniques (a derivative-based method, the polytope method, and the GA) applied to the optimization of a networked production system by varying parameters such as separator pressure, diameters of tubing, and pipeline vs. surface choke. Calculations were restricted to relatively simple production rate equations because the use of a reservoir simulator was judged to be too time-consuming. Later, Bittencourt and Horne (1997) used a GA combined with economics and simulation to determine the optimal relocation of wells in a proposed 33-well layout and the best platform location. Their experiences reinforce the motivation behind the ANN–GA/SA approach: that the advantages of optimization techniques will not be fully exploited until some method is found to reduce the computational burden imposed by the reservoir simulator.

These recent advances suggest that the petroleum engineering field is beginning to pay attention to more structured approaches to the optimization of development strategies. Use of the ANN–GA/SA approach can promote further interest in this process by alleviating the computational bottleneck created by the reservoir simulator. This is accomplished not by eliminating the simulator from the optimization but by capturing simulator predictions in the weights of artificial neural networks. In this way, the results of the optimization continue to benefit from the increased accuracy of predictions that a reservoir simulator can provide without having to pay the full price in computational time. The critical role played by the simulator is reinforced when the best-performing scenarios generated by the search are submitted to it for validation.

2. RESERVOIR DESCRIPTION

The Pompano Field in the deep water Gulf of Mexico is the test site for this project. BP and Kerr-McGee are joint operators of this field, which has been in production since April, 1995. They have developed and calibrated a reservoir model, using Landmark's VIP® simulator, for the Miocene section.

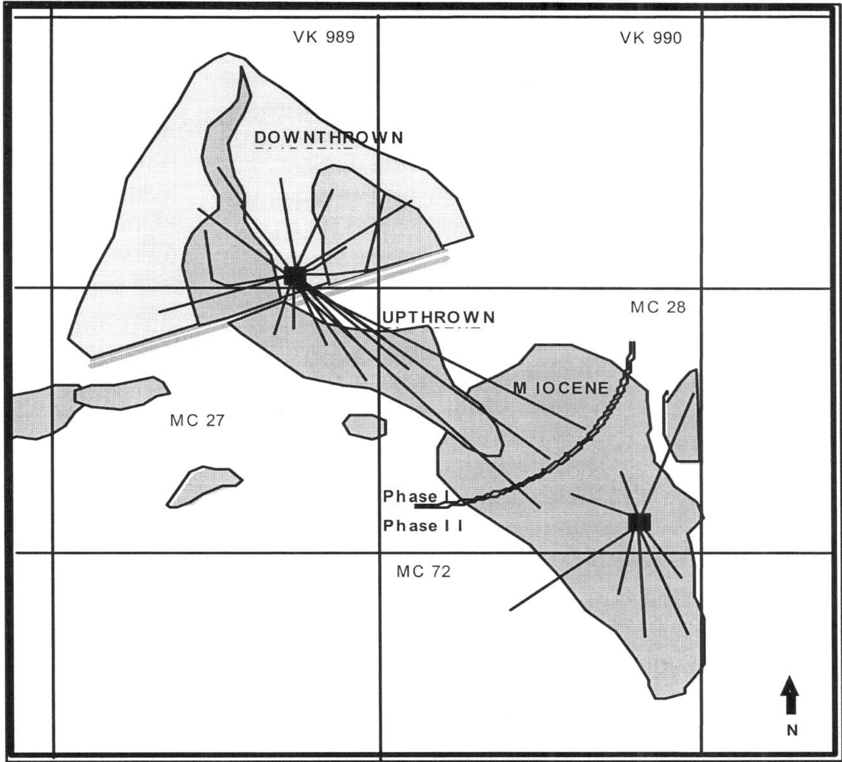

Fig. 2. Pompano Field.

The Pompano field consists of multiple turbidite reservoirs in a variety of structural traps and settings. An intrusive salt body and a large counter-regional growth fault are important structural features in the field. The field is divided into three areas, shown in Fig. 2. To the north and northwest of the fault is the downthrown Pliocene which consists of 10 independent, stacked reservoirs. It is generally underlain by the more sheet-like part of the salt body. The upthrown Pliocene is south of the salt and growth fault. Its reservoirs are a group of related channel sand deposits. An older Miocene channel complex lies to the southeast, in the syncline between the Pompano and Mickey salt bodies. This Miocene complex consists of an interconnected group of turbidite sands. Since more than two thirds of the total recoverable reserves are estimated to be in the Miocene portion of the field, it became the focus for field development planning.

The Miocene reservoir sands were deposited as mid-slope turbidites in a large, aggradational channel complex. There is significant connectivity between channels as younger channels frequently eroded into previously deposited ones. Pressure depletion in successively drilled wells suggests that most of the reservoir complex is in pressure and fluid continuity. Grain size ranges from very fine to medium, with the bulk being fine grained. The average thickness of the Miocene sand is 50 net ft of oil in a vertical target interval of 300 ft to 400 ft, and the thickest sand penetrated is 110 net ft of oil in a single sand.

A north–south trending channel system draped over an east–west trending structural nose forms the trap. The channel sands are laterally confined by the shales and silty shales of the overbank deposits. An oil–water contact at −10,200 ft true vertical depth subsea (TVDSS) has been drilled on the southern edge of the field and is implicated on the north/northwest end by seismic interpretation and water production. Maximum hydrocarbon column height is approximately 600 ft. The large aquifer system below, estimated to be three-fold larger than oil-in-place, is judged to be an advantage to help offset pressure losses during reservoir depletion.

The Miocene oil has very favorable properties which help in achieving high production rates. API gravity is 32°, viscosity is 0.38 cp, and the gas–oil ratio (GOR) was initially 1037 scf/stbo and is climbing with increased production. The very restricted range of variability in the producing wells emphasizes the connectivity in the Miocene reservoirs. There are 12 production wells in operation, five drilled from the platform to the north during Phase I, and seven drilled from a subsea template to the south during Phase II (see Fig. 3). The average initial flow rate was 788 stb/day for the five Phase I wells and 6343 stb/day for the seven Phase II wells. The gas and oil production decline

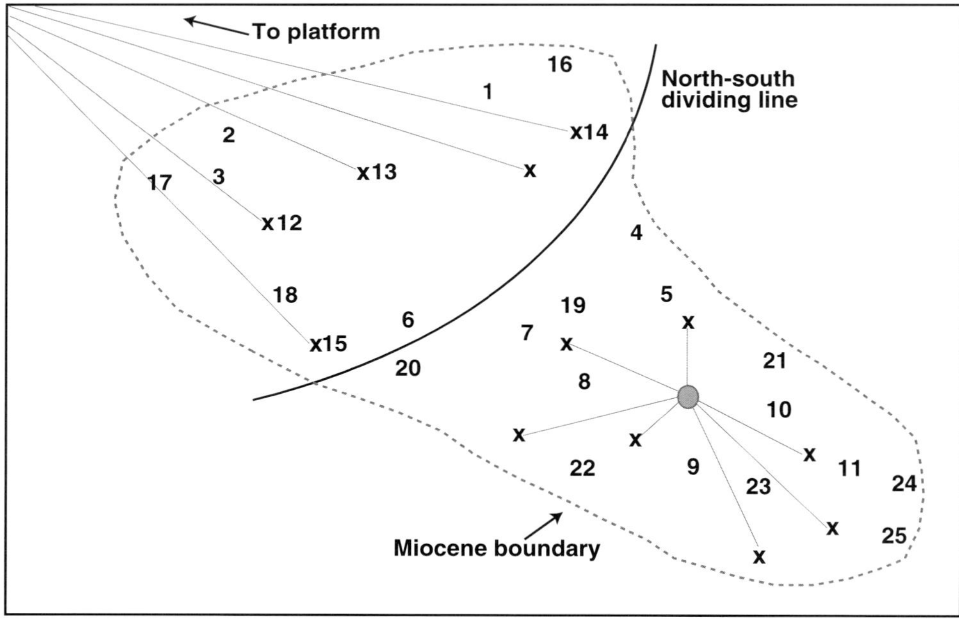

Fig. 3. Production and injection well locations in the northern and southern portions of the Miocene. The 12 existing producers are identified by '×'. Numbered locations represent either new injection locations (1–11, 16–25) or producers tagged for potential conversion to injectors (12–15). Wells in the north portion of the field would be drilled from the existing platform. Wells in the southern portion would be drilled from a second subsea template which would have to be installed to support them.

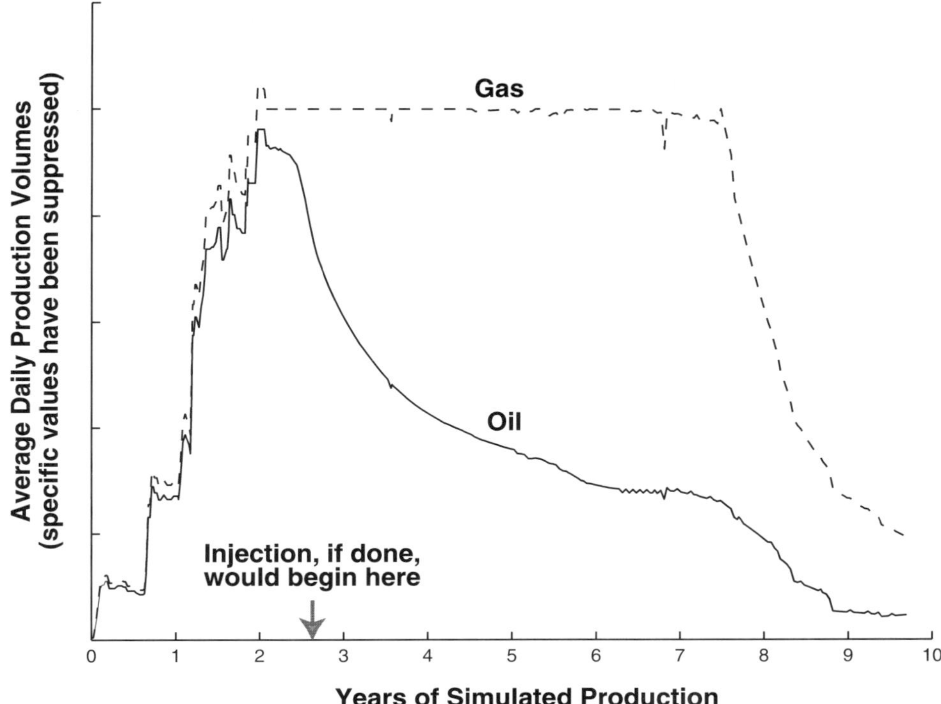

Fig. 4. Baseline oil and gas production decline curves. Production to the left of the arrow is a history match to actual production over the first 2.67 years of the field's operation. Production to the right of the arrow forecasts the decline to be expected if the existing producers continue to operate over the next 7 years.

curves for 2.67 years of production (from April, 1995 through December, 1997) and seven additional years of simulated production without injection are shown in Fig. 4. The cap on gas production is a function of surface facility limitations.

The heterogeneous anticlinal turbidite reservoir was discretized first into an approximately three million cell block model at seismic resolution. It was then scaled up to a 40,000 cell block simulation model, implemented in VIP®, with dimensions of 40×40 in plan view and 25 layers. Seven-year simulations of the existing producers plus one to four injectors required an average of 3.5 hours to complete on a dedicated Sun UltraSparc 2 workstation.

3. MANAGEMENT QUESTION

The planning question posed by BP is whether a water injection program will improve production from the Miocene. A pool of 25 prospective locations for injector wells was created, based on high transmissivity, spatial coverage, and economics (see the methodology section below for details). The optimization problem was then formulated to search for the combinations of one to four injector wells which maximize simple net profit, subject to facility constraints. Although only one or two injectors were being considered by the Pompano asset team at the time this project was initiated, the problem scope was expanded to include the possibility of a more aggressive program. Management time horizons of both three and seven years were examined.

3.1. Assumptions and constraints

For this management formulation, the following assumptions and constraints were operative:

(1) The maximum time period over which alternative water flood scenarios would be assessed was January 1, 1998 to January 1, 2005.

(2) A candidate pool of 25 injection sites, including both existing production wells and newly drilled injectors, would be developed. Given this candidate pool, the search would identify the particular subsets, which could vary in size from one to four wells, which maximize some measure of economic performance.

(3) The Phase I producers located in the northern portion of the field would be considered for conversion to injectors; but, for engineering reasons, the Phase II producers would not.

(4) Due to limitations on unused slots, only two new injectors could be drilled from the northern platform. Drilling any injectors in the southern section would require the emplacement of a new subsea template at substantial cost, from which up to four injectors could be drilled.

(5) The 12 production wells would continue to operate, except for any that were converted to injectors.

(6) The implementation of the water flood would be kept simple. On January 1, 1998, all sites in the well combination would commence injection and would continue to do so for the duration of simulation. Individual flow rates would be capped at 20,000 bbl/day; but actual rates would be determined by the reservoir simulator's own algorithms. No phasing of either injection or production wells would be considered.

(7) Existing surface facilities constraints would be maintained. However, if a well combination's total demand for sea water to inject were to exceed the existing limit of 40,000 bbl/day, upgrading would be permitted for additional cost.

3.2. Cost estimates

Estimates for costs are given below. The dollar amounts represent general industry figures and are not reflective of actual costs incurred at Pompano.

(1) Conversion of producers: Conversion of a Phase I producer into an injector is estimated at \$7 million for the first injector and \$3 million for each subsequent injector.

(2) New injector wells: New locations are considered in two cost categories. A north–south dividing line is drawn to separate wells which can be drilled from the platform and those which must be drilled from the new subsea template in the south (see

Fig. 3). Locations within reach of the platform can be drilled for \$13 million each. Locations in the southern portion of the field require an up-front investment of \$25 million to move the drilling platform into place and install the template. Each well would then cost \$13 million to drill.

(3) Seawater pumping facilities: Well combinations whose combined peak injection rates exceed 40,000 bbl/day will necessitate upgrading the facilities at an estimated cost of \$2 million for each additional 30,000 barrels pumped.

(4) Maintenance and operation (M&O): M&O costs associated with the 12 existing producers are estimated at \$182,000/well per year. M&O costs for injectors are estimated at \$1 million/year for the first injector and \$500,000/year for each additional injector.

(5) Value of produced oil/gas: The oil price used in the net profit calculations was \$15.50/bbl. The gas price was \$2.50/mcf. These values were based on New York Mercantile Exchange quotes from May, 1998.

3.3. Performance measure (objective function)

The measure used to evaluate the performance of individual well combinations and serve as the objective function to be optimized is simple net profit (SNP). This measure is the sum of all revenues from sale of the produced oil and gas over the time period being evaluated minus the sum of the capital and M&O costs detailed above for the same period. No discounting or inflation factors were taken into account. By using a simple measure such as this, the number of individual estimates of oil and gas production required for the calculations could be kept to a minimum: one estimate each of cumulative oil and cumulative gas production over the desired time-frame. For this particular problem, nothing was lost by optimizing on the basis of the simpler formulation. For the 550 well combinations comprising the knowledge base, the squared correlation over seven years between SNP and net present value (which was calculated using a 0.10 discount factor and a 0.03 inflation factor) was $r^2 = 0.99$. However, this correlation may not stand over longer periods of simulation. The form of the objective function is given below:

$$\max F(\boldsymbol{x}), \tag{1}$$

where $\boldsymbol{x}$ = vector of 25 binary variables, each representing a prospective injection location, subject to various constraints (see Section 3.1. above)

$$F = \mathrm{SNP} = R - C \tag{2}$$

$$R = V_{\mathrm{O}} \times P_{\mathrm{O}} + V_{\mathrm{G}} \times P_{\mathrm{G}} \tag{3}$$

V_{O} = cumulative volume of oil produced by the field over some fixed time-frame; P_{O} = some fixed price per unit of oil; V_{G} = cumulative volume of gas produced by the field over some fixed time-frame; P_{G} = some fixed price per unit of gas

$$C = U + D + O_{\mathrm{I}} + O_{\mathrm{P}} \tag{4}$$

U = initial cost of upgrading seawater pumping facilities for injection, if required; D = initial cost of drilling new injection wells and/or converting existing producers, as dictated by the particular injection locations selected for evaluation; O_I = total cost of operating and maintaining all injectors over some fixed time-frame; O_P = total cost of operating and maintaining all producers over some fixed time-frame.

The reader will be spared further details of the cost calculations. They are defined in Section 3.2. above.

4. APPLICATION OF THE ANN–GA/SA METHODOLOGY

Fig. 1 shows the general flow of the methodology. The application of each component to the Pompano water flood problem is discussed below.

4.1. Create a knowledge base of simulations

This is the most critical component in the entire process and consists of several steps.

4.1.1. Define the problem scope

In this step, the boundaries of the problem to be optimized are determined. The decisions made in this step will guide the sampling of representative runs for the reservoir simulations and, as a result, will set the limits within which management questions can be asked. Most of the decisions that are made at this time are embodied in the assumptions and constraints detailed in section 3.1 above. One critical issue is the maximum time-frame over which performance will be evaluated. The maximum time-frame, seven years in this case, determines the simulation period for the reservoir simulation runs. By saving intermediate yearly results, this time-frame can be shortened, if desired. But it cannot be extended without further simulation. For the Pompano problem, the seven-year time-frame was selected for practical reasons concerning the limited number (2) of workstations licensed to run the simulator.

Another set of issues involves separating factors in the problem which will be held constant from those that will be allowed to vary (i.e. the 'decision variables', in optimization terminology). For example, one of the assumptions listed earlier is that production at the existing wells will continue as before, except for any that are converted to injectors. This means that field development scenarios that involve the drilling of additional production wells cannot be considered later on because that option will not have been included in the sampling plan from which the knowledge base is created. For the Pompano problem, the only variables are (1) the size of the well combinations (from one to four) and (2) which particular wells, from the candidate pool of 25, will compose the combination.

Finally, it is necessary to identify the output variables that will go into the calculation of objective functions, such as the SNP measure defined earlier. At this stage, it is most important to define the performance measures (e.g. gas/oil production) and parameters (e.g. water injection volumes) that must be calculated by the simulator, since these decisions will determine the type and timing of output saved from each run. VIP® provides a wealth of information at each time step, ranging from production figures at the well-, region-, and field-level to updated 40,000-cell arrays of pertinent physical properties. While all that information can be archived for later exploitation, only information pertinent to the management questions likely to be posed needs to be saved. For the Pompano problem, it was anticipated that only production-related objective functions would be of interest. No spatial information, such as the distribution of pressures or oil-in-place, was archived.

4.1.2. Select the candidate pool of well locations

In theory, injection could occur at any of the 40,000 cell blocks comprising the reservoir model. In practice, there will be geological and engineering constraints on the siting and completion of wells. Furthermore, it is desirable to restrict consideration to some manageable number of locations, to avoid wasting simulation and search time on unprofitable scenarios. For the Pompano problem, this manageable number was set at 25, largely based on past experience with the groundwater examples cited in the literature review. The production criteria described below should be considered only suggestive of those which could be applied.

The initial candidate pool consisted of all five Phase I producers, included because conversion is less expensive than drilling a new well and because more is known about the reservoir at those points, and 50 tentative new locations. The new locations were selected as follows: each of the 1600 (40×40 in plan view) columns in the model grid was examined to locate those columns with five or more (of the 25 possible) layers having either an x or y-transmissivity greater than 1.0. A 10×22 block in the southeastern corner of the grid was removed from consideration because the high transmissivities in that area were due to intersection with the aquifer. Of the 302 columns meeting these criteria, 50 were chosen, randomly but with some manual adjustment to improve spatial dispersion, for evaluation. All 55 initial candidates were submitted to the simulator as 1-well injection scenarios, ranked by the total hydrocarbon production (i.e. oil plus gas in oil-equivalent units) after seven years of injection, and compared to the no-injection baseline production case. The final 25 locations shown in Fig. 4 consist of the top-ranked 21 locations and four of the five Phase I wells that at least performed better than the baseline case.

4.1.3. Sample over the decision variables

This process begins by setting an overall target size for the knowledge base, 550 in this case, and sampling over the decision variables until that size is achieved. There is an approximate relationship between the number of decision variables and the number of examples required for ANN training and testing; but this relationship is also affected by the complexity of the physical relationships being modeled by the ANNs. An earlier 2D groundwater remediation problem having 30 prospective well locations had successfully employed a total knowledge base of 400 examples (300 for training and 100 for testing). For the 3D Pompano problem, targets of 400 training and 150 testing examples were set. The adequacy of these targets will be discussed in later sections.

The examples in the knowledge base set aside for ANN training contained the no-injection baseline case and all 25 1-well injection combinations. The remaining 374 training examples were generated in a three-step process:

(1) randomly select the size, from 2–4, of the combination,

(2) randomly select specific well locations, from the set of 25, to fill out the combination, and
(3) delete duplicates and those violating certain facility constraints (e.g. no more than two new wells could be drilled from the northern platform).

The 150 examples set aside for testing the ANNs' generalization performance were generated in the same fashion, except that sampling proceeded until exactly 50 2-, 3-, and 4-well combinations were obtained. This balancing by size is intended to avoid inadvertently biasing the test set in favor of any particular size, which can occur when random methods are applied to a relatively small sample.

By the standard formula for combinations of n elements taken r at a time, the total possible combinations of 1-, 2-, 3- and 4-well combinations are 25, 300, 2300, and 12,650, respectively. The entire knowledge base, including both training and testing examples, contained 25, 158, 184 and 182 combinations, respectively. The rate of inclusion of each of the 25 locations ranged from 11% to 14%.

4.1.4. Carry out the simulations

A key feature of the collection of examples generated in the sampling step is that they are independent of each other. The input to example B is not dependent on the outcomes of example A. Consequently, they can be farmed out, either manually or in an automated fashion, to as many processors as the simulator's license allows. At the time the knowledge base simulations were conducted for this project, only two single-user licenses were available. So, the simulation step required about six weeks to complete. Given additional licenses, this task could have been completed more quickly.

In contrast to typical reservoir modeling studies where detailed attention is paid to setting the simulation parameters and the analysis of outcomes on a case-by-case basis, both the creation of input files and the analysis of output is automated. General rules for assigning skin factors to injection locations, determining the layers in which a well would be completed, and setting facility constraints, together with appropriate simulation parameters, were obtained from members of the Pompano asset team who had been closely involved in the development and use of the numerical model. Given these rules and the list of well combinations to simulate, Perl scripts tailored input files for each run, launched the simulation, and extracted and saved information from each run's output.

4.2. Train ANNs to predict reservoir performance

The architecture used for all ANNs in the Pompano project was a feedforward network, trained by the familiar backpropagation learning algorithm (Rumelhart et al., 1986). In this paradigm, a network is initialized with small random weights, as is illustrated in Fig. 5. Training consists of presenting example inputs to the network and calculating the corresponding outputs, given the current values of the connection weights. The calculated output values are compared to the target values from the examples; and the connection weights are updated according to any of several learning algorithms to minimize the difference between calculated and target values on the next iteration. Over time, the connection weights associated with important relationships grow large and those associated with trivial relationships decay to zero. In the particular

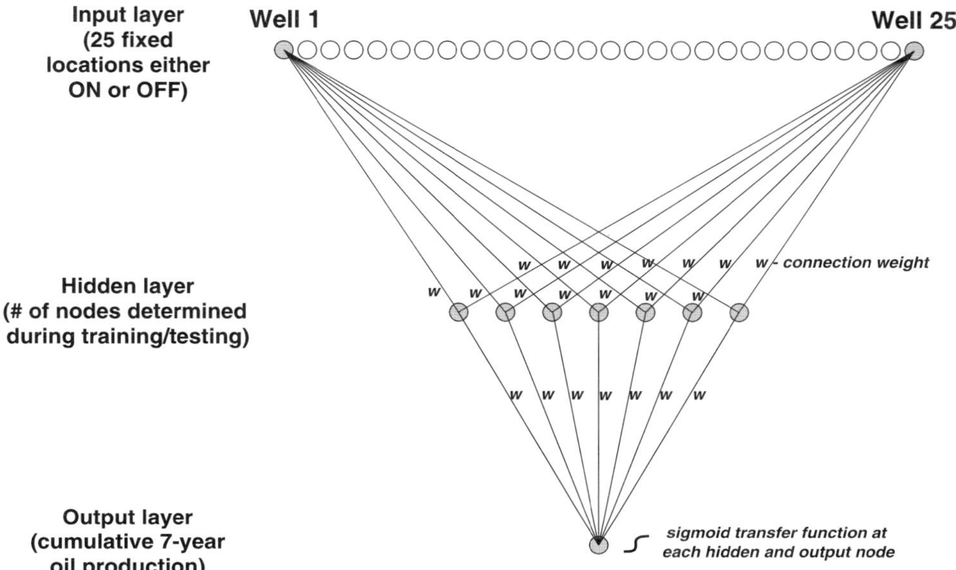

Fig. 5. Diagram of an ANN to predict cumulative 7-year oil production given a well combination as input. Connection weights between all input nodes except the first and last have been left out for visual simplicity.

implementation used for the Pompano project, a conjugate gradient optimization method (Johansson et al., 1992), employing the Polak–Ribiere weight update rule, was used to speed convergence and reduce the likelihood of becoming trapped in local minima. A sigmoid was used as the transfer function. To avoid overfitting of the network weights to idiosyncratic features of the training examples, batch updating of weights and a relatively short number (300) of training epochs was employed.

The goal of training is to construct a network with maximal capacity to accurately generalize its predictions to previously unseen combinations. Accuracy is defined here as the square of the Pearson product–moment correlation, r^2, between the ANN's and the simulator's predictions for a given attribute on some set of examples. Training accuracy, then, is the r^2 between the ANN and simulator predictions on the examples in the training set. Testing or generalization accuracy is this same measure on the examples in the test set. Factors that are known to contribute to generalization include the complexity of the network as reflected in the number of connection weights, the size and composition of the training set, and the degree of noise in the training/testing sets (Atiya and Ji, 1997). In the current study, noise in the usual sense of the term is not at issue since the examples are generated by mathematics, not nature. This is probably the main reason why all ANNs in the Pompano problem achieved very high levels of *training set* accuracy ($r^2 \geq 0.95$), a necessary but not sufficient condition for generalization accuracy.

The issue of training set size, on the other hand, is much more problematic. The allocation of 400 combinations to the training set and 150 to the testing set was based mainly on experience gained in two prior optimization studies conducted on a groundwater remediation problem. Although these numbers were thought to be low, given the greater degree of nonlinearity in 3D multiphase flow, it was thought preferable to proceed with a manageable number of simulations and leave the question of the relationship between training/testing set sizes and predictive accuracy to later research efforts.

The third factor, network complexity, is addressed by the manner in which variations on a given network are constructed and tested. As illustrated by the simplified network in Fig. 5, the size of the input and output layers are fixed at 25 nodes and one node, respectively, these dimensions having been established as the minimum necessary to adequately represent the Pompano problem. Earlier efforts to express well locations in x–y coordinates to permit a network to make spatial interpolations produced greatly degraded predictive accuracy. So, the convention of employing a set of preselected locations that constitutes the domain about which questions can be asked has been followed in this work. To keep the architecture similarly stream-lined, networks are constructed to predict only one attribute at a time: 7-year cumulative oil production, 7-year cumulative gas production, and peak injection volume. The results of searches optimizing SNP over three years, which required 3-year versions of cumulative oil and gas production, proved to be less interesting, because there was little performance spread between well combinations over such a short period of time. Consequently, the 3-year ANNs will not be discussed, except to illustrate an occasional point about neural network training and testing. One such point is that, since the knowledge base contained yearly performance data, it could be used to train ANNs over any desired time-frame up to the maximum of seven years.

The only variable architectural element is the number of nodes in the hidden layer. The value of this attribute which best promotes generalization is determined empirically by training variant networks with anywhere from 1 to 10 hidden nodes and selecting the variant with the best test set (i.e. generalization) accuracy. The protocol for selecting the best possible ANN for a given predictive task cannot end there, however. Backpropagation training is, itself, a nonlinear optimization problem and suffers from vulnerability to entrapment in local minima in the error-surface, depending on the randomly assigned initial values of the connection weights. The variance caused by those initial values is partly a function of the complexity of the input–output relationships being mapped and can be reduced by increasing the size of the training set. However, with the relatively small training/testing set sizes in the Pompano problem, some other procedure had to be developed to confront the initial-weights issue.

Fig. 6 illustrates the kind of initial-weights analysis that was performed. The graphs show mean test set accuracy, $\pm$ one standard deviation, over 25 different weight initializations for each hidden layer size from 1 to 10. The complete training of variant networks for each attribute (e.g. 3-year cumulative gas) required 250 training/testing cycles. The task was performed by a batch process that required a total of about one hour to complete, per attribute. The purpose of this exercise was to select a size for the hidden layer with not only the highest mean but also the smallest standard deviation, in an effort to identify the network architecture with the *best and most stable* generalization. Having narrowed the number of variants being considered to 25 by selecting the size of the hidden layer, the network chosen to participate in the searches was simply the variant with the highest test set accuracy.

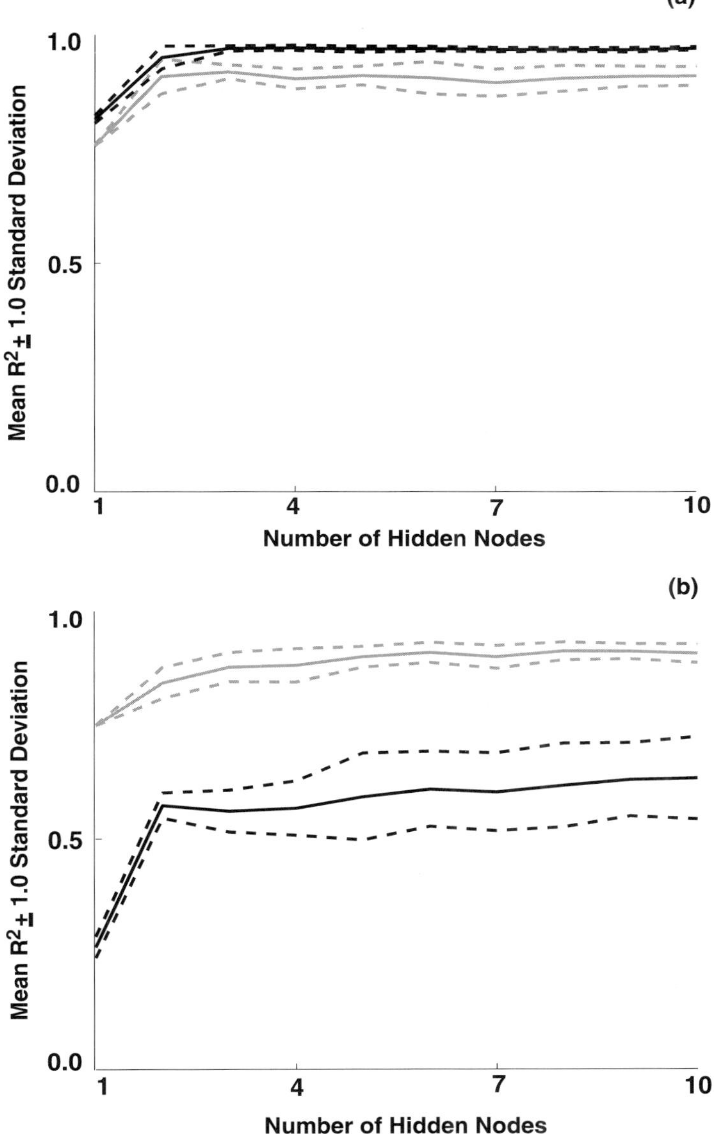

Fig. 6. Mean ANN generalization accuracy (shown by solid lines), ± one standard deviation (indicated by broken lines), as a function of the number of hidden nodes in the network, averaged over 25 random initializations of the network connection weights. Gray lines indicate 3-year predictions, with black used for 7-year predictions. (a) Cumulative gas production. (b) Cumulative oil production.

Fig. 6 also shows how the various attributes being predicted can differ from each other. The easiest attribute to accurately predict is 7-year cumulative gas, as shown by its very high means and tiny standard deviations. Defying the usual rule-of-thumb that predictive accuracy declines with increasing time, 3-year cumulative gas shows slightly lower accuracies. The situation reverts to expectations with cumulative oil, however. Three-year performance is considerably easier to predict accurately than 7-year performance. These results underscore the critical point that the mapping of inputs to outputs by the ANNs is an empirical procedure. The complexities and outcomes of the mapping is a function of the particular examples in the training and test sets and do not necessarily reflect more general physical principles.

4.3. Search for optimal well combinations

Although the trained nets can be coupled with a variety of search techniques, the genetic algorithm (GA) and simulated annealing (SA) methods were selected for their robustness and flexibility. Like all optimization drivers, these techniques are highly sensitive to some of the parameters guiding their search and relatively insensitive to others. The parameter settings used in the Pompano project have been determined by extensive trial-and-error experimentation. A discussion of these methods is given below.

4.3.1. Genetic algorithm

Given the attention that has been paid to GA applications in recent years, readers are probably familiar with the basic mechanisms of and rationale for this family of search techniques. Consequently, this section will mainly address the specific procedures chosen for implementation in the current study. Excellent introductions can be found in Goldberg (1989) and Michalewicz (1992). Goldberg is the source for all information concerning the GA presented below, unless otherwise noted.

The upper portion of Table 1 presents a summary of parameters and procedures used in the current GA. The 25 well locations which form the decision variables are represented in the GA as a string of 25 bits, each of which can either be on or off. The spatial location of each well is fixed and implicit in the representation. The order of the well locations in the string is indicated by their identification numbers in Fig. 3. That numbering is arbitrary, as is their location in the bit-string.

The search is initialized with a set of 100 well combinations. In fact, this initial population is simply a random subset of the 150 cases in the ANN testing set. The population size of 100 chosen for the current study is a fairly small value. In water resources applications, values have ranged from 64 (McKinney and Lin, 1994) to 300 (Ritzel et al., 1994) and even up to 1,000 (Cieniawski et al., 1995). A larger population helps maintain greater diversity but does so at considerable computational cost when the full model is being used to generate performance predictions.

The basic cycle of the GA is as follows. The initial population of 100 well combinations is evaluated according to an objective function, SNP in this case. A new generation of 100 combinations is created from the old population by means of three mechanisms: selection, reproduction, and mutation. The new population is then evaluated according to the objective function; and the entire process is repeated until some termination criterion is reached. The manner in which the three mechanisms have been implemented is as follows:

TABLE 1

Heuristic search parameters

Genetic Algorithm (GA)	
Population size	100
Initial population	Randomly generated patterns
String length	25 (one bit per well)
Selection for mating:	
Method	Rank order
Selection bias factor	1.5
Crossover:	
Method	Uniform
Exchange Probability	0.5
Mutation rate	0.01
Termination criterion	5 generations without exceeding peak mean fitness, 25 generations maximum
Simulated Annealing (SA)	
Initial pattern	All 25 wells off
Cooling schedule:	
Iterations/temperature	50
Decrement	0.9 (i.e. $t_{n+1} = t_n - 0.9t_n$)
Starting temperature	0.25
Neighborhood search:	
Perturbations at t_1	Up to 10 randomly selected wells
Perturbations at t_{25}	1 randomly selected well
Reduction Schedule	Same as cooling schedule
Termination Criterion	5 temperatures without exceeding peak mean fitness, 16 temperatures maximum

(1) *Selection* – This mechanism determines which members of the current generation will be selected for carry-over, in one form or another, to the new generation. To make sure that the highest-ranking combinations are not lost to the population through accidents of selection and crossover, the top three combinations are copied over to the new generation intact. The remaining 97 slots in the new population are filled by a form of sexual reproduction, a process for which parents must be selected.

The most popular method of selection is the roulette wheel, in which each member's likelihood of being selected for reproduction is the ratio of its own performance score to the total performance score of the population. The larger a given member's score is in relation to the other members', the larger portion of the roulette wheel it occupies, increasing the odds that the member will be selected one or more times for reproduction. When large discrepancies exist in the scores of individual members, the members with the higher scores come to dominate the population too quickly. Conversely, when differences between members become very small, the selection process becomes random. To avoid these cases, the current GA employs selection based on the combinations' rank order (Whitley, 1989) rather than their proportional scores. Combinations are selected by sampling from a uniform distribution over the ranks, with a bias factor of 1.5 serving to favor high-ranking combinations over lower-ranked combinations.

Selections for reproduction are made, two at a time, to obtain parent combinations from which a child combination will be formed. This process is repeated until 97 children have been generated. The same combination may constitute both members of the pair, in which case the child is simply a clone of the parent.

(2) *Reproduction (crossover)* – The most common form of reproduction is single-point crossover. Child combinations are constructed by breaking the parent combinations apart at some randomly selected crossover position in the bit-string and joining segments from each parent. For example, given two parents in a 5-bit problem (0 1 0 0 0 and *1 1 0 1 1*) and a crossover point of 2, two different children could be constructed (0 1 *0 1 1* and *1 1* 0 0 0).

Creating new combinations from 'chunks' of old ones makes the most sense when proximity in the bit-string is important. That is, the proximity of wells in the bit-string should reflect one or more dimensions of relatedness in the physical problem it represents. This is not necessarily the case in the Pompano problem. In fact, the earlier groundwater studies employing the GA had discovered a 'sticky' well problem. That is, particular wells kept appearing in the optimal solutions sets whose individual contributions to the efficiency of remediation were minimal but which were adjacent in the bit-string to wells making major contributions. To break up these spurious associations, a different reproductive mechanism, uniform crossover, is used (Syswerda, 1989). In this method, the value of each bit in the child string is set independently of every other bit. A coin-toss at each bit-position determines from which parent the child will inherit the value for that particular bit. The exchange probability can be biased to favor the fitter parent, if any; but in this study the exchange probability is kept at an impartial 0.5.

(3) *Mutation* – Mutation is a way to maintain diversity in a population by arbitrarily changing the values of bits in the child combinations according to some rate, often the inverse of the population size. A high mutation rate can undermine the effects of crossover; a low one limits the introduction of 'novelty' into the population. For this study, the inverse rule yields a mutation rate of 0.01.

4.3.2. Simulated annealing

Like the GA, SA techniques are based on an analogy to a natural process. Instead of Darwinian concepts of evolution, which are the foundation of the GA, SA is based on an analogy to the cooling of materials in a heat bath. The fundamental idea is that if the amount of energy in a system is reduced very slowly, the system will come to rest in a more perfect state than if the energy is reduced quickly. When translated into terms pertinent to optimization, the energy in the system refers to the tolerance for pursuing apparently poorer solutions in an effort to avoid being trapped in local minima. As the search proceeds, this tolerance is slowly reduced until the search converges to a final optimal solution. SA algorithms have appeared in several water resources optimization applications (Dougherty and Marryott, 1992; Christakos and Killam, 1993; Marryott et al., 1993; Rizzo and Dougherty, 1996). A highly readable introduction to the subject can be found in Dowsland (1993), which is also the source for the material discussed below, unless otherwise noted.

The SA parameters employed in the current study are given in the lower portion of Table 1. SA represents a return to single-point search, in contrast to the multiple-point or population-based search of the GA. At every step, there is only one new well combination being compared to the current combination. The initial combination represents the starting point for search. In this implementation, the initial combination is the no-injection case. Trial and error experimentation with the algorithm has shown that the initial starting point has only a small effect on the duration of search. The current study's annealing algorithm, adapted from the standard algorithm as presented in Dowsland (1993), proceeds as follows:

```
Set the current combination c = initial combination
Set the current energy in the system t = initial temperature
Select a temperature decrement function α
Repeat
   Repeat
      Generate a new combination n in the neighborhood of c
      δ = fitness(n) − fitness(c)
      if δ > 0 then c = n
      else
         generate a random value x uniformly in the range (0, 1)
         if x < exp(−δ/t) then c = n
   Until the iteration counter = iterations per temperature
   Set t = α(t)
Until termination criteria are met
```

The purpose of the temperature parameter in the algorithm is to control the tolerance for accepting a newly generated combination n as the current combination c, even when its performance score is lower than the current combination's score. If the new combination's score is greater than the current combination's, it is always accepted as the new current combination. If not, there is a probability of accepting it anyway that is a function of the current temperature t in the system, leavened by the magnitude of the difference δ between the two scores. The initial temperature and the range over which it is allowed to vary are empirically determined parameters. The experimenter decides, in advance, what overall percentages of poorer combinations it is desirable to accept in the initial and final stages of search and adjusts the temperature range until those percentages are achieved.

On the other hand, the temperature decrement or cooling function and the number of iterations per temperature are parameters that have received more attention in the literature. As was mentioned earlier, the rate of cooling has considerable impact on the likelihood of converging to an optimal solution. The function chosen for the current study, a geometric decrement function with a decrement factor of 0.9, is one of the two most widely used approaches. The issue of how many iterations to perform at a given temperature level has been the subject of considerable analysis in certain applications (Dougherty and Marryott, 1992; Marryott et al., 1993). While theory suggests that extremely large values for this parameter should be used to guarantee that the algorithm is given an adequate opportunity to sample the search space, experimentation with this parameter indicates that much smaller values, 10–100 times the number of decision variables, can be employed. At a minimum, this rule of thumb would imply that iterations per temperature should be set to 250 for the current study. Instead, a very small value, 50, has been selected, mainly to permit more timely comparisons between ANN- and simulator-based searches.

The algorithm listed above does not indicate how to generate a new combination from the neighborhood of the current combination. This is another domain-dependent decision because the manner in which valid new combinations can be constructed from old ones is a function of the problem representation. In the current implementation, the temperature parameter is used again to control the extent to which the new combination can vary from the current combination. This is equivalent to controlling the size of the local neighborhood being searched at a specified temperature level. Initially, the number of well locations in the current combination that will be switched is determined by randomly selecting an integer from 1 to 10. The particular locations to alter are then selected at random from the available 25 locations, subject to the usual facility constraints described earlier, until the pre-specified number of locations in the current combination have had their status changed from *on* to *off* or vice versa. As temperature decreases, the maximum number of locations that can be potentially changed is reduced from 10 to 1.

A small departure, also not shown above, from the serial nature of the standard algorithm has been implemented. According to the standard algorithm, the current combination at the end of processing at a given temperature level is not necessarily the highest-scoring combination encountered during the 50 iterations at that level, because there is a certain probability that an inferior new combination will replace the current combination. However, the algorithm implemented in the current study remembers the best combination ever encountered and makes it the current combination before proceeding to the next temperature level. This is somewhat akin to the practice in the GA of preserving the top combinations from one generation to the next so that they are not lost through the vicissitudes of selection and crossover.

4.3.3. Procedures common to both GA and SA searches

Termination criteria in optimization are usually based on some notion of convergence to a single best solution. In keeping with the philosophy of heuristic search, however, the current study is more interested in generating *sets* of near-optimal solutions rather than a single best solution. This goal is achieved by tying termination criteria to the performance score of the population (in the case of the GA) or the temperature (in the case of SA) rather than the performance of the highest-ranking individual combination. Search terminates when either (a) the mean population or temperature performance score fails to improve over five consecutive generations or temperatures, or (b) some maximum number of generations or temperatures have elapsed, whichever comes first. The maximum number of the GA generations is 25; the maximum number of SA temperatures was reduced to 16, to prevent over-long searches when the ANN–SA vs. VIP®–SA comparison was being conducted. At the end of every generation or temperature, combinations with scores above a predetermined cut-off are saved to a file. The top-ranked unique combinations in this file become the set of near-optimal solutions.

The outcome of search in both the GA and SA is influenced by the particular random choices that are made. To improve the stability of the outcome, the results of each search in the current study (with one exception, which will be noted below) actually consist of combined results from 10 searches, each with a different seed initializing the pseudo-random number generator.

4.4. *Verify optimal combinations with the simulator*

In an actual engineering application of the ANN–GA/SA methodology, the asset team may choose to only submit a handful of well combinations to the simulator. For this demonstration project, however, the top 25 well combinations from the near-optimal set were submitted for verification. The resulting simulator predictions of 7-year oil and gas production and peak injection volume are used to recalculate the SNP. The *updated* SNPs become the measure for subsequent analysis and decision-making.

5. SEARCH RESULTS

The results of various efforts to identify optimal well combinations to maximize SNP over seven years are shown in Tables 2–6. Throughout, the production figures for the no-injection baseline case serve as the standard against which alternative scenarios are judged. All values appearing in the tables are reported in increments/decrements of the appropriate unit (e.g. dollars, mmcf). SNP is calculated according to the cost estimates and definitions described earlier. Scenarios are designated by a list of the identification numbers (see Fig. 3) of the wells making up the combination.

5.1. *Context scenarios*

It is useful to begin by considering the performance of some simple conversions of existing producers in the northern section to injection locations. There is considerable appeal to pursuing such scenarios, in part because knowledge of the reservoir is much greater in the vicinity of an existing producer than around the new injection locations. The effects of converting the four northern producers which survived the initial screening of locations (according to the criterion that their effect, singly, on total hydrocarbon production over seven years must exceed the no-injection baseline) are

TABLE 2

Performance measures of selected single-well conversion scenarios relative to the no-injection case

Scenario	Simple net profit (millions)	Oil (mstb)	Gas (mmcf)	Costs (millions)
No Injection	0.00	0	0	0.00
Well 12	−5.75	+3,227	−17,223	+12.72
Well 13	+14.68	+5,696	−24,353	+12.72
Well 14	+3.59	+3,859	−17,400	+12.72
Well 15	−10.18	+688	−3,250	+12.72

TABLE 3

Top ten well combinations from the 550-case knowledge base ranked by their improvement over baseline on simple net profit

Scenario	Simple net profit (millions)	Oil (mstb)	Gas (mmcf)	Costs (millions)
7–9	+95.08	+11,899	−11,139	+61.5
7–24	+94.95	+11,715	−11,057	+61.5
6–19–21–24	+91.62	+19,007	−12,362	+94.5
7–16–18–24	+91.20	+15,152	−18,863	+96.5
6–7–10–16	+87.67	+14,307	−15,832	+94.5
7–11–15–16	+86.31	+14,176	−17,675	+89.2
7–25	+86.10	+11,233	−10,600	+61.5
6–7–10	+85.44	+12,275	−10,728	+78.0
6–7–9	+81.22	+12,676	−14,103	+80.0
7–11–21–23	+78.90	+13,666	−14,567	+96.5

shown in Table 2. When the more complex SNP performance measure is used, two of the wells now show a negative impact on total field productivity and the positive influence of the other two is minimal. The best conversion, well 13, produces only a 1.72% improvement over baseline performance. This result illustrates the sensitivity of outcomes to the particular performance measure being used and suggests that many different measures should be used to evaluate scenarios for field development.

5.2. Best in knowledge base

The next most obvious tactic is to query the 550-case knowledge base to identify the well combinations which yield the highest SNPs. The attraction of this tactic is that the oil production, gas production, and peak injection volume inputs to the SNP calculations come directly from the simulator, without any estimation errors introduced by the ANNs. The drawback is that results are limited to well combinations already in the knowledge base. As Table 4 shows, the information in the knowledge base alone makes a considerable improvement in expected performance of the field over the simple single-well conversions of Table 2. The best combination, consisting of wells 7 and 9, shows an 11.11% improvement over the baseline SNP.

5.3. ANN–GA search results

The reason for going to the extra effort of implementing an actual search for optimal well combinations is that there may be combinations not sampled in the knowledge base which have superior performance characteristics. A directed search technique can usually identify peak performers which a random sampling may miss. Since the time required to train ANNs and conduct the searches is small (at least once the methodologies are mastered) relative to the time required to create the knowledge base, there is ample reason to proceed.

TABLE 4

VIP®-verified simple net profit (SNP) of well combinations from the ANN–GA search which exceed the SNP of the best combination in the knowledge base

Scenario	Simple net profit (millions)
7–16–24	+115.54
7–16–23	+114.85
7–11–16	+113.89
1–7–24	+109.76
7–19–24	+109.76
6–7–24	+108.53
6–7–23	+107.96
7–16–25	+107.88
7–11–19	+106.94
1–7–11	+105.66
6–7–11	+104.02
7–20–24	+101.05
7–11–20	+99.14
7–9–20	+97.66
6–7–25	+96.83
7–9–16	+95.42

The entire 10-cycle GA search required less than an hour on the same class of workstation used to perform the simulations. All well combinations with estimated SNPs above a certain cut-off were saved and combined for post-processing. The top 25 well combinations from this pool were submitted to the simulator to verify the oil, gas, and peak injection numbers and calculate an updated SNP. The 16 combinations whose updated SNP exceeded the best in the knowledge base are shown in Table 4. The best combination, 7–16–24, yields a 13.5% improvement over baseline.

In addition to fulfilling the final step in the ANN–GA/SA process (i.e. verifying the optimal set of well combinations so that engineering decisions can be made on the best-available information), the data generated by the verification runs provide an opportunity to assess the final-stage accuracy of the ANNs. The generalization accuracies, expressed as the squared correlation between ANN and simulator predictions, of the 7-year oil, 7-year gas, and peak injection volume ANNs on the 150-case test set were 0.81, 0.98, and 0.99, respectively. To the extent that the test set is a good, if low-resolution, representation of the total space over which the search might roam, these numbers indicate excellent generalization for the gas and peak injection ANNs and borderline-acceptable generalization for the oil ANN. This does not mean, however, that the ANNs' level of accuracy will be maintained during the final stages when small subregions of the search space are being searched at high resolution. It is to be expected that ANNs trained and tested on a coarse sampling will lose accuracy when required to make fine-grained distinctions. And, in fact, the squared correlations between ANN and simulator predictions on the top 25 well combinations generated by the ANN–GA search on 7-year oil, 7-year gas, and peak injection volume were 0.39, 0.38, and 0.96, respectively. Furthermore, the correlation between the SNP estimates based on the ANN predictions

TABLE 5

VIP®-verified simple net profit (SNP) of well combinations from the ANN–SA search which exceed the SNP of the best combination in the knowledge base

Scenario [a]	Simple net profit (millions)
7–16–24	+115.54
7–16–23	+114.85
7–11–16	+113.89
1–7–24	+109.76
7–19–24	+109.76
6–7–24	+108.53
7–16–25	+107.88
7–11–19	+106.94
1–7–11	+105.66
6–7–11	+104.02
6–11–19	+102.68
7–20–24	+101.05
7–11–20	+99.14
7–9–20	+97.66
7–9–16	+95.42

[a] Bold face scenarios were also located by the ANN–GA search (see Table 4).

and the *updated* SNP figures based on the simulator-verified numbers were virtually zero. What this suggests is that the GA, using SNP calculations based on the relatively coarse-grained ANN predictions, is able to locate approximate regions where optimal combinations lie and to identify several near-optimal candidates. However, at the current level of investment in training/testing examples, only the simulator itself can sort out the relative ranking among that final set of candidates.

5.4. *ANN–SA search results*

One issue that needs to be explored is the extent to which the results in Table 4 are a function of the ANNs or a function of the GA. To address this question, an ANN–SA search was conducted, holding all procedures used in the previous search constant except for the substitution of the SA search method. The results, shown in Table 5, are a very clear indication that the final set of well combinations is a reflection of ANN prediction and not on the specific search engine. The shift to a different search method made almost no difference to the final set of well combinations.

5.5. *VIP®–SA search results*

Given that the ANNs' final stage accuracy does deteriorate, one last question is whether superior well combinations are being missed because the ANNs, rather than the simulator, are supplying the predictions which are influencing the direction of search. The only way to answer this question is to conduct a search in which the simulator is called each time a new well combination is being evaluated by the search engine.

TABLE 6

Simple net profit (SNP) of well combinations from the VIP®–SA search which exceed the SNP of the best combination in the knowledge base

Scenario [a]	Simple net profit (millions)
7–16–24	+115.54
1–7–24	+109.76
7–19–24	+109.76
1–7–23	+108.73
6–7–23	+107.96
7–16–25	+107.88
7–16–20–24	+103.73
7–20–24	+101.05
7–9–20	+97.66
1–7–23–24	+97.61
1–7–11–23	+97.28
1–7–20–24	+97.28
6–7–10–23	+97.02
6–7–25	+96.83
7–16–20–25	+95.83
1–7–10–24	+95.27

[a] Bold face scenarios were also located by the ANN–SA search (see Table 5).

A VIP®–SA search, the results of which are shown in Table 6, was conducted in the following manner: because the simulator was called to supply the oil, gas, and peak injection data needed for the SNP calculations and each call required an average of 3.5 hours, only three rather than 10 repetitions of the search were performed. Three workstations ran in parallel, sharing a common cache of results so that no time would be wasted on duplicate calls for the same well combination. The SA engine was chosen over the GA because it tends to converge more quickly. Even so, the three searches required several weeks to complete on each workstation, involved 936 unique calls to the simulator (or 3276 total computational hours) and matched but did not beat the best well combination (7–16–24) found by the ANN–SA search. In other words, using the simulator directly in the search did not improve the quality of results for this particular problem; it merely took an inordinate amount of computational time. Furthermore, results saved from these runs are not likely to be re-usable in new searches. The well combinations which dominate for the current definition of SNP will not necessarily appear at all if another set of cost estimates is used.

6. SUMMARY AND CONCLUSIONS

The purpose of the Pompano project was to apply a methodology originally developed to optimize the placement of wells in groundwater remediation to a problem in reservoir management. The project sought to illustrate the improvements in decision-making which can be achieved with only minimal adaptation of methods from the earlier work. The following conclusions seem to be warranted by the results shown in Tables 2–6.

Current practice in the industry is to treat the reservoir simulator as a tool for detailed analysis of the reservoir. Members of the asset team propose a small number of scenarios for well-field development based on the available reservoir characterization data, any existing production data, and their own knowledge and experience. These scenarios are submitted to the simulator, with results confirming or refuting the team's proposals and possibly suggesting new design variations to explore. Because the emphasis is on detailed examination of results, the total number of scenarios that are likely to be considered in this approach is small: on the order of 'tens'.

Using the archive of simulations as a knowledge base, the project has introduced a change of perspective, expanding the scope of study from tens to hundreds of scenarios. The simulator is now viewed as a tool for providing rapid answers to a variety of engineering and management questions. Querying the knowledge base has highlighted the considerable increase in performance that may possibly be achieved by switching from an approach to injection based on converting one or more existing producers to one involving the drilling of three to four new injectors, despite the increased capital and operating expenses associated with this latter approach.

Even greater value is mined from the reservoir simulator when the archive of simulations is used in its second capacity: as a source of examples for training and testing ANNs. ANNs were trained to predict peak injection volumes and volumes of produced oil and gas over seven years of injection. The rapid estimates of these quantities provided by the ANNs were fed into simple net profit calculations, which in turn were used by the GA or SA to evaluate the effectiveness of different well-field scenarios and generate improved scenarios. The search engine explored scenarios not contained in the original archive of simulations, expanding the scope of study from hundreds to thousands of scenarios. This expansion has enabled the identification of new scenarios which exceed the simple net profits of the best scenarios found by simply querying the knowledge base.

6.1. Outstanding issues

Both substantive and methodological issues have been raised in the course of this project.

6.1.1. Substantive interpretation of results

In the discussion of results in Tables 2–6, emphasis was placed on the best performing scenario located by each method. However, the results actually consist of *sets* of near-optimal scenarios which can be analyzed in an effort to better understand the underlying physical reasons why these scenarios are optimal answers to a particular management question. For example, an examination of the top 25 well combinations from the ANN–GA search found that well 7 figured in 100% of the combinations, followed at a distance by well 24 (32%), well 11 (28%), and well 16 (24%). One might speculate that well 7 has a larger sweep of neighboring producers that are important to production over the seven year time-frame. The other popular wells may be reflective of more conventional wisdom regarding the desirability of raising pressures near the boundaries of the reservoir. Given results from several searches addressing different management questions (e.g. varying economic parameters and time-frames or narrowing the focus to wells of special interest), the asset team has the opportunity to build a body of operating principles about the field, some of which may transfer to other fields, as well. The process of developing feedback loops between asset team hypotheses and optimization analysis could facilitate integrating the reservoir modeling with geologic/geophysical interpretations.

6.1.2. ANN accuracy issues

An ANN's generalization accuracy has to be examined in two ways. First, there is the question of how accurately it makes predictions over the entire space in which predictions might be called for by the search engine. In the Pompano problem, these initial accuracies were estimated by correlating ANN and simulator predictions on the 150 cases in the test set. Accuracy varied considerably by both the attribute to be predicted (oil vs. gas, for example) and the time-frame over which the prediction was being made (three vs. seven years). It appears, however, that despite errors of estimation, the search engines still gravitated to the specific regions where optimal well combinations were to be found and generated several near-optimal candidates. It was then the job of the simulator to sort out the proper rankings among the 'finalists'. One possible drawback to using an ANN as a proxy for the simulator, namely that ANN-introduced errors would cause the search to completely miss the best combinations, proved not to be true for this particular problem. In a comparison search where the simulator itself was called upon to provide predictions as demanded by the search engine, the best combination located in the ANN-based searches was matched but not beaten. Still, this outcome might not be born out on other management questions or over longer time-frames. Consequently, a critical research task is to identify strategies for improving both initial and final accuracy. An example of a technical strategy which might improve accuracy involves simplifying the prediction task for the ANNs by training separate nets for the 2-, 3-, and 4-well combinations. A more knowledge-based approach to the problem would involve supplying more information about the reservoir to the nets, in the form of more inputs. For example, instead of describing a well combination as a set of binary inputs, the local average permeability of each well which will be turned on in the combination could serve as the inputs.

6.1.3. Uncertainties in the underlying model

A third concern that has not yet been addressed involves uncertainties associated with the reservoir simulator itself. So far, a single model of the reservoir has been taken as a kind of norm or best-bet on which to base reservoir management decisions. However, since there are likely to be reasonable alternatives to the normative model which may greatly affect the optimal solutions to management questions, decision-makers are better served if they are presented with at least some indication of how great a variation is introduced by considering these alternatives.

The problem of estimating and managing model uncertainties is huge and will not be solved anytime in the near future. There are, however, incremental strategies for incorporating aspects of uncertainty analysis into the ANN–GA/SA methodology at different stages of the optimization process. A very simple strategy is to rank each well location by the relative certainty of the physical properties in its vicinity. The objective function being optimized would contain a penalty term based on that rank, which will reflect the informational-risk associated with including that well in the scenario. A much more laborious approach would be to create separate knowledge bases for a small set (e.g. three) of geologically-reasonable alternative models and carry out the entire process separately for each one, comparing results for common locations.

ACKNOWLEDGEMENTS

This work has been funded by DeepLook, a consortium of major oil and service companies, dedicated to collaborative research in the area of locating and exploiting bypassed hydrocarbons. Special thanks are due to Landmark Graphics, Inc., for use of their reservoir simulator VIP®, and to Ed Stoessel, for his review of the paper. This work has been performed under the auspices of the U.S. Department of Energy by Lawrence Livermore National Laboratory under contract W-7405-ENG-408.

REFERENCES

Aadland, A., Dyrnes, O., Olsen, S.R. and Dronen, O.M., 1994. Statfjord field: field and reservoir management perspectives. *SPE Reservoir Eng.*, 9(3): 157–161.

Aanonsen, S.I., Eide, A.L., Holden, L. and Aasen, J.O., 1995. Optimizing reservoir performance under uncertainty with application to well location. In: *Proc. SPE Annual Technical Conference and Exhibition: Reservoir Engineering*. Society of Petroleum Engineers, Tulsa, OK, pp. 67–76.

Atiya, A. and Ji, C., 1997. How initial conditions affect generalization performance in large networks. *IEEE Trans. Neural Networks*, 8(2): 448–451.

Behrenbruch, P., 1993. Offshore oilfield development planning. *J. Pet. Technol.*, 45(8): 735–743.

Bittencourt, A.C. and Horne, R.N., 1997. Reservoir development and design optimization. In: *Proc. SPE Annual Technical Conference and Exhibition: Reservoir Engineering*. Society of Petroleum Engineers, Tulsa, OK, pp. 554–558.

Breitenbach, E.A., 1991. Reservoir simulation: state of the art. *J. Pet. Technol.*, 43(9): 1033–1036.

Christakos, G. and Killam, B.R., 1993. Sampling design for classifying contaminant level using annealing search algorithms. *Water Resour. Res.*, 29(12): 4063–4076.

Cieniawski, S.E., Eheart, J.W. and Ranjithan, S., 1995. Using genetic algorithms to solve a multiobjective groundwater monitoring problem. *Water Resour. Res.*, 31(2): 399–409.

Coskuner, G. and Lutes, B., 1996. Optimizing field development through infill drilling coupled with surface network: a case study of low permeability gas reservoir. In: *Proc. SPE Annual Technical Conference and Exhibition: Reservoir Engineering*. Society of Petroleum Engineers, Tulsa, OK, pp. 273–284.

Dougherty, D.E., 1991. Hydrologic applications of the connection machine CM-2. *Water Resour. Res.*, 27(12): 3137–3147.

Dougherty, D.E. and Marryott, R.A., 1992. Markov chain length effects on optimization in groundwater management by simulated annealing. In: Fitzgibbon, W.E. and Wheeler, M.F. (Eds.), *Computational Methods in Geoscience*. Society for Industrial and Applied Mathematics, Philadelphia, PA, pp. 53–65.

Dowsland, K.A., 1993. Simulated annealing. In: Reeves, C.R. (Ed.), *Modern Heuristic Techniques for Combinatorial Problems*. Halstead Press, New York, NY, pp. 20–69.

Fujii, H. and Horne, R.N., 1994. Multivariate optimization of networked production systems. *Soc. Pet. Eng.*,

Tulsa, OK, SPE 27617.

Goldberg, D.E., 1989. *Genetic Algorithms in Search, Optimization, and Machine Learning*. Addison-Wesley, Reading, MA.

Johansson, E.M., Dowla, F.U. and Goodman, D.M., 1992. Backpropagation learning for multi-layer feed-forward neural networks using the conjugate gradient method. *Int. J. Neural Systems*, 2(4): 291–301.

Karatzas, G.P. and Pinder, G.F., 1993. Groundwater management using numerical simulation and the outer approximation method for global optimization. *Water Resour. Res.*, 29(10): 3371–3378.

Kikani, J. and Smith, T.D., 1996. Recovery optimization by modeling depletion and fault block differential pressures at Green Canyon 110. In: *Proc. SPE Annual Technical Conference and Exhibition: Reservoir Engineering*. Society of Petroleum Engineers, Tulsa, OK, pp. 157–170.

Kirkpatrick, S., Gelatt, C.D. and Vecchi, M.P., 1983. Optimization by simulated annealing. *Science*, 220(4598): 671–680.

Kumar, M. and Ziegler, V.M., 1993. Injection schedules and production strategies for optimizing steamflood performance. *SPE Reservoir Eng.*, 8(2): 101–108.

Marryott, R.A., Dougherty, D.E. and Stollar, R.L., 1993. Optimal groundwater management. II. Application of simulated annealing to a field-scale contamination site. *Water Resour. Res.*, 29(4): 847–860.

McKinney, D.C. and Lin, M.D., 1994. Genetic algorithm solution of groundwater management models. *Water Resour. Res.*, 30(6): 1897–1906.

Michalewicz, Z., 1992. *Genetic Algorithms + Data Structures = Evolution Programs, 2nd ed.* Springer-Verlag, New York, NY.

Reeves, C.R., 1993. Introduction. In: Reeves, C.R. (Ed.), *Modern Heuristic Techniques for Combinatorial Problems*. Halstead Press, New York, NY, pp. 1–19.

Ritzel, B.J., Eheart, J.W. and Ranjithan, S., 1994. Using genetic algorithms to solve a multiple objective groundwater pollution containment problem. *Water Resour. Res.*, 30(5): 1589–1603.

Rizzo, D.M. and Dougherty, D.E., 1996. Design optimization for multiple management period groundwater remediation. *Water Resour. Res.*, 32(8): 2549–2561.

Rogers, L.L. and Dowla, F.U., 1994. Optimization of groundwater remediation using artificial neural networks and parallel solute transport modeling. *Water Resour. Res.*, 30(2): 457–481.

Rogers, L.L., Dowla, F.U. and Johnson, V.M., 1995. Optimal field-scale groundwater remediation using neural networks and the genetic algorithm. *Environ. Sci. Technol.*, 29(5): 1145-1155.

Rumelhart, D.E., Hinton, G.E. and Williams, R.J., 1986. Learning internal representations by error propagation. In: Rumelhart, D.E. and McClelland, J.L. (Eds.), *Parallel Distributed Processing: Explorations in the Microstructure of Cognition, Vol. 1: Foundations*. MIT Press, Cambridge, MA, pp. 318–362.

Syswerda, G., 1989. Uniform crossover in genetic algorithms. In: Schaffer, J.D. (Ed.), *Proc. 3rd Int. Conf. Genetic Algorithms*. Morgan Kaufman, San Mateo, CA, pp. 2–9.

Tompson, A.F.B., Ashby, S.F., Falgout, R.D., Smith, S.G., Fogwell, T.W. and Loosmore, G.A., 1994. Use of high performance computing to examine the effectiveness of aquifer remediation. In: Peters, A. (Ed.), *Computational Methods in Water Resources X, Volume 2*. Kluwer Academic Publishers, Boston, MA, pp. 875–882.

Wackowski, R.K., Stevens, C.E., Masoner, L.O., Attanucci, V., Larson, J.L. and Aslesen, K.S., 1992. Applying rigorous decision analysis methodology to optimization of a tertiary recovery project: Rangely Weber sand unit, Colorado. *Soc. Pet. Eng.*, Tulsa, OK, SPE 24234.

Wagner, B.J., 1995. Recent advances in simulation–optimization groundwater management modeling. In: *Review of Geophysics, Supplement.* U.S. National Report to International Union of Geodesy and Geophysics, 1991–1994, pp. 1021–1028.

Whitley, D., 1989. The GENITOR algorithm and selection pressure: Why rank-based allocation of reproductive trials is best. In: Schaffer, J.D. (Ed.), *Proc., 3rd Int. Conf. on Genetic Algorithms*. Morgan Kaufman, San Mateo, CA, pp. 116–123.

York, S.D., Peng, C.P. and Joslin, T.H., 1992. Reservoir management of Valhall field, Norway. *J. Pet. Technol.*, 44(8): 918–923.

Developments in Petroleum Science, 51
Editors: M. Nikravesh, F. Aminzadeh and L.A. Zadeh

Chapter 19

NEURAL NETWORK PREDICTION OF PERMEABILITY IN THE EL GARIA FORMATION, ASHTART OILFIELD, OFFSHORE TUNISIA

J.H. LIGTENBERG and A.G. WANSINK [1]

dGB Earth Sciences, Boulevard-1945 24, 7511 AE, Enschede, The Netherlands

ABSTRACT

The Lower Eocene El Garia Formation forms the reservoir rock at the Ashtart oilfield, offshore Tunisia. It comprises a thick package of mainly nummulitic packstones and grainstones, with variable reservoir quality. Although porosity is moderate to high, permeability is often poor to fair with some high permeability streaks.

The aim of this study was to establish the relationships between log-derived data and core data, and to apply these relationships in a predictive sense to un-cored intervals. The main objective was to predict from measured logs and core data the limestone depositional texture (as indicated by the Dunham classification), as well as porosity and permeability. A total of nine wells with complete logging suites, multiple cored intervals with core plug measurements together with detailed core interpretations were available.

We used a fully connected multi-layer perceptrons network (MLP, a type of neural network) to establish possible non-linear relationships. Detailed analyses revealed that no relationship exists between log response and limestone texture (Dunham class). The initial idea to predict Dunham class and subsequently use the classification results to predict permeability could therefore not be pursued. However, further analyses revealed that it was feasible to predict permeability without using the depositional fabric, but using a combination of wireline logs and measured core porosity. Careful preparation of the training set for the neural network proved to be very important. Early experiments showed that low to fair permeability (1–35 mD) could be predicted with confidence, but that the network failed to predict the high permeability streaks. 'Balancing' the data set solved this problem. Balancing is a technique in which the training set is increased by adding more examples to the under-sampled part of the data space. Examples are created by random selection from the training set and white noise is added. After balancing, the neural network's performance improved significantly. Testing the neural network on two wells indicated that this method is capable of predicting the entire range of permeability with confidence.

---

[1] Present address: Wintershall Noordzee B.V., Eisenhowerlaan 142–146, 2517 KN, The Hague, The Netherlands.

1. INTRODUCTION

The reservoir rock in the Ashtart oilfield, offshore Tunisia, is composed of the Lower Eocene El Garia Formation of the Metlaoui Carbonate Group. The quality of this carbonate reservoir is variable. Although it has a moderate to high porosity, the permeability is poor to fair, generally below 10 mD (Loucks et al., 1998, fig. 6). The best reservoir intervals are characterized by high permeability streaks, which occur where interparticle porosity is preserved or at strongly dolomitized or fractured zones. This is predominantly dependent on a number of diagenetic processes such as mechanical compaction and dolomitization. According to Loucks et al. (1998), permeability in this formation is also controlled by factors such as the abundances of lime mud and nummulithoclastic debris.

Although there is no standard method of estimating permeability in carbonate rocks, a simple model like the 'best-fit curve' technique is too inaccurate to predict high-quality reservoirs. This method draws a curve through measured core plug data by simple interpolation. The quality of the result between two core plugs strongly depends on the sampling distance of the plugs and the homogeneity of the rock. Thin intervals with high permeabilities will not be detected if they were not included in the core plug analyses. Possibly the largest limitation of this technique is that it can only be applied to wells where cores and plugs were taken. Another technique that is sometimes applied in carbonate settings is the so-called 'field-wide layer' method. A detailed sequence-stratigraphic interpretation of the field is made, and porosity–permeability relationships are derived layer by layer. The method does deliver permeability predictions in non-cored wells but the predictive power of the derived relationship is often dubious in nature. Moreover, the sequence stratigraphic interpretation is a difficult and time consuming task.

Loucks et al. (1998) noted that core data is essential for the analysis of reservoir quality and that it is not possible to predict permeability from wireline logs. However, in this study, we will demonstrate that neural networks can predict permeability from wireline data successfully. We also attempt to show if there is any relationship between the measured logs, depositional fabric (Dunham class) and porosity.

Our approach to predicting reservoir quality is deterministic in nature. We aim to find subtle relationships between measured logs and core data that can be captured by a supervised neural network. The trained network is subsequently applied to predict relevant reservoir properties in sections lacking core information.

2. GEOLOGICAL SETTING

The Ypresian El Garia Formation is part of the Metlaoui Carbonate Group. It consists mainly of thick nummulitic packstones and grainstones (Fig. 1), and was deposited in a carbonate ramp environment (Loucks et al., 1998; Macaulay et al., 2001; Racey et al., 2001). The formation forms the reservoir in the Ashtart oilfield in the Gulf of Gabes, offshore Tunisia, and in several other gas and oilfields in Tunisia and Libya (Racey et al., 2001). The main source rock in this region is the organic-rich Bou Dabbous

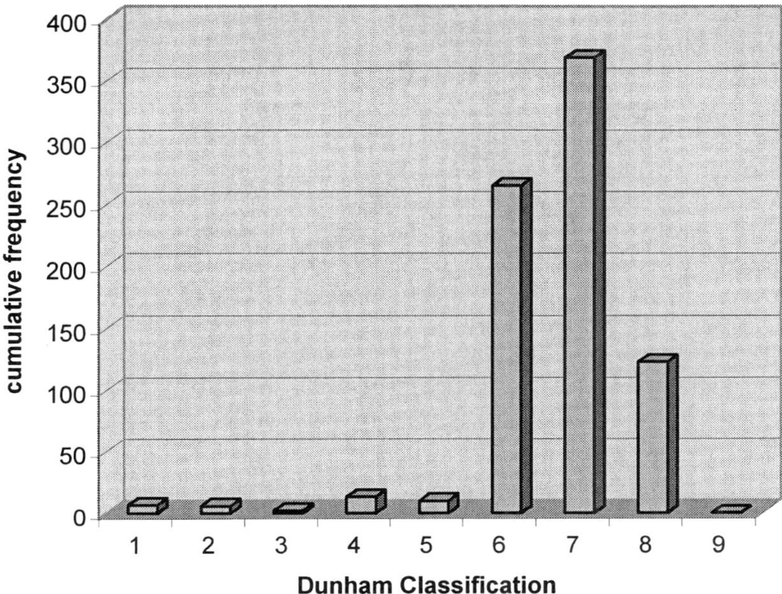

Fig. 1. Distribution of limestone textures in the El Garia Formation at the Ashtart oilfield, based on core interpretations (see Table 1 for Dunham classes).

Formation, the deep marine equivalent of the El Garia Formation (Racey et al., 2001). The Compact Micrite Member forms the top seal.

The quality of the El Garia reservoir is variable. Porosity is generally moderate to high, mainly in the form of intraparticle and micro-porosity, whilst permeability is dominantly poor to fair and generally below 20 mD. The best quality reservoir is associated with some high permeability streaks, which are often difficult to detect. In such intervals, the primary interparticle porosity has either been preserved or reflects significant dolomitization resulting in high intercrystalline porosity (Loucks et al., 1998). Reservoir quality depends on many factors, such as grain type and size, cementation and compaction, which make it difficult to predict.

3. NEURAL NETWORKS

Artificial neural networks belong to a group of mathematical algorithms which in general are inspired by the 'brain metaphor', meaning that they try to emulate the internal processes of the human brain. They usually consist of many processing nodes that are connected by weights. Neural networks are used in many industries today to solve a range of problems including, (but not limited to) pattern recognition, regression analysis and data clustering. In the oil industry, neural networks are now routinely used in seismic reservoir characterization and seismic pattern analysis (Wong et al., 1995; Mohaghegh et al., 1996; de Groot, 1999) and in general for solving complicated data problems.

Reservoir quality prediction in the El Garia Formation is just such a complicated problem. The high permeability zones in the reservoir are extremely difficult to detect since no apparent visible correlation exits between permeability and the measured/ interpreted data. To tackle this problem, we choose to use a fully connected multi-layer perceptron (MLP) neural network. MLPs are the most common type of neural networks and are sometimes (mistakenly) referred to as 'back-propagation' networks after the popular training algorithm used in the learning process. MLPs are supervised neural networks, i.e. they learn by example. The processing elements in MLPs are called perceptrons and are organized in layers: an input layer, (usually one) hidden layer, and an output layer. In our networks all nodes are connected to all nodes in the next layer and all connections have associated weights. During the training phase, the weights are updated such that the error in the output layer is minimized. The nodes of the networks depicted in this article are colour coded to indicate the relative importance. A black node has more weights attached, hence is more important in predicting the output, than white or light grey nodes.

Supervised learning approaches require an independent test set to stop training before overfitting occurs. 'Overfitting' is the process in which the neural network loses its predictive power because it starts to memorize individual examples from the training set. The error on the training set continues to decrease with prolonged training but the error on the test set starts to increase. When working with geological data it is important to realize that geographical variations can significantly influence the training results. To avoid this potential problem, we decided to merge data from all wells to create training and test sets by random selection from the merged file.

It is important to realize that input attributes used in training a network must also be present when we apply the network. For example let us assume that density is one of the input attributes that is used to train the network. To apply this network, the density (log) must also be present. If not, the network will output an 'undefined value'. It is also important to select relevant attributes; i.e. attributes that will contribute to the training result and that are not fully correlated with each other. This requires detailed study of the data and the derived attributes. Finally the data set may need special preparation to increase the network's performance. One such technique is 'balancing' the data, which is required when the data space is not sampled evenly. Balancing proved to be an important issue in this study and will be discussed later.

4. AVAILABLE DATA

Data from thirteen wells were available for the lithology, porosity and permeability analyses. Four wells had incomplete logging suites and had to be discarded, leaving nine wells for further analyses. Available wireline logs in these wells are gamma-ray, sonic, density, neutron porosity, sonic porosity and laterologs (deep and shallow). The wells had been cored over large sections of the El Garia Formation. Detailed core interpretations, including the depositional fabric, described in terms of the Dunham classification, dolomite percentage, presence of baroque dolomite, fossil percentage, debris percentage, abundance of lime mud and the types of porosity (interparticle, intraparticle, vug/mouldic, fracture and micro-porosity) were available. Porosity and permeability were measured on core plugs. No fracture data was available for this study.

5. DATA ANALYSIS

The data was analyzed to establish relevant relationships between log and core data with the aim of predicting depositional texture (Dunham classification), porosity and permeability. Cross-plots were used to increase our understanding of the data and to detect (linear) relationships between the parameters. We evaluated whether different correlations exist when the data is analyzed per stratigraphic group, member or sub-member. The results were used to select the input attributes for neural network training.

Supervised neural networks established whether non-linear relationships exist between the wireline and core parameters. This produced trained neural networks that predict Dunham classification, porosity or permeability. The prediction results were compared with the results from conventional methods to determine if the neural networks were significantly better in the prediction of porosity and permeability.

5.1. Dunham classification

The Dunham classification (Dunham, 1962) is a widely used system of classifying limestones based on their depositional texture, and five classes are recognized (e.g. Tucker and Wright, 1990, p. 20): mudstone, wackestone, packstone, grainstone and boundstone. These have been extended to nine in the present study by recognizing intermediate categories (e.g. mud/wackestone) (Table 1). The presence of these rock types is not evenly distributed in our database. Packstone/grainstone, grainstone, and grain/boundstone are the dominant classes in the El Garia Formation, whilst the other types are rarely present (Fig. 1).

Since the depositional fabric can only be described from core data, it would be useful to be able to predict Dunham class from wireline data to assign a Dunham value at each logged sample along the well track. Resistivity logs are often used to discriminate

TABLE 1

Extended Dunham classification scheme used in this study

No.	Lithology type
1	Mudstone
2	Mud/Wackestone
3	Wackestone
4	Wacke/Packstone
5	Packstone
6	Pack/Grainstone
7	Grainstone
8	Grain/Boundstone
9	Boundstone

between different carbonate rock types and facies (Keith and Pittman, 1983; Serra, 1986). However, our cross-plots for the El Garia Formation showed no relationship between any of the log types and the Dunham classification. The lack of correlation is supported by the poor training performance of the (MLP) neural networks. Various combinations of logs (density, sonic, impedance, gamma-ray, neutron porosity, sonic porosity, deep laterolog, shallow laterolog and resistivity difference (LLD–LLS)) were used as input to predict the Dunham type, but no relationship was found. Nor did the correlation improve by using input data from a smaller stratigraphic interval rather than from the entire El Garia Formation (for example, from a single section or member), since it reduces the size of the database, causing a negative effect on the training performance.

Since the Dunham classes could not be predicted from wireline data, the initial idea to use Dunham class as a basis for estimating the reservoir quality or for predicting permeability could not be pursued.

5.2. Core porosity

Porosity was measured directly from core data and is plotted as a frequency histogram in Fig. 2. The Dunham class and core plug measurements, like dolomite percentage, fossil content and abundance of lime mud, did not show any relationship with the core porosity. These results confirm the observations made by Macaulay et al. (2001), who stated that a limited relationship exits between the depositional fabric (i.e. Dunham class) and the reservoir quality in the El Garia Formation. It is likely that later changes in the rock texture by compaction and diagenesis have resulted in this poor correlation. From the poor training of the neural network it can be deduced that even a combination of diagenetic features is not sufficient to predict porosity.

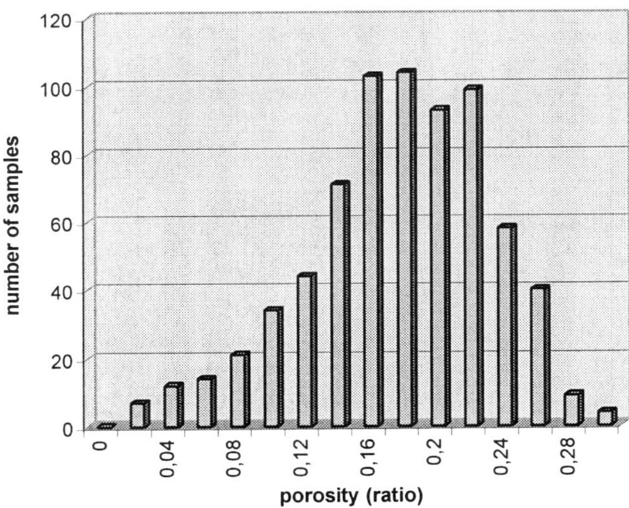

Fig. 2. Frequency distribution of porosity in the El Garia Formation, measured on core plugs.

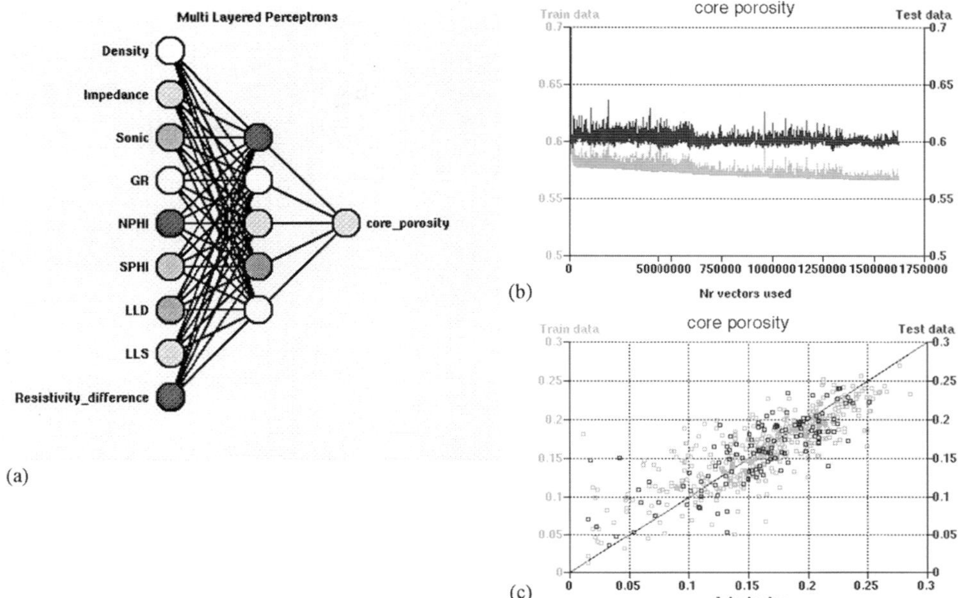

Fig. 3. Results of the neural network training to predict porosity from log data. Normalized RMS: training data 0.57, test data 0.60. (a) Neural network topology (black = strong weights attached, white = low weights attached); (b) Normalized RMS (root-mean-square) of the error between the actual porosity value and the predicted porosity (Y-axis) plotted during network training (X-axis, number of vectors) for training (grey) and test data sets (black); (c) Scatter-plot of actual core porosity versus predicted porosity, for training data (grey) and test set data (black).

For comparison, the neural network was also trained using wireline data instead of core data as an input. Cross-plots showed that the core porosity has strong correlations with wireline data, especially with the density, sonic, neutron porosity and sonic porosity logs. A moderate relationship was found with the gamma-ray values, but no correlation appears to exist with the resistivity and induction logs.

The MLP neural networks were trained using all types of logs, including the impedance and the resistivity difference (LLD–LLS) (Fig. 3a). From the colour coded input nodes it is derived that the neutron porosity and the resistivity difference logs (coloured black), are the most important in predicting the porosity value. The weights that are attached to these input nodes are higher than to any of the other input nodes (coloured white or light grey). The plot of the normalized root-mean-square (RMS) value of the error (Fig. 3b) shows that during training of the neural network the normalized RMS value of the error between the real porosity value and the predicted porosity values reduces. The lower the normalized RMS, the better the neural network is trained. The data that is used for neural network training (= grey) and the data in the independent test set (= black), follow the same trend, but the performance of the neural network is slightly better for the training dataset (RMS = 0.57) than for the test data set (RMS = 0.60). The scatter-plot (Fig. 3c) shows the actual porosity values (X-axis)

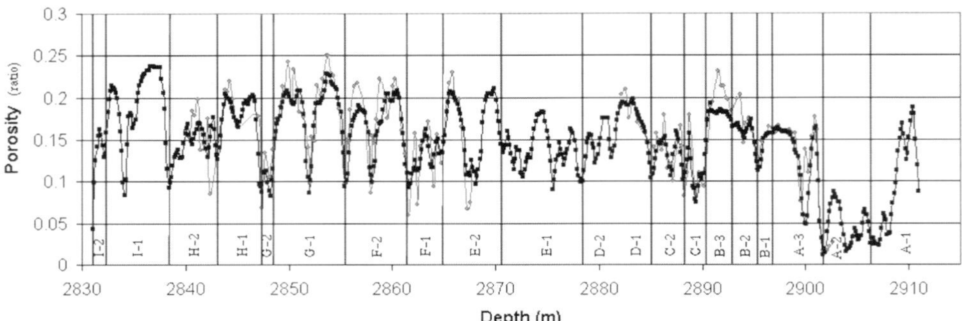

Fig. 4. Porosity prediction results on well Ashtart-A. Grey dots are the porosity values measured on core plugs; black line is the core porosity predicted by the neural network. The stratigraphic units are annotated from A-1 to I-2.

versus the predicted values for the training data (= grey) and the independent test data set (= black). Ideally the data are clustered around the diagonal line where the predicted value is the same as the actual value. The plot shows that the entire porosity range, from low to high values, is predicted well.

The trained neural network was applied to one of the wells to evaluate its predictive quality. The result is shown in Fig. 4. The black line represents the porosity predicted by the neural network, while the core plug porosity measurements are shown in grey. The correlation between the predicted porosity and the porosity measured on the cores is

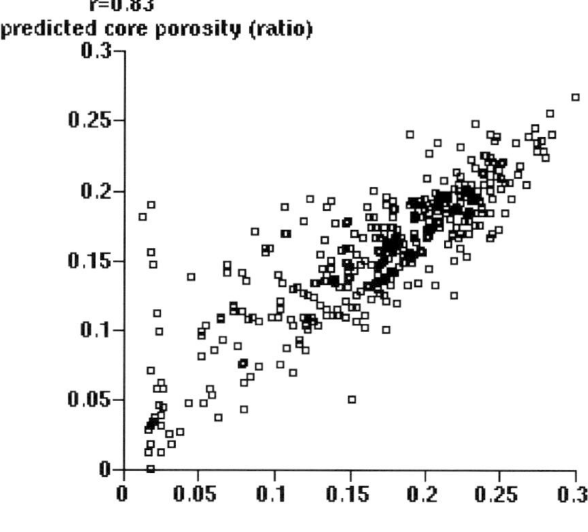

Fig. 5. Correlation between porosity measured on core plugs and predicted core porosity. Correlation coefficient $r = 0.83$.

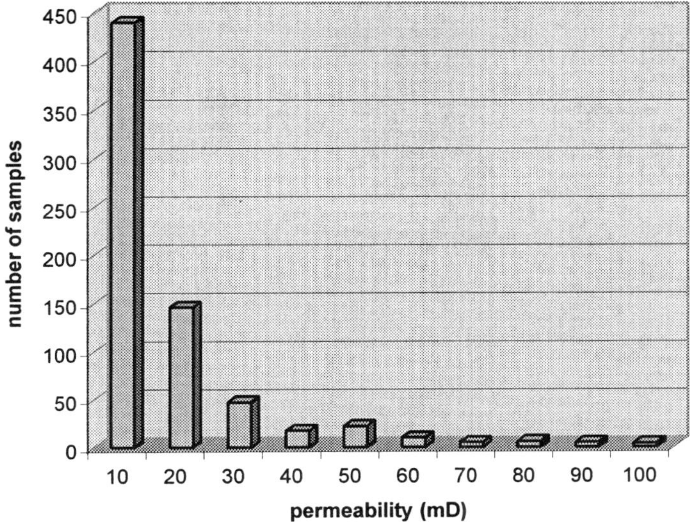

Fig. 6. Frequency distribution of permeability (mD) in the El Garia Formation measured on the core plugs.

illustrated by the cross-plot in Fig. 5. In general, the prediction is good and the predicted porosity follows the same trend as the measured core plug data.

5.3. Core permeability

Permeability values in the El Garia Formation of the Ashtart Field range from 0 mD to several Darcies, but most permeability values lie below 10 mD (Loucks et al., 1998), as illustrated in Fig. 6. Good reservoir intervals in the El Garia Formation are characterized by their preserved interparticle porosity or are dolomitized. The zones with high permeability are most often related to significant dolomitization during later diagenesis (Loucks et al., 1998; Macaulay et al., 2001). Permeability in these zones ranges from approximately 50 mD to several Darcies. However, the high permeability zones are very thin (a few meters) and are usually hard to detect. We investigated if the neural network approach could be applied to find these high permeability zones, using wireline and/or core data as an input to the neural network. Because of the distribution of the permeability data, logarithmic values were used in cross-plots and as input for the neural networks.

Loucks et al. (1998) proposed that high permeabilities are associated in particular with (1) low abundance of lime mud, (2) a low abundance of nummulithoclastic debris, and echinoderm fragments, (3) moderate sorting, (4) minor precipitation of late burial cements and (5) dolomitization. An MLP neural network was trained using the Dunham classification and other core measurements to verify the validity of this proposal for the El Garia Formation. It was obvious that the dominant factor was core porosity, followed by the dolomite percentage and fossil content. The abundance of lime mud and debris are less important than suggested by Loucks et al.

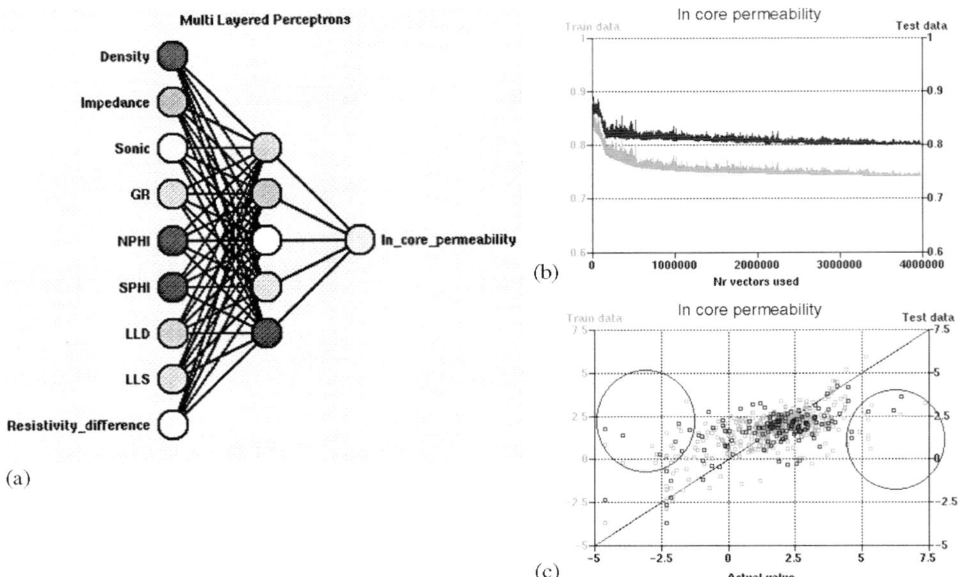

Fig. 7. Results of the neural network training to predict the logarithmic value of permeability from log data. Normalized RMS: training data 0.74, test data 0.80. (a) Neural network topology (black = strong weights attached, white = low weights attached); (b) Normalized RMS (root-mean-square) of the error between the actual logarithmic permeability value and the predicted value (Y-axis) plotted during network training (X-axis, number of vectors) for training (grey) and test data sets (black); (c) Scatter-plot of actual core permeability versus predicted permeability, for training data (grey) and test set data (black). Circles indicate data points that are predicted far too low (right circle) or too high (left circle).

Stronger relationships were found between the permeability values and the wireline data. Cross-plots show a good correlation between permeability and the density, sonic, neutron porosity and sonic porosity logs. A moderate relationship exists with the gamma-ray log and a poor correlation was found with the resistivity and induction logs. All logs formed input data for the MLP neural network, including impedance and the resistivity difference (LLD–LLS) logs, but the density, neutron and sonic porosity logs dominated the prediction of permeability (Fig. 7). The overall training performance, however, illustrates that the quality of the prediction is low. The normalized RMS is 0.74 for the training data and 0.8 for the independent test set (Fig. 7b). The scatter plot in Fig. 7c, shows that in particular the prediction of high and low permeability values, indicated by the two circles, was found to be difficult.

Since it was noted that core porosity correlates well with core permeability, it was added as input node to the neural network. The consequence of using core porosity is that a complete core porosity log must be present in order to be able to apply the neural network. Where core porosity does not exist, it must first be predicted. The training of the neural network should be based purely on the original measured core porosity to ensure that the prediction will not inherit errors from the porosity prediction.

The combination of wireline data and core porosity improved the training perfor-

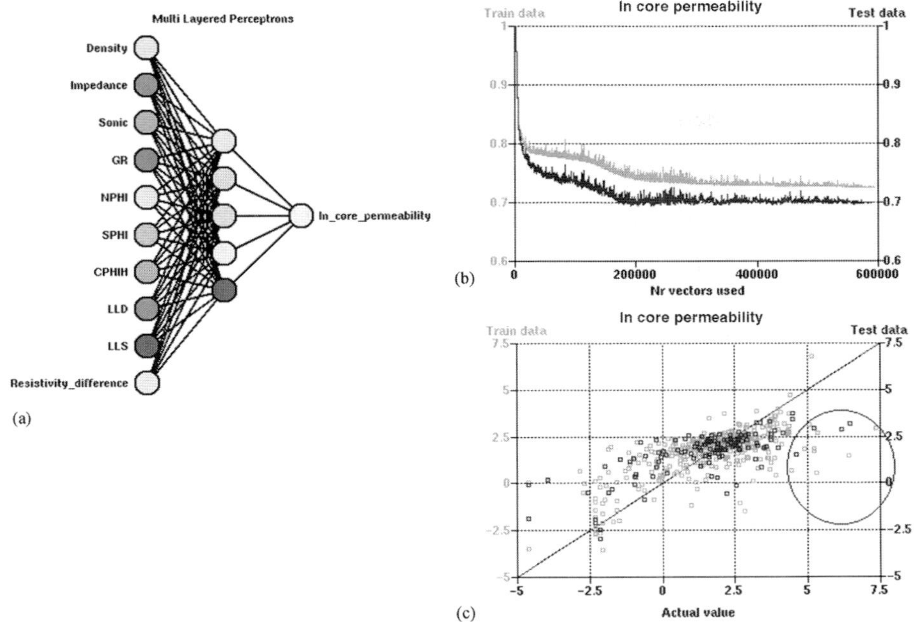

Fig. 8. Results of the neural network training to predict the logarithmic value of permeability from log data and core porosity data. Normalized RMS: training data 0.72, test data 0.70. (a) Neural network topology (black = strong weights attached, white = low weights attached); (b) Normalized RMS (root-mean-square) of the error between the actual logarithmic permeability value and the predicted value (Y-axis) plotted during network training (X-axis, number of vectors) for training (grey) and test data sets (black); (c) Scatter plot of actual core permeability versus predicted permeability, for training data (grey) and test set data (black). Circle indicates data points that are predicted far too low.

mance, and the normalized RMS of the training set decreased from 0.74 to 0.72 (Fig. 8). The scatter plot indicates that the lower permeability values are predicted better, though the high permeability values remain difficult to predict with confidence (plotted in the circle) (Fig. 8c). This was expected because only a few examples of high permeability were fed to the neural network and it was therefore incapable of making accurate predictions in this range. It was concluded that a balanced data set was required in which an even distribution of examples exists throughout the logarithmic permeability range.

A balanced data set was created using density, sonic, impedance, gamma-ray, neutron porosity, sonic porosity, deep and shallow laterolog, resistivity difference (LLD–LLS) logs together with the measured core porosity and permeability values from nine wells. The data set was extracted from the entire El Garia Formation and originally contained 1463 samples. Fig. 6 displays the uneven distribution of the permeability measurements, nearly 80% of which were between 0 and 20 mD, and with a peak below 10 mD (933 measurements, i.e. 64%). To create a more even distribution, the data was subdivided into twelve sets based on the logarithmic permeability values. When the original number of samples per set was higher than 50, the set was reduced to 50 by deleting lines

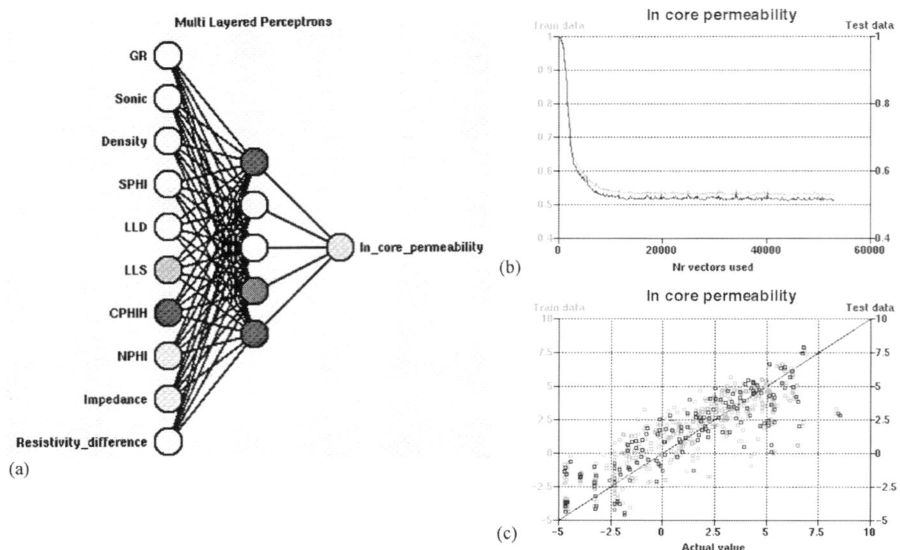

Fig. 9. Results of the neural network training to predict the logarithmic value of permeability from log data and core porosity data, using a balanced data set. Normalized RMS: training data 0.53, test data 0.52. (a) Neural network topology (black = strong weights attached, white = low weights attached); (b) Normalized RMS (root-mean-square) of the error between the actual logarithmic permeability value and the predicted value (Y-axis) plotted during network training (X-axis, number of vectors) for training (grey) and test data sets (black); (c) Scatter-plot of actual core permeability versus predicted permeability, for training data (grey) and test set data (black).

at random. If the number of samples per set was lower than 50, the set was enlarged to approximately 50 by duplicating lines. A consequent randomized filter (± 2%), a so-called 'white-noise' filter, was applied to all duplicated log and permeability data, to avoid exactly the same examples being fed to the neural network. After balancing, the total number of examples was 619. The data set was randomly split into a training and a test set to control the training of the neural network and to avoid overtraining.

The training results using this balanced data set are given in Fig. 9. The performance of the neural network was significantly improved compared to the previously trained neural networks that use unbalanced data sets. The normalized RMS-error dropped from 0.72 to 0.53 for the training set, and from 0.70 to 0.52 for the test data (Fig. 9b). The prediction improvement in the high permeability range is striking (Fig. 9c).

It was concluded that MLP neural networks can predict the permeability in the El Garia Formation successfully from a combination of wireline log data and core porosity measurements, provided that a balanced data set is used with an equal distribution of samples over the entire permeability range.

The trained neural network was applied to two wells. The permeability prediction for well Ashtart-A is good to fair and values range from 0 to 60 mD (Fig. 10a). The predicted permeability (black curve) clearly follows the trend of the measured core plug permeabilities (grey curve). The neural network picked up the thin intervals with high

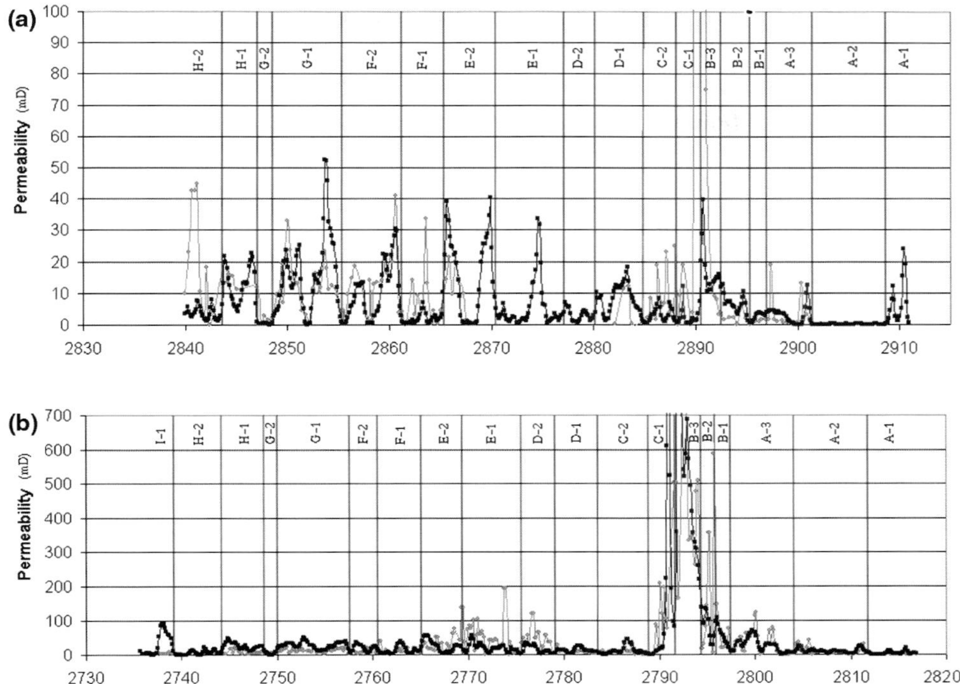

Fig. 10. Permeability prediction results. The grey curves are the permeability measured on core plugs; the black curves are the predicted core permeability by the neural network. The stratigraphic units are plotted from A-1 to I-1. (a) Well Ashtart-A; (b) Well Ashtart-B.

permeability values. In stratigraphic sections where no core plug measurements were taken, e.g. E2 to D2, the neural network detected some high permeability zones. Due to the absence of core plug data, it was not possible to check if the prediction is correct. However, the high permeability values seem realistic since they correspond to high predicted porosity values that resemble the measured porosity values from the logs.

A high permeability zone in this well at approximately 2,892 m, the so-called 'drain' (Loucks et al., 1998), is detected, but the predicted permeability value is much lower than the core value. It should be noted that in this well, the 'drain' is very thin and comprises only one core plug sample (700 mD). The surrounding values are below 50 mD. The interval is so thin that the logs might have missed the entire interval due to their sampling rate. Alternatively, the neural network cannot predict these extreme values, or the core measurement is unrealistic; it could be affected, for example, by laboratory conditions or core plug preparation.

The predicted result for the second well (Ashtart-B) is displayed in Fig. 10b. In both the low and high permeability range it is very close to the measured values. The majority of the data is predicted to be below 100 mD, but the B3-member ('drain') stands out in particular. It has a much higher permeability, with values up to 700 mD. The interval is thicker than in the previous well, and it is likely that the logs have sampled the interval

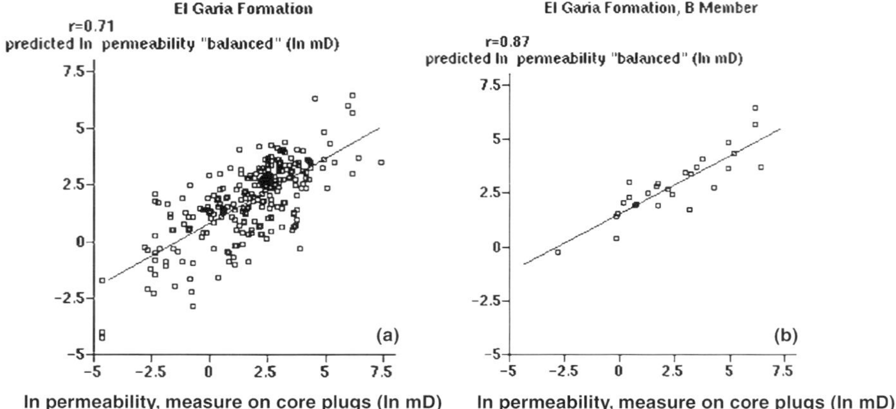

Fig. 11. Correlation between (logarithmic) permeability measured on core plugs and predicted (logarithmic) permeability. (a) in the complete El Garia Formation ($r = 0.71$); (b) in the B-member of the El Garia Formation ($r = 0.87$).

better, resulting in a more accurate neural network prediction for the B2 and B3 intervals in this well.

The permeability prediction for the E2 to D2 members is less than in the other sections of this well. This could be related to the fact that less core data is available from the nine wells in this interval. The neural network was not specifically trained for these members.

From the cross-plot of the predicted permeability against the core plug permeability in wells Ashtart-A and Ashtart-B for the entire El Garia Formation (Fig. 11a), and the objective interval B (Fig. 11b), it can be concluded that the prediction of permeability is accurate (correlation coefficients 0.71 and 0.87 respectively).

6. CONCLUSIONS

Neural networks have proven to be successful in the prediction of permeability in carbonate rocks. Although correlation with single parameters is weak, a reliable prediction can be made from the combination of wireline and core (or predicted) porosity data. To achieve a high neural network training performance, it is necessary to use a well-balanced data set that has an even distribution of samples over the entire permeability range. Since core porosity is included as an input parameter for the neural network, a porosity log must be predicted prior to the prediction of permeability. The core porosity is most closely correlated with neutron porosity and the difference between the deep and shallow resistivity logs. The entire porosity range was predicted with a high degree confidence.

The initial idea to use (predicted) Dunham classes for the prediction of permeability was discarded because the rock type could not be predicted from wireline data. As it does not correlate with porosity or permeability, it cannot be used as an indicator of reservoir quality.

The neural network method is a robust technique. Provided that enough and representative data is given to the neural network for training, then reservoir quality can be reasonably predicted in un-cored intervals located in the same geological setting as wells from which training data were derived. Thin-bedded sections with alternations of, for example, high and low permeability zones, are detected as long as they were properly logged.

ACKNOWLEDGEMENTS

We thank Preussag Energie and ETAP for permission to publish this paper. We also thank Dr. P.F.M. de Groot for his useful comments. The evaluations and conclusions made in this paper are for the account of the authors and are not necessarily supported by other parties.

REFERENCES

Dunham, R.J., 1962. Classification of carbonate rocks according to depositional texture. In: Ham, W.E. (Ed.), *Classification of Carbonate rocks. Am. Assoc. Pet. Geol. Mem.*, 1: 108–121.

de Groot, P.F.M., 1999. Seismic reservoir characterisation using artificial neural networks. *19th Mintrop Seminar*, Münster, 16–19 May 1999.

Keith, B.D. and Pittman, E.D., 1983. Bimodal porosity in oolitic reservoir – effect on productivity and log response, Rodessa Limestone (Lower Cretaceous), East Texas Basin. *Am. Assoc. Pet. Geol. Bull.*, 67(9): 1391–1399.

Loucks, R.G., Moody, R.T.J., Bellis, J.K. and Brown, A.A., 1998. Regional depositional setting and pore network systems of the El Garia Formation (Metlaoui Group, Lower Eocene), offshore Tunisia. In: MacGregor, D.S., Moody, R.T.J. and Clark-Lowes, D.D. (Eds.), 1998. *Petroleum Geology of North Africa. Geol. Soc., London, Spec. Publ.*, 132, 355–374.

Macaulay, C.I., Beckett, D., Braithaite, K., Bliefnick, D. and Philps, B., 2001. Constraints on diagenesis and reservoir quality in the fractured Hasdrubal field, offshore Tunisia. *J. Pet. Geol.*, 24(1): 55–78.

Mohaghegh, S., Areti, R., Ameri, S., Aminiand, K. and Nutter, R., 1996. Petroleum reservoir characterisation with the aid of artificial neural networks. *J. Pet. Sci. Eng.*, 16: 263–274.

Racey, A., Bailey, H.W., Beckett, D., Gallagher, L.T., Hampton, M.J. and McQuilken, 2001. The petroleum geology of the Early Eocene El Garia Formation, Hasdrubal field, offshore Tunisia. *J. Pet. Geol.*, 24(1): 29–53.

Serra, O., 1986. *Advanced Interpretation of Wireline Logs*. Schlumberger, Special Publication.

Tucker, M.E. and Wright, V.P., 1990. *Carbonate Sedimentology*. Blackwell Scientific Publications, 482 pp.

Wong, P.M., Jian, F.X. and Taggart, I.J., 1995. A critical comparison of neural networks and discriminant analysis in lithofacies, porosity and permeability predictions. *J. Pet. Geol.*, 18(2): 191–206.

Developments in Petroleum Science, 51
Editors: M. Nikravesh, F. Aminzadeh and L.A. Zadeh

Chapter 20

USING RBF NETWORK TO MODEL THE RESERVOIR FLUID BEHAVIOR OF BLACK OIL SYSTEMS

ADEL M. ELSHARKAWY

Petroleum Engineering Department, Kuwait University, P.O. Box 5969, Safat 13060, Kuwait

ABSTRACT

This paper presents a new technique to model the behavior of crude oil systems. The proposed technique is using a radial basis function neural network model (RBFNM). The model predicts oil formation volume factor, solution gas–oil ratio, oil viscosity, saturated oil density, undersaturated oil compressibility, and evolved gas gravity. Input data to the RBFNM are reservoir pressure, temperature, stock tank oil gravity, and separator gas gravity. The model is trained using differential PVT analysis of numerous black-oil samples collected from various oil fields. The proposed RBFNM is tested using PVT properties of other samples that have not been used during the training process. Accuracy of the proposed network model to predict PVT properties of black-oils systems is compared to the accuracy of numerous published PVT correlations.

1. INTRODUCTION

Accurate estimation of PVT properties of black oil systems is essential for all petroleum engineering calculations such as material balance calculations, reservoir simulation, fluid flow in porous media, evaluation of new formation for potential development, design of production equipment, and planing future enhanced oil recovery projects. The PVT properties can be obtained from experimental tests conducted on available bottom hole samples or recombined surface sample early in the production life of the reservoir. They are usually measured at reservoir temperature. However, if the representative bottom hole or recombined surface sample can not be obtained, the PVT properties of oil and gas systems are estimated from the equations of state or PVT correlations. PVT correlation and the equations of state are also used to estimate PVT data at temperatures other than reservoir temperature, which is needed for production equipment design or planning thermal oil recovery. The equations of state are very poor predictive tools unless they are tuned using initial fluid composition and some experimentally measured properties, e.g. saturation pressure of a given hydrocarbon system. Once the equation of state is adjusted to match the experimentally measured data, it can then be used to predict the PVT properties of such system. If the initial composition of the hydrocarbon system, and some experimentally measured properties are not available then the equation of state can not be used. The equation of state involves numerous numeric computations and their estimation of the viscosity of crude oil is not accurate. On the contrary, PVT correlations do not need compositional data, no tuning process is required, and involve simple mathematical computations. Nevertheless, PVT correlations have limited accuracy. These PVT correlations were developed using simple regression or graphical techniques. Some of these correlations do not properly describe the physical behavior of PVT properties as a function of pressure, temperature, stock tank oil gravity, and gas gravity. Katz (1942) developed the first correlation for predicting oil formation volume factor using crude oils from Mid-Continent. Standing (1947) presented correlation for estimating bubble point pressure and oil formation volume factor as a function of solution gas–oil ratio, gas gravity, stock tank oil gravity, and reservoir temperature. Since Standing presented his correlations, several PVT correlations from various locations all over the world have been published in the literature and new promising ones are coming every day. A list of the published PVT correlations presented from various parts of the world is shown in Table 1. The use of

TABLE 1

PVT correlations from various geographical locations in the world

Author(s)	Year(s)	Location
Katz	1942	Mid Continent and North America
Standing	1947, 1962, 1977, 1981	California, USA
Borden and Rzasa	1950	Rock Mountain
Elam	1957	Texas, USA
Trube	1957	
Lasater	1958	North and South America
Abdus Sattar	1959	Rock Mountain, USA
Knopp and Ramsey	1960	Venezuela
Niakan	1967	Iran
Tahrani	1968	Iran
Cronquist	1973	Gulf Coast, USA
Beggs and Robinson	1975	Collected data
Caixerio	1976	Brazil
Vasquez and Beggs	1977, 1980	Data bank
Galsϕ	1980	North Sea
Labedi	1982, 1990, 1992	North Africa
Owolabi	1983	Alaska, USA
Obomanu and Okpobiri	1987	Nigeria
Abdul Majeed and salman	1988	Unknown
Al-Marhoun	1988	Saudi Arabia
Asgarpour et al.	1988	Canada
Santamaria and hernamdez	1989	Mexico
Kartoatmodjo and Schmidt	1990, 1994	Data bank
Dokla and Osman	1992	United Arab Emirates
Farshad et al	1992	Colombia
Petrosky and Farshad	1993, 1998	Gulf of Mexico
McCain and Hill	1995	Collected data
Elsharkawy et al.	1996, 1997, 1999a, 1999b	Kuwait
Gharbi and Elsharkawy	1996, 1999	Data bank

these correlations for new locations should be taken with caution. Numerous studies have been published to assess the accuracy of this correlations (Ostermann et al., 1983; Abdul Majeed, 1985, 1988; Saleh et al., 1987; Ahmed, 1989; Sutton and Farshad, 1990; Ghetto et al., 1994; Elsharkawy et al., 1995; and Mohamood and Al-Marhoun, 1996). The general conclusion of these studies is that their performance is poor outside their range of application.

Recent years have witnessed a steady increase in the literature concerning the application of neural network models in petroleum engineering (Al-kaabi and Lee, 1990; Osborne, 1992; Accarain and Desbrandes, 1993; Juniardi and Ershaghi, 1993; Zhou, 1993; Ali, 1994; Briones, 1994; Kumoluyi, 1994; Mohaghegh, 1994, 1996; Gharbi and Elsharkawy, 1996, 1999; Habiballah, 1996; and Sung, 1996). Neural networks are computing systems composed of a number of highly interconnected simple neuron-like processing units, which process information by their dynamic response to external inputs. They attempt, in greatly simplified way, to mimic the behavior of their biologic counterparts and are used in different artificial intelligence problems such as machine vision, pattern recognition, classifications, and association memory; to mention only a few. Once trained, these network models can predict output values for novel sets of input. In these network models, the entire network collectively performs computations with knowledge represented as distributed pattern of activity over all processing elements. The collective operations result in high degree of parallelism, which enable the network to solve complex problem in a relatively small time. The distributed representation leads to greater fault tolerance and to degradation immunity when problem encountered is beyond the experience of the network. Moreover, the network can adjust itself dynamically to changes, infer general rules from specific examples, and recognize invariance from complex multidimensional data. Several papers about application of neural network have been presented to the oil industry. Almost all these papers have used back propagation network model. This paper presents the first attempt to use the radial basis function to model the entire PVT properties of crude oils. The conventional approach to develop PVT correlations was based on classical regression techniques (CRT). However, PVT models based on successfully trained neural network regression techniques, (NNRT) can be excellent and reliable tools for reservoir engineering studies. The massive interconnection in the network design produces large degree of freedom that allows the network to capture the system nonlinearity better than the CRT. Another advantage of the NNRT is its ability to learn and adapt itself to new situations in which the network performance is inadequate. Finally, unlike CRT, NNRT can be used to map multiple-input, multiple-output systems. This will be illustrated in this paper.

This work is aimed at the following:

(1) Presenting a new technique of neural network that is the radial basis function (RBF).
(2) Show that this RBF once properly trained it can be comprehensive and reliable to predict all the properties of crude oil and gas systems using differential PVT.
(3) Check the physical performance of the new technique (RBF) by comparing its prediction to experimental data.
(4) Compare the accuracy of the proposed RBF model to the accuracy of several published PVT correlations.

2. PRESENT STUDY

PVT properties of one hundred crude oil and gas systems collected from different oil fields were available for this study. The data include measurements of solution gas–oil ratio (SGOR), oil formation volume factor (OFVF), oil viscosity, and oil density at seven to ten pressure steps above and fifteen to thirteen pressure steps below the bubble point pressure for every system. The data also include undersaturated oil compressibility above bubble point pressure and evolved gas gravity below bubble point pressure. These properties were measured isothermally at reservoir temperature by differential liberation. The data was divided into two groups. The first group comprised properties of ninety samples that are used for training the proposed model. The second group comprised properties of another ten samples, which were selected at random for testing the performance of the proposed model. Description of the PVT properties of these reservoir fluids is given in Table 2. Supplementary material containing details of the experimental data for the black oil systems used in this study are available.

2.1. Development of the RBFNM

Several different neural network architectures were tried to model the PVT properties of the crude oil systems described in this study. These networks are:
(1) Back propagation with momentum.
(2) Accelerated back propagation.
(3) Levenberg–Marquardt BP network.
(4) Elman recurrent network.
(5) General regression network with radial basis function.

The first four network models failed to recognized the PVT pattern of the data. The last network, the general regression network with radial basis function (RBF) was found to successfully match the PVT behavior of the reservoir fluids. Most neural network applications in the petroleum literature have used the back propagation neural network paradigm. This study presents the first attempt to use RBF to model all PVT properties

TABLE 2

Description of the PVT data used in this study

	Minimum	Maximum
Pressure, psi	0	9900
Temperature, F	130	245
API (water = 10)	20	45
Separator gas gravity, g/cc	0.807	1.234
Solution gas–oil ratio, scf/stb	0	1568
OFVF, bbl/stb	1.033	1.932
Oil viscosity, cP	0.24	16.00
Oil gravity, (air = 1)	0.583	0.867
Undersaturated oil compressibility, 10^{-6} 1/psi	3.055	29.81
Evolved gas gravity (air = 1)	0.630	1.195

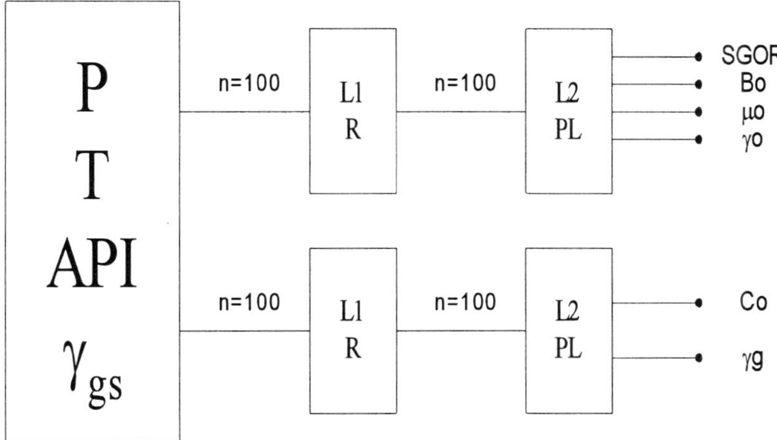

Fig. 1. RBFNM architecture.

of black oil systems. RBF of the general regression network (GRN) may require more neurons than the standard feed-forward or back propagation network, but the RBF are faster to train. The GRN usually goes through a large training set of the data and reaches a stable state in short period. Depending on data, the RBF network model may generate more accurate results than the BP network model, which proved to be the case in this paper. The architecture of the radial basis function of the general regression model used in this study is shown in Fig. 1. The processing units are arranged in four layers: an input layer, two hidden layers, and an output layer.

Two models are described here for predicting the PVT properties of the crude oil systems. The input layer for the two models contains four neurons: reservoir pressures, reservoir temperature, stock tank oil gravity, and separator gas gravity. The output layer for model #1 contains solution gas–oil ratio (SGOR), oil formation volume factor (OFVF), oil viscosity, and oil density. However, The output layer for model #2 contains undersaturated oil compressibility, and evolved gas gravity. The reason for having two models is explained in the training section. Each hidden layers has one hundred neurons. The hidden layers processing elements construct the mapping process that generalizes the input/output map for the given information. Details of the RBF are given in the Appendix A.

2.2. *Training the model*

PVT properties of ninety black-oil samples was used in the training process. It was found that one network paradigm design can not accurately match all the PVT properties simultaneously. Therefore, attempts were made to train the radial basis function network to match only the solution gas–oil ratio. Unfortunately, the radial basis function network that was successfully trained to match gas–oil ratio could not match the oil formation volume factor. After investigation, it was found that the reason for failure of the network to match the gas–oil ratio and the formation volume factor simultaneously using one paradigm was the large variations in the data. Gas–oil ratio varies from 0 to 1568 scf/stb whereas oil formation volume factor from 1.033 to 1.932 bbl/stb. Normalizing each parameter of the PVT properties solved this problem. The normalized variable has the following form:

$$X_{\text{new}} = \frac{X_{\text{old}} - X_{\min}}{X_{\max} - X_{\min}} \tag{1}$$

Where X denotes the PVT parameters such as gas–oil ratio, oil formation volume factor, crude oil viscosity, or density. The subscript old refer to the actual value, max refer to the maximum and min to the minimum value of the variable. The new normalized variable takes the range from zero to 1 for all the parameters. Using the normalized variable, one architecture of the radial basis function network successfully matched the gas–oil ratio, oil formation volume factor, oil density, and oil viscosity, but failed to satisfactorily match the undersaturated oil compressibility and evolved gas gravity of the training data. The reason was that the first four parameters are measured above and below the bubble point pressure. However, the undersaturated oil compressibility is measured above the bubble point pressure and evolved gas gravity is measured below the bubble point pressure. Therefore both the oil compressibility and gas gravity are separated into another RBF network.

2.3. Testing the model

Calculating the errors and standard deviation for the values estimated by the model tests the accuracy of the proposed models. These errors and standard deviations are compared to those occurred form the PVT correlations. The behavior of the RBFNM in matching the PVT parameters of black oil system is also checked via comparing plots of the predicted parameters versus pressure to the measured parameters for the test sample.

3. ACCURACY OF THE MODEL

3.1. Solution gas–oil ratio

Several correlations have been published to estimate the solution gas–oil ratio (SGOR) for different crude oil systems. Most of these correlations expressed SGOR as a function of reservoir pressure, temperature, stock tank oil gravity, and gas gravity. The following correlations for estimating SGOR are considered in this section: Standing (1962), Vasquez and Beggs (1980), Glasø (1980), Obomanu and Okpobiri (1987), Al-Marhoun (1988), Dokla and Osman (1992), Farshad et al. (1992), Petrosky and Farshad (1993), Kartoatmodjo and Schmidt (1994), and Elsharkawy and Alikhan (1997). The equations for these correlations are shown in Appendix B. Table 3 shows the accuracy of the proposed radial basis function network model as compared to that of the before mentioned PVT correlations. It is clear from this table that the RBFNM is more accurate than the PVT correlations. The RBFNM has an average relative error (E_r) of 0.61%, average absolute error (E_a) of 5.22%, standard deviation (SD) of 7.83% and correlation coefficient (R) of 99.28% for the training samples and 0.69%, 4.53%, 7.37%, and 98.97%, respectively

TABLE 3

Accuracy of RBFNM and correlations in estimating SGOR for training and testing samples

	Training samples						Testing samples					
	$E_{a,min}$ %	$E_{a,max}$ %	E_r %	E_a %	SD %	R %	$E_{a,min}$ %	$E_{a,max}$ %	E_r %	E_a %	SD %	R %
RBFNM	0.00	55.00	0.61	5.22	7.83	99.28	0.00	50.34	0.69	4.53	7.37	98.97
Standing	0.04	82.14	23.78	24.31	19.88	96.53	1.21	83.26	18.96	19.94	21.42	97.65
Vasquez	0.86	209.95	28.57	36.10	31.24	87.57	0.80	239.19	−2.81	48.79	67.35	66.41
Glasø	7.53	71.77	32.22	32.22	11.85	98.38	8.90	74.04	30.67	30.67	12.25	97.45
Al-Marhoun	0.07	415.30	−92.87	110.11	114.93	97.06	0.55	125.39	−40.92	62.30	55.71	97.38
Dokla	0.03	93.32	17.88	27.80	31.01	98.68	0.09	95.69	21.07	23.40	27.05	97.56
Petrosky	0.09	87.99	12.31	16.90	15.59	97.27	0.30	133.54	5.61	10.90	14.80	97.15
Farshad	0.97	88.32	28.60	28.78	18.62	98.98	5.59	91.09	31.91	31.91	18.16	97.44
Kartoatmodjo	2.51	85.61	31.90	31.60	16.33	96.15	2.79	88.53	21.74	21.74	17.79	96.34
Obomanu	0.36	208.54	−61.63	62.74	38.95	92.47	1.05	129.02	−68.06	69.75	31.40	96.63
Elsharkawy	0.01	81.96	9.01	12.62	17.23	98.90	0.44	82.94	12.50	15.58	21.44	98.03

for the testing samples. On the other hand, the PVT correlations have higher errors, higher standard deviations and lower coefficient of correlations. Thus, the proposed model performed better than the PVT correlations. Fig. 2a and 2b show cross plots of measured SGOR versus that predicted by the RBFNM for both the training and testing samples. These plots illustrate that all plotted points fall closely to the unit slope line. Thus, the RBFNM closely matches the points for the training and testing samples.

3.2. Oil formation volume factor

Oil formation volume factor (OFVF) correlations use solution gas–oil ratio, reservoir temperature, and stock-tank oil gravity and gas gravity as input data. It is important to note here that when differential PVT data becomes unavailable, the SGOR is estimated from correlation. Thus, error in estimated SGOR would result in erratic estimation of OFVF. The correlation presented by Standing (1977), Vasquez and Beggs (1980), Glasø (1980), Obomanu and Okpobiri (1987), Abdul Majeed (1985), Al-Marhoun (1980), Dokla and Osman (1992), Labedi (1992), Farshad et al. (1992), Petrosky and Farshad (1993), and Kartoatmodjo and Schmidt (1994) are used in this section to estimate the oil formation volume factor for the crude oil and gas systems. These equations are shown in Appendix C. The accuracy of the proposed RBF model as compared to the above-mentioned correlations is shown in Table 4. The RBFNM has a better accuracy than all the PVT correlations. It predicts OFVF with an E_r of −0.06%, E_a of 0.86%, SD of 1.28% and *R* of 99.46% for training samples and 0.08%, 0.53%, 0.57%, 98.24% for the testing samples. The PVT correlations considered in this study have E_a in the range of 1 to 3% and SD in the range of 1 to 6%. Again the proposed RBFN model has a better accuracy than the PVT correlations. Fig. 3a and 3b show the cross plots for the training and testing samples. These plots show excellent behavior of the RBFNM in predicting OFVF for both the testing and training samples.

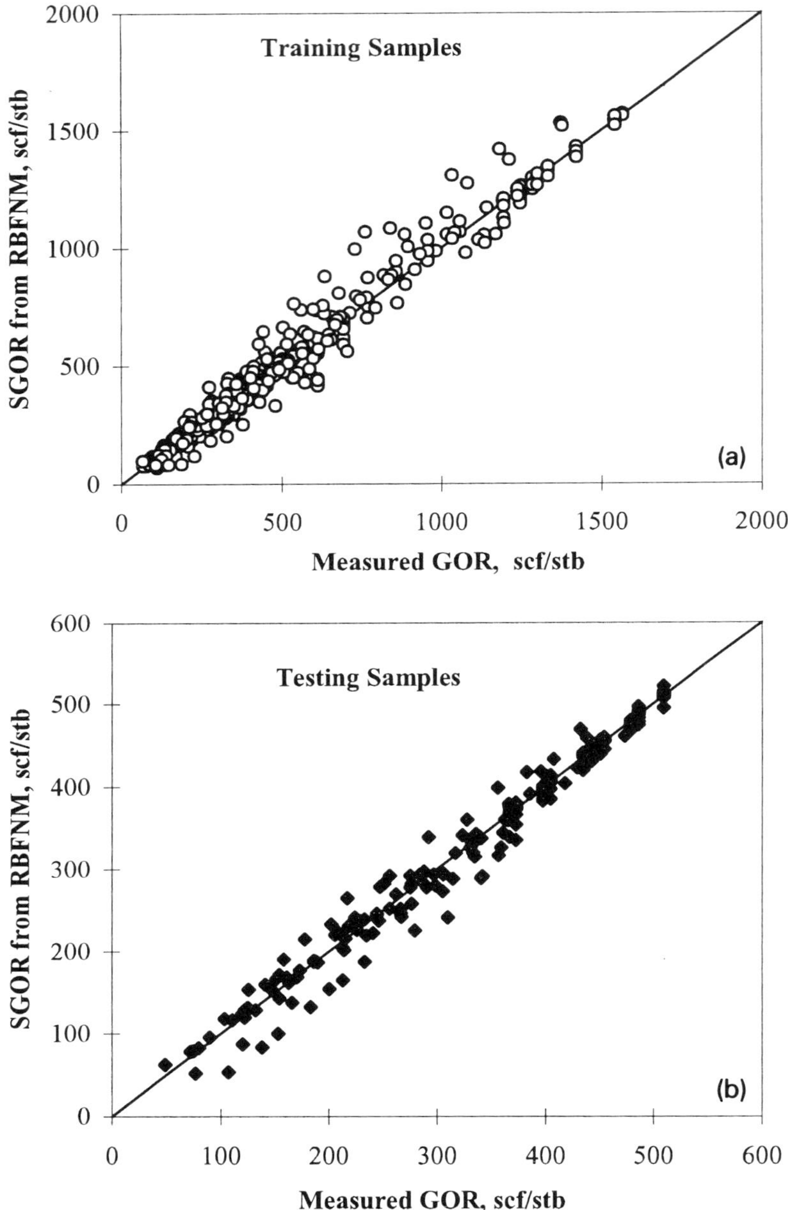

Fig. 2. (a) Crossplot of GOR for training samples. (b) Crossplot of GOR for testing samples.

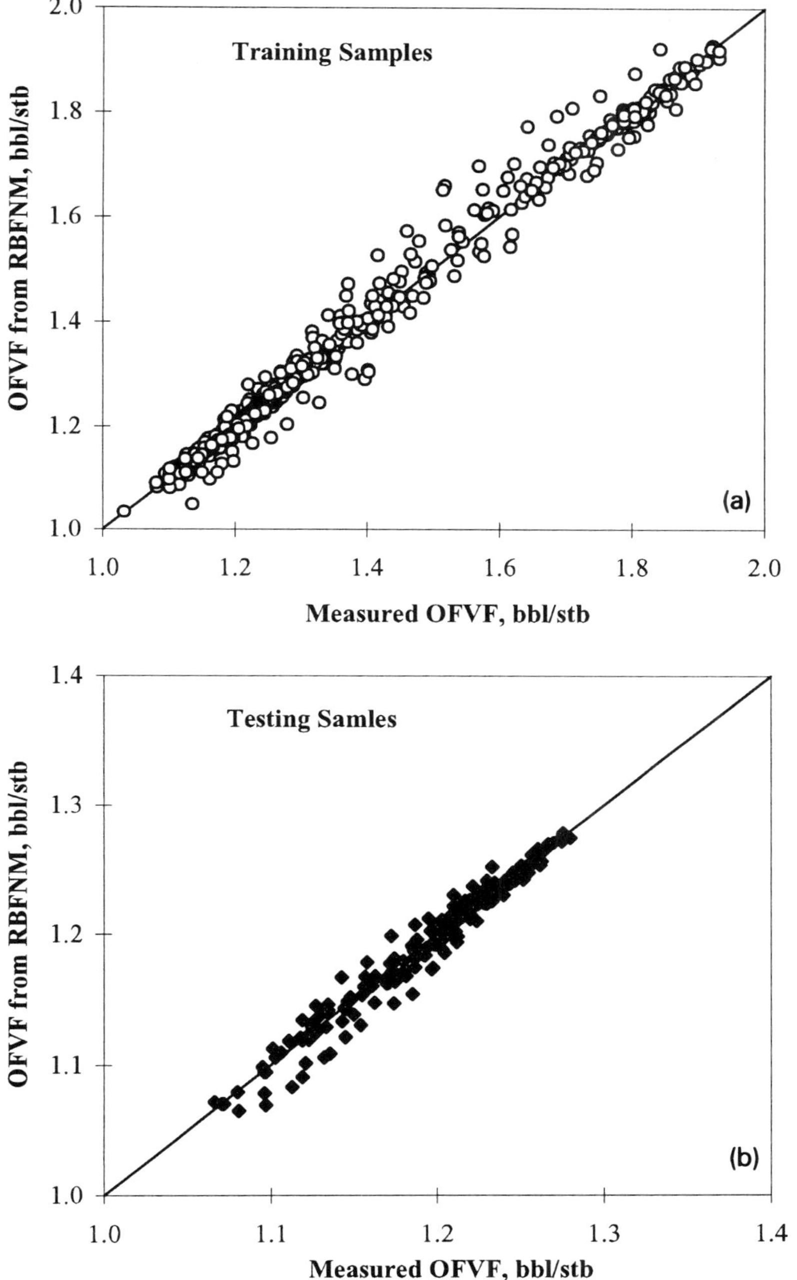

Fig. 3. (a) Crossplot of OFVF for training samples. (b) Crossplot of OFVF for testing samples.

TABLE 4

Accuracy of RBFNM and correlations in estimating OFVF for training and testing samples

	Training samples						Testing samples					
	$E_{a,min}$ %	$E_{a,max}$ %	E_r %	E_a %	SD %	R %	$E_{a,min}$ %	$E_{a,max}$ %	E_r %	E_a %	SD %	R %
RBFNM	0.00	9.24	−0.06	0.86	1.28	99.46	0.00	2.69	0.08	0.53	0.57	98.24
Standing	0.00	12.52	0.76	2.45	3.06	98.5	0.00	4.00	0.58	1.53	1.75	95.51
Vasquez	0.00	15.84	1.43	2.98	3.96	96.29	0.00	3.76	0.02	1.10	1.39	96.28
Glasø	0.00	15.02	1.74	3.31	3.78	97.86	0.00	5.65	1.43	2.39	2.56	93.19
Al-Marhoun	0.00	17.00	−0.67	1.93	2.81	98.87	0.00	3.18	−0.07	0.96	1.21	97.72
Dokla	0.00	20.42	0.76	3.56	4.51	98.15	0.00	7.36	2.00	2.95	3.12	93.17
Labedi	0.00	12.02	0.52	2.16	2.82	98.51	0.00	2.96	0.15	1.13	1.38	96.16
Petrosky	0.00	24.76	−0.91	2.17	3.52	98.2	0.00	3.15	−0.08	0.94	1.18	97.35
Farshad	0.00	35.17	−1.52	2.47	4.68	96.72	0.00	2.54	−0.25	0.84	1.04	97.90
Abdul Majeed	0.00	59.24	−3.37	4.06	6.42	91.07	0.00	7.59	−1.55	1.59	1.56	93.53
Kartoatmodjo	0.00	26.07	−1.28	2.29	3.85	97.83	0.00	2.54	−0.18	0.87	1.09	97.77
Obomanu	0.00	14.10	−0.3	2.32	3.09	98.78	0.00	4.81	0.72	1.68	1.98	94.31
Elsharkawy	0.00	11.07	−0.15	1.66	2.43	98.71	0.00	2.54	−0.53	0.88	0.93	97.76

3.3. Oil viscosity

Most PVT correlations use oil API gravity and reservoir temperature to estimate dead oil viscosity (μ_{od}). These correlations also use solution gas–oil ratio (SGOR) and μ_{od} to estimate saturated oil viscosity (μ_{ob}). The PVT correlations use μ_{ob} and differential pressure above bubble point pressure to estimate undersaturated oil viscosity (μ_{oa}). Thus, the estimated values of μ_{oa} and μ_{ob} are dependent on accuracy of both SGOR and μ_{od}. Contrary to the PVT correlations, the estimated value of μ_{ob} by the RBFNM is independent of SGOR and μ_{od}, and estimate of μ_{oa} is independent of μ_{ob}. The following correlations have been used to estimate crude oil viscosity and compared to the proposed RBFNM: Beggs and Robinson (1975), Vasquez and Beggs (1980), Labedi (1992), Kartoatmodjo and Schmidt (1994), and Elsharkawy and Alikhan (1999). These equations are given in Appendix D. The accuracy of the RBFNM as well as the before-mentioned correlations is shown in Table 5. This table shows that the model has better accuracy than all the viscosity correlations. The RBF model has an E_r of −0.23%, E_a of 7.58%, SD of 10.94%, and R of 97.56% for the training samples and −5.89%, 8.72%, 10.11%, 97.8% for the testing samples. The viscosity correlations have E_a in the range of 10 to 40% and SD in the range of 10 to 60%. It is clear from Table 5 that the RBF model has a better predictability than the viscosity correlations for crude oils under study. Crossplot of oil viscosity for training and testing samples are shown in Fig. 4a and 4b. These figures show that the data points are evenly distributed around the unit slope line.

3.4. Oil density

Table 6 shows the accuracy of the RBFNM in estimating the oil density (γ_o) as compared to the accuracy of McCain's correlations (1995) and Elsharkawy's correlations

TABLE 5

Accuracy of RBFNM and correlations in estimating oil viscosity for training and testing samples

	Training samples						Testing samples					
	$E_{a,min}$ %	$E_{a,max}$ %	E_r %	E_a %	SD %	R %	$E_{a,min}$ %	$E_{a,max}$ %	E_r %	E_a %	SD %	R %
RBFNM	0.00	88.31	−0.23	7.58	10.94	97.56	0.02	54.66	−5.89	8.72	10.11	94.88
Elsharkawy	0.09	120.09	10.51	10.84	22.14	88.32	1.28	70.74	30.40	30.57	15.25	78.70
Beggs-R&V	0.00	82.60	5.41	9.81	20.88	89.52	0.10	55.33	24.39	24.80	16.31	79.29
Labedi	0.02	341.01	−37.38	38.00	60.38	74.61	0.09	133.57	−43.13	48.53	39.85	61.64
Kartoatmodjo	0.99	82.23	20.81	20.82	19.63	79.08	12.37	82.05	49.19	49.19	14.32	56.51

TABLE 6

Accuracy of RBFNM and correlations in estimating oil gravity for training and testing samples

	Training samples						Testing samples					
	$E_{a,min}$ %	$E_{a,max}$ %	E_r %	E_a %	SD %	R %	$E_{a,min}$ %	$E_{a,max}$ %	E_r %	E_a %	SD %	R %
RBFNM	0.00	5.59	0.07	0.56	0.71	99.41	0.00	1.99	0.00	0.40	0.34	97.26
Elsharkawy	0.00	7.49	−0.11	0.82	1.51	98.97	0.00	1.68	0.0	0.45	0.56	97.51
McCain	1.31	26.05	−6.68	6.68	3.69	93.37	1.04	6.28	−4.28	4.24	1.14	89.92

(1999) for the training and testing samples. Correlations for estimating saturated oil density are given in Appendix E. The model has an E_a of 0.56%, SD of 0.71%, and R of 99.41% for the training samples and 0.40%, 0.34% and 97.62%, respectively for the testing samples. The PVT correlations have E_a in the range of E_a of 4 to 7% and SD of from 1 to 4%. 6.8%. Thus, the network model is highly accurate than the PVT correlations in estimating the density of all crude oils samples. Fig. 5a and 5b show cross plot of the training and testing data. These crossplots show that all points are evenly distributed around the 45-degree line, which is an indication of excellent performance of the RBF network model.

3.5. Undersaturated oil compressibility

The accuracy of the RBFNM in estimating undersaturated oil compressibility (C_o) is shown in Table 7 for both training and testing samples. Table 7 also shows the accuracy of the correlations by Vasquez and Beggs (1980), Farshad et al. (1992), Petrosky and Farshad (1993), Kartoatmodjo and Schmidt (1994), Ahmed (1989), and Elsharkawy and Alikhan (1997). These PVT correlations are given in Appendix F. The model has an E_r of −2.26%, E_a of 8.68%, SD of 13.33% and R of 92.28% for the training samples, and −1.09%, 5.98%, 7%, and 86.34% respectively for the testing samples. The PVT correlations considered in the comparison have errors in the range of 20 to 70% and standard deviation in the range of 20 to 50%. Cross plots of estimated compressibility

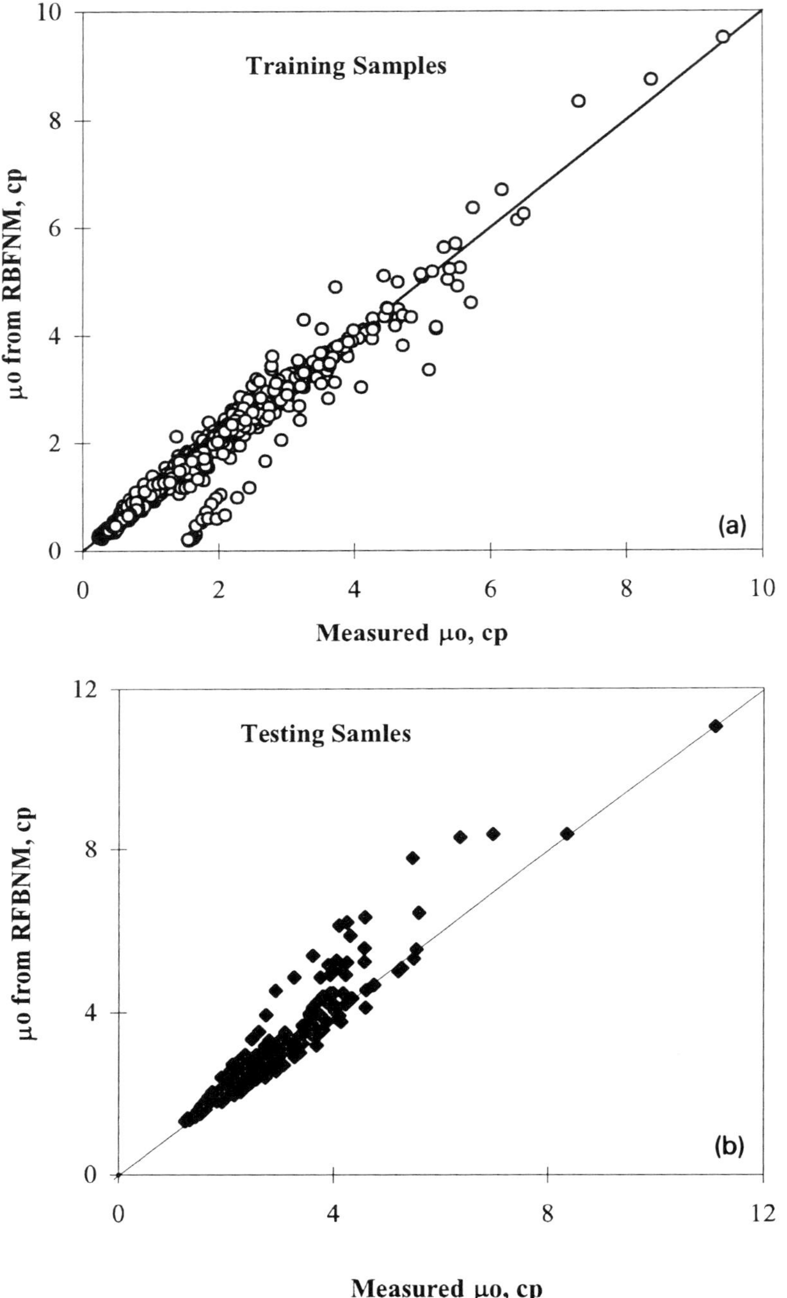

Fig. 4. (a) Crossplot of oil viscosity for training samples. (b) Crossplot of oil viscosity for testing samples.

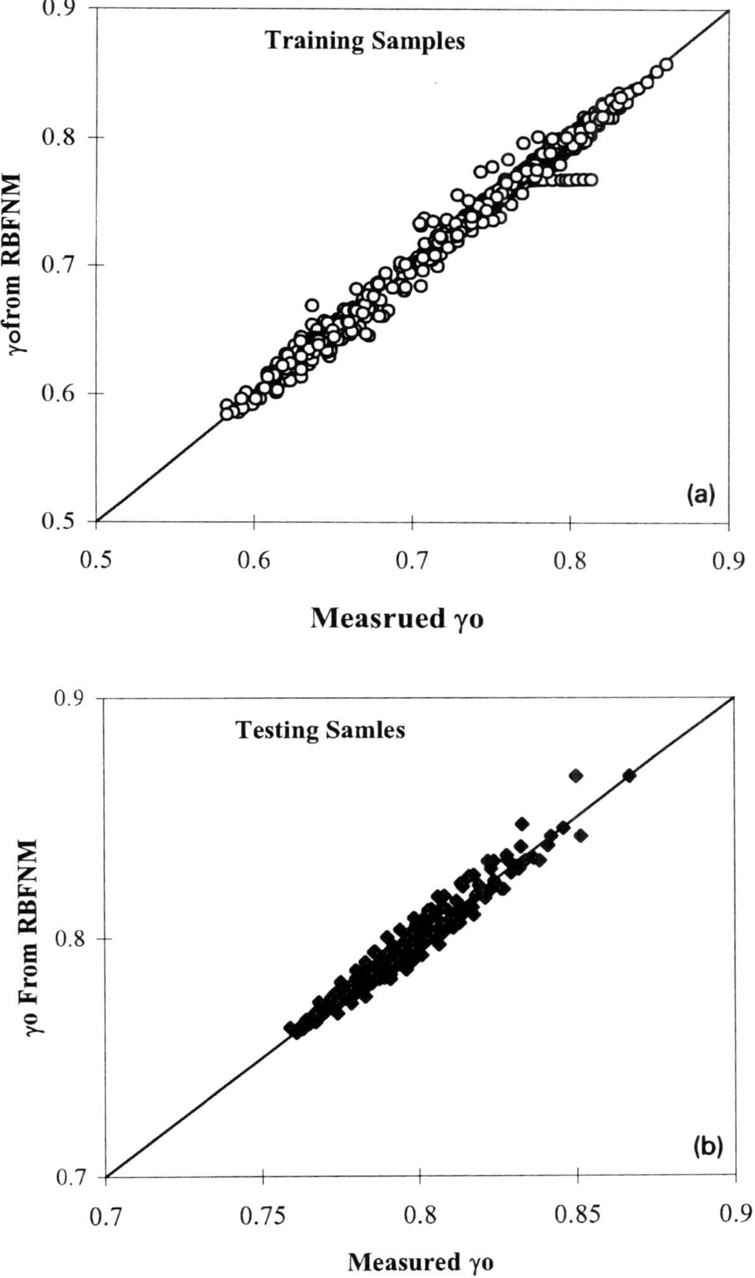

Fig. 5. (a) Crossplot of oil gravity for training samples. (b) Crossplot of oil gravity for testing samples.

TABLE 7

Accuracy of RBFNM and correlations in estimating undersaturated oil compressibility for training and testing samples

	Training samples						Testing samples					
	$E_{a,min}$ %	$E_{a,max}$ %	E_r %	E_a %	SD %	R %	$E_{a,min}$ %	$E_{a,max}$ %	E_r %	E_a %	SD %	R %
RBFNM	0.00	102.89	−2.26	8.86	13.33	92.82	0.05	44.29	−1.09	5.98	7.00	86.34
Vasquez	0.43	108.75	−25.90	30.54	28.42	85.39	0.93	105.07	−23.12	30.15	31.35	60.47
Farshad	0.17	141.79	−14.31	21.47	22.77	79.73	0.48	60.13	−14.23	21.11	20.63	34.49
Kartoatmodjo	0.11	111.63	−5.18	23.27	28.55	76.44	0.51	102.37	−12.09	24.28	28.81	59.73
Ahmed	0.20	157.98	−67.62	71.84	54.05	49.78	14.94	182.42	−102.6	102.9	35.05	53.32
Petrosky	0.13	127.93	−30.62	33.05	21.87	86.30	0.14	71.50	−22.11	25.28	21.50	57.18
Elsharkawy	0.33	75.06	−5.75	18.71	23.55	85.72	0.19	72.06	−2.17	21.19	26.06	60.61

by the network model versus measured ones is shown in Fig. 6a for training samples and 6b for testing samples. Even though these plots show some scatter around the unit slope line, the RBFNM is much more accurate than all the PVT correlations in estimating undersaturated oil compressibility for all the samples.

3.6. Gas gravity

Table 8 shows the accuracy of the RBFNM compared to other PVT correlations in predicting the evolved gas gravity (γ_g) for the training and testing samples. The correlations considered in the comparison are Labedi (1990), McCain and Hill (1995), and Elsharkawy and Alikhan (1999). These PVT correlations are given in Appendix G. Crossplots of the training samples, Fig. 7a, shows that all the plotted points fall around the unit slope line. However, Fig. 7b shows some scatter around the unit slope line for the testing samples. Nevertheless, The RBFNM shows better accuracy than all the PVT correlations considered in the comparison. The Radial basis function model predicts evolved gas gravity with an E_r of −0.12%, E_a of 1.23%, SD of 1.41%, and R of 97.97% for the training samples and −1.57%, 3.03%, 4.80%, 91.64%, respectively for the testing samples.

TABLE 8

Accuracy of RBFNM and correlations in estimating oil gravity for training and testing samples

	Training samples						Testing samples					
	$E_{a,min}$ %	$E_{a,max}$ %	E_r %	E_a %	SD %	R %	$E_{a,min}$ %	$E_{a,max}$ %	E_r %	E_a %	SD %	R %
RBFNM	0.01	14.99	−0.12	1.23	1.41	97.97	0.07	23.50	−1.57	3.03	4.80	91.64
Labedi	0.01	40.68	1.37	4.87	6.22	76.94	0.07	48.32	0.46	3.31	6.16	85.96
McCain	0.83	55.61	−10.85	11.06	6.19	85.54	0.16	78.01	−6.87	8.47	9.14	56.17
Elsharkawy	0.01	32.15	−0.87	2.96	4.62	87.43	0.01	64.28	−0.37	2.57	7.12	82.48

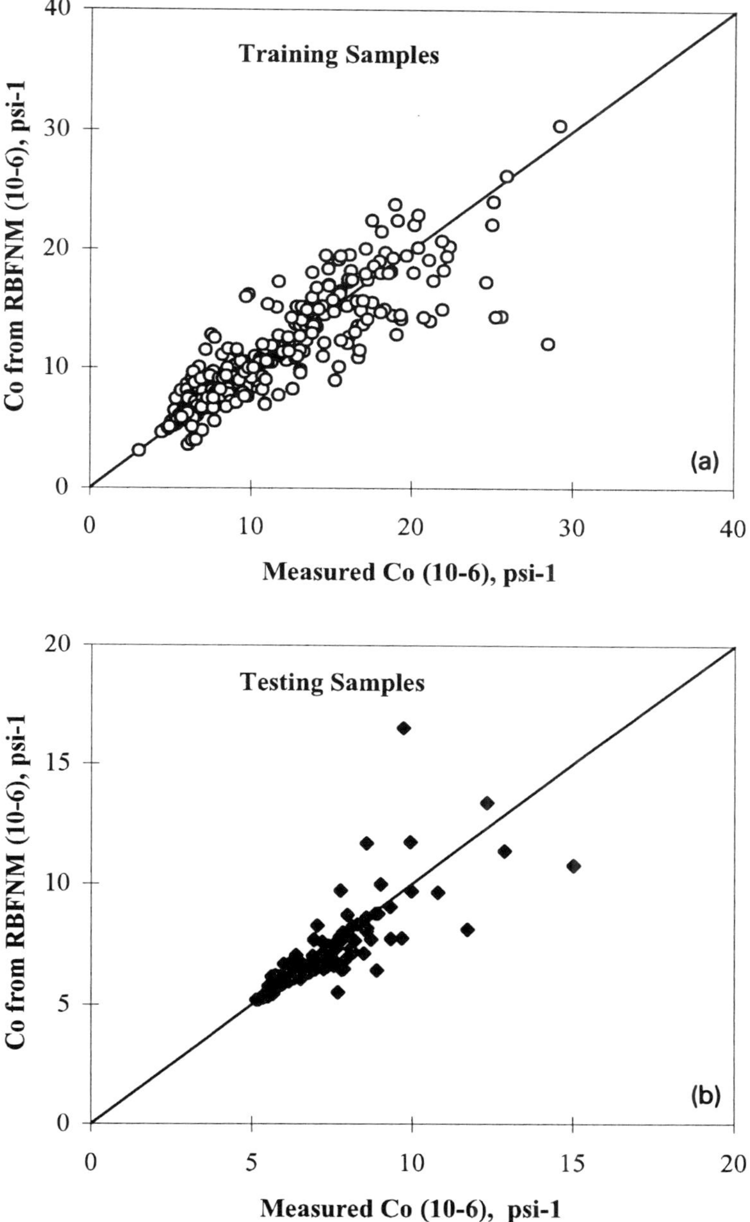

Fig. 6. (a) Crossplot of undersaturated oil compressibility for training samples. (b) Crossplot of undersaturated oil compressibility for testing samples.

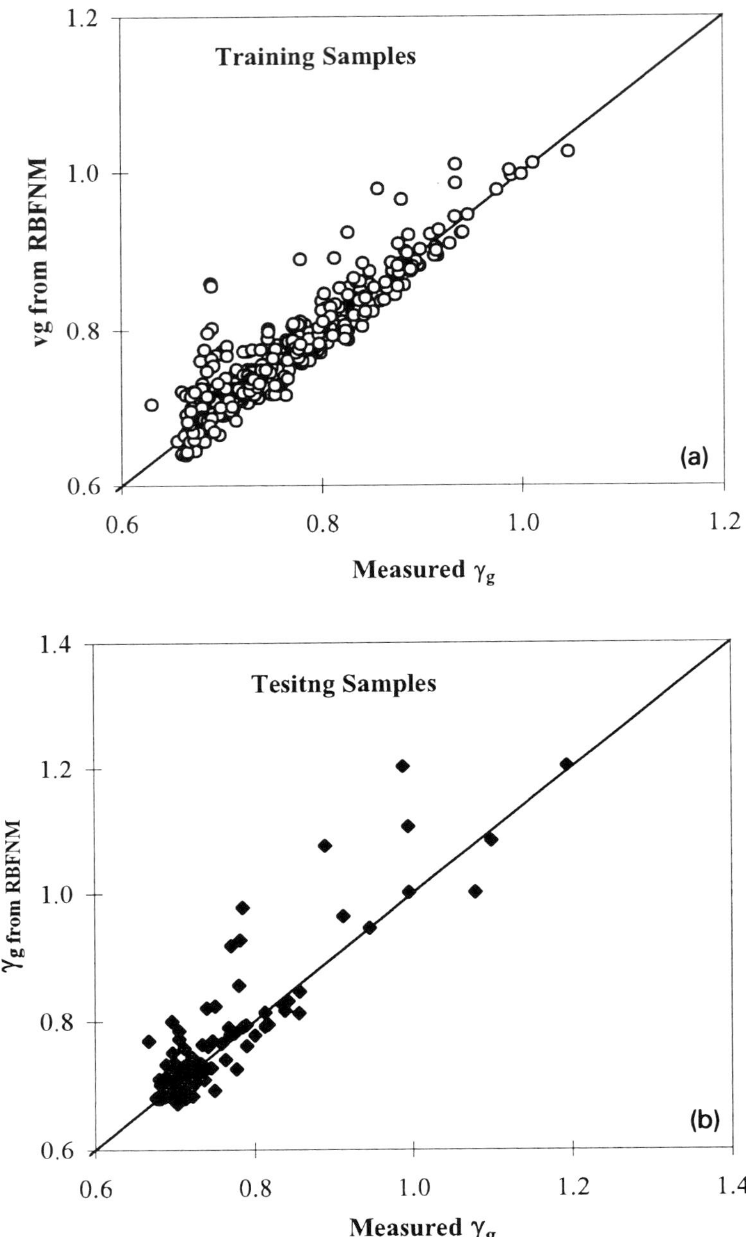

Fig. 7. (a) Crossplot of gas gravity for training samples. (b) Crossplot of gas gravity for testing samples.

4. BEHAVIOR OF THE MODEL

Fig. 8a through 8f show the behavior of the RBFNM in matching the experimentally measured solution gas–oil ratio, oil formation volume factor, oil viscosity, oil density, undersaturated oil compressibility, and evolved gas gravity for one of the test samples. These plots illustrate that no over fitting occur in the behavior of the model and that the model captures the physical trend of the measured PVT properties. Thus the model is reliable, stable, and physically correct.

5. CONCLUSION

A radial basis function network model (RBFNM) has been developed to predict the properties of the black oil systems. The model has been trained using differential PVT data of ninety samples and tested using another ten novel samples. Input data to the RBFNM are reservoir pressure, temperature, stock tank oil gravity, and separator gas gravity. The accuracy of the model in predicting the solution gas oil ratio, oil formation volume factor, oil viscosity, oil density, undersaturated oil compressibility and evolved gas gravity has been compared for training and testing samples to several PVT correlations. The comparison shows that the proposed model is much more accurate than published correlations in predicting the properties of the crude oils under study. The behavior of the model in capturing the physical trend of the PVT data has also been checked against experimentally measured PVT properties of the test samples. The proposed model proved to be stable and reliable. The RBFNM can be used in reservoir simulation studies and to check accuracy of future PVT reports. Although, this model was developed using some crude oil systems, the idea of using neural network to model behavior of reservoir fluid can be extended to other crude oils as a substitute to PVT correlations that were developed by conventional regression techniques.

6. NOTATIONS

γ_g	Separator gas gravity
γ_{gr}	Reservoir gas gravity
γ_o	Specified oil gravity
API	Stock tank oil gravity
B_{ob}	Oil formation volume factor, scf/stb
Co	Undersaturated oil compressibility, 1/psi
P	Reservoir pressure, psi
R_s	Solution gas–oil ratio, scf/stb
T	Reservoir temperature, °F
T_k	Reservoir temperature, °K
T_r	Reservoir temperature, R
μ_{oa}	Undersaturated oil viscosity, cP
μ_{ob}	Saturated oil viscosity, cP

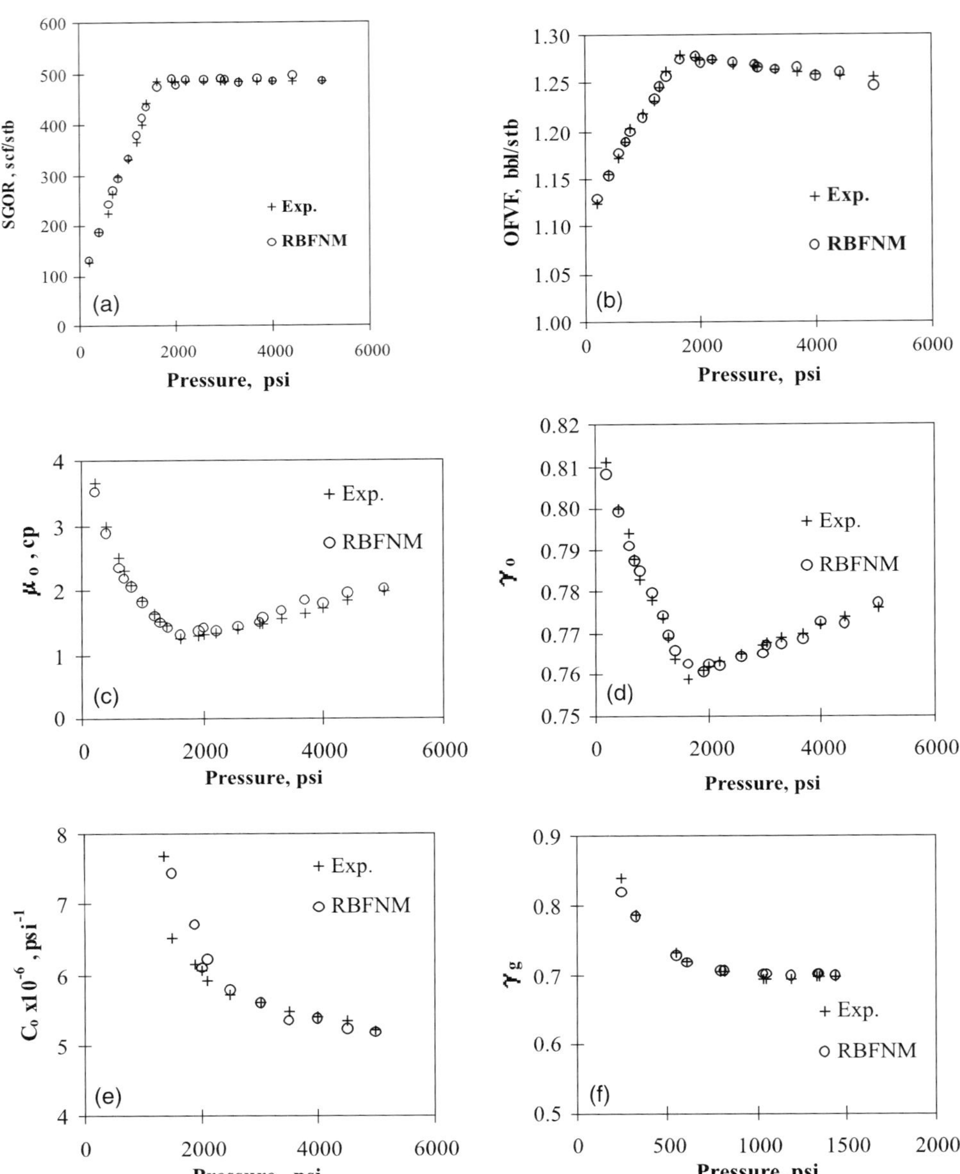

Fig. 8. (a) Measured and predicted SGOR. (b) Measured and predicted OFVF. (c) Measured and predicted oil viscosity. (d) Measured and predicted oil gravity. (e) Measured and predicted undersaturated oil compressibility. (f) Measured and predicted evolved gravity.

μ_{od} Dead crude oil viscosity, cP
$E_{a,min}$ Minimum absolute error
$E_{a,max}$ Maximum absolute error

E_r	Average relative error
E_a	Average absolute error
SD	Standard deviation
R	Coefficient of correlation

7. SI METRIC CONVERSION FACTORS

Degree API 141.5/(131.5 + API°)	= g/cm^3
bbl × 1.589,873 E−01	= m^3
Degree F, (F° − 32)/1.8	= °C
Degree R, F° + 460	= °R
PSI × 6.894,757 Eoo	= kPa
scf/stb × 1.801,175 E−01	= stdm^3/m^3

ACKNOWLEDGEMENTS

The author thanks Kuwaiti Foundation for Advancement of Sciences for the financial supports for this study, research grant No. KFAS 960904.

APPENDIX A. RADIAL BASIS FUNCTIONS

In recent years a special class of artificial neural networks, the radial basis function (RBF) networks have received considerable attention (Krzyzak, 1988). Radial basis function networks may require more training (more neurons) than standard feed-forward back propagation, but often they can be designed and trained in fraction of the time it takes to train standard feed-forward back propagation networks. They work best when many training vectors are available (Chen et al., 1991; and Park and Sandberg, 1991). Radial basis function networks have been applied to function approximation and pattern recognition problems. They have the capability to represent arbitrary functions (Leonard et al., 1992).

RBF networks consist of two layers: a hidden radbas layer of **S1** neuron and O/P purelin layer of **S2**. In the architecture of RBF, **P** represents the input vector, **W** the weight vector, **b** the bias, and **a** the output vector. RBF networks can be designed with the function *solverbe* from MATLAB that can actually designed a network with zero error on training vectors.

$$[\mathbf{W1}, \mathbf{b1}, \mathbf{b2}] = \mathit{solverbe}(\mathbf{P}, \mathbf{T}, \mathbf{\mathit{SC}})$$

The function takes matrix of input vector **P** and target vector **T**, and a spread constant for the radial basis layer **SC**, and returns the weight and biases for the network which outputs exactly **T** given **P**. This function works by creating as many *radbas* neurons as there are input vector in **P**. Each *radbas* neuron's weight vector in **W1** is set to the transpose of different input vector. Thus, we now have a layer of *radbas* neurons such that each neuron acts as a detector for a different input vector (Demuth and Beale,

1994). Radial basis functions (RBF) are a particular choice of basis Q_i such that our input/output map can be written as:

$$f(x) = \sum_{i=1}^{N} C_i Q_i(x) \tag{A-1}$$

where $\{C_i\}$ is a set of constants and x is an input data vector. RBFs can be described by

$$\phi_i(x) = h(\Pi x - \varsigma_i \Pi) = 1/\sigma\sqrt{2\Pi}\mathrm{e}^{-\Pi x - \varsigma_i \Pi^2/2\sigma^2} \tag{A-2}$$

$$\phi_i(x) = h(\Pi x - \varsigma_i \Pi) = 1/\sigma\sqrt{2\Pi}\mathrm{e}^{-\Pi x - \varsigma_i \Pi^2/2\sigma^2} \tag{A-3}$$

which is a Gaussian function where σ is spread constant. The name radial basis comes from the fact that the value of $Q_i(x)$ depends on the radial distance of x from the center value ξ_i of the basis function. Therefore, as the input moves away a given center, the neuron output drops off rapidly to zero. That is to say that the neurons in the RBFN have localized receptive fields because they only respond to inputs that are very close to their centers. The output layer is linear and produces a weighted sum of the outputs of the RB layer (Powell, 1987; Moody and Darken, 1989; Poggio and Girosi, 1990). The architecture of the radial basis function of the general regression model used in this study is shown in Fig. 1. The processing units are arranged in four layers: an input layer, two hidden layers, and an output layer. Two models are described here for predicting the PVT properties of the crude oil and gas systems. The input layer for the two models contains four neurons: reservoir pressures, reservoir temperature, stock tank oil gravity, and separator gas gravity. The output layer for model #1 contains solution gas–oil ratio (SGOR), oil formation volume factor (OFVF), oil viscosity, and oil density. However, The output layer for model #2 contains undersaturated oil compressibility, and evolved gas gravity. Corresponding weight and bias for the proposed models are shown in Tables A-1 through A-4.

APPENDIX B. SOLUTION GAS OIL RATIO CORRELATIONS

Standing (1977)

$$R_s = \gamma_g[\{(pb/18.2) + 1.4\}10^x]$$
$$x = 7.916 \times 10^{-4}\mathrm{API}^{1.5410} - 4.561 \times 10^{-5}T^{1.3911}$$

Vasquez and Beggs (1980)

$$R_s = C_1\gamma_g P^{C_2}\exp\{C_3(\mathrm{API}/(T+460))]$$

$\mathrm{API} \leq 30$	$\mathrm{API} > 30$
$C_1 = 0.0362$	0.0178
$C_2 = 1.9037$	1.178
$C_3 = 25.7240$	23.9310

TABLE A-1

Connection weight for the B_o, R_s, μ_o and γ_o

	First layer weight W_{ij}				Second layer weight W_{ij}			
i/j	1	2	3	4	1	2	3	4
1	0.92	28.03	140.0	300.0	1.0	34.19	230	1600
2	0.95	30.40	135.0	656.0	0.985	35.96	240	1100
3	0.88	27.49	134.0	1110	1.009	36.15	240	2100
4	0.92	28.03	140.0	1450	1.009	36.15	240	2600
5	0.92	30.40	134.0	1912	1.029	36.35	241	3558
6	0.95	30.40	135.0	60.0	1.009	36.15	240	3109
7	0.92	28.03	140.0	900.0	0.988	37.15	241	700
8	0.92	30.40	134.0	2940	1.0043	33.80	241	3925
9	0.91	31.89	135.0	2480	1.009	36.55	243	5000
10	0.91	26.42	166.0	500.0	0.997	35.96	241	390
11	0.92	28.93	132.0	3945	1.0702	36.15	243	4500
12	0.92	28.93	132	3445	0.95	30.40	135	1350
13	0.88	32.84	130	5000	1.0043	33.80	241	6100
14	0.91	27.85	134	1680	1.234	30.77	208	1877
15	0.88	32.84	130	4500	0.985	35.96	240	7060
16	0.92	31.71	133	1270	1.0702	36.15	243	8000
17	0.84	28.93	135	2140	1.000	34.19	230	2360
18	1.23	30.77	208	700	0.988	36.55	235	2820
19	1.01	36.55	243	200	0.985	35.96	240	4250
20	0.92	30.40	134	3675	1.234	30.77	208	900
21	0.98	35.96	240	1100	1.0702	36.15	243	3325
22	0.92	28.03	140	3200	1.029	36.35	241	5500
23	1.00	28.39	132	2625	1.009	36.55	243	200
24	0.91	27.85	134	2325	1.0043	33.80	241	4760
25	1.00	34.19	230	9000	1.00	34.19	230	9575
26	1.23	30.77	208	7000	1.009	36.15	240	550
27	0.96	30.21	135	718	1.029	36.35	241	3750
28	1.00	33.80	241	3925	0.988	36.55	235	9205
29	1.01	36.15	240	1565	0.988	37.15	241	8505
30	1.03	36.35	241	3558	1.009	36.55	243	1300
31	1.00	35.96	241	3218	1.234	30.77	208	3000
32	1.00	34.19	230	9575	1.0043	33.80	241	5325
33	1.23	30.77	208	400	0.988	36.55	235	7500
34	1.04	34.77	134	4500	0.988	37.15	241	7815
35	1.00	34.19	230	6000	0.988	37.15	241	5650
36	1.00	34.19	230	8000	0.988	36.55	235	1450
37	1.23	30.77	208	1000	0.988	36.55	235	1700
38	1.23	30.77	208	9900	1.00	34.19	230	9000
39	1.02	39.81	132	700	1.0702	36.15	243	3950
40	1.23	30.77	208	7500	1.0702	36.15	243	5000
41	0.92	38.16	135	5000	1.009	36.55	243	6000
42	0.93	24.51	135	300	0.988	37.15	241	6250
43	0.92	28.93	132	2800	1.009	36.55	243	7200
44	0.91	26.42	166	2873	0.925	38.16	135	2500
45	1.23	30.77	208	5500	0.957	26.78	134	2500
46	1.00	33.80	241	8000	1.009	36.55	243	2300
47	1.01	36.55	243	1800	1.009	36.55	243	2000
48	0.92	28.03	140	2065	1.0702	36.15	243	2000
49	1.00	35.96	241	1220	1.009	36.15	240	1100
50	1.00	33.80	241	7000	0.883	27.49	134	1110

TABLE A-1 (continued)

	First layer weight W_{ij}				Second layer weight W_{ij}			
i/j	1	2	3	4	1	2	3	4
51	0.92	30.40	134	4410	1.0043	33.80	241	6960
52	1.04	34.77	134	2625	0.923	28.03	140	3605
53	0.99	36.55	235	6000	0.985	35.96	240	2400
54	1.00	33.80	241	5500	0.921	28.93	132	2035
55	1.23	30.77	208	300	1.009	36.55	243	2700
56	0.85	30.96	135	1554	1.234	30.77	208	9900
57	0.92	31.71	133	1788	1.00	34.19	230	4000
58	1.00	34.19	230	1400	1.234	30.77	208	7000
59	0.99	37.15	241	3700	1.0043	33.80	241	4500
60	1.01	36.55	243	8550	0.988	37.15	241	5000
61	1.00	34.19	230	200	1.0043	33.80	241	1400
62	1.19	34.14	130	100	0.997	35.96	241	4500
63	0.92	30.96	132	108	1.234	30.77	208	5500
64	0.96	26.78	134	1110	1.009	36.55	243	900
65	1.01	36.15	240	9260	1.0043	33.80	241	900
66	1.19	34.14	130	700	1.0043	33.80	241	8000
67	0.99	36.55	235	7500	1.194	34.14	130	1616
68	0.92	30.96	132	2106	0.91	26.42	166	1615
69	0.99	37.15	241	9100	0.997	35.96	241	3218
70	0.92	38.16	135	700	0.923	28.03	140	3064
71	0.81	31.52	133	700	0.988	37.15	241	3500
72	0.92	30.96	132	308	1.029	36.35	241	3500
73	0.93	24.51	135	1400	0.997	35.96	241	3500
74	1.00	34.19	230	7000	1.234	30.77	208	7500
75	0.93	33.42	136	1400	1.0043	33.80	241	6000
76	1.01	36.55	243	7200	1.021	39.81	132	3000
77	1.00	34.19	230	100	0.962	31.52	135	3000
78	1.07	36.15	243	4715	0.924	30.96	132	725
79	1.01	36.55	243	3500	1.0702	36.15	243	2700
80	0.98	35.96	240	3750	1.0378	34.77	134	3375
81	1.01	36.15	240	1100	0.924	30.96	132	35
82	0.98	35.96	240	7060	1.009	36.15	240	9500
83	0.91	27.85	134	1930	0.997	35.96	241	9600
84	1.00	33.80	241	5325	1.021	39.81	132	900
85	0.99	37.15	241	700	0.988	37.15	241	9100
86	0.92	30.40	134	700	0.921	28.93	132	4535
87	0.93	31.14	134	2940	0.957	26.78	134	1110
88	0.93	31.14	134	3308	0.988	37.15	241	3069
89	1.19	34.14	130	3000	1.0702	36.15	243	3620
90	0.93	24.51	135	4000	0.988	36.55	235	4000
91	1.01	36.55	243	9600	1.0043	33.80	241	4000
92	1.01	36.15	240	9500	1.001	28.39	132	2625
93	0.92	28.93	132	4535	0.908	27.85	134	2325
94	0.88	32.84	130	2500	0.985	35.96	240	3750
95	0.92	30.40	134	2205	0.997	35.96	241	1220
96	0.88	32.84	130	1000	0.985	35.96	240	1200
97	0.92	38.16	135	900	0.997	35.96	241	790
98	0.97	33.03	132	1000	0.988	37.15	241	2600
99	0.96	30.58	134	1440	1.029	36.35	241	4500
100	1.01	36.15	240	550	1.0043	33.80	241	3605

TABLE A-2

Bias for the B_o, R_s, μ_o and γ_o ($b_{11} = 0.008325546$, $b_{21} = 0.008325546$)

	First layer B_{ij}		Second layer B_{ij}	
i/j	1	2	1	2
1	0.036453132	0.034312861	0.05290036	0.004033181

Glasø (1980)

$$R_s = \gamma_g [P_b^* \gamma \, \mathrm{API}^{0.989}/(T^{0.172})]1.2255$$
$$P_b^* = 10\{2.8869 - [14.1811 - 3.3093 \log(p)]^{0.5}\}$$

Obomanu and Okpobiri (1987)

$$R_s = (0.03008 p^{0.927} \gamma_g^{2.15} \mathrm{API}^{1.27})/\{10^{0.811}(1.8T_k - 460)^{0.497}\}$$

(where reservoir temperature is degree K)

Al-Marhoun (1988)

$$R_s = [185.843 p \gamma_g^{1.877840} \gamma_o^{-3.1437} T_r^{-1.32657}]^{1.398441}$$

(T_r is the reservoir temperature degree °R)

Dokla and Osman (1992)

$$R_s = [1.195620 \times 10^{-4} p \gamma_g 1.0104 \gamma_o^{-0.107991} T_r^{-0.952584}]^{1.3811258}$$

Farshad et al. (1992)

$$R_s = (0.0156p + 7.282)^{1.577} \gamma_g - 1.81410^{(0.0125\mathrm{API} - 0.000528T)}$$

Petrosky and Farshad (1993)

$$R_s = [\{(pb/112.727) + 12.340\} \gamma_g^{0.8439} 10^x]$$
$$x = 7.916 \times 10^{-4} \mathrm{API}^{1.5410} - 4.56 \times 10^{-5} T^{1.3911}$$

Kartoatmodjo and Schmidt (1994)

API ≤ 30 $\quad R_s = 0.05958 \gamma_g^{0.7972} p^{1.0014} 10^{(13.1405\mathrm{API}/T + 460)}$

API > 30 $\quad R_s = 0.03150 \gamma_g^{0.7587} p^{1.0937} 10^{(11.2895\mathrm{API}/T + 460)}$

Elsharkawy and Alikhan (1997)

API ≤ 30 $\quad R_s = \gamma_g (p^{1.18026} 10^{[-1.2179 + 0.4636\mathrm{API}/T]}$

API > 30 $\quad R_s = p^{94776} \gamma_g 0.0443_{\mathrm{API}} 1.139410^{(-2.188 + 0.0008392T)}$

TABLE A-3

Connection weight for C_o and γ_g

	First layer weight W_{ij}		Second layer weight W_{ij}	
i/j	1	2	1	2
1	35.74598	24.11257	2.554852	1.808335
2	0.7501365	1.262608	334.2763	1417.378
3	166.2872	0.9254646	0.4988829	0.3703352
4	1.575462	3.637014	35.57034	96.11373
5	0.8765853	1.222365	6.360369	0.1074685
6	2.992908	2.127222	0.7286772	1.035021
7	3.071836	3.668724	0.447443	0.3732625
8	51.38244	89.57623	9.653893	6.085307
9	0.6396314	0.5398248	962.0249	869.0175
10	0.3837543	0.5425555	0.5139371	0.353178
11	0.3091691	0.6434867	41.10466	16.75800
12	0.3894748	0.8111666	0.1714241	0.08066349
13	3.165264	5.233759	0.2703956	0.2968597
14	0.0841218	0.2049584	0.5495678	0.3475373
15	5.066246	8.627882	0.8478541	1.362599
16	0.2722602	0.5711312	39.28868	83.97691
17	1.118174	0.6885659	2.098851	2.087063
18	48.14442	45.02184	0.7500231	0.7220706
19	5.532377	6.652916	0.8260019	0.8622047
20	0.3938701	0.8872098	13.98876	13.75005
21	593.55600	771.10220	0.479329	0.4419861
22	0.4262171	0.8474149	1.493771	2.800643
23	2.95439	10.17063	0.534676	0.3287948
24	0.4391009	0.697665	0.8362356	0.8420284
25	0.3135694	0.7751689	7.438658	8.292005
26	56.53442	93.68191	0.5108836	0.314311
27	1.709999	2.891292	11.7368	15.3991
28	0.2051371	0.3917754	0.9139244	0.5629787
29	0.1014746	0.1124389	0.853117	0.777676
30	0.2278181	0.6551849	0.9514196	0.69291
31	0.2181234	0.457091	0.0956961	0.1880189
32	3.999123	7.675135	0.6890489	0.6608059
33	0.06431021	0.09032439	3.802651	4.064953
34	5.055288	9.154837	0.7124249	0.7581441
35	17.91358	36.36773	0.7352077	0.774722
36	12.47615	24.57039	0.1203896	0.08465432
37	0.5546461	1.172782	0.7571558	0.6615387
38	0.1992272	0.5255603	0.623407	0.2417958
39	4557.721	3148.911	12.5561	7.7632
40	1.896941	3.587235	971.212	870.5399
41	2.859627	4.56319	29.39777	63.43624
42	30.82611	18.56475	0.7456571	0.7959032
43	0.5256966	0.9414822	0.7005025	0.4312372
44	0.5248454	0.834724	12.00972	6.62184
45	1.423359	2.822799	11.86984	6.452496
46	12.3622	24.22985	1.500569	1.437428
47	0.1000056	0.123639	886.1668	751.5815
48	0.05785572	0.2068581	885.8085	751.2066
49	0.08333781	0.01793729	335.1615	1418.088
50	120.0949	200.0505	326.4867	79.9624

TABLE A-3 (continued)

	First layer weight W_{ij}		Second layer weight W_{ij}	
i/j	1	2	1	2
51	0.1936124	0.5998847	1.436652	1.966807
52	2.90994	9.896642	0.2821151	0.0138668
53	17.78917	36.02798	1.738015	1.696771
54	1.326303	2.548804	0.1638018	0.07351729
55	2.65522	1.820181	2186.03	1936.268
56	0.2372554	0.5400339	0.4620528	0.2674904
57	0.2493352	0.4797075	92.8816	64.72498
58	0.2843839	0.4875086	1.327223	2.350113
59	0.5012234	1.328083	33755.62	37266.85
60	0.03422742	0.1830561	10.01284	2.290607
61	5.781283	7.142961	0.9179431	0.4940004
62	10.01845	5.666465	219889	243186.7
63	7.650318	4.085574	1.053188	2.457481
64	166.5802	0.5804538	189.8188	161.945
65	0.6838361	0.3135628	201.5181	173.4944
66	6627.874	3898.726	38.5876	83.35072
67	1.793459	3.258126	4.512933	3.204951
68	0.6301887	0.1950276	5.998674	4.341583
69	0.1446443	0.3144072	0.64472	0.7608127
70	4434.74	3050.938	0.77072	0.54781
71	14602.32	7732.133	22087.04	67838.63
72	3.341804	4.618771	66643.42	206176.8
73	7.096934	9.796459	44557.09	138343.8
74	174.0688	289.6285	3.100269	3.542542
75	7.466223	10.59856	28.73851	62.85073
76	0.08344062	0.267207	15.96804	13.07846
77	0.3796794	0.3100336	15.4336	12.70484
78	0.00695941	0.6744896	0.04643679	0.01147644
79	0.2882149	0.7212626	2185.764	1936.067
80	0.3094192	0.9037049	0.1743809	0.1076471
81	593.6657	770.9344	0.07005464	0.064095
82	0.2211114	0.3832149	2.567914	2.559042
83	1.108793	1.72827	6.581244	7.277481
84	0.00368648	0.08614754	1.332567	1.736476
85	18.46776	19.41026	0.5392409	0.3428158
86	8125.052	3956.354	0.63682	0.37422
87	50.89776	88.73341	326.4461	79.85439
88	0.0145092	0.1235268	1.795002	2.115715
89	0.0384226	0.1543936	16.87714	4.10311
90	0.1182365	0.2602295	176.0443	120.5635
91	3.244775	6.172774	78.69241	52.90435
92	1.040548	1.86203	0.6206906	0.8980576
93	0.2114291	0.9453539	0.2101218	0.07396833
94	0.869427	1.005106	11.21115	15.25902
95	0.408061	0.2015472	4.938832	4.005344
96	22.80864	53.37302	4.154425	3.418799
97	2.792938	3.200247	0.1625593	0.1204214
98	23.00587	53.86826	35.82872	96.26394
99	1.961104	4.463388	186104.1	205916.4
100	0.0528698	0.04505904	20.95917	3.661544

APPENDIX C. OIL FORMATION VALUE FACTOR CORRELATIONS

Standing (1977)

$$B_{\mathrm{ob}} = 0.972 + 1.47 \times 10^{-4}[R_{\mathrm{s}}(\gamma_{\mathrm{g}}\gamma_{\mathrm{o}})^{0.5} + 1.25T]^{1.175}$$

Vasquez and Beggs (1980)

$$B_{\mathrm{ob}} = 1.0 + C_1 R_{\mathrm{s}} + C_2(T - 60)(\gamma\,\mathrm{API}/\gamma\,\mathrm{gs}) + C_3 R_{\mathrm{s}}(T - 60)(\gamma\,\mathrm{API}/\gamma\,\mathrm{gs})$$

API ≤ 30	API > 30
$C_1 = 4.677 \times 10^{-4}$	$C_1 = 4.670 \times 10^{-4}$
$C_2 = 1.751 \times 10^{-5}$	$C_2 = 1.100 \times 10^{-5}$
$C_3 = 1.8106 \times 10^{-8}$	$C_3 = 1.337 \times 10^{-9}$

Glasø (1980)

$$B_{\mathrm{ob}} = 1.0 + 10^{[-6.58511 + 0.91329 \log(B^*_{\mathrm{ob}}) - 0.27683(\log B^*_{\mathrm{ob}})^2]},$$

$$B^*_{\mathrm{ob}} = R_{\mathrm{s}}(\gamma_{\mathrm{g}}/\gamma_{\mathrm{o}})^{0.526} + 0.968T$$

Obomanu and Okpobiri (1987)

API > 30

$$B_{\mathrm{ob}} = 0.3321 + 7.88374 \times 10^{-4} R_{\mathrm{s}}^{-2.335} \times 10^{-3} R_{\mathrm{s}}(\gamma_{\mathrm{g}}/\gamma_{\mathrm{o}}) + 2.0855 \times 10^{-3} T_{\mathrm{k}}$$

(R_{s} in $\mathrm{m^3/m^3}$ and reservoir temperature in degree K)

API ≤ 30

$$B_{\mathrm{ob}} = 1.0232 + 1.065 \times 10^{-4}[R_{\mathrm{s}}(\gamma_{\mathrm{g}}/\gamma_{\mathrm{o}}) + 1.8T_{\mathrm{k}} - 460]^{0.79}$$

Al-Marhoun (1988)

$$B_{\mathrm{ob}} = 0.497069 + 0.862963 \times 10^{-3} T_{\mathrm{r}} + 0.182594 \times 10^{-2} F + 0.318099 \times 10^{-5} F^2$$

(where T_{r} is the reservoir temperature degree R)

$$F = R_{\mathrm{s}}^{0.742390} \gamma_{\mathrm{g}}^{0.323294} \gamma_{\mathrm{o}}^{-1.202040}$$

Abdul Majeed (1988)

$$B_{\mathrm{ob}} = 0.9657876 + 4.8141 \times 10^{5} F - 6.8987 \times 10^{-10} F^2 + 7.73 \times 10^{-4} T F$$

$$= R_{\mathrm{s}}^{1.2} \gamma_{\mathrm{g}}^{0.147} \gamma_{\mathrm{o}}^{-5.222}$$

Labedi (1990)

$$B_{\mathrm{ob}} = 0.9897 + 0.0001364[R_{\mathrm{s}}(\gamma_{\mathrm{g}}/\gamma_{\mathrm{o}})^{0.5} + 1.25T]^{1.175}$$

Dokla and Osman (1992)

$$B_{\mathrm{ob}} = 0.0431935 + 0.156667 \times 10^{-2} T_{\mathrm{r}} + 0.139775 \times 10^{-2} F + 0.380525 \times 10^{-5} F^2$$

$$F = R_{\mathrm{s}} 0.773572 \gamma_{\mathrm{g}} 0.404020 \gamma_{\mathrm{o}} - 0.882605$$

Farshad et al. (1992)

$$B_{\mathrm{ob}} = 1 + 10^{[-2.6541+0.557\log(A)+0.333(\log A)^2]}$$
$$A = R_{\mathrm{s}}^{0.5956}\gamma_{\mathrm{g}}^{0.2369}\gamma_{\mathrm{o}}^{-1.3282} + 0.0976T$$

Petrosky and Farshad (1993)

$$B_{\mathrm{ob}} = 1.0113 + 7.2046\times 10^{-5}[R_{\mathrm{s}}^{0.37738}(\gamma_{\mathrm{g}}^{0.2914}/\gamma_{\mathrm{o}}^{0.6265}) + 0.24626T]^{3.0936}$$

Kartoatmodjo and Schmidt (1994)

$$B_{\mathrm{ob}} = 0.98496 + 0.0001[R_{\mathrm{s}}^{0.755}\gamma_{\mathrm{g}}^{0.25}\gamma_{\mathrm{o}})^{-1.50} + 0.45T]$$

Elsharkawy and Alikhan (1997)

$$B_{\mathrm{ob}} = 1.0 + 40.428\times 10^{-5}R_{\mathrm{s}} + 63.802\times 10^{-5}T + 0.0780\times 10^{-5}[R_{\mathrm{s}}(T-60)\gamma_{\mathrm{g}}/\gamma_{\mathrm{o}}]$$

APPENDIX D. OIL VISCOSITY CORRELATIONS

Beggs and Robinson (1975)

$$u_{\mathrm{od}} = 10^{x} - 1$$
$$x = y(T-460)^{-1.163}$$
$$y = 10^{z}$$
$$z = 3.0324 - 0.0203\mathrm{API}$$
$$u_{\mathrm{ob}} = (10)^{a}(uod)^{b}$$
$$a = 10.715(R_{\mathrm{s}}+100)^{-0.515}$$
$$b = 5.440(R_{\mathrm{s}}+150)^{-0.338}$$

Vasquez and Beggs (1980)

$$u_{\mathrm{oa}} = u_{\mathrm{ob}}(p/pb)^{m}$$
$$m = 2.6(p^{1.187})(10^{a})$$
$$a = -3.9(10^{-5})p - 5$$

Labedi (1992)

$$\mu_{\mathrm{od}} = 10^{9.224}/\mathrm{API}^{4.7013}T_{f}^{0.6739}$$

At the bubble point

$$u_{\mathrm{ob1}} = (10^{2.344-0.03542\mathrm{API}})(u_{\mathrm{od}}^{0.6447})/(pb^{0.426})$$

Below the bubblepoint (saturated)

$$u_{\mathrm{ob2}} = u_{\mathrm{ob1}}/1 - M_{\mathrm{ub}}[1-(p/p_b)]$$
$$M_{\mathrm{ub}} = 10^{-3.876}p_b^{0.5423}\mathrm{API}^{1.1302}$$

Above bubble point (undersaturated)

$$u_{\text{oa}} = u_{\text{ob}} - M_{\text{ua}}[1 - p/pb)]$$
$$M_{\text{ua}} = 10^{-2.488} u_{\text{od}}^{0.9036} p_b^{0.6151} / 10^{0.01976\text{API}}$$

Kartoatmodjo and Schmidt (1994)

$$uod = 16(10^8) T_f^{-2.8177} (\log \text{API})^x$$
$$x = 5.7526 \log(T_f) - 26.9718$$
$$u_{\text{ob}} = -0.06821 + 0.9824 f + 0.000403 f^2$$
$$f = [0.2001 + 0.8428(10^{-0.000845 R_S})]x$$
$$x = u_{\text{od}}^{(0.43+0.5165y)}$$
$$y = 10^{-0.00081 R_S}$$
$$u_{\text{oa}} = 1.00081 u_{\text{ob}} + 0.001127(p - p_b)(-0.006517 u_{\text{ob}}^{1.8148} + 0.038 u_{\text{ob}}^{1.590})$$

Elsharkawy and Alikhan (1999)

$$\log_{10}\{\log_{10}(u_{\text{od}} + 1)\} = 2.16924 - 0.02525\text{API} - 0.68875 \log_{10}(T)$$
$$u_{\text{ob}} = A(u_{\text{od}})^B$$
$$A = 1241.932(R_s + 641.026)^{-1.12410}$$
$$B = 178.841(R_s + 1180.335)^{-1.06622}$$
$$u_{\text{oa}} = u_{\text{ob}} + 10^{-2.0771}(p - pb)(u_{\text{od}}^{1.19279} u_{\text{ob}}^{-0.40712} p_b^{-0.7941})$$

APPENDIX E. SATURATED OIL DENSITY CORRELATIONS

McCain and Hill (1995)

McCain method needs iteration for ρa. The first iteration can be obtained from the following equation.

$$\rho a = 52.8 - 0.01 R_{\text{sb}}$$

For the next iteration, use the following equation;

$$\rho a = a_0 + a_1 \gamma_{\text{gsp}} + a_2 \gamma_{\text{gsp}} \rho_{\text{po}} + a_3 \gamma_{\text{gsp}}^{\rho_{\text{po}}^2} + a_4 \rho_{\text{po}} + a_5 \rho_{\text{po}}^2$$

where a_0, a_1, a_2, a_3, a_4 and a_5 are the following constants:

$$a_0 = -49.8930 \qquad a_1 = 85.0149$$
$$a_2 = -3.70373 \qquad a_3 = 0.0479818$$
$$a_4 = 2.98914 \qquad a_5 = -0.0356888$$

Elsharkawy and Alikhan (1997)

$$\gamma_{\text{o}} = 10^{0.18671} R_s 0.061307 T^{-0.008061} p^{-0.030472} \text{API}^{-0.189797} \gamma_{\text{gsp}} 0.006447 B_{\text{o}}^{-0.6675}$$

APPENDIX F. UNDERSATURATED OIL COMPRESSIBILITY CORRELATIONS

Vasquez and Beggs (1980)

$$C_o = (-1433 + 5R_s + 17.2T - 1180\gamma_g + 12.61\text{API})/10^5 p$$

Farshad et al. (1992)

$$C_o = 1.705 \times 10^{-7} R_s^{0.69357} \gamma_g 0.1885\text{API}^{0.3272} T^{0.6729} p^{0.5906}$$

Petrosky and Farshad (1993)

$$C_o = 1.705 \times 10^{-7} R_s 0.69357 \gamma_g 0.1885\text{API}^{0.3272} T^{0.6729} p^{0.5906}$$

Kartoatmodjo and Schmidt (1994)

$$C_o = (6.8257 \times 10^{-6} R_s^{0.5002} \text{API}^{0.3613} T^{0.76606} \gamma_g^{0.35505})/p$$

Elsharkawy and Alikhan (1997)

$$C_o = (-27321 + 33.784R_s + 238.81T)/10^6 p$$

APPENDIX G. EVOLVED GAS GRAVITY CORRELATIONS

Labedi (1982)

$$\gamma_g = A_g + M_g(P_b/P - 1)$$
$$A_g = 0.7176(\gamma_{gsp})^{0.5672} T_r^{0.0003}$$
$$M_g = (10^{-3.9778}(\gamma_{gsp})^{1.5781}(\gamma_o)^{1.9167})/(10^{(0.0001942P)})$$

McCain and Hill (1995)

$$1/\gamma_g = a_1/p + a_2/p^2 + a_3 \cdot P + a_4/\sqrt{T} + a_5 \cdot T + a_6 \cdot R_s + a_7 \cdot \text{API} + a_8/\gamma_{gsp} + a_9 \cdot \gamma_{gsp}^2$$

where a_1 through a_9 are the coefficients used with reservoir pressure above 300 psig:

$a_1 = -208.0797$ $\quad a_2 = 22885$ $\quad a_3 = -0.000063642$

$a_4 = 3.38346$ $\quad a_5 = -0.000992$ $\quad a_6 = -0.000081147$

$a_7 = -0.001956$ $\quad a_8 = 1.081956$ $\quad a_9 = 0.394035$

Elsharkawy and Alikhan (1997)

$$\gamma_{gr} = 0.53853 + 39.944(\tfrac{1}{p}) + 0.44696(T_R/460)$$
$$-35.29 \times 10^6 R_s - 1.0956\gamma_{gsp} + 0.66211\gamma_{gsp}^2$$

REFERENCES

Abdul Majeed, G.H., 1985. Evaluation of PVT correlations. *Soc. Pet. Eng.*, SPE Paper 14478 (unsolicited).

Abdul Majeed, G.H., 1988. Statistical evaluation of PVT correlation solution gas–oil ratio. *J. Can. Pet. Technol.*, 27(4): 95–101.

Abdul Majeed, G.H. and Salman, N.H., 1988. An empirical correlation for oil FVF prediction. *J. Can. Pet. Technol.*, 27(6): 118–122.

Abdus Sattar, A., 1959. *Correlation technique for evaluating formation volume factor and gas solubility of crude oil in Rocky Mountain regions.* Pet. Eng. Dept., Colorado School of Mines.

Accarain, P. and Desbrandes, R., 1993. Neuro-computing help pore pressure determination. *Pet. Eng. Int.*, Feb., 39–42.

Ahmed, T., 1989. *Hydrocarbon Phase Behavior.* Gulf Publishing Co., Houston, TX, USA, pp. 163–164.

Al-kaabi, A.W. and Lee, J.W., 1990. An artificial neural network approach to identify the well test interpretation model: application. *65th Ann. Tech. Meet.*, New Orleans, LA, Sept. 23–26, SPE Paper 2055.

Al-Marhoun, M.A., 1988. PVT Correlations for Middle East Crude Oils. *J. Pet. Technol.*, May: 650–66; *Trans. AIME*, 285.

Ali, J.K., 1994. Neural networks: A new tool for the petroleum industry. *European Per Conf.*, Aberdeen, Mar. 15–18, Paper SPE 27561.

Asgarpour, S., Mclauchlin, L., Womg, D. and Cheung, V., 1988. Pressure–volume temperature correlations for western Canadian gases and Oil. *39th Ann. Tech. Meet.*, Calgary, Jun. 12–16, CIM Paper No. 88-39-62.

Beggs, H.D. and Robinson, J.R., 1975. Estimating the viscosity of crude oil systems. *J. Pet. Technol.*, Sept., 1140–1141.

Borden, G. and Rzasa, M.J., 1950. Correlation of bottom hole sample data. *Trans. AIME*, 189: 345–348.

Briones, 1994. Application of neural network in the prediction of reservoir hydrocarbon mixture composition from production data. *69th Ann. Tech. Meet.*, New Orleans, Sept. 25–28, SPE Paper 28598.

Caixerio, E., 1976. *Correlation of Reservoir Properties, Miranga field, Brazil.* M.Sc. Report, Stanford University.

Chen, S., Cowan, C.F.N. and Grant, P.M., 1991. Orthogonal Least Square learning algorithm for radial basis function networks. *IEEE Trans. Neural Networks*, 2(2): 302–309.

Cronquist, C., 1973. Dimensionless PVT behavior of gulf coast crude oils. *J. Pet. Technol.*, May, 1–8.

Demuth, H. and Beale, M., 1994. *Neural Network Toolbox for User with MATLAB.* Math Works Inc.

Dokla, M.E. and Osman, M.E., 1992. Correlation of PVT properties for UAE crudes. *SPE Form. Eval.*, March, 41–46.

Elam, F.M., 1957. *Prediction of Bubblepoint Pressure and Formation Volume Factors from Field Data.* M.Sc. Thesis, Texas A&M University.

Elsharkawy, A.M., 1998. Changes in gas and oil gravity during depletion of oil reservoirs. *Fuel*, 77(8): 837–845.

Elsharkawy, A.M. and Al-Matter, D., 1996. Geographic location considered in PVT calculation program. *Oil Gas J.*, 22: 36–39.

Elsharkawy, A.M. and Alikhan, A.A., 1997. Correlations for predicting solution gas–oil ratio, oil formation volume factor, and undersaturated oil compressibility. *J. Petrol. Sci. Eng.*, 17(3/4): 291–302.

Elsharkawy, A.M. and Alikhan, A.A., 1999. Predicting the viscosity of Middle East crude oils. *Fuel*, 78: 891–903.

Elsharkawy, A.M., Elgibly, A.A. and Alikhan, A.A., 1995. Assessment of the PVT correlations for predicting the properties of Kuwaiti crude oils. *J. Pet. Sci. Eng.*, 13: 219–232.

Farshad, F.F., LeBlance, J.L., Garbeer, J.O. and Osorio, J.G., 1992. Empirical PVT correlations for Colombian crudes. *Soc. Pet. Eng.*, SPE Paper 24538 (unsolicited).

Gharbi, R.B. and Elsharkawy, A.M., 1996. Neural network model for estimating the PVT properties of Middle East crude oil systems. *In Situ*, 20(4): 367–394.

Gharbi, R.B. and Elsharkawy, A.M., 1999. Universal neural network based model for estimating the PVT properties of crude oil systems. *Energy Fuel*, 13: 454–458.

Ghetto, G., Panone, F. and Villa, M., 1994. Reliability analysis of PVT correlations. *European Petroleum Conference*, London, Oct. 25–27, SPE Paper 28904.

Glasø, O., 1980. Generalized pressure–volume–temperature correlations. *J. Pet. Technol.*, May, 785–795.

Habiballah, 1996. Use of neural networks for the prediction of vapor/liquid equilibrium K Values for light hydrocarbon mixtures. *Soc. Pet. Eng. Res. Eng.*, May, 121–125.

Juniardi, I.R. and Ershaghi, I., 1993. Complexities of using neural network in well test analysis of faulted reservoir. *West Reg. Meet.*, Alaska, May 26–28, SPE Paper 26106.

Kartoatmodjo, T., 1990. *New Correlations for Estimating Hydrocarbon Liquid Properties.* M.Sc. Thesis, Univ. of Tulsa.

Kartoatmodjo, T. and Schmidt, Z., 1994. Large databank improves crude oil physical property correlations. *Oil Gas J.*, July, 51–55.

Katz, D.L., 1942. Prediction of the shrinkage of crude oils. *API Drill. Prod. Pract.*, 137.

Knopp, C.R. and Ramsey, L.A., 1960. Correlation for oil formation volume factor and solution gas–oil ratio. *J. Pet. Technol.*, August, 27–29.

Krzyzak, A., 1988. Radial basis function networks and complexity regularization in function learning. *IEEE Trans. Neural Networks*, 9(2): 247–256.

Kumoluyi, 1994. Identification of well test models using high order neural net. *European Comp. Conf.*, Aberdeen, Mar. 15–17, SPE Paper 27558.

Labedi, R., 1982. *PVT Correlations of the African Crudes.* Ph.D. Thesis, Colorado School of Mines.

Labedi, R., 1990. Use of production data to estimate volume factor density and compressibility of reservoir fluids. *J. Pet. Sci. Eng.*, 4: 357–390.

Labedi, R., 1992. Improved correlations for predicting the viscosity of light crudes. *J. Pet. Sci. Eng.*, 8: 221–234.

Lasater, J.A., 1958. Bubble point pressure correlation. *Trans. AIME*, 213: 379–381.

Leonard, J.A., Kramer, M.A. and Ungar, L.H., 1992. Using Radial basis function networks to approximate a function and its error. *IEEE Trans. Neural Networks*, 3(4): 624–627.

McCain, W.D. and Hill, N.C., 1995. Correlations for liquid densities and evolved gas specific gravity for black oil during pressure depletion. *Soc. Pet. Eng.*, SPE Paper 30773.

Mohaghegh, S., 1994. Design and development of an artificial Neural network for estimation of formation permeability. *SPE Comp. Conf.*, Dallas, TX, Jul. 31–Aug. 3, SPE Paper 28237.

Mohaghegh, S., 1996. Petroleum reservoir characterization with the aid of artificial neural networks. *J. Pet. Sci. Eng.*, 16: 26–274.

Mohamood, M.A. and Al-Marhoun, M.A., 1996. Evaluation of empirically derived PVT properties for Pakistani crude oils. *J. Pet. Sci. Eng.*, 16: 275–290.

Moody, J.E. and Darken, C.J., 1989. Fast Learning in Network of Locally-Tuned Processing Units. *Neural Comput.*, 1: 281–294.

Niakan, M.R., 1967. *Correlation of Oil Formation Volume Factors for Asmari and Bangsten Crudes within the Agreement Area.* Iranian Oil Co., Tech. Note No. 1115.

Obomanu, D.A. and Okpobiri, G.A., 1987. Correlating the PVT properties of Nigerian crudes. *J. Energy Resour. Tech., Trans. ASME*, 109: 214–217.

Osborne, O.A., 1992. Neural networks provide more accurate reservoir permeability. *Oil Gas J.*, 28: 80–83.

Ostermann, R.D., Ehlig-Economides, C.A. and Owolabi, O.O., 1983. Correlations for the reservoir fluid properties of Alaskan crudes. *Soc. Pet. Eng.*, SPE Paper 11703.

Owolabi, O., 1983. *Reservoir fluid properties of Alaskan crudes.* M.Sc. Thesis, Univ. of Alaska.

Park, J. and Sandberg, I.W., 1991. Universal approximation using radial-basis function networks. *Neural Comput.*, 3: 246–257.

Petrosky, G.E. and Farshad, F.F., 1993. Pressure volume temperature correlations for Gulf of Mexico crude oils. *Soc. Pet. Eng.*, SPE 26644.

Poggio, T. and Girosi, F., 1990. Networks for approximation and learning. *Proc. IEEE*, 78: 1481–1497.

Powell, M.J.D., 1987. Radial basis functions for multivariable interpolation: a review. In: Mason, J.C. and Cox, M.G. (Eds.), *Algorithms for the Approximates of Functions and Data.* Clarendon Press, Oxford.

Saleh, A.M., Mahgoub, I.S. and Assad, Y., 1987. Evaluation of empirically derived PVT properties for Egyptian crudes. *Soc. Pet. Eng.*, SPE Paper 15721.

Santamaria, G.N.E. and Hernandez, P.M.A., 1989. Development of empirical PVT correlations for Mexican crude oils. *Rev. Inst. Mex. Pet.*, 21(1): 60–79.

Standing, M.B., 1947. A pressure–volume–temperature correlation for mixtures of California oils and gases. *API Drill. Prod. Pract.*, pp. 275–287.

Standing, M.B., 1962. Oil-system correlations: In: Frick, T.C. (Ed.), *Petroleum Production Handbook., Vol. 2.* SPE, Richardson, TX, chapter 19.

Standing, M.B., 1977. *Volumetric and Phase Behavior of Oil Field Hydrocarbon Systems.* Society of Petroleum Engineers, Richardson, TX, 124 pp.

Standing, M.B., 1981. *Volumetric Phase Behavior of Oil Field Hydrocarbon Systems, 9th ed.* Society of Petroleum Engineers, Dallas, TX.

Sung, 1996. Development of HT-BP neural network system for the identification of well test interpretation model. *SPE Comp. Appl.*, August, 102–105.

Sutton, R.P. and Farshad, F.F., 1990. Evaluation of empirically derived PVT properties for Gulf of Mexico crude oils. *SPE Res. Eng.*, February, 79–86.

Tahrani, H.O., 1968. Bubblepoint pressure correlation for reservoir crudes of Southwest Iran. *Second AIME Reg. Tech. Symp.*, Saudi Arabia, Mar. 27–29.

Trube, A.S., 1957. Compressibility of hydrocarbon reservoir fluids. *Trans. AIME*, 210: 341–344.

Vasquez, M.E., 1976. *Correlations for Fluid Physical Property Prediction.* M.Sc. Thesis, Univ. of Tulsa, Tulsa, OK.

Vasquez, M.E. and Beggs, H.D., 1980. Correlations for fluid physical property prediction. *J. Pet. Technol.*, June, 968–970.

Zhou, 1993. Determining reservoir properties in reservoir studies using a fuzzy neural network. *68th Ann. Tech. Meet.*, Houston, TX, Oct. 3–6, SPE Paper 26430, .

Developments in Petroleum Science, 51
Editors: M. Nikravesh, F. Aminzadeh and L.A. Zadeh

Chapter 21

ENHANCING GAS STORAGE WELLS DELIVERABILITY USING INTELLIGENT SYSTEMS

SHAHAB D. MOHAGHEGH [1]

West Virginia University, 345E Mineral Resources Building, Morgantown, WV 26506, USA

1. INTRODUCTION

Gas storage fields have numerous wells that are used for both injection during low demand periods and withdrawal during high demand periods. As these wells age, their deliverability declines due to several factors. Stimulation treatments (hydraulic fracturing of the formation) are routinely used in gas industry to improve gas well productivity.

This study was conducted on a large natural gas storage field located in Northeastern Ohio. The formation is tight gas sandstone and is called the Clinton Sand. All of the storage wells were initially stimulated by hydraulic fracturing. Restimulation is considered a last resort method of deliverability enhancement in this storage field. However, some wells are selected to be restimulated each year based on maintenance history, past fracture response, years since previous stimulation and overall deliverability potential. Since 1970, an average of twenty-five wells have been refractured (restimulated) each year for a total of around 600 refracturing treatments. Since most wells in the field have been refractured (restimulated), some up to three times, the need for post stimulation well performance estimates and optimal fracture design is very important to maximize deliverability gains. The experience with the Clinton Sandstone indicates that hydraulic fractures grow vertically out of the zone, regardless of rate and fluid viscosity. Therefore, it appears critical to use high proppant concentrations in a viscous fluid to create a conductive fracture in the pay interval. Treatment designs for the storage field currently include a 25 to 30 pound linear gel with maximum sand concentrations from 3 to 4 pounds per gallon (ppg) (McVay et al., 1994).

Several well testing methods are available for predicting hydraulically fractured well performance including type curve matching and computer simulation (Millheim and Cichowicz, 1968; Gringarten et al., 1975; Cinco-Ley et al., 1978; Agarwal et al., 1979; Hopkins and Gatens, 1991). In addition, two- and three-dimensional computer simulators are frequently used for fracture design. Use of these tools, however, requires access to several types of reservoir data. Reservoir data necessary for hydraulic fracture simulation include porosity, permeability, thickness and stress profiles of the formation. Experience has shown that given the aforementioned data and assuming availability of a good geologic and structural definition of the reservoir, hydraulic fracturing simulators can predict the outcome of the hydraulic fracturing process with reasonable accuracy.

[1] Tel.: +1 (304) 293-7682 ext. 3405, fax: (304) 293-5708, E-mail: shahab@wvu.edu

When dealing with storage fields that are old (this is true for most of the storage fields since they are usually old, depleted gas fields that have been converted to storage fields), the aforementioned information is not available. Acquiring these types of information on an old reservoir is usually very expensive. It involves massive coring of the reservoir, where pieces of the rock from the target formation are brought to the surface and tested in the laboratory under simulated field conditions to measure the reservoir's porosity and permeability. It also involves elaborate well testing procedures and subsequent analysis of well test data. This article introduces a new and novel method for predicting the outcome of hydraulic fracture treatments in gas storage fields, with minimal cost.

Another important factor that must be considered is that fundamentally different stimulation jobs such as refracturing versus chemical treatments have been historically practiced in the Clinton Sandstone. Each of these restimulation jobs must be treated differently during the model building process. Moreover, economic considerations play an important role in restimulation projects.

During a stimulation/restimulation program the engineers face several challenging questions. The hydraulic fractures cost four to five times as much as a chemical treatment, and yet some wells respond reasonably well to chemical treatments. Given the economic parameters involved, should a well be refractured or chemically treated? What would be the maximum potential post-treatment deliverability if the wells were refractured as oppose to chemically treated? Would the decline behavior be different? Would extra cost of the refrac job justify the extra deliverability gains? These are not simple questions to be answered. Considering the fact that every year the engineers must select a handful of wells for restimulation from a total of more than 700 wells emphasizes the complexity of the problem.

In order to address this problem and expect reasonable result it is obvious that many factors must be taken into account. These factors include the history of the well. How it has responded to different hydraulic fractures and refrac processes in the past? Have chemical treatments been performed on the well? If yes, then how did the well responded to those treatments? If the well has been through several fracs, refracs and chemical treatments, do the sequence of these jobs have any significance on the post-treatment deliverability? Has the decline in post-treatment deliverability been sharper in the case of refracs or chemical treatments? These and many other technical questions may be posed.

In addition to the above technical questions many economical considerations also need to be addresses. It is a fact that refracs cost much more than chemical treatments yet many wells have shown that a well-designed and implemented chemical treatment may provide the same kind of post-treatment deliverability. Economic parameters other than the cost of the treatment may include the price of the gas and the total budget for the year's stimulation/restimulation program.

The objective of this study is to provide a methodology – and build a software tool based on this methodology – to address the above questions. The ultimate output of the software tool is a list of the restimulation candidates for each year. The list will contain the selected candidates and specifies whether that particular candidate should be refractured or chemically treated. In either case the software tool would provide recommendation on the parameters used in the refrac or the number and amount of chemical used for the chemical treatment.

It is not hard to see that the problem that has been described here is one of process modeling and optimization, and a challenging one. The software tool will take into account all the economic as well as technical concerns that were mentioned here through the use of virtual intelligence techniques. In a nut shell, virtual intelligence – also known as computational intelligence and soft computing – is an attempt to mimic life in solving highly complex and non-linear problems that are either impossible or unfeasible to solve using conventional methods.

In this study author uses a series of artificial neural networks and genetic algorithm routines, integrated with an extensive relational database – specifically developed for this study – to achieve the goals of the project. Since introductory discussions about neural networks and genetic algorithms have been published in the many previous SPE papers by the authors (Mohaghegh et al., 1996a,b; 1997) and other researchers in this area, further discussion on the nature of these sciences will not be included here.

2. METHODOLOGY

Fig. 1 is a schematic diagram of the flow of the information through the software application that was developed for this study. As it is shown in this figure the input data that resides in a relational database is fed into the application. The input data includes general well information, such as well ID number, well location, and some wellbore characteristics, some historical deliverability indicators such as pre-treatment

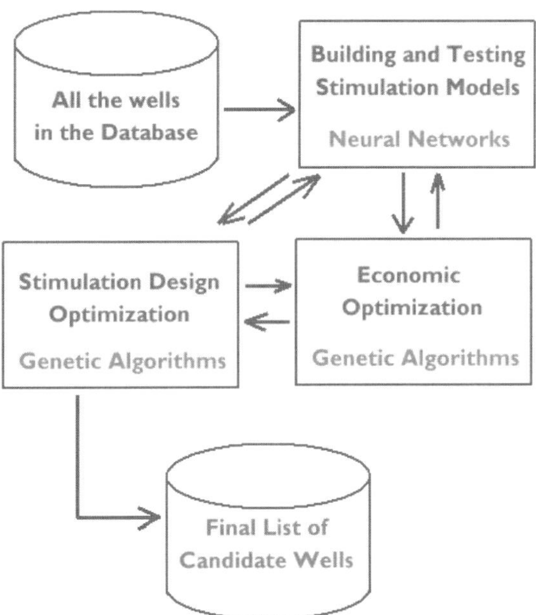

Fig. 1. Flow chart of the process used to build the software application.

deliverability, maximum and minimum deliverability for the life of the well and average deliverability for the past twenty years, as well as stimulation parameters.

It is worthwhile to mention that an extensive relational database was created for this particular gas storage field. This database is used as input data source for the application as well as a valuable source during the training of the several neural networks that are used as the main engines of the application. The database also provides an excellent manual rapid screening tool for the engineers. Several useful queries are built into the database in order to simplify visualization of the extensive amount of data that reside in the database.

The software application includes three separate modules. The first module includes the rapid screening neural networks – one network for the refracs and one for the chemical treatments. These networks use general well information and the historical data as input and attempts to predict the post-treatment deliverability. The rapid screening module works as follows: The user specifies a minimum post-deliverability (threshold). The rapid screening module – using the two neural networks – will identify all the wells that have the potential to meet this minimum. These are the wells that are used into the next modules.

To construct the second module, four neural networks – one for refracs and three for chemical treatments – are trained to work as fitness functions for the optimization process using genetic algorithms. The input data into these neural networks include the same data as rapid screening networks plus detail stimulation data. Second module includes four genetic algorithm routines, one for each neural network. Working in the batch mode this module optimizes all possible stimulation treatments for each well and ranks them. The user at this time has the option to inspect the results one well at a time or he/she may continue with the batch mode. Fig. 2 provides a schematic diagram of the module two.

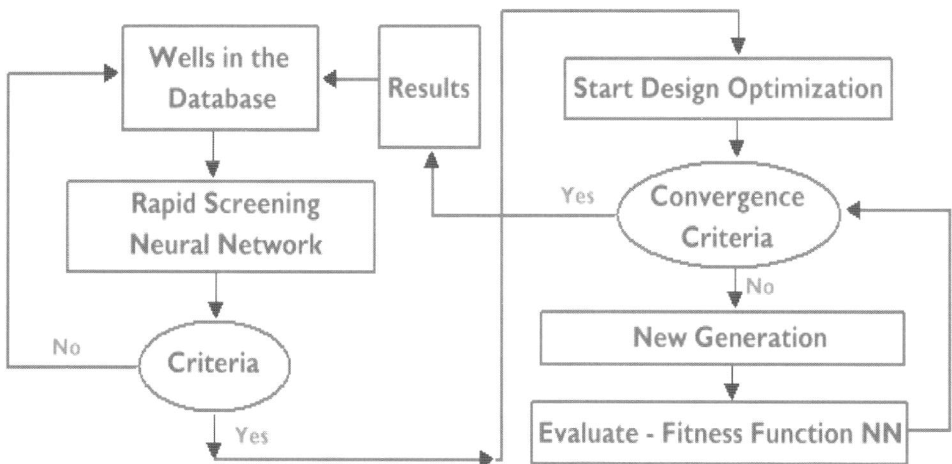

Fig. 2. The schematic diagram for the module two process.

In the batch mode, after completion of the second module the third module takes over. This module is a specialized genetic algorithm routine that uses the provided economic parameters to optimize the return on investment. The outcome of this module is a list of candidate wells and the type of the stimulation job that should be performed on each well.

During the second module design optimization is applied to a number of wells that have passed the rapid screening process. At this point the best candidates are identified. But there still remains one question to be answered. Given the cost of frac jobs and chemical treatments – which are inputs to the third module as economic parameters – what is the optimum combination of frac jobs and chemical treatments to be performed? And which wells are to be fractured or treated, in order to maximize the return on the investment?

Module three is an attempt to answer this question. The economic optimization genetic algorithm in this module is designed to provide the optimum list of candidate wells and the corresponding stimulation jobs that should be performed on each well. The function to be optimized in this module is what we call the profit function. As can be seen from Eq. (1), the profit function at this time is a simplified function. Increasing the complexity of this function will not in any shape or form change the applicability of this module. The profit function to be optimized is given by the following equation:

$$P = \$g \times \left[\sum_{1}^{n_{\mathrm{F}}} \Delta Q_{\mathrm{F}} + \sum_{1}^{n_{\mathrm{T}}} \Delta Q_{\mathrm{T}}\right] - \left[n_{\mathrm{F}} \times \$F + n_{\mathrm{T}} \times \$T\right] \tag{1}$$

where P = profit; $\$g$ = market gas price; ΔQ_{F} = total deliverability increase due to applied frac jobs; ΔQ_{T} = total deliverability increase due to applied chemical treatments; n_{F} = number of frac jobs performed; n_{T} = number of chemical treatments performed; $\$F$ = average cost of a frac job; and $\$T$ = average cost of a chemical treatment

In the above equation the only variables are the numbers of frac jobs and chemical treatments to be performed. The goal is to find the optimum combination of these two values such that it would maximize P. There is, however, a constraint that has to be imposed, namely the capital investment available for each stimulation program, i.e. annual stimulation budget. Eq. (2) provides the constrain:

$$\left[n_{\mathrm{F}} \times \$F + n_{\mathrm{T}} \times \$T\right] \leqslant \$Total \tag{2}$$

where $\$Total$ = annual stimulation program budget.

While the above equation provides assurance that we are not overspending, Eq. (3) is imposed as another constrain so the budget is spent in a fashion that most of it is used:

$$\$Total - \left[n_{\mathrm{F}} \times \$F + n_{\mathrm{T}} \times \$T\right] = \min \tag{3}$$

In module three the genetic algorithm maximizes the profit function – as the fitness function – using the two constrain equations. It should be noted that during this optimization process the genetic algorithm uses the list of the wells that have already gone through the stimulation optimization process in module two.

The total number of candidate wells and their optimum stimulation job are identified at this point. The data is stored in a database and can be viewed. A recommended set

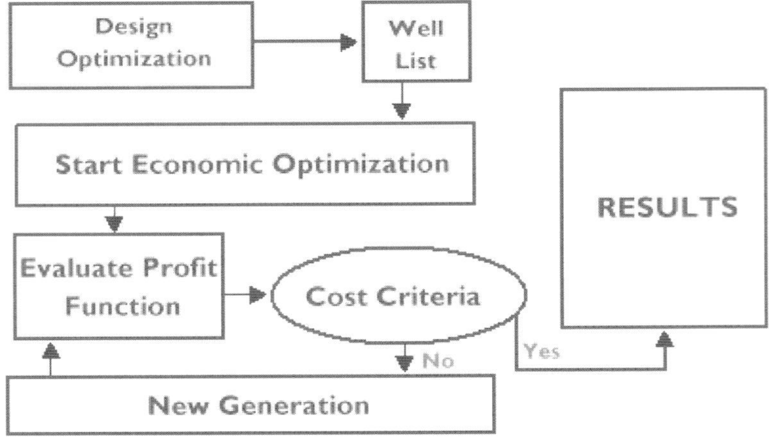

Fig. 3. The schematic diagram for the module three process.

of parameters for the frac jobs or the chemical treatments – whatever the case may be – accompanies the candidate. Fig. 3 shows a schematic diagram of the module three.

3. RESULTS AND DISCUSSION

In this section the results of neural model built for the refracturing process is presented first. The application of the genetic algorithms to optimize the refracturing design will follow the first set of results. The quality of the neural model that has been constructed and trained for the chemical treatment will then be presented.

For the refracturing process, available data covers the 1968–1991 period. Table 1 is the list of parameters that was available for all the hydraulic fracturing jobs. The data has been divided into two parts, input and output. The input, as can be seen from Table 1

TABLE 1

Data used in this study

Data type	Parameters
INPUT	Well number, Year well was fractured, Date well was fractured (A relative pressure function), Number of times well was fractured, Type of fracture (water, gelled water, foam), Fluid viscosity, Total water used, Nitrogen used per barrel of water, Total sand used, Sand Concentration, Sand type, Acid volume and type, Chemicals (Iron control, bacterial control...), Treatment injection rate, Occurrence of screen out, Contractor, Hole size, Completion type, Well type (Injection, Production), Date of completion, Date converted to storage, Well group number, Sand thickness, Minimum 20 year flow test value, Maximum 20 year flow test value, Average 20 year flow test value, and Flow test before refracture.
OUTPUT	Maximum Post-Fracture Deliverability

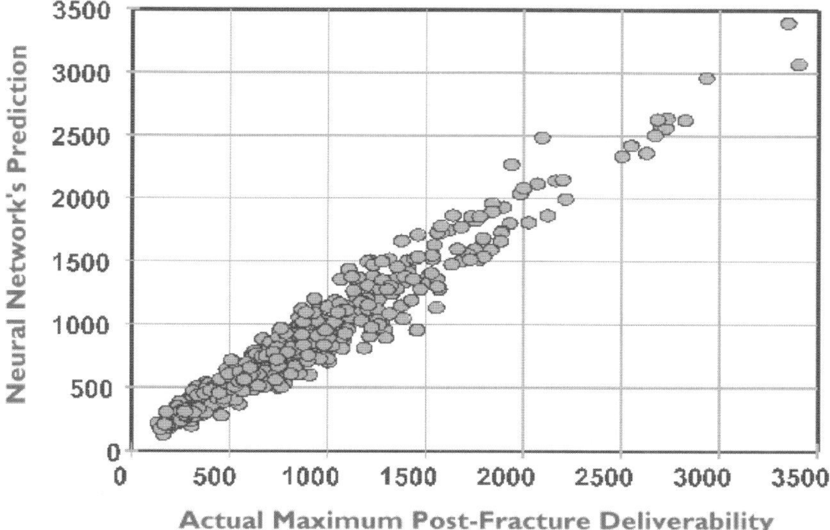

Fig. 4. Neural network's prediction for all the available data.

contains all the parameters but the one parameter that has been designated as the output. As a first attempt, 13% of the entire data were randomly selected and put aside as the test set and the network was trained on the remaining 87% of the data. Once the network reached convergence, network's predictions were compared against actual field results. This comparison provided a coefficient of correlation of 0.97 for training data and 0.98 for the test set. Fig. 4 shows the result of this part of the study for the entire data.

In order to simulate a practical situation, the data from 1968 to the end of 1988 were used to train the network. The input–output pairs were presented to the network for training. Once the network reached a stable state, the network's prediction on the training data was compared to the output that was provided to the network. It produced a coefficient of correlation of above 0.96. To test the generalization capabilities of the developed network, the input data from the years 1989, 1990, and 1991 were given to the network to forecast the post-fracture well deliverability. This would be the result upon which candidate wells for fracturing would be selected. Figs. 5–7 show the comparison between network's predictions with actual flow test data. The result is quite satisfactory. Using flow test indicator of 500 as the cut off point, the figures show that a total of 9 wells (five wells from 1988, 1 well from 1990, and 3 wells from 1991) would not have been chosen for refracturing if this tool was available at that time. Using this tool, the cost involved in fracturing these 9 wells would have been saved and also 9 other wells with acceptable post-fracture deliverability would have been fractured, which means even more economic benefits. The time for a well to reach its peak deliverability after a hydraulic fracture (the indicator that is the target of this study) is between two to three years. At the time this study was being completed, results from 1991 were the most recent peak post-fracture deliverabilities that were available. At a later time, the

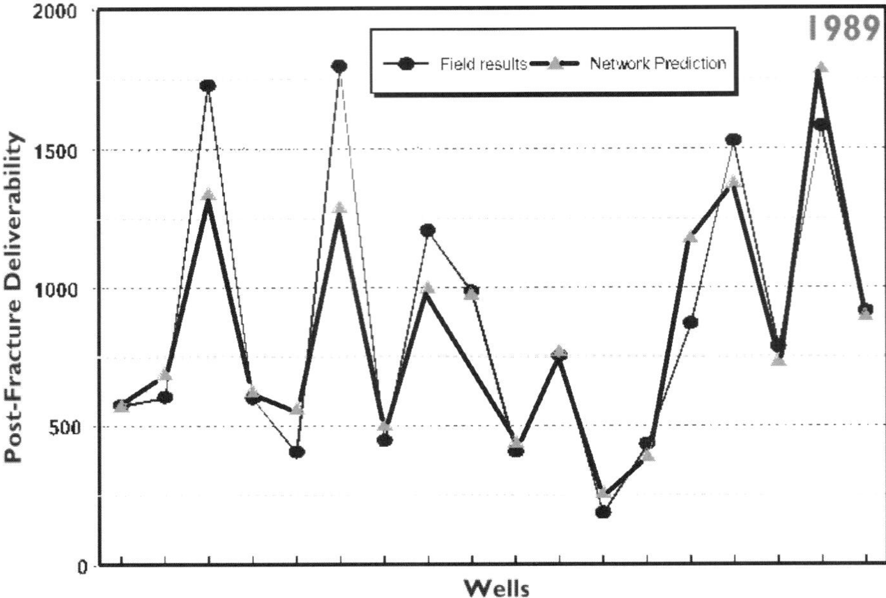

Fig. 5. Comparison of network's prediction and actual field results for the year 1989. Network was trained using data from 1968 to 1988.

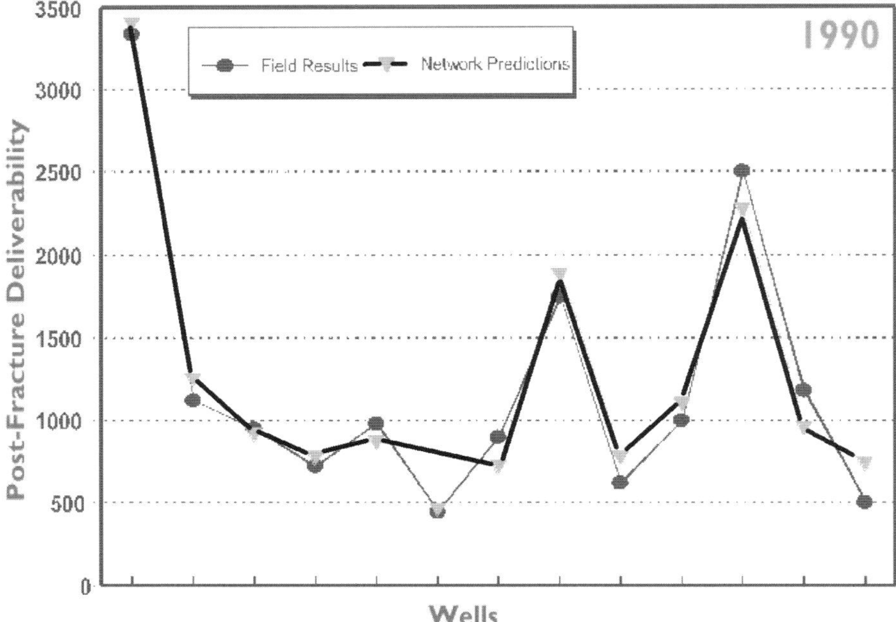

Fig. 6. Comparison of network's prediction and actual field results for the year 1990. Network was trained using data from 1968 to 1988.

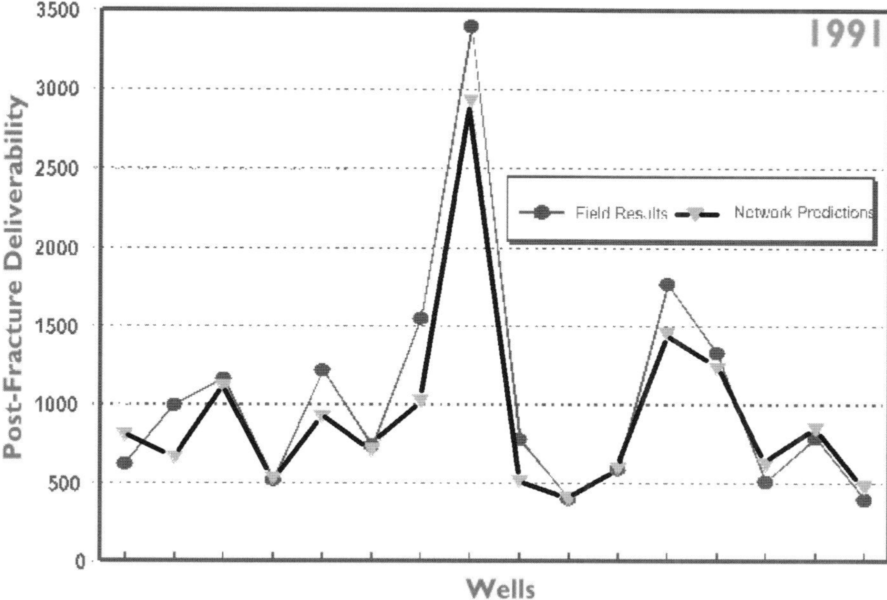

Fig. 7. Comparison of network's prediction and actual field results for the year 1991.Network was trained using data from 1968 to 1988.

developed network was used to predict the peak post-fracture deliverabilities for the wells stimulated in 1992. During 1992, nineteen wells were stimulated. By the study was being completed, results from 11 wells had become available. Fig. 8 shows the comparison between network's prediction and actual field results for the year 1992. As can be seen, network made quite accurate predictions for all but one well, which is the first well in Fig. 8. For this well, neural network predicted a post-fracture deliverability of 1400 mscfd, while the actual deliverability peaked at about 900 mscfd. Since 500 mscfd was used as the cut-off point, neural network's prediction (1400 mscfd) would have suggested that hydraulic fracturing be performed on this well. In retrospect, this would have been a good suggestion since the actual deliverability was above 500 mscfd.

In a separate attempt to demonstrate the power and robustness of this new methodology, the network was trained with data from 1968 to 1974. The coefficient of correlation at this point was almost 0.98. In 1975, a new fracturing fluid was used for the first time (foam). When data from 1975 was introduced to network, the performance of the network degraded and its prediction accuracy dropped to 0.88. This performance bounced back up by the year 1980, when network observed and learned the new behavior that was displayed by the new fracturing fluid. This process was repeated two more times, when new fluids were introduces in 1981 and 1989. Fig. 9 shows the neural network's prediction capabilities as new information is added to the network. This further proves the notion that systems developed, based on neural network, do not break down when new situations are encountered, rather, they degrade gracefully.

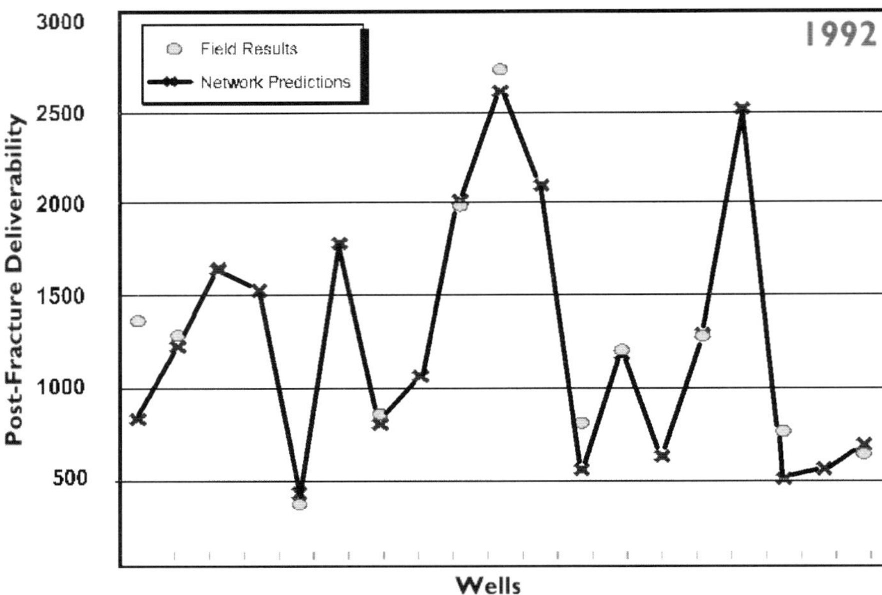

Fig. 8. Comparison of network's prediction and actual field results for the year 1992. Network was trained using data from 1968 to 1988.

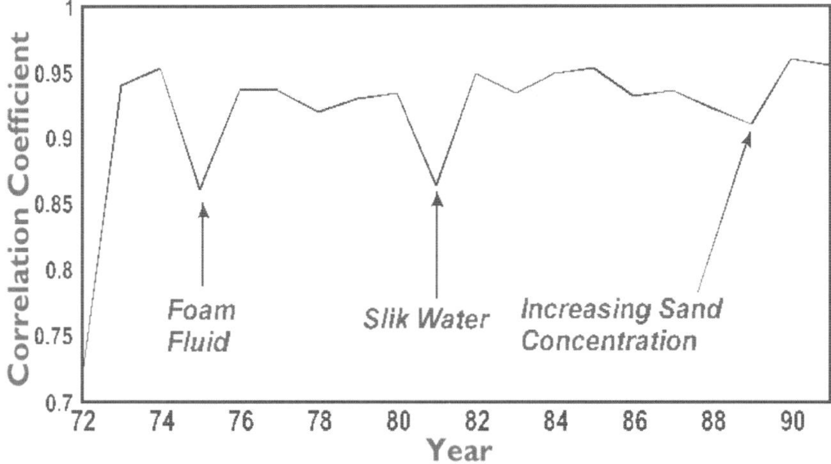

Fig. 9. Neural networks robustness demonstrated by its reaction to addition of new information to the dataset.

It is also important to note that the so-called 'conventional engineering wisdom' (whatever it may mean) about the available data may not be quite applicable here. In other words, a piece of data that might look very unimportant in terms of its information content about the reservoir, the well or the fracturing process, may actually contain valuable *implicit* information useful for the neural network. An example on our experience may clarify this point. During our analysis, it was noted that the well ID number played a role in the overall pattern that has been established between inputs and the post-fracture deliverability. It prompted us to look further into the conventions that might have been used in numbering these wells. It was later determined that these wells were numbered according to (a) their date of completion, and (b) their relative geographic location in the field. Although this information was not explicit and was unknown to us at the time of analysis, the network was able to deduce it from the data. It was also interesting to note that, although no information regarding the physics of the problem was provided to the network during the training, once the network was trained, it provided us with information that made perfect engineering sense (McVay et al., 1994).

3.1. Genetic optimization

Once the neural model for the hydraulic fracturing was constructed and tested and it was concluded that its performance is satisfactory, the next step was to incorporate this neural model into a genetic optimization routine to identify optimum refracturing design. This neural network (neural module #2) would be the fitness function for the genetic algorithms. A two-stage process is now developed to optimize the frac design in Clinton Sandstone. A detail, step by step procedure will be covered in the following section. Fig. 10 presents a schematic diagram of the procedure. For the first stage a new neural network (neural module #1) is designed and trained. As it was mentioned earlier this neural network is not given any information on the frac design parameters. The only data available to this neural net is basic well information and production history. After all this will be all the information that will be available in each well that is being considered for a frac job. This neural network is trained to accept the aforementioned information as input data and estimate a post-frac deliverability as output. The post-frac deliverability predicted by this neural net is the same as an average (generic) frac job within a certain degree of accuracy.

This neural net is used only as a screening tool. It will identify and put aside the so-called 'dog wells' that would not be enhanced considerably even after a frac job. The wells that have passed the screening test will enter the second stage that is the actual frac design stage. A second neural net (neural module #2) has been trained for this stage. This neural net has been trained with more than 570 different frac jobs that have been performed on Clinton Sandstone. This network is capable of providing post-frac deliverability with high accuracy given well information, historical data and frac design parameters. This neural net will play the role of fitness function or the environment in the genetic algorithm part of the methodology. Fig. 11 is an elaboration on how this neural network is being used in conjunction with the genetic algorithm. The output of the genetic algorithm portion of this methodology is the optimized frac design for each well. The tool will also provide the engineer with expected post-frac deliverability once the suggested design is used for a frac treatment. This result may be saved and printed. The design parameters can then be given to any service company for implementation.

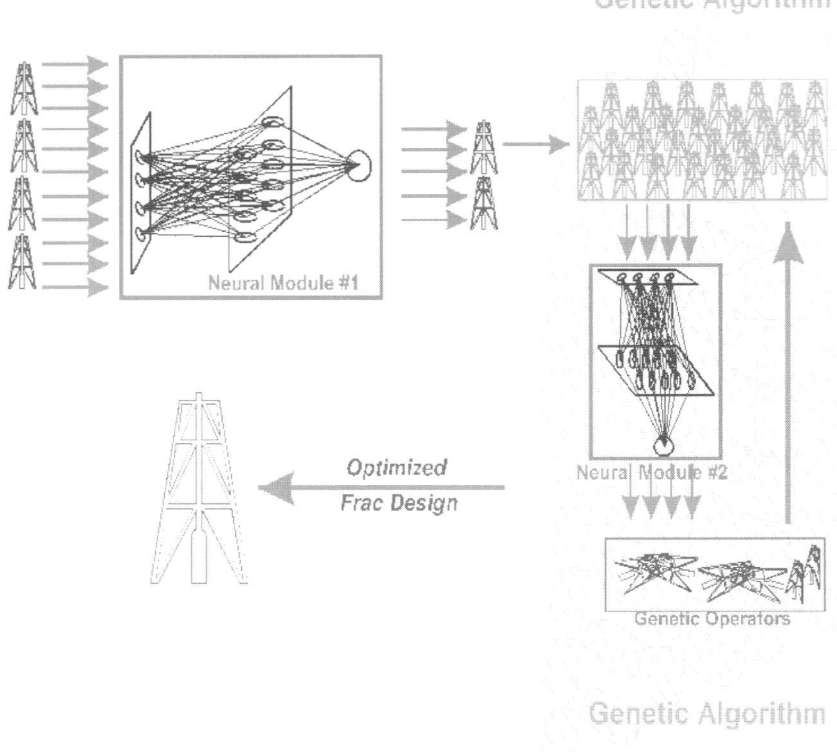

Fig. 10. Schematic diagram of the genetic optimization procedure.

3.2. Procedure

The well selection and hydraulic fracture design take place in two stages:

3.2.1. Stage 1: Screening

In this stage a criteria is set for screening the candidate wells. Neural module #1 that has been trained on well completion and production history is used to screen the candidate wells, and selects those wells that meet a certain post-frac deliverability, set by design engineer as threshold. In other words, well completion and production history for all candidate wells are provided to the software with a threshold value for post-frac deliverability. Those wells that meet or exceed the threshold will be identified and prepared for further analysis and hydraulic fracture design. A preliminary post-frac deliverability for each well will be calculated and displayed. The post-frac deliverability that is presented at this stage is what is expected if a generic frac is designed for this well, i.e. with no optimization.

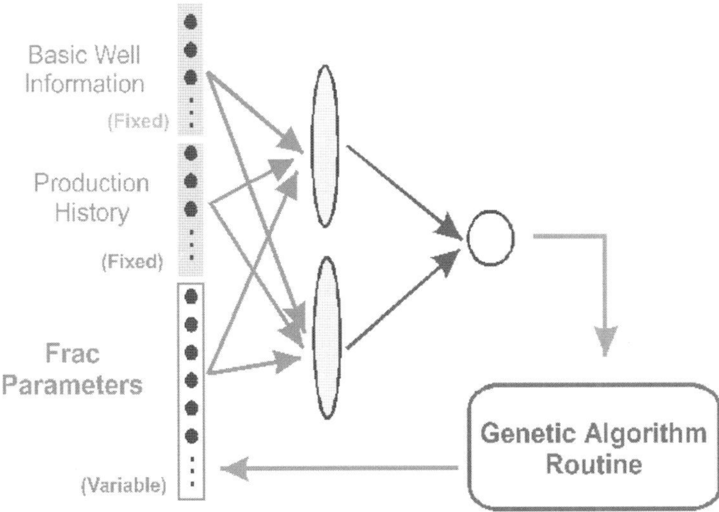

Fig. 11. Schematic diagram of the neural module #2.

It must be noted that if actual threshold is for example 500 mcfd, then 400 mcfd should be used at this point. This is due to the fact that optimization process has an average post-frac deliverability enhancement of 20% in this field.

At this point design engineer is presented with a list of candidate wells that meet and or exceed the post-frac deliverability threshold set previously. He/she will select one well at a time and enters the second stage for optimization.

3.3. Stage 2: Optimization

In this stage following steps will be taken:

Step 1: One out of four frac fluids (water, gel, foam, foam and water) is selected. Please note that these four frac procedures were chosen because they have been routinely performed in the aforementioned field in the past.

Step 2: One hundred random combinations of input variables (frac parameters) are generated. This is called the original population.

Step 3: Neural module #2 that has been proven to have higher than 95% accuracy in predicting post-frac deliverability for this particular field is used to forecast post-frac deliverability for 100 cases generated in step 1.

Step 4: The outcome of neural module #2 will be ranked from 1 to 100, 1 being the highest post-frac deliverability.

Step 5: The highest-ranking frac parameters combination (design) is compared with the last highest-ranking design and the better of the two is saved in the memory as optimum design.

Step 6: Top 25 designs of step 4 will be selected for the next step and rest will be discarded.

Step 7: Crossover, mutation, and inversion operators are used on the top 25 designs of step 6 and a new population of 100 designs is generated.
Step 8: Procedure is repeated from step 3.

In order to demonstrate the application of this optimization methodology it was decided to perform design optimization on wells that were treated during 1989, 1990, and 1991. Since the actual results of frac treatments on these wells were available, it would provide a good comparison. We used the software to

(a) Predict the frac treatment results (please be reminded that these results were not seen by the software in advance and they are as new to the software as any other set of input values) and compare it with the actual field results, and
(b) See how much enhancement would have been made if this software were used to design the treatment.

Neural module #2 in the software is responsible for prediction of output (frac treatment results) from new sets of input data (frac designs for particular wells). It would be reasonable to expect that if this module predicts frac treatment results within a certain degree of accuracy for one set of the input values, it should predict the results of another set of input values approximately within the same degree of accuracy.

Figs. 12–14 show the results of this demonstration. In these figures actual field results are shown (Field Results) as well as software's prediction (Predicted). It is obvious that the software does a fine job predicting frac treatment results from frac design parameters, however this had already been established. Frac treatment parameters that have been generated by the software itself using the combined neuro-genetic procedure resulted

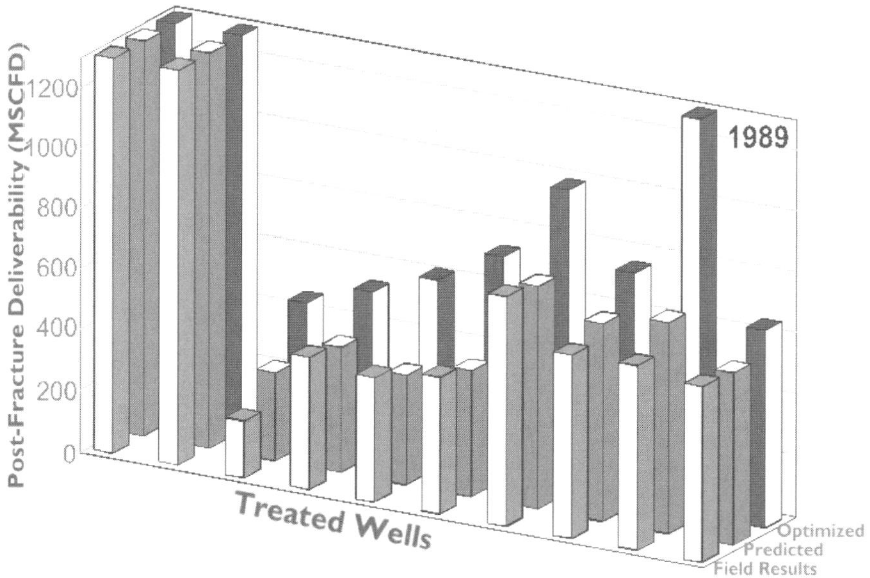

Fig. 12. Enhancement that could have been achieved on wells treated in 1989.

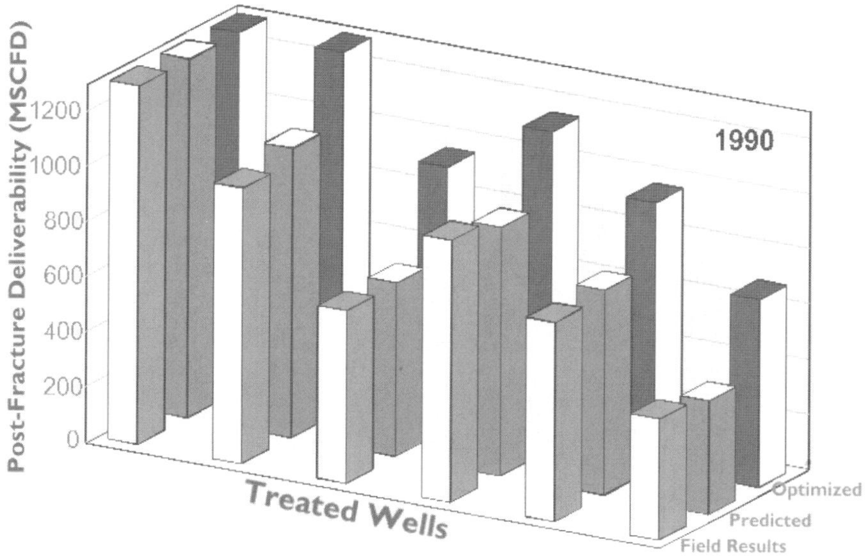

Fig. 13. Enhancement that could have been achieved on wells treated in 1990.

in the frac treatment results shown in the figures as 'Optimized'. Again, please note that the same module in the software that has produced the triangles has produced the crosses, and in both cases from new set of input data (new to the module).

From these figures it can be seen that by using this software to design a frac treatment for this field, one can enhance treatment results by an average of 20 percent. It should also be noted that these wells were chosen from among 100 candidate wells each year. If this software was available at the time the selection process a different set of wells might have been used as restimulation candidates.

Table 2 shows the result of this process on particular well. Well #1166 was treated and its post-frac deliverability was determined to be 918 mcfd. The software predicted this well's post-frac deliverability to be 968.6 mcfd, which is within 5.5% of the actual value. Using the neuro-genetic optimization process introduced here the software predicts a post-frac deliverability of 1507.5 mcfd. Using the 5.5% tolerance for the software's

TABLE 2

Optimization results from well #1166

Well number	1166
Actual, mscfd	918
Prediction, mscfd	968.6
Percent difference, %	5.5
After optimization, mscfd	1507.5
Within the 5.5% difference, mscfd	1590–1425
Enhancement, mscfd	672–507

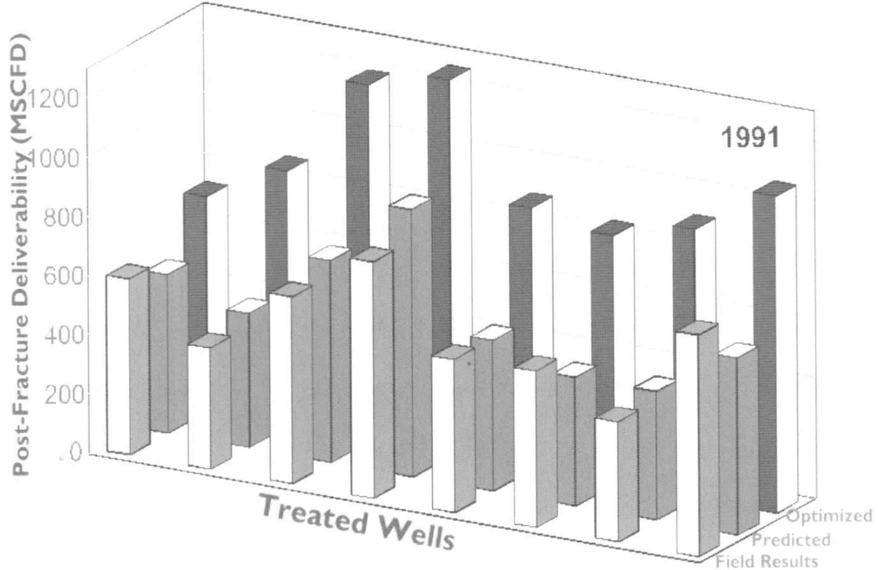

Fig. 14. Enhancement that could have been achieved on wells treated in 1990.

accuracy this methodology could have enhanced this well's post-frac deliverability by 55 to 73%.

3.4. Chemical treatments

As was mentioned before historical data in this field included many frac and refrac jobs as well as a variety of different chemical treatments. Upon a closer inspection of the data it was possible to classify the chemical treatments into three categories. The classification was made based on the number of chemicals used in the treatments. They were divided into one, two and three components chemical treatments. Table 3 shows the chemicals used in each category.

For chemical treatments, similar to the refracturing procedure, module one of the software application includes the rapid screening neural nets. These nets are constructed and trained to look at the general information of the well and the historical data to estimate a post-stimulation deliverability. The only information provided to the network about the stimulation job at this point is the type of the stimulation jobs i.e. refrac or chemical treatment.

A separate set of neural networks were constructed and trained for module two. These networks are trained using all available data that includes detail stimulation parameters. These are the networks that are used as fitness functions in the genetic algorithm routines.

Fig. 15 shows the accuracy of the module one neural networks for the chemical treatments. Figs. 16–18 are the plots of the actual post-treatment deliverabilities versus neural network predictions for the second module of the chemical treatment

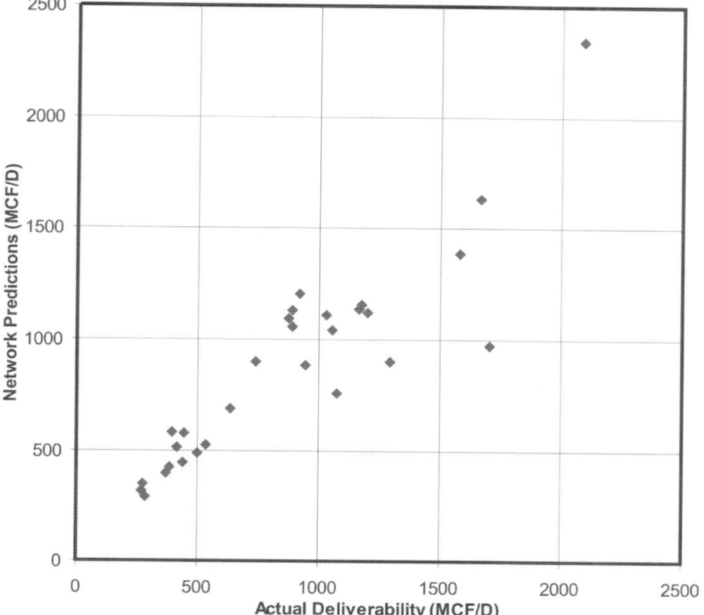

Fig. 15. Module one neural net for chemical treatments; rapid Screening.

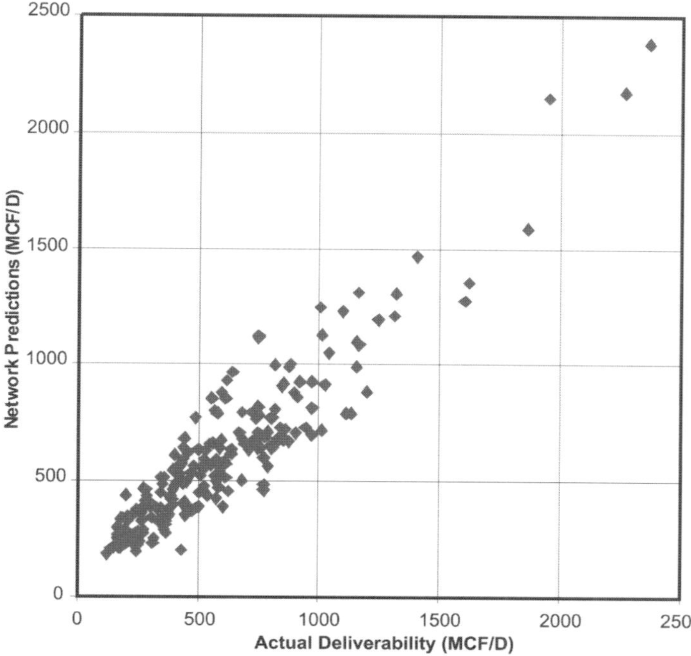

Fig. 16. Module two neural net for one component chemical treatments; optimization.

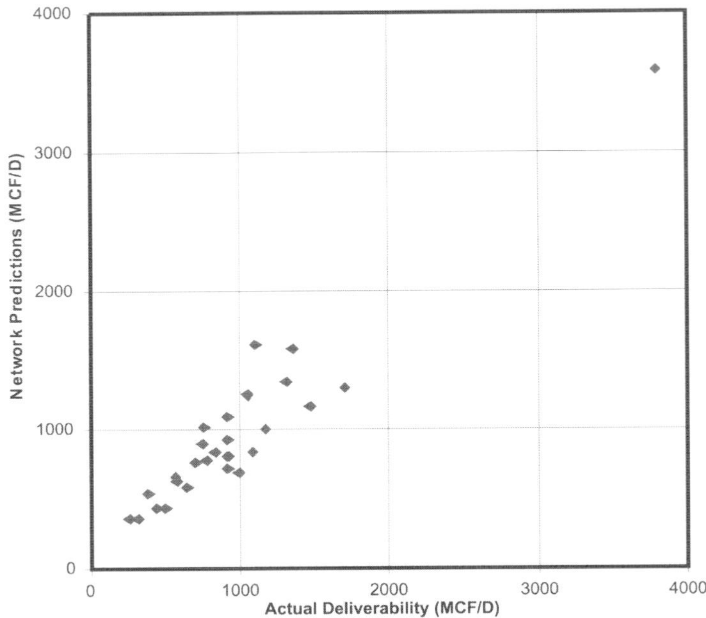

Fig. 17. Module two neural net for two component chemical treatments; optimization.

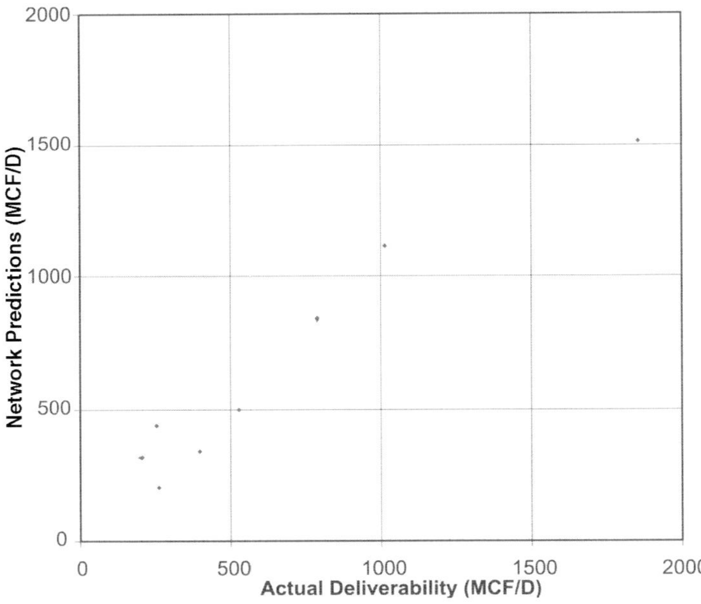

Fig. 18. Module two neural net for three component chemical treatments; optimization.

TABLE 3

Chemical treatment classification

Treatment type	Types of chemicals used	
One component		Kerosene Solvent Surfactant Water
Two components	Cold water-based	Paraffin dispersant PEN-5 VISCO 4750
	Hot water-based	B-11 Drill foam Nalco Paraffin dispersant PEN-5 Surflo S-24 Tretolite VISCO W-121
Three components	Acid-based	Methanol + water
	Water-based	Methanol + B-11 Methanol + SEM-7 Methanol + W-121

portion of the software application. Three different networks were trained for this module. Figs. 16–18 are graphs of network predictions versus actual post-treatment deliverabilities for one-, two- and three-component chemical treatments. These graphs show how well these networks have been trained.

To clearly demonstrate their generalization capabilities correlation coefficients for these neural networks are provided in Table 4. In this table two separate correlation coefficients are provided for each network, one correlation coefficient for the training data set and one for verification data set. The verification data set includes data that have

TABLE 4

Quality of the neural networks that were trained for this study

Neural Networks	Modules in the software application		Training set		Verification set	
			Cor. coef.	Rec.	Cor. coef.	Rec.
Chemical treatments	Rapid screening		96%	1830	92%	783
	Optimization	1 component	97%	370	91%	157
	Optimization	2 components	95%	1492	91%	637
	Optimization	3 components	97%	63	94%	25

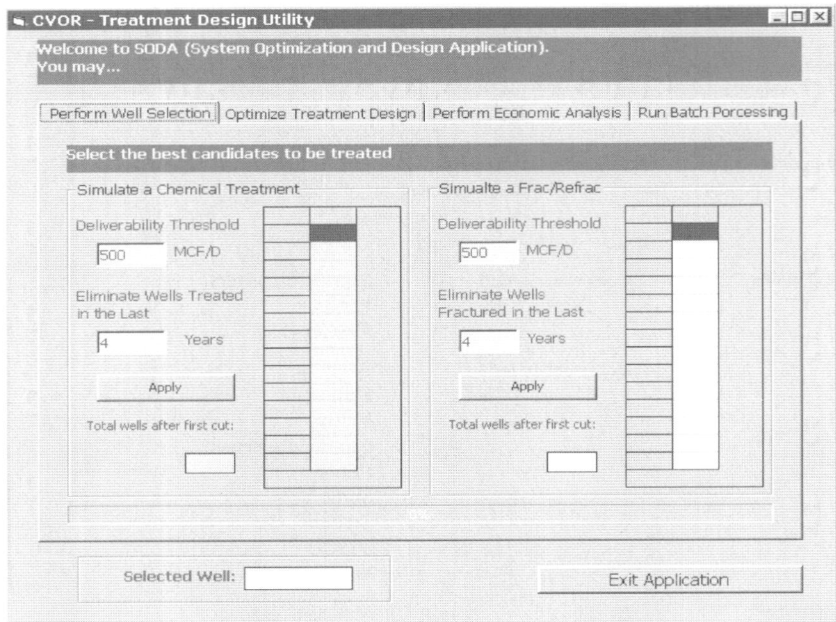

Fig. 19. Software application interface for module one.

not been used during the model construction and therefore the networks had not seen them before.

Figs. 19–21 show screen shots from the software application that was developed for this study.

4. APPLICATION TO OTHER FIELDS

This methodology can be applied not only to gas storage operation but to other types of operations as well. This is true as long as production history for some wells and results of some prior treatment are available. With some modifications this methodology can also be applied to new fields where no hydraulic fractures are performed in the past. It should be noted that in such cases (no prior frac jobs) it is necessary that some reservoir data be available. This data may be in the form of well logs with corresponding core data as well as some stress profiles from several wells in the fields (Cinco-Ley et al., 1978). The reason a specific number of wells are not suggested (for logs, cores and stress profiles) is due to the fact that it is a function of the size of the field under investigation.

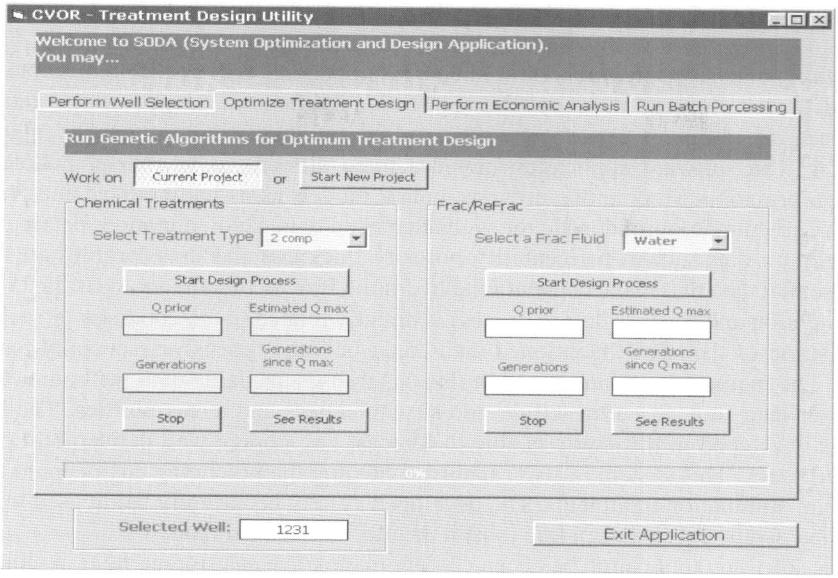

Fig. 20. Software application interface for module two.

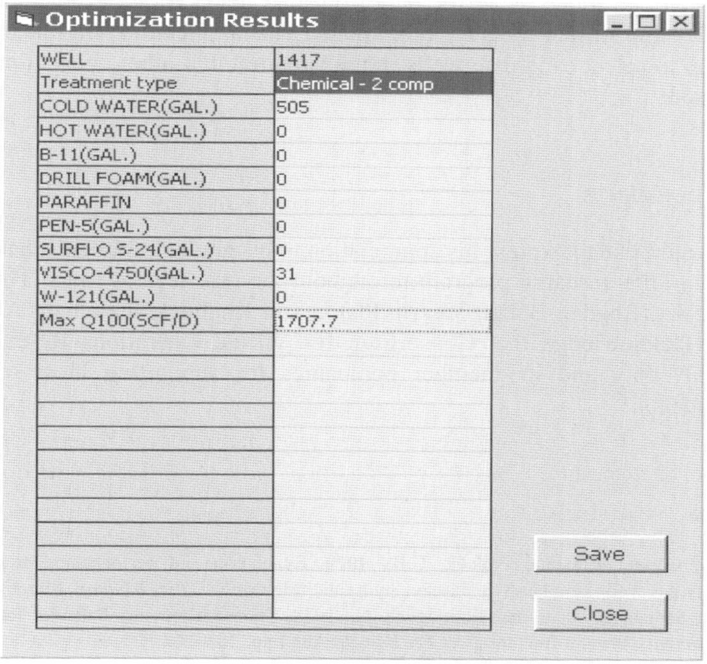

WELL	1417
Treatment type	Chemical - 2 comp
COLD WATER(GAL.)	505
HOT WATER(GAL.)	0
B-11(GAL.)	0
DRILL FOAM(GAL.)	0
PARAFFIN	0
PEN-5(GAL.)	0
SURFLO S-24(GAL.)	0
VISCO-4750(GAL.)	31
W-121(GAL.)	0
Max Q100(SCF/D)	1707.7

Fig. 21. Software application interface.

5. CONCLUSIONS

A comprehensive software tool has been developed that will assist engineers to select candidate wells for restimulation. The application has been developed using a relational database, six different neural network models and five different genetic algorithm routines. The software application includes three different independent modules that share information. Module one uses two neural models as its main engine and provides a rapid screening tool to identify the wells that need to be studied in more detail.

Reservoir data such as permeability, porosity, thickness and stress profiles are among the essential data that make conventional hydraulic fracture simulation possible. Success of simulation and fracture design process is directly related to the goodness of such data. Acquisition of the above mentioned data could be very expensive especially for older fields. The methodology introduced in this paper, uses available data, without access to reservoir data such as permeability, porosity, thickness and stress profiles. The hybrid system developed in this study is able to forecast gas storage well deliverabilities with high accuracy. This system is also capable of helping the practicing engineers to design optimum stimulation treatments. The developed system is currently being used to select candidate wells and to design frac jobs in the aforementioned field.

This software application has been custom made for a gas storage field in Ohio. The customization of the application is related to the neural network models and the genetic algorithm routines. These models and routines are specific to this storage field since they have been developed using the data from this field. The same methodology may be used to develop similar tools for other fields. This application will make it easier for the engineers to select candidate wells in the situation that other conventional methods cannot be used.

ACKNOWLEDGEMENTS

Author would like to express his appreciation to his graduate students that contributed significantly to this research program throughout the past several years. These students are Dan McVey, Bogdan Balan, Valeriu Platon and Khalid Mohammad. Author would also like to acknowledge that Consortium for Virtual Operations Research at West Virginia University and its member companies for supporting most of the above graduate students.

REFERENCES

Agarwal, R.G. Carter, R.D. and Pollock, C.B., 1979. Evaluation and performance prediction of low permeability gas wells simulated by massive hydraulic fracturing. *J. Pet. Technol.*, March, pp. 362–372.

Cinco-Ley, H., Samaniego-V., F. and Dominguez, N., 1978. Transient pressure behavior for a well with a finite conductivity vertical fracture. *Soc. Pet Eng. J.*, Aug., pp. 253–264.

Gringarten, A.C., Ramey, H.J., Jr. and Raghavan, R., 1975. Applied pressure analysis for fractured wells. *J. Pet. Technol.*, July, pp. 887–892.

Hopkins, C.W. and Gatens, J.M., 1991. Stimulation optimization in a low-permeability, Upper Devonian sandstone reservoir: a case history. *SPE Gas Technology Symposium*, Houston, TX, Jan. 23–25, SPE Paper 21499.

McVay, D., Mohaghegh, S. and Aminian, K., 1994. Identification of parameters influencing the response of gas storage wells to hydraulic fracturing with the aid of a neural network. *Proc. of 1994 SPE Eastern Regional Conference and Exhibition*, Charleston, WV, SPE Paper 29159.

Millheim, K.K. and Cichowicz, L., 1968. Testing and analyzing low-permeability fractured gas wells. *J. Pet. Technol.*, Feb., pp. 193–198.

Mohaghegh, S., Aminian, K., Ameri, S. and McVey, D.S., 1995. Predicting well stimulation results in a gas storage field in the absence of reservoir data, using neural networks. *Soc. Pet. Eng.*, SPE Paper 31159.

Mohaghegh, S., Arefi, R. and Ameri, S., 1996a. Petroleum reservoir characterization with the aid of artificial neural networks. *J. Pet. Sci. Eng.*, 16: 263–274.

Mohaghegh, S., Arefi, R. and Ameri, S., 1996b. Virtual measurement of heterogeneous formation permeability using geophysical well log responses. *Log Analyst*, March–April, pp. 32–39.

Mohaghegh, S., Arefi, R. and Ameri, S., 1997. Determination of permeability from well log data. *SPE Formation Eval. J.*, Sept., pp. 263–274.

PART 5.

INTEGRATED FIELD STUDIES

Developments in Petroleum Science, 51
Editors: M. Nikravesh, F. Aminzadeh and L.A. Zadeh

Chapter 22

SOFT COMPUTING: TOOLS FOR INTELLIGENT RESERVOIR CHARACTERIZATION AND OPTIMUM WELL PLACEMENT

MASOUD NIKRAVESH [a,b,1], ROY D. ADAMS [b] and RAYMOND A. LEVEY [b]

[a] *BISC Program Department of Electrical Engineering and Computer Sciences University of California, Berkeley, CA 94720, USA*
[b] *Zadeh Institute for Information Technology (ZIFIT)*
[c] *Energy and Geoscience Institute, University of Utah, Salt Lake City, Utah 84108, USA*

ABSTRACT

An integrated methodology has been developed to identify nonlinear relationships and mapping between 3D seismic data and production log data. This methodology has been applied to a producing field. The method uses conventional techniques such as geostatistical and classical pattern recognition in conjunction with modern techniques such as soft computing (neuro computing, fuzzy logic, genetic computing, and probabilistic reasoning). An important goal of our research is to use clustering techniques to recognize the optimal location of a new well based on 3D seismic data and available production-log data. The classification task was accomplished in three ways; (1) k-mean clustering, (2) fuzzy c-means clustering, and (3) neural network clustering to recognize similarity cubes. Relationships between each cluster and production-log data can be recognized around the well bore and the results used to reconstruct and extrapolate production-log data away from the well bore. This advanced technique for analysis and interpretation of 3D seismic and log data can be used to predict: (1) mapping between production data and seismic data, (2) reservoir connectivity based on multi-attribute analysis, (3) pay zone estimation, and (4) optimum well placement.

1. INTRODUCTION

In reservoir engineering, it is important to characterize how 3D seismic information is related to production, lithology, geology, and logs (e.g. porosity, density, gamma ray, etc.) (Aminzadeh and Chatterjee, 1984/85; Yoshioka et al., 1996; Boadu, 1997; Chawathe et al., 1997; Monson and Pita, 1997; Schuelke et al., 1997; Nikravesh, 1998a,b; Nikravesh et al., 1998). Knowledge of 3D seismic data will help to reconstruct the 3D volume of relevant reservoir information away from the well bore. However, data from well logs and 3D seismic attributes are often difficult to analyze because of their complexity and our limited ability to understand and use the intensive information content of these data. Unfortunately, only linear and simple nonlinear information can

[1] Fax: (510) 642-5775; E-mail: Nikravesh@cs.berkeley.edu

be extracted from these data by standard statistical methods such as ordinary least squares, partial least squares, and nonlinear quadratic partial least-squares. However, if *a priori* information regarding nonlinear input–output mapping is available, these methods become more useful.

Simple mathematical models may become inaccurate because several assumptions are made to simplify the models in order to solve the problem. On the other hand, complex models may become inaccurate if additional equations, involving a more or less approximate description of phenomena, are included. In most cases, these models require a number of parameters that are not physically measurable.

Neural networks (Hecht-Nielsen, 1989) and fuzzy logic (Zadeh, 1965) offer a third alternative and have the potential to establish a model from nonlinear, complex, and multi-dimensional data. They have found wide application in analyzing experimental, industrial, and field data (Baldwin et al., 1989, 1990; Rogers et al., 1992; Wong et al., 1995a,b; Nikravesh et al., 1996; Pezeshk et al., 1996; Nikravesh and Aminzadeh, 1997). In recent years, the utility of neural network and fuzzy logic analysis has stimulated growing interest among reservoir engineers, geologists, and geophysicists (Klimentos and McCann, 1990; Aminzadeh et al., 1994; Yoshioka et al., 1996; Boadu, 1997; Chawathe et al., 1997; Monson and Pita, 1997; Schuelke et al., 1997; Nikravesh, 1998a,b; Nikravesh and Aminzadeh, 1999; Nikravesh et al., 1998). Boadu (1997) and Nikravesh et al. (1998) applied artificial neural networks and neuro-fuzzy successfully to find relationships between seismic data and rock properties of sandstone. In a recent study, Nikravesh and Aminzadeh (1999), Nikravesh et al. (1998) and Nikravesh (1998b) used an artificial neural network to further analyze data published by Klimentos and McCann (1990) and analyzed by Boadu (1997). It was concluded that to find nonlinear relationships, a neural network model provides better performance than does a multiple linear regression model. Neural network, neuro-fuzzy, and knowledge-based models have been successfully used to model rock properties based on well log databases (Nikravesh, 1998b).

Monson and Pita (1997), Chawathe et al. (1997) and Nikravesh (1998b) applied artificial neural networks and neuro-fuzzy techniques successfully to find the relationships between 3D seismic attributes and well logs and to extrapolate mapping away from the well bore to reconstruct log responses.

In this study, we analyzed 3D seismic data to recognize the most important patterns, structures, relationships, and characteristics based on classical pattern-recognition techniques, neural networks and fuzzy logic models. Nonlinear mapping between production data and 3D seismic was identified. Finally, based on integrated clustering techniques, optimal locations to drill new wells were predicted.

1.1. Neural networks

During the last decade, application of neural networks with back propagation for modeling complex multi-dimensional field data has greatly increased (Rumelhart et al., 1986; Hecht-Nielsen, 1989; Widrow and Lehr, 1990; Aminzadeh et al., 1994; Horikawa et al., 1996) (see Appendix A). This widespread usage has been due to several attractive features of neural networks: (1) they do not require specification of structural relationships between input and output data; (2) they can extract and recognize underlying patterns, structures, and relationships between data; and (3) they can be used for parallel processing. However, developing a proper neural network model that is an 'accurate' representation of the data may be an arduous task that requires sufficient experience with the qualitative effects of structural parameters of neural network models, scaling techniques for input–output data, and a minimum insight into the physical behavior of the model. In addition, neural network models are frequently complex, need a large amount of precise data, and the underlying patterns and structure are not easily visible. Unlike statistical methods, conventional neural network models cannot deal with probability. In addition, conventional neural network models cannot deal with uncertainty in data due to fuzziness (see Appendix A).

A typical neural network has an input layer in which input data is presented to the network, an output layer in which output data is presented to the network (network prediction), and at least one hidden layer (see Appendix A). Several techniques have been proposed for training neural network models. The most common technique is the back propagation approach (Rumelhart et al., 1986; Hecht-Nielsen, 1989). The objective of the learning process is to minimize global error in the output nodes by adjusting the weights (see Appendix A). This minimization is usually set up as an optimization problem. The Levenberg–Marquardt algorithm (The Math Works™, 1995) can be used, which is faster and more robust than conventional algorithms but requires more memory.

1.2. Fuzzy logic

Zadeh (1965) first introduced the basic theory of fuzzy sets (see Appendix A). Unlike classical logic which is based on crisp sets who's members are either 'True' or 'False', fuzzy logic views problems as having a degree of 'Truth.' Fuzzy logic is based on the concept of fuzzy sets who's members may be 'True' or 'False' or any number of gradations between 'True' and 'False.' Another way of expressing this is that a member of a fuzzy set may have varying amounts of both 'True' and 'False.' In classical or crisp sets, the transition between membership and non-membership in a given set for an element in the universe is abrupt (crisp). For an element in a universe that contains fuzzy sets, this transition can be gradual rather than abrupt. Therefore, 'fuzzy' and 'fuzziness' can be defined as having the fuzzy set characteristic. Mapping and sets in fuzzy theory are described and characterized as membership functions. Membership functions assign to each element (object) a membership value between zero and one.

Fuzzy logic is considered to be appropriate to deal with the nature of uncertainty in human error and in systems, neither of which are included in current reliability theories. Despite the common meaning of the word 'fuzzy', fuzzy set theory does not permit vagueness. It is a methodology that was developed to obtain an approximate solution where problems are subject to vague descriptions. In addition, it can help engineers and researchers tackle uncertainty and handle imprecise information in a complex situation (Zadeh, 1973, 1976). During the past several years, successful application of fuzzy logic for solving complex problems subject to uncertainty has greatly increased. Today, in numerous engineering disciplines, fuzzy logic plays an important role in various procedures ranging from conceptual reasoning to practical implementation.

In recent years, considerable attention has been devoted to the use of hybrid neural-network/fuzzy-logic approaches (Jang, 1991; Horikawa et al., 1992) as an alternative for pattern recognition, clustering, and statistical and mathematical modeling (Kaufmann and Gupta, 1988). It has been shown that neural network models can be used to construct internal models, which recognize fuzzy rules.

1.3. Pattern recognition

In the 1960s and 1970s, pattern recognition techniques were used only by statisticians and were based on statistical theories. Due to recent advances in computer systems and technology, artificial neural networks and fuzzy logic models have been used in many pattern recognition applications ranging from simple character recognition, interpolation, and extrapolation between specific patterns to the most sophisticated robotic applications. To recognize a pattern, one can use the standard multi-layer perceptron with a back-propagation learning algorithm or simpler models such as self-organizing networks (Kohonen, 1997) or fuzzy c-means techniques (Bezdek, 1981; Jang and Gulley, 1995) (see Appendix A). Self-organizing networks and fuzzy c-means techniques can easily learn to recognize the topology, patterns, and distribution in a specific set of information.

1.4. Clustering

Cluster analysis encompasses a number of different classification algorithms that can be used to organize observed data into meaningful structures. For example, k-means (see Appendix A) is an algorithm to assign a specific number of centers, k, to represent the clustering of N points ($k < N$). These points are iteratively adjusted so that each point is assigned to one cluster, and the centroid of each cluster is the mean of its assigned points. In general, the k-means technique will produce exactly k different clusters of the greatest possible distinction.

Alternatively, fuzzy techniques can be used as a method for clustering. Fuzzy clustering partitions a data set into fuzzy clusters such that each data point can belong to multiple clusters. Fuzzy c-means (FCM) (see Appendix A) is a well-known fuzzy clustering technique that generalizes the classical (hard) c-means algorithm and can be used where it is unclear how many clusters there should be for a given set of data.

Subtractive clustering is a fast, one-pass algorithm for estimating the number of clusters and the cluster centers in a set of data. The cluster estimates obtained from subtractive clustering can be used to initialize iterative optimization-based clustering methods and model identification methods.

In addition, the self-organizing map technique (see Appendix A) known as Kohonen's self-organizing feature map (Kohonen, 1997) can be used as an alternative for clustering purposes. This technique converts patterns of arbitrary dimensionality (the pattern space) into the response of one- or two-dimensional arrays of neurons (the feature space). This unsupervised learning model can discover any relationship of interest such as patterns, features, correlations, or regularities in the input data, and translate the discovered relationship into outputs.

Our neural networks and fuzzy c-means techniques are implemented on a personal computer using Matlab™ software, a technical computing environment combining computation, numerical analysis, and graphics (Bemuth and Beal, 1994; Jang and Gulley, 1995; The Math Works™, 1995).

2. RESERVOIR CHARACTERIZATION

Fig. 1 shows schematically the flow of information and techniques to be used for intelligent reservoir characterization (IRESC). The main goal is to integrate soft data such as geological data with hard data such as 3D seismic, production data, etc. to build reservoir and stratigraphic models. In this case study, we analyzed 3D seismic attributes to find similarity cubes and clusters using three different techniques: (1) k-means, (2) neural network (self-organizing map), and (3) fuzzy c-means. The clusters can be interpreted as lithofacies, homogeneous classes, or similar patterns that exist in the data. The relationship between each cluster and production-log data was recognized around the well bore and the results were used to reconstruct and extrapolate production-log data away from the well bore. The results from clustering were superimposed on the reconstructed production-log data and optimal locations to drill new wells were determined.

2.1. Examples

Our example are from fields that produce from the Ellenburger Group. The Ellenburger is one of the most prolific gas producers in the conterminous United States, with greater than 13 tcf of production from fields in west Texas. The Ellenburger Group was deposited on an Early Ordovician passive margin in shallow subtidal to intertidal envi-

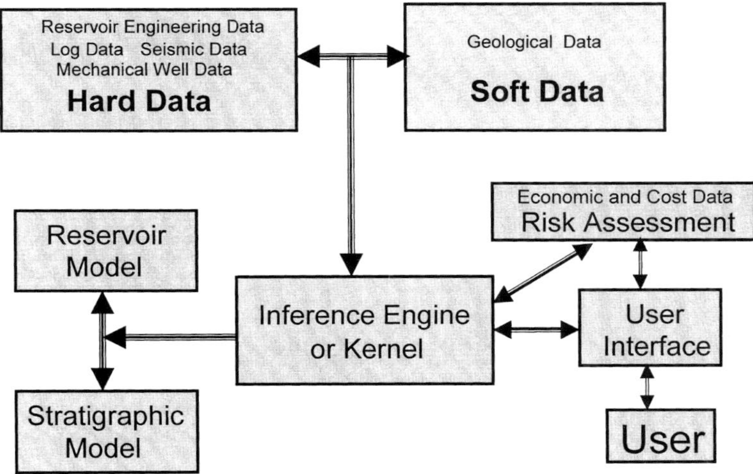

Fig. 1. Integrated Reservoir Characterization (IRESC).

ronments. Reservoir description indicates the study area is affected by a karst-related, collapsed paleocave system that acts as the primary reservoir in the field studied (Adams et al., 1999; Levey et al., 1999).

2.1.1. Area 1

The 3D seismic volume used for this study has 3,178,500 data points (Table 1). Two hundred, seventy-four well-log data points intersect the seismic traces. Eighty-nine production log data points are available for analysis (19 production and 70 non-production). A representative subset of the 3D seismic cube, production log data, and an area of interest were selected in the training phase for clustering and mapping purposes. The subset (150 samples, with each sample equal to 2 msec of seismic data or approximately 20 feet of Ellenburger dolomite) was designed as a section (670 seismic traces) passing through all the wells as shown in Fig. 2 and has 100,500 (670×150) data points. However, only 34,170 (670×51) data points were selected for clustering purposes, representing the main Ellenburger focus area. This subset covers the horizontal boreholes of producing wells, and starts approximately 15 samples (300 feet) above the Ellenburger, and ends 20 samples (400 feet) below the locations of the horizontal wells. In addition, the horizontal wells are present in a 16-sample interval, for a total interval of 51 samples (102 msec or 1020 feet). Table 1 shows typical statistics for this case study. Fig. 3 shows a schematic diagram of how the well path intersects the seismic traces. For clustering and mapping, there are two windows that must be optimized, the seismic window and the well log window. Optimal numbers of seismic attributes and clusters need to be determined, depending on the nature of the problem. Fig. 4 shows the iterative technique that has been used to select an optimal number of clusters, seismic attributes, and optimal processing windows for the seismic section shown in Fig. 2. Expert knowledge regarding geological parameters has also been used to constrain the maximum number of clusters to be selected. In this study,

TABLE 1

Typical statistics for main focus area, Area 1, and Ellenburger

Data			
Cube		*Section*	
InLine:	163	Total number of traces:	670
Xline:	130	Time sample:	150
Time sample:	150	Total number of points:	100,500
Total number of points:	3,178,500	Used for clustering:	34,170
		Section/cube:	3.16%
		For clustering: 1.08%	
Well data		*Production data*	
Total number of points:	274	Total number of points:	89
Well data/section:	0.80%	Production:	19
Well data/cube:	0.009%	No production:	70
		Production data/section:	0.26%
		Production data/cube:	0.003%

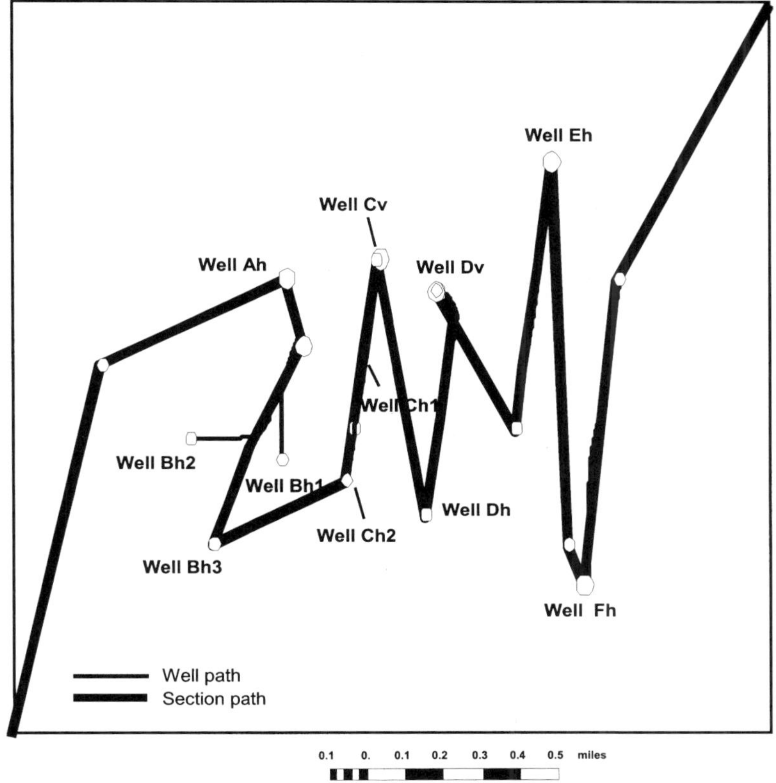

Fig. 2. Seismic section passing through all the wells, area 1.

six attributes have been selected (Raw Seismic, Instantaneous Amplitude, Instantaneous Phase, Cosine Instantaneous Phase, Instantaneous Frequency, and Integrate Absolute Amplitude) out of 17 attributes calculated (Table 2).

The following functions and equations have been used to calculate the seismic attributes:

- Raw Seismic Trace Amplitude: $s(t)$
- Hilbert Transform of the Raw Seismic Trace: $s'(t)$
- The Complex Trace Signal:

$$a(t) = s(t) + js'(t)$$

- Instantaneous Amplitude or Envelope Amplitude:

$$\mathrm{AMP}(t) = |a(t)| = \sqrt{s^2(t) + s'^{(2)}(t)}$$

- Instantaneous Phase:

$$\phi(t) = \cos^{-1}\left[s(t)/\mathrm{AMP}(t)\right]$$

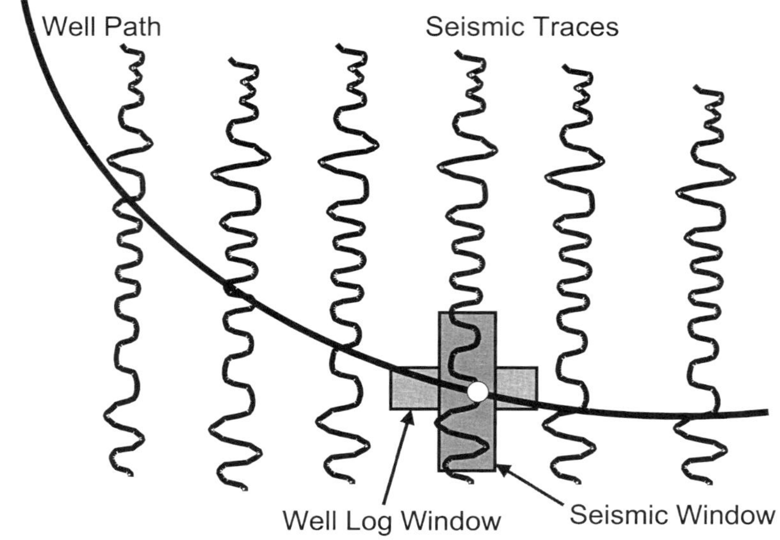

Fig. 3. Schematic diagram of how the well path intersects the seismic traces.

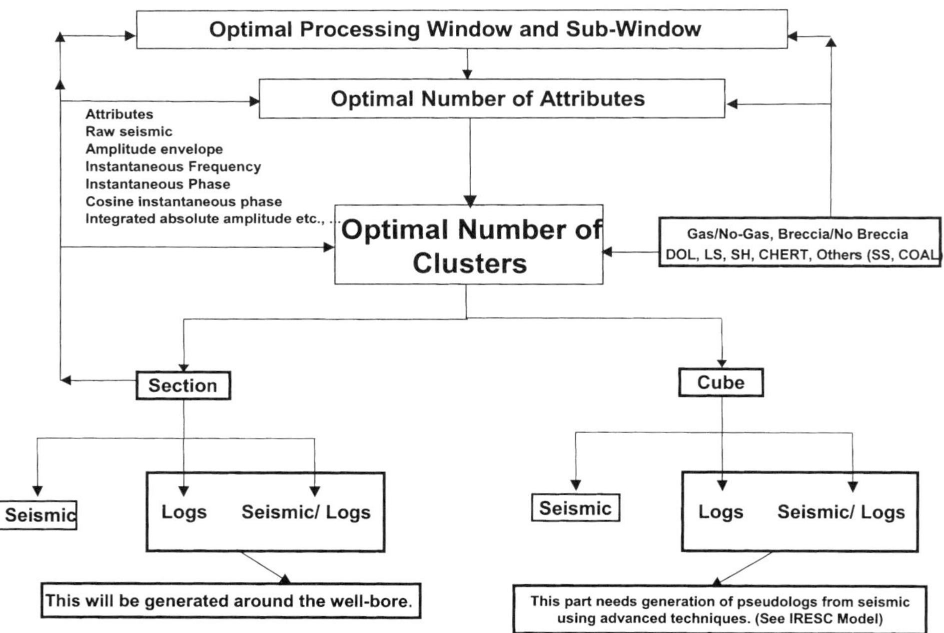

Fig. 4. Iterative technique to select an optimal number of clusters, seismic attributes, and optimal processing windows.

TABLE 2

List of the attributes calculated in this study

Attribute No.	Abbreviation	Attribute
1.	ampenv	Amplitude envelope
2.	ampwcp	Amplitude weighted cosine phase
3.	ampwfr	Amplitude weighted frequency
4.	ampwph	Amplitude weighted phase
5.	aprpol	Apparent polarity
6.	avgfre	Average frequency
7.	cosiph	Cosine instantaneous phase
8.	deriamp	Derivative instantaneous amplitude
9.	deriv	Derivative
10.	domfre	Dominant frequency
11.	insfre	Instantaneous frequency
12.	inspha	Instantaneous phase
13.	intaamp	Integrated absolute amplitude
14.	integ	Integrate
15.	raw	Raw seismic
16.	sdinam	Second derivative instantaneous amplitude
17.	secdev	Second derivative

- Cosine Instantaneous Phase:

$$\cos(\phi(t)) = \left[s(t)/\mathrm{AMP}(t)\right]$$

- Instantaneous Frequency:

$$f(t) = (1/2\pi)\cdot\left[\mathrm{d}\phi(t)/\mathrm{d}t\right] \times 360$$

- Integrate Absolute Amplitude:

$$\mathrm{IAAAI}(t) = \left(\sum_{i=1}^{i=t}\mathrm{AMP}(t)\right) - \overline{\mathrm{AMP}(t)}$$

where $\overline{\mathrm{AMP}(t)}$ = The Moving Window Smoothed Amplitude Envelope.

Figs. 5 through 10 show typical representations of these attributes in our case study. The following criteria are used to generate Figs. 5–10.

- **Rule 1**: Define the upper and lower range of the color map for each attribute based on the following criteria:
  - Lower Value = Mean − 3 × Standard Deviation
  - Upper Value = Mean + 3 × Standard Deviation
  - IF the Lower Value is less than Min of data THEN Lower Value is equal to Min of data.
  - IF the Upper Value is greater than Max of data THEN Upper Value is equal to Max of data.
- The values outside the range are assigned to the first or last color map color.

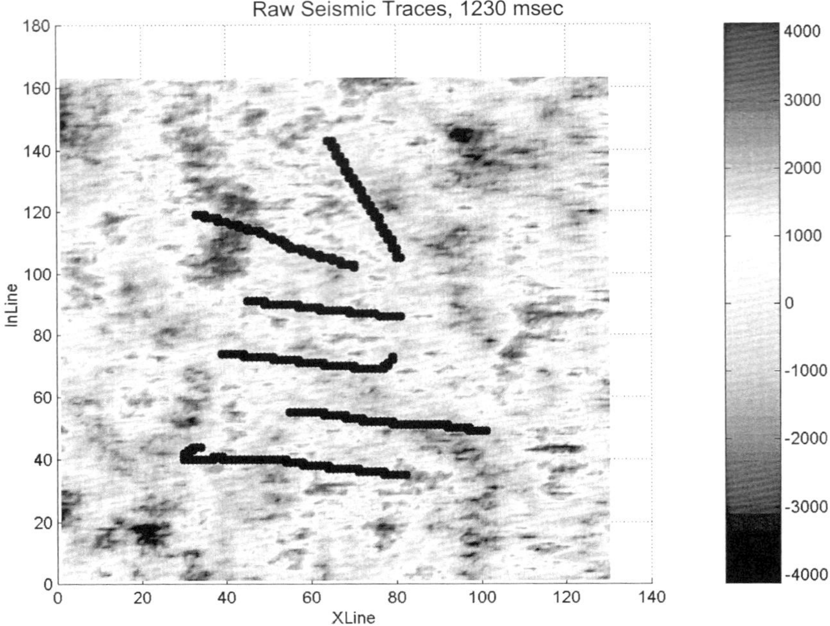

Fig. 5. Typical time slice of raw seismic in area 1 with rule 1.

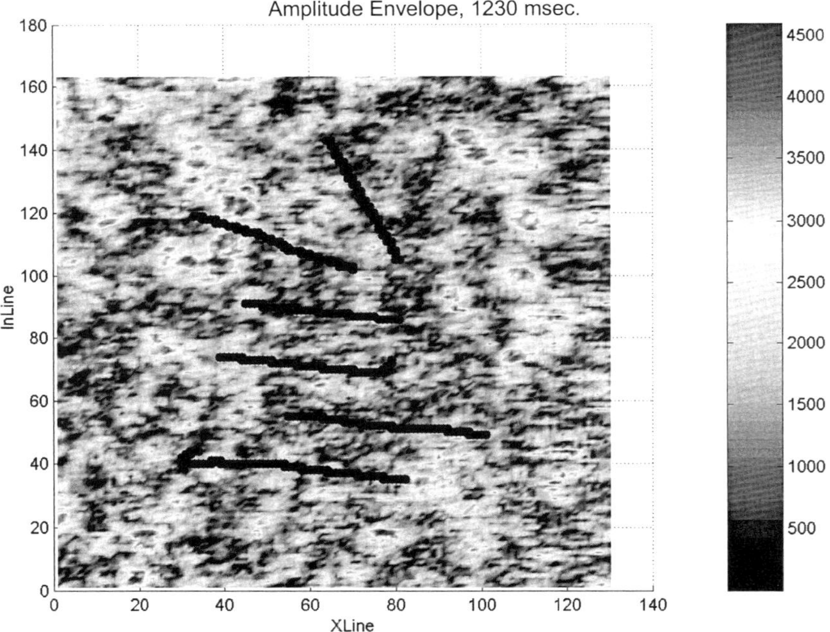

Fig. 6. Typical time slice of amplitude envelope in area 1 with Rule 1.

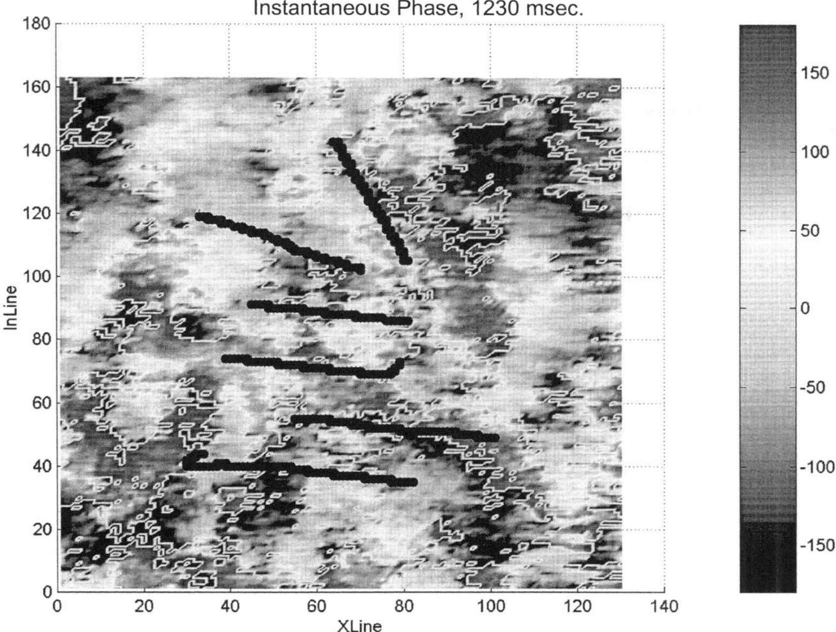

Fig. 7. Typical time slice of instantaneous phase in area 1 with Rule 1.

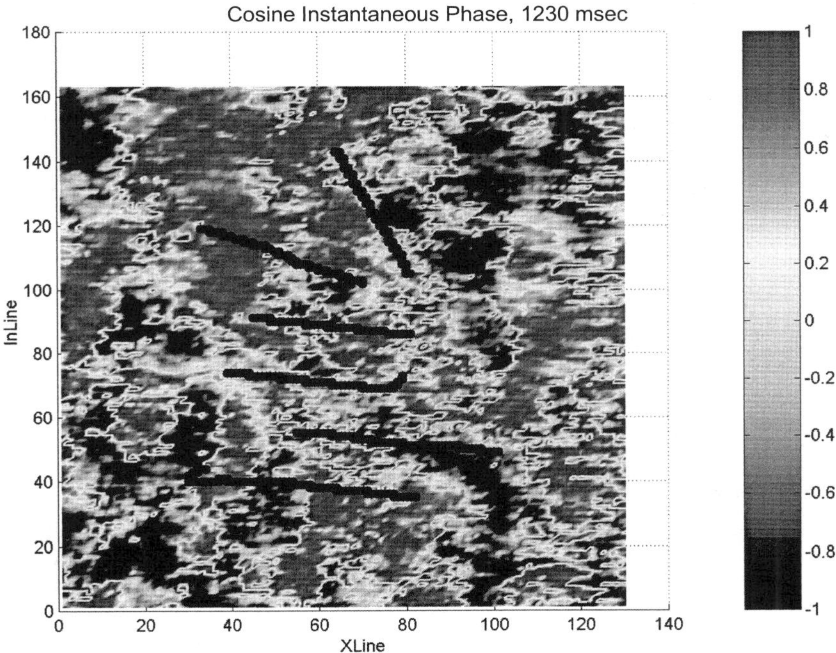

Fig. 8. Typical time slice of cosine instantaneous phase in area 1 with rule 1.

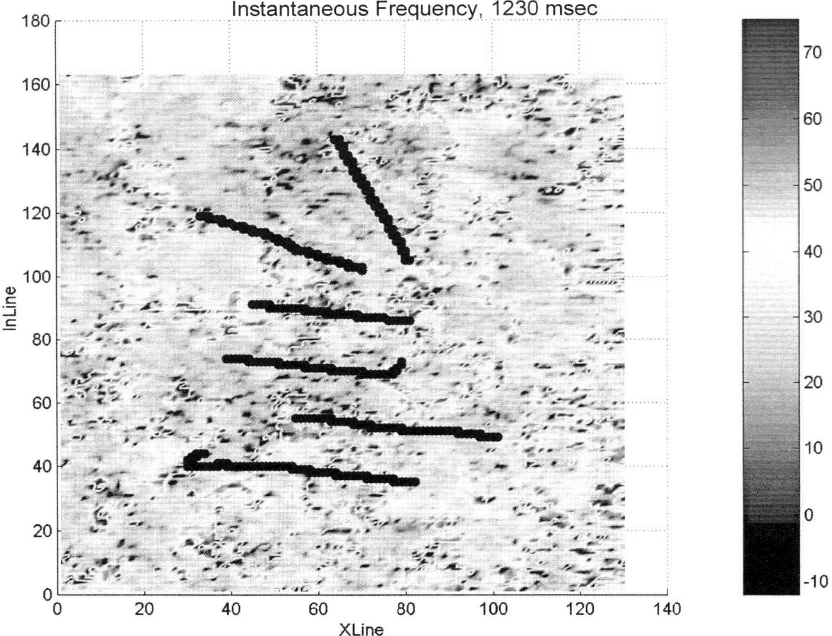

Fig. 9. Typical time slice of instantaneous frequency in area 1 with rule 1.

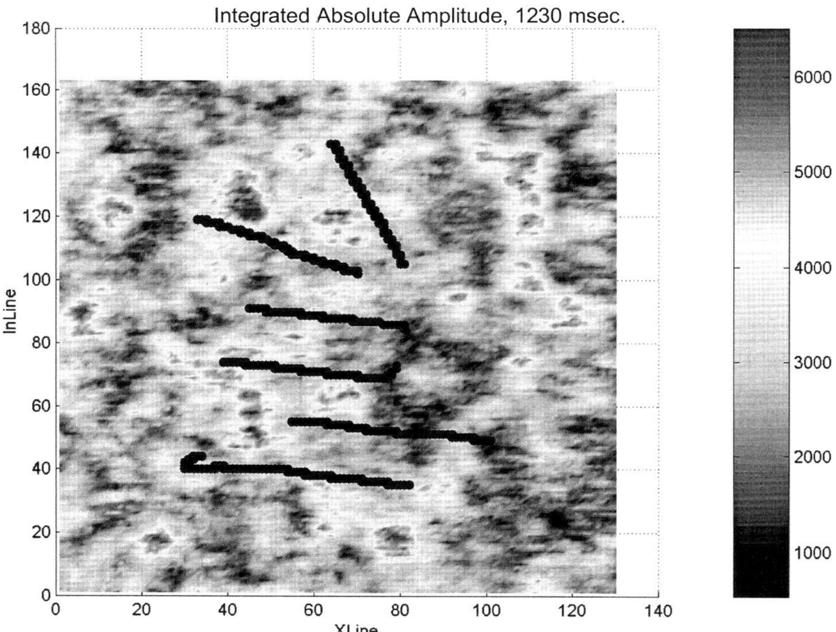

Fig. 10. Typical time slice of integrated absolute amplitude in area 1 with rule 1.

Ten clusters were recognized, a window of one sample was used as the optimal window size for the seismic, and a window of three samples was used for the production log data. Based on qualitative analysis, specific clusters with the potential to be in producing zones were selected. Software was developed to do the qualitative analysis and run on a personal computer using Matlab™ software. Fig. 11 shows typical windows and parameters of this software. Clustering was based on three different techniques, k-means (statistical), neural network, and fuzzy c-means clustering. Different techniques recognized different cluster patterns as shown by the cluster distributions (Fig. 12A through 14). Figs. 12–14 show the distribution of clusters in the section passing through the wells as shown in Fig. 2. By comparing k-mean (Fig. 12A) and neural network clusters (Fig. 13) with fuzzy clusters (Fig. 14), one can conclude that the neural network predicted a different structure and patterns than did the other techniques. Fig. 12B and 12C show a typical time-slice from the 3D seismic cube that has been reconstructed with the extrapolated k-means cluster data. Finally, based on a qualitative analysis, specific clusters that have the potential to include producing zones were selected. Each clustering technique produced two clusters that included most of the production data. Each of these three pairs of clusters is equivalent.

To confirm such a conclusion, cluster patterns were generated for the section passing through the wells as shown in Fig. 2. Fig. 15 through 17 show the two clusters from each technique that correlate with production: clusters one and four from k-means clustering (Fig. 15); clusters one and six from neural network clustering (Fig. 16); and clusters six and ten from fuzzy c-means clustering (Fig. 17). By comparing these three cross-sections, one can conclude that, in the present study, all three techniques predicted the same pair of clusters based on the objective of predicting potential producing zones. However, this may not always be the case because information that can be extracted by the different techniques may be different. For example, clusters using classical techniques will have sharp boundaries whereas those generated using the fuzzy technique will have fuzzy boundaries.

Based on the clusters recognized in Fig. 15 through 17 and the production log data, a subset of the clusters has been selected and assigned as cluster 11 as shown in Figs. 18 and 19. In this sub-cluster, the relationship between production-log data and clusters has been recognized and the production-log data has been reconstructed and extrapolated away from the well bore. Finally, the production-log data and the cluster data were superimposed at each point in the 3D seismic cube. Fig. 20A and 20B show a typical time-slice of a 3D seismic cube that has been reconstructed with the extrapolated production-log data and cluster data. The color scale in Fig. 20A and 20B is divided into two indices, Cluster Index and Production Index.

Criteria used to define Cluster Indices for each point are expressed as a series of dependent IF–THEN statements. To determine the Cluster Index of a point, the program starts with the first IF–THEN statement. Whenever an IF-clause is false, the program moves to the next IF–THEN statement. Whenever an IF-clause is true, the value of the Cluster Index is determined. The following criteria were used to define Cluster Indices for each point:

IF the values of the six attributes at each point are between plus and minus one standard deviation from the center of the specified cluster, THEN Cluster Index = 5,

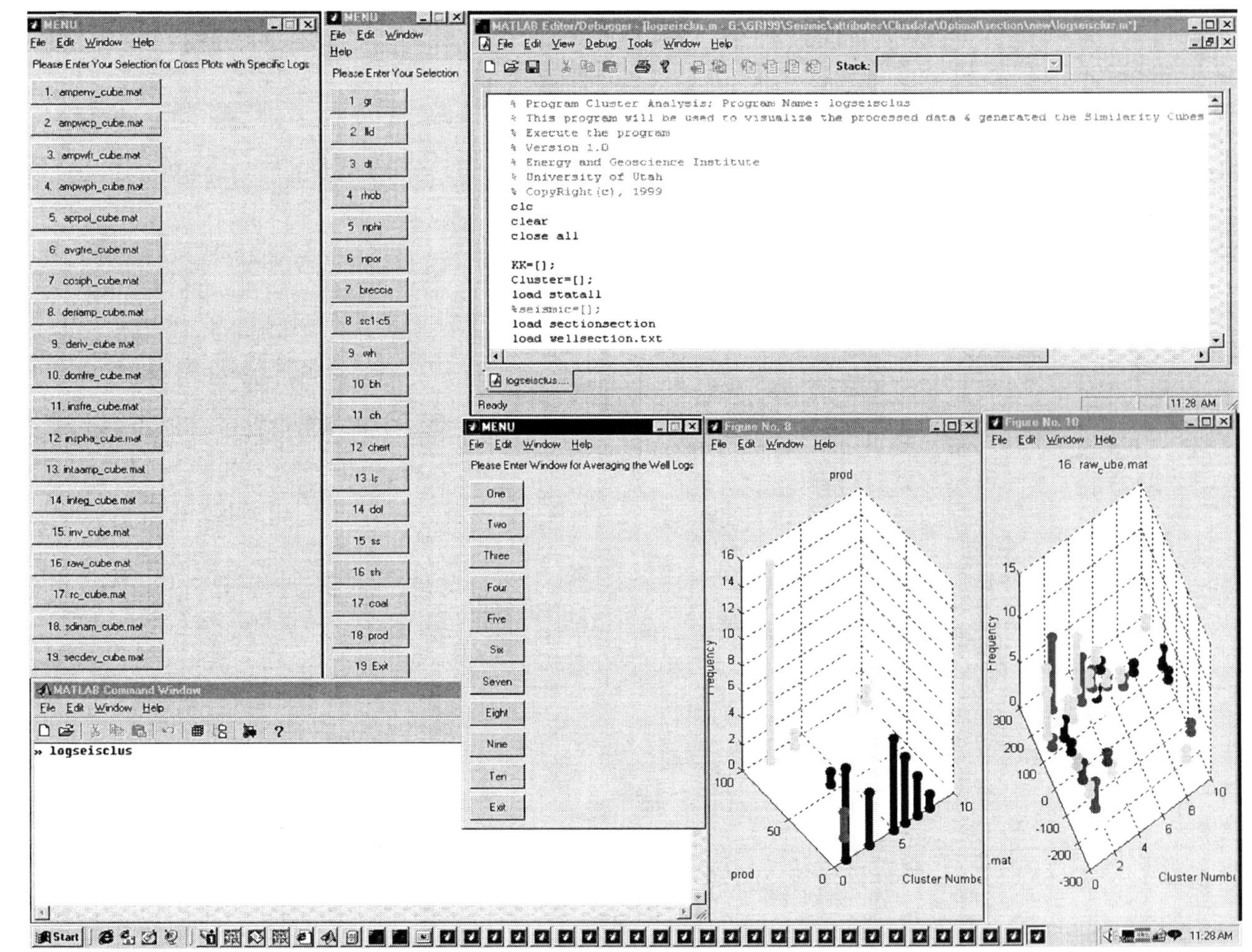

Fig. 11. MatLab software to do the qualitative analysis.

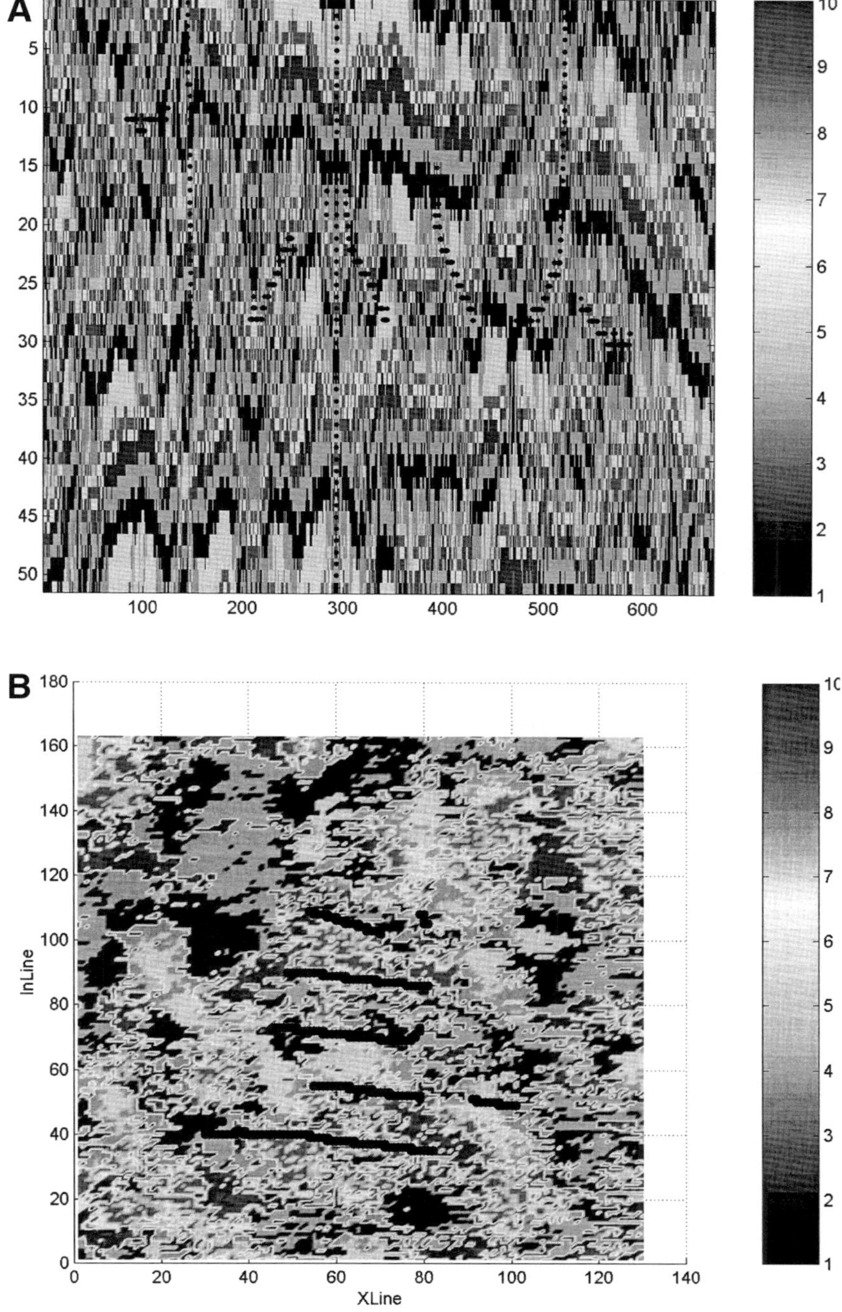

Fig. 12. (A) Typical k-means distribution of clusters in the section passing through the wells as shown in Fig. 2. (B) Typical 2D presentation of time-slice of k-means distribution of clusters.

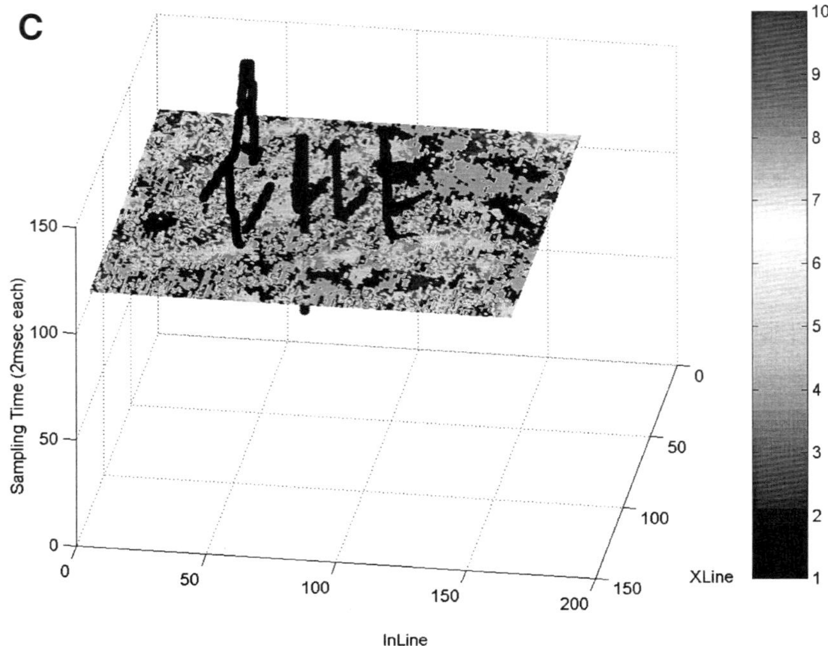

Fig. 12 (continued). (C) Typical 3D presentation of time-slice of k-means distribution of clusters.

IF the values of the six attributes at each point are between plus and minus two standard deviations from the center of the specified cluster and it is not Cluster Index 5, THEN Cluster Index = 4,

IF the values of the six attributes at each point are between plus and minus three standard deviations from the center of the specified cluster and it is not Cluster Index 4, THEN Cluster Index = 3,

IF the values of the six attributes at each point are between plus and minus four standard deviations from the center of the specified cluster and it is not Cluster Index 3, THEN Cluster Index = 2,

IF the values of the six attributes at each point are between plus and minus five standard deviations from the center of the specified cluster and it is not Cluster Index 2, THEN Cluster Index = 1,

IF the point does not belong to the specified cluster, THEN Cluster Index = 0.

Before defining Production Indices for each point within a specified cluster, a new cluster must first be defined based only on the seismic data that represents production-log data with averaged values greater than 0.50. Averaged values are determined by assigning a value to each sample of a horizontal borehole (two feet/sample). Sample intervals that are producing gas are assigned values of one and non-producing sample

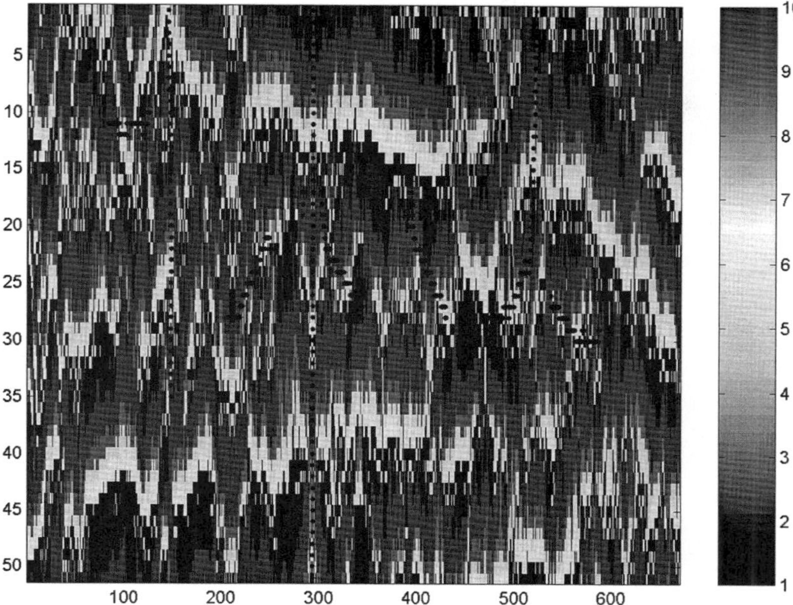

Fig. 13. Typical neural network distribution of clusters in the section passing through the wells as shown in Fig. 2.

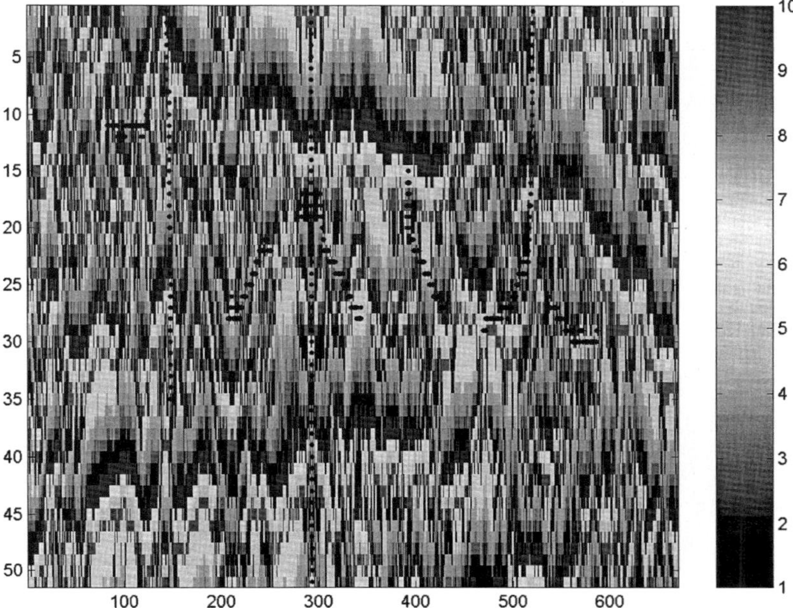

Fig. 14. Typical fuzzy c-means distribution of clusters in the section passing through the wells as shown in Fig. 2.

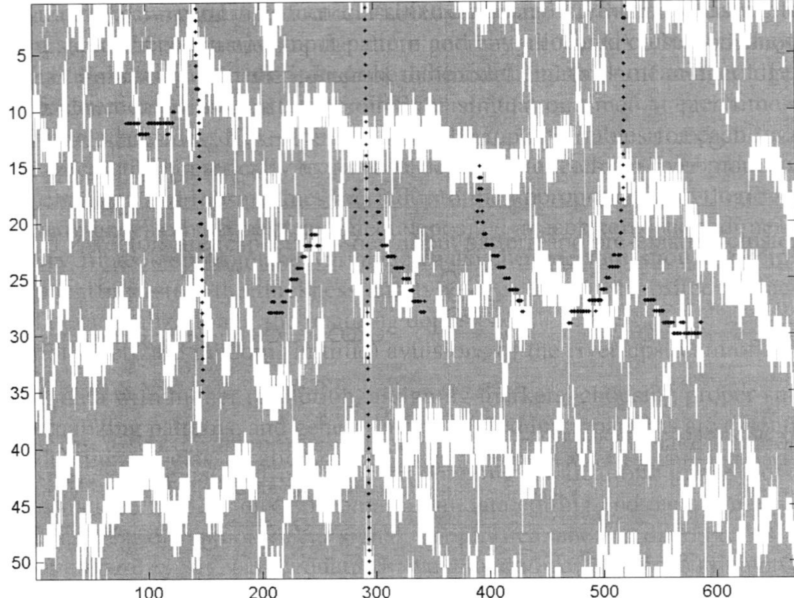

Fig. 15. Clusters one and four that correlate with production using the k-means clustering technique on the section passing through the wells as shown in Fig. 2.

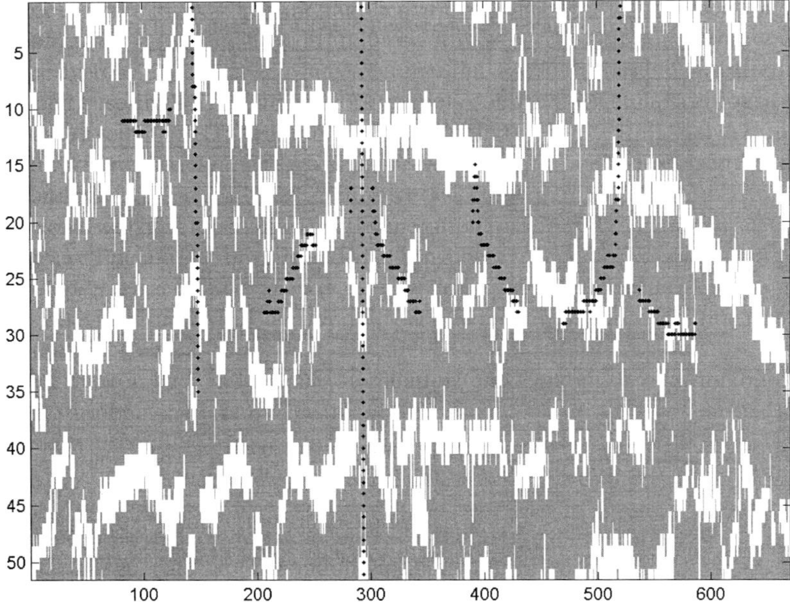

Fig. 16. Clusters one and six that correlate with production using the neural network clustering technique on the section passing through the wells as shown in Fig. 2.

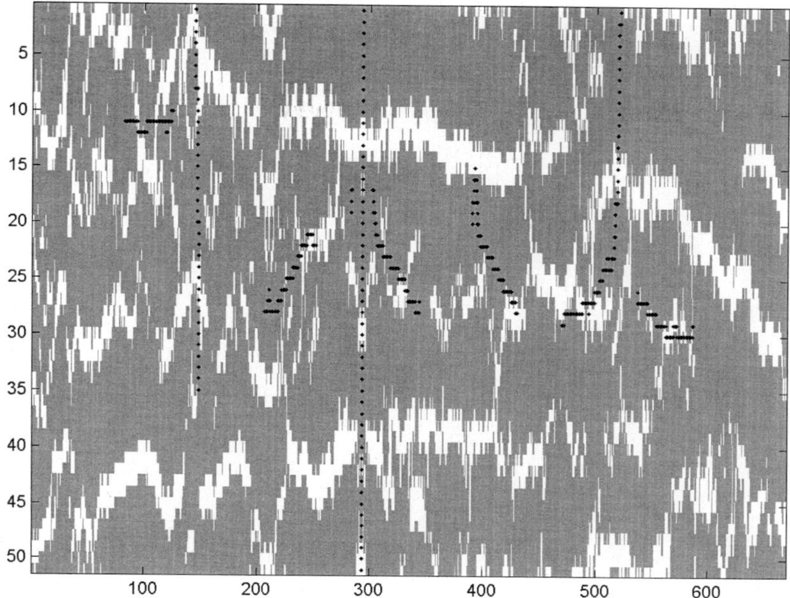

Fig. 17. Clusters six and ten that correlate with production using the fuzzy c-means clustering technique on the section passing through the wells as shown in Fig. 2.

intervals are assigned values of zero. The optimum window size for production-log data is three samples and the averaged value at any point is the average of the samples in the surrounding window. After the new cluster is determined, a series of IF–THEN statements is used to define the Production Indices. Again, these are dependent IF–THEN statements and to determine the Production Index, the series is gone through until an IF-clause is true. The following criteria were used to define the Production Indices:

IF the values of the six attributes at each point are between plus and minus one standard deviation from the center of the specified cluster, THEN Production Index = 10,

IF the values of the six attributes at each point are between plus and minus two standard deviations from the center of the specified cluster and it is not Production Index 10, THEN Production Index = 9,

IF the values of the six attributes at each point are between plus and minus three standard deviations from the center of the specified cluster and it is not Production Index 9, THEN Production Index = 8,

IF the values of the six attributes at each point are between plus and minus four standard deviations from the center of the specified cluster and it is not Production Index 8, THEN Production Index = 7,

IF the values of the six attributes at each point are between plus and minus five standard deviations from the center of the specified cluster and it is not Production Index 7, THEN Production Index = 6,

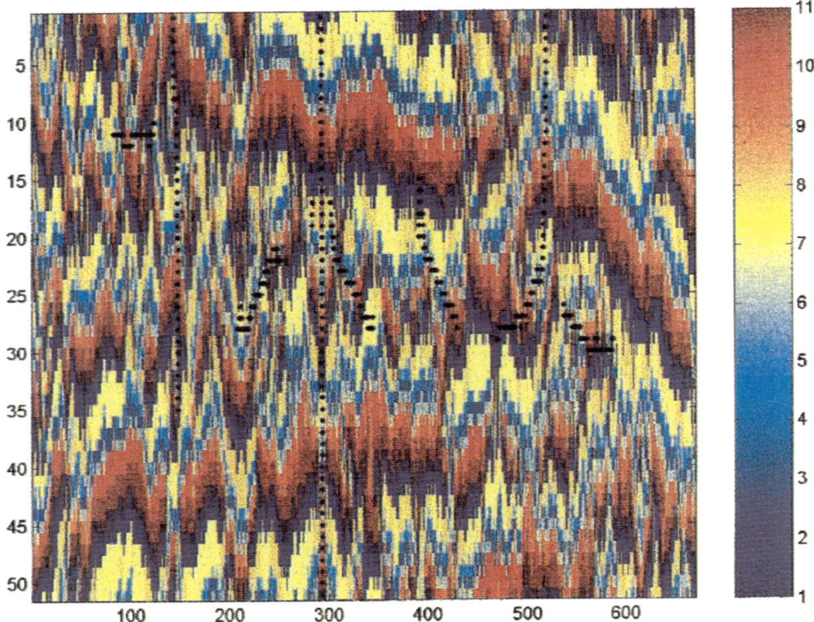

Fig. 18. The section passing through all wells showing a typical distribution of clusters generated by combining the three sets of 10 clusters created by each clustering technique (k-means, neural network, and fuzzy c-means). Note that eleven clusters are displayed rather than the ten clusters originally generated. The eleventh cluster is a subset of the three pairs of clusters that contain most of the production.

> IF the point does not belong to the specified cluster, THEN Production Index = 0.

Three criteria have been used to select potential locations for infill drilling or recompletion: (1) continuity of the selected cluster, (2) size and shape of the cluster, and (3) existence of high Production-Index values inside a selected cluster with high Cluster-Index values. Based on these criteria, locations of the new wells were selected and two such locations are shown in Fig. 20B, one with high continuity and potential for high production and one with low continuity and potential for low production. The neighboring wells that are already in production confirm such a prediction as shown in Fig. 20B.

3. CONCLUSIONS

In this study, a new integrated methodology was developed to identify a nonlinear relationship and mapping between 3D seismic data and production-log data and the technique was applied to a producing field. This advanced data analysis and interpretation methodology for 3D seismic and production-log data uses conventional statistical

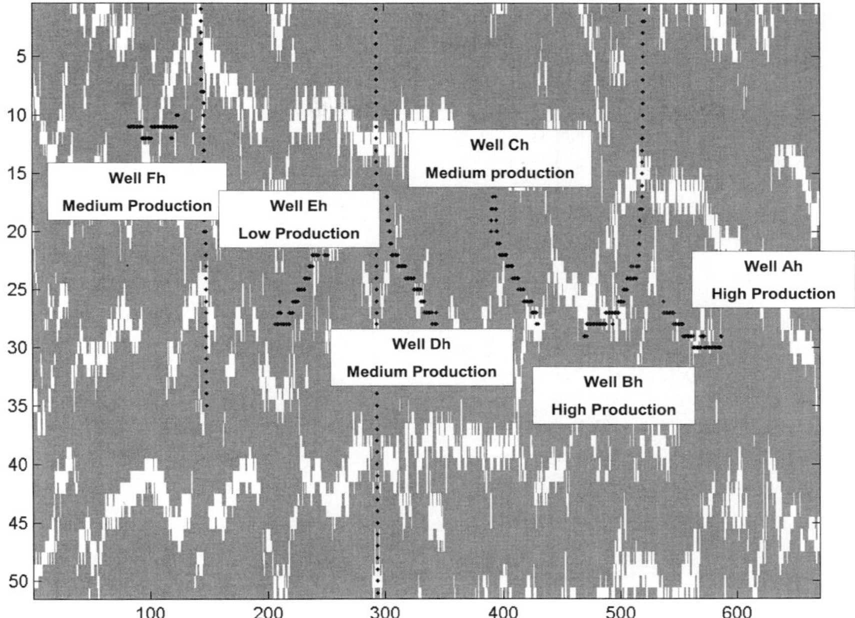

Fig. 19. The section passing through all wells showing only the eleventh cluster, a subset of the three pairs of clusters that contains most of the production.

techniques combined with modern soft-computing techniques. It can be used to predict: (1) mapping between production-log data and seismic data, (2) reservoir connectivity based on multi-attribute analysis, (3) pay zone recognition, and (4) optimum well placement. It is important to note that these optimal well locations are based on static data (3D seismic data and original well logs) and can be considered only as potential locations for new wells to be drilled. Therefore, final decisions for selecting locations of new drilling sites should be confirmed by dynamic reservoir modeling and simulation and by future-performance prediction through reservoir simulation and history-matching techniques. Although these methodologies have limitations, the usefulness of the techniques will be for fast screening of production zones with reasonable accuracy.

4. POTENTIAL RESEARCH OPPORTUNITIES IN THE FUTURE

4.1. Quantitative 3D reconstruction of well logs and prediction of pay zone thickness

This new methodology, combined with techniques presented by Nikravesh (1998a,b), Nikravesh and Aminzadeh (1998), and Nikravesh et al. (1998), can be used to reconstruct well logs such as porosity, density, resistivity, etc. away from the well bore. By doing so, net-pay-zone thickness, reservoir models, and geological representations will

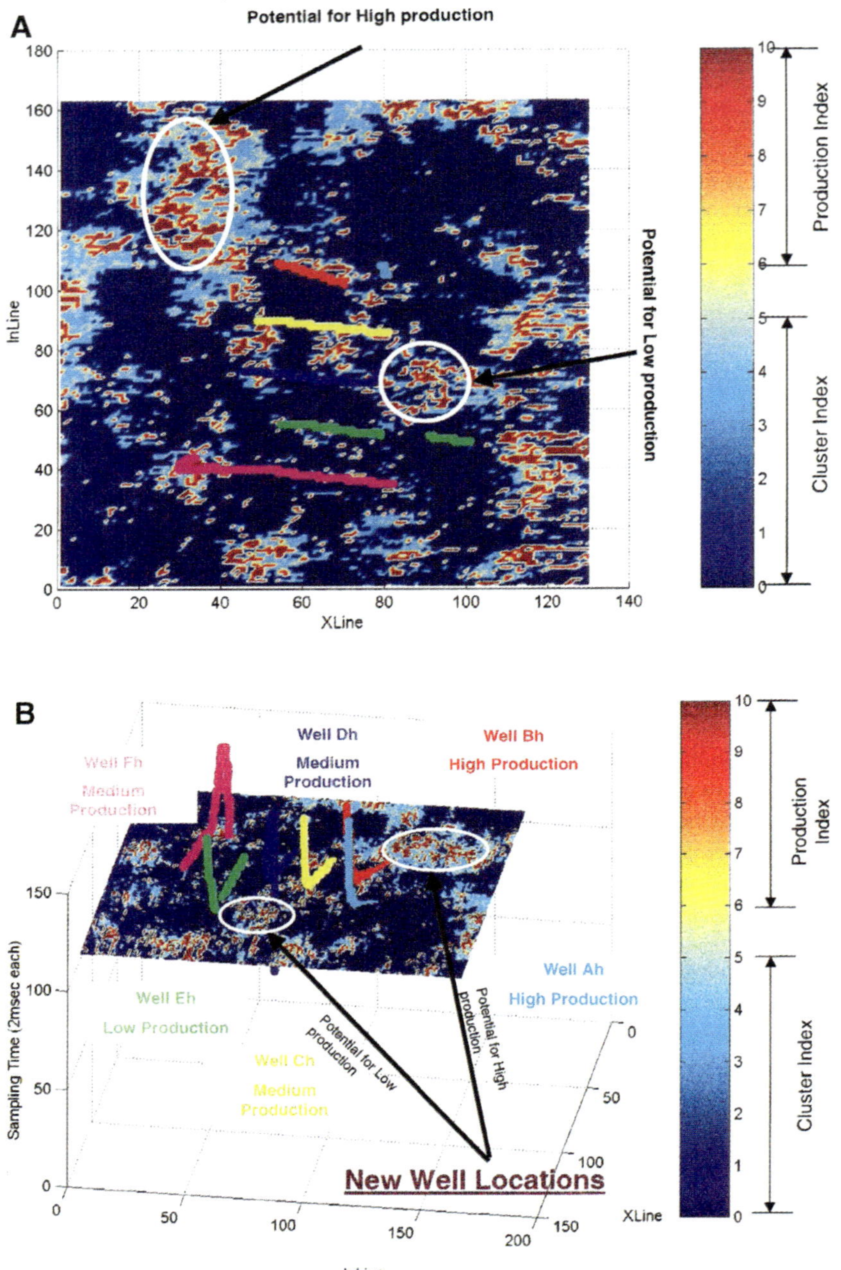

Fig. 20. (A) A 2D time slice showing clusters and production, with areas indicated for optimal well placement. (B) An oblique view of the 2D time slice in (A) showing clusters and production, with areas indicated for optimal well placement.

be accurately identified. Accurate reservoir characterization through data integration is an essential step in reservoir modeling, management, and production optimization.

4.2. IRESC model

Fig. 1 shows schematically the flow of information and techniques to be used for intelligent reservoir characterization (IRESC). The main goal will be to integrate soft data such as geological data with hard data such as 3D seismic, production data, etc. to build a reservoir and stratigraphic model.

4.3. Neuro-fuzzy techniques

In recent years, considerable attention has been devoted to the use of hybrid neural-network/fuzzy-logic approaches as an alternative for pattern recognition, clustering, and statistical and mathematical modeling. It has been shown that neural network models can be used to construct internal models that recognize fuzzy rules.

Neuro-fuzzy modeling is a technique for describing the behavior of a system using fuzzy inference rules within a neural network structure. The model has a unique feature in that it can express linguistically the characteristics of a complex nonlinear system. As a part of any future research opportunities, we will use the neuro-fuzzy model originally presented by Sugeno and Yasukawa (1993). The neuro-fuzzy model is characterized by a set of rules. The rules are expressed as follows:

$$\begin{aligned} R^i: \quad & \text{if } x_1 \text{ is } A_1^i \text{ and } x_2 \text{ is } A_2^i \ldots \text{ and } x_n \text{ is } A_n^i \quad && \text{(Antecedent)} \\ & \text{then } y^* = f_i(x_1, x_2, \ldots, x_n) && \text{(Consequent)} \end{aligned} \tag{1}$$

where $f_i(x_1,x_2,\ldots,x_n)$ can be constant, linear, or a fuzzy set.

$$\text{For the linear case:} \quad f_i(x_1,x_2,\ldots,x_n) = a_{i0} + a_{i1}x_1 + a_{i2}x_2 + \ldots + a_{in}x_n \tag{2}$$

Therefore, the predicted value for output y is given by:

$$y = \sum_i \mu_i f_i(x_1,x_2,\ldots,x_n) \Big/ \sum \mu_i \tag{3}$$

with

$$\mu_i = \prod_j A_j^i(x_j) \tag{4}$$

where R_i is the ith rule, x_j are input variables, y is output, A_j^i are fuzzy membership functions (fuzzy variables), and a_{ij} constant values.

As a part of any future research opportunities, we will use the adaptive neuro-fuzzy inference system (ANFIS) technique originally presented by Jang (1992). The model uses neuro-adaptive learning techniques, which are similar to those of neural networks. Given an input/output data set, the ANFIS can construct a fuzzy inference system (FIS) whose membership function parameters are adjusted using the back-propagation algorithm or other similar optimization techniques. This allows fuzzy systems to learn from the data they are modeling.

ACKNOWLEDGEMENTS

Funding for this research was provided by the Gas Research Institute under GRI contract No. 5097-21-4101, project manager Robert W. Siegfried. The authors express their thanks to Dr. Fred Aminzadeh from Fact Inc. and Zadeh Institute for Information Technology (ZIFIT) for his many useful comments and suggestions.

APPENDIX A

A.1. K-means clustering

An early paper on k-means clustering was written by MacQueen (1967). K-means is an algorithm to assign a specific number of centers, k, to represent the clustering of N points ($k < N$). These points are iteratively adjusted so that each point is assigned to one cluster, and the centroid of each cluster is the mean of its assigned points. In general, the k-means technique will produce exactly k different clusters of the greatest possible distinction.

The algorithm is summarized in the following:

(1) Consider each cluster consisting of a set of M samples that are similar to each other: $x_1, x_2, x_3, \ldots, x_m$
(2) Choose a set of clusters $\{y_1, y_2, y_3, \ldots, y_k\}$
(3) Assign the M samples to the clusters using the minimum Euclidean distance rule
(4) Compute a new cluster so as to minimize the cost function
(5) If any cluster changes, return to step 3; otherwise stop.
(6) End

A.2. Fuzzy c-means clustering

Bezdek (1981) presents comprehensive coverage of the use of fuzzy logic in pattern recognition. Fuzzy techniques can be used as an alternative method for clustering. Fuzzy clustering partitions a data set into fuzzy clusters such that each data point can belong to multiple clusters. Fuzzy c-means (FCM) is a well-known fuzzy clustering technique that generalizes the classical (hard) c-means algorithm, and can be used where it is unclear how many clusters there should be for a given set of data.

Subtractive clustering is a fast, one-pass algorithm for estimating the number of clusters and the cluster centers in a set of data. The cluster estimates obtained from subtractive clustering can be used to initialize iterative optimization-based clustering methods and model identification methods.

The algorithm is summarized in the following:

(1) Consider a finite set of elements $X = \{x_1, x_2, x_3, \ldots, x_n\}$ or x_j, $j = 1, 2, \ldots, n$
(2) Select a number of clusters c
(3) Choose an initial partition matrix, $U^{(0)}$

$$U = [u_{ij}] \qquad i = 1, 2, \ldots, c;\ j = 1, 2, \ldots, n$$

where u_{ij} express the degree to which the element of x_j belongs to the ith cluster

$$\sum u_{ij} = 1 \qquad \text{for all } j = 1,2,\ldots,n$$
$$0 < \sum u_{ij} < 1 \qquad \text{for all } i = 1,2,\ldots,c$$

(4) Choose a termination criteria, ε
(5) Set the iteration index l to 0
(6) Calculate the fuzzy cluster centers using $U^{(l)}$ and an objective function
(7) Calculate the new partition matrix $U^{(l+1)}$ using an objective function
(8) Calculate $\Delta = \|U^{(l+1)} - U^{(l)}\| = \max_{i,j} |u_{ij}^{(l+1)} - u_{ij}^{(l)}|$
(9) If $\Delta > \varepsilon$, then set $l = l + 1$ return to step 6, otherwise stop.
(10)End

A.3. Neural network clustering

Kohonen (1987, 1997) wrote two fundamental books on neural network clustering. The self-organizing map technique known as Kohonen's self-organizing feature map (Kohonen, 1997) can be used as an alternative for clustering purposes (Fig. A.1). This technique converts patterns of arbitrary dimensionality (the pattern space) into the response of one- or two-dimensional arrays of neurons (the feature space). This unsupervised learning model can discover any relationship of interest such as patterns,

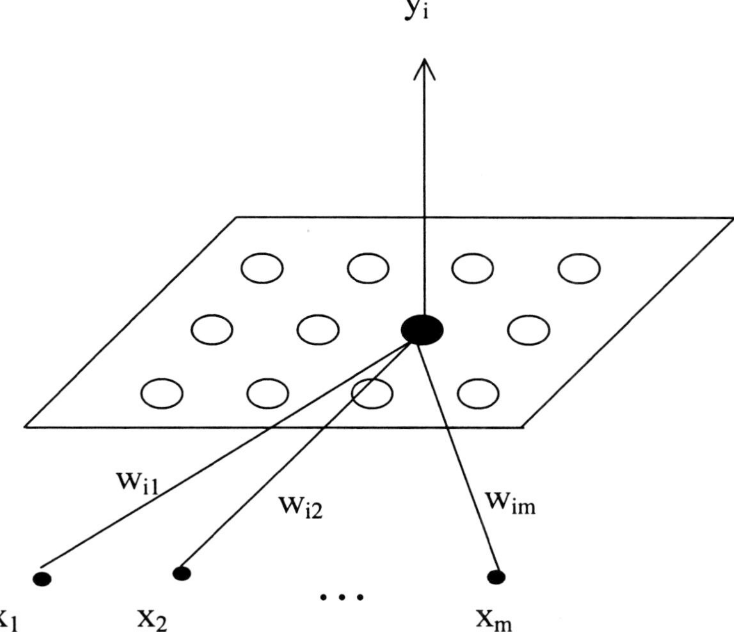

Fig. A.1. Basic Kohonen self-organizing map.

features, correlations, or regularities in the input data, and translate the discovered relationship into outputs.The algorithm is summarized in the following:

(1) Consider the network structure as shown in Fig. A.1.

(2) The learning rule is defined as:

- The similarity match is defined as:

$$\|\boldsymbol{x} - \boldsymbol{w}_{i^*}\| = \min_j \{\|\boldsymbol{x} - \boldsymbol{w}_j\|$$

- The learning rule is defined as:

$$w_{ij}^{(k+1)} = \begin{cases} w_{ij}^{(k)} + \alpha^{(k)}\left[x_j^{(k)} - w_{ij}^{(k)}\right], & \text{in the neighborhood set of the winner node } i^* \text{ at the time step } k. \\ w_{ij}^{(k)}, & \text{otherwise.} \end{cases}$$

with α defined as a learning constant.

REFERENCES

Adams, R.D., Collister, J.W., Ekart, D.D., Levey, R.A. and Nikravesh, M., 1999. Evaluation of gas reservoirs in collapsed paleocave systems. Ellenburger Group, Permian Basin, Texas. *Am. Assoc. Pet. Geol. Annu. Meet.*, San Antonio, TX, 11–14 April.

Aminzadeh, F. and Chatterjee, S., 1984/85. Applications of clustering in exploration seismology. *Geoexploration*, 23: 147–159.

Aminzadeh, F., Katz, S. and Aki, K., 1994. Adaptive neural network for generation of artificial earthquake precursors. *IEEE Trans. Geosci. Remote Sensing*, 32(6).

Baldwin, J.L., Otte, D.N. and Wheatley, C.L., 1989. Computer emulation of human mental process: Application of neural network simulations to problems in well log interpretation. *Soc. Pet. Eng.*, SPE Paper #19619, 481.

Baldwin, J.L., Bateman, A.R.M. and Wheatley, C.L., 1990. Application of neural network to the problem of mineral identification from well logs. *Log Analysis*, 3: 279.

Bemuth, H. and Beal, M., 1994. *Neural Network Toolbox*. The Math Works Inc., Natick, MA.

Bezdek, J.C., 1981. *Pattern Recognition with Fuzzy Objective Function Algorithm*. Plenum Press, New York, NY.

Bezdek, J.C. and Pal, S.K. (Eds.), 1992. *Fuzzy Models for Pattern Recognition*. IEEE Press, 539 pp.

Boadu, F.K., 1997. Rock properties and seismic attenuation: Neural network analysis. *Pure Appl. Geophys.*, 149: 507–524.

Chawathe, A., Quenes, A. and Weiss, W.W., 1997. Interwell property mapping using crosswell seismic attributes. *SPE Annual Technical Conference and Exhibition*, San Antonio, TX, 5–8 Oct., SPE Paper #38747.

Hecht-Nielsen, R., 1989. Theory of backpropagation neural networks. *IEEE Proc., Int. Conf. Neural Network*, Washington DC.

Horikawa, S.I., Furushashi, T. and Uchkawa, Y., 1992. On fuzzy modeling using neural networks with the backpropagation algorithm. *IEEE Trans. Neural Networks*, 3: 801–806.

Horikawa, S., Furuhashi, T., Kuromiya, A., Yamaoka, M. and Uchikawa, Y., 1996. Determination of antecedent structure for fuzzy modeling using genetic algorithm. *Proc. ICEC 1996, IEEE International Conference on Evolutionary Computation*, Nagoya, Japan, 20–22 May.

Jang, J.S.R., 1991. Fuzzy modeling using generalized neural networks and Kalman filter algorithm. *Proc. Ninth Natl. Conf. Artificial Intelligence*, pp. 762–767.

Jang, J.S.R., 1992. Self-learning fuzzy controllers based on temporal backpropagation. *IEEE Trans. Neural Networks*, 3(5).

Jang, J.S.R. and Gulley, N., 1995. *Fuzzy Logic Toolbox*. The Math Works Inc., Natick, MA.

Kaufmann, A. and Gupta, M.M., 1988. *Fuzzy Mathematical Models in Engineering and Management Science*. Elsevier, Amsterdam.

Klimentos, T. and McCann, C., 1990. Relationship among compressional wave attenuation, porosity, clay content and permeability in sandstones. *Geophysics*, 55: 991014.

Kohonen, T., 1987. *Self-Organization and Associate Memory, 2nd Edition*. Springer-Verlag, Berlin.

Kohonen, T., 1997. *Self-Organizing Maps, Second Edition*. Springer-Verlag, Berlin.

Levey, R., Nikravesh, M., Adams, R., Ekart, D., Livnat, Y., Snelgrove, S. and Collister, J., 1999. Evaluation of fractured and paleocave carbonate reservoirs. *Am. Assoc. Pet. Geol. Annual Meeting*, San Antonio, TX, 11–14 April.

MacQueen, J., 1967. Some methods for classification and analysis of multivariate observation. In: LeCun, L.M. and Neyman, J. (Eds.), T*he Proceedings of the 5th Berkeley Symposium on Mathematical Statistics and Probability*. University of California Press, 1: 281–297.

Monson, G.D. and Pita, J.A., 1997. Neural network prediction of pseudo-logs for net pay and reservoir property interpretation: Greater Zafiro field area, Equatorial Guinea. *SEG 1997 Meet.*, Dallas, TX.

Nikravesh, M., 1998a. *Mining and Fusion of Petroleum Data with Fuzzy Logic and Neural Network Agents.* CopyRight© Report, LBNL-DOE, ORNL-DOE, and DeepLook Industry Consortium.

Nikravesh, M., 1998b. Neural network knowledge-based modeling of rock properties based on well log databases. *1998 SPE Western Regional Meet.*, Bakersfield, CA, 10–13 May, SPE Paper #46206.

Nikravesh, M. and Aminzadeh, F., 1997. *Knowledge Discovery from Data Bases: Intelligent Data Mining.* FACT Inc. and LBNL Proposal, submitted to SBIR-NASA.

Nikravesh, M. and Aminzadeh, F., 1999. Opportunities to apply intelligent reservoir characterizations to the oil fields in Iran: A tutorial and lecture notes for intelligent reservoir characterization. *The 4th IAA Annual Conference, Oil, Petrochemicals, Energy and the Environment*, The City University of New York, New York, September 18–19.

Nikravesh, M., Farell, A.E. and Stanford, T.G., 1996. Model identification of nonlinear time-variant processes via artificial neural network. *Comput. Chem. Eng.*, 20(11): 1277.

Nikravesh, M., Novak, B. and Aminzadeh, F., 1998. Data mining and fusion with integrated neuro-fuzzy agents: rock properties and seismic attenuation. *JCIS 1998, The Fourth Joint Conference on Information Sciences*, NC, 23–28 October.

Pezeshk, S., Camp, C.C. and Karprapu, S., 1996. Geophysical log interpretation using neural network. *J. Comput. Civil Eng.*, 10: 136.

Rogers, S.J., Fang, J.H., Karr, C.L. and Stanley, D.A., 1992. Determination of Lithology, from well logs using a neural network. *Am. Assoc. Pet. Geol. Bull.*, 76: 731.

Rumelhart, D.E., Hinton, G.E. and Williams, R.J., 1986. Learning internal representations by error propagation. In: Rumelhart, D. and McClelland, J. (Ed.), *Parallel Data Processing*. MIT Press, Cambridge, MA.

Schuelke, J.S., Quirein, J.A., Sarg, J.F., Altany, D.A. and Hunt, P.E., 1997. Reservoir architecture and porosity distribution, Pegasus field, West Texas – an integrated sequence stratigraphic-seismic attribute study using neural networks. *SEG 1997 Meet.*, Dallas, TX.

Sugeno, M. and Yasukawa, T., 1993. A fuzzy-logic-based approach to qualitative modeling. *IEEE Trans. Fuzzy Systems*, 1(1).

The Math Works™, 1995. Natick.

Widrow, B. and Lehr, M.A., 1990. 30 years of adaptive neural networks: perceptron, madaline, and backpropagation. *Proc. IEEE*, 78(9): 1414.

Wong, P.M., Jiang, F.X. and Taggart, I.J., 1995a. A critical comparison of neural networks and discrimination analysis in lithofacies, porosity and permeability prediction. *J. Pet. Geol.*, 18: 191.

Wong, P.M., Gedeon, T.D. and Taggart, I.J., 1995b. An improved technique in prediction: a neural network approach. *IEEE Trans. Geosci. Remote Sensing*, 33: 971.

Yoshioka, K., Shimada, N. and Ishii, Y. 1996. Application of neural networks and co-kriging for predicting reservoir porosity-thickness. *GeoArabia*, 1(3).

Zadeh, L.A., 1965. Fuzzy sets. *Inf. Control*, 8: 33353.

Zadeh, L.A., 1973. Outline of a new approach to the analysis of complex systems and decision processes. *IEEE Trans. Systems, Man Cybern.*, 3: 244.

Zadeh, L.A., 1976. A fuzzy-algorithm approach to the definition of complex or imprecise concepts. *Int. J. Man-Machine Studies*, 8: 249–291.

Developments in Petroleum Science, 51
Editors: M. Nikravesh, F. Aminzadeh and L.A. Zadeh

Chapter 23

COMBINING GEOLOGICAL INFORMATION WITH SEISMIC AND PRODUCTION DATA

JEF CAERS [a,1] and SANJAY SRINIVASAN [b,2]

[a] *Stanford University, Department of Petroleum Engineering, Stanford, CA 94305-2220, USA*
[b] *University of Calgary, Department of Chemical and Petroleum Engineering, 2500 University Drive, N.W., Calgary, AB T2N 1N4, Canada*

ABSTRACT

The traditional practice of geostatistics for reservoir characterization is limited by the variogram which, as a measure of geological continuity, can capture only two-point statistics. Important curvi-linear geological information, beyond the modelling capabilities of the variogram, can be taken from training images and later used in model construction. Training images can provide multiple-point statistics which describe the statistical relation between multiple spatial locations considered jointly. Stochastic reservoir simulation then consists of anchoring the borrowed geo-structures in the form of multiple-point statistics to the actual subsurface well, seismic and production data.

1. INTRODUCTION

Extensive outcrop data, photographs of present day depositions or even single drawings from expert geologists contain important structural information about geological curvi-linear continuity of the subsurface reservoir that is beyond the modelling capability of the variogram. The *training image* is coined as a term for images that depict in 3D or in a series of 2D sections, the believed geological continuity of the reservoir. Such images are only conceptual, they need not be conditioned to any subsurface data. Multiple training images can be used each depicting a different scale of geological variation or each carrying a different geological interpretation (geological scenarios).

Typical to petroleum geostatistics is the scarcity of hard data, particularly in the horizontal directions. For the purpose of geostatistical modelling, horizontal variograms are borrowed from training data sets such as outcrops. The same training sets, because they are exhaustive, could be used to extract *multiple-point statistics* rather than mere two-point correlation or variogram between two spatial locations only. Ignoring multiple-point information from training sets is often regarded as safe practice because such information cannot be checked from the limited well data, or because its selection and quantification calls for a subjective interpretation. Yet any single mapping algo-

[1] Tel: 1 650 723 1774, fax: 1 650 725 2099; e-mail: jef@pangea.stanford.edu
[2] E-mail: ssriniva@ucalgary.ca

rithm, be it stochastic or deterministic, calls for the same amount of prior structural information, i.e. multiple-point statistics. Hence, when explicit modelling is limited to two-point statistics only, the missing higher order statistics is provided by simplistic mathematical models, such as the Gaussian random field which is determined completely by its mean and variogram. One should then question the geological relevance of such purely mathematical models beyond their analytical convenience.

In this paper, we propose to combine established geostatistical methods with pattern recognition tools such as neural networks in order to borrow geological structures from training images in the form of multiple-point statistics (image analysis, pattern extraction and pattern recognition). The resulting structural model (that can be visualized in an unconditional simulation) consists of geo-patterns identified through multiple-point statistics instead of a simple variogram function.

Stochastic reservoir simulation then consists of anchoring these geo-patterns to the actual subsurface hard and soft data (pattern reproduction). Hard data are frozen at their data location; the information content of soft data is calibrated by forward simulation of the physical process (seismic/flow) onto the same training image(s). The pair of corresponding hard and soft training images provides the non-linear multipoint relationship between hard (rock properties) and soft (physical response) variables. It will be shown that instead of calling for the solution of a difficult inverse problem, integration of both seismic and production data can now proceed in an easier forward mode.

2. A DEMONSTRATION

Fig. 1A displays an horizontal section of a fluvial reservoir, which will be considered as the "true" reference reservoir. This section of size 100×150 pixels was generated using an object-based algorithm fluvsim (Deutsch and Wang, 1996).

The variable describing the sand facies is the indicator defined as $I(\mathbf{u}) = 1$ if sand is observed at location $\mathbf{u}$, $I(\mathbf{u}) = 0$ otherwise. The overall true proportion of sand is $p = 0.5$.

Only two wells are extracted from the true reservoir (Fig. 1C) each containing 38 regularly sampled observations of the sand indicator: the sample proportion is $\hat{p} = 0.5$.

Next we construct a set of indirect data (soft data) which could have been provided from a seismic survey (Fig. 1B). A linear averaging of the true sand indicators was performed, followed by a non-linear transformation. Again, any other algorithm could have been used for this purpose. First, at any location $\mathbf{u}$ a linear average $b(\mathbf{u})$ is constructed as follows

$$b(\mathbf{u}) = \frac{1}{25} \sum_{\alpha=1}^{25} i(\mathbf{u} + \boldsymbol{h}_\alpha) \tag{1}$$

where the $\boldsymbol{h}_\alpha$'s are the 24 vectors of the square 5×5 template shown in Fig. 2. The linear averages $b(\mathbf{u})$ are then transformed into their standard ranks $v(\mathbf{u})$ uniformly distributed in $[0, 1]$. These ranks $v(\mathbf{u})$ are the soft data shown in Fig. 1B. These soft data do not provide the location of individual channels, they do however indicate areas of high channel concentration, as one could expect from an actual seismic survey.

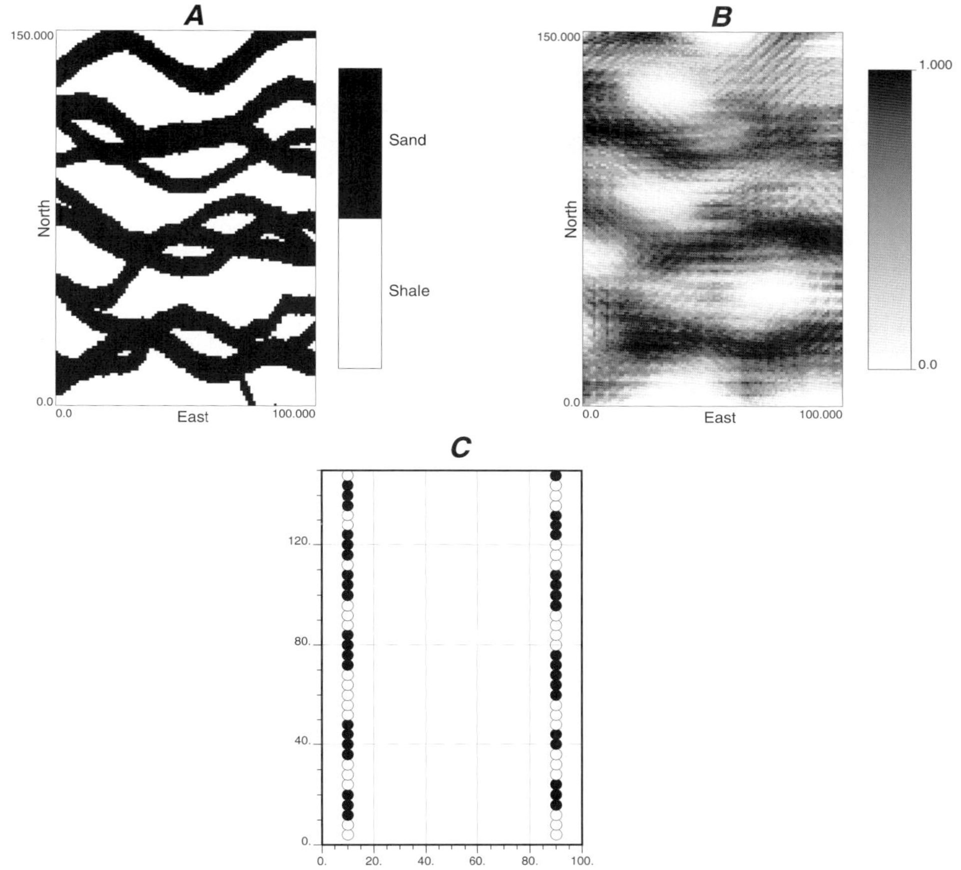

Fig. 1. (A) True reference reservoir, (B) soft or indirect data, (C) hard data observed at wells.

3. BORROWING STRUCTURES FROM TRAINING IMAGES

A training image was generated using the same object-based simulation program fluvsim, see Fig. 3A. One can rightfully argue that the similarity of that training image with the "true" image of Fig. 1A is an extremely favorable situation, since in reality one never knows perfectly the geology and sand continuity. In reality the conceptual geological model is more uncertain, in which case one could provide a set of training images (different geological scenarios) that span the believed uncertainty about the patterns of sand continuity (different channel thickness, sinuousity, orientation, etc. ...). The proposed methodology allows using such multiple training images.

The extraction of multiple-point statistics from this training image and their reproduction in a stochastic reservoir model proceeds in three steps: pattern extraction, pattern recognition and pattern reproduction.

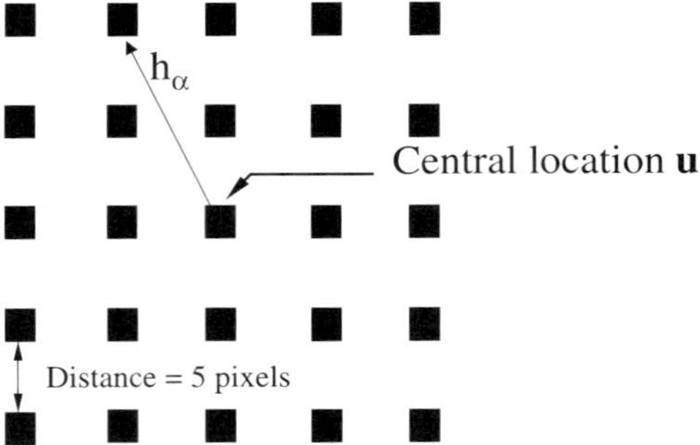

Fig. 2. Template geometry used for obtaining the soft data through an averaging process.

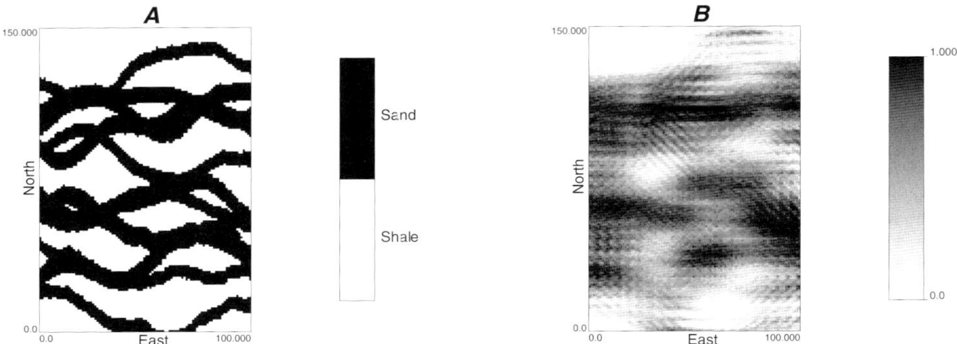

Fig. 3. (A) Training image, (B) soft data training image.

3.1. Pattern extraction

The training image of Fig. 3A is scanned using a *template* t constituted by n_t locations $\mathbf{u}_\alpha$ and a central location $\mathbf{u}$:

$$\mathbf{u}_\alpha = \mathbf{u} + \boldsymbol{h}_\alpha \quad \alpha = 1, \ldots, n_t$$

Fig. 4A displays such a template with $n_t = 12$. The template is used to scan the training image and collect at each location $\mathbf{u}$ the *d*ata *ev*ent:

$$dev(\mathbf{u}) = \{i(\mathbf{u})\,;\, i(\mathbf{u} + \boldsymbol{h}_\alpha),\ \alpha = 1, \ldots, n_t\} \tag{2}$$

We will also use the notations

$$\{i(\mathbf{u} + \boldsymbol{h}_\alpha),\ \alpha = 1, \ldots, n_t\} \equiv \mathbf{i}_t(\mathbf{u}) \equiv (n_t)$$

The set of all data events scanned from the training image results in a *training data set*

$$S_t = \{dev(\mathbf{u}_j),\quad j = 1, \ldots, N_t\}$$

where S_t refers to the training data set constructed with template t. N_t is the number of different center locations of template t over the training image.

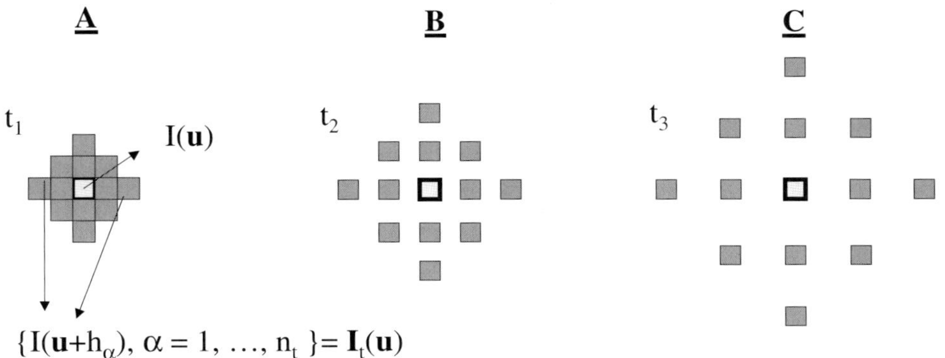

Fig. 4. (A) Template geometry for scanning the training image at the small scale, (B) at the intermediate scale, (C) at the coarse scale.

The choice of the template geometry is important and one should try various template sizes (n_t) and geometries in order to observe (in the resulting simulations) which one reproduces best the structures deemed important in the training image. However, experience (Guardiano and Srivastava, 1992; Wang, 1996) has already shown that isotropic templates perform well in reproducing even complex anisotropic structures. Our own experience has shown that the star-shape templates of Fig. 4 perform well overall.

In presence of large scale structures, the use of one single limited size template, as in Fig. 4A, would not suffice to model the large-scale EW channeling observed in the training image of Fig. 3A. Therefore the three different templates of Fig. 4 were used to scan that training image, resulting in three different data sets S_{t_1}, S_{t_2} and S_{t_3}. Larger scale templates can simply be expanded from the small scale template as shown in Figs. 4B and 4C. As shown in the next two sections, the multiple data sets (at multiple scales) will result in a multigrid simulation (Tran, 1994). Recall that in a multigrid simulation, a simulation is first performed on the coarsest grid. Once that first simulation is finished, the simulated values are assigned to the correct grid locations on the finer grid, and are used as conditioning data on the finer grid and so on. The procedure terminates when the finest grid is simulated. Using the templates of Fig. 4B simulation will therefore be performed on a coarse grid comprising 25 × 37 nodes, next on a 50 × 75 grid, finally on a 100 × 150

3.2. Pattern recognition

Indicator kriging (Journel, 1983) can be viewed as a probabilistic classification, hence it is a pattern recognition tool based on two-point statistics. Indicator kriging provides an estimate of the probability $\Pr\{I(\mathbf{u}) = 1|(n)\}$ that the facies at location **u** be sand given the neighboring well information (n). More precisely, that estimate is provided by the simple kriging expression

$$i^*(\mathbf{u}) = \hat{p} + \sum_{\alpha=1}^{n} \lambda_\alpha(\mathbf{u})[i(\mathbf{u}_\alpha) - \hat{p}] \tag{3}$$

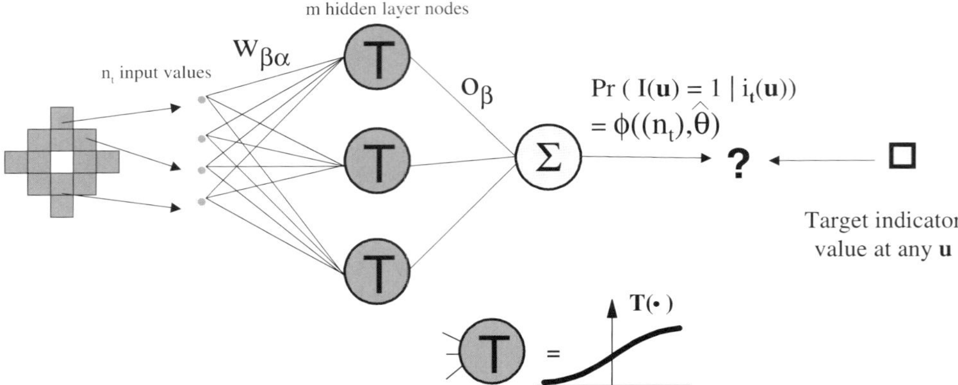

Fig. 5. One-hidden layer neural network. Inputs are the neigboring facies values $\mathbf{i}_t(\mathbf{u}) = (n_t)$ at any location u, target outputs is the facies value $i(\mathbf{u})$. After training the neural network provides the conditional probability $\Pr\{I(\mathbf{u}) = 1|(n_t)\}$.

$i^*(\mathbf{u}|(n))$ is an estimate for the conditional probability $\Pr\{I(\mathbf{u}) = 1|(n)\}$ of having sand at $\mathbf{u}$. $(n) = i(\mathbf{u}_\alpha)$, $\alpha = 1,\ldots,n$ is the set of local well data at $\mathbf{u}$.

The limitation of kriging lies in the representation (3) of the probability. In this paper we propose to use neural networks (Bishop, 1995) to provide an improved model for the local probability model as they are well established classification tools in the field of statistical pattern recognition. We will establish a neural network that provides a general non-linear mapping between any set of neighboring data and the unknown $I(\mathbf{u})$; this mapping will consider all data (n_t) jointly instead of weighting them one by one as done in the kriging expression (3). The neural network, denoted as ϕ_t, is determined by a set of parameters $\boldsymbol{\theta}$ and can be used as a model for the probability of having sand at location $\mathbf{u}$

$$\Pr\{I(\mathbf{u}) = 1|(n_t)\} = \phi_t(\boldsymbol{\theta}) \tag{4}$$

If a one-hidden-layer neural network is used (Fig. 5), the function ϕ_t is given by the following expression (Bishop, 1995)

$$\phi_t(\boldsymbol{\theta}) = \sum_{\beta=1}^{m} o_\beta T\left(\sum_{\alpha=1}^{n_t} w_{\beta\alpha} i(\mathbf{u}+\boldsymbol{h}_\alpha)\right), \tag{5}$$

with

$$\boldsymbol{\theta} = \{o_\beta,\ w_{\beta\alpha},\ \beta = 1,\ldots,m,\ \alpha = 1,\ldots,n_t\}$$

T is the logistic sigmoidal activation function typically used for neural network models (Bishop, 1995, p. 83). The parameters o_β and $w_{\beta\alpha}$ need to be estimated, or in neural network language, the neural network needs to be trained. m is the number of nodes in the hidden layer of the neural network (Fig. 5). Training is performed using a training data set S_t obtained from the training image. Typically neural networks are trained by minimizing a sum-of-square error between a target output and the output ϕ provided by the network for a given input. Thus, the input into the neural network is the set of (n_t) neighboring facies categories at any location **u**, $\mathbf{i}_t(\mathbf{u}) = \{i(\mathbf{u}+\boldsymbol{h}_\alpha),\ \alpha = 1,\ldots,n_t\}$, the target output is the sand indicator at that location **u** as provided by a training data set S_t. In this case, the neural network has only one single output that aims at predicting the indicator of sand at any location **u**. The neural network notation $\phi_t(\boldsymbol{\theta})$ is therefore extended to $\phi_t(\mathbf{i}_t(\mathbf{u});\boldsymbol{\theta})$ to accentuate the explicit dependency to the input, namely $\mathbf{i}_t(\mathbf{u})$. In general when K facies categories need to be modeled the network should have $K-1$ outputs. The network parameters are determined by minimizing the following sum-of squares error

$$Ef = \frac{1}{2N_t}\sum_{j=1}^{N_t}(\phi(\mathbf{i}_t(\mathbf{u}_j);\boldsymbol{\theta}) - i(\mathbf{u}_j))^2 \tag{6}$$

The trained neural network is then denoted as $\phi(\mathbf{i}_t(\mathbf{u}),\hat{\boldsymbol{\theta}})$ and can be evaluated for any arbitrary input vector $\mathbf{i}_t(\mathbf{u})$. In Bishop (1995, p. 201–202) it is shown that the output of a traditional one-layer neural network trained with a sum-of-squares error function trained can be interpreted as a conditional expectation, namely Bishop shows that

$$\phi(\mathbf{i}_t(\mathbf{u}),\hat{\boldsymbol{\theta}}) = \mathrm{E}[I(\mathbf{u})|\mathbf{i}_t(\mathbf{u})]$$

Since the expectation of an indicator variable is a probability, i.e.

$$\mathrm{E}[I(\mathbf{u})|\mathbf{i}_t(\mathbf{u}_j)] = \Pr\{I(\mathbf{u}) = 1|\mathbf{i}_t(\mathbf{u})\} \tag{7}$$

the neural network output verifies the conditional probability of $I(\mathbf{u}) = 1$ given its neighbors $\mathbf{i}_t(\mathbf{u})$. The least-square criterion of Eq. (6) however does not ensure that this probability is within $[0,1]$. However, experience has shown that the recorded violations are very small (in the order of 0.01).

If a multigrid simulation is warranted, then a conditional probability distribution (7) for each grid needs to be established. Therefore, for each grid, a different neural network has to be trained using the data S_{t_i}. In the case study of Fig. 1, three grids are used, and three data sets S_{t_i} were obtained through scanning. For each of the data sets S_{t_i}, a neural network was constructed with 30 hidden-layer nodes, i.e. 30 functions T in Eq. (5). The number (30) of hidden layer nodes seems rather arbitrary in this case, but some sensitivity analysis showed that increasing the number of nodes beyond 30 did not improve the resulting simulated realizations. A cross-validation, typically used in neural network training is applied while training (see Bishop, 1995, p. 343–344).

3.3. Pattern reproduction

Once various networks have been trained (one for each template), they can be used to perform unconditional simulation or conditional simulation to local hard data. Markov chain Monte Carlo (McMC) sampling with a Metropolis–Hastings sampler (see Hegstad et al., 1993 for a general account) is applied. The method proceeds as follows:

First, an initial arbitrary (e.g. purely random) image is constructed, i.e. all simulation grid nodes are assigned a random value. Any local hard data from wells are assigned to the nearest grid node locations. The McMC sampler will update that random image by iteratively visiting each node, except the hard data locations, using a random path and by performing a two-step operation at each node: proposal and acceptance. First, at each nodal location $\mathbf{u}$, a new facies $i_{\text{new}}(\mathbf{u})$ is proposed to replace the current category $i_{\text{current}}(\mathbf{u})$, that new facies is drawn from the sample prior distribution $\hat{p}$. If the new facies is the same as the current, one moves to another node. Else, the new facies value is accepted with a probability given by

$$\alpha(i_{\text{current}}(\mathbf{u}), i_{\text{new}}(\mathbf{u})) = \min\left\{1, \frac{\Pr\{\textit{new} \text{ proposed facies at } \mathbf{u} \mid \mathbf{i}_t(\mathbf{u})\}}{\Pr\{\textit{current} \text{ existing facies at } \mathbf{u} \mid \mathbf{i}_t(\mathbf{u})\}} \frac{p_{\text{current}}}{p_{\text{new}}}\right\} \quad (8)$$

Since there exists only two facies, the acceptance probability (8) has only two case: in the case where $i_{\text{new}}(\mathbf{u}) = 1$ (sand facies), Eq. (8) becomes

$$\alpha(i_{\text{current}}(\mathbf{u}) = 0, i_{\text{new}}(\mathbf{u}) = 1) = \min\left\{1, \frac{\phi(\mathbf{i}_t(\mathbf{u}), \hat{\boldsymbol{\theta}})}{1 - \phi(\mathbf{i}_t(\mathbf{u}), \hat{\boldsymbol{\theta}})} \frac{1 - \hat{p}}{\hat{p}}\right\}$$

where $\phi(\mathbf{i}_t(\mathbf{u}), \hat{\boldsymbol{\theta}})$ is the neural network output evaluated with $\mathbf{i}_t(\mathbf{u})$ (the neural net input). In the case where the new proposed facies $i_{\text{new}}(\mathbf{u}) = 0$ (mud), Eq. (8) becomes

$$\alpha(i_{\text{current}}(\mathbf{u}) = 1, i_{\text{new}}(\mathbf{u}) = 0) = \min\left\{1, \frac{1 - \phi(\mathbf{i}_t(\mathbf{u}), \hat{\boldsymbol{\theta}})}{\phi(\mathbf{i}_t(\mathbf{u}), \hat{\boldsymbol{\theta}})} \frac{\hat{p}}{1 - \hat{p}}\right\}$$

In order to simulate the long range structure of the channels a multigrid simulation is performed. First, simulation is performed on a coarse grid comprising 25×37 nodes. For this simulation, we use the neural network trained with the data set S_{t_3}. Once that first simulation is finished, the simulated values are assigned to the correct grid locations on the finer grid, in this case the 50×75 grid. Next, conditional simulation with the neural network trained on the S_{t_2} dataset is performed on the finer grid. Finally the same procedure is repeated for the final grid of 100×150. For this grid we use the neural network trained with the S_{t_1} data set.

Fig. 6A shows an unconditional simulation using the described methodology. From a visual inspection, the channel characteristics of the training image (Fig. 3A) are well reproduced. Fig. 6B shows a simulation, conditional to the well observations. The channel location is now much more confined due to the conditioning to the two wells. Consider next in Fig. 6C a conditional simulation with a traditional indicator method (sisim, see Deutsch and Journel, 1998) using the exhaustive variogram of the training image as a model. Clearly the use of two-point statistics only fails to reproduce the channeling behavior.

4. CONDITIONING TO INDIRECT DATA

Stochastic reservoir models, and more generally the task of stochastic image construction, must be conditioned to an abundance of indirect data measured at different scales and with different precision. In our case the above developed method for conditional simulation needs to be extended to be able to condition to the indirect data of

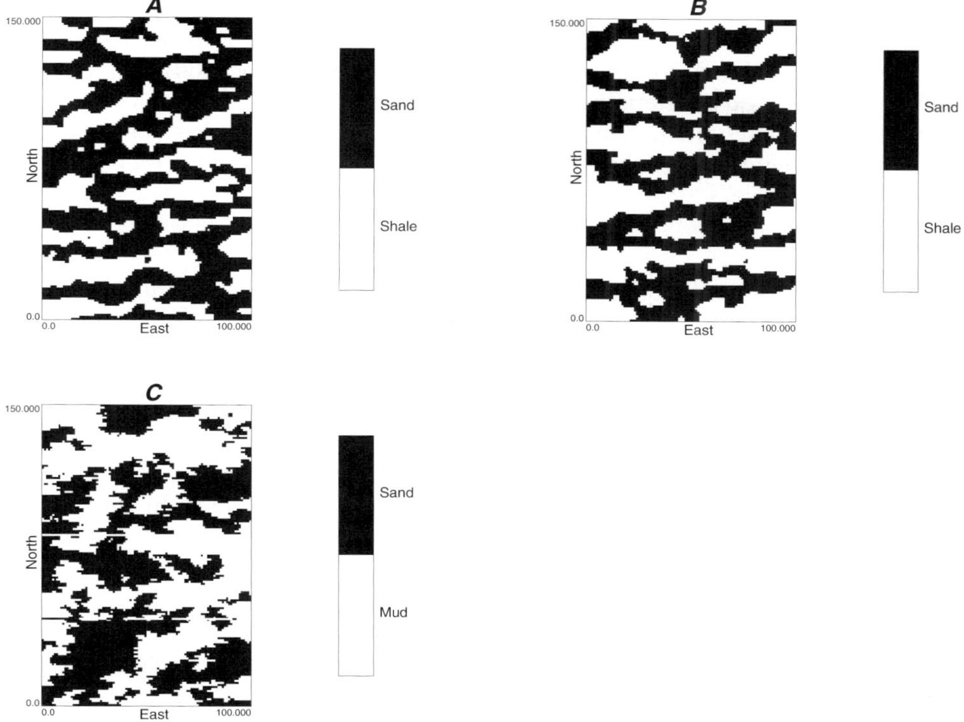

Fig. 6. (A) unconditional simulation (representing a sample from the prior structural model), (B) conditional simulation, (C) conditional simulation based on traditional indicator model variogram only.

Fig. 1B. This problem has in other disciplines been formulated as an inverse problem (Tarantola, 1987). The inverse problem consists in finding the unknown lithology at each location (sand or mud) given the seismic data. Such inverse problem is generally ill-posed, i.e. the solution to the inverse problem is not unique. To solve such inverse problems, Tarantola proposes to construct a forward model, i.e. a model that provides the seismic, given the lithology at each location. The forward model is unique in the sense that given the lithology, there exists only one unique seismic response (up to measurement errors). Tarantola's method however relies on the classical Gaussian assumptions to invert the lithology from the seismic using the forward model and does not deal with spatially correlated phenomena. We extend his idea to condition the fluvial channel reservoir (clearly non-Gaussian) to the seismic data.

In order to be able to construct such forward model, the same averaging process as in Eq. (1) is performed but now on the (hard-data) training image of Fig. 3A to obtain a soft-data training image (Fig. 4A). Again, this is a rather favorable situation since in reality, one would never know the exact physical transfer function between facies types and seismic response. In an actual case study one could obtain synthetic seismic by applying a forward seismic simulation model to the hard data training image. This forward modelling can be performed using techniques described in Mavko et al. (1998). The pair of hard and soft data training image is used to obtain a model relating the seismic information $y(\mathbf{u})$ to facies values at $\mathbf{u}$ and neigboring locations. Again neural networks will be used to determine such relation.

Next, we propose an algorithm to draw stochastic reservoir models conditioned to both hard and indirect data using a similar three steps procedure:

4.1. Pattern extraction

The pair of hard and soft training images is used to extract information on the relationship between the soft and the hard variable, hence evaluating the "softness" of the indirect data. In this case, we decided to relate a single seismic datum at any location $\mathbf{u}$ in the soft data training image to collocated and to neigboring facies information at the corresponding $\mathbf{u}$ in the hard data training image. Again, we construct a data set (to be used for regression in next section), now by scanning both training images. In the hard data training image facies information at each location $\mathbf{u}$ and within a neighboorhood is retained. The neigborhood is again defined using the template in Fig. 4. In the soft data training image, we retain only a single seismic data at each $\mathbf{u}$. The data set constructed then consists of a number of data events, namely

$$dev(\mathbf{u}) = \{i(\mathbf{u})\,;\, i(\mathbf{u}+\boldsymbol{h}_\alpha),\ \alpha = 1,\dots,n_t;\ y(\mathbf{u})\} \tag{9}$$

$y(\mathbf{u})$ is the seismic event at $\mathbf{u}$ in the training image. The training data set then consist of all data scanned from both training hard and training soft image.

$$S_t = \{dev(\mathbf{u}_j),\ j = 1,\dots,N_t\} \tag{10}$$

As in the case of borrowing structure from training images, we establish three data sets, S_{t_1}, S_{t_2}, S_{t_3}, each with a template defined on a different scale.

4.2. Pattern recognition

Neural networks are extremely powerful tools for recognizing the possible non-linear multipoint relation existing between soft data and hard variables. CPU-cheap neural network computations have already proven to adequately model difficult seismic inversion tasks (Röth and Tarantola, 1994) or replace computationally demanding flow simulators (Srinivasan, 1999). The training data (9) can be used to train so-called *calibration neural networks* ϕ_{cal}, see Fig. 7. The task of such trained calibration neural network is to build a regression model between the multiple facies values and the single seismic event, based on the training data (10):

$$\mathrm{E}[Y(\mathbf{u})|i(\mathbf{u}),\mathbf{i}_t(\mathbf{u})] = \phi_{\text{cal}}(i(\mathbf{u}),\mathbf{i}_t(\mathbf{u});\boldsymbol{\theta})$$

Again we rely on the fact that the neural network output can be interpreted as an expectation (Bishop, 1995). The calibration neural network serves as a forward model because it models the seismic $Y(\mathbf{u})$ based on the lithology in the neighborhood, namely $i(\mathbf{u}),\mathbf{i}_t(\mathbf{u})$. Once the network is trained, it can predict any seismic $y(\mathbf{u})$ from any given

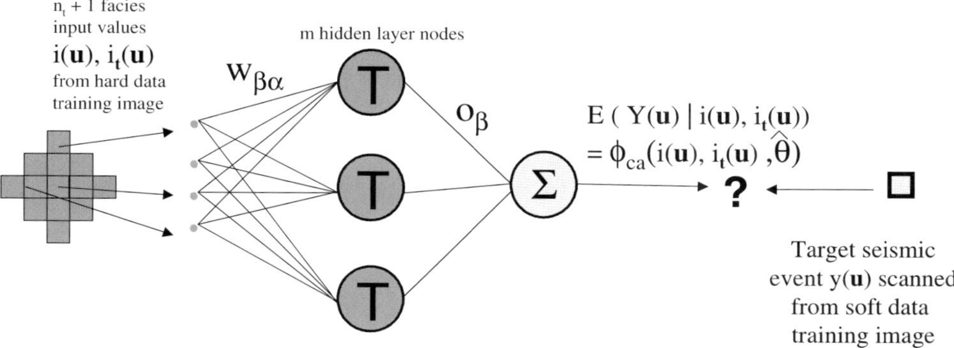

Fig. 7. Inputs to the calibration neural network are the facies values $i(\mathbf{u})$ and $\mathbf{i}_t(\mathbf{u}) = (n_t)$ at any location u, target outputs is the seismic datum $y(\mathbf{u})$. After training the neural network provides the conditional expectation $\mathrm{E}[Y(\mathbf{u})|i(\mathbf{u}_j), \mathbf{i}_t(\mathbf{u}_j)]$.

neigboring facies information. The training is again performed by minimizing a sum-of-squares error function between the target seismic value observed from the training image and the output of a one-hidden-layer network (30 hidden layer nodes)

$$Ef_{\text{soft}} = \frac{1}{2N_t} \sum_{j=1}^{N_t} (\phi_{\text{cal}}(i(\mathbf{u}_j), \mathbf{i}_t(\mathbf{u}_j); \boldsymbol{\theta}) - y(\mathbf{u}_j))^2$$

After the network parameters are determined, the performance of the training must be evaluated. The trained neural net output is compared to the target data $y(\mathbf{u})$ on a scatterplot in Fig. 8. The scatterplot shows the difference between what the neural network has learned from the input and what it should have learned, i.e. the target output data. The correlation coefficient, which should be 1 for perfect training, is 0.86.

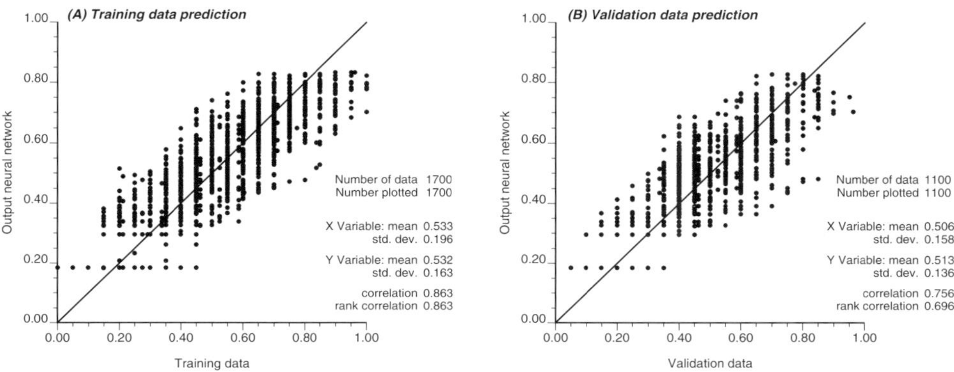

Fig. 8. Scatterplots of the neural network predicted seismic versus training seismic scanned from soft data training image.

Such scatterplot can then be used (program scatsmth, see Deutsch and Journel, 1998, p. 218–221) to model the conditional probability of obtaining a seismic event at any location **u** given a prediction of the seismic event using the network. We denote this conditional probability as

$$\Pr\{Y(\mathbf{u}) = y(\mathbf{u}) | \phi_{\text{cal}}(i(\mathbf{u}); \mathbf{i}_t(\mathbf{u}), \hat{\boldsymbol{\theta}})\}$$

This conditional distribution will be used in the following step to condition to the seismic data.

4.3. Pattern reproduction

Using the trained calibration neural network, the channel reservoir can now be conditioned to the soft information. The aim is to construct simulations such that, when the procedure of Eq. (1) is evaluated on these simulations one retrieves back the soft data.

The Metropolis–Hastings sampler can be adapted to account for the soft data by changing the acceptance criterion as follows

$$\alpha(i_{\text{current}}(\mathbf{u}), i_{\text{new}}(\mathbf{u})) = \min\{1, C_1 \times C_2\} \tag{11}$$

where C_1 refers to the term in the original criterion in Eq. (8) without the soft data. The term C_2 accounts for conditioning to the soft data as follows (see appendix A how this term is established)

$$C_2 = \frac{\Pr\{Y(\mathbf{u}) = y(\mathbf{u}) | \phi_{\text{cal}}(i_{\text{new}}(\mathbf{u}); \mathbf{i}_t(\mathbf{u}), \hat{\boldsymbol{\theta}})\}}{\Pr\{Y(\mathbf{u}) = y(\mathbf{u}) | \phi_{\text{cal}}(i_{\text{current}}(\mathbf{u}); \mathbf{i}_t(\mathbf{u}), \hat{\boldsymbol{\theta}})\}}$$

$y(\mathbf{u})$ is now the *observed* soft data of Fig. 1B.

As before, a multigrid procedure is used to obtain the simulated realizations. Fig. 9A shows a conditional simulation constrained to the seismic data of Fig. 1B. Fig. 9B shows the forward simulated seismic (using the averaging process of Eq. (1)) on that simulation and when visually compared to Fig. 1B shows that the seismic is approximately reproduced. The reproduction is only approximative because, as shown in the previous section, the network training was not perfect (correlation was only 0.86).

The aim of using geostatistics is to provide a probability statement about the reservoir given the available information. In this binary case (sand/mud), the probability of having sand is quantified by a map showing the average over a number realizations (termed E-type estimate, Deutsch and Journel, 1998). An E-type estimate of a binary random variable indicates the probability of obtaining sand at any location **u**. Fig. 10-A shows the E-type estimates over 25 realization for a conditional simulation with hard well data only and Fig. 10-B a conditional simulation with hard and soft data. Fig. 10 shows that the seismic reduces considerably the uncertainty about the lithology between the wells.

5. PRODUCTION DATA INTEGRATION

Spatial variations exhibited by the reservoir permeability field have an immediate impact on the fluid producing characteristics of the reservoir. The relationship between

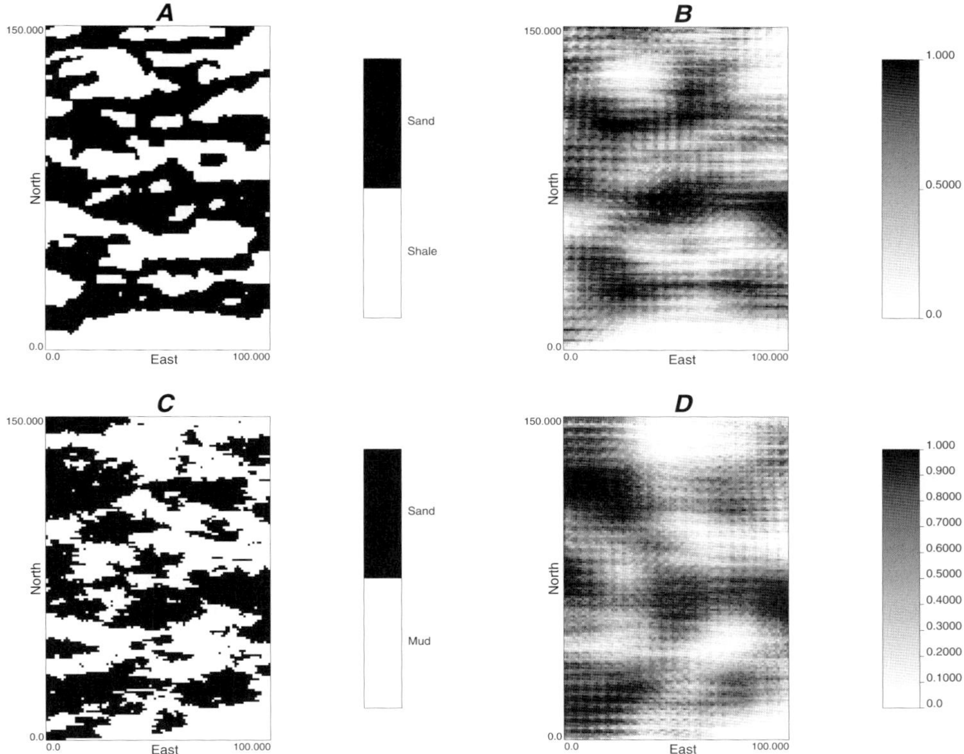

Fig. 9. (A) conditional simulation with the soft data (B) forward simulated seismic (C) single sisim realization constrained to the soft data, (D) corresponding forward simulated seismic.

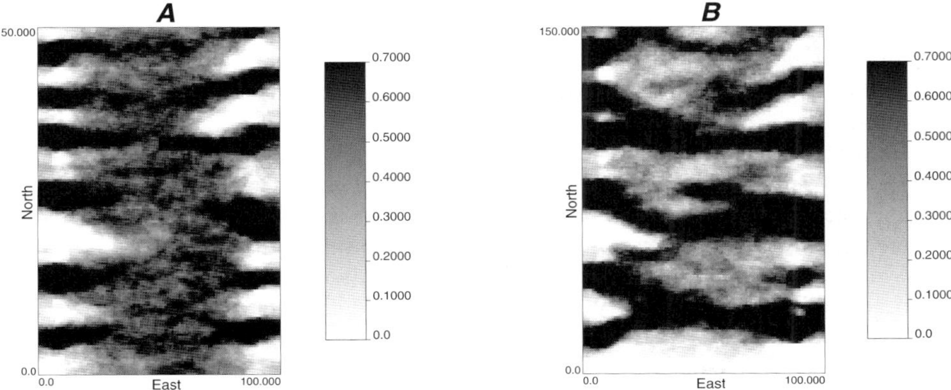

Fig. 10. (A) E-type estimate for conditional simulation with hard data only, (B) E-type estimate for conditional simulation with hard and soft data.

a reservoir response, say the well pressure $p_w(\mathbf{u}',t)$ at location $\mathbf{u}'$ and time t, and the permeability field $k(\mathbf{u}), \mathbf{u} \in$ Reservoir is complex and varies with both location $\mathbf{u}'$ and time t. Over the years, considerable time and effort has been spent on understanding this complex relationship and has resulted in the development of numerous, numerical flow simulators.

If the permeability field is not fully known, it is modeled with a suite of L equi-probable realizations $k^{\ell}(\mathbf{u}),\ \ell = 1,\ldots L$. All these realizations should be such that application of the flow simulator would yield a simulated well response close to the observed value:

$$p_w^{\ell}(\mathbf{u}',t) \simeq p_w(\mathbf{u}',t), \quad \forall \mathbf{u}', t, \text{ and } \forall \ell = 1,\ldots,L$$

Since flow and future productivity is controlled by the spatial connectivity of permeability in the reservoir, the flow related data $p_w(\mathbf{u}',t)$ is particularly valuable for idenifying patterns of spatial variability of the reservoir. By identifying and replicating these patterns accurately, one hopes that the resultant permeability models $k^{\ell}(\mathbf{u}), \mathbf{u} \in$ Reservoir could be used to predict the future reservoir performance accurately.

There are various avenues to constrain reservoir permeability models to available historic production data:

(1) Use traditional geostatistical algorithms to condition permeability fields to well log data and a variogram model $\gamma(\mathbf{h})$. Subsequently perform a history matching step, i.e. process each geostatistical model through the flow model to get the response $p_w^{\ell}(\mathbf{u}',t)$. Retain only those geostatistical models that exhibit minimum deviation of p_w^{ℓ} from the target p_w. The probability of drawing one or more models (ℓ) that will match acceptably the target p_w is usually extremely low, rendering that brute force approach unpractical.

(2) In the second approach the relationship between the stochastic permeability value $k(\mathbf{u}), \mathbf{u} \in$ Reservoir and the response p_w is expressed in the form of a cross-correlation $C_{KP}(\mathbf{h})$. Once that correlation is computed, fast geostatistical algorithms exist for simulating permeability field $k^{\ell}(\mathbf{u})$ under the additional covariance constraint $C_{KP}(\mathbf{h})$ (Kitanidis and Vomvoris, 1983; Dagan 1985). However, since the relation TF_w between P_w and K is complex, significant assumptions are needed to compute the correlation C_{KP}.

(3) A third more practical approach calls for calculating the sensitivity of the well response p_w to permeability perturbations in the reservoir and then utilize this sensitivity to systematically perturb an initial model of the permeability field until it matches the desired response p_w (Ramarao et al., 1995; Gomez et al., 1997). This approach requires multiple runs of the flow simulator TF_w in order to establish the sensitivity of the flow response to permeability perturbations. The issue of which of the very large number of permeability values $k^{\ell}(\mathbf{u}), \mathbf{u} \in Reservoir$ should be perturbed is resolved by some authors by limiting the perturbation to a few master locations; the perturbation at these master locations are then spread to other locations by some interpolation process (Gomez et al., 1997).

Application of these techniques to invert production data and obtain a representation of the permeability field in 3-D is difficult and computationally expensive. An alternate, quicker methodology for integrating production data into reservoir models is presented in this section. This proposed methodology utilizes well pressure history data and the master point approach mentioned above, to arrive at an areal proportion map indicative of the high pay regions of the reservoir. Once such a coarse determination of areal pay proportions is made, that information is then used to constrain geostatistical models that additionally honor other available information such as well logs and seismic data. In fact, the novel multiple point simulation procedure outlined earlier is used to integrate the areal proportion map obtained by inverting well pressure data. The resultant reservoir models retain the correct pattern of spatial variability of the reservoir in addition to being history matched.

5.1. Information in production data

History matching is a critical step in most reservoir simulation studies. The historic production data is used for fine-tuning the flow simulator and adjust critical parameters such as the spatial distribution of reservoir permeability, other flow related functions such as relative permeabilities and fluid contacts. The tuned reservoir model is then used to predict the future performanc of the reservoir. Frequently, in order to arrive at a history match, the reservoir is segregated into zones and the average permeability within the zones is perturbed. This facilitates history matching, but more importantly, it underscores the influence of the mean permeability in different regions of the reservoir on the observed production response. While the connectivity of the reservoir (in terms of the multiple point characteristics of the permeability field) has an important influence on the areal sweep efficiency corresponding to displacement proceses and responses such as breakthrough time and water-cuts, other reservoir responses such as average pressure, fluid rates, cummulative recoveries are more likely impacted by mean levels of permeability in different regions of the reservoir. In the limit, it can be postulated that the available production data such as the well pressure profiles are indicative of the areal proportions of pay at various locations of the reservoir. For a 3-D flow model, this amounts to achieving a history match by perturbing the vertically averaged permeability at different locations within the reservoir. This is the underlying paradigm of the proposed methodology outlined below.

The updating of a vertically averaged permeability field such that it reproduces the observed well response can be viewed in an optimization context. The objective of the optimization procedure is to update that prior permeability model such that applying the flow simulator on the final optimized model would yield the observed response p_w. Numerous techniques for efficiently solving this optimization problem have been put forth. The sequential self-calibration (SSC) technique, see Gomez et al. (1997) is one such technique.

Practical implementation of an optimization algorithm for iteratively updating a prior vertically averaged permeability model using the flow response data requires an efficient scheme for perturbing the permeability field. Clearly, perturbing permeability values one location at a time and then computing the corresponding well response p_w^i at each iteration step i using the transfer function flow simulator is computationally inefficient. The SSC algorithm utilizes a set of master points for perturbing jointly the permeability values at a few select locations within the reservoir and, then subsequently spreading the perturbations to all locations within the reservoir in such a way that the spatial continuity of the permeability field is preserved. Master points are located on a pseudo-regular grid at the start of the algorithm. The optimum perturbation at the master point location is one that results in the closest match between the computed flow response and the target value p_w. In order to calculate this optimal perturbation, an iterative procedure is implemented in which trial perturbations are proposed at the master point locations and spread to all locations using kriging. The flow response corresponding to the perturbed permeability field is computed using a linearized flow model. A gradient based optimization scheme then yields the optimal perturbation based on the linearized flow model. After convergence of this iterative optimization procedure using the linearized flow model, the full complex, non-linear flow simulator is applied on the resultant permeability field. The mismatch between the computed flow response and the target is calculated. If this mismatch is deemed significant, a new set of perturbations are determined and the inner optimization loop is again repeated. The algorithm derives the name 'sequential self-calibration' due to the calibration process achieved by this outer loop using the full flow simulator.

Although, significant speed-up of the inversion process is realized due to the concept of master points and fast optimization using linearized flow models, the methodology is *cpu* demanding, when applied to a full 3-D field simulation scenario. The proposal to utilize SSC to condition the vertically averaged permeability field (2-D) to production data reduces the *cpu* requirements significantly.

As a demonstration of the SSC based inversion procedure, consider the following reference 3-D reservoir characterized by ellipsoids of shale embedded in sandstone as depicted in Fig. 11. The proportion of shale in the reservoir is assumed to be 40%. In order to clearly differentiate between pay and non-pay, the shale is assigned a constant perme-

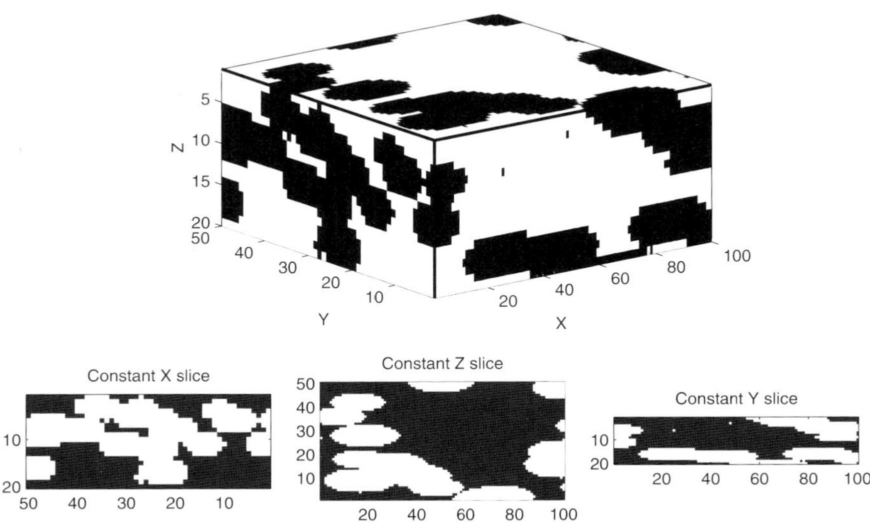

Fig. 11. Reference reservoir comprising of shale bodies embedded in sandstone.

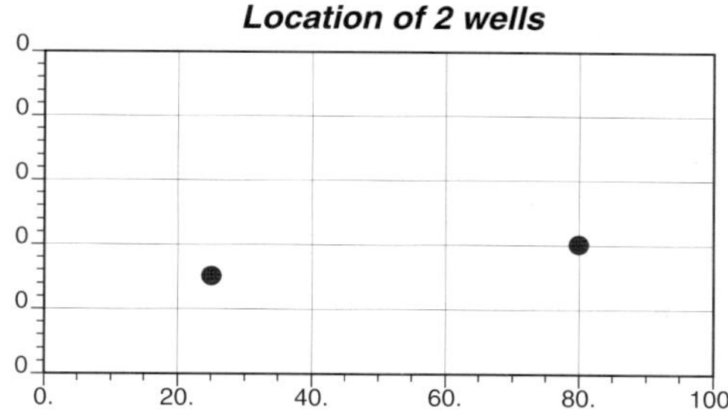

Fig. 12. Location of the two wells where production data is available.

ability of 3 md. while the sand is assumed to be of medium quality and hence assigned a uniform permeability of 700 md. A pair of wells are assumed for production from this reservoir and their locations are shown in Fig. 12. The reservoir is discretized using 100 blocks in the E-W direction, 50 blocks in the N-S direction and 20 blocks in the vertical. The block size is assumed to be 40 ft. in the horizontal directions and 2 ft. in height. Porosity is assumed to be constant throughout the reservoir and equal to 30%. The reference flow response at the two wells due to primary production is obtained by flow simulation and is shown in Fig. 13. In order to mimic a real production scenario, one of the wells is assumed to cycle and the interference response at the other well is recorded.

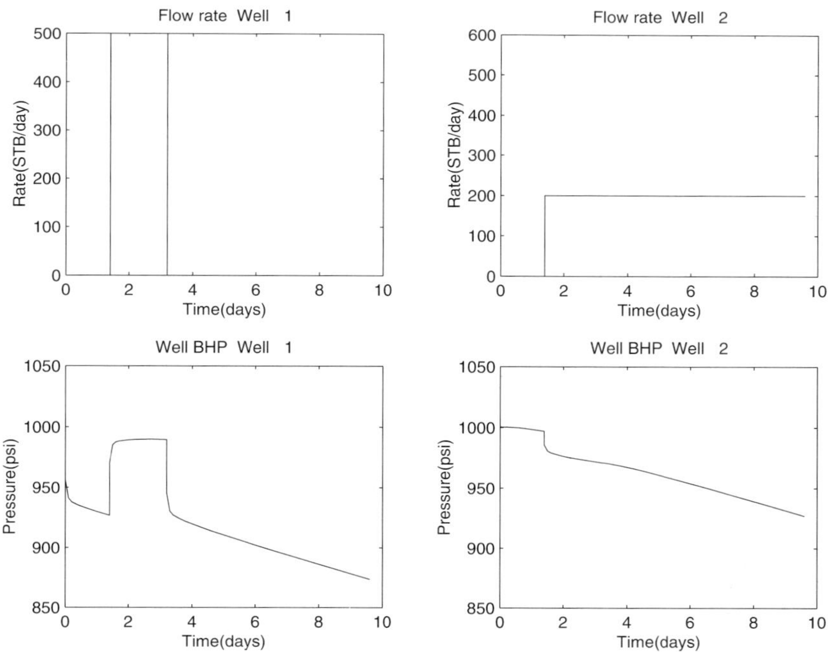

Fig. 13. Historic flow rate data and corresponding pressure response at the wells.

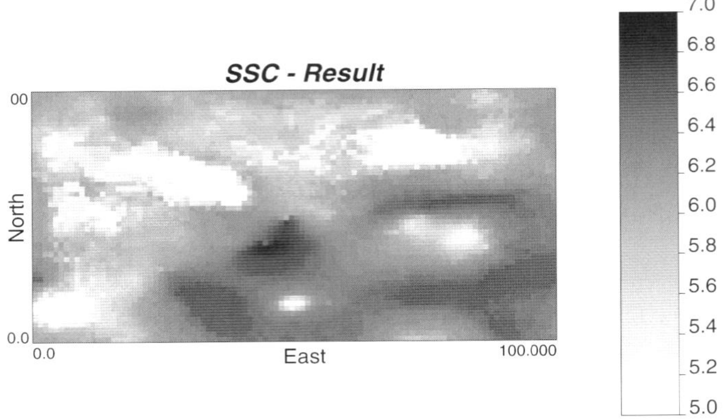

Fig. 14. One realization of the vertically averaged permeability field obtained at the end of the SSC procedure.

It is postulated that the observed production response is influenced by the areal variation of permeability in the reservoir. The SSC procedure outlined above is used to perturb an initial uniform vertically averaged permeability field into a final map reflecting the variations in average permeability in the reservoir. The fluid properties and saturation functions are assumed to be exactly known for the SSC inversion procedure as are the location of the no flow boundary. The pressure response at the wells corresponding to known flow rate changes is used to guide the optimization procedure. Fig. 14 is a realization of the vertically averaged permeability map obtained at the end of the SSC procedure. The initial (corresponding to the initial uniform field) match as well as the final match of well pressures is shown in Fig. 15. For comparison, the corresponding vertically avareged permeability map corresponding to the reference reservoir is shown in Fig. 16. The comparison reveals that the interference patterns in the production data at the two wells serves to constrain the average permeability in the region between the two wells. The high permeability regions in the middle of the reservoir observed in the reference map is reproduced quite well in the SSC derived realization.

In order to confirm that the production data recorded at wells is influenced by the areal variations in the average permeability and that the SSC procedure does indeed detect patterns of variations in production data influenced by the permeability field, the procedure was repeated using data from four wells. The locations of the four wells is shown in Fig. 17. The reference production data at the four wells is shown in Fig. 18. The SSC procedure was repeated utilizing the available production information and the resultant areal variations in average permeability is shown in Fig. 19. The convergence characteristics of the SSC procedure in terms of the pressure match is shown in Fig. 20.

Comparing the average permeability variations indicated by the two well production data and the four well production data, it can be observed that the two well interference data causes permeability features to be simulated that extend in the direction of the wells. In the four well case, on the other hand, the interaction between the wells causes

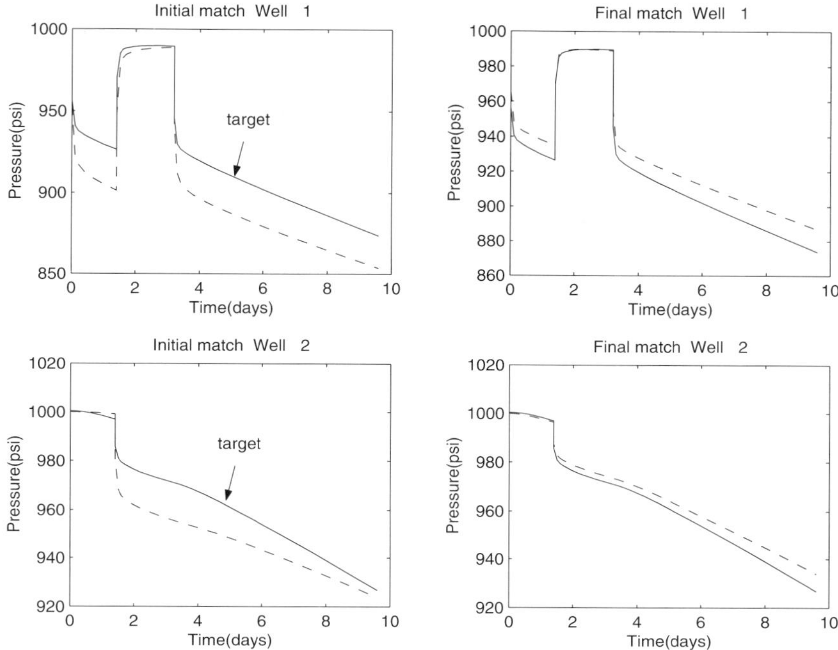

Fig. 15. The history match corresponding to the initial uniform permeability field and the final match at the end of the SSC procedure.

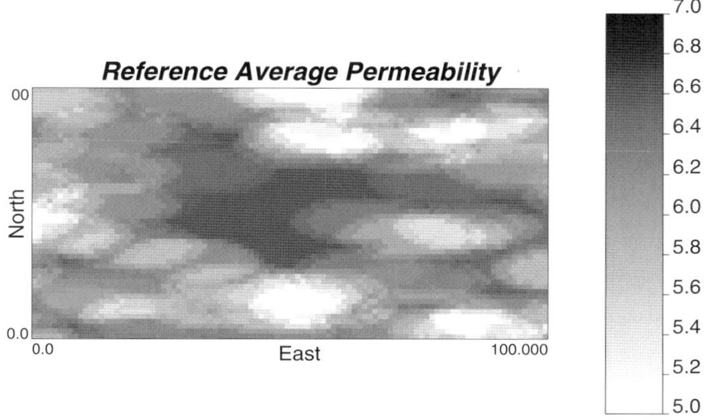

Fig. 16. Vertically averaged permeability corresponding to the reference reservoir.

a high permeability zone to be simulated in the middle of the reservoir. There is no preferential directionality of the simulated high average permeability regions. The same variogram model, inferred from the reference average permeability map is utilized for both the cases. The results suggest that it is more difficult to resolve subtle variations in

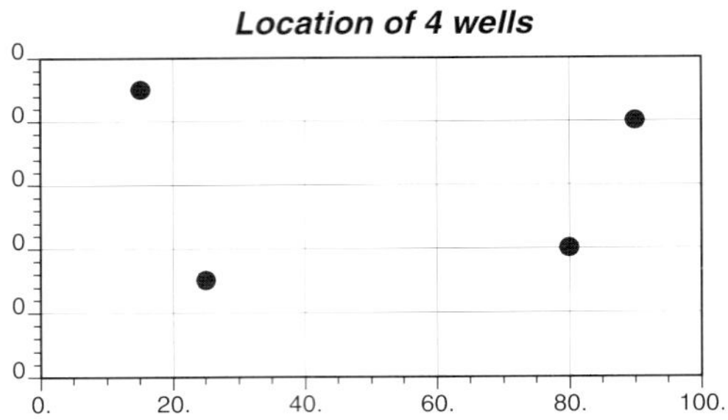

Fig. 17. Location of the four production wells.

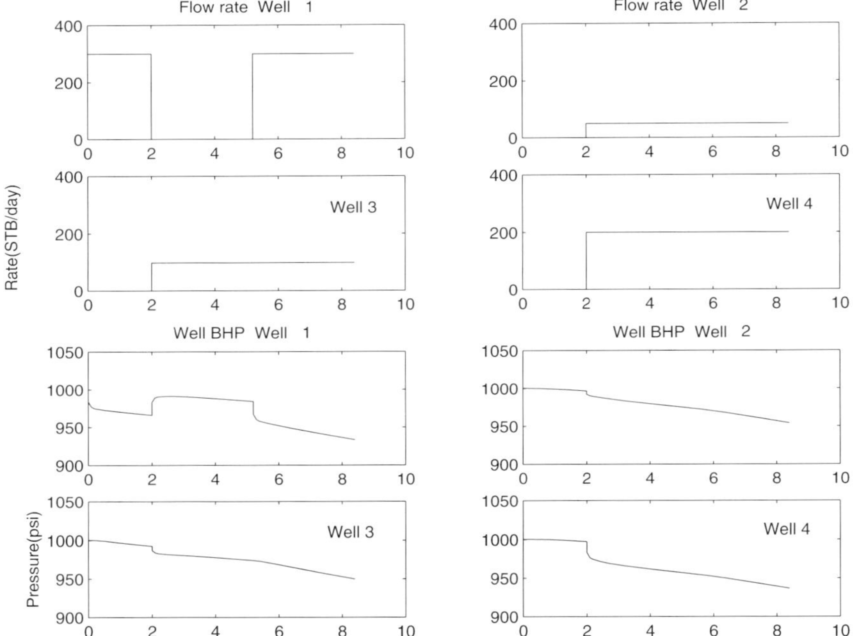

Fig. 18. Production history at the four wells. The top four figures are the production rate history while the bottom figures are the correspoding well pressures

the permeability field using data at multiple wells due to complex interactions between wells. Nevertheless, the SSC procedure does result in fairly accurate identification of regions of the reservoir that have higher vertically averaged permeability.

5.2. Integrating production based data into reservoir models

The vertical average permeability variations obtained using SSC can be converted to the corresponding areal proportion map. Thus if k_{shale} and k_{sand} are known and $p(\mathbf{u})$ is

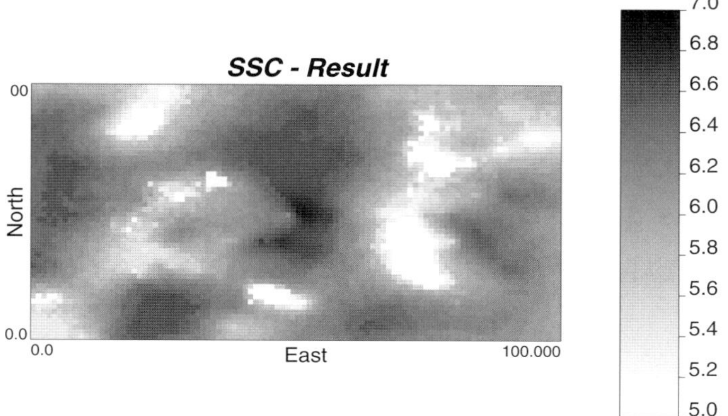

Fig. 19. SSC realization of vertically averaged permeability, constrained to the four well production data.

the unknown proportion of shale corresponding to the 2-D location $\mathbf{u}$, then:

$$k_{\mathrm{SSC}}(\mathbf{u}) = p(\mathbf{u}) \cdot k_{\mathrm{shale}} + (1 - p(\mathbf{u})) \cdot k_{\mathrm{sand}}$$

from which the proportion of shale $p(\mathbf{u})$ can be determined.

A training image containing ellipsoidal shapes similar to the true reservoir is generated using a Boolean technique. Recall that the 3D training image is not constrained by any well or production data, hence is fast and easy to generate. Instead of the neural net approach outlined above, we use an alternative method sequential simulation method, termed snesim (Single Normal Equation SIMulation, Strebelle, 2000) to generate multiple realizations, constrained to the 2D areal proportion map and the 3D well data. The aim of the snesim methodology is the same as for the neural net approach: it renders geological structure constrained to hard and soft data. The snesim method is a non-iterative conditional simulation technique as opposed to teh neural net technique which is iterative. The snesim technique allows constraining to vertical or areal proportion data.

The results obtained by implementing the simulation procedure accounting for areal proportions from SSC are shown in Fig. 21 corresponding to the 2 well case and Fig. 22 for the 4 well case. These figures correspond to one realization of the reservoir obtained by the simulation procedure. The ellipsoidal shape of the shale objects is reproduced remarkably well. The elongated structure of the permeability field in the direction of the two wells observed for the two well case manifests itself in the form of arrangement of shale objects in a pattern between the wells.

5.3. Results and discussion

The combination of the SSC procedure followed by the *snesim* algorithm accounting for areal proportions from production data results in reservoir models that are consistent with the pattern of variability of the target reservoir. However verification of the flow performance of the simulated reservoir models is necessary before any statement about

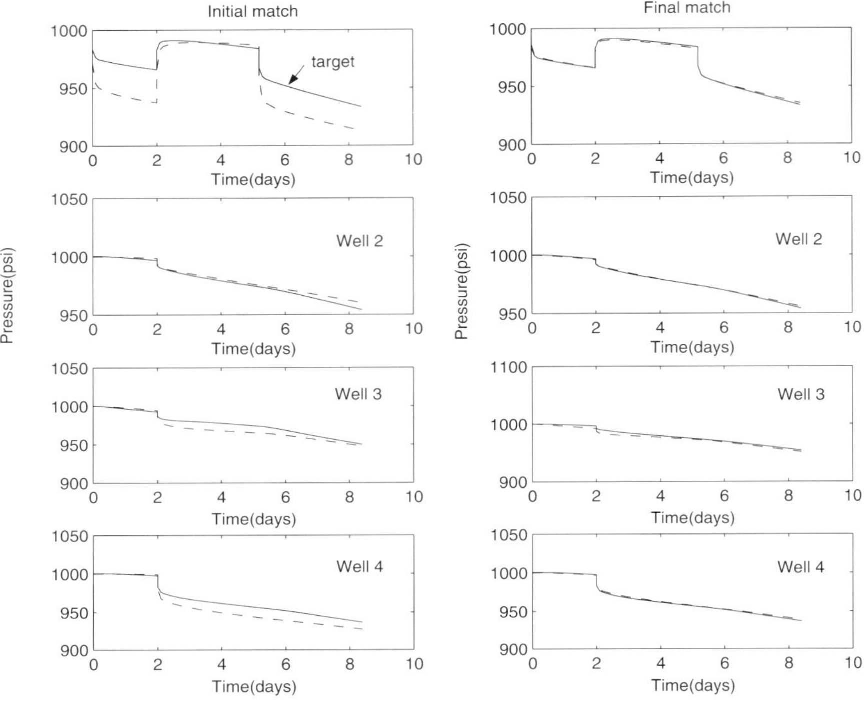

Fig. 20. Initial match corresponding to the initial uniform field and the final match at the end of the SSC simulation.

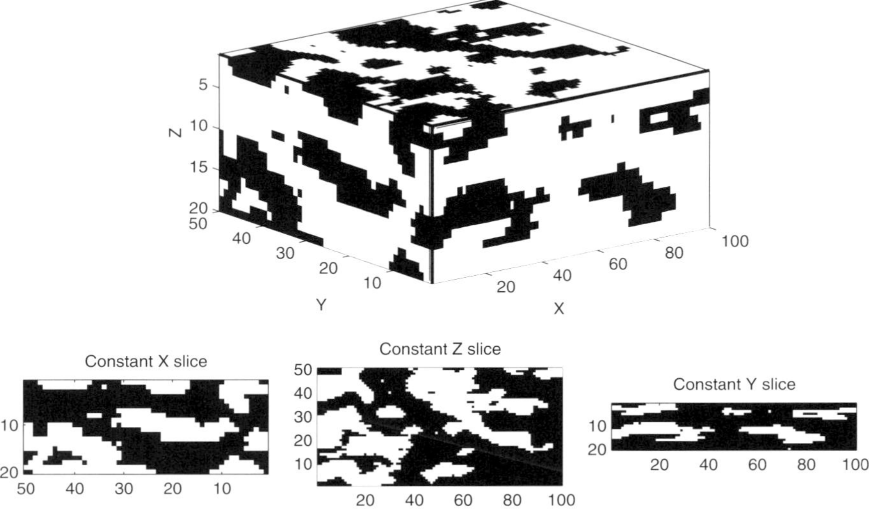

Fig. 21. A 3-D reservoir model constrained to multiple point statistics and production data from two wells.

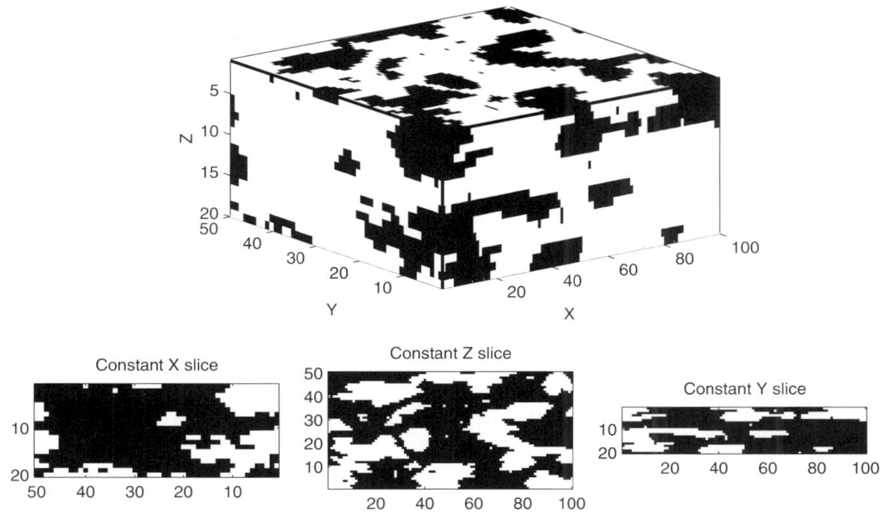

Fig. 22. One realization of the 3-D permeability field constrained to multiple point statistics and production data from four wells.

the validity of the proposed methodology can be made. For this purpose, ten equi-probable models of the reservoir are considered, each constrained only to the correct multiple point statistics inferred from a suitable training reservoir model. The flow response at the same four well locations as before are obtained by flow simulation. The resultant spread in flow responses is shown in Fig. 23. The flow responses are significantly dispersed, despite the correct pattern of spatial variability exhibited by the reservoir models.

Another set of 10 realizations is generated, constrained to the production data at two wells, well 1 and 2. The pressure response at these wells for the ten realizations is shown in Fig. 24. The reduction in spread of the simulated responses is remarkable. Utilizing the same ten realizations the pressure response at well 3 and 4 is checked. Recall that in the two-well case well 3 and 4 are *not* used to constrain the permeability variation, yet, the results in Fig. 25 shows that the information on well 3 and 4 influences the overall flow characteristics of the reservoirs in regions away from the wells. The spread in response values observed at the two additional wells is reduced even though no production data at these wells are considered.

A final set of 10 realizations are considered that are constrained to the production information on all four wells. The corresponding flow simulation responses are shown in Fig. 26. These responses exhibit match the reference response excellently and exhibit little spread, thereby validating the proposed procedure for integrating production information.

6. DISCUSSION AND CONCLUSIONS

The aim of this paper is to outline a novel methodology for stochastic reservoir modelling that combines two important fields within the statistical sciences: geostatistics

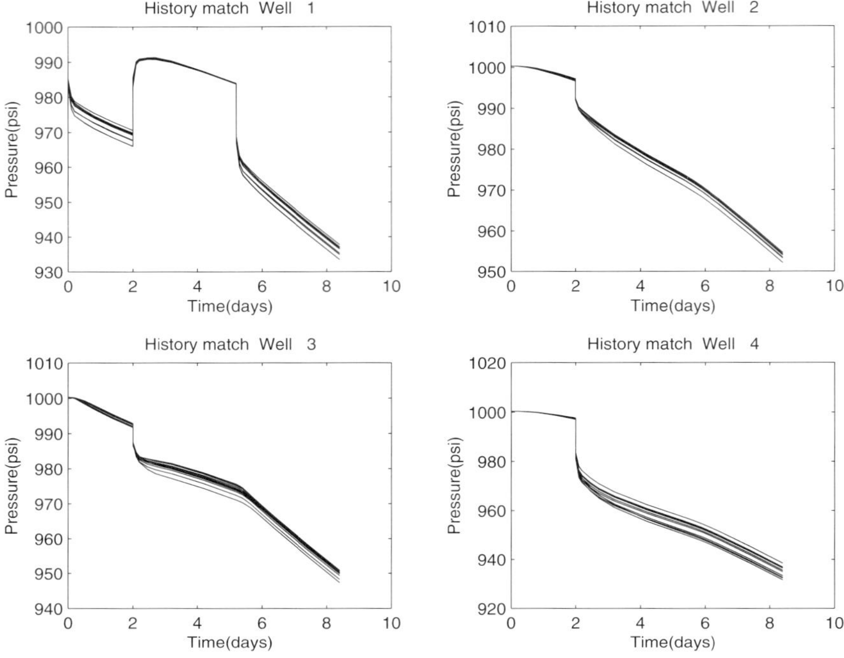

Fig. 23. History match corresponding to realizations constrained only to the correct multiple point statistics but not to the production data.

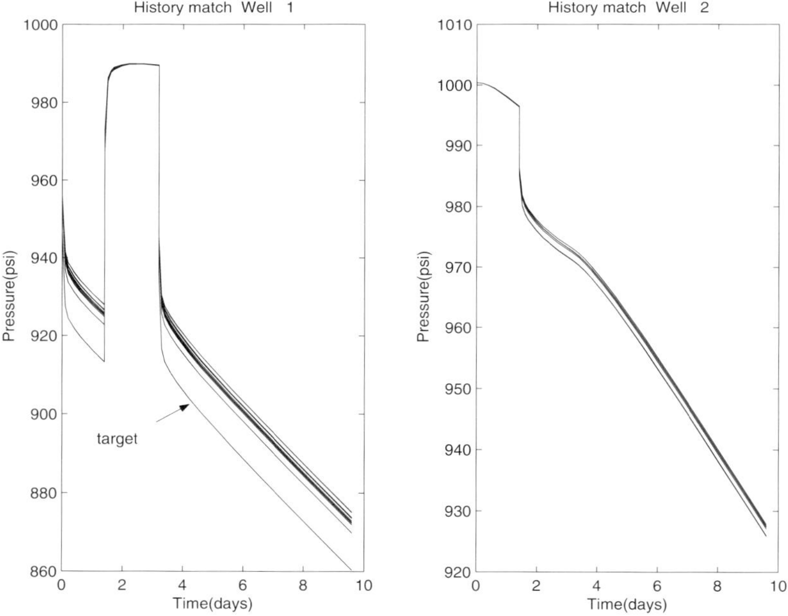

Fig. 24. Match of the well pressure history at the two wells where data is available.

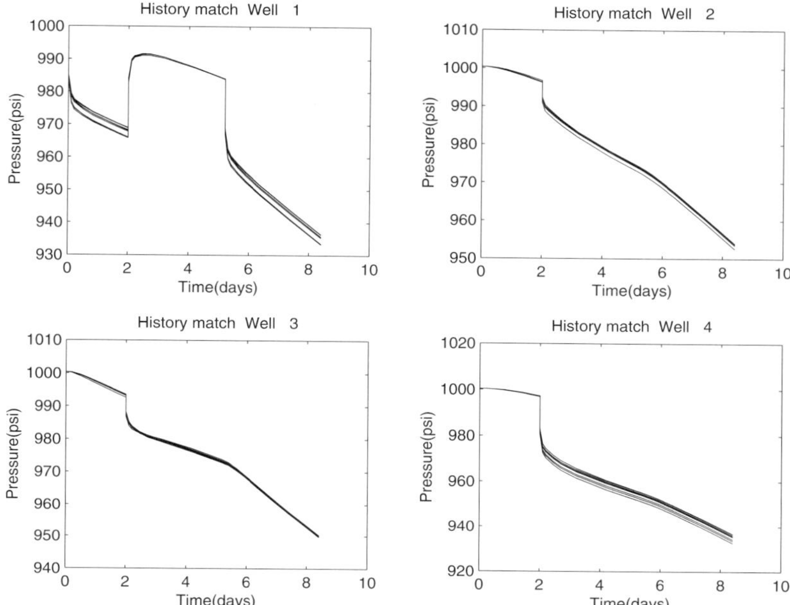

Fig. 25. Match of the well history at two wells (marked 3 and 4) where no production data was available for constraining the permeability field.

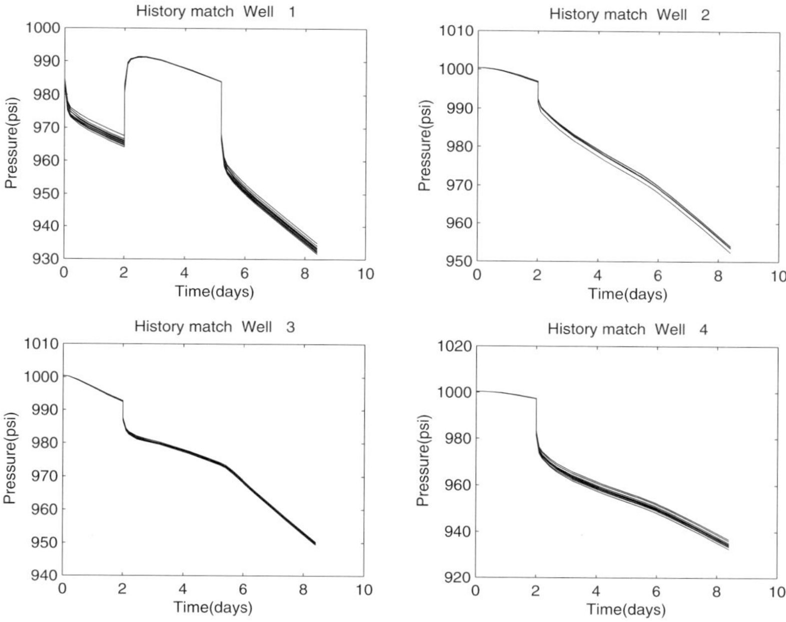

Fig. 26. Pressure history match when the production data from all four wells are used for constraining the permeability field.

and statistical pattern recognition. For a full case history of this methodology applied to actual reservoir data we refer to Caers, Srinivasan and Journel (1999). The method works equally well in 3D.

The systematic use of training images is certainly new to geostatistics but certainly not to reservoir characterization. In the latter field there has always been a need to borrow structural data from analogs such as outcrops since the hard observations at wells provide only a partial insight in vertical geological variation. Seldomly does the well data provide enough information on the important detailed geological continuity in horizontal directions.

It is often argued that borrowing higher order statistics from training images leads to models that are too subjective compared to models that borrow only covariances. Nevertheless, the fact that the structural model is not made explicit, does not mean it will be less constraining in the resulting simulated realizations. A multivariate-Gaussian random function model imposes onto the simulation a maximum entropy characteristic that goes beyond the variogram. In the Gaussian case, the higher order statistics are therefore implicit to the mathematical model, (and thus not controllable), their correctness with regard to the actual geological continuity is never checked. Training images allow an explicit visualization of this geological continuity and are easier to refute, i.e. it is never taken for granted.

Nevertheless, these are some important research question that need to be addressed:

- Statistical pattern recognition captures the patterns present in training images in probability terms (local conditional distributions). The proposed methodology does not provide a check on which patterns will actually be reproduced. Therefore, it is not yet clear what essential patterns are actually captured by these probability distributions.
- It might be difficult to construct 3D training images, but multiple two dimensional cross-section (training cross-sections) in various planes might be available, e.g. from outcrop modelling. The neural network methodology needs to be extended to be able to include the structural information from these various 2D cross-sections to build full 3D reservoir models.

APPENDIX A

The Metropolis–Hastings sampler can be extend to include soft information. Consider the following abbreviated notation:

A^* = the new proposed class value $i_{\text{new}}(\mathbf{u})$ at location $\mathbf{u}$

A = the current class value $i_{\text{current}}(\mathbf{u})$ at location $\mathbf{u}$

B = the neighboring hard data class values $\mathbf{i}_t(\mathbf{u})$ within a template around $\mathbf{u}$

C = the soft information $y(\mathbf{u})$ at location $\mathbf{u}$

Here C is collocated information only, although the method can be extended to non-collocated information. Then the following development leads automatically to the

Metropolis criterion in the text (Eq. (11))

$$\begin{aligned}\Pr\{A|B \cap C) &= \frac{\Pr\{A \cap B \cap C\}}{\Pr\{B \cap C\}} \\ &= \frac{\Pr\{C|A \cap B\}\Pr\{A \cap B\}}{\Pr\{B \cap C\}} \\ &= \frac{\Pr\{C|A \cap B\}\Pr\{A|B\}\Pr\{B\}}{\Pr\{B \cap C\}}\end{aligned}$$

hence

$$\frac{\Pr\{A^*|B \cap C\}}{\Pr\{A|B \cap C\}} = \frac{\Pr\{A^*|B\}}{\Pr\{A|B\}}\frac{\Pr\{C|A^* \cap B\}}{\Pr\{C|A \cap B\}}$$

REFERENCES

Bishop, C.M., 1995. *Neural Networks for Pattern Recognition*. Oxford University Press, Oxford.

Caers, J., and Journel, A.G. 1998. Stochastic reservoir simulation using neural networks, in: Proc. SPE Annu. Tech. Conf. and Exhibition, New Orleans, Sept. 321-336. SPE # 49026

Caers, J., Srinivasan, S. and Journel, A.G., 1998. Geostatistical quantification of geological information in a North-Sea reservoir. *SPE Res. Eval. Eng.*, 3: 457–467.

Dagan, G., 1985. Stochastic modeling of groundwater flow by unconditional and conditional probabilities: The inverse problem. *Water Resources Res.*, 21(1): 65–72.

Deutsch, C.V. and Wang. L., 1996. Hierarchical Object-Based Stochastic Modeling of Fluvial Reservoir *Math. Geol.*, 28: 857–880.

Deutsch, C.V. and Journel, A.G., 1998. *GSLIB: Geostatistical software library*. Oxford University Press, Oxford

Gomez-Hernandez, J., Sahuquillo, A. and Capilla, J.E., 1997. Stochastic simulation of transmissivity fields conditional to both transmissivity and piezometric data - 1. Theory. *J. Hydrol.*, 203: 162–174.

Guardiano, F. and Srivastava, M., 1992. Multivariate geostatistics: Beyond bivariate moments, in: A. Soares (Ed.), Geostatistics-Troia, Kluwer, Dordrecht, Holland, pp. 133–144

Hegstad, B.K., Omre, H., Tjelmeland, H. and Tyler, K., 1993. Stochastic simulation and conditioning by annaling in reservoir description, in: Armstrong, M. and Dowd, P.A. (Eds.), Geostatisical Simulation, Kluwer Academic Publisher, pp. 43–56.

Journel, A.G., 1983. Non-parametric estimation of spatial distributions. *Math. Geol.*, 15: 445–468.

Kitanidis, P.K. and Vomvoris, E.G., 1983. A geostatistical approach to the inverse problem in groundwater modeling (steady state) and one-dimensional simulations. *Water Resources Res.*, 19(3): 677–690.

Mavko, G., Mukerji, T., Dvorkin, J., 1998. *The Rock Physics Handbook*. Cambridge University Press, Cambridge.

RamaRao, B.S., LaVenue, A.M., de Marsily, G. and Marietta, M.G., 1995. Pilot point methodology for automated calibration of an ensemble of conditionally simulated transmissivity fields, 1. Theory and computational experiments. *Water Resources Res.*, 31(3): 475–493.

Röth, G. and Tarantola, A., 1994. Neural networks and inversion of seismic data. *J. Geophys. Res.*, 99(B4): 6753–6768.

Srinivasan, S., 1999. Integration of production data into reservoir models: a forward modeling perspective. Ph.D. Dissertation, Stanford University, Stanford, California, USA.

Strebelle, S., 2000. Sequential simulation drawing structure from training images. Ph.D. Dissertation, Stanford University, Stanford, CA, USA.

Tarantola, A., 1987. *Inverse Problem Theory*. Elsevier, Amsterdam.

Tran, T., 1994. Improving variogram reproduction on dense simulation grids. *Computers Geosci.*, 20: 1161–1168.

Wang, L., 1996. Modeling complex reservoir geometries with multipoint statistics. *Math. Geol.*, 28: 895–908.

Developments in Petroleum Science, 51
Editors: M. Nikravesh, F. Aminzadeh and L.A. Zadeh

Chapter 24

INTERPRETING BIOSTRATIGRAPHICAL DATA USING FUZZY LOGIC: THE IDENTIFICATION OF REGIONAL MUDSTONES WITHIN THE FLEMING FIELD, UK NORTH SEA

M.I. WAKEFIELD [1], R.J. COOK, H. JACKSON and P. THOMPSON

BG Group, 100 Thames Valley Park Drive, Reading RG6 1PT, UK

ABSTRACT

The Fleming gas-condensate Field, located on the eastern flank of the Central Graben, UK North Sea, is an elongate stratigraphic pinch-out whose reservoir is composed of stacked turbidite sandstones of the lower Palaeogene Maureen Formation. The sandstones have a sheet-like geometry with each sandstone lobe being partially offset into the swales of the preceding lobes. In such depositional environments, the understanding of lateral and vertical sandstone connectivity, a major uncertainty in reservoir modelling and well planning and production strategies, depends upon the choice of depositional model that is applied and the lateral continuity of pelagic mudstones.

Previously published work defined a model, based on variations in the composition of agglutinated foraminiferal populations, that could be used to derive a qualitative measure of the level of pelagic influence within mudstones interbedded with turbidite sandstones. It was considered that mudstones with a high pelagic influence are likely to be more laterally extensive than those with a low pelagic index. A fuzzy logic workflow was constructed using this model and was applied to the Fleming Field in order to identify laterally persistent mudstones. This approach was combined with high-resolution correlation of bioevents using graphic correlation. A detailed layering scheme for the Fleming Field was defined and this predicted the presence of a field-wide mudstone.

Initial attempts at history matching during reservoir simulation using a simple six-layer lithostraigraphical scheme were not successful. A revised layering scheme defined by biofacies modelling and graphic correlation was used to produce a 13-layer model; this was later simplified by combination with the six-layer model to produce a ten-layer model. This layering scheme is shown to provide a better understanding of both net-to-gross distribution and the dynamic behaviour of the field, and also improved history matching against production data. The biostratigraphical model applied using a fuzzy logic approach is authenticated by the reservoir simulation (fluid flow) and pre- and post-maintenance well pressure tests of well 22/5b-A3 that showed the perforated interval in that well to be isolated from the perforated intervals in the other producing wells. While history matching during reservoir simulation is important, the predictive capability of the fuzzy logic model proved to be critical to our understanding of the field.

[1] E-mail: matthew.wakefield@bg-group.com

1. INTRODUCTION

While taxonomic identification remains *the* fundamental building block of biostratigraphical data generation, variations in population composition and in the abundance of individual species provide the basis for straigraphical and palaeoenvironmental interpretations. Biostratigraphical data has sometimes been considered to be inferior with respect to more quantifiable datasets such as those provided by geophysics and petrophysics, because of the taxonomic subjectivity inherent in its collection. Furthermore, the biostratigraphical data collected, particularly by oil companies, may be sparse, having a sample spacing of 30 ft (10 m) or more; and it is often representative of an interval of time (cuttings) rather than a pinpoint of time (conventional core or side-wall core). The interpretation of biostratigraphical data is also complicated by the effects of post mortem transport (particularly upon palynological assemblages), the reworking of older taxa into younger sediments, and the caving of younger taxa into cuttings from older sediments that can occur during drilling. All these factors can result in the 'smearing' of stratigraphical and ecological signals. Provided that such problems are understood and effectively accounted for, this apparent degradation does not mean that the information obtained is either poor or of limited use. However, statistical techniques such as time series analysis, which are applicable to more data-rich disciplines, may not always be suitable for biostratigraphical data interpretation, since they may produce ambiguous results and an increased but false level of confidence which can lead to inaccurate inferences.

In general, the analysis of biostratigraphical data in the oil industry has been undertaken subjectively, for example the use of palynological assemblages for delineating sea level cycles in tropical deltaic successions by Poumot (1989). Subjective interpretations use 'case-based' reasoning (problem solving by analogy) which can lead to a low measure of interpretational repeatability even by the most rigorous scientist. Fuzzy logic provides a means of interpreting biostratigraphical data that removes subjectivity and increases repeatability. Fuzzy sets are generalistic and contain natural variations such as those that are present in species abundance gradients between neighbouring environments. By contrast, 'crisp' or 'binary logic' (case-based reasoning) has hard threshold boundaries that do not lend themselves to the modelling of environmental gradients. Gradients of this type are readily identifiable by the specialist but are often difficult to characterize consistently or to recognise in adjacent wells.

Fuzzy logic has not been widely used by geoscientists; however, in recent years the technique has been applied to the prediction of permeability (Fang and Chen, 1997), to the understanding of geological uncertainty (Fang and Chen, 1990; Foley et al., 1997) and to stratigraphical modelling (Norland, 1996, 1999).

2. THE FUNDAMENTALS OF FUZZY LOGIC

We generally communicate with each other using lexically imprecise descriptors or dimensionally relative terms such as 'the man is *tall*' or 'the man is *old*'. Such relative terms are difficult to quantify. A similar problem exists with the interpretation of biostratigraphical data – for example, the determination of water depth using the Planktonic : Benthonic foraminiferal ratio (P : B; Fig. 1A). In this example, a P : B ratio of 20% corresponds to depths between 70 m and 1000 m. Although some relationship exists between water depth and the P : B ratio, it is not precise and therefore the ratio is only of limited value.

Fuzzy Logic was developed by Zadeh (1965) and permits approximate reasoning with imprecise terms such as 'shallow', 'deep', 'weak', 'strong', 'short', 'tall', 'light' and 'heavy'. Fuzzy sets attempt to contain natural variations, transitions and gradations, such as those illustrated by the P : B – water-depth relationship in Fig. 1A. A variable (in this instance a value of P : B) can belong to several fuzzy sets (depth zones), its relationship to each being defined by a '*grade of membership value*'. This is different from 'binary' or 'crisp' logic where a variable can belong only to a single group. Crisp logic by definition cannot model gradational boundaries (Fig. 1B) and is therefore a poor method for interpreting ecological data where gradational boundaries often exist. Membership grades are determined from functions that represent the relationship of the variable (in the instance of biostratigraphical data, a 'taxon') to each of the fuzzy sets to which it can belong. Kosko (1994) provides an excellent introduction to the application

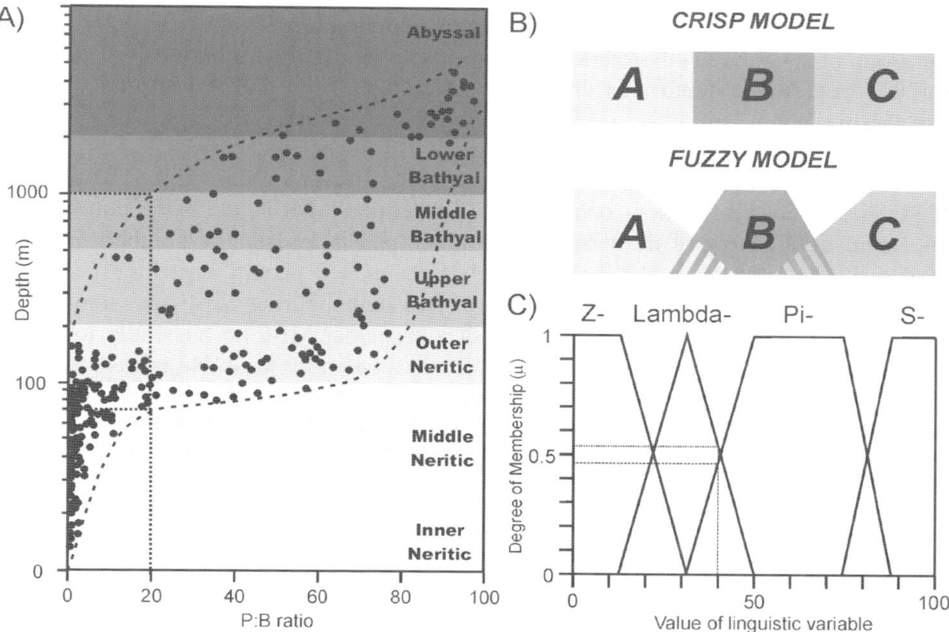

Fig. 1. (A) Graph of planktonic : benthonic foraminiferal ratio (P : B ratio) against water depth (m) for the Atlantic Coast of the US (data from Gibson, 1989). Heavy dashed lines indicate maximum and minimum depths. A P : B ratio of 20% can equate to a depth range of 70 m to 1000 m, or the depth zones Middle Neritic through Upper Bathyal inclusive. (B) The P : B ratio data does not lend itself to modelling with crisp boundaries where a value can only belong to a single set of outcomes (or in this case a single depth zone), but more to modelling where a value can belong to more than one set of outcomes (or depth zones) and has 'degrees of membership' to each outcome set to which it can belong. (C) Types of Membership function (MBF) used to model membership sets. A data point with a value of 40 has a membership value of 0.46 to the Pi-shaped linguistic variable and 0.54 to the Lambda-shaped linguistic variable. See text for explanation.

of fuzzy logic for the layman using everyday examples. A brief outline of a typical fuzzy logic workflow is given in the following paragraphs.

2.1. Linguistic variables

The primary building block within any fuzzy logic system is a '*linguistic variable*' such as temperature or water depth. For the definition of water depth, an oceanographer can use *linguistic terms* such as 'shallow' and 'deep'; these represent possible values of the *linguistic variable* 'water depth'. Fuzzy systems usually have at least three linguistic variables because people generally consider the two extremes of a system as well as the mid-point, although this number is not prescriptive. Models are normally symmetrical so most systems have an odd number of terms – three, five or seven terms being the most common.

2.2. Membership functions

Membership functions (MBFs) express a continuous variable in terms of a degree of membership – for example, the variation in specimen numbers of a particular species (or group of species) as environmental conditions alter. MBFs for all terms of the same variable (e.g. water depth) are drawn on a single diagram and are normalized such that the maximum is always $\mu = 1$ and the minimum is always $\mu = 0$. The degree of membership is sometimes likened to the 'degree of truth' of a statement.

There are four basic types of MBF: Z, Lambda, P and S types (Fig. 1C), although MBF shapes can be far more complicated. To define an MBF, the value that best fits the linguistic meaning of the term is determined and is assigned a membership value of $\mu = 1$. For each term, the membership degree $\mu = 0$ occurs where the next term has its most typical value, i.e. the overlap of two neighbouring MBFs is symmetrical. The membership values $\mu = 1$ and $\mu = 0$ of a particular MBF are generally joined by a straight line, though curved and stepped lines can also be used. No terms exist next to the outer edge of the data set for MBFs at the left-hand and right-hand extremes of the plot. Therefore, every value outside $\mu = 1$ is considered to belong to the MBF term being considered i.e. $\mu = 1$. Data points are plotted on these MBF plots and their membership to each linguistic variable are calculated, that is they are attributed a fuzzy value. For example, consider a value of 40 on Fig. 1C; this has a membership of 0.46 to the Pi-shaped *linguistic variable*, and 0.54 to the Lambda-shaped linguistic variable.

2.3. Fuzzy logic rules

The knowledge and behaviour of any fuzzy logic system is represented by rules using linguistic variables as vocabulary. Fuzzy logic rules, consisting of a condition (*If* . . .) and a conclusion (*then* . . .), represent the relationship between different linguistic variables in a system and define actions or results produced by those relationships. The *If* . . . condition can contain several *linguistic variables* joined by *linguistic conjunctions* such as '*and* . . . ' and '*or* . . . '. An example would be: *if* P : B is high *and* agglutinated foraminifera are common *then* water depth is deep. System behaviour is modified using *Degree of Support* (DoS) weightings between 0 and 1 for each rule. Redundant rules, i.e. those that have no impact on the system or those that never operate, should be removed, simplifying the system. With three inputs and one output each having four terms, there will be 64 rules ($4 \times 4 \times 4 = 64$). Rules can be automatically defined by most commercial software. However, the user must determine DoS weightings, which in a large system can be time consuming. DoS weightings can be varied during testing/optimisation of the fuzzy logic system in order to prevent the over- or under-emphasis of certain rules.

2.4. *Defuzzification*

Once the rule-base has been defined the workflow is completed by 'defuzzification', i.e. the derivation of a crisp value that describes the fuzzy value. This derivation is non-unique as different values of a linguistic variable can map to the same defuzzified value. Therefore, it is important to evaluate the fuzzy values as well as the defuzzified result. There are two basic approaches to defuzzification: 'best compromise' and 'most plausible result'.

Best compromise approaches include Centre of Maximum (CoM), Centre of Area (CoA) or Centre of Gravity (CoG) defuzzification. CoM determines the most typical value for each term and computes the best compromise of the fuzzy logic inference result by considering inference results as weights, the balancing of which produce the 'crisp' result. Sometimes this approach does not work; for example when two results are equally valid, the result will be that no evidence exists. With single MBFs, the CoA and CoM approaches produce identical results.

In the *most plausible result* approach, the Mean of Maximum (MoM) method selects the typical value of the most valid term, i.e. the term with the highest resulting DoS. However, when the maximum is non-unique (i.e. a Pi-shaped MBF: Fig. 1C), then the mean of the maximizing interval, or plateau width, is calculated. A drawback of this method of defuzzification is that it will reach a point at which a small input change will result in a large change in the output variable which over-emphasizes changes. Therefore, it has a similar outcome to 'crisp' systems with their hard boundaries. Pattern-recognition applications use MoM defuzzification.

2.5. *Previous application of fuzzy logic to palaeontological data analysis*

Gary and Filewicz (1997) quantified palaeoenvironmental changes in Neogene estuarine sediments from offshore China using fuzzy logic. As input variables, they used abundance variations of freshwater dinoflagellates, marine dinoflagellates, benthonic foraminifera and calcareous nannofossils. Simple MBFs were constructed and these related the abundance of each microfossil group to four palaeoenvironmental zones that were used as *linguistic variables*: freshwater, inner estuary, middle estuary and outer estuary. To accommodate offshore transport, a rule was invoked such that the freshwater dinoflagellate function was only activated when membership to middle or outer estuarine palaeoenvironments was zero. In this way, each sample was assigned a membership value to each of the four palaeoenvironments and the results were not defuzzified. Plotting these membership values in the form of adjacent logs highlighted palaeoenvironmental gradients that could be correlated between wells. Gary and Filewicz (1997) considered that their results provided a higher degree of detail than was achievable using a 'crisp' representation of the palaeoenvironments and that the uncertainties inherent in both the data set and the interpretation of the palaeoenvironment were quantified.

3. APPLICATION OF FUZZY LOGIC MODELLING IN THE FLEMING FIELD

3.1. Geological setting

The Fleming gas-condensate Field, operated by BG Group as part of the Armada complex (Fleming, Drake and Hawkins fields), is located on the eastern flank of the Central Graben, UK North Sea (Fig. 2) and has estimated reserves of 900 bcf of gas and 40 MM bbl of condensate (O'Connor and Walker, 1993). The field is an elongate stratigraphic pinch-out, some 20 km from north to south and 2–3 km wide. The reservoir is composed of stacked turbidite sandstones of the lower Palaeogene Maureen Formation, which pinch out onto the Jaeren High where they are replaced by condensed mudstones. The Maureen Formation sandstones was deposited from a number of high density turbidite flows sourced from the East Shetland Platform to the north and northwest, which combined to form a lowstand fan/wedge complex (Stewart, 1987; Vining et al., 1993). Overall net-to-gross is high, generally between 40 and 60%. The sandstone lobes have a sheet-like geometry and are 8–12 km wide; they thin rapidly onto the basin margins, and each lobe is partially offset into the swales of the preceding lobes. Reworking and slumping of the underlying Ekofisk Formation on the flanks of the Jaeren High, particularly evident in the south of the field, supplied allochthonous carbonates into the basin prior to and during deposition of the Maureen Formation. Dip closure is provided by drape over Zechstein salt swells.

In this depositional environment, one of the main uncertainties in reservoir modelling is lateral and vertical sandstone connectivity. This can depend upon the choice of depositional model that is applied to the field, for example sheet sands versus nested lobes, the degree of scouring, and the continuity of the pelagic mudstones. The identification of laterally persistent mudstones has important implications for well planning and production strategies in such a depositional system.

3.2. Stratigraphical data

In this paper we use the lithostratigraphical subdivision and nomenclature of the Palaeogene compiled by Knox and Holloway (1992), which is essentially that of Mudge and Copestake (1992a). We use the timescale of Berggren et al. (1995) that subdivides the Palaeocene series into three stages – Danian, Selandian and Thanetian. The Selandian spans the time between the end of the Danian (ca. 61 Ma) and the beginning of the Thanetian (57.9 Ma). The Danian/Selandian boundary represents a major change in the geotectonic evolution of the NE Atlantic, marked by marginal uplift and basinal subsidence prior to late Palaeogene sea-floor spreading in the Norwegian–Greenland Sea during Chrons C25–26. It coincides with uplift of the East Shetland

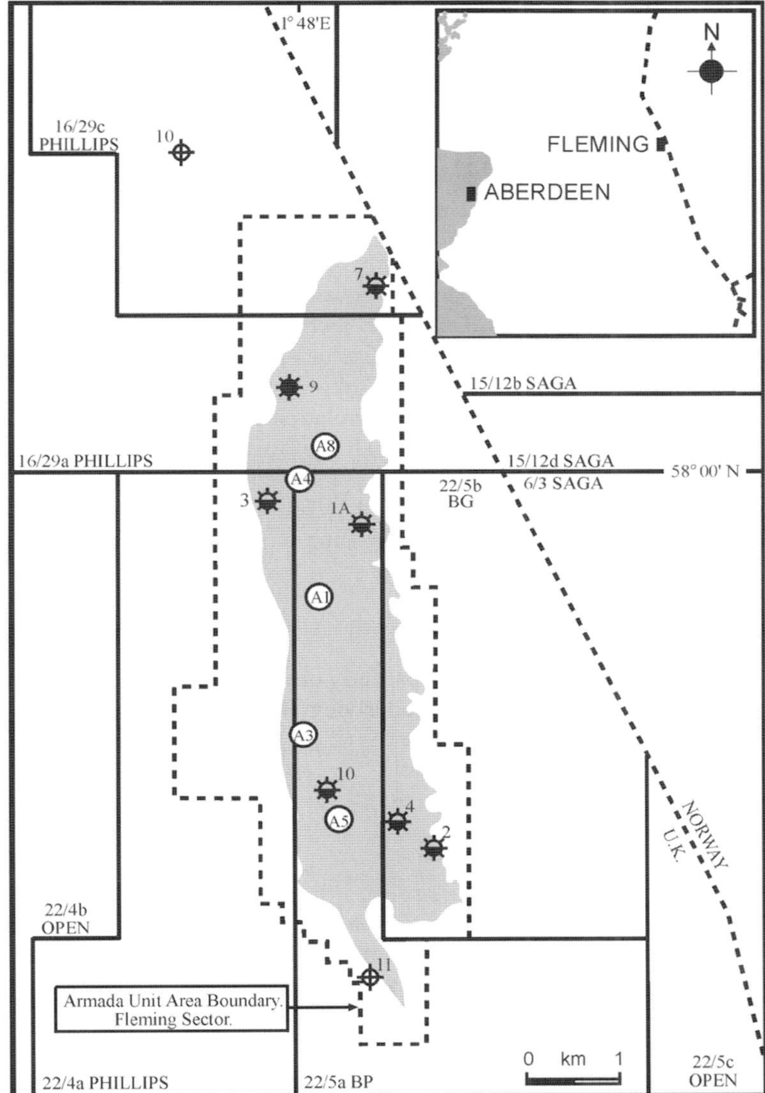

Fig. 2. Location of the Fleming Field, UK North Sea, showing the position of the studied wells. Production wells are circled. Note position of wells is given where they penetrate the top of the Maureen Formation.

Platform, erosion of the Danian Chalk Group and the initiation of clastic sedimentation (Maureen Formation).

The Lexis Group generated the foraminiferal and dinoflagellate data from high-density sampling of Fleming Field wells. Bioevent data from wells *16/29a-9, 16/29c-7, 16/29c-10, 22/4-3, 22/5a-10, 22/5a-11, 22/5a-1A, 22/5b-2, 22/5b-4, 22/5b-A1,*

22/5b-A1z, *22/5b-A2*, *22/5b-A3*, and *22/5b-A5* were used (see Fig. 2 for locations). The majority of the bioevents used in this study are species extinctions (first downhole occurrences). The dinoflagellate bioevents used are those of Mudge and Bujak (1996a,b), which were updated from Mudge and Copestake (1992a,b). All the depths cited are measured depths in feet. As the production wells are not vertical, the depths do not represent true vertical thickness.

3.3. Graphic correlation of bioevent data

Graphic correlation (Shaw, 1964) is a powerful semi-quantitative technique for the correlation of straigraphical data. The technique is not limited to biostratigraphical data, and other stratigraphically useful single horizons or events that are synchronous/isochronous and discrete in nature (e.g. maximum flooding surfaces, volcanic tuffs) can also be included. Graphic correlation produces a Composite Standard Reference Section (CSRS), which contains all the 'event' data from an area, against which detailed correlations can be made. The graphic correlation approach plots unitary event data in a pairwise fashion i.e. an *xy* graph, where events that are common between two wells are displayed as points (Fig. 3). A Line of Correlation (LOC) based on the

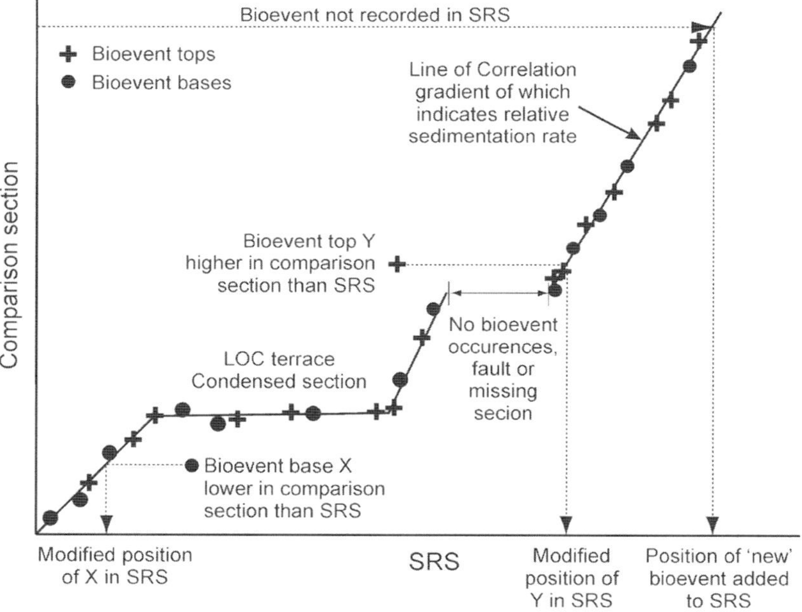

Fig. 3. Fundamentals of graphic correlation. The relative positions of bioevents in the comparison well and the Standard Reference Section (SRS) defines the Line of Correlation (LOC). The SRS is developed as each successive comparision section is plotted. New bioevents are included in the SRS via the LOC. The position of bioevents that already occur in the SRS may be modified: bioevent tops (e.g. Y on the figure) that occur *above* the LOC and bioevent bases (e.g. X) that occur below the LOC can be repositioned via the LOC as indicated.

trend of these correlation points is drawn and represents the best estimate of correlation between the well sections. Changes in gradient of the LOC represent differing relative sedimentation rates between the sections, but do not account for differential compaction.

Construction of a CSRS begins with the selection of the most stratigraphically complete and fossiliferous section of those studied (the Standard Reference Section: SRS), which is plotted on the x-axis. The comparison section is plotted on the y-axis, and a LOC is drawn using the positions of the correlation points of bioevents common to both sections as a guide. Taxon tops that plot above the LOC are lower in the SRS than in the comparison succession (this may also be due to reworking which should be checked), and are projected across to the LOC and then perpendicularly to the SRS (e.g. bioevent top Y, Fig. 3). Taxon tops that plot below the LOC are already higher in the SRS and no adjustment is made. Conversely, taxon bases that plot below the LOC are higher in the SRS, and are adjusted in a similar manner to give the maximum range, although the data should be checked for evidence of caving (e.g. bioevent base X, Fig. 3). Taxon bases that plot above the line are already lower in the SRS and no adjustment is made. Any bioevents not already in the SRS are added in a similar way. Projections to the LOC are always made parallel to the SRS axis. This process is repeated until all successions have been studied, at which point the SRS becomes the CSRS and contains the maximum ranges for all taxa. The final correlation is obtained by plotting each well against the CSRS, and comparing the results with one another.

The composite plotting of LOCs from each well on a single figure (Neal et al., 1994, 1995) enable stratigraphically equivalent intervals of low sedimentation ('LOC terraces') to be identified. Neal et al. (1994, 1995) interpreted LOC terraces in a sequence stratigraphic context: on the shelf, LOC terraces represent missing time either in the form of an unconformity or sediment bypass resulting in a sequence boundary; whereas in the basin, LOC terraces represent condensation at the maximum flooding surface and the subsequent highstand systems tract when sediments are trapped on the shelf and basin sedimentation rates are consequently very low. However, in the Fleming Field, LOC terraces were found to coincide with mudstone deposition i.e. consistent with low sedimentation rates. Therefore, any overlapping LOC terraces are interpreted to represent field-wide mudstones.

This process was applied to bioevent data from the Fleming Field, permitting a high-resolution correlation of mudstone intervals to be made which was independent of biofacies modelling. It was used in combination with biofacies modelling to identify laterally continuous mudstones. Neal et al. (1994, 1995) and Stein et al. (1995) published graphic correlation studies of the North Sea Palaeogene, although there is limited detail relating to the Maureen Formation in those papers.

3.4. Agglutinated foraminiferal community structure and mudstone continuity

Holmes (1999) used 'morphogroups' of agglutinated foraminifera (Jones and Charnock, 1985) to derive a qualitative measure of the level of pelagic influence in mudstones interbedded with turbidite sandstones. Holmes' model used the inferred life habits of agglutinated foraminifera to deduce likely conditions at the sediment–water interface. Thus, mudstones deposited in short intervals of time between successive turbidites will have a simple agglutinated foraminiferal assemblage dominated by suspension feeders of Morphogroup A (*sensu* Jones and Charnock, 1985). Mudstones deposited under quiescent pelagic conditions will have a more complex community structure dominated by detritivores of Morphogroups B and C. The latter mudstones are considered likely to be more laterally extensive, while the former are considered more likely to be relatively impersistent. There is a gradation between these two end members. An increase in taxonomic diversity has been used in a similar way to identify critical surfaces in bio-sequence stratigraphical studies (Armentrout, 1987, 1996; Armentrout and Clement, 1990; Armentrout et al., 1990, 1999; Mitchum et al., 1993; Pacht et al., 1990; and Shaffer, 1987, 1990).

Holmes (1999) used fluctuations in the relative abundance of these morphogroups to define a qualitative 'Pelagic Index' (see his Fig. 3). He assessed the lateral persistence of mudstones that were considered to be important as flow baffles and barriers in the Palaeocene turbidite reservoirs of the North Sea Joanne and Andrew fields. This approach has also recently been applied at the nearby Palaeocene Blenheim Field during pre-drilling appraisal (Dickenson et al., 2001).

Kaminski et al. (1988a); p. 170) described agglutinated foraminifera community successions in similar deep-water settings that support Holmes' model. They summarised the findings of several previous authors as follows:

> "at the contact between turbidites and hemipelagic layers, assemblages . . . of primitive tubular varieties occur [Morphogroup A], whereas . . . in the hemipelagic layer a more diverse fauna is generally reported. This pattern has been interpreted as evidence for gradual recolonisation of the sea floor after deposition by a turbidity current."

However, Kaminski et al. (1988a) went on to discuss additional work suggesting that concentrations of tubular species occurring directly above fine-grained sandstones are evidence of current winnowing. In addition, Kaminski et al. (1988b) showed that tubular agglutinated foraminifera may not be particularly good colonisers. Concentrations of tubular species were considered to be the result of concentration by turbidity currents that entrained the surface layer of sediment. Tubular species would therefore be found in unit E of the classic Bouma sequence, or in any mudstone with a significant redeposited component. Further, Jones and Charnock (1985) reported that Morphogroup A agglutinated foraminifera are found in greatest abundance and diversity in quiescent present-day environments. However, we consider the model of Holmes (1999) to be correct. Although the reasoning behind the order of morphotypes recovered may not be consistent with some of the authors above, unpublished work that calibrates the Pelagic Index against palynofacies, ichnofacies and sedimentological data strengthen the validity of the model (Nicholas Holmes, personal communication, 2001; see also Dickinson et al., 2001).

The application of Holmes' foraminiferal model is dependent upon accurate correlation, as it does not take into account the erosional effects of succeeding turbidites that may locally remove the finer-grained, more pelagic biofacies. The biofacies model must therefore be integrated with other quantitative biostratigraphical correlation techniques (e.g. graphic correlation) and where possible with other data (e.g. from wireline logs, sedimentology or ichnofacies analysis) to ensure that coeval mudstones are compared.

Both Holmes (1999) and Dickinson et al. (2001) use detailed sedimentological data and, where available, key marker taxa to enable confident correlation. A modified version of the Pelagic Index model of Holmes (1999) is used here in conjunction with graphic correlation to define a layering scheme within the Fleming Field that can be used for reservoir modelling.

Holmes (1999) used a number of variables to determine a 'Pelagic Index'. These can be broken down such that changes in morphotype can be expressed as changes in lifestyle and include:

(1) Number of species (simple diversity).
(2) Population composition using agglutinated foraminiferal morphotypes.
(3) Relative dominance of agglutinated foraminiferal morphotypes.

Dominance has commonly been calculated as the abundance of the two most abundant taxa in a sample divided by the total number of specimens in that sample; however, this is simplistic. Therefore, the Simpson (D) diversity index (Simpson, 1949), which takes into account the relative abundance of species within a sample population, is used here. The Simpson diversity index is calculated as:

$$D = \sum_{i-1}^{S} \left[\frac{n_i(n_i - 1)}{N(N - 1)} \right]$$

where n_i = the number of individuals in the ith taxon; N = the total number of individuals; S = the number of taxa.

D increases as diversity decreases. The index is very sensitive to the most abundant species and is less sensitive to rare species. D depends solely upon the distribution of species abundance, and when all individuals belong to a single species $D = 1$. This measure is the inverse of those determined by other diversity indices which increase as species richness increases, and D is therefore often reported as $1 - D$ or $1/D$.

Membership functions were defined for five input variables that describe Holmes' model:

(1) Total number of agglutinated foraminiferal specimens.
(2) Simple diversity.
(3) Dominance Index (Simpson's diversity index; D).
(4) Epifaunal : Infaunal ratio (E : I).
(5) Detritivore : Suspension feeder ratio (D : S).

The ratios Epifauna : Infauna and Detritivore:Suspension feeders were used to represent the variation in agglutinate morphotypes used by Holmes (1999) and their relative dominance. Morphogroup A, which dominates the agglutinate community in mudstones immediately above turbidite sandstones, are suspension feeders. As the agglutinate community evolves, the numbers the detritivorous agglutinates (Morphotypes B and C) increases. Therefore, the Detritivore : Suspension feeder ratio will increase with time following the cessation of sandstone deposition.

3.5. Calibration of the fuzzy model

The values for each membership function were calibrated against Palaeogene turbidite reservoir data sets that were subjectively interpreted using the modified pelagic

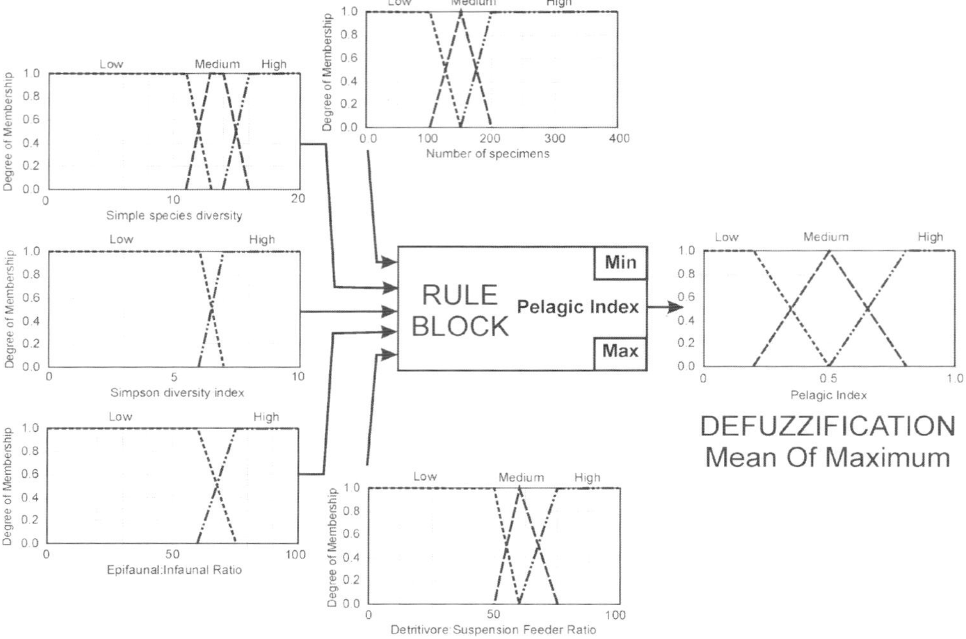

Fig. 4. Fuzzy Logic workflow illustrating modelling of the Pelagic Index (Holmes, 1999) for mudstones interbedded with turbiditic sandstones in the Fleming Field using agglutinated foraminifera. Membership functions (MBFs) are defined for each of the five input variables for the linguistic terms 'low', 'medium' and 'high pelagic index'. Rules are generated from these MBFs, for example: *If* number of specimens is high *and* simple species diversity is high *and* Simpson diversity index is high *and* Epifaunal : Infaunal ratio is high *and* Detritivore : Suspension feeder ratio is high *then* Pelagic Index is high. A fuzzy value for each sample run through the model is then defuzzified using the Mean of Maximum approach (see text for explanation) to provide a simple means of recognising changes in the Pelagic Index through time.

index model of Holmes (1999). Histograms and simple statistics (max–min range, mean and mode) for each input variable against each level of pelagic index were generated. These were used to define the shape of the MBF for each input variable (Fig. 4). In general, three linguistic terms were used for each variable equating to values attributable to low, moderate and high pelagic index. Three variables were considered to be sufficient to highlight trends within the data without over-complicating the model. The MBFs for the inputs 'total agglutinates', 'simple species diversity' and 'Detritivore : Suspension feeder ratio' had three linguistic terms each – low, medium and high pelagic index (Fig. 4). However, data for the input variables 'Simpson diversity' and 'Epifaunal : Infaunal ratio' were less well structured, particularly in mudstones deposited immediately above turbiditic sandstones. Therefore, only two linguistic variables were used – 'low' which probably equates to the low and medium values in the other three input variable MBFs, and 'high'.

Based on the membership functions defined for the five input variables (Fig. 4), 108 rules were generated: this is based on the number of linguistic terms per rule, which with the model in Fig. 4 is $3 \times 3 \times 2 \times 2 \times 3 = 108$. Each rule was given a specific outcome. For example, the rule "*if* total agglutinates is high *and* simple diversity is high *and* Simpson diversity is high *and* the Epifaunal : Infaunal ratio is high *and* the Detritivore : Suspension feeder ratio is high *then* the pelagic index is high". Rules considered redundant were given a DoS value of 0 and these rules did not operate. Other rules that included all input variables with the same linguistic value, for example the rule described above, were all set at a DoS of 1, a value that remained constant throughout the calibration process. Although the number of rules originally generated was high, careful identification of redundant and high probability rules (i.e. those where all input variables had the same linguistic variable) reduced the number of rules requiring calibration, such that manual editing of the DoS weightings was achievable. Weightings for the remaining rules were initially set to 0.5 and were iteratively altered by running the model against a calibration well that had previously been subjectively interpreted. Each iteration of the model produced fuzzy values for the low, medium and high pelagic index, which were plotted against the subjective interpretation. Weightings were applied to rules one at a time and the results were examined. This process was run until an acceptable answer was achieved. The model was then compared with a second well and final modifications were undertaken. Once calibrated against wells with known answers, the model can be applied consistently against other well data sets and the resulting interpretations can be confidently compared.

Defuzzification used three values of mudstone persistence – low, moderate and high pelagic index, which can be interpreted in terms of the likely lateral persistence of the mudstone under investigation (Fig. 4). Mean of maximum defuzzification was used in order to recognise the three degrees of persistence. The defuzzification process was again calibrated against the subjective interpretation of two test wells.

3.6. Data handling

The Integrated Paleontological System (IPS) software (Technical Alliance for Computational Stratigraphy: *http://tacs.egi.utah.edu*) was used to calculate the input variables used in the model from the foraminiferal data (Fig. 5). Data was then exported to the *FuzzyTECH* software (Inform Gmbh: *http://www.fuzzytech.com*) for fuzzy logic modelling. Results from *FuzzyTECH* were exported to and displayed in IPS together with the input data and other data including palaeobathymetry and gamma-ray log signature (Fig. 5).

IPS palaeobathymetry uses a depth framework of eight zones: transitional (e.g. lagoonal); Inner Neritic (0–20 m/0–60 ft); Middle Neritic (20–100 m/60–300 ft); Outer Neritic (100–200 m/300–600 ft); Upper Bathyal (200–500 m/600–1500 ft); Middle Bathyal (500–1000 m/1500–3000 ft); Lower Bathyal (1000–2000 m/3000–6000 ft) and Abyssal (2000 m+/6000 ft+). Each benthonic foraminiferal species in a sample was cross-referenced against an in-house database of species with their upper and lower water-depth limits. Bathymetric ranges of Recent taxa are readily available (e.g. Culver, 1988) and can be used with confidence. Substantive uniformitarianism is applicable throughout the Tertiary to define palaeobathymetric ranges for benthonic foraminifera (e.g. Charnock and Jones, 1990), and these can therefore be used with some confidence.

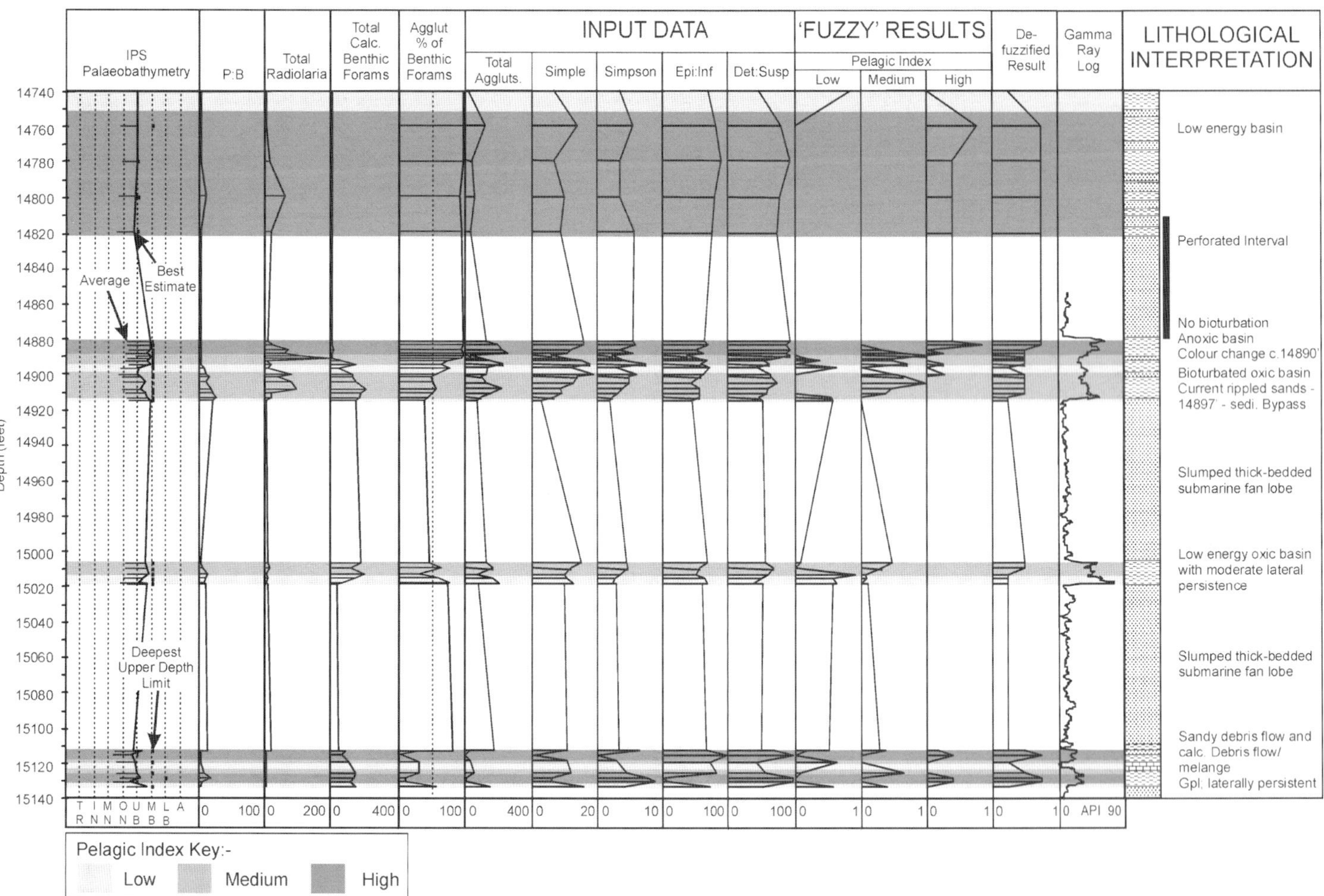
IPS Palaeobathymetry
P:B
Total Radiolaria
Total Calc. Benthic Forams
Agglut % of Benthic Forams
INPUT DATA
Total Aggluts.
Simple
Simpson
Epi:Inf
Det:Susp
'FUZZY' RESULTS
Pelagic Index
Low
Medium
High
De-fuzzified Result
Gamma Ray Log
LITHOLOGICAL INTERPRETATION
Low energy basin
Perforated Interval
No bioturbation
Anoxic basin
Colour change c.14890'
Bioturbated oxic basin
Current rippled sands - 14897' - sedi. Bypass
Slumped thick-bedded submarine fan lobe
Low energy oxic basin with moderate lateral persistence
Slumped thick-bedded submarine fan lobe
Sandy debris flow and calc. Debris flow/ melange
Gpl. laterally persistent
Average
Best Estimate
Deepest Upper Depth Limit
T I M O U M L A
R N N N B B B
0 100
0 200
0 400
0 100
0 400
0 20
0 10
0 100
0 100
0 1
0 1
0 1
0 1
0 API 90
Depth (feet)
14740
14760
14780
14800
14820
14840
14860
14880
14900
14920
14940
14960
14980
15000
15020
15040
15060
15080
15100
15120
15140
Pelagic Index Key:-
Low
Medium
High

IPS calculates an 'average paleobathymetry' using a weighted average of upper depth limits based on all taxa present in a sample that have a match in the paleobathymetry database. Empirical analysis of modern box-core samples from the present-day Gulf of Mexico indicates that the majority of the benthonic foraminiferal population for a given species begins about one standard deviation deeper than its upper depth limit (A.C. Gary, personal communication, 2000). Therefore, IPS calculates a fraction of the standard deviation and appends that to the weighted average paleobathymetry, thereby producing a shift towards deeper water. This measure is termed the 'bathymetric best estimate' (Fig. 5). IPS also indicates if taxa in a sample have upper depth limits deeper than the 'bathymetric best estimate' (Fig. 5).

3.7. Results of the fuzzy logic modelling

Both fuzzy and defuzzified (MoM) results are given in detail for well *22/5b-A3* (Fig. 5). CoM was not used, as the aim was to identify different types of mudstone rather than trends.

The mudstones sampled were deposited in upper to middle bathyal waters, probably in excess of 1500 ft (Fig. 5). The lower mudstone interval (ca. 15,110 ft and 15,030 ft) was tentatively identified during graphic correlation analysis (overlapping LOC terraces) as being laterally persistent, and was labelled by the Gpl bioevent (*Globorotalia planocompressa* extinction). Gpl was recorded immediately beneath a thin allochthonous limestone at ca. 15,120 ft (also recorded in well *22/5a-10*). There is a palaeobathymetric peak immediately beneath that limestone that correlates with dominance by detritivorous agglutinates (Det : Susp: Fig. 5). The mudstones above the thin limestone contain an increased proportion of suspension feeders (Morphogroup A) and consistently high levels of reworked Ekofisk Formation nannofossils. The allochthonous limestone is developed in the south of the field and its presence suggests that the mudstone at this level could have been 'breached'.

The mudstone interval between ca. 15,010 ft and 15,020 ft records an up-hole increase in the proportion of detritivores (Det : Susp column; Fig. 5). The base of this mudstone interval is considered to represent the suspension fallout from the preceding turbidite and is dominated by Morphogroup A. This may be supported by the relatively shallower palaeobathymetry, indicating a possible allochthonous component within the foraminiferal fauna. The remaining portion of the interval represents the gradual onset of pelagic conditions with a slight bathymetric deepening. Fully pelagic conditions are not represented in this interval and the mudstone is probably laterally impersistent.

The mudstone interval at ca. 14,880 ft and 14,910 ft contains variations in Det : Susp, Simple and Epi : Inf (Fig. 5) indicating a complex depositional history. There is a

---

Fig. 5. Micropalaeontological signals from well *22/5b-A3*, Fleming Field. All core samples. Input data, fuzzified and defuzzified output are shown with the interpreted 'Pelagic Index'. Column 1 displays the IPS palaeobathymetry curve where, TR = Transitional; IN = Inner Neritic; MN = Middle Neritic; ON = Outer Neritic; UB = Upper Bathyal; MB = Middle Bathyal; LB = Lower Bathyal and AB = Abyssal (depth ranges of these bathymetric zones are given in the text).

palaeobathymetric increase to the top of the interval, which coincides with dominance of detritivores and of agglutinate over calcareous benthonic foraminifera. Fully pelagic conditions are interpreted between 14,880 ft and 14,900 ft. This level coincides with a marked mudstone colour change from green-grey to dark-grey/black, reflecting a change from oxic to dysoxic/anoxic conditions that is also marked by a total loss of ichnofauna (*Chondrites, Zoophycus, Planolites* and *Rhizocorallian*). The colour change is accompanied by major changes within the recovered microfauna, including the loss of planktonic foraminifera, the loss of calcareous benthonic foraminifera and a sudden increase in radiolaria such as *Cenodiscus*. This colour change was noted in a number of wells (*16/29a-9, 22/4-3, 22/5a-10, 22/5b-4, 22/5b-A4* and *22/5b-A5*). These findings indicate a laterally persistent mudstone between 14,880 ft and 14,900 ft that is highlighted by the distinctive agglutinate biofacies in wells where the colour change was not noted. The agglutinate benthonic foraminifera from the remaining five samples between 14,740 ft and 14,820 ft indicate an increase in pelagic content to a peak between 14,760 ft and 14,080 ft that coincides with the lithostratigraphical pick for the top of the Maureen Formation. The wide spacing of these samples precluded recognition of primary colonisation above the several thin sandstones.

3.8. Integration of graphic correlation and mudstone continuity modelling

Fuzzy logic modelling reveals changes in mudstone persistence. A small number of intervals containing a poor agglutinate fauna (<50% Det : Susp ratio; Fig. 5) occur immediately after turbidite sandstone deposition and are interpreted as being laterally impersistent. However, the majority of these mudstones form only parts of thicker mudstone intervals that contain a more diverse agglutinated fauna. Therefore, these intervals represent heterogeneity within moderately to highly persistent mudstone units rather than flow baffles. In general, the majority of the agglutinated foraminiferal biofacies recovered from the mudstone intervals are of an intermediate character (50–70% Det : Susp ratio; generally 7–12 species; Fig. 5), and probably indicate a moderate likelihood of lateral persistence. Significantly, for reservoir compartmentalisation, it is possible that these mudstones may not be field-wide and could have been breached by subsequent turbidite flows. Mudstone persistence is interpreted to increase with vertical distance or depositional time above an underlying turbidite sandstone. However, a reduction in the pelagic index occurs between 14,820 ft and 14,740 ft (Fig. 5), probably as a result of the widely spaced samples. Other fine-scale variations are also evident such as that recorded at ca. 14,897 ft. This interval coincides with a ca. 1 cm thick rippled sandstone, interpreted as evidence of sediment bypass, which we now suggest could have killed the existing agglutinated foraminifera in the immediate vicinity of well *22/5b-A3*. The mudstone interval between 14,897 ft and 14,880 ft records the gradual re-establishment of an agglutinate foraminiferal community with time after the 'sterilizing' effect of currents associated with sediment bypass.

In general, an increase in mudstone persistence appears to correlate positively with IPS palaeobathymetry. This is particularly evident between 14,880 ft and 14,915 ft. A positive correlation is also present between mudstone persistence and increasing gamma-ray signature.

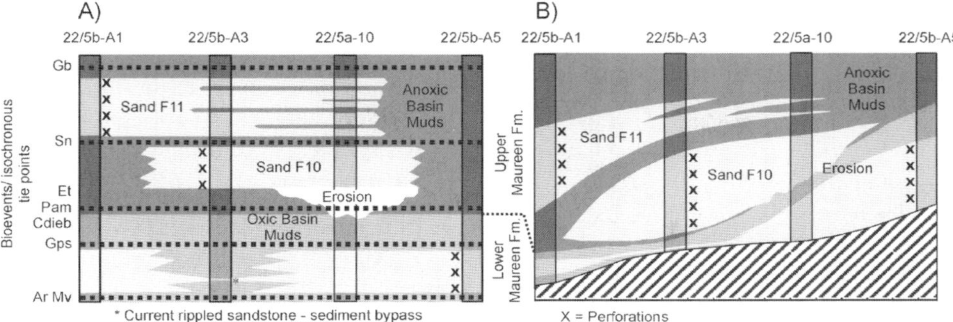

Fig. 6. (A) Schematic 'Wheeler' type diagram showing sediment type and proposed correlation for four wells in the south of the Fleming Field (see Fig. 2 for location of wells). Bioevent positions in each well are determined by graphic correlation. Gb = *Gavelinella beccariformis*, Sn = *Spiroplectammina naibensis*, Et = *Eoglobigerina trivialis*, Pam = *Pullenia americana*, Cdieb = *Cerodinium diebelii*, Gps = *Globorotalia pseudobulloides*, Ar = *Alisocysta reticulata* and Mv = *Matanzia varians*. (Not to scale). (B) Schematic cross-section showing successive turbidites banking onto the Jaeren High and earlier sands. Note the erosion of 'anoxic' mudstones in well *22/5a-10*. (Not to scale).

In Fig. 6, we illustrate a model to explain sandstone/mudstone connectivity in the south of the Fleming Field around wells *22/5b-A1, 22/5b-A3, 22/5a-10* and *22/5b-A5*. Indications of moderate to high continuity of the mudstones suggests that the sandstones where deposited as discrete events across spatially distinct sections of the field. These sands appear to bank against each other and are separated by mudstone intervals (Fig. 6A and 6B), a depositional model consistent with experimental analyses of turbidite flows (Kneller and McCaffrey, 1999). Where a turbidite flow is reflected off a linear high such as the Jaeren High, flow deceleration and possible flow stripping can result in deposition of the bulk of the coarse sediment load adjacent to the high (Kneller and McCaffrey, 1999). Subsequent flows would be expected to build out from the high.

The biofacies-derived model for the upper Maureen Formation interval shows that the perforated sand in well *22/5b-A3* is separated from the perforated intervals in well *22/5b-A5* by the laterally persistent Et/Pam mudstone (Fig. 6B). The model predicts that the perforated interval in well *22/5b-A3* will be isolated from the rest of the field by a laterally persistent mudstone. Support for the model was provided by data gathered during maintenance of well *22/5b-A3* during February–May 1999. The pressure within well *22/5b-A3* did not alter when the well was shut in, but production continued from the other wells.

4. THE USE OF BIOSTRATIGRAPHICAL CORRELATION IN RESERVOIR MODELLING

Following phase one development drilling a six-layer reservoir model for the Fleming Field was constructed in 1997 using a simple lithostratigraphical correlation scheme (Fig. 7A). A 13-layer model was constructed in early 1999 (Fig. 7B) based on the available biostratigraphical data. Fluid flow simulations in the reservoir were run for

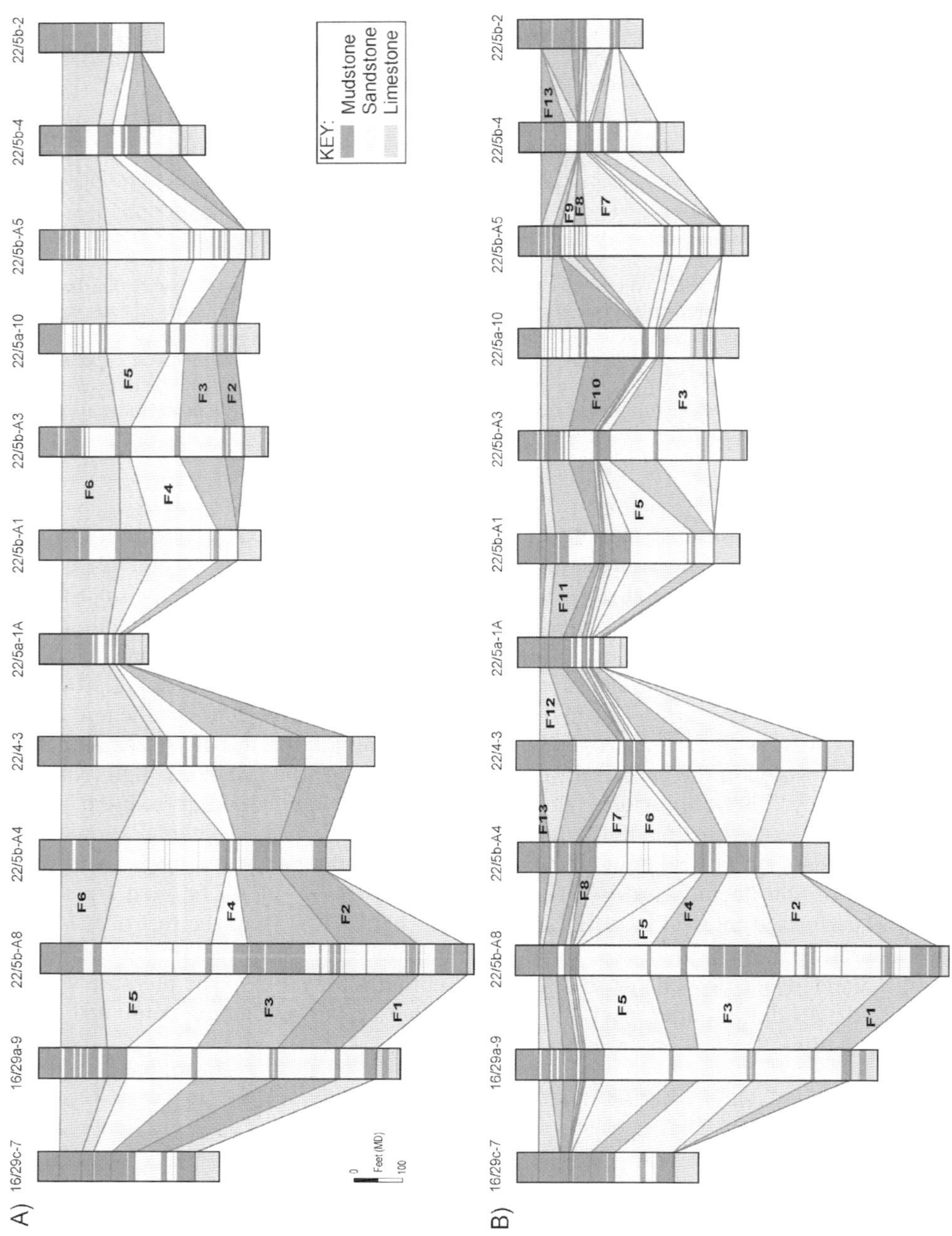
A)
16/29c-7
16/29a-9
22/5b-A8
22/5b-A4
22/4-3
22/5a-1A
22/5b-A1
22/5b-A3
22/5a-10
22/5b-A5
22/5b-4
22/5b-2
F1
F2
F3
F4
F5
F6
KEY:
Mudstone
Sandstone
Limestone
0
Feet (MD)
100
B)
16/29c-7
16/29a-9
22/5b-A8
22/5b-A4
22/4-3
22/5a-1A
22/5b-A1
22/5b-A3
22/5a-10
22/5b-A5
22/5b-4
22/5b-2
F1
F2
F3
F4
F5
F6
F7
F8
F9
F10
F11
F12
F13

both of the models and history-matching exercises, a process in which modelled data is compared to the real field production data, were carried out. Providing the modelled data matches the real data, history matching increases confidence that the geological model captures reservoir behaviour, thereby enabling predictions to be made regarding future production. If the model does not match the existing production data, there is increased uncertainty and risk in field behaviour prediction and in decisions relating to the drilling of additional production wells.

History matching exercises using the six- and 13-layer Fleming models were not able fully to describe the gas distribution encountered. The six-layer model could reproduce the gas volumes but not the distribution of the gas around the producing wells. The 13-layer model, by contrast, reproduced the separation of the wells and well groups apparent in the field but not the gas volumes. A reservoir model rebuild was therefore undertaken with the specific aim of explaining the apparent separations between the perforated intervals in wells *22/5b-A3, 22/5b-A5* and the group of wells *22/5b-A1, 22/5b-A4* and *22/5b-A8* (Fig. 2), and to match the gas volumes in contact with these wells and well groups indicated in the history matching.

Close re-investigation of available 3D seismic data indicated no significant levels of faulting that could explain the well separations. Core and thin-section analysis revealed no porosity-occluding or permeability-reducing cements. It was concluded that the porosity and permeability distribution in Fleming is purely a reflection of the net-to-gross distribution. It was felt, therefore, that the key to understanding the production behaviour of the Fleming Field was to define a suitable stratigraphical layering model, and to determine the net-to-gross distribution (and hence gas volume) in the reservoir. Two approaches were taken to address these aims:

(1) A ten-layer reservoir model based on the 13-layer model. This layering scheme honoured the biostratigraphical data and made the isolation of particular wells or well groups possible. In passing, we note that it is easier to construct a simplified reservoir model from a complicated model than vice versa.

(2) A 'hybrid' model based on both the 13-layer and the six-layer model. The upper Maureen Formation was modelled using the 13-layer model (layers F5 through to F13/F12 inclusive) as this made the isolation of particular wells or well groups possible; the lower Maureen Formation was modelled using the six-layer model, as this had previously given the best history match.

4.1. The ten-layer model

The original 13-layer model was reduced to a ten-layer model with the same top- and base-reservoir horizons (top Maureen and top Chalk, respectively). Reduction in the layering (i) provided a better understanding and modelling of reservoir sand distribution; (ii) explained the apparent separation of the wells; and (iii) reduced simulation times

---

Fig. 7. North–south section across the Fleming Field flattened at top Maureen Formation (see Fig. 2 for location of wells). (A) Ten-layer model. (B) 13-layer model. See Table 1 for detailed comparison of the two layering schemes.

TABLE 1

Comparison of the 13- and ten-layer schemes (see Fig. 7 for graphical representation of the layering scheme)

13-layer model	10-layer model
F13 F12	F13-12
F11	F11
F10	F10
F9 F8	F9–8
F7	F7
F6	F6
F5	F5
F4	F4
F3	F3
F2 F1	F2–1

during history matching by reducing the number of cells required in the flow simulation model. By combining layers the overall number of layers was reduced; layers were combined or were kept separate based on the information provided by the fuzzy logic modelling of mudstone continuity. The new layering scheme and its relation to the original 13-layer model is shown in Table 1 (see also Fig. 7B).

A flow simulation grid was constructed using this new layering scheme. Each layer was modelled as a single cell thickness, giving an overall count of 52,170 cells compared to 67,821 cells in the original 13-layer model. After initial history matching, the F10 layer was subdivided into three layers in order to improve the match at well *22/5b-A3*. This gave a final model of 62,604 cells.

4.2. The hybrid model

This model was a combination of the 13-layer model (upper Maureen Formation) and the six-layer model (lower Maureen Formation) as summarized in Table 2 (see also Fig. 7A and 7B). The flow simulation grid constructed for this model, again with a single cell thickness, gave an overall total of 57,387 cells.

4.3. Parameter grids

Net-to-gross distribution was considered critical to the understanding of the dynamic behaviour of the reservoir. The original net-to-gross distribution maps from the 13-layer model were thought to be pessimistic, with too rapid sanding-out onto the Jaeren High (J. Clark: personal communication, 1999). The maps were therefore fine-tuned to produce a result more compatible with well data and previous history matching exercises. In all, four criteria were used to edit the net-to-gross maps. First, the maps had to match the net-to-gross at each well, and secondly, they had to match the overall net-to-gross trend in Fleming as was indicated by the wells. Thirdly, the new layering scheme would

TABLE 2

Construction of the layering scheme used in the hybrid model and its relationship with the six- and 13-layer models

Model layer	ECLIPSE layer	Surface origin
F13/12	1	These surfaces are from the 13-layer model.
F11	2	
F10	3	
F9/8	4	
F7	5	
F6	6	
F5	7	
F4	8	These surfaces are from the six-layer model and were edited to fit.
F3	9	
F2	10	
F1	11	

be used to understand how successive turbidite flows would build upon the underlying topography (Fig. 6), and how this would relate to the net-to-gross distribution. The final criterion was to use the previous history matching data and the indications it gave as to the likely volumes of gas that were connected to the wells or well groups.

After editing, the net-to-gross trends were converted into 3D parameter grids for inclusion in the flow simulation grid. The procedure was repeated for each layer in turn to produce a 3D net-to-gross parameter grid for the entire reservoir. The process of editing the net-to-gross maps for both the ten-layer and hybrid models was iterative: this involved producing the maps, generating the 3D parameters and running a flow simulation to check the history match (Fig. 8). The methods for producing the net-to-gross grids were the same for both the ten-layer model and the hybrid model.

Porosity grids for each layer were produced using standard contouring techniques upon average porosity values for each layer at all the well points. Permeability grids were defined by applying a porosity–permeability transform to the porosity grids.

4.4. History matching results

Flow simulation was undertaken on both ten-layer and hybrid models. Bottom hole pressure results were compared to actual field production data, and results from the six-

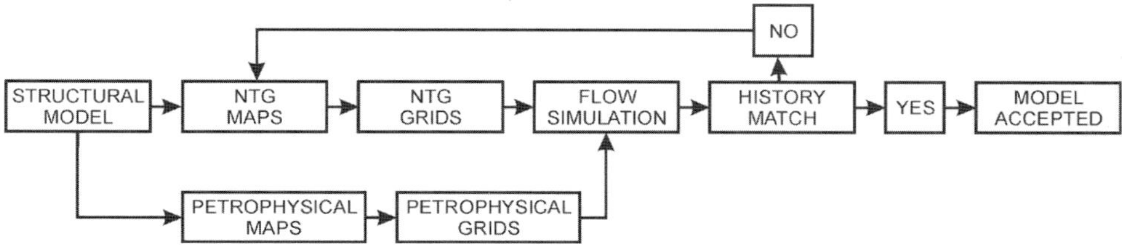

Fig. 8. A flow diagram of the iterative process used to achieve a history match.

and 13-layer models. For the ten-layer model, a good history match was produced after approximately eight iterations of editing the net-to-gross maps. For the hybrid model, a reasonable history match was produced after approximately ten iterations of editing the net-to-gross maps. As a good history match had already been produced for the ten-layer model, history matching of the hybrid model proceeded until a reasonable match had been produced but no further.

In Fig. 9A, 9B and 9C, we illustrate history matches for wells *22/5b-A3*, *22/5b-A4* and *22/5b-A5*. These three wells were chosen as they illustrate problems associated with the earlier six- and 13-layer models. The biostratigraphical data indicates that the perforated interval in well *22/5b-A3* is isolated from the other production wells in the Fleming Field, and when this is included in the reservoir model (as it is in the hybrid, 13- and ten-layer models) a good history match results (Fig. 9A). The *22/5b-A3* well is isolated in the six-layer model, however, and the history match is a poor one (Fig. 9A). In wells *22/5b-A4* and *22/5b-A5*, the 13-layer model gives a poor match (Fig. 9B and 9C), whereas the six- and ten-layer models give good history matches. Only the ten-layer and hybrid models give good history matches for the entire field and could therefore be used to predict future field performance. Of these two models, the ten-layer model gives a slightly better history match (Fig. 9A, 9B and 9C).

4.5. Discussion

We consider that fuzzy logic provides a powerful interpretation tool when applied to biostratigraphical data where simple models can communicate powerful results. Comparisons between individual wells will be more robust when interpreted using the fuzzy model as compared with the subjective interpretations of the biostratigrapher, i.e. the interpretation is calculated objectively and consistently against a stable model. In addition, every sample will have been analysed, whereas subjective manual interpretations often look at trends of data points rather than every single data point. Therefore, fuzzy logic model results are likely to be more detailed. Further, models can be run repeatedly and consistently and will give comparable results for different wells making correlation work easier. Though model construction may be a time consuming process, particularly the definition of MBFs and iterative tuning of the model via DoS weightings, the benefits gained in repeatability outweigh this possible disadvantage.

5. CONCLUSIONS

We show that a combination of graphic correlation and fuzzy modelling results enable high-resolution correlation of reservoir sandstones and interbedded mudstones

---

Fig. 9. History match (bottom hole pressure) for Fleming Field wells illustrating differences between the six-, ten-, 13-layer and hybrid models. Note that the ten-layer model (green line) gives the more consistent match to the measured bottom hole pressure in pounds per square inch (PSI) to all the wells. (A) *22/5b-A3*, (B) *22/5b-A4* and (C) *22/5b-A5*. (see Fig. 2 for location of wells).

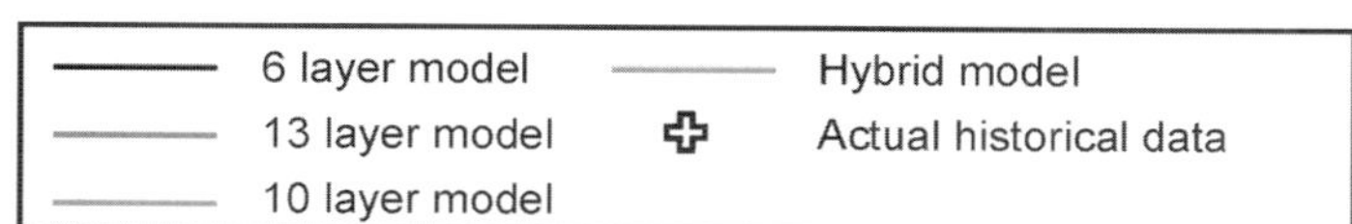
6 layer model
13 layer model
10 layer model
Hybrid model
Actual historical data

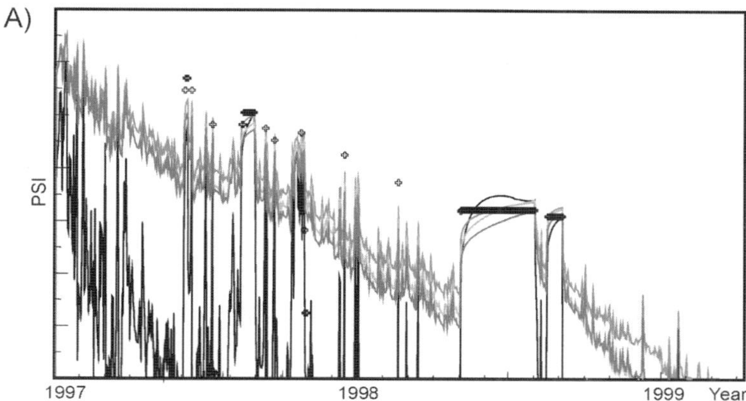
A)
PSI
1997
1998
1999
Year

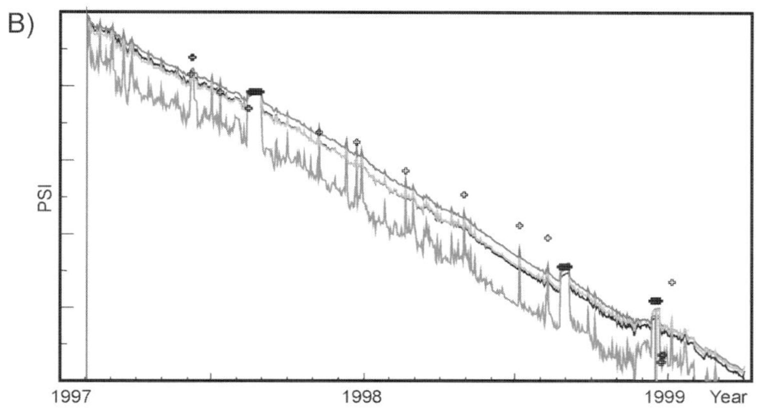
B)
PSI
1997
1998
1999
Year

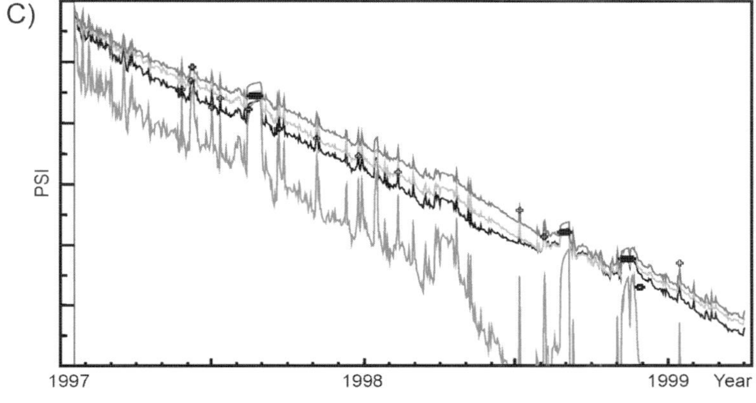
C)
PSI
1997
1998
1999
Year

at the Fleming Field. Detailed correlation of wells *22/5b-A1*, *22/5b-A3*, *22/5a-10* and *22/5b-A5* enabled identification of sandstones either side of a field-wide mudstone. A correlation model was produced that predicts that the perforated sand in well *22/5b-A3* was not in communication with the perforated intervals in wells *22/5b-A1* or *22/5b-A5*. This separation is supported by pressure data taken during maintenance of well *22/5b-A3* and subsequent fluid flow reservoir simulation.

Both ten-layer and hybrid models produced good history matches, indicating that the subdivision of the upper Maureen Formation based on the biostratigraphical work presented here is crucial to understanding connectivity in the reservoir, but less critical in the higher net-to-gross lower Maureen Formation. It may be that the mudstones in the lower Maureen Formation, identified in the fuzzy logic modelling as being likely to be continuous across the field, have been locally breached by subsequent sandy turbidite flows. The history matching process, however, was generally more difficult with the hybrid model compared to the ten-layer model. This in turn suggests that although the laterally continuous mudstones in the lower Maureen Formation may have been breached, they still have an influence on the dynamic behaviour of the reservoir, perhaps acting as baffles rather than barriers. While history matching is important, it is the predictive capability of the fuzzy logic model used that has proved to be critical in this case.

ACKNOWLEDGEMENTS

We thank BG Group and its partners in the Armada complex (Agip, BP, ConocoPhillips, TotalFinaElf and Yorkshire Energy) for permission to publish. We also wish to thank Jim Brown, Stuart Burley, Greig Cowan, Ross Garden, Phil Highton, Andy Samuel, Ian Stuart (all BG Group) and Nicholas Holmes (Ichron Ltd.) for constructive comments on an earlier draft of the manuscript. JPG reviewers Steve Cuddy (Petro-Innovations), Stewart Molyneux (British Geological Survey), Massoud Nikravesh (Berkeley) and D. Tamhane (University of New South Wales) are thanked for their constructive criticisms and suggestions. The Technical Alliance for Computational Stratigraphy is thanked for access to IPS.

REFERENCES

Armentrout, J.M., 1987. Integration of biostratigraphy and seismic stratigraphy: Pliocene–Pleistocene, Gulf of Mexico. In: *Innovative Biostratigraphic Approaches to Sequence Analysis: New Exploration Opportunities*. CGS-SEPM, 8th Annual Research Conference, pp. 6–14.

Armentrout, J.M., 1996. High resolution sequence biostratigraphy: examples from the Gulf of Mexico Plio-Pleistocene. In: Howell, J.A. and Aitken, J.F. (Eds.), *High Resolution Sequence Stratigraphy: Innovations and Applications. Geol. Soc. London, Spec. Publ.*, 104: 65–86.

Armentrout, J.M. and Clement, J.F., 1990. Biostratigraphic calibration of depositional cycles: a case study in High Island–Galveston–East Banks areas, offshore Texas. In: Armentrout, J.M. and Perkins, B.F. (Eds.), *Sequence Stratigraphy as an Exploration Tool: Concepts and Practices in the Gulf Coast*. CGS-SEPM, 11th Annual Research Conference, pp. 21–51.

Armentrout, J.M., Echols, R.C., and Lee, T.D., 1990. Patterns of foraminiferal abundance and diversity:

implications for sequence stratigraphic analysis. In: Armentrout, J.M. and Perkins, B.F. (Eds.), *Sequence Stratigraphy as an Exploration Tool: Concepts and Practices in the Gulf Coast*. CGS-SEPM, 11th Annual Research Conference, pp. 53–59.

Armentrout, J.M., Fearn, L.B., Rodgers, K., Root, S., Lyle, W.D., Herrick, D.C., Bloch, R.B., Snedden, J.W. and Nwnakwo, B., 1999. High-resolution sequence biostratigraphy of a lowstand prograding deltaic wedge: Oso Field (late Miocene), Nigeria. In: Jones, R.W. and Simmonds, M.D. (Eds.), *Biostratigraphy in Production and Development Geology. Geol. Soc. London, Spec. Publ.*, 152: 259–290.

Berggren, W.A., Kent, D.V., Swisher, C.C. and Aubry, M.-P., 1995. A revised cenozoic geochronology and chronostratigraphy. In: Berggren, W.A., Kent, D.V., Aubry, M.-P. and Hardenbol, J. (Eds.), *Geochronology, Time Scales and Global Stratigraphic Correlation. Soc. Econ. Paleontol. Mineral, Spec. Publ.*, 54: 129–212.

Charnock, M.A. and Jones, R.W., 1990. Agglutinated foraminifera from the Palaeogene of the North Sea. In: Hemelben, C., Kaminski, M.A., Khunt, W. and Scott, D.B. (Eds.), *Paleoecology, Biostratigraphy, Paleoecology and Taxonomy of Agglutinating Foraminifera*. Kluwer Academic Publishers, Dordrecht, pp. 139–144.

Culver, S.J., 1988. New foraminiferal depth zonation of the northwestern Gulf of Mexico. *Palaios*, 3: 69–85.

Dickenson, B., Waterhouse, M., Goodall, J. and Holmes, N., 2001. Blenheim Field: the appraisal of a small oil field with a horizontal well. *Pet. Geosci.*, 7: 81–95.

Fang, J.H. and Chen, H.C., 1990. Uncertainties are better handled by fuzzy arithmetic. *Am. Assoc. Pet. Geol. Bull.*, 74: 1228–1233.

Fang, J.H. and Chen, H.C., 1997. Fuzzy modelling and the prediction of porosity and permeability from the compositional and textural attributes of sandstone. *J. Pet. Geol.*, 20(2): 185–204.

Foley, L., Ball, L., Hurst, A., Davies, J. and Blockley, D., 1997. Fuzziness, incompleteness and randomness: classification of uncertainty in reservoir appraisal. *Pet. Geosci.*, 3: 203–209.

Gary, A.C. and Filewicz, M., 1997. Quantifying Descriptive Paleoenvironmental Models using Fuzzy Logic. In: *Biostratigraphy in Production and Development Geology*. Geol. Soc. London, Petroleum Group Conference (Aberdeen). Abstract.

Gibson, T.G., 1989. Planktonic benthonic foraminifera ratios: modern patterns and Tertiary applicability. *Mar. Micropalaeontol.*, 15: 29–52.

Holmes, N.A., 1999. The Andrew Formation and 'biosteering' – different reservoirs, different approaches. In: Jones, R.W. and Simmons, M.D. (Eds.), *Biostratigraphy in Production and Development Geology. Geol. Soc. London, Spec. Publ.*, 152: 155–166.

Jones, R.W. and Charnock, M.A., 1985. 'Morphogroups' of agglutinating foraminifera. Their life positions and feeding habits and potential applicability in (paleo)ecological studies. *Rev. Paleobiol.*, 4: 311–320.

Kaminski, M.A., Gradstein, F.M., Berggren, W.A., Geroch, S. and Beckmann, J.P., 1988a. Flysch-type agglutinated foraminiferal assemblages from Trinidad: taxonomy, stratigraphy and paleobathymetry. In: Gradstein, F.M. and Rogl, F. (Eds.), *2nd International Workshop on Agglutinated Foraminifera, Vienna 1986, Proceedings. Abh. Geol. Bundesanst.*, 41: 155–227.

Kaminski, M.A., Grassle, J.F. and Whitlatch, R.B., 1988b. Life history and recolonisation among agglutinated foraminifera in the Panama Basin. In: Gradstein, F.M. and Rogl, F. (Eds.), *2nd International Workshop on Agglutinated Foraminifera, Vienna 1986, Proceedings. Abh. Geol. Bundesanst.*, 41: 229–244.

Kneller, B. and McCaffrey, W., 1999. Depositional effects of flow non-uniformity and stratification within turbidity currents approaching a bounding slope: deflection, reflection, and facies variation. *J. Sediment. Res.*, 69: 980–991.

Knox, R.W.O'B. and Holloway, S., 1992. Palaeogene of the Central and Northern North Sea. In: Knox, R.W.O'B. and Cordey, W.G. (Eds.), *Lithostratigraphic Nomenclature of the UK North Sea*. British Geological Survey, Nottingham, 133 pp.

Kosko, B., 1994. *Fuzzy Thinking – The New Science of Fuzzy Logic*. Flamingo, New York, 336 pp.

Mitchum, R.M., Sangree, J.B., Vail, P.R. and Wornardt, W.W., 1993. Recognizing sequences and systems tracts from well logs, seismic data, and biostratigraphy: examples from the Late Cenozoic of the Gulf of Mexico. In: Weimer, P. and Posamentier, H.W. (Eds.), *Siliciclastic Sequence Stratigraphy – Recent Developments and Applications. Am. Assoc. Pet. Geol. Mem.*, 58: 163–197.

Mudge, D.C. and Bujak, J.P., 1996a. An integrated stratigraphy for the Paleocene and Eocene of the North

Sea. In: Knox, R.W.O'B., Corfield, R.M. and Dunay, R.E. (Eds.), *Correlation of the Early Palaeogene in Northwest Europe. Geol. Soc. London, Spec. Publ.*, 101: 91–113.

Mudge, D.C. and Bujak, J.P., 1996b. Palaeocene biostratigraphy and sequence stratigraphy of the UK Central North Sea. *Mar. Pet. Geol.*, 13: 295–312.

Mudge, D.C. and Copestake, P., 1992a. Revised Lower Palaeogene lithostratigraphy for the Outer Moray Firth, North Sea. *Mar. Pet. Geol.*, 9: 53–69.

Mudge, D.C. and Copestake, P., 1992b. Lower Palaeogene stratigraphy of the northern North Sea. *Mar. Pet. Geol.*, 9: 287–301.

Neal, J.E., Stein, J.A. and Gamber, J.H., 1994. Graphic correlation and sequence stratigraphy in the Palaeogene of NW Europe. *J. Micropaleontol.*, 13: 55–80.

Neal, J.E., Stein, J.A. and Gamber, J.H., 1995. Integration of the graphic correlation methodology in a sequence stratigraphic study: examples from the North Sea Palaeogene sections. In: Mann, K.O. and Lane, H.R. (Eds), *Graphic Correlation. Soc. Econ. Paleontol. Mineral. Spec. Publ.*, 53: 95–113.

Norland, U., 1996. Formalizing geological knowledge with an example of modeling stratigraphy using fuzzy logic. *J. Sediment. Petrol.*, 66: 689–698.

Norland, U., 1999. Stratigraphic modeling using common sense rules. In: Harbaugh, J.W., Watney, W.L., Rankey, E.C., Slingerland, R., Goldstein, R.H. and Franseen, E.K. (Eds.), *Numerical Experiments in Stratigraphy: Recent Advances in Stratigraphic and Sedimentologic Computer Simulations. Soc. Econ. Paleontol. Mineral, Spec. Publ.*, 62: 245–251.

O'Connor, S.J. and Walker, D., 1993. Paleocene reservoirs of the Everest Trend. In: Parker, J.R. (Ed.), *Petroleum Geology of Northwest Europe*. Proceedings of the 4th Conference, pp. 145–160.

Pacht, J.A., Bowen, B.E., Shaffer, B.L and Pottorf, B.R., 1990. Sequence stratigraphy of Plio-Pleistocene strata in the offshore Louisiana Gulf Coast: applications to hydrocarbon exploration. In: Armentrout, J.M. and Perkins, B.F. (Eds.), *Sequence Stratigraphy as an Exploration Tool: Concepts and Practices in the Gulf Coast*. CGS-SEPM, 11th Annual Research Conference, pp. 269–285.

Poumot, C., 1989. Palynological evidence for eustatic events in the tropical Neogene. *Bull. Cent. Rech. Explor.-Prod. Elf-Aquitaine*, 13(2): 437–453.

Shaffer, B.L., 1987. The potential of calcareous nannofossils for recognizing Plio-Pleistocene climatic cycles and sequence boundaries on the Shelf. In: *Innovative Biostratigraphic Approaches to Sequence Analysis: New Exploration Opportunities*. GCS-SEPM, 8th Annual Research Conference, pp. 142–145.

Shaffer, B.L., 1990. The nature and significance of condensed sections in Gulf Coast late Neogene sequence stratigraphy. *Trans., Gulf Coast Assoc. Geol. Soc.*, XL: 767–776.

Shaw, A.B., 1964. *Time in Stratigraphy*. McGraw-Hill, New York, NY, 365 pp.

Simpson, E.H., 1949. Measurement of diversity. *Nature*, 189, 732–735.

Stein, J.A., Gamber, J.H., Krebs, W.N. and LA Coe, M.K., 1995. A composite standard approach to biostratigraphic evaluation of the North Sea Palaeogene. In: Steel, R.J., Felt, V.L., Johannesson, E.P. and Mathieu, C. (Eds.), *Sequence Stratigraphy of the Northwest European Margin. NPF Special Publication, 5*. Elsevier, Amsterdam, pp. 401–414.

Stewart, I.J., 1987. A revised stratigraphic Interpretation of the early Palaeogene of the central North Sea. In: Brooks, J. and Glennie, K. (Eds.), *Petroleum Geology of North West Europe*. pp. 557–576.

Vining, B.A., Ioannides, N.S. and Pickering, K.T., 1993. Stratigraphic relationships of some Tertiary lowstand depositional systems in the central North Sea. In: Parker, J.R. (Ed.), *Petroleum Geology of Northwest Europe*. Proceedings of the 4th Conference. pp. 17–30.

Zadeh, L.A., 1965. Fuzzy sets. *Inf. Control*, 8: 338–353.

Developments in Petroleum Science, 51
Editors: M. Nikravesh, F. Aminzadeh and L.A. Zadeh

Chapter 25

GEOSTATISTICAL CHARACTERIZATION OF THE CARPINTERIA FIELD, CALIFORNIA

RAJESH J. PAWAR [a], EDWIN B. EDWARDS [b] and EARL M. WHITNEY [b]

[a] *Los Alamos National Laboratories, Los Alamos, NM, USA*
[b] *Pacific Operators Offshore Inc., Santa Barbara, CA, USA*

ABSTRACT

This paper describes application of geostatistical techniques for detailed characterization of a geologically complex oil reservoir. The reservoir is mature and the characterization is aimed at facilitating its further development. A geological structure and stratigraphy model of the field has already been developed. Stochastic distributions of important reservoir rock properties were subsequently generated. Different techniques were applied to generate these distributions. A variety of data sources were integrated for each property. Multiple realizations were combined to assess uncertainty in property distributions. These measures of uncertainty and the detailed property realizations will be helpful for the field operator for assessing future field development scenarios.

1. INTRODUCTION

The Carpinteria Offshore Field in the Santa Barbara channel is a mature oil field. Since its discovery in 1964, the field has produced over 100 million barrels of oil through approximately 200 deviated wells from five platforms. The field is part of a geologically complex reservoir, currently operated by Pacific Operators Offshore Inc. (POOI), a small, independent producer in California. Several companies have operated various parts of the field at various times since its discovery. The resulting disparate production practices have led to a variety of interpretations of the geologic and engineering data. No field-wide reservoir interpretation integrating the geological, geophysical, engineering and production data was available when POOI purchased the field. In order to efficiently manage this mature field over a long term, it was necessary to understand and model the complex geological environment. Developing a conceptual model of the field was difficult because of the multiple prior interpretations for the extensive geological and engineering data. Careful reconciliation of existing interpretations and reinterpretation of the legacy data have led to an integrated, self-consistent geological model for the structure of the reservoir. Distributions of reservoir rock properties in the initial version of the geologic model were generated using deterministic techniques.

Attempts were made to validate the geologic model with a history match exercise, but these attempts were unsuccessful. Because production in the field has been commingled among dozens of pay zones and even among fault blocks, production data, which was monitored only at the surface, provides insufficient information to describe the historical movement of fluids in the field. Additional data that might help to reduce this uncertainty either do not exist, or are very limited in scope.

It was thought that in the absence of a history match exercise, confidence in the geologic model could be increased through improved characterization of the reservoir using advanced stochastic techniques. A variety of geostatistical techniques were applied to generate distributions of reservoir rock properties. Primary as well as multiple secondary data, wherever available, were integrated in the process. Multiple realizations were generated for each property. These realizations were subsequently used to quantify the uncertainty in the given property data. This detailed characterization and the carefully quantified uncertainty will be useful for the field operator for long term reservoir management.

2. RESERVOIR GEOLOGY AND GEOLOGIC STRUCTURE MODELING

The Carpinteria field is part of a geologically complex, turbidite reservoir. The productive structure is an east plunging asymmetric anticline, which has been modified by several generations of folding and faulting. Current productive zones lie within the Repettian foraminiferal stage of the Pliocene Pico formation. The sequence consists of alternating sandstone and shale beds. The depositional styles include turbidites and associated channelized sands and overbank deposits. The presence of these depositional types indicates a deep-water fan environment where the exact position relative to individual fan elements has varied through time. Geologic depositional studies suggest that the source for the sediments has been the Santa Ynez mountains to the north. The lowest reservoir zones show a tendency for porosity and thickness isolines to trend approximately east–west. This suggests deposition in a paleoenvironment with a gradient at right angles to the modern structure, probably an outer mid fan area or outer fan margin. In the higher zones depositional patterns show more variability along the axis of the structure, suggesting some sort of channelized flow environment, such as that found in the distributary channel section of suprafan lobe in the mid fan position.

Thirty productive zones, identified from well log analysis, are present in the field. These zones, ranging in thickness from 10 to 200 feet, represent major depositional events that can, for the most part, be correlated field wide. Each zone is made up of multiple individual sedimentary events, but there are not enough appropriate data to sufficiently define the facies. The productive zones have been classified into five groups designated C, D, E, F and G. Each group is further divided by numbers and, in some cases, yet again using letters, e.g., G6C. The tops and bottoms of each zone have been correlated throughout the field.

The Carpinteria structure has been cut by a major reverse fault, the Hobson fault. The amount of the displacement varies from place to place, but is typically on the order of about 800 to 1000 feet. Hobson divides the field in two sub-fields, the upper or supra-thrust part and the lower or sub-thrust part. In addition to the Hobson, a number of other faults of lesser displacement are present in the reservoir. Previous interpretations of the Carpinteria field have shown it as being divided into numerous individual fault

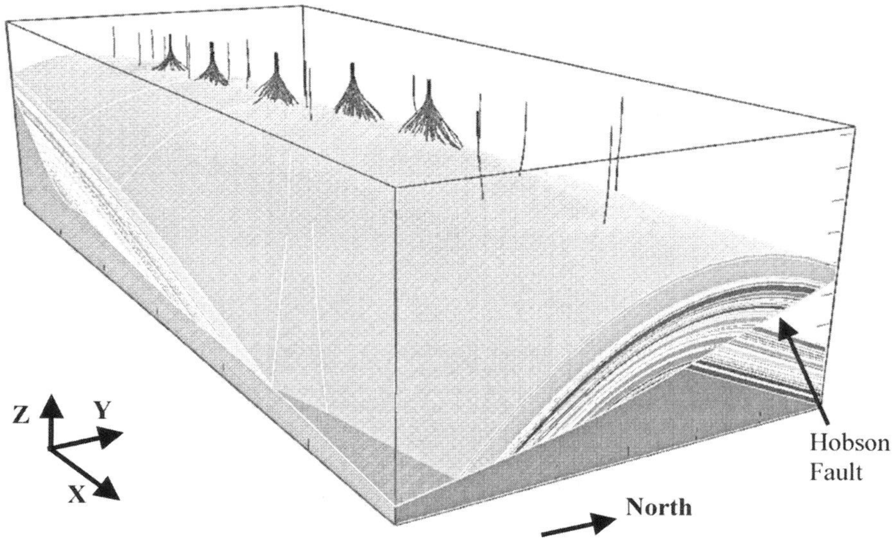

Fig. 1. A three-dimensional view of the model for Carpinteria Field along with the wells. (The wells are indicated by yellow lines.)

blocks by a multitude of approximately north-trending cross faults. These faults were interpreted largely because of a need to account for an extremely complicated distribution of oil/water contacts observed in the wells. Oil/water contacts approximating the original (pre-development) condition of each of the zones can be obtained from the resistivity logs for the set of initial development wells. Such studies by the current operator showed a gradual rather than step-wise inclination of oil/water contacts for all zones, and showed that these inclinations are congruent over multiple zones. Thus, it is not a simple matter to relate oil/water contact elevations to faulting at Carpinteria. Furthermore, seismic data is sparse and of such coarse resolution as to preclude its use in the delineation of secondary faults. Therefore, fault recognition became a matter of detailed log correlation, with assistance from sparse dipmeter data. More than 80 fault intercepts have been identified to date and 14 fault planes delineated. Several of the small-displacement faults (10 to 30 feet) have been verified by subsequent drilling.

The information on zone boundaries and faults was used to generate a model for the geologic structure of the field. Figs. 1 and 2 show two different views of that model. The geostatistical analysis of the field presented in this paper is constrained by the volumetric boundaries defined by this model.

3. AVAILABLE DATA

Geostatistical distributions of porosity, permeability, and shale volume fraction were generated within the structural model. Various sources of data were available to interpret

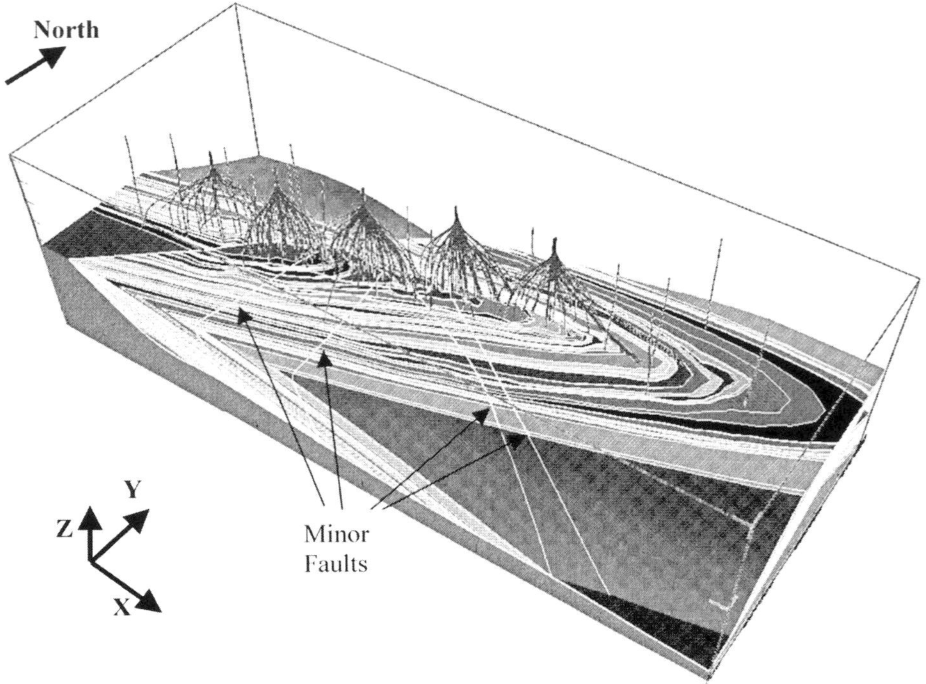

Fig. 2. A cross-sectional view of the model for Carpinteria Field along with the wells. (The wells are indicated by yellow lines.)

the values of these properties. These sources included a suite of geophysical logs such as neutron, density, gamma ray, and traditional spontaneous potential and resistivity electric logs that were collected in approximately 200 wells on the field. Laboratory derived permeability and porosity data were available from core samples taken from various wells. A few well test derived permeability values were also available.

Values of porosity were calculated from core as well as density and neutron log data. Gamma ray and spontaneous potential logs were used to infer the values of volume shale fraction. Permeability values were calculated from core derived porosity–permeability correlations, and from porosity and shale fraction data. Not all of the above mentioned sources of data were available for each well, but calculations took advantage of all available data at any given geospatial point.

The spatial tracks of the well bores were available from deviation surveys. Information on stratigraphic markers defining the tops and bottoms of each zone was available for each well. The surfaces representing the distribution of these top and bottom markers were imported from the geologic structure model.

The distributions of reservoir rock properties were generated for a grid with dimensions of $74 \times 31 \times 10$ in the x, y and z directions respectively. The cell dimensions were 400 feet each in the x and y direction. The approximate average thickness of the

F1 zone is 200 feet, and most of the log data was available at the resolution of one foot. Generating distributions at the resolution of the available data would have resulted in models with an enormous number of cells. The model size was reduced to only 10 cells in the vertical direction by reducing the data resolution. As mentioned earlier, it is unfortunate that no facies data were available for the reservoir. The rock property distributions were therefore generated using logs for the entire zone.

Although rock property distributions were generated for all 30 zones present in the reservoir, results are presented only for the F1 zone in this paper. This zone is one of the most productive in the reservoir. The same techniques used to generate property distributions for this zone were applied to all the other zones. All these distributions were limited to the supra-thrust volume of the field.

4. POROSITY DISTRIBUTION

Multiple distributions of porosity were generated from the available porosity data using a sequential simulation technique. These simulations were conditioned not only to the porosity data, but to the available density and neutron log data as well. The first step was to determine the parameters defining the spatial continuity in the observed data.

4.1. Semivariogram for porosity

A three-dimensional semivariogram was calculated to ascertain the spatial continuity of the porosity data in the horizontal and vertical direction. It is important to determine the principle directions of anisotropy in the spatial continuity of the data. Multiple variograms were therefore calculated, with different principle axes in the horizontal direction. Analyses of these variograms always resulted in the N79°E as the major direction of continuity. However, this principle direction of continuity is likely an artifact of the preferential spatial availability of data due to the placement of the wells. Because the Carpinteria anticline is oriented approximately in the N79°E direction, the five platforms from which most of the wells originate are also aligned in this general direction. This has resulted in placement of most of the wells and hence availability of the observed data in the N79°E direction. As mentioned earlier, the source for the deposition for the field has been to the north. Some of the data have also shown a gradient in the general principle direction of the anticline.

Although various attempts were made to determine the principle direction for deposition by limiting the range of the data used to calculate the semivariogram, analysis of these directional variograms never revealed a prominent direction for continuity. And although principle directions consistent with the general depositional nature of the field were observed in some subsets of the data, these trends were not repeated throughout the field. In view of this bias of the spatial placement of the observed data, isotropic semivariograms were used in all subsequent analyses. The semivariogram and its corresponding spherical model are shown in Fig. 3a. A semivariogram was also calculated in the vertical direction. The vertical variogram is

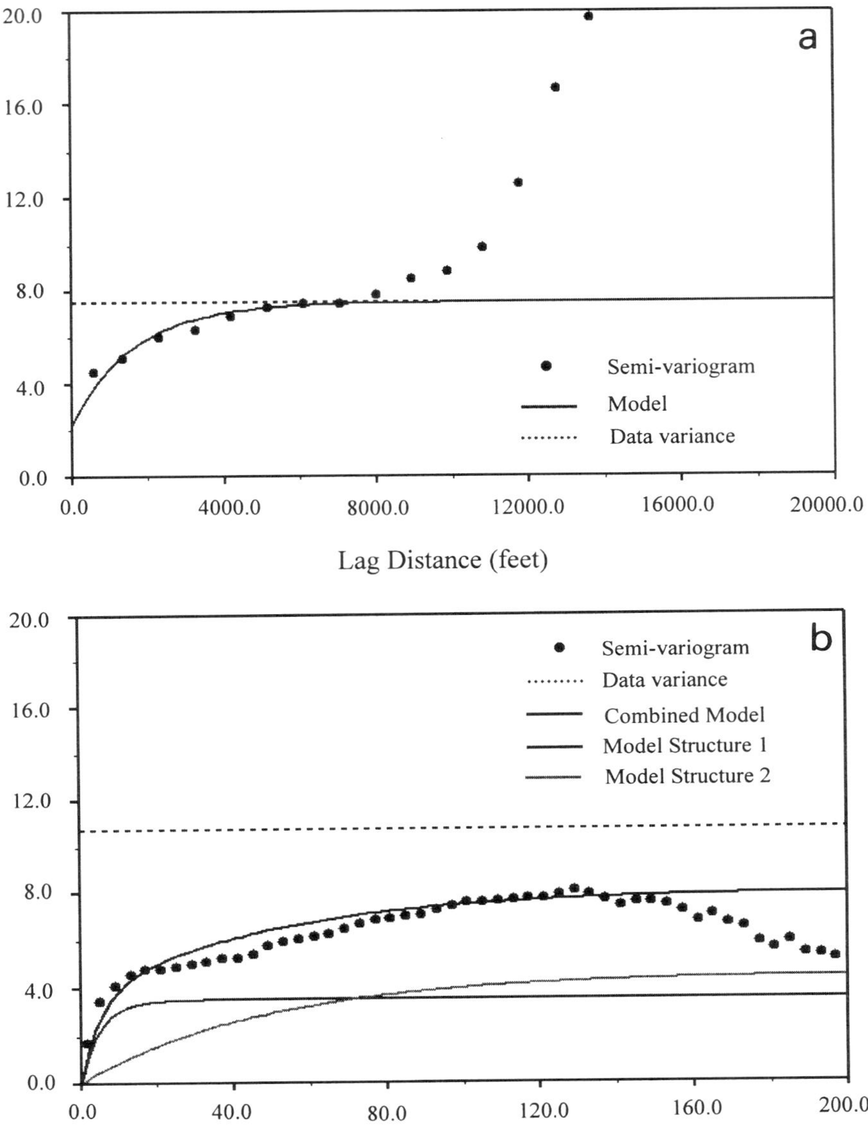

Fig. 3. (a) Isotropic semivariogram for porosity in the horizontal direction and corresponding model. (b) Semivariogram for porosity in the vertical direction and corresponding model.

shown in Fig. 3b. The equations for the two variogram models are shown below.

$$\gamma_{\text{horizontal}}(h) = 2.25 + 5.25\,\text{Sph}\left(\sqrt{\frac{h_x^2}{4866.67} + \frac{h_y^2}{4866.67}}\right)$$

$$\gamma_{\text{vertical}}(h) = 3.5\,\text{Sph}\left(\sqrt{\frac{h_z^2}{15}}\right) + 4.5\,\text{Sph}\left(\sqrt{\frac{h_z^2}{140}}\right)$$

Similarly, parameters of spatial continuity were calculated for neutron and density porosity log data.

4.2. Porosity realizations

The porosity realizations were calculated using a co-collocated co-kriging approach. This co-kriging approach is based on the Markov hypothesis of collocated co-kriging (Deutsch and Journel, 1982). This approach allows use of additional or secondary variables without performing a full-scale co-kriging. It approximates the cross-covariance between two variables by the product of the covariance of the secondary variable and correlation coefficient between the primary and secondary variables. Thus, the secondary variable is retained in the calculation but the inference process is simplified by eliminating a need for a full co-kriging approach. Application of this technique allowed use of both the density and neutron log data values as the secondary or soft data. Multiple porosity realizations were generated using a sequential Gaussian simulation method based on the co-collocated co-kriging technique. The simulation procedure is described in detail here.

(1) In order to use the sequential Gaussian simulation algorithm, it is necessary that the variable being simulated be multivariate normal. Not only should the univariate cumulative distribution function (cdf) of the primary variable be normal, but it is also necessary to check for the normality of the bivariate cdf. The input porosity data was transformed to follow a normal distribution satisfying the first condition. The normality of the bivariate cdf was checked by comparing the indicator variograms of the data calculated at different cutoffs (Deutsch and Journel, 1982). Indicator variograms were calculated at seven different cdf cutoffs, namely, 0.05, 0.1, 0.3, 0.5, 0.7, 0.9 and 0.95. The semivariograms for each cutoff are shown in Fig. 4a through 4g. As can be seen from the figure, the semivariograms continuously and symmetrically decrease as the cdf cutoffs decrease towards zero or one. The semivariograms for the lower and upper cutoffs, 0.05 and 0.95 are essentially flat indicating pure nugget effects. On the other hand, the semivariograms for the remaining cutoffs are symmetric around that for a cutoff of 0.5. This validates the bi-gaussian nature of the input data.

(2) In order to perform collocated co-kriging it was necessary to have the secondary variable present at all the locations where the primary variable was being estimated. The density porosity data was available at about 145 well locations while the neutron porosity data was available at only 69 well locations. For this reason, the density porosity data was chosen as the principle secondary variable. The density well log data were first used to generate a three-dimensional realization of the density porosity through ordinary kriging, and then this realization was used as input for the secondary data.

(3) The correlation coefficients between the primary variable (porosity) and each of the two secondary variables (density and neutron porosity) were calculated. In order to account for the spatial variability of the correlation coefficients, they were calculated at each well location. These different correlation coefficients were then used to generate a kriged surface of the coefficients for the entire calculation grid.

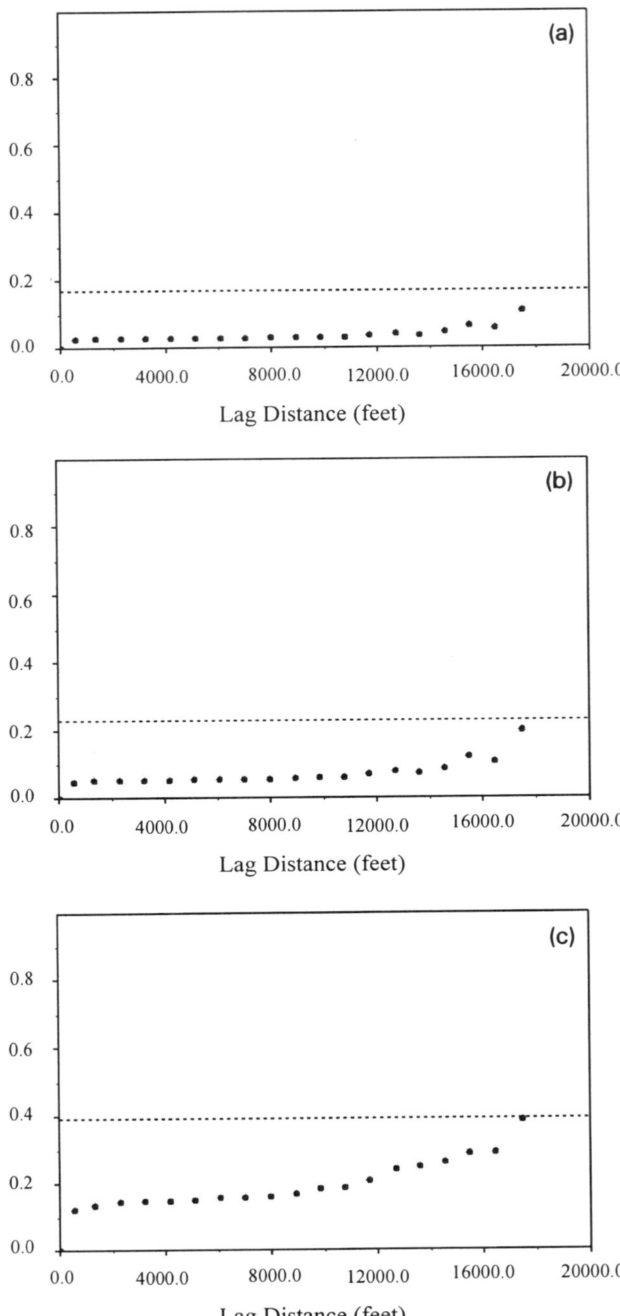

Fig. 4. Indicator variograms for seven cutoffs of input porosity data: (a) cutoff 0.05, (b) cutoff 0.1, (c) cutoff 0.3.

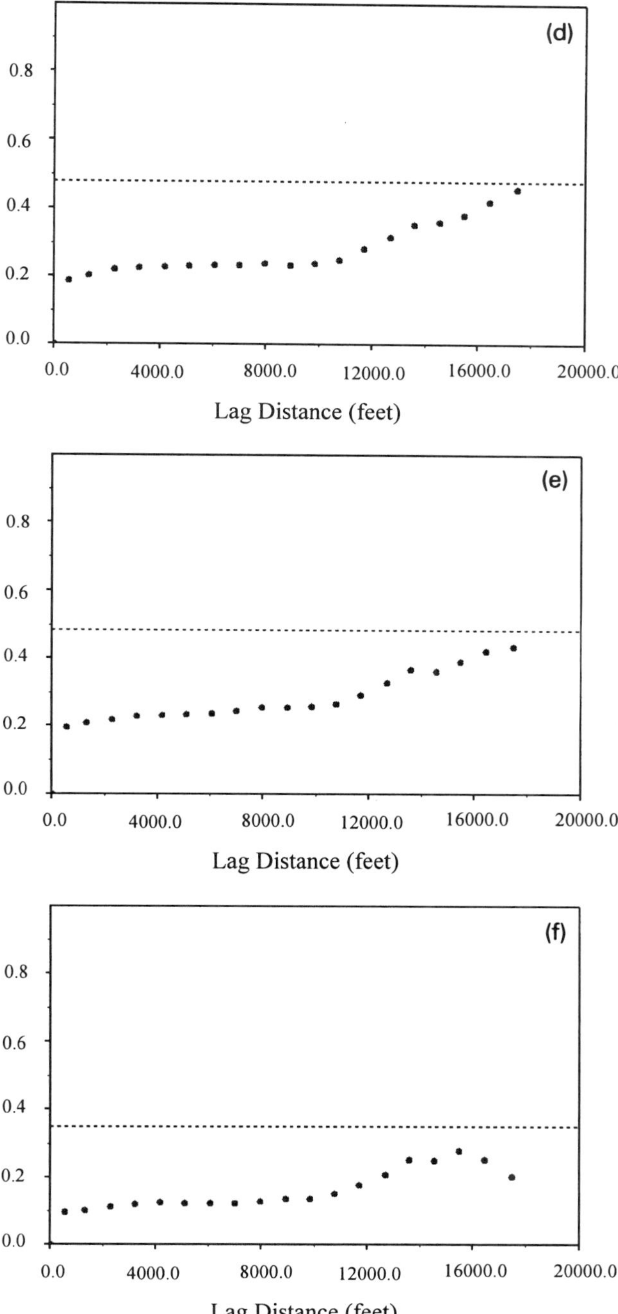

Fig. 4 (continued). Indicator variograms for seven cutoffs of input porosity data: (d) cutoff 0.5, (e) cutoff 0.7, (f) cutoff 0.9.

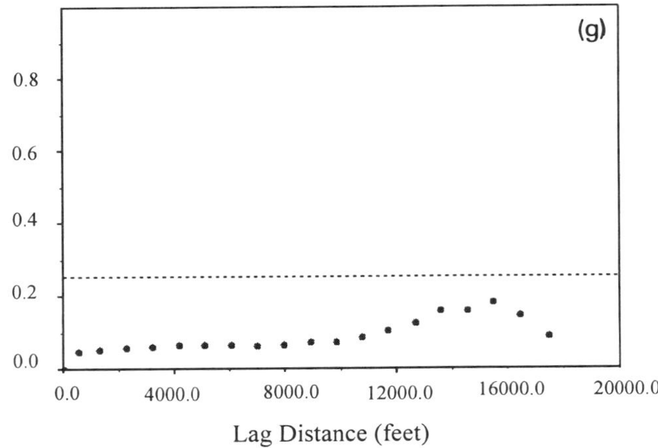

Fig. 4 (continued). Indicator variograms for seven cutoffs of input porosity data: (g) cutoff 0.95.

As mentioned earlier, realizations of porosity distributions were conditioned to the primary porosity data and secondary density and neutron log data. The 3D model for the density porosity and the kriged surfaces for the primary/secondary correlation coefficients were used for secondary conditioning of porosity. Twenty five realizations of porosity were generated with the co-collocated co-kriging-based sequential simulation approach. Fig. 5 shows a three-dimensional view of one such realization. In Table 1, the statistics of 25 simulated porosity realizations are compared with the statistics of the observed porosity data. Analysis of the table shows that the output realizations are statistically comparable with the input data. The most noticeable difference is in the minimum and maximum values. It should be noted that the statistics for the output data are actually averages for the 25 realizations. The minimum values actually range between 0.03 to 8.6, while the maximum values range between 37.5 to 49.3 for all realizations. It should also be noted that the actual data used for the simulations is thickness-averaged data, generated for the ten vertical layers used in the geostatistical model. The averaged values will have slightly different minimum and maximum values.

TABLE 1

Comparison of statistics of input data and the simulated realizations for porosity

Statistics	Input data	Average values for 25 realizations
Minimum	0.00	4.80
Maximum	49.30	47.08
Average	24.34	24.44
Lower quartile	22.40	22.82
Median	24.30	24.35
Upper quartile	26.40	26.11

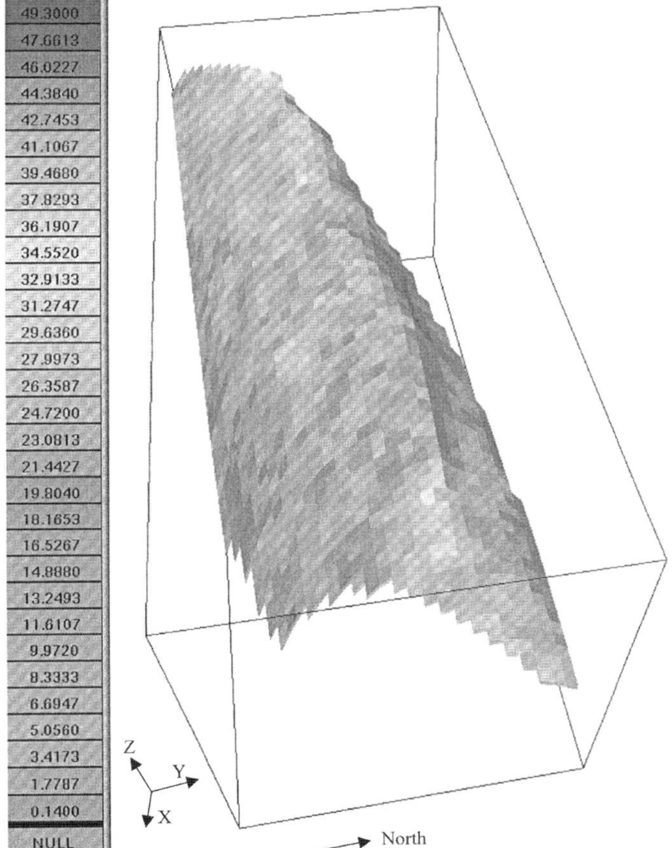

Fig. 5. A three-dimensional view of one of the simulated porosity realizations.

5. SHALE VOLUME FRACTION REALIZATION

The data for shale volume fraction were calculated mainly from the gamma ray and spontaneous potential data, but resistivity logs were used where they were available. A different approach was used to generate distributions for shale fraction. Because the secondary support data for shale fraction was not as extensively available as that for porosity, a simple sequential Gaussian simulation (sGs) approach was used to generate multiple realizations that were conditioned only to shale fraction data.

5.1. Spatial correlation

The spatial continuity parameters for shale fraction were calculated from a three-dimensional semivariogram. Independent semivariograms were calculated to determine the parameters in the horizontal and vertical directions. For the reasons mentioned

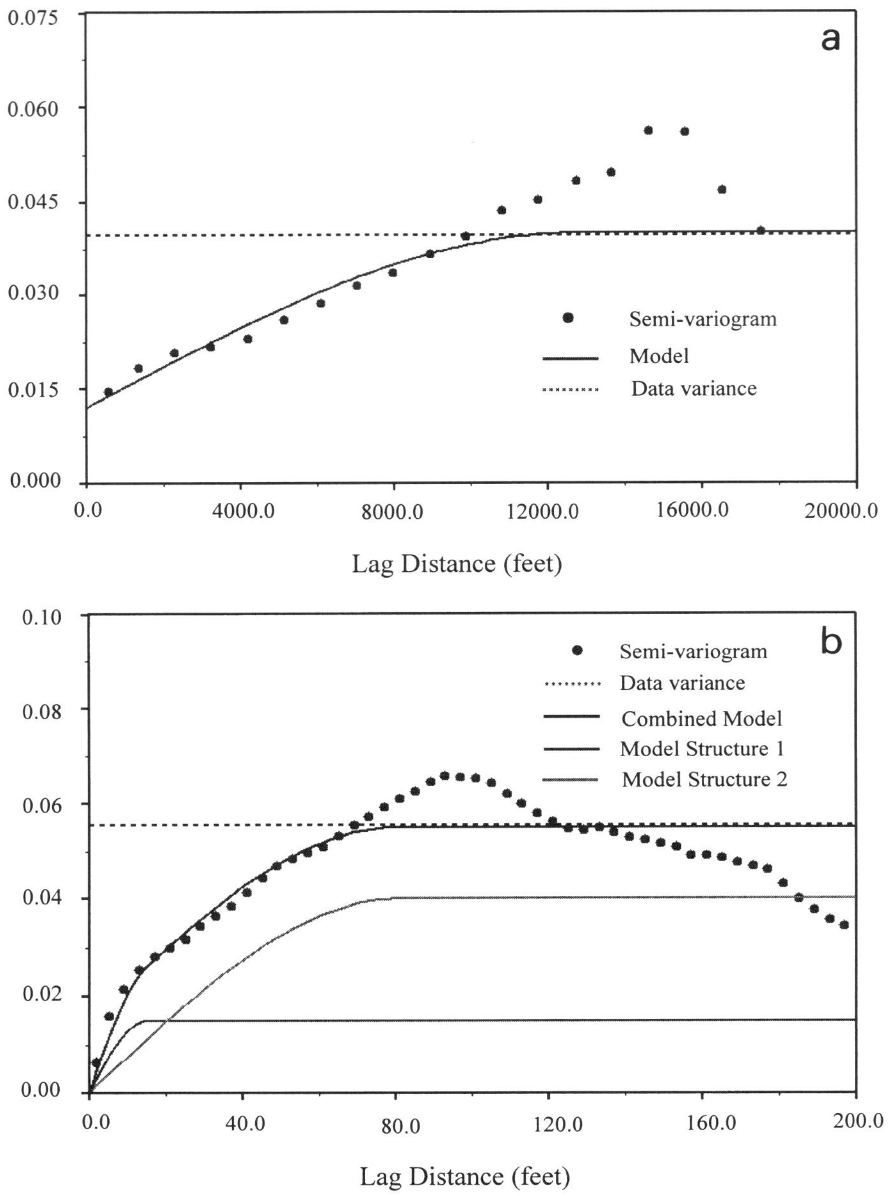

Fig. 6. (a) Isotropic semivariogram for shale fraction in the horizontal direction and corresponding model. (b) Semivariogram for shale fraction in the vertical direction and corresponding model.

earlier, only an isotropic semivariogram was calculated in the horizontal direction. Fig. 6a shows the calculated semivariogram for the horizontal direction and its model, and the semivariogram for the vertical direction and its model fit are shown in Fig. 6b.

The models for the two semivariograms are shown below.

$$\gamma_{\text{horizontal}}(h) = 0.012 + 0.028\,\text{Sph}\left(\sqrt{\frac{h_x^2}{12666.67} + \frac{h_y^2}{12666.67}}\right)$$

$$\gamma_{\text{vertical}}(h) = 0.015\,\text{Sph}\left(\sqrt{\frac{h_z^2}{15}}\right) + 0.04\,\text{Sph}\left(\sqrt{\frac{h_z^2}{80}}\right)$$

5.2. Realizations of shale fraction

As described earlier, applying a gaussian-based algorithm requires that the input data exhibit multivariate Gaussian behavior. For this reason, the input data were examined to determine whether they are bi-gaussian. The input data were transformed to follow a normal distribution. The bi-gaussian nature of the input data was checked by examining indicator variograms that were calculated at seven cdf cutoffs of 0.05, 0.1, 0.3, 0.5, 0.7, 0.9 and 0.95. These variograms are shown in Fig. 7a through 7g. The two features may be noted in the figures – first, high and low cutoff values exhibit a pure nugget structure, and second, all variograms exhibit a symmetric behavior around the cdf cutoff of 0.5. Satisfaction of these two conditions implies that applying the sequential Gaussian simulation algorithm is warranted.

Using a sequential Gaussian simulation approach 25 different realizations of shale fraction distribution were generated. Fig. 8 shows a three-dimensional view of one of these realizations. Statistics for the 25 realizations are compared with the statistics for the input data in Table 2. Examination of the table shows that the statistics for the output realizations are comparable to the input data. Another important aspect of the input data was the bimodal nature of its histogram as shown in Fig. 9a. In order to assure the proper statistical distribution of the simulated values, it was important to reproduce this observed nature. Fig. 9b shows the histogram for one of the realizations. As can be seen from the figure, the overall trends in the two histograms are similar.

TABLE 2

Comparison of statistics of input data and the simulated realizations for shale fraction

Statistics	Input data	Average values for 25 realizations
Minimum	0.000	0.000
Maximum	1.000	0.974
Average	0.327	0.370
Lower quartile	0.147	0.159
Median	0.217	0.255
Upper quartile	0.518	0.579

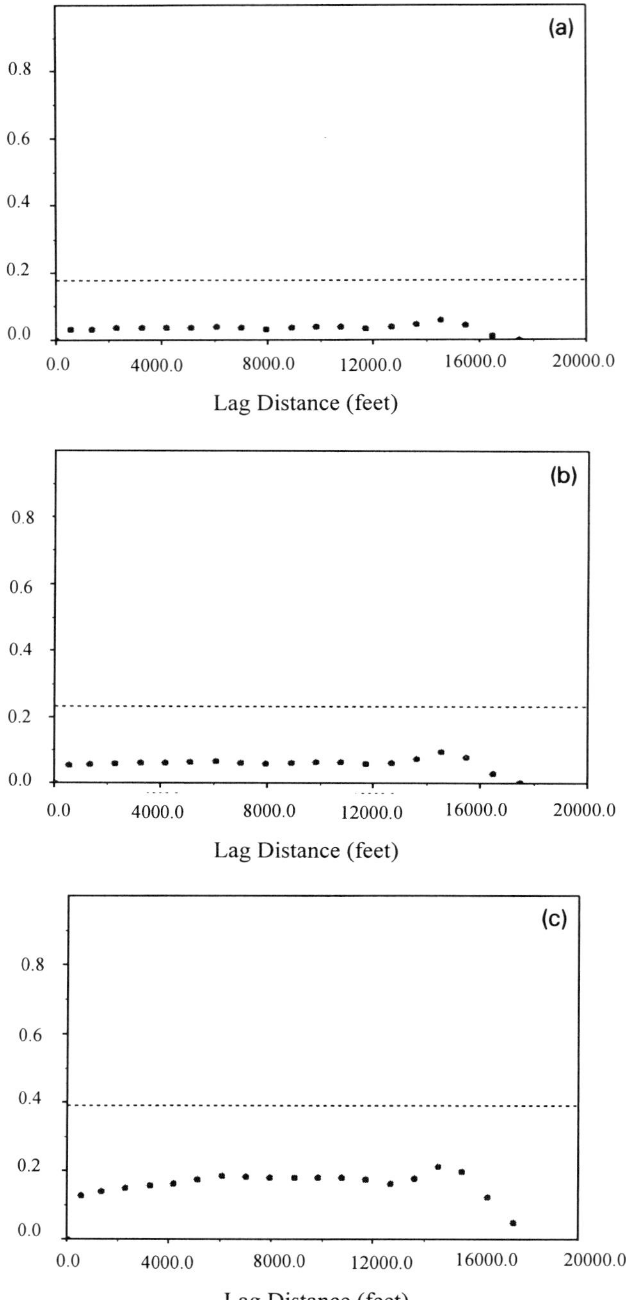

Fig. 7. Indicator variograms for seven cutoffs for input shale fraction data: (a) cutoff 0.05, (b) cutoff 0.1, (c) cutoff 0.3.

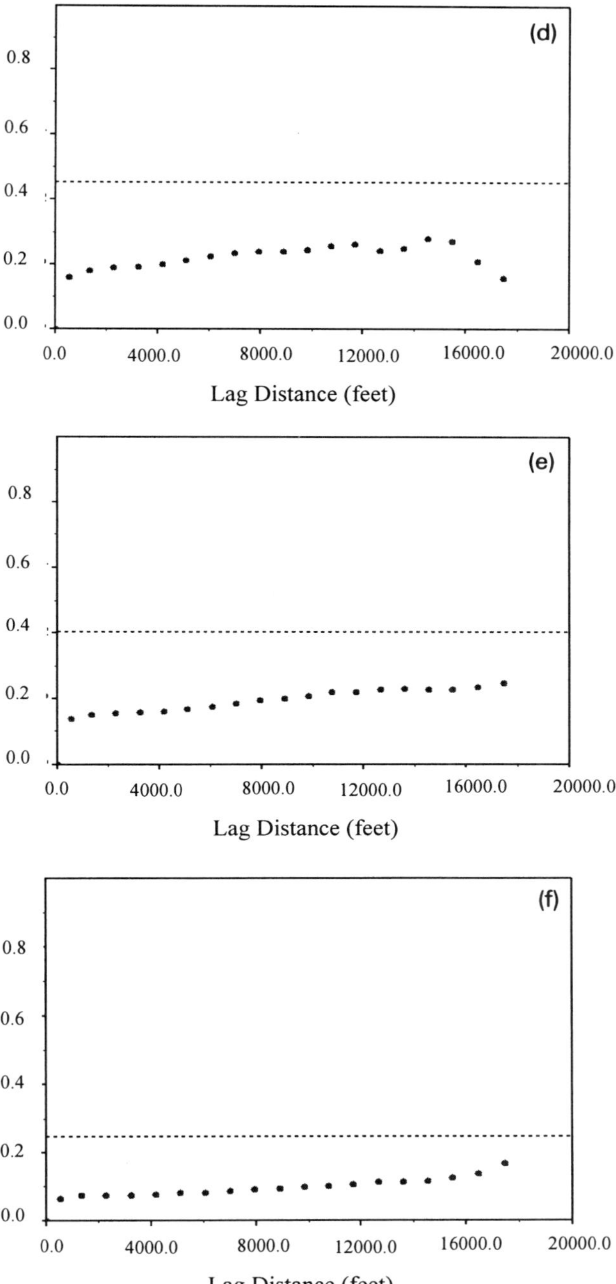

Fig. 7 (continued). Indicator variograms for seven cutoffs for input shale fraction data: (d) cutoff 0.5, (e) cutoff 0.7, (f) cutoff 0.9.

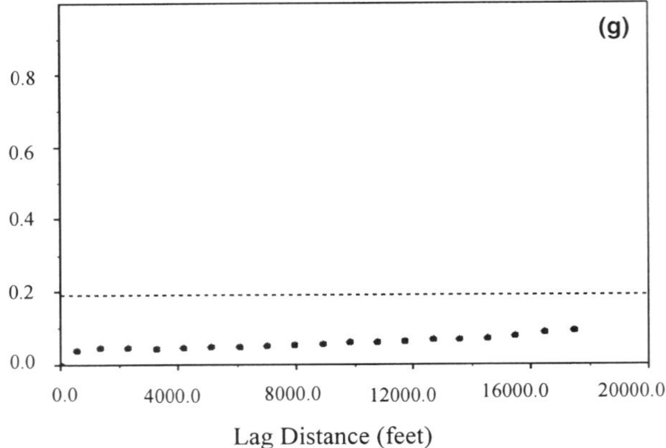

Fig. 7 (continued). Indicator variograms for seven cutoffs for input shale fraction data: (g) cutoff 0.95.

6. PERMEABILITY DISTRIBUTION

Porosity and shale content both have an effect on the intrinsic value of permeability in a sand. There are differing relationships among these three variables. In most of the sands, permeability is directly proportional to porosity. Permeability, on the other hand, has a semi-inverse relationship with shale fraction. That is to say, high values of shale fraction generally indicate lower values of permeability. However, a low value of shale fraction may not necessarily correspond to high permeability. In fields where facies data is available, the porosity and permeability correlations are conditioned to individual facies. These facies specific correlations are subsequently used to generate the permeability values from porosity realizations. Different facies have different shale content. Thus, conditioning to facies can take into account the effect of shaliness on the permeability. Because no facies data were available for the Carpinteria field, both porosity and shale fraction values were used to generate the permeability values.

Before utilizing the porosity and shale fraction data, their relationship with permeability was studied. Fig. 10a to 10c show scatterplots among these three variables. As can be seen from the figures, both porosity and shale fraction have a complex relationship with permeability. The high porosity values correspond to high as well as low permeability values. High values of shale fraction tend to correspond to low permeability values, but low shale fraction values correspond to high as well as low permeability values. It should be noted that high porosity corresponds to high permeability only for low values of shale fraction and that low values of shale faction correspond to high permeability only for high porosity. It can be concluded from these figures that the three variables must be considered together when generating permeability distributions.

In the absence of facies data, we defined artificial facies data that accounts for the complex relationships among these three properties. Two different approaches were used two define these artificial facies data. In the first approach, facies were defined based on the value of shale fraction at a location. Five different facies were defined based on the shale fraction cutoffs of 0.2, 0.4, 0.6 and 0.8. In the second approach,

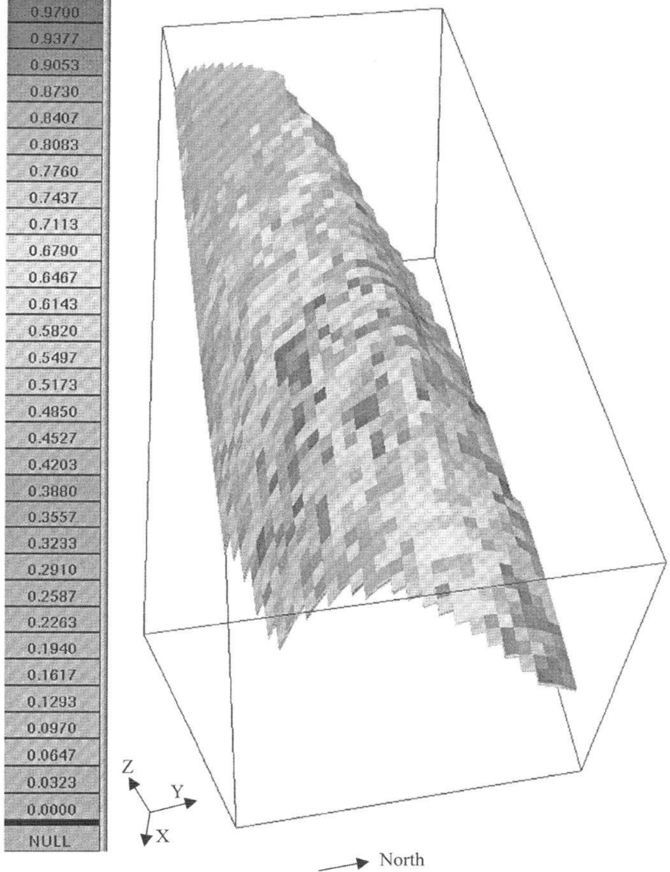

Fig. 8. A three-dimensional view of one of the simulated shale fraction realizations.

both shale fraction and porosity values were combined to generate facies data. A value for the product of porosity and shale fraction was calculated at each well location. The relationship between this shale fraction–porosity product and permeability was studied. Fig. 10d shows a plot of permeability versus the shale fraction–porosity product data. This plot exhibits a similar relationship to the one that exists between shale fraction and permeability. However, the spread of the data in Fig. 10d is much smaller than that in Fig. 10b. This shale fraction–porosity product was used to define facies at 2 cutoff values, 10 and 20. By examining Fig. 10a and 10d, it can be seen that the cutoffs correspond to different ranges of permeability. These cutoffs were chosen by visual inspection of the scatterplot in Fig. 10d. Two different artificial logs were calculated based on these two different approaches. These artificial logs indicated the presence of a particular facies depending on the range corresponding to the shale fraction or the shale fraction–porosity product value at that location.

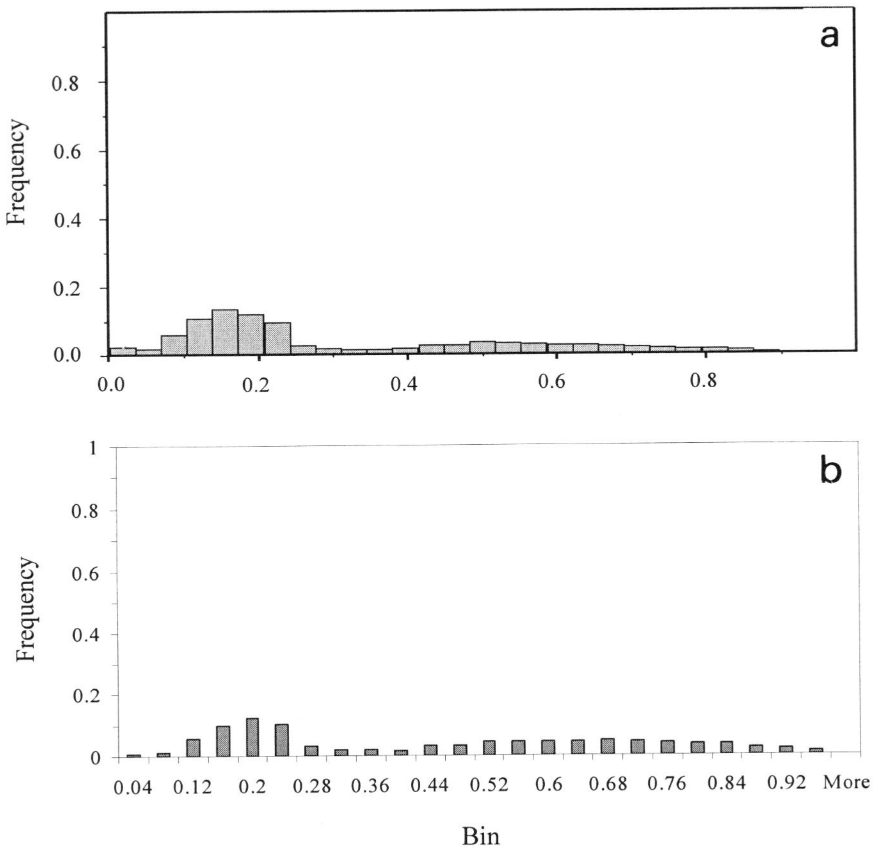

Fig. 9. (a) A histogram for input shale fraction data. (b) A histogram for one of the simulation realizations of shale fraction.

As mentioned earlier, the purpose of this effort was to utilize the relationships between porosity, permeability and shale fraction data as well as the character of variability in distributions of porosity and shale fraction to generate permeability realizations. A facies-based cloud transform approach was chosen. A cloud transform facilitates the transformation of a model of one property into a model for another related property. Usually, the transformation is performed using the crossplot between the two properties. A correlated probability field model is calculated from this crossplot. The range and distribution of a dependent variable such as permeability is calculated, with the aid of the crossplot, from an independent variable such as porosity. This approach seeks to reproduce the original scatter in the data rather than a simple correlation such as linear regression. In the facies-based approach the cloud transformation is further constrained by the facies data. At a particular location, a permeability value is calculated from the porosity–permeability crossplot for the facies identified at that location.

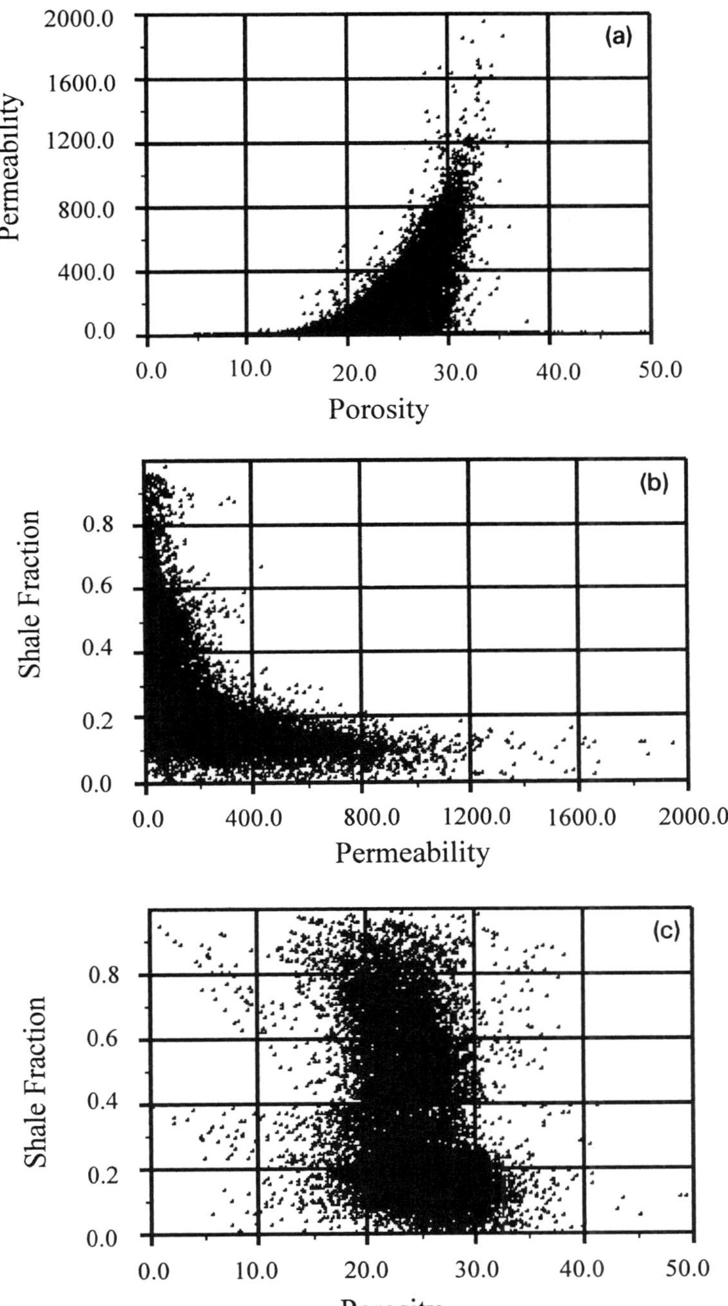

Fig. 10. (a) A scatterplot of input porosity versus permeability data. (b) A scatterplot of input permeability versus shale fraction data. (c) A scatterplot of input porosity versus shale fraction data.

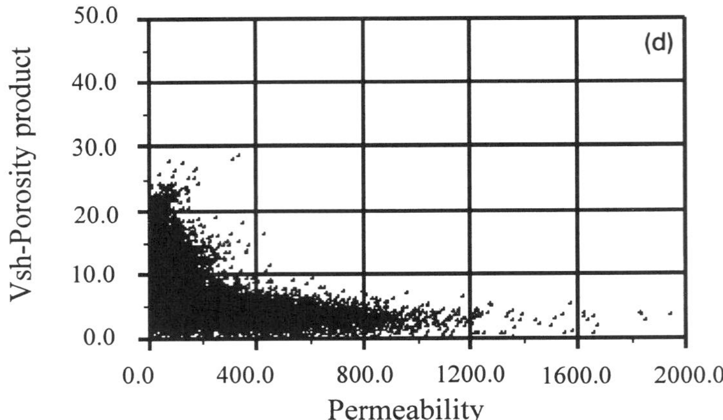

Fig. 10 (continued). (d) A scatterplot of input permeability versus shale fraction–porosity product data.

6.1. Input data

Porosity and permeability logs were used to calculate the cloud transforms specific to each facies. The simulation realizations for porosity and shale fraction were used to define the facies distribution. Twenty five different facies models were generated combining the 25 porosity and shale fraction models. These facies models and the porosity models from which they were derived were used in turn to generate the permeability model.

6.2. Results

Twenty five different realizations of permeability distributions were generated with each of the two facies-based approaches. Fig. 11 shows one of the realizations. This particular realization was based on the first approach of defining facies from shale fraction data alone. The statistics of these realizations from both approaches are compared with the statistics of the input data in Table 3. It should be noted that the statistics presented are an accumulation of 25 realizations. Statistically, the outputs from the two approaches do not differ significantly. The statistics of the input data are

TABLE 3

Comparison of overall permeability statistics from the two facies-based approaches

Statistics	Input data	Facies approach 1	Facies approach 2
Minimum	0.00	0.00	0.00
Maximum	1944.30	1628.29	1674.58
Average	165.81	113.66	113.45
Lower quartile	17.40	23.46	22.97
Median	97.50	77.79	76.88
Upper quartile	241.80	153.87	154.45

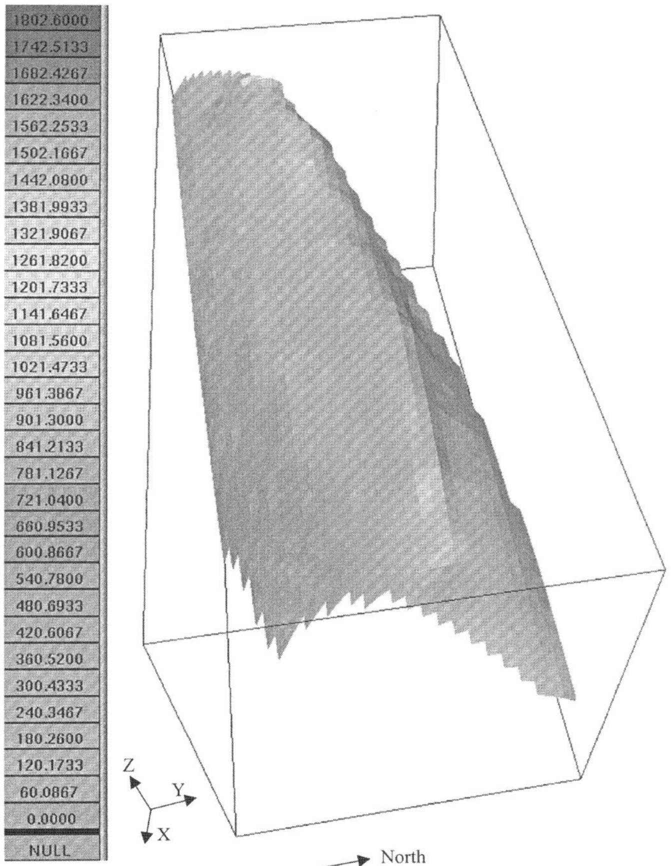

Fig. 11. A three-dimensional view of one of the simulated permeability realizations.

not reproduced as well for permeability realizations as they are in the realizations of shale fraction or porosity. The statistics for individual facies are compared in Tables 4 and 5 for the 2 approaches. It should be repeated that these facies were defined from porosity and shale fraction data, and that each of the two approaches honors porosity and shale fractions data at any given location. The two methods produce distributions of facies with different statistics. The second approach reproduces the range of input data better than the first approach, but quartile binning of the data is not reproduced well. There are two reasons for this result. First, we are comparing statistics of input data that is available at a high resolution to the simulated data, which is calculated at a more coarse resolution. The decrease in resolution results in averaged values. Second, because realizations of permeability are generated from realizations of porosity and shale fraction, it is possible that significant error propagation has occurred.

Even with these observations, this approach does seem to reproduce the overall trends in permeability data based on shale fraction and porosity data. Figs. 12 and 13 show

TABLE 4

Comparison of statistics for permeability for individual facies-based on facies approach 1

Statistics	Facies 1		Facies 2		Facies 3		Facies 4		Facies 5	
	Input data	Results	Input data	Results	Input data	Results	Input data	Results	Input data	Results
Minimum	0.00	0.00	0.00	0.00	0.00	0.00	0.00	0.00	0.00	0.01
Maximum	333.00	337.01	431.40	274.31	368.60	346.92	957.60	978.11	1944.30	1598.76
Average	11.72	50.06	15.71	33.16	54.28	54.22	123.25	149.17	262.36	167.61
Lower quartile	0.10	0.78	2.70	4.04	23.00	19.05	51.90	72.42	90.00	66.77
Median	0.70	32.94	8.80	21.59	42.70	44.79	103.15	126.45	217.85	128.72
Upper quartile	2.80	76.48	18.80	47.30	72.50	75.99	169.50	197.99	379.30	222.62

TABLE 5

Comparison of statistics for permeability for individual facies- based on facies approach 2

Statistics	Facies 1		Facies 2		Facies 3	
	Input data	Results	Input data	Results	Input data	Results
Minimum	0.00	0.00	0.00	0.00	0.00	0.00
Maximum	333.00	329.22	446.30	491.17	1944.30	1674.58
Average	28.20	49.27	36.68	46.97	214.46	156.76
Lower quartile	0.15	0.90	4.20	8.38	59.50	82.53
Median	2.95	30.68	19.60	32.74	159.50	121.50
Upper quartile	50.00	77.14	51.00	68.72	307.80	210.16

scatterplots of simulated permeability values versus the conditioning porosity and shale fraction values, for the two approaches. The figures show that the overall relationships between these three properties are qualitatively similar to the input data (Fig. 10a–c).

7. UNCERTAINTY ANALYSIS

As mentioned in the introduction, the purpose of this study was not only to generate distributions of simulated reservoir rock properties but also to quantify the uncertainty of the original and the simulated data. We combined different realizations of porosity, shale fraction and permeability for the uncertainty analysis.

7.1. Pore volume

Different realizations of porosity and shale fraction were combined to estimate the total pore volume in the F1 zone. Pore volume was defined as the product of rock volume, porosity, and shale fraction. The individual porosity and shale fraction realizations were generated independent of each other, so a total of 625 values of pore volume were calculated. The results of the pore volume calculations are shown

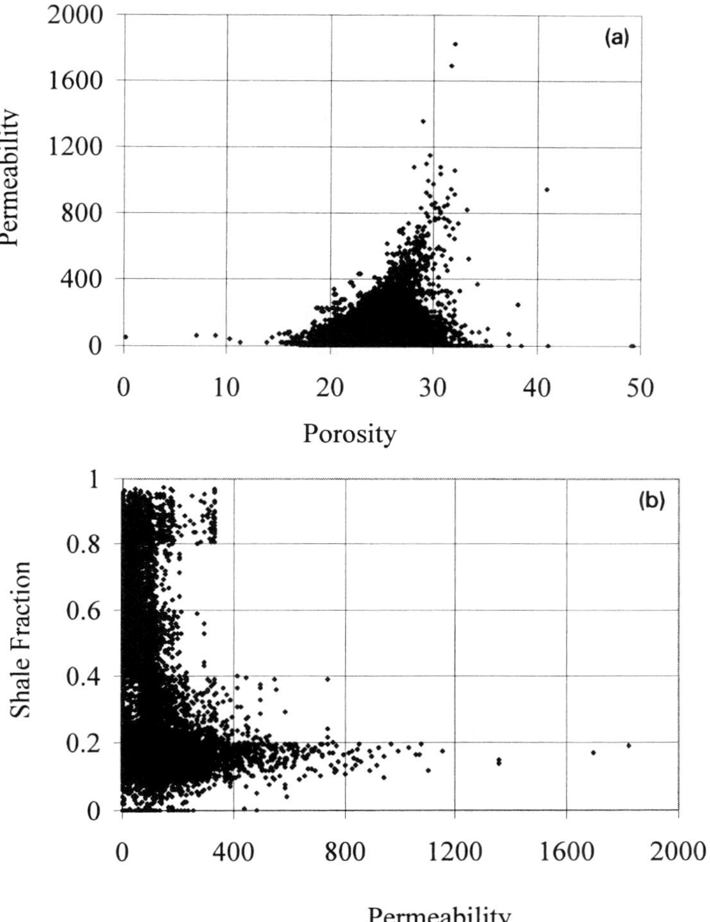

Fig. 12. (a) A scatterplot of conditioning porosity values versus corresponding simulated permeability values for one of the permeability simulation realizations (based on facies approach 1). (b) A scatterplot of conditioning shale fraction values versus corresponding simulated permeability values for one of the permeability simulation realizations (based on facies approach 1).

in Table 6. As can be seen from the table, the average pore volume in the F1 zone is 267×10^7 ft^3. The standard deviation in the pore volume calculation is 17.6×10^7 ft^3, or 6% of the average volume.

7.2. Uncertainty in porosity and shale fraction

In order to quantify the variability in simulations of porosity and shale fraction, simulated data were accumulated from every grid point, and from all 25 realizations. Figs. 14 and 15 show three-dimensional distributions of the standard deviation in the

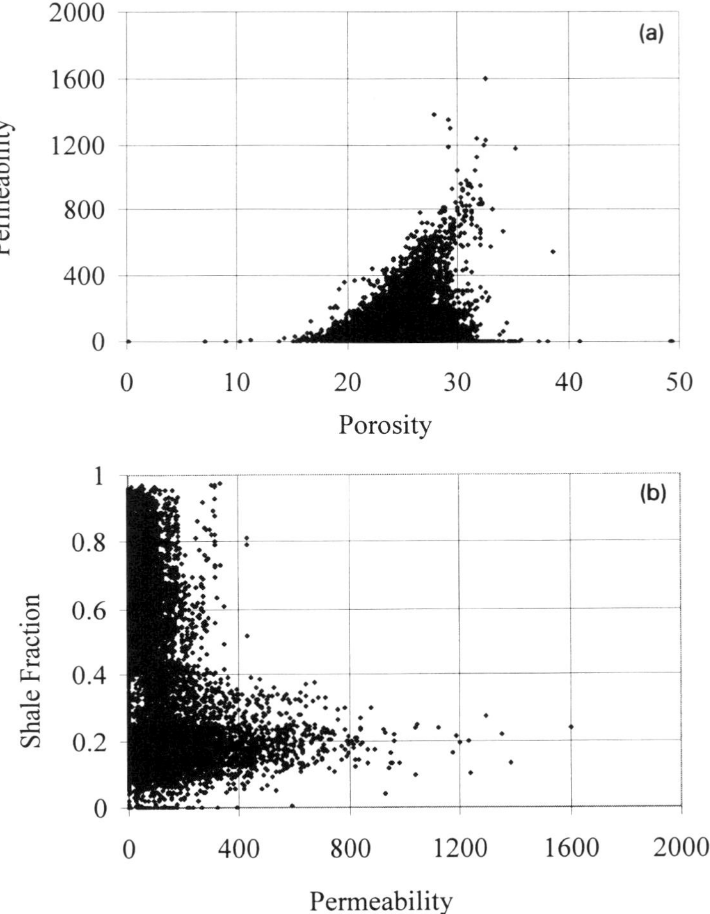

Fig. 13. (a) A scatterplot of conditioning porosity values versus corresponding simulated permeability values for one of the permeability simulation realizations (based on facies approach 2). (b) A scatterplot of conditioning shale fraction values versus corresponding simulated permeability values for one of the permeability simulation realizations (based on facies approach 2).

TABLE 6

Statistics of pore volume calculation

Statistics	Pore volume ($\times 10^7$)
Minimum	232
Maximum	323
Average	267
Standard deviation	17.6

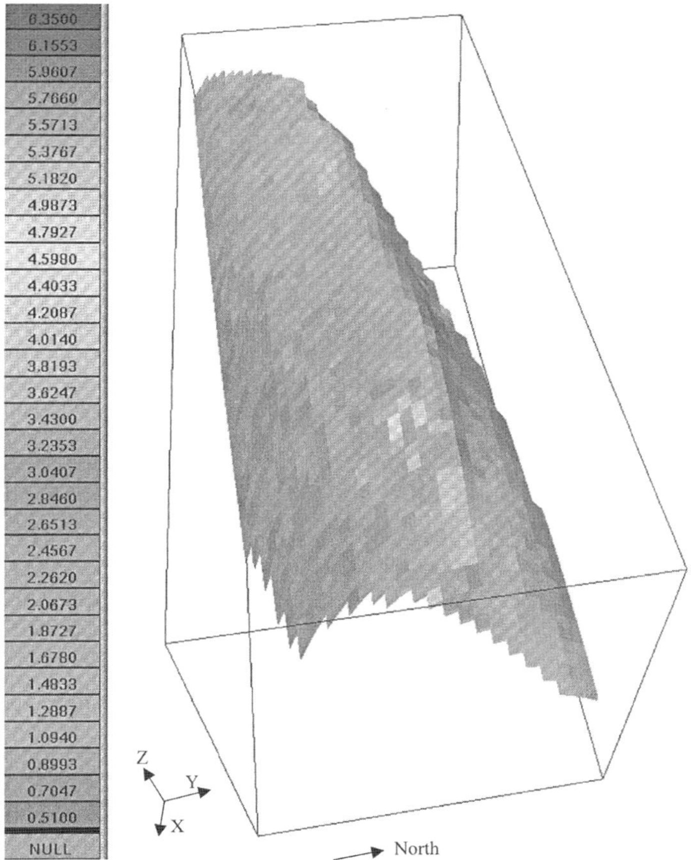

Fig. 14. A three-dimensional view of standard deviation in simulated porosity values.

porosity and shale fraction, respectively. It should be noted that the standard deviations are low in the area near the crest of the anticline. As the distance from the crest increases, standard deviations increase. As mentioned earlier, the wells in the field are preferentially located near the crest. This results in greater availability of conditioning data near the crest and reduces the variation in the simulated values. The porosity values show less variation than values of shale fraction. Support for shale fraction data is not as extensively available as that for porosity data. Because simulations of permeability were conditioned to simulated porosity and shale fraction data, we have not calculated the standard deviation in permeability data.

7.3. Variation in productive volume around wells

The productive volume is a non-physical variable, calculated to show the general quality of reservoir near a given well. This variable was calculated for every well by

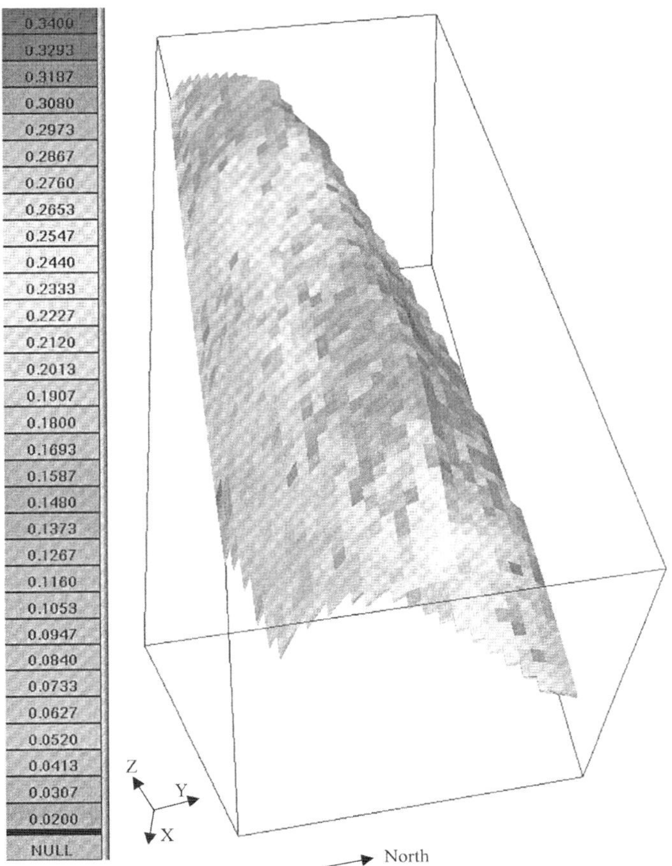

Fig. 15. A three-dimensional view of standard deviation in simulated shale fraction values.

summing the product of pore volume and permeability for each grid block within a given distance of the well. The 25 realizations for porosity, shale fraction and permeability were used as primary data, and were combined in 25 groups for the analysis. These 25 groups of data were slightly correlated, because corresponding realizations of porosity and shale fraction were generated with the same random number seed (thereby dictating the same sequential path in the simulation), and because permeabilities were derived from these same porosity and shale fraction data. Results for these calculations are displayed in Fig. 16 for four different wells in the field. The figure shows the average from 25 data sets and the variation in the productive volume. There is considerable variation for the four wells. Well HU16B38 has significant productive volume and favorable permeability compared to well HI1517. Similar products were calculated for other wells in the field. These comparisons will be useful to target the wells for further examination.

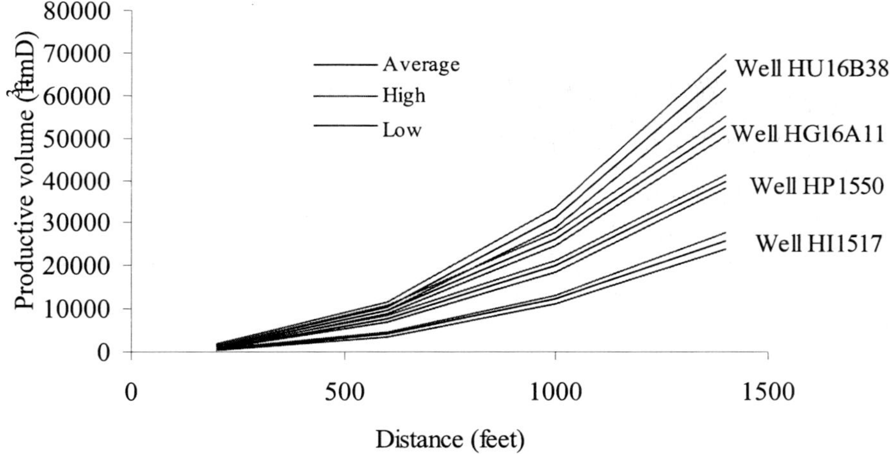

Fig. 16. Comparison of cumulative productive volumes for four wells in the field.

8. DISCUSSION OF RESULTS

The geological depositional environment of the Carpinteria Field appears to be spatially continuous. Analysis of the spatial continuity parameters for porosity and shale fraction reveals that properties can be correlated for length scales of the order of 1000s of feet. The calculated major axis of continuity for the reservoir sand seems to support the general belief that the field was deposited from a source to the north. However, this behavior is exhibited by limited data, and not observed over the entire field. The spatial availability of the data appears to be limited to verify the general depositional pattern.

The F1 zone appears to be well characterized in terms of porosity. The value for third quartile for standard deviation in the simulated porosity values was 2.46, while that for the 90th percentile was 2.74. The maximum observed standard deviation in the porosity values was 6.47. The 90th percentile value is about 6% of the average porosity in the field. The low values of standard deviation show that porosity data is well characterized. The zone appears to have significant porous rock volume. The value for the lower quartile of the porosity is about 23% while that for the upper quartile is about 25%. Overall, the observed porosity values show why historically this zone has been one of the significant producers.

The longer correlation length for shale fraction indicates presence of significantly continuous shale bodies in F1 zone. The zone appears to be not as well characterized for shale fraction as porosity. The maximum value for standard deviation in the simulated shale fraction values is 0.34. The value for the 90th percentile of standard deviation in simulated shale fraction values is 0.26. Percentage wise these values are significantly high. In general, the uncertainty in the shale fraction data is high away from the crest of the anticline. Even with higher uncertainty in the shale fraction data, the rock pore volume exhibits lower uncertainty. This indicates that the effect of higher uncertainty associated with shale fraction is masked by the lower uncertainty in porosity.

Permeability in the F1 zone is strongly influenced by porosity as well as shale fraction. Even in the absence of facies data, available porosity and shale fraction data were combined to condition permeability distributions. Overall, F1 zone has significant permeability, with a median value of 98 mD.

The different realizations of important reservoir rock properties, namely, porosity, permeability and shale fraction can be used together to study the reservoir quality around each well. A non-physical parameter such as the productive volume defined in this paper, can be used to get a quantitative idea of the volumes surrounding different wells. This parameter gives an idea about storage capacity as well as fluid flow capability by combining the three properties. The F1 zone, near wells HU16B38 and HG16A11 appears to have significant reservoir quality even up to distances of 1500 feet from the wells. On the other hand, well HI1517 does not appear to have as significant reservoir quality. The uncertainty associated with the productive volume increases with the distance from the well. This is because the data away from the well is stochastically generated. The uncertainty in the data increases with the distance from the well bores and manifests in the productive volume calculations.

9. CONCLUSIONS

Extensive data, available from over 200 wells and coreholes, in the Carpinteria Field were integrated to generate stochastic realizations of important reservoir rock properties. Multiple geostatistical techniques were applied to generate property distributions. For porosity data, secondary sources such as density and neutron porosity logs were utilized. Both the porosity and shale fraction values have significant effect on the permeability values. Permeability values were simulated, and conditioned to both the porosity and shale fraction data. Two different approaches were used to condition permeability data. These approaches utilized some artificially defined facies distributions based on the porosity and shale fraction data, in the absence of any real facies data. The simulated realizations for porosity and shale fraction reproduced the statistics of the original input data. The statistics of simulated permeability data, however, did not exhibit the variability of the original data as well. The simulated permeability data exhibited same relationship with porosity and shale fraction, as the observed data.

Geostatistical analysis suggests that the F1 zone is well characterized in terms of porosity. However, the zone appears to have higher uncertainty associated with the shale fraction data. The shale bodies in the zone appear to be continuos over large length scales. In general, the uncertainty in the simulated property values increased as the distance from the crest increased. The zone also appears to have significant permeability values.

Cumulative productive volumes were defined around each well location, in order to understand the combined effect of variations in the porosity, shale fraction and permeability in the area. The quality of reservoir surrounding all wells in the field was compared on the basis of these volumes. Of the four wells compared, the reservoir quality is significantly good around well HU16B38 compared to one around well HI1517.

The uncertainty in the productive volumes increased with the distance because of the uncertainty associated with the stochastically simulated values used in the calculations.

10. LIST OF SYMBOLS

$\gamma(h)$ Semivariance at lag distance 'h'
h_x Lag distance in the x direction
h_y Lag distance in the y direction
h_z Lag distance in the z direction
Sph Spherical model for semivariogram
Exp Exponential model for semivariogram

ACKNOWLEDGEMENTS

We thank the Pacific Operators Offshore Inc. for providing the data for the study. The geologic insights of Stan Coombs of Pacific Operators were valuable in defining the approach. This study was supported by the U.S. Department of Energy.

REFERENCES

Deutsch, C.V. and Journel, A.G., 1982 *GSLIB: Geostatistical Software Library and User's Guide*. Oxford University Press, New York, NY.

Developments in Petroleum Science, 51
Editors: M. Nikravesh, F. Aminzadeh and L.A. Zadeh

Chapter 26

INTEGRATED FRACTURED RESERVOIR CHARACTERIZATION USING NEURAL NETWORKS AND FUZZY LOGIC: THREE CASE STUDIES

A.M. ZELLOU [1] and A. OUENES [2]

Reservoir Characterization, Research & Consulting (RC)2, a subsidiary of Veritas DGC, 13 rue Pierre Loti, 92340 Bourg-La-Reine, France

ABSTRACT

The characterization of naturally fractured reservoirs is a challenging task for geologists and petroleum engineers. A good understanding of the fracture network in the subsurface requires a knowledge of fracture genesis. Natural fracture systems consist of two major categories: regional orthogonal fractures, and structure-related fractures (tectonic fractures). During the past eight years or so, a methodology using neural networks and fuzzy logic has been developed to assist geologists in the characterization of these complex reservoirs. Conventional methods use only one or two geological parameters (i.e. structure or curvature, which is the second derivative of the structure) to characterize a naturally fractured reservoir. However, an integrated approach that utilizes all the information available (including lithology, thickness, state of stress and fault patterns) is necessary. Our approach uses fuzzy logic to quantify and rank the importance of each geological parameter on fracturing, and neural networks to find the complex, non-linear relationship between these geological parameters and the fracture index.

Three case studies are described in this paper to illustrate this methodology. They report on a faulted limestone (oil reservoir), showing structure-related fractures associated with faults; a slope carbonate (oil reservoir), showing structure-related fractures associated with folds; and a sandstone (gas reservoir) in a large basin showing regional orthogonal fractures.

These three case studies illustrate fracture identification and prediction using static information (i.e. image logs) and/or dynamic information (i.e. well performance) as a fracture index. The results of these studies in terms of infill drilling, understanding the fractured reservoir and improved reservoir simulation will be described.

1. INTRODUCTION

Naturally fractured reservoirs (NFR) are by definition complex, and the challenges faced by geologists, geophysicists and reservoir engineers when studying them are two-fold: first, to understand the origin of the fracturing at a specific location; and second,

[1] E-mail: azellou@spemail.org
[2] E-mail: ouenes@att.net

to model fracture intensity over an entire field. The petroleum industry has focused for many years on team integration, with geologists, geophysicists and reservoir engineers working closely together. This integration of people is now being followed slowly by data integration, i.e. the combined use of seismic, geological and dynamic data. In this paper, we show how data integration and the use of artificial intelligence (fuzzy logic and neural networks) can help geoscientists in the study of naturally fractured reservoirs. We attempt to demonstrate through three very different case-history examples the practical application of this methodology.

The first case history focuses on a faulted oil reservoir in North Africa, where image logs from four horizontal wells have been used to assess the degree of fracturing. The integration of seismic, log and core data, together with the image logs allows a prediction of fracturing over the entire field. The second case study reports on a folded carbonate reservoir in SE New Mexico, and illustrates how modeling can assist with understanding a naturally fractured reservoir in terms of shear and conjugate fractures. The last case focuses on a tight sandstone reservoir located in NW New Mexico. The main concern of the operator of the field was to locate 'sweetspots', directly related to fractures.

2. FRACTURED RESERVOIR MODELING USING AI TOOLS

When using a fractured reservoir model to help understand an interwell region (by utilizing the known 3D distribution of the geologic drivers), it is imperative to find a fracture intensity predictor which can 'see' further than a few inches around the wellbore. Production-based indicators such as the productivity index (PI) and transmissibility (kh) estimated from well tests, and the estimated ultimate recovery (EUR), can be used to provide an average fracture intensity over an area as big as the drainage radius and integrates the fracture and matrix contribution. Production-based fracture intensities, which only seem appropriate for 2D models, have successfully been used in 3D modeling by simply allocating single measurements along the wellbore. Different methods can be used to transform a single kh or PI value into a vertical log: the most commonly used being the multiplication of the production log with the kh or PI. Our methodology preferably uses a production-based fracture intensity, although fracture counts from image logs are used in the case-history pertaining to the faulted limestone reservoir. Once the fracture intensity is chosen, it represents the output of the neural network. Inputs which can be related to this fracture intensity are geological drivers (porosity, permeability, lithology, facies proportion, bed thickness, proximity to faults, etc.) and present-day indicators (seismic, stress, strain, etc), all of which could be made available to the entire reservoir volume by using appropriate geostatistical methods. The search for possible relationships between the drivers and the chosen fracture intensity is a three-stage process, as described below.

2.1. Ranking the drivers

Prior to any modeling, appropriate ranking methods must be used to analyze the effects of each driver on the chosen fracture intensity. The engineer or geologist must check at this stage for the validity of the ranking over the entire area of the study and on specific zones. For example, if the reservoir was under extensional strain, one could expect that curvature would have a high rank.

There are many benefits which can be derived from this ranking exercise; the most notable is to achieve a better understanding of what the primary drivers are. There are also computational benefits; for example, low-ranked drivers may indicate that they have no effect on the fracture intensity, and therefore they can be dropped from the inputs. These examples show that AI tools can be used for geologic interpretation; we emphasize that a neural network exercise does not simply consist of feeding an input into a black box and then deriving some number through a mysterious data-driven relationship.

2.2. Training and testing the models

Once the user has decided on which drivers to use in the modeling process, the set of available data is then divided into two subsets: a training set and a testing set. Since this approach is done in a stochastic framework, many realizations are needed. Each realization can be derived by selecting the training set either randomly or according to some rule. The neural network modeling process consists of adjusting the weights until the actual fracture intensity matches the estimated ones. Once the matching process is completed, we can assume that the model can be used for testing and cross-validation. Depending on the ability of the model to predict the fracture intensity at test locations, the model can be kept for further use or discarded.

2.3. Simulation process

The stochastic aspect of this approach is related to uncertainty in the above regression models. Therefore, several realizations or simulations are generated cell-by-cell over the entire field in which the fracturing drivers are known to predict the fracture intensity FI. With a large number of realizations, a probability and average model is estimated over the entire field.

2.4. Transforming fractured models into 3D effective permeabilities

Since all the drivers are available over the entire 3D reservoir volume and a relationship has been established between the drivers and the fracture intensity, the application of the neural network to all the gridblocks in the reservoir will lead to a 3D distribution of fracture intensity. As the fracture intensity is derived from a production-based indicator such as PI or kh, the 3D fracture intensity model can easily be converted to an effective 3D permeability model using simple reservoir engineering relationships.

3. CASE STUDY 1: FAULTED LIMESTONE RESERVOIR, NORTH AFRICA

3.1. *Field geology*

In this oil-bearing reservoir, the porosity ranges from 10 to 35% (average 20%). The matrix permeability measured from core plugs ranges from 0 to 1000 mD (several tens of millidarcies on average), whereas the well test permeability ranges from 10 to 110 mD. The ratio between test and plug permeability, which is generally interpreted as an indicator of fracturing effects, varies from 1 to 10. Therefore, the reservoir cannot be considered strictly as a fractured reservoir since a minimum ratio of 10 between these two types of permeability measurements is considered as an indicator of dual permeability behaviour. Single-porosity single-permeability reservoir simulations generally give good well history matches. However, bad matches observed at some locations could be due to high permeability streaks related to small-scale faults and/or fracture swarms. In other locations, fracturing is believed to improve the matrix performance homogeneously.

In this field, a carbonate sequence of Late Paleocene age separates an oil-producing upper reservoir unit from a mostly water-bearing lower unit. Non-reservoir shales and carbonates separate these two units. The upper reservoir unit is subdivided into four layers, each 5 to 40 ft thick. These consist of tight limestone (variably argillaceous), to high porosity limestone (locally interbedded with dolomitic streaks) and calcareous shale. Our study focuses on the high porosity limestone layers.

The structural history of the field is quite complex but is generally interpreted as reflecting its location in a transtensional basin, which originated in the Cretaceous and which was reactivated during the Tertiary. However, the tectonic phase(s) post-dating reservoir rock deposition is (are) poorly documented. Fig. 1 shows a structural map of the whole field at top-reservoir level overlain with the fault pattern at a deeper level. Note the change in the general strike of the seismically defined faults with depth. Thus at top reservoir level, the faults are mainly oriented N120°–130°. This fault network is developed above a deeper trend which strikes preferentially N170°. The field structure can be explained by the reactivation of deep basement faults within an oblique extensional regime. This structural style results at top reservoir in the dominant oblique NW–SE normal faults and in secondary N170° faults and flexures. From the fault pattern in the top reservoir, the deformation seems to be concentrated on the flanks of the structure just above the deep faults. Indeed, the top fault density and continuity increase in the southern part of the field where the two principal deep faults merge.

3.2. *Factors affecting fracturing at this field*

Nelson (1985) documented the geological parameters controlling the fracture frequency in a layered sedimentary rock: composition, grain size, porosity, bed thickness and structural position. To incorporate all these parameters (except grain size), and in

---

Fig. 1. Structural map at top reservoir with deep faults.

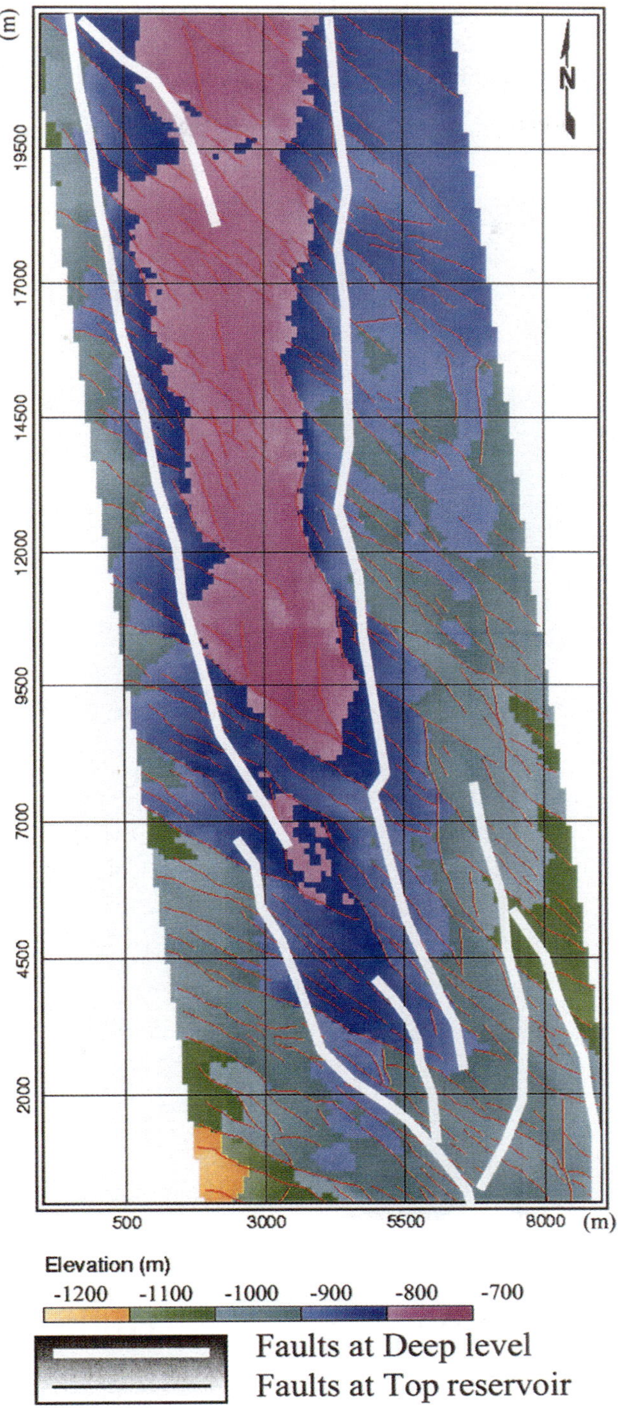
(m)
19500
17000
14500
12000
9500
7000
4500
2000
N
500
3000
5500
8000
(m)
Elevation (m)
-1200
-1100
-1000
-900
-800
-700
Faults at Deep level
Faults at Top reservoir

order to achieve the most comprehensive (i.e. most geologically validated) result, four main sources of data were used here:

(1) Seismically derived attributes, i.e. seismic amplitude and coherence, top and deep fault maps;

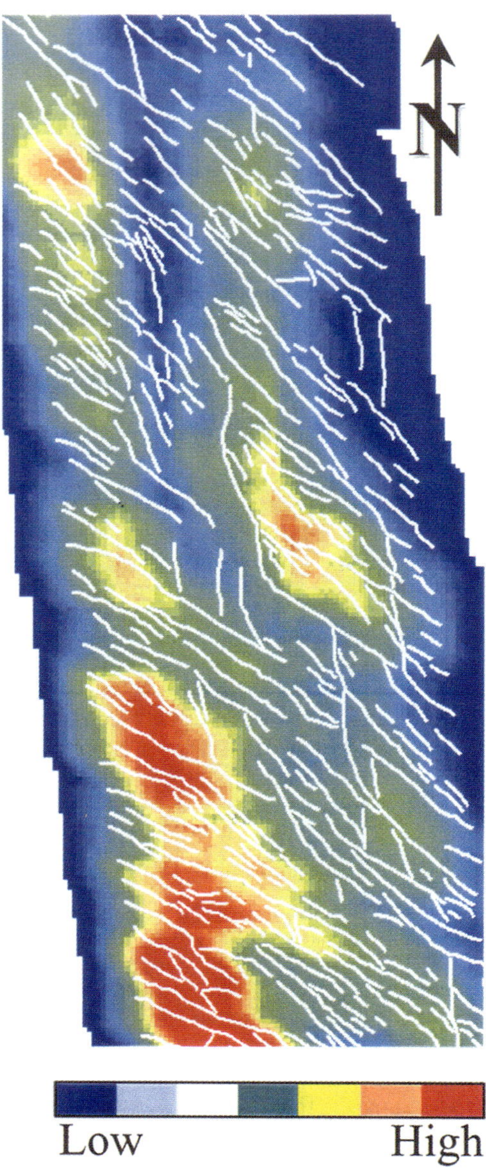

Fig. 2. Strain map. This map is one of the fracturing drivers used in the modeling.

(2) Log-derived data, i.e. porosity and gross thicknesses;
(3) Core-related lithological data – dolomite, calcite and illite content;
(4) Computed data: strain, maximum structural curvature, and distance to top faults.

All these fracturing drivers were mapped over the entire field, driver values being assigned or calculated at the center of a 100×100 m cell (with a total of 16,534 cells). Fig. 2 shows the strain map: it is one of the fracturing driver maps used in the modeling.

This paper only focuses on the modeling of the dominant NW–SE fracture set interpreted from four horizontal wells.

3.3. Application of the fractured reservoir modeling using AI tools

On each of the 24 cells crossed by the four horizontal wells, the fracture intensity FI is calculated as the average fracture frequency over the section of the well drain crossing that cell.

3.3.1. Ranking the fracturing drivers

Using the initial 24 control cells where the drivers (the input) and the FI (the output) are defined, a fuzzy logic tool was used to rank the fracturing drivers. A prior elimination process identifying irrelevant drivers, based on a knowledge of the field, was performed using this tool. Fig. 3 shows the final result of this ranking, which confirms that the two main controls on fracturing in this field are the lithology (i.e. the dolomite and calcite content) and the tectonic intensity (i.e. influence of strain and the deep faults). This ranking process allowed a better understanding of the fracturing and has the benefit of improving the full field fracture modeling by keeping only the most relevant drivers. In this case, the first eight were selected based on their weighting factor, to accelerate the modeling.

3.3.2. Training and testing

Different models were derived starting with a high number of training cells (20 out of 24) to test the modeling limit of the technique. The final minimum number of cells used for training and needed to build a consistent model was 16. Eight tested cells were hidden to validate the model. Once the predicted and the actual FI values for the training set and the testing set are higher than a user-defined correlation coefficient (0.8 and 0.7, respectively) the realization was kept.

3.3.3. Simulation results

75 realizations of FI were performed and a first qualitative screening was done to select the best realizations based on the correlation coefficient and the realization map. After this first screening, 40 realizations were kept to compute average and probability maps over the entire field. Fig. 4 shows the probability map at a given threshold, which is 0.5 fractures/m.

3.4. Conclusions and recommendations

The following conclusions and recommendations were derived from the results of this case study:

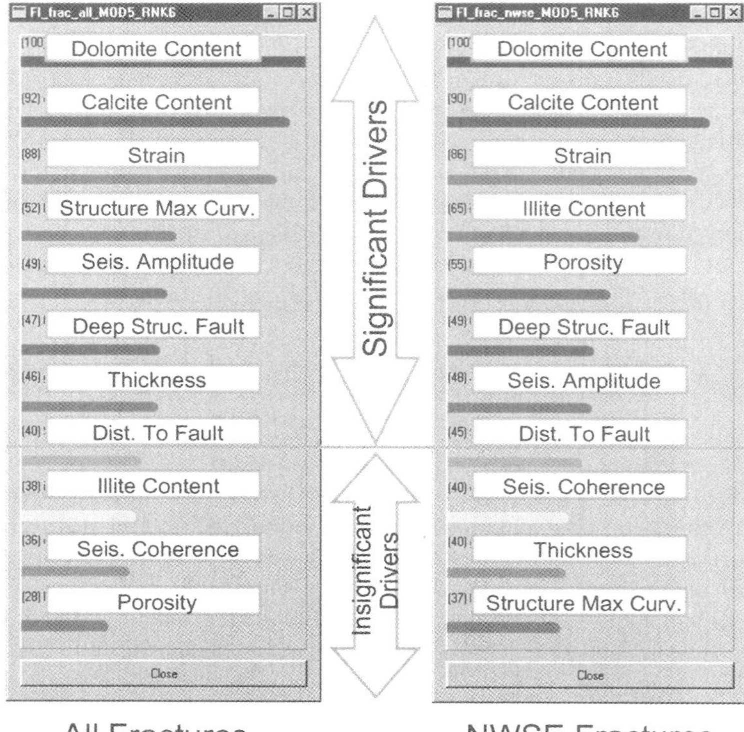

Fig. 3. Ranking of the drivers. Here the case of all the fractures combined is also considered. It can be noted that both rankings are consistent.

(1) The use of fuzzy logic allowed us to define the drivers controlling fracturing, and to gain a better understanding of the origin of fracturing, i.e. tectonics and sedimentology;
(2) Using these drivers, the fracture frequency was modeled over the entire field.
(3) The maps derived from this analysis can be used in infill drilling and to improve the success of horizontal wells;
(4) These maps can be used to control the local discrete fracture networks (DFN) involved in well test simulations at any location.

4. CASE STUDY 2: SLOPE CARBONATE OIL RESERVOIR, SE NEW MEXICO

4.1. Background

The intensity and direction of fractures in a reservoir are a function of the state of stress, which is strongly related to reservoir depth, thickness, tectonic history, and structure. Murray (1968) and Lisle (1994) took into account a fourth parameter by

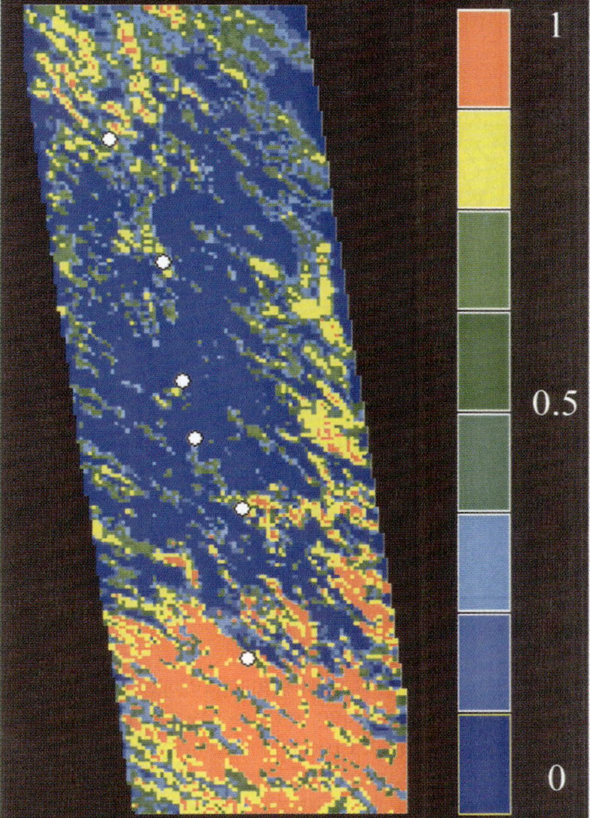

Fig. 4. Probability map of having fracture frequency larger than 0.5 fractures/m.

introducing the use of curvatures. Murray's model included the unrealistic assumption that fractures only occur in a direction perpendicular to the radius of folding, meaning that the shear properties of the state of stress are not considered. The idea of using the curvature of structures to estimate fracture intensity is a sound approach, as the stress in the rock depends on the curvature. However, the stress state is described by a tensor and has therefore directional properties. Ouenes et al. (1994) and Richardson (1995) generalized the curvature method to include curvature in many directions. Curvature values in four directions can be obtained from an interpolated structure map, created by using well log data and a mapping method (Ouenes et al., 1995). These curvature values represent one set of inputs for a neural network. The other inputs for this application are the depth, the thickness, and the cumulative production of the reservoir. A fracture intensity map was obtained after training the neural network by using all the known geomechanical parameters, the structure, and the bed thickness.

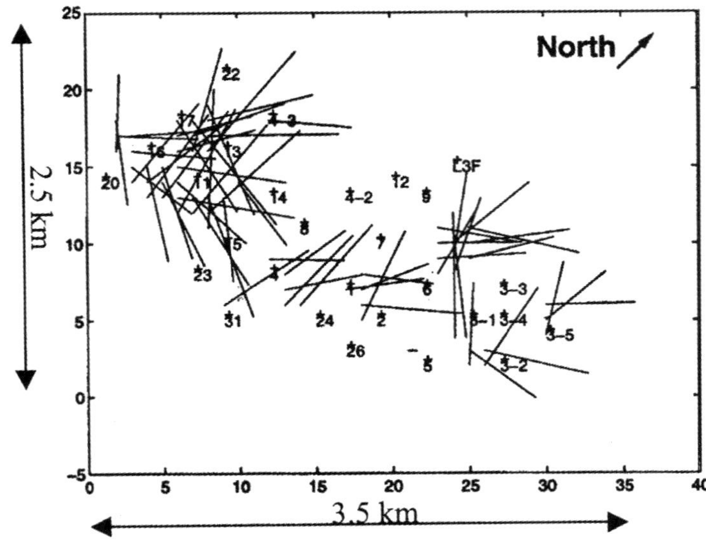

Fig. 5. YDU fracture network map.

4.2. Results and discussions

The reservoir for this case study is a carbonate referred to as the Young Deep Unit (YDU). The YDU is one of the eight Bone Springs reservoirs located at the margin of the NW Shelf in SE New Mexico (USA). Production history and log data for 30 wells were available at the YDU which spans an area of about 2 sections (33×23 gridblocks of 350 ft). The well density per section is 15 wells/sq. mile. Fig. 5 shows a fracture intensity map of the YDU and depth, curvatures, and cumulative oil production were inputs for this model. A neural network provides a value for the fracture intensity for each grid block. This intensity is a combination of the length, the aperture, and the population of the fractures in the gridblock considered. Using a 'weighting method' described by Zellou et al. (1995), a fracture network map (Fig. 5) was created from the fracture intensity map (Fig. 6). Analysis of this map clearly shows that the fracture system is structure-related (Fig. 7). Note that the YDU is in the nose of a fold. One can see the fracture strip, from the west through the center back to the northeast, following the shape of the fold nose, and the conjugate shear fractures making an angle of approximately 60°. An illustration of this type of fold-related fracturing is given in Fig. 8.

4.3. Conclusions

Based on the results of this case study, we arrive at the following conclusions:

(1) The '*weighting method*' has been developed to describe the subsurface fracture network.

(2) This fracture network provides geologic information on the type of fracture (structure related fractures associated to folds).

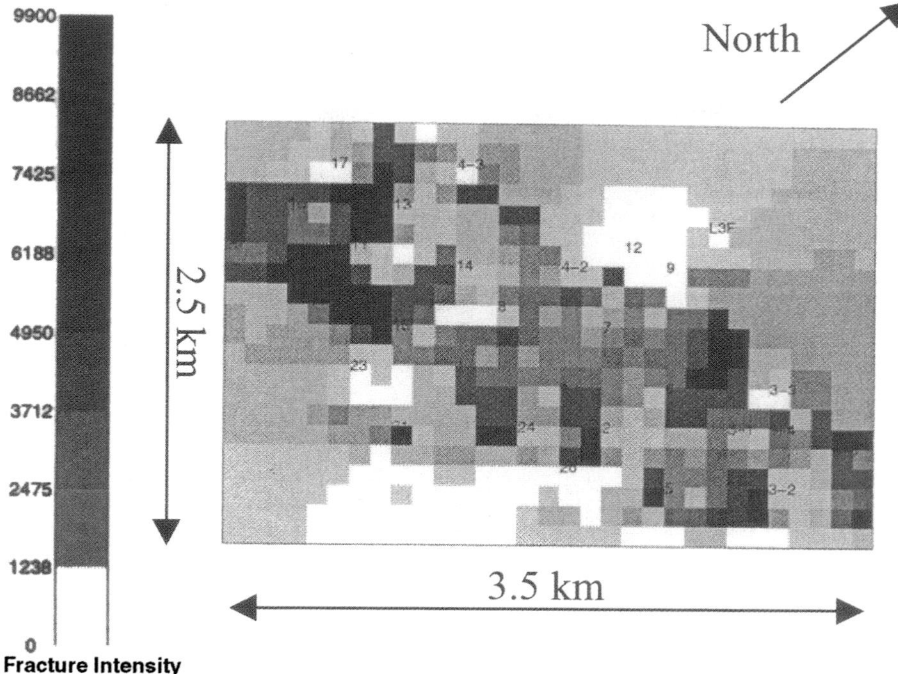

Fig. 6. YDU fracture intensity map.

(3) The location and the direction of natural fractures is obtained from both the fracture intensity map and the fracture network map.

5. CASE STUDY 3: A SANDSTONE GAS RESERVOIR, NW NEW MEXICO

5.1. Dakota production and geology

The San Juan Basin of NW New Mexico contains some of the largest gas fields in the United States. Gas is produced from three Upper Cretaceous sandstone formations: the Dakota, Mesaverde, and Pictured Cliffs as well as from the Fruitland coal. Fassett (1991) suggested that the gas produced from the Dakota, Mesaverde, and Pictured Cliffs reservoirs are structurally enhanced by natural fracturing. Unfortunately, little has been published on the subsurface pattern of these fractures.

In excess of 4.8 trillion cubic feet (tcf) of gas (135.9 billion m^3) as well as substantial volumes of condensate have been produced from three primary sandstone complexes in the upper Dakota Formation. This production comes mainly from the Basin Dakota field located in the south-central part of the San Juan Basin. The productive upper Dakota sandstones comprise a complex of marine and fluvial–marine sandstone reservoirs. The upper Dakota is a tight gas reservoir with porosity generally between 5 to 10% and per-

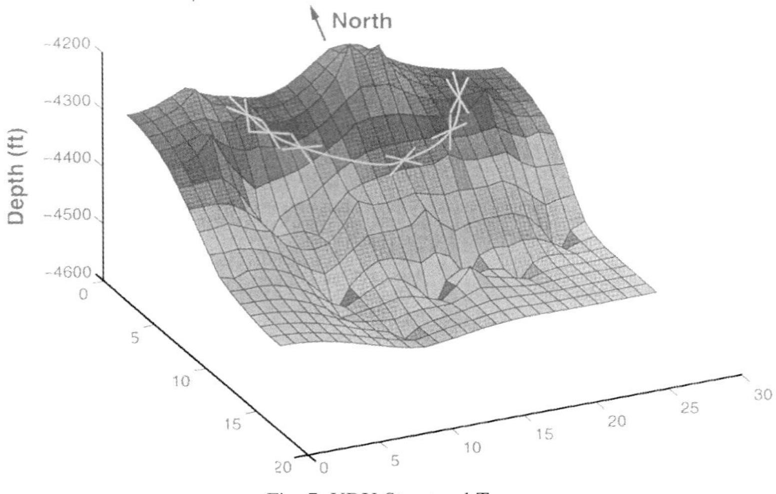

Fig. 7. YDU Structural Top.

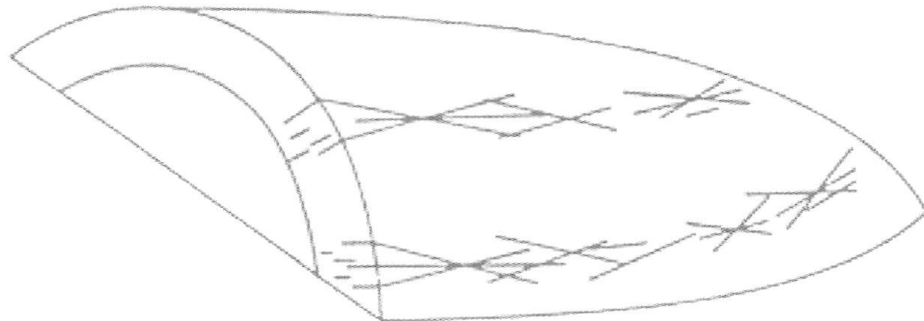
Fig. 8. Schematic illustration of fractures associated with the nose of a fold (after Stearn and Friedman, 1972).

meability ranging from less than 0.1 to 0.25 mD. Consequently, these Dakota reservoirs generally cannot produce commercial flow rates without natural or induced fracturing.

The key for optimal exploration and development of upper Dakota reservoirs resides in the ability to predict the spatial distribution of the fracture intensity. However, when considering a productive interval in excess of 200 ft (60 m) thick covering over 4300 sq. miles – (11,130 km^2), the use of simple analytical tools focused on one or two fracture parameters, such as the structural curvature, will not result in a comprehensive understanding of the spatial distribution of the fracture intensity. Hence, numerous fracture drivers must be characterized and integrated into the analysis.

5.2. Factors affecting fracturing in the Dakota

The study area covering some 870 sq. miles (2250 km^2), has been divided into grid blocks which measure 0.5 miles × 0.5 miles (0.8 km × 0.8 km). A total of 3478 grid

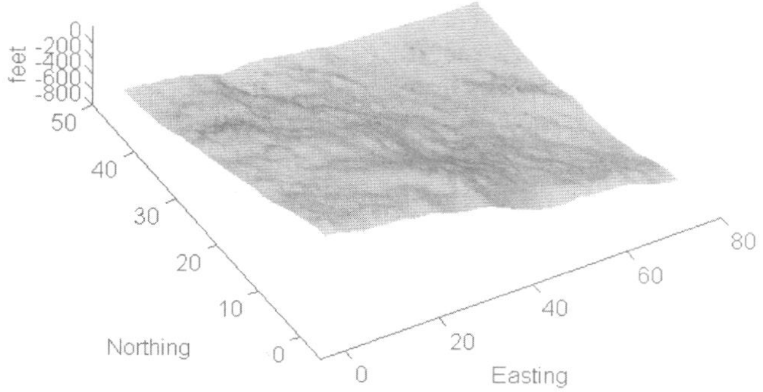

Fig. 9. Dakota structure in the 24 township area.

blocks was used. For each fracturing factor, a two-dimensional map was drawn over the entire grid.

The structure of the Dakota (Fig. 9) was computed using Graneros reservoir log top in 2108 wells and the Laplacian interpolation method developed by Ouenes et al. (1994). The same interpolation method was used to draw a map of estimated ultimate recovery (EUR) (Fig. 10) using 1304 'parent' wells drilled on 320 acres spacing. The EURs vary from 0.1 to 14.7BCF (2.83 to 416.1 million m^3) and four major, highly productive 'sweetspots' are present in the eastern and western parts of the study area.

Extensive geological and engineering analysis of the Dakota Formation (by Burlington Resources) established that the primary driver to reservoir performance is the degree of natural fracturing. The EUR is considered to be a measure of fracture density.

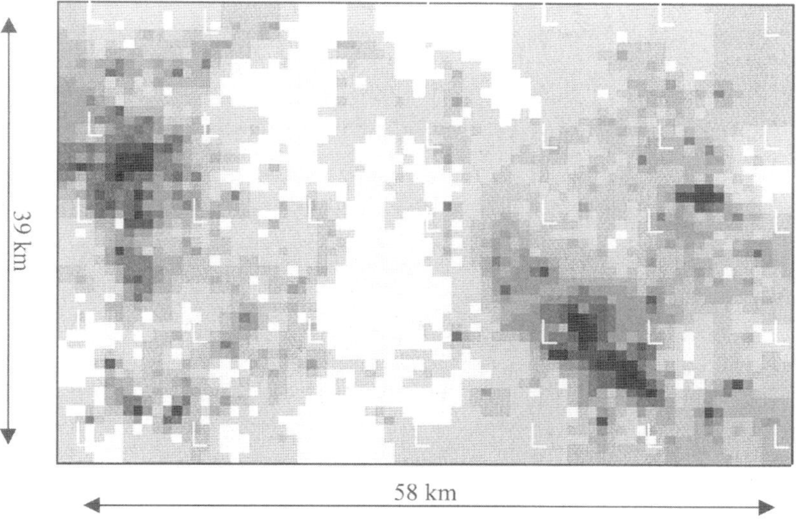

Fig. 10. EUR map for the 24 township area. The dark areas represent 'sweetspots'.

The superposition of the EUR map on the structure shows clearly that some of the 'sweetspots' are associated with structural features. However, other 'sweetspots' are present in areas with little structural variation. Some poorly productive areas have major structural variations but poor EUR. This suggests that other parameters are required to characterize natural fracture development.

In previous studies, structural effects have been accounted for by using curvatures taken in one or two directions. Ouenes et al. (1995) introduced the use of curvatures (second derivatives) and first derivatives of the structure in multiple directions, which better reflects stress states which can be described by a tensor that has directional properties. Another reason for the use of multiple directions is that structural effects could have different directions over the study area. Therefore, in each gridblock, structural effects are represented by a set of four curvatures and slopes computed in four directions: north–south, east–west, NW–SE and NE–SW. In addition to these structural effects, this study considered thickness and degree of 'shaliness' as parameters driving Dakota natural fracturing.

The upper Dakota is composed of three major productive sandstone units which in ascending order are the Oak Canyon/Cubero, Paguate and TwoWells Formations, these are referred to in this paper as Sands A, B, and C, respectively. Gross isopach maps of these sands (Figs. 11–13) illustrate the significant pay thickness variation across the study area. This thickness variation dramatically influences the intensity of natural fracture development and is considered in the analysis.

As mentioned above, EUR from these three sand complexes is used as a fracture intensity indicator. This analysis was prompted by the results of a previous fracture prediction study in the San Juan Basin Mesaverde Formation by Basinski et al. (1997). In addition to pay thickness, facies maps are needed to delineate the third fracturing factor, lithology.

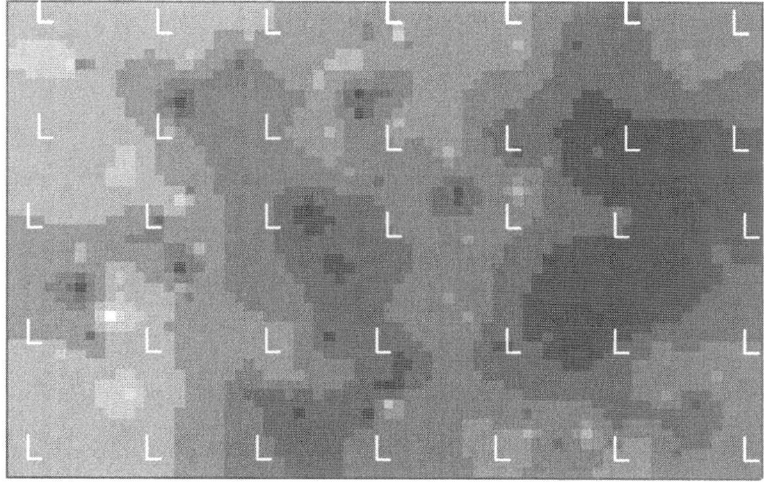

Fig. 11. Thickness of Sand A over the 24 townships. The dark parts represent the thicker sand.

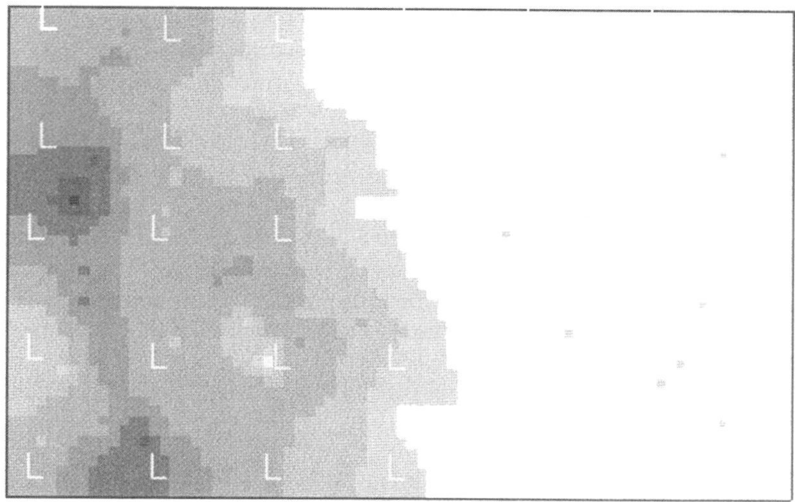

Fig. 12. Thickness of Sand B over the 24 townships. The dark parts represent the thicker sand.

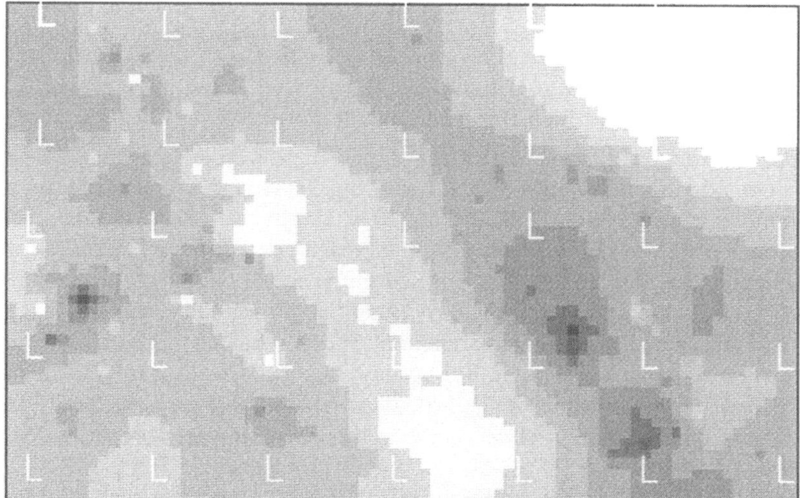

Fig. 13. Thickness of Sand C over the 24 townships. The dark parts represent the thicker sand.

Whole core analysis indicates that there was very strong correlation between resistivity and fractures. In the study area, resistivity is as a rule a function of facies and not hydrocarbon saturation. Highly resistive intervals represent facies with minimal clay content which were subsequently diagenetically silicified and rendered brittle. Consequently, where resistivity is high, abundant natural fractures are developed in the core. Although resistivity may not be the ideal lithological indicator, in the study area it contains valuable information about the facies changes and degree of fracturing in Sands A and B. In this study, the average pay thickness and resistivity for Sands A, B, and C over the entire net pay was used. This data was mapped using information from less than 200 wells in the study area.

At each gridblock, thirteen possible parameters which may have played a role in rock fracturing are used as inputs for the neural network analysis: four first derivatives (slopes), four second derivatives (curvatures), three thicknesses, and two resistivities. These reservoir properties were easy to derive and were adequate to characterize the reservoir fracture intensity robustly. For any set of chosen reservoir properties, the objective is to design a geological model that can explain the observed production data. In exploration areas, the structural properties and thicknesses could be estimated from 3D seismic travel time data (Zellou et al., 1995).

5.3. Building a geologic model

Different techniques are available to build quantitative geologic models. For example, geostatistics is commonly used in conventional reservoirs to build geologic models that account for existing spatial correlations of a given reservoir property. Some geostatistical algorithms are even able to provide improved models by simultaneously using two reservoir properties such as a seismic attribute and porosity. These techniques are sophisticated mapping methods but they are of little help when it comes to understanding complex fractured reservoirs. In such problems, the first issue is to identify the geologic parameters that control rock fracturing and to establish their relative importance.

5.3.1. Ranking fracturing factors

When considering thirteen possible fracturing factors, the relative importance of each parameter and the impact on Dakota EUR is critical. Given the data set, many different techniques could be used to rank the thirteen parameters. The most commonly used techniques are statistical methods based on principal components analysis. In our methodology, we employ a fuzzy logic tool that has proved very efficient and quick in finding the ranking of the geologic parameters. When considering the entire study area, the fuzzy logic ranking reveals that the six most important factors are: (i) the thickness of Sand B; (ii) the inverse thickness of Sand A; (iii) the thickness of Sand C; (iv) the resistivity of Sand B; (v) the first derivative in the NW–SE direction; and (vi) the Sand A resistivity.

The most important factor, the thickness of Sand B, suggests that a large portion of gas produced is from the Sand B and that thicker reservoir equates to higher EUR. Consequently, Sand B behaves more like a conventional gas reservoir in this ranking where the production is mostly controlled by original gas-in-place. The inverse thickness of Sand A implies typical fractured sandstone behaviour, where the thinner intervals contain the highest fracture intensity. Thickness of Sand C, the third-most important parameter, indicates an analogous relationship to Sand B. Very significantly, the first item of structural information, the NW–SE orientation of the first derivative (or slope) of the structure is ranked fifth and illustrates that structural changes are not the main fracturing control over the entire study area. However, this orientation is consistent with the major axis of the basin and is parallel with many Dakota producing trends, indicating that structural changes cannot be neglected in fracture analysis.

The first second-derivative parameter, or curvature, in the rankings is seventh in overall importance and illustrates that structural changes can sometimes be better represented by slopes rather than curvatures. Armed with this knowledge, a geologic model can be built to relate all the thirteen factors to EUR. Indeed, this ranking was confirmed for the various productive sands after the parameters were empirically tied back to core analysis, petrophysics, and production.

5.3.2. Neural network analysis

Neural network inputs for this study were the first and second structural derivatives in four directions, the thicknesses of Sands B and C, the inverse thickness of Sand, and the resistivity of Sands A and B; the network output was the EUR of the 'parent' wells. The study area encompasses 1304 'parent' wells, and the intensity of the fracturing is represented by the EUR of these wells. The level of confidence in these EURs is high: most of the 'parent' wells have over 40 years of production history. Only the 'parent' well EUR was used to represent reservoir fracture intensity because 'infill' wells are somewhat depleted in highly fractured areas.

Many geologic models were constructed using a very limited number of training wells ranging from 70 to 150. In each model, a large area that contains a known 'sweetspot' was considered without any training well. The objective of the geologic model was to test the ability of the neural network to predict the 'sweetspot' without providing any EUR indication in the training set. This exercise is illustrated with the following example:

Out of the 1304 'parent' wells in the 2250 km^2 study area (2250 km^2), only 100 wells were used to train the neural network. Using the Laplacian mapping technique and the 100 training wells, an EUR map was drawn (Fig. 14). Note that two major 'sweetspots' are absent in the highlighted windows. This model was then applied to the rest of the area to ascertain its predictive strength. Therefore, in the context of this study, the parts of the study area highlighted by the two windows can be considered exploratory. Using the neural network model, an estimated EUR map over the entire study area is shown in Fig. 15. The two 'sweetspots' absent in the training set appear in the east and west windows, showing the ability of the model to detect 'sweetspots' and model the complex interplay and competition between the thirteen factors affecting Dakota fractures.

5.4. Conclusions

The application of an innovative prospecting technique in the naturally fractured upper Dakota gas reservoir was demonstrated in a 2250 km^2 in the San Juan Basin, New Mexico. The following conclusions can be derived:

(1) Numerous parameters control production from the upper Dakota. In order of relative importance, they include pay thickness, resistivity (geological facies), and structure of the three productive sand complexes present in the study area.
(2) The relationships between the thirteen parameters used in this study to predict EUR

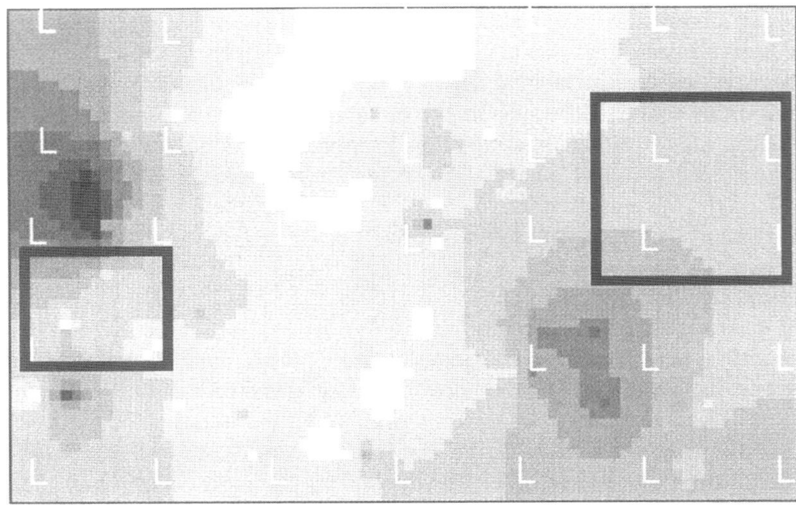

Fig. 14. EUR map drawn with 100 wells used for training.

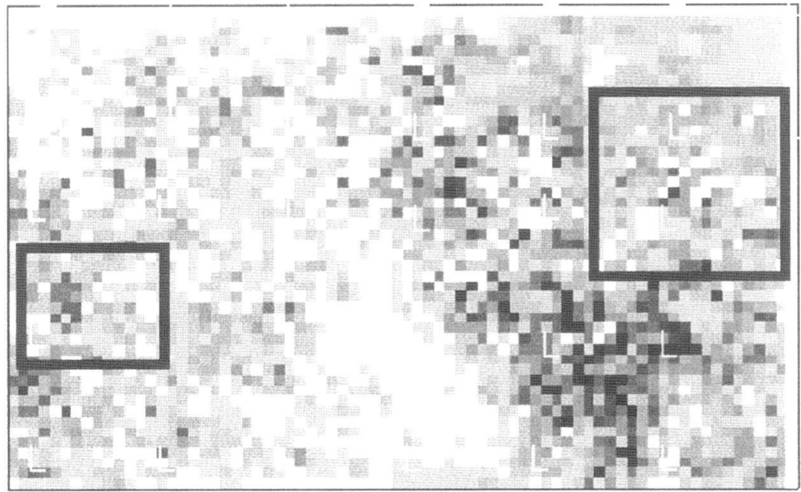

Fig. 15. EUR map estimated by the neural network.

and natural fracturing is not straightforward. The use of neural network analysis provides a link between the fracture drivers and production.

(3) A neural network was able to delineate highly productive 'sweetspot' areas with minimal input data, and had significant potential to high-grade new drilling opportunities after a model is derived.

(4) This methodology is currently used in the San Juan Basin and has allowed the average EUR of new drilled wells to more than double.

6. CONCLUSIONS

From these three case studies, we can draw the following conclusions:

- With the use of fuzzy logic and neural networks to characterize fractured reservoirs, it becomes possible to classify fractures and different fractured reservoirs, following the classifications of Stearn and Friedman (1972) and Nelson (1885).
- The use of fuzzy logic allows us to rank the importance of each driver towards fracturing. The case studies illustrate explicitly and quantitatively the well-known fact that each fractured reservoir is specific in terms of the origin of its fracturing.
- Conversely to numerous simulations methods (i.e. sequential simulation, geostatistics), the prediction of the fracture intensity networks on each gridblock using a neural network is computed independently of the neighbouring gridblock. Hence, the full field fracture intensity prediction, as shown in the case studies, becomes more realistic: a high fracturing zone neighbouring a low one.
- This static modeling of fractured reservoir has been and is currently used for the design of infill drilling and reservoir simulation, i.e. water, oil and pressure history matching.
- These three examples illustrate how AI technology fits into fractured reservoir characterization.

ACKNOWLEDGEMENTS

This paper is based on three different applications of neurocomputing in fractured reservoir characterization. The authors would like to thank Paul Basinski (El Paso, Houston) for being a precursor in the oil and gas industry in the use of this innovative technology. We would also like to thank Bertrand Gauthier (TotalFinaElf, France) for his support.

REFERENCES

Basinski, P., Zellou, A. and Ouenes, A., 1997. Prediction of Mesaverde estimated ultimate recovery using structural curvature and neural network analysis, San Juan Basin, NM, USA. *1997 AAPG Rocky Mountain Section*, Denver, Co., USA.

Fassett, J.E., 1991. Oil and gas resources of the San Juan Basin, New Mexico and Colorado. In: Gluskoter, H., Rice, D. and Taylor, R. (Eds.), *Geology of North America. Vol. P-2, Economic Geology*. 357 pp.

Lisle, J.L., 1994. Detection of zones of abnormal strains in structures using Gaussian curvature analysis. *Am. Assoc. Pet. Geol. Bull.*, 78(12): 1811–1819.

Murray Jr., G.H., 1968. Quantitative fracture study – Sanish Pool, McKenzie County, North Dakota. *Am. Assoc. Pet. Geol. Bull.*, 52(1): 57–65.

Nelson, R., 1985. *Geologic Analysis of Naturally Fractured Reservoirs*. Gulf Publishing Co., Houston, TX, 320 pp.

Ouenes, A., Weiss, W.W., Richardson, S., Sultan, A.J., Gum, T. and Brooks, L.C., 1994. A new method to characterize fracture reservoirs: application to infill drilling. *SPE/DOE Symposium on Improved Oil Recovery*, Tulsa, OK, Paper *SPE/DOE 27799*.

Ouenes, A., Richardson, S. and Weiss, W.W., 1995. Fractured reservoir characterization and performance forecasting using geomechanics and artificial intelligence. *1995 SPE Annual Technical Conference and Exhibition*, Dallas, TX, Paper *SPE 30572*.

Richardson, S., 1995. *Generalization of the Curvature Method and Application of Neural Networks to Fractured Reservoir Characterization*. Master Thesis, New Mexico Institute of Mining and Technology, Socorro, NM.

Stearn, D.W. and Friedman, M., 1972. *Reservoirs in Fractured Rock. Am. Assoc. Pet. Geol. Mem.*, 16: 82–106.

Zellou, A.M., Ouenes, A. and Banik, A., 1995. Improved Fractured Reservoir Characterization Using Neural Network, Geomechanics and 3-D Seismic. *1995 SPE Annual Technical Conference and Exhibition*, Dallas, TX, Paper SPE 30722.

PART 6.
GENERAL APPLICATIONS

Developments in Petroleum Science, 51
Editors: M. Nikravesh, F. Aminzadeh and L.A. Zadeh

Chapter 27

VIRTUAL MAGNETIC RESONANCE LOGS, A LOW COST RESERVOIR DESCRIPTION TOOL

SHAHAB D. MOHAGHEGH [1]

West Virginia University, 345E Mineral Resources Building, Morgantown, WV 26506, USA

ABSTRACT

Magnetic resonance imaging (MRI) logs are well logs that use nuclear magnetic resonance to accurately measure free fluid, irreducible water (MBVI), and effective porosity (MPHI). Permeability is then calculated using a mathematical function that incorporates these measured properties. This paper describes the methodology developed to generate synthetic magnetic resonance imaging logs using data obtained by conventional well logs such as spontaneous potential (SP), gamma-ray, caliper, and resistivity. The synthetically generated logs are named virtual magnetic resonance logs or 'VMRL' for short.

Magnetic resonance logs provide the capability of in-situ measurement of reservoir characteristics. The study also examines and provides alternatives for situations in which all required conventional logs are unavailable for a particular well. Synthetic magnetic resonance logs for wells with an incomplete suite of conventional logs are generated and compared with actual magnetic resonance logs for the same well.

In order to demonstrate the feasibility of the concept being introduced here, the methodology is applied in two different fashions. First, it is applied to four wells; each from a different part of the country. These wells are located in East Texas, Gulf of Mexico, Utah, and New Mexico. Since only one well from each region is available, the model is developed using a segment of the pay zone and consequently is applied to the rest of the pay zone. In a second approach, the technique is applied to a set of wells in a highly heterogeneous reservoir in East Texas. Here the model was developed using a set of wells and then was verified by applying it to a well away from the wells used during the development process. This technique is capable of providing a better reservoir description (effective porosity, fluid saturation, and permeability) and more realistic reserve estimation at a much lower cost.

1. INTRODUCTION

Austin and Faulkner (1993) published a paper in August 1993 in 'The American Oil and Gas Reporter' providing some valuable information about the Magnetic Resonance Logging technique and its benefits to low resistivity reservoirs.

[1] Tel.: (304) 293-7682 ext. 3405, fax: (304) 293-5708. E-mail: Shahab@wvu.edu

The MR log measures effective porosity – total porosity minus the clay bound porosity – as well as irreducible water saturation. The irreducible water saturation is the combinations of clay bound water and water held due to the surface tension of the matrix material. The difference between effective porosity (MPHI) and the irreducible water saturation (MBVI) is called the free fluid index. This is the producible fluid in the reservoir. This demonstrates how valuable MR log is to low resistivity reservoirs. In many low resistivity reservoirs, matrix irreducible water rather than producible water may cause a drop in resistivity. While producible water can seriously hamper production and make the pay quite unattractive, the same cannot be said for the irreducible water. Therefore, a reservoir that seems to be a poor candidate for further development – looking only at the conventional logs – may prove to be an attractive prospect once MR log is employed.

MRI logs may provide information that results in an increase in the recoverable reserves. This takes place simply by including the portions of the pay zone into the recoverable reserve calculations that were excluded during the analysis using only the conventional well logs. General background about neural networks has been published in numerous papers and will not be repeated in this paper. An example of such publications is included in the references Mohaghegh et al. (1995; 1996a,b; 1997).

As was mentioned earlier, in this study the developed technique is applied in two different ways. In the first attempt the author will show that it is possible to generate virtual magnetic resonance logs using conventional wireline logs. The concept is tested on several wells from different locations in the United States and the Gulf of Mexico. It is demonstrated that using artificial neural networks, it is possible to generate accurate virtual magnetic resonance logs. In this segment of the study part of the pay zone is used for the model development and then the model is tested on the rest of the pay zone. It is further demonstrated that using the virtual magnetic resonance logs for reserve calculation provides very accurate estimations (within 3%) when compared to reserve estimation obtained by actual magnetic resonance logs.

In the second attempt, which is considered to be the ultimate test for this methodology, it is tested in a manner that would simulate its actual use. This time data from several wells in a particular field is available. This methodology would work best when conventional logs are available from most of the wells in the field and magnetic resonance logs are performed only on a handful of wells (these wells should also have the conventional logs). The wells with magnetic resonance log are used for model development and consequent testing and verification of the model. Then the developed (and verified) model is applied to all the wells in the field. This would generate a much better and more realistic picture of the reservoir characteristics for the entire field. Having such an accurate picture of reservoir characteristics would be a valuable asset for any study that requires accurate reservoir description such as, reservoir simulation, modeling, and reservoir management.

In the second part of this article, the methodology is applied to a field in East Texas (Cotton Valley formation) that is known for its heterogeneity as well as for the fact that the well logs and reservoir characteristics are non-correlatable from well to well. A recently published paper (Mohaghegh et al., 1999) demonstrated the non-correlatable nature of formation characteristics and well logs in this formation.

2. METHODOLOGY

In this section the procedure and methodology for completing this study is explained. This section is divided into two parts. First, the methodology for the intra-well virtual magnetic resonance logs is covered. This is the first part of the study, where wells from different parts of the country are used to show the robustness of the methodology with respect to the type of the formation it is being applied to. The second part of this study is concentrated on one particular formation, Cotton Valley in East Texas. In this part, it is demonstrated that the methodology can be applied to a particular field and can considerably reduce the cost of reservoir characterization.

2.1. *Wells from different formations*

Four wells from different locations in the United States are used to demonstrate the development of virtual MRI logs. These wells are from East Texas, New Mexico, Utah, and Gulf of Mexico. These wells are from different formations and since the virtual MR methodology is a formation specific process, there was no option but to test the methodology using single wells. In this section, part of the pay zone will be used to develop the model and then the model is verified by using the remainder of the pay. The ideal way to show the actual potential of this methodology is to use several wells from the same formation (which also is presented in this article). The prerequisite is that both conventional and MRI logs for the wells should be available. In such a situation, a few of the wells would be used to develop the model and the remaining wells would be used for verification purposes.

For each well in this study, gamma-ray, spontaneous potential (SP), caliper, and resistivity logs were available. These logs were digitized with a resolution of six inches for the entire pay zone. Thirty percent of the data were chosen randomly for the model development and the remaining 70% of the pay were used for verification. In all four cases, the model was able to generate synthetic MR logs with correlation coefficients of up to 0.97 for data that was not used during the model development process.

The model development process was implemented using a fully connected; five layer neural network architecture with different activation functions in each layer. These layers included one input layer, three hidden layers and one output layer. Each of the hidden layers has been designed to detect distinct features of the model. A schematic diagram of the network architecture is shown in Fig. 1.

Please note that in this figure all of the neurons and/or connections are not shown. The purpose of the figure is to show the general architecture of the network used for this study. A supervised gradient descent backpropagation of error method was used to train the neural networks. The input layer has six neurons representing depth, gamma-ray, SP, caliper, medium and deep resistivity. Each hidden layer included five neurons. Upon completion of the training process, each neural network contained six weight matrices. Three of the weight matrices had 30 elements while the remaining matrices each had five. In most cases, acceptable generalization was achieved in less than 500 visits to the entire training data. Once the network was trained, it was used to generate the virtual MPHI, MBVI and MPERM logs. The MPHI and MBVI logs were then used to estimate

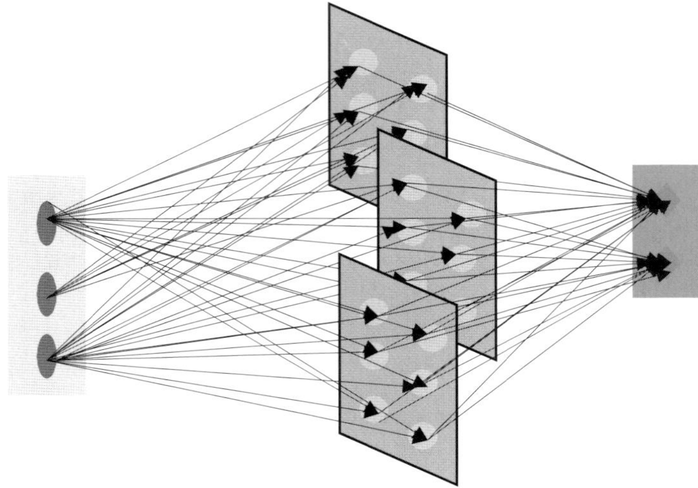

Fig. 1. Schematic diagram of the neural networks used in this study.

the reserves. In one case – the well in New Mexico – that was a tight reservoir, the resolution of the permeability data made it impossible to train an adequate network.

It should be noted that it might be more effective not to use a neural network to replicate the MPERM log. Since this log is derived from MPHI and MBVI, it would be better to calculate the virtual MPERM log from the virtual MPHI and MBVI.

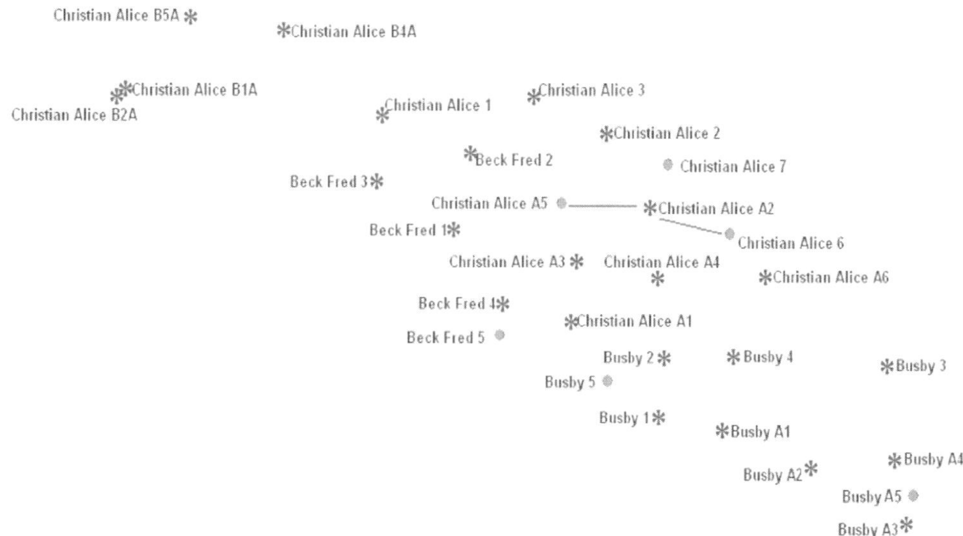

Fig. 2. Relative location of wells in the Cotton Valley, East Texas.

2.2. *Wells from the same formation*

The study area includes a total of 26 wells. There is magnetic resonance logs available from only six wells. The other 20 wells have conventional logs but no magnetic resonance logs. Fig. 2 demonstrates the relative location of the wells. In this figure wells with magnetic resonance logs are shown with circles and wells that have no magnetic resonance logs are shown with asterisks. Also no conventional porosity logs are available for wells with magnetic resonance logs. This could simply be due to the fact that magnetic resonance logs provides effective porosity values that are much more accurate than their conventional counterparts such as neutron porosity, density porosity, and bulk density logs, so not running these logs was an economic decision. Table 1 provides a complete list of the wells and logs that were available for each well in this study.

During the intra-well study, it was observed that existence of porosity indicator logs such as neutron porosity, density porosity, and bulk density, is helpful during the model building process. Therefore, it was decided to generate the virtual version of these logs for the wells with magnetic resonance logs prior to attempting to develop the

TABLE 1

List of the wells in this study and available logs for each well

Well ID	CALI	SP	GR	ILD	ILM	SFL	NPHI	DPHI	RHOB	MBVI	MPERM	MPHI
Beck Fred 5	×	×	×	×	×					×	×	×
Chr. Alice A5	×	×	×	×	×	×				×	×	×
Chr. Alice A7	×	×	×	×	×	×				×	×	×
Chr. Alice 6	×	×	×	×	×	×				×	×	×
Busby 5	×	×	×	×	×					×	×	×
Busby A5	×	×	×	×	×					×	×	×
Chr. Alice B5A	×	×	×	×	×		×	×	×			
Chr. Alice B2A	×	×	×	×	×	×	×	×	×			
Chr. Alice B4A	×	×	×	×	×	×	×	×	×			
Beck Fred 1	×	×	×	×	×	×	×	×	×			
Chr. Alice 3	×	×	×	×	×	×	×	×				
Chr. Alice 2	×	×	×	×	×		×	×	×			
Beck Fred 4		×	×	×		×	×	×				
Chr. Alice A3		×	×	×			×	×				
Chr. Alice A1	×	×	×	×	×		×	×				
Chr. Alice A2	×	×	×	×	×	×	×	×	×			
Chr. Alice A4	×	×	×	×		×	×	×	×			
Busby 2	×	×	×	×	×	×						
Busby 1	×	×	×	×	×	×	×	×	×			
Busby A1	×	×	×	×	×	×	×	×				
Busby 4	×	×	×	×	×		×	×	×			
Chr. Alice A6	×	×	×	×	×	×		×				
Busby 2	×	×	×	×	×		×	×	×			
Busby 3	×	×	×	×	×	×	×	×	×			
Busby A4	×	×	×				×	×	×			
Busby A3	×	×	×	×	×	×	×	×	×			

virtual magnetic resonance logs. Therefore the process of generating virtual magnetic resonance logs becomes a two-step process. A set of neural networks is trained in order to provide input for another set of neural networks. This may sound counter-intuitive from a neural network theoretical point of view. A strong and theoretically sound argument can be made that since neural networks are model free function estimators, and since they are part of an armament of tools that are capable of deducing implicit information form the available data, then adding a set of input values that are essentially a function of other inputs (since they have been generated using the same inputs) should not provide any additional information.

Actually, the opposite of this approach is usually practiced. In cases that there are many input parameters but not as many training records, several analysis including principal component analysis are used to identify the co-dependency of input parameters to one another and removing those input parameters that are a function of others inputs.

A possible respond to such an argument would be as follows. Theoretically there is an ideal neural network structure that when coupled with the ideal training algorithm and ideal neural network parameters will be able to generate the same result with the original inputs and there will be no need for supplemental inputs generated by another set of neural networks, which are the porosity indicator logs such as neutron porosity, density porosity and bulk density in this case. But since such a network is not available, certain (not any) functional relationships (that can be based on domain expertise) between input parameters can indeed help the training and learning process by explicitly revealing some valuable information. A schematic diagram of the two-step process used for the development of virtual magnetic resonance logs are presented in Fig. 3.

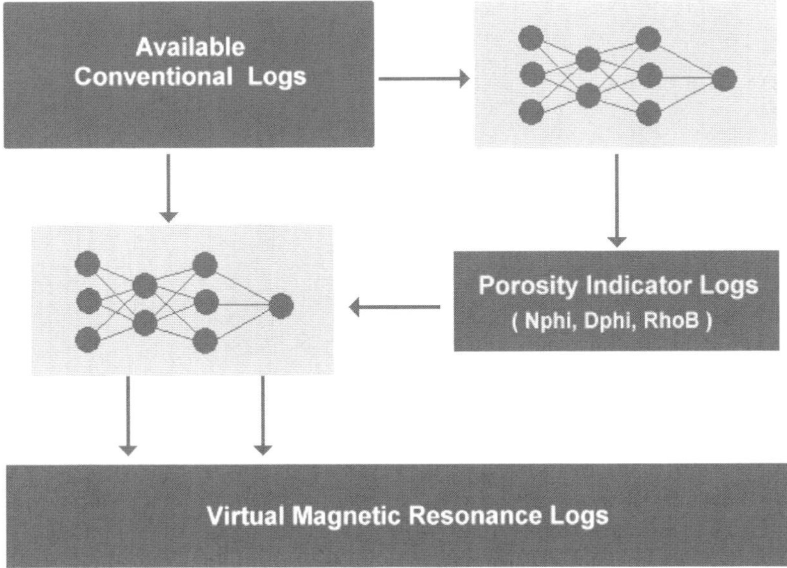

Fig. 3. Schematic diagram of the process for developing virtual magnetic resonance logs.

2.3. Synthetic conventional logs

As was mentioned in the previous section the process of developing virtual magnetic resonance logs starts by generation of synthetic conventional logs. Therefore the wells that had a complete suite of conventional logs were used to develop a neural network model that is capable of replicating the conventional logs such as neutron porosity, density porosity and bulk density for the wells with magnetic resonance logs that lack these logs.

In order to make sure that the neural network model that we are building provides accurate suite of porosity indicator logs, well Christian Alice A2 was used as a test well. This simply means that the data from this well was not used during the training and model building process; rather it was put aside so the capabilities of the trained neural network or neural model can be tested and verified. Fig. 4 shows the actual and virtual versions of all three logs (neutron porosity, density porosity, and bulk density) for the well Christian Alice A2.

As can be seen in this figure, we have been successful in building a representative model that is capable of generating virtual porosity indicator logs for this field. The virtual (synthetic) logs closely follow the trend of the actual logs. All these porosity indicator logs were generated for the wells with magnetic resonance logs. In order to further demonstrate the validity of the virtual (synthetic) porosity indicator logs, neutron porosity logs of three wells were plotted on the same graph. This is shown in Fig. 5. These wells (Christian Alice A5, Christian Alice 6 and Christian Alice A2) are in the proximity of each other. Christian Alice A5 and Christian Alice 6 did not have any conventional porosity indicator logs and well Christian Alice A2's conventional porosity indicator logs were not used during the model development process. Fig. 5 shows the virtual neutron porosity for all three wells as well as the actual neutron porosity for well Christian Alice A2. Formation signatures are easily detectable from all these wells. The distance between these wells Christian Alice A2 and each of the wells Christian Alice A5 and Christian Alice 6 is about 7000 ft. These distances are indicated with a line in Fig. 2.

The methodology explained in this section can be used in many different situations where a complete suite of logs is required for all wells but cannot be accessed due to the fact that some wells lack some of the logs.

3. RESULTS AND DISCUSSION

Similar to the last section, results and discussions will also be presented separately for intra-well study and the study of the Cotton Valley field.

Figs. 6 and 7 show the actual and virtual MPHI, MBVI, and MPERM logs for the well in East Texas. Fig. 6 shows only the verification data set – data never seen by the network before – while Fig. 7 contains the virtual and actual logs for the entire pay zone. Virtual effective porosity log (MPHI) has a correlation coefficient of 0.941 for the verification data set and a 0.967 correlation coefficient for the entire pay zone. The values for virtual MBVI log are 0.853 and 0.894, respectively. The virtual permeability

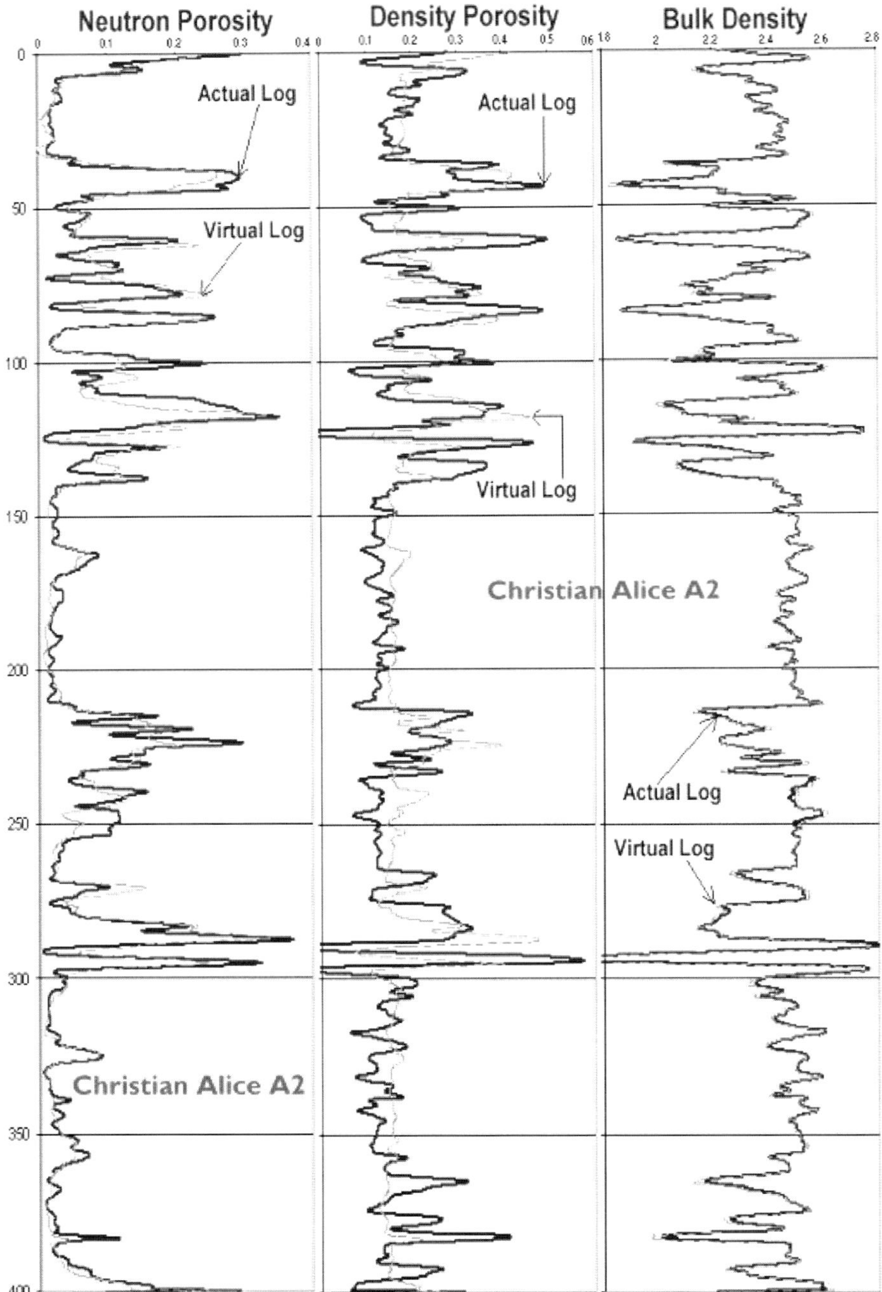

Fig. 4. Actual and virtual porosity indicator logs for well Christian Alice A2.

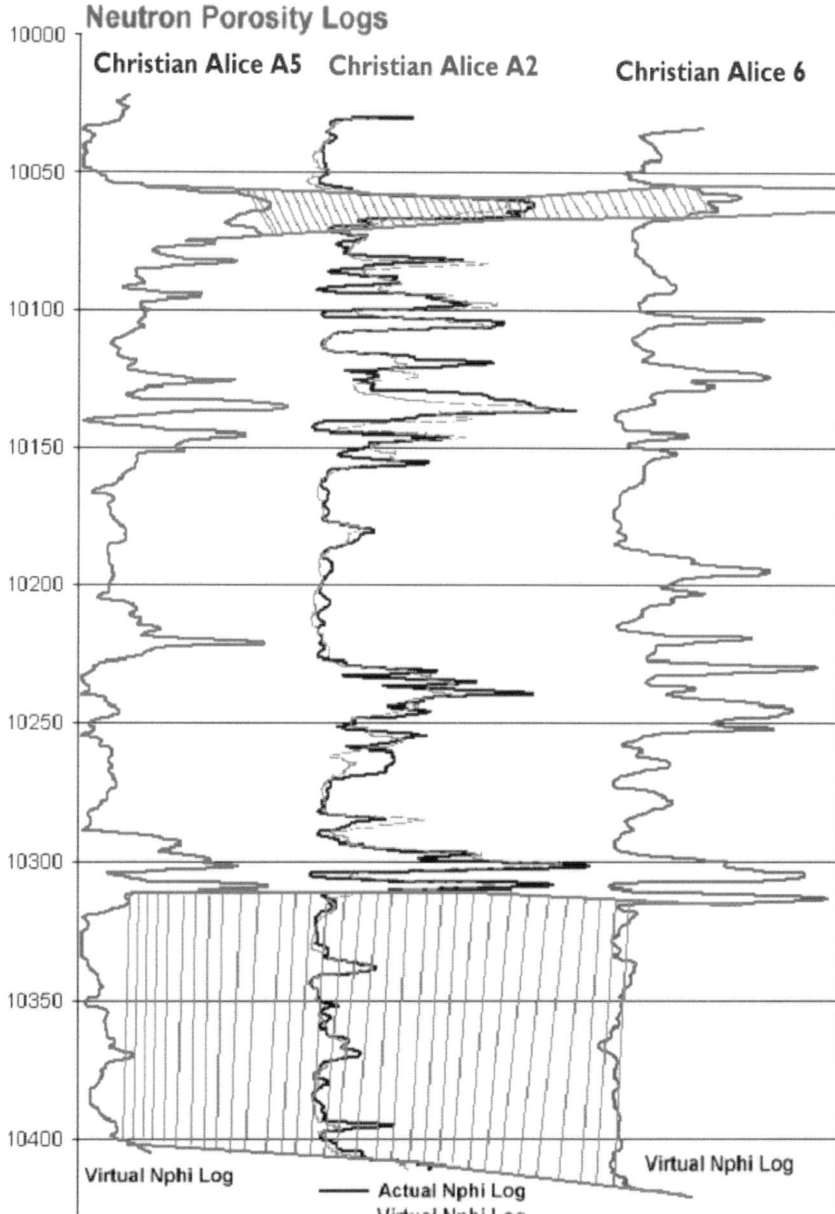

Fig. 5. Actual and virtual neutron porosity logs for well Christian Alice A2 along with virtual Nphi for wells Christian Alice A5 and Christian Alice 6.

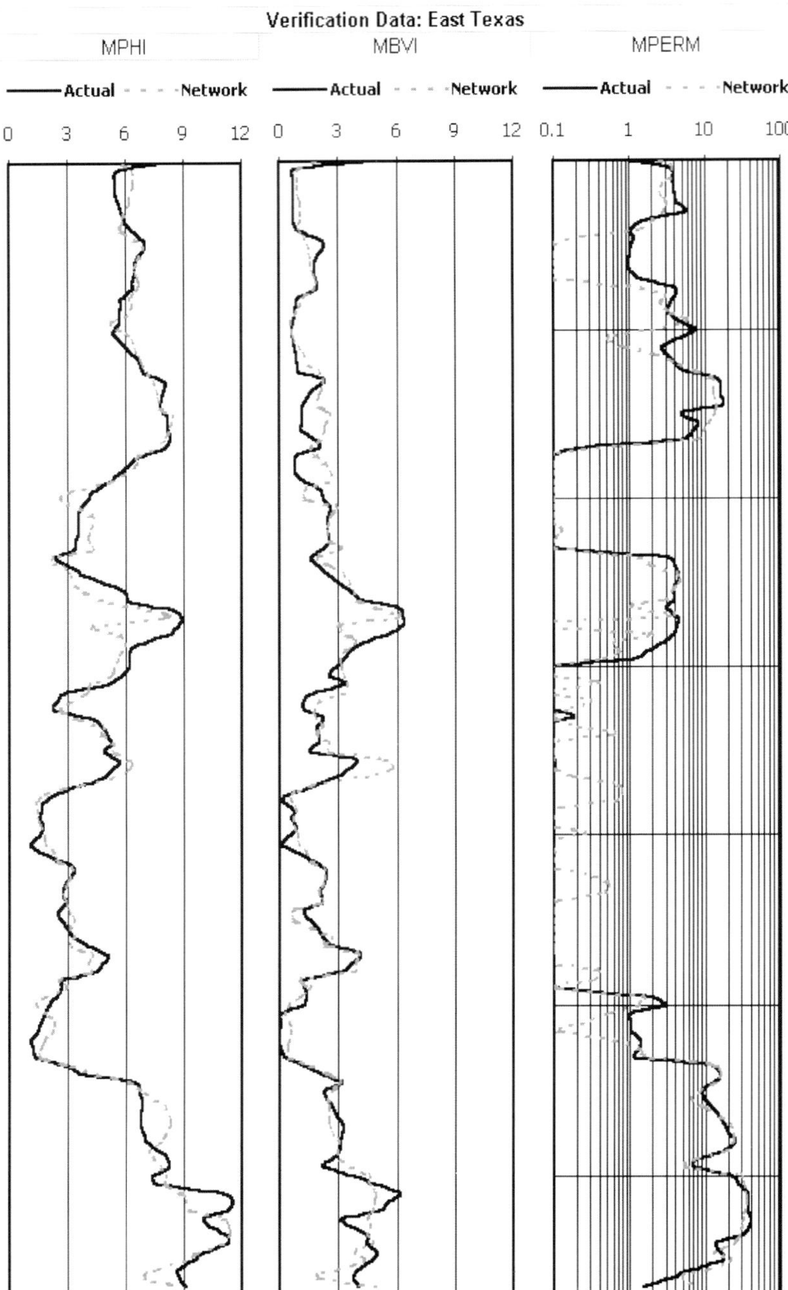

Fig. 6. Virtual and actual MRI logs for the verification data set for the well in East Texas.

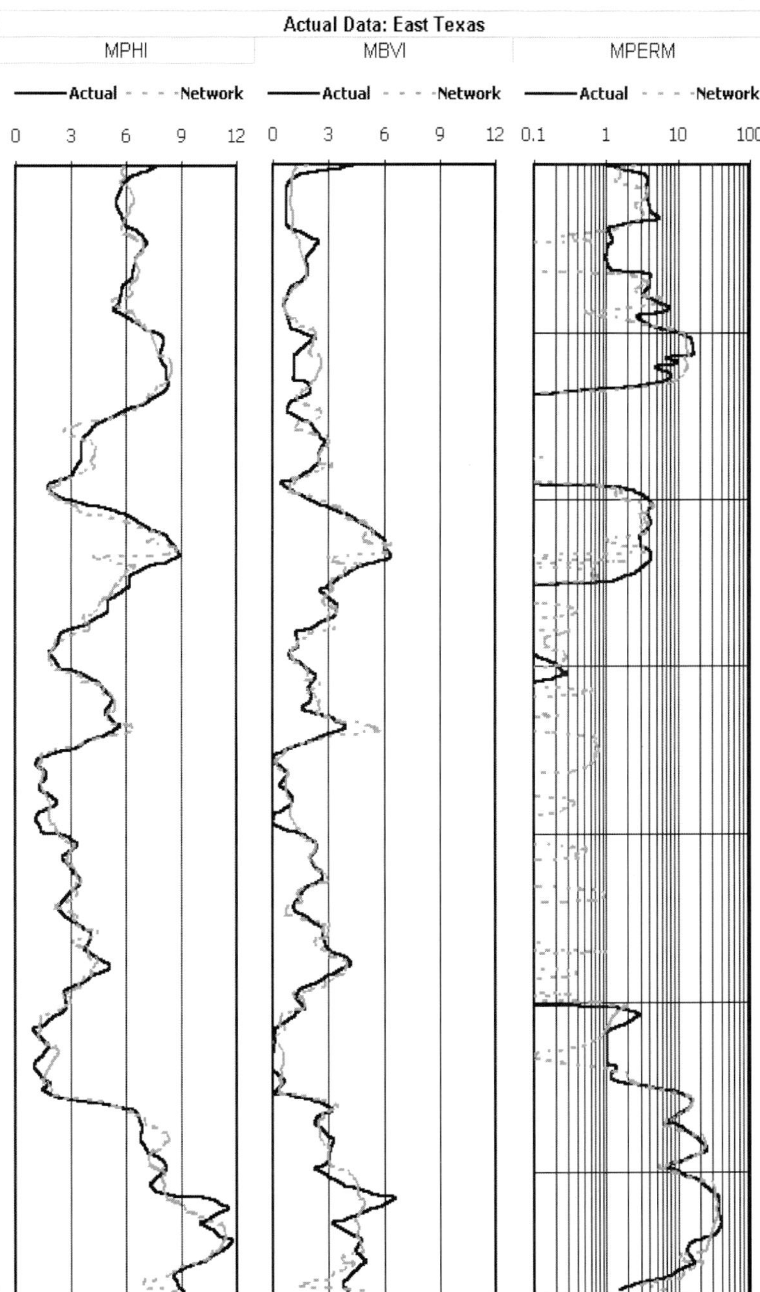

Fig. 7. Virtual and actual MRI logs for the entire pay zone for the well in East Texas.

log for this well also shows a strong correlation, 0.966 for the verification data set and 0.967 for the entire pay zone.

Figs. 8–13 show similar results for the wells from Utah, Gulf of Mexico, and New Mexico respectively. In all cases shown in these figures, virtual MRI logs closely follow the general trends of the actual MRI logs. Please note that MPERM logs are shown in logarithmic scale and therefore the difference in the lower values of the permeability can be misleading. The correlation coefficient provides a more realistic mathematical measure of closeness of these curves to one another.

Table 2 is a summary of the analysis done on all four wells. This Table contains the correlation coefficients for all the logs that were generated. This table shows the accuracy of the virtual MR log methodology on wells from different locations in the United States. The lowest correlation coefficient belongs to virtual MPHI log for the well located in Utah – 0.800 – while the best correlation coefficient belongs to virtual MPERM log for the well located in East Texas – 0.966.

Although the correlation coefficients for all the virtual logs are quite satisfactory, it should be noted that once these logs are used to calculate estimated recoverable reserves, the results are even more promising. This is due to the fact that many times the effective porosity and saturation is averaged. After all, MRI logs are used in two different ways.

TABLE 2

Correlation coefficient between actual and virtual MR logs for four wells in the United States

Well location	MR log type	Data set	Corr. coeff.
Texas	MPHI	Verification	0.941
		Entire Well	0.967
	MBVI	Verification	0.853
		Entire Well	0.894
	MPERM	Verification	0.966
		Entire Well	0.967
Utah	MPHI	Verification	0.800
		Entire Well	0.831
	MBVI	Verification	0.887
		Entire Well	0.914
	MPERM	Verification	0.952
		Entire Well	0.963
Gulf of Mexico	MPHI	Verification	0.858
		Entire Well	0.893
	MBVI	Verification	0.930
		Entire Well	0.940
	MPERM	Verification	0.945
		Entire Well	0.947
New Mexico	MPHI	Verification	0.957
		Entire Well	0.960
	MBVI	Verification	0.884
		Entire Well	0.926

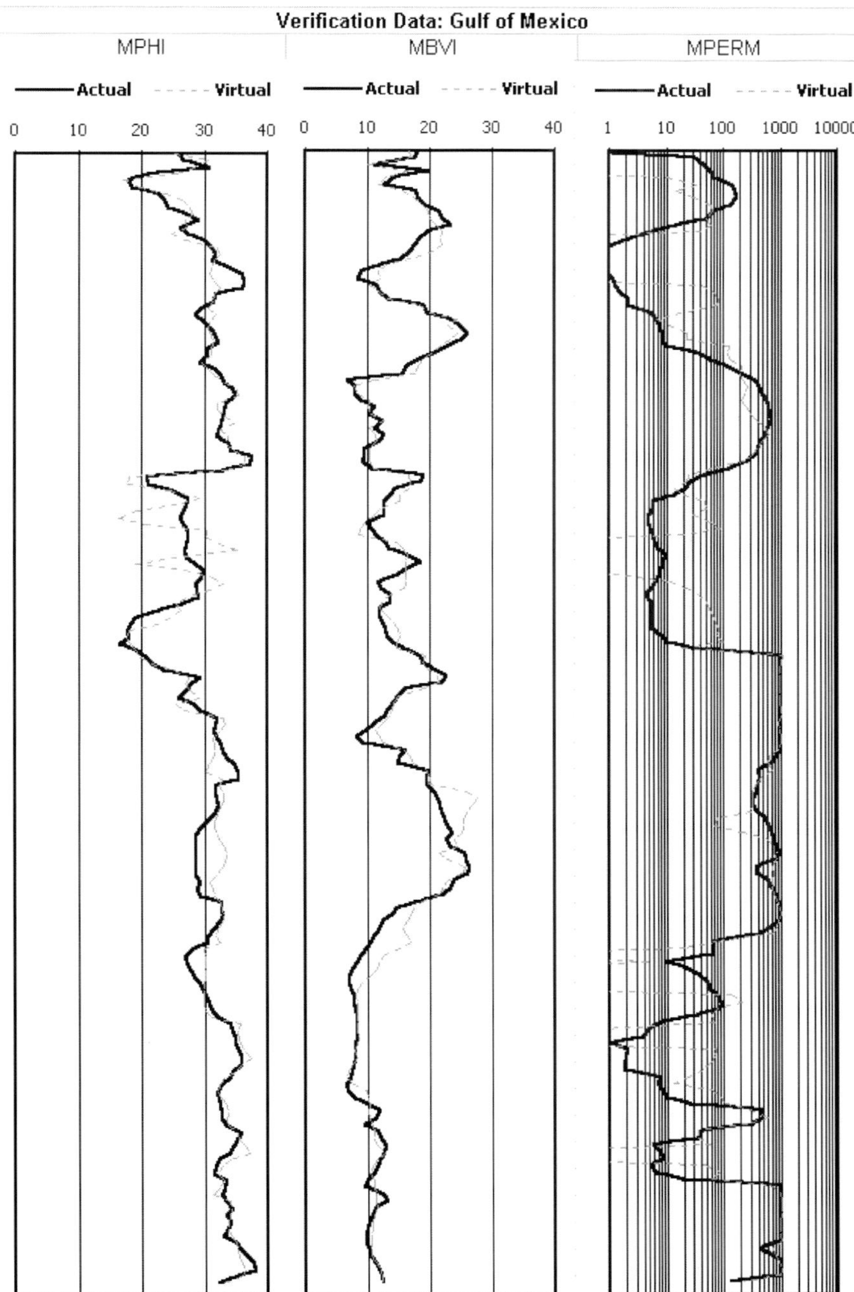

Fig. 8. Virtual and actual MR logs for the verification data set for the well in Gulf of Mexico.

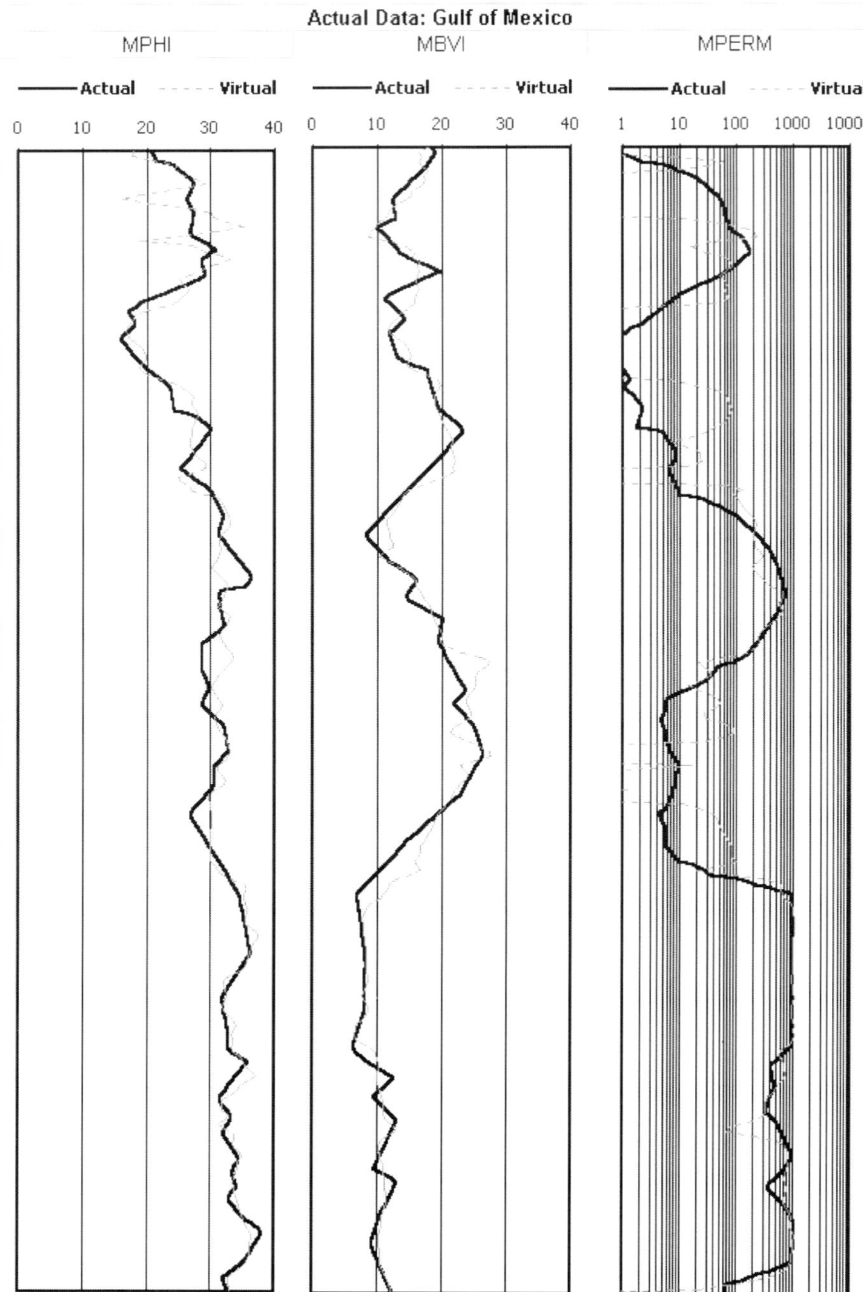

Fig. 9. Virtual and actual MRI logs for the entire pay zone for the well in Gulf of Mexico.

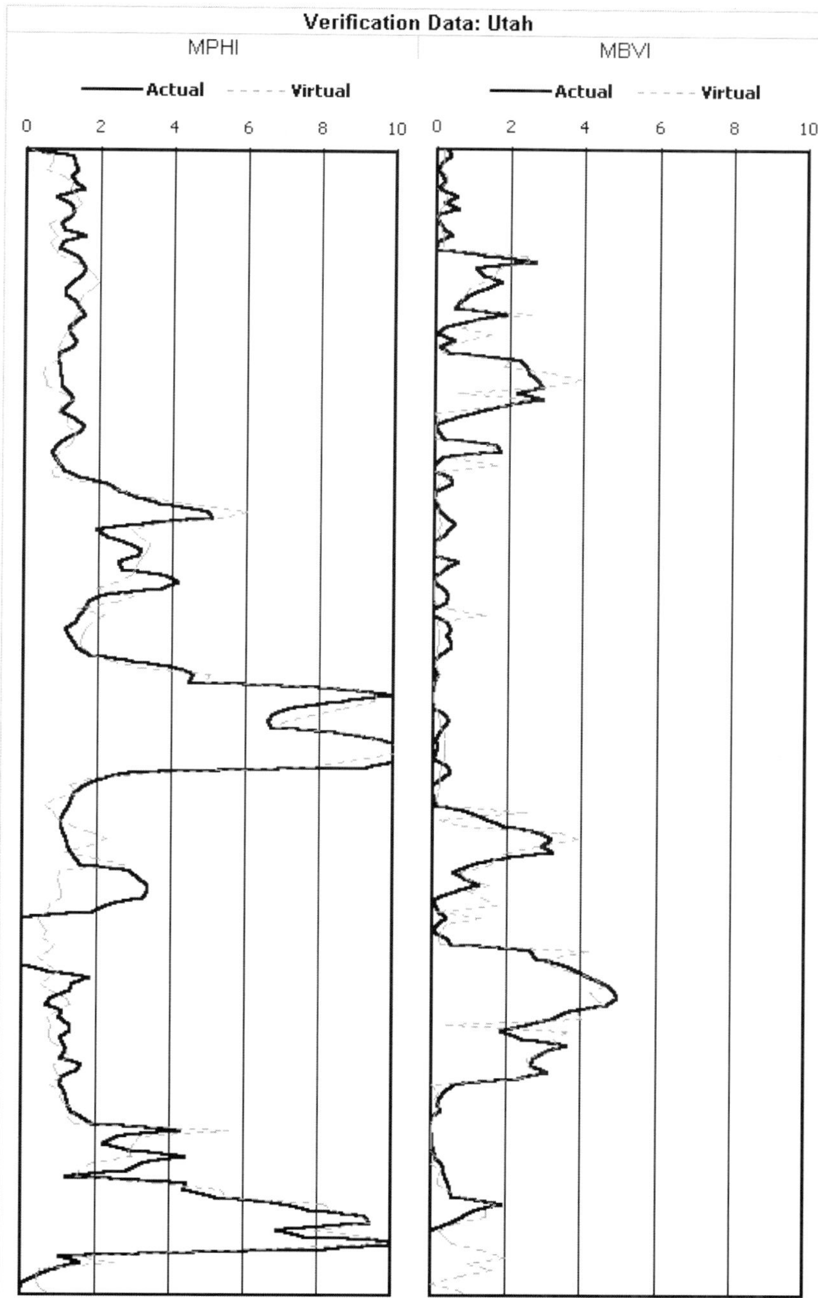

Fig. 10. Virtual and actual MRI logs for the verification data set for the well in Utah.

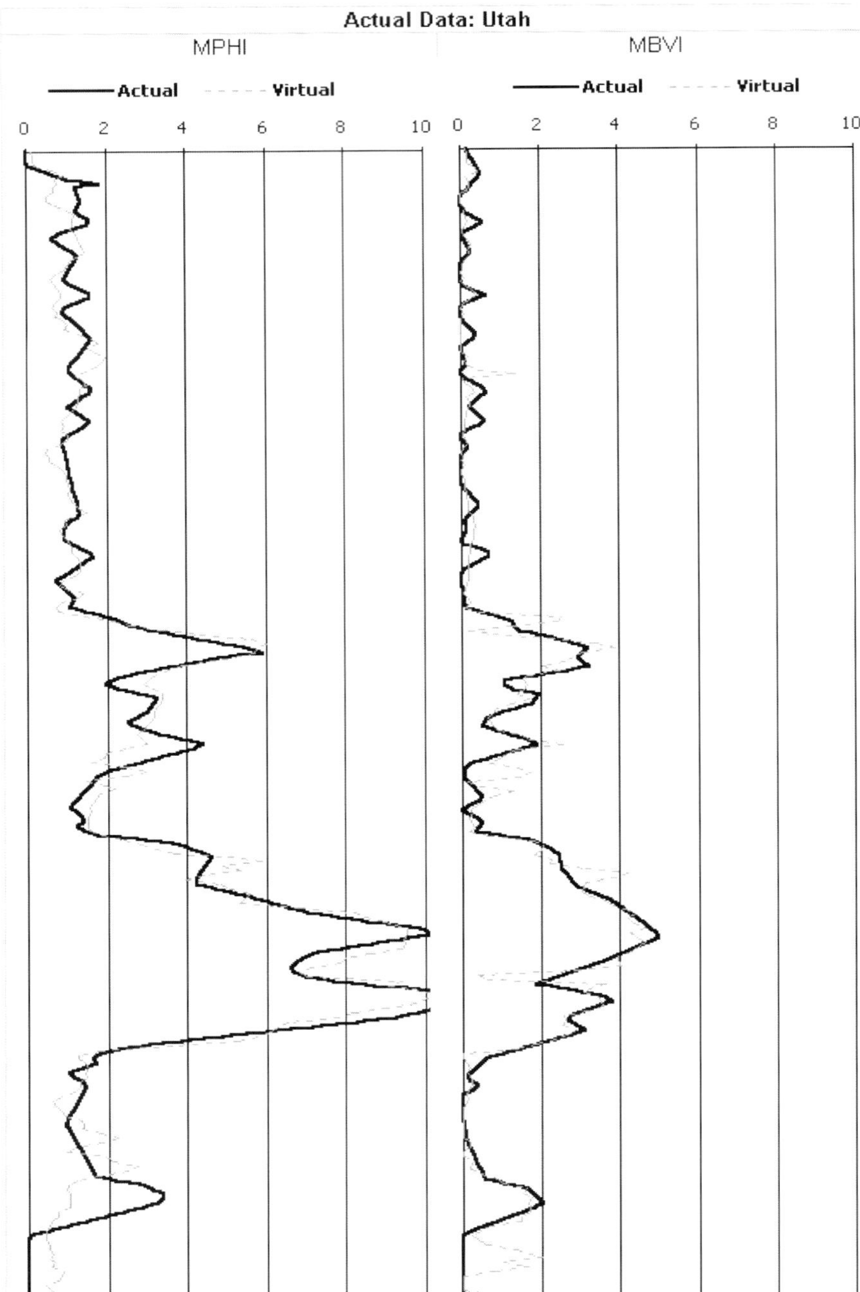

Fig. 11. Virtual and actual MRI logs for the entire pay zone for the well in Utah.

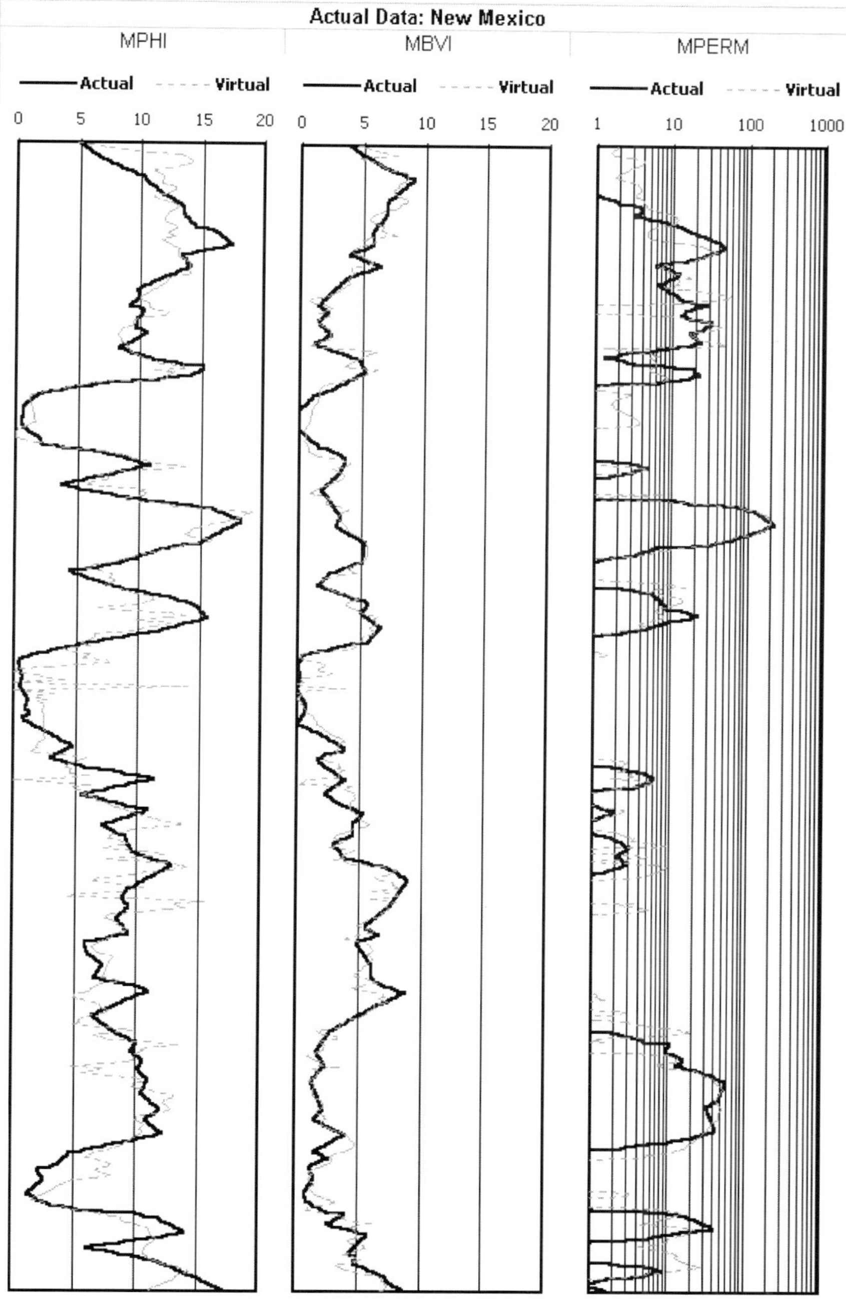

Fig. 12. Virtual and actual MRI logs for the verification data set for the well in New Mexico.

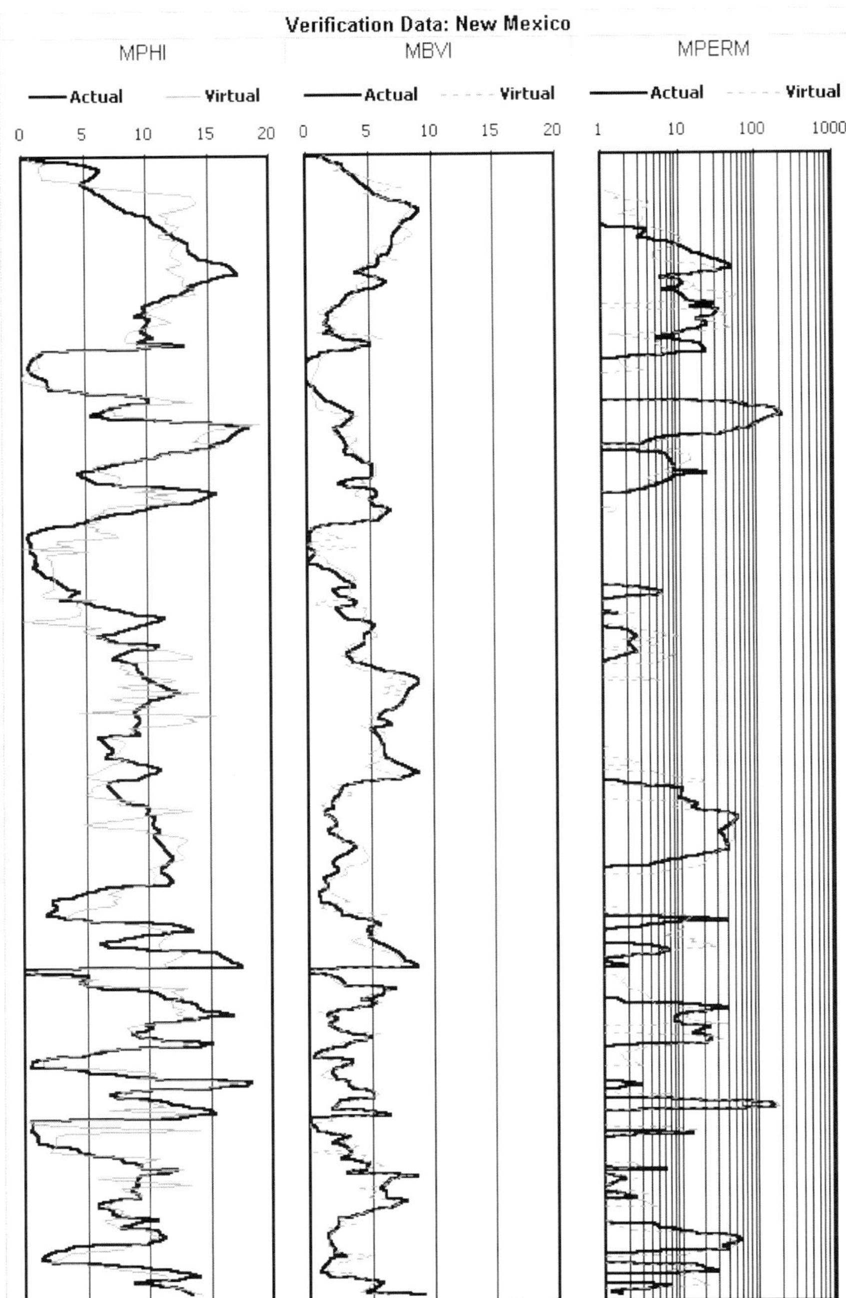

Fig. 13. Virtual and actual MRI logs for the entire pay zone for the well in New Mexico.

TABLE 3

A per-acre estimate of the recoverable reserves using actual and virtual MR logs for four wells in the United States

Well location	MR log type	Reserve Bbls/acre	Percent difference
Texas	Actual	52,368	−1.4
	Virtual	51,529	
New Mexico	Actual	24,346	−1.9
	Virtual	23,876	
Gulf of Mexico	Actual	240,616	+0.3
	Virtual	241,345	
Utah	Actual	172,295	−1.8
	Virtual	169,194	

One-way is to locate and complete portions of the pay zone that have been missed due to the conventional log analysis. This is more a qualitative analysis than a quantitative one since the engineer will look for an increase in the difference between MBVI and MPHI that correspond to a high permeability interval. The second use of these logs is to estimate the recoverable reserves more realistically.

The reserve estimates calculated using virtual MRI logs when compared to estimates calculated using actual MRI logs were quite accurate. As shown in Table 3, the reserve estimates using virtual MRI logs ranged from underestimating the recoverable reserves by 1.8% to over estimating it by 0.3%.

Figs. 14–17 show the virtual and actual MR logs for wells in East Texas and the Gulf of Mexico. These logs are shown in the fashion that MRI logs are usually presented. These logs clearly show the free fluid index – difference between MBVI and MPHI logs – and the corresponding permeability values. This particular representation of the MRI logs is very useful to locate the portions of the pay zone that should be completed. The parts of the pay that has a high free fluid index and corresponds to a reasonably high permeability value are excellent candidates for completion.

So far it was demonstrated that this methodology presented here is a viable tool for generating virtual magnetic resonance logs for different formations. As was mentioned before the objective of this study is to develop a methodology that significantly decreases the cost of field-wide reservoir characterization by generating virtual magnetic resonance logs for all the wells in the field. This will be done through selecting a few wells in the field to be logged using the magnetic resonance logging tools and using this data to develop an intelligent model that can replicate the magnetic resonance logs for other wells in the field.

If a company decides to use this methodology on one of its fields it would be desirable to start by some planning prior to performing any magnetic resonance logging in the field. This would have an important impact on the modeling process. During the planning process the number of the wells that should be logged using the magnetic resonance tools and the location of these well with respect to the rest of the wells in the

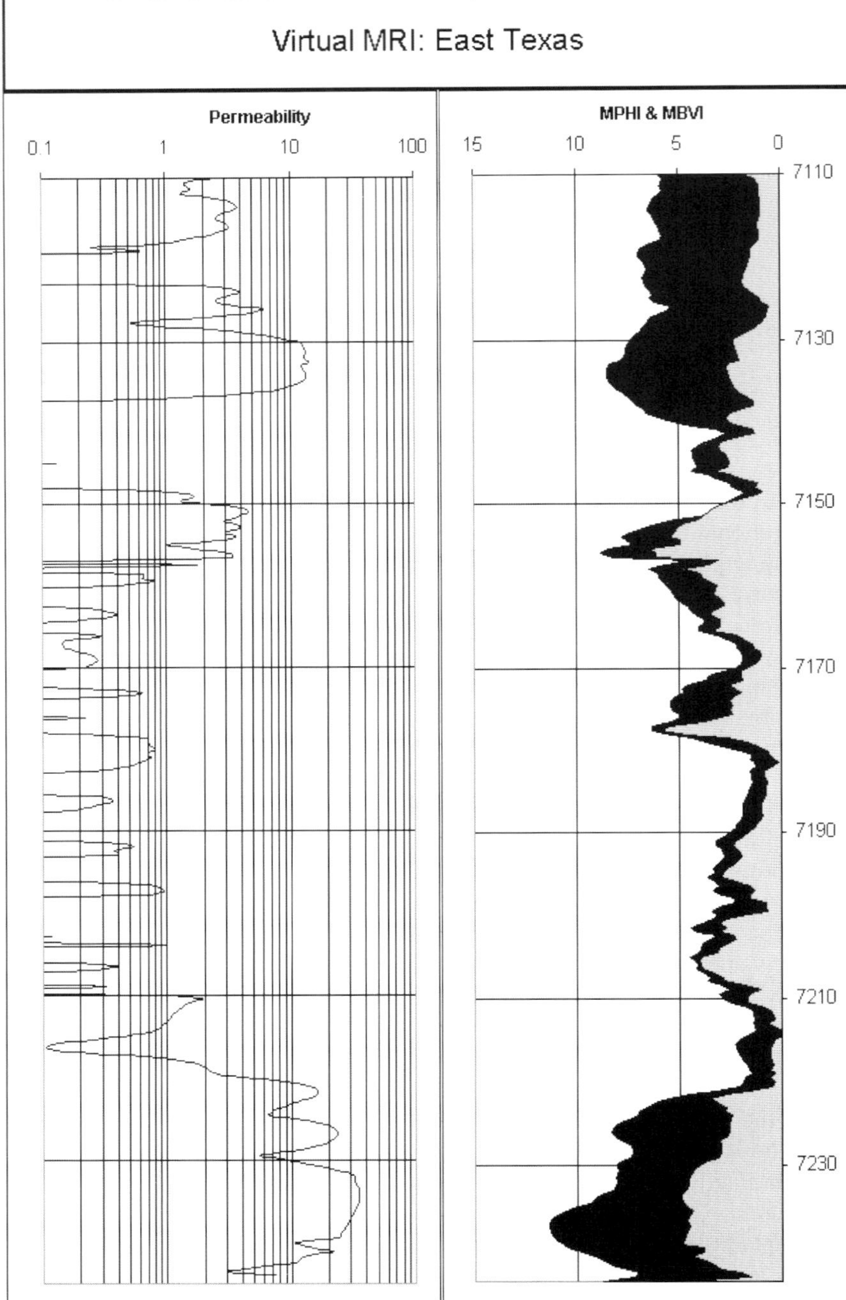

Fig. 14. Virtual MR logs for the well in East Texas

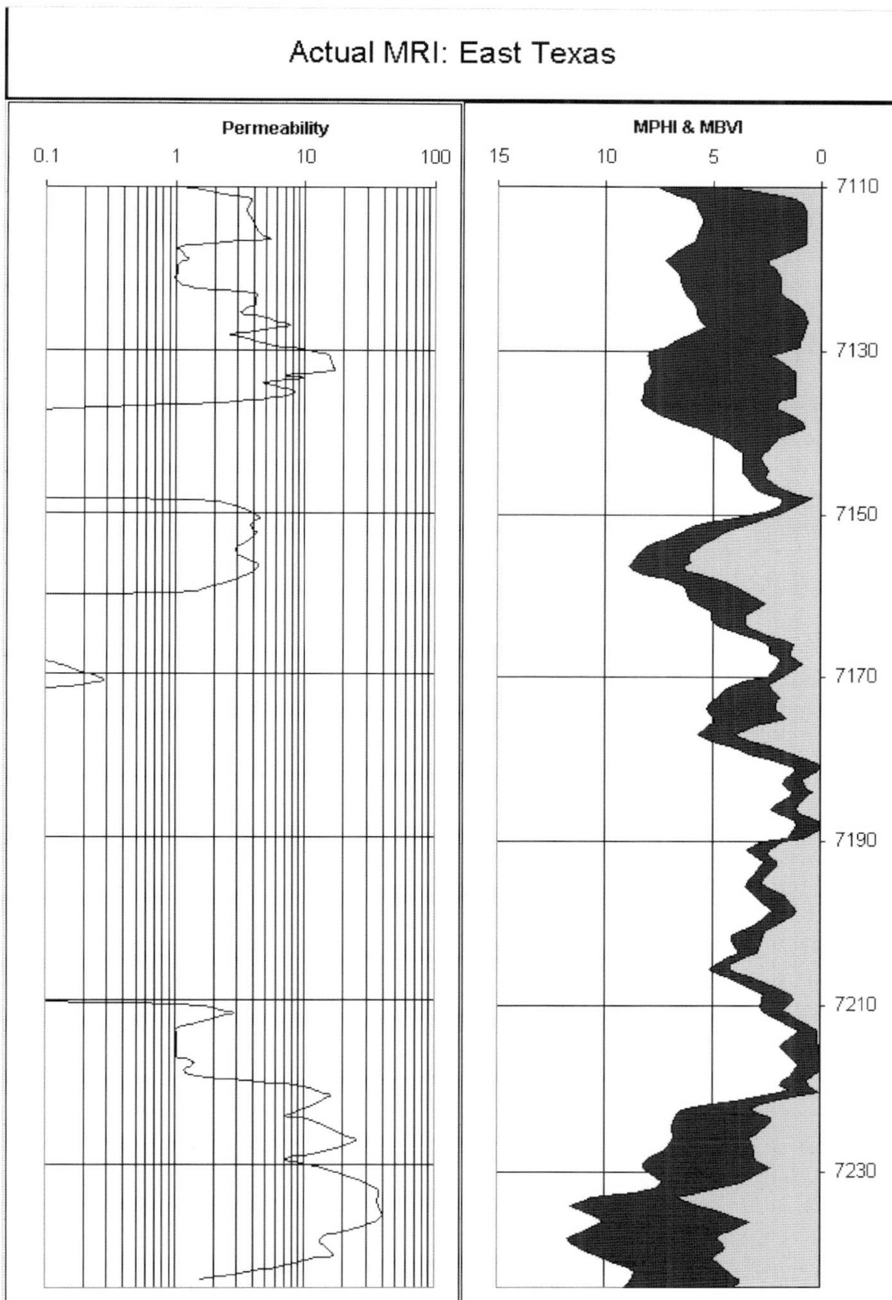

Fig. 15. Actual MR logs for well in East Texas

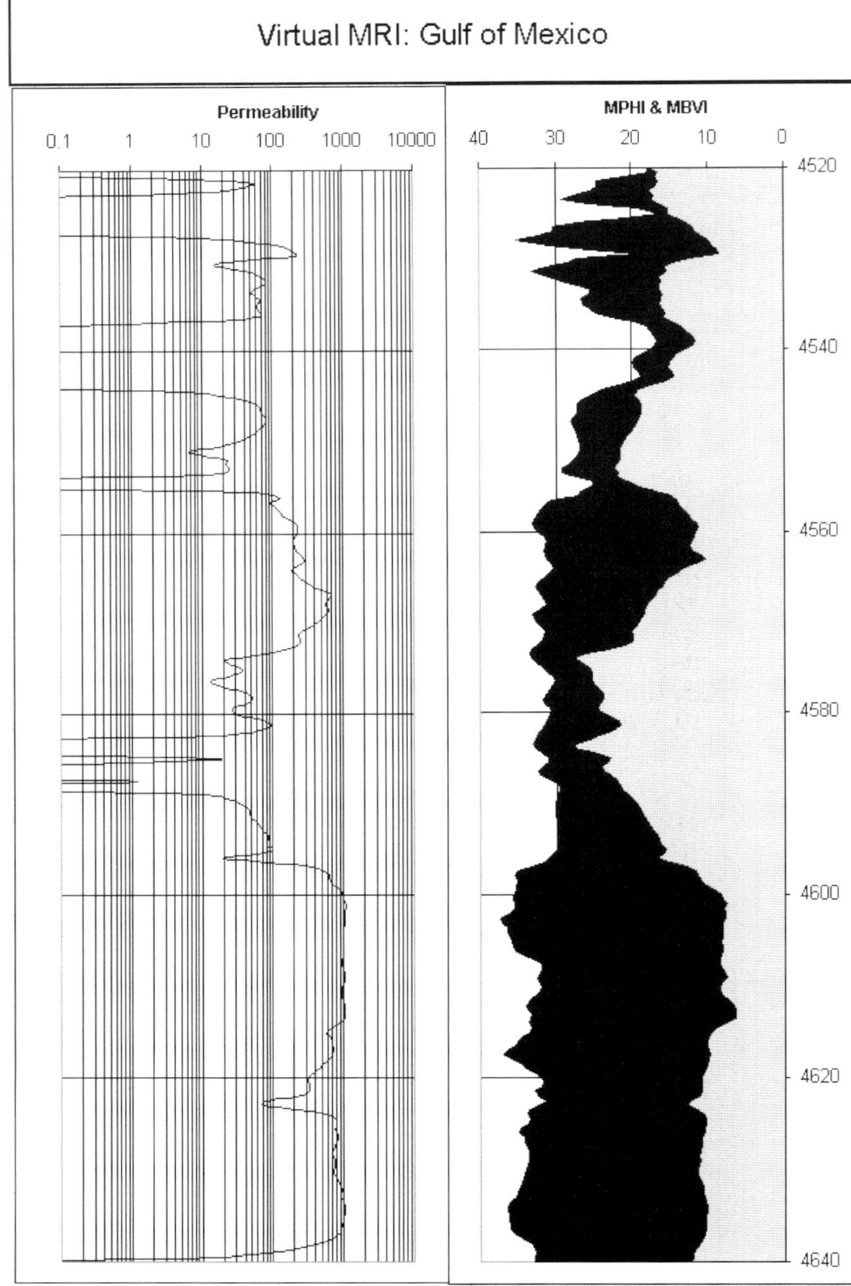

Fig. 16. Virtual MR logs for the well in Gulf of Mexico.

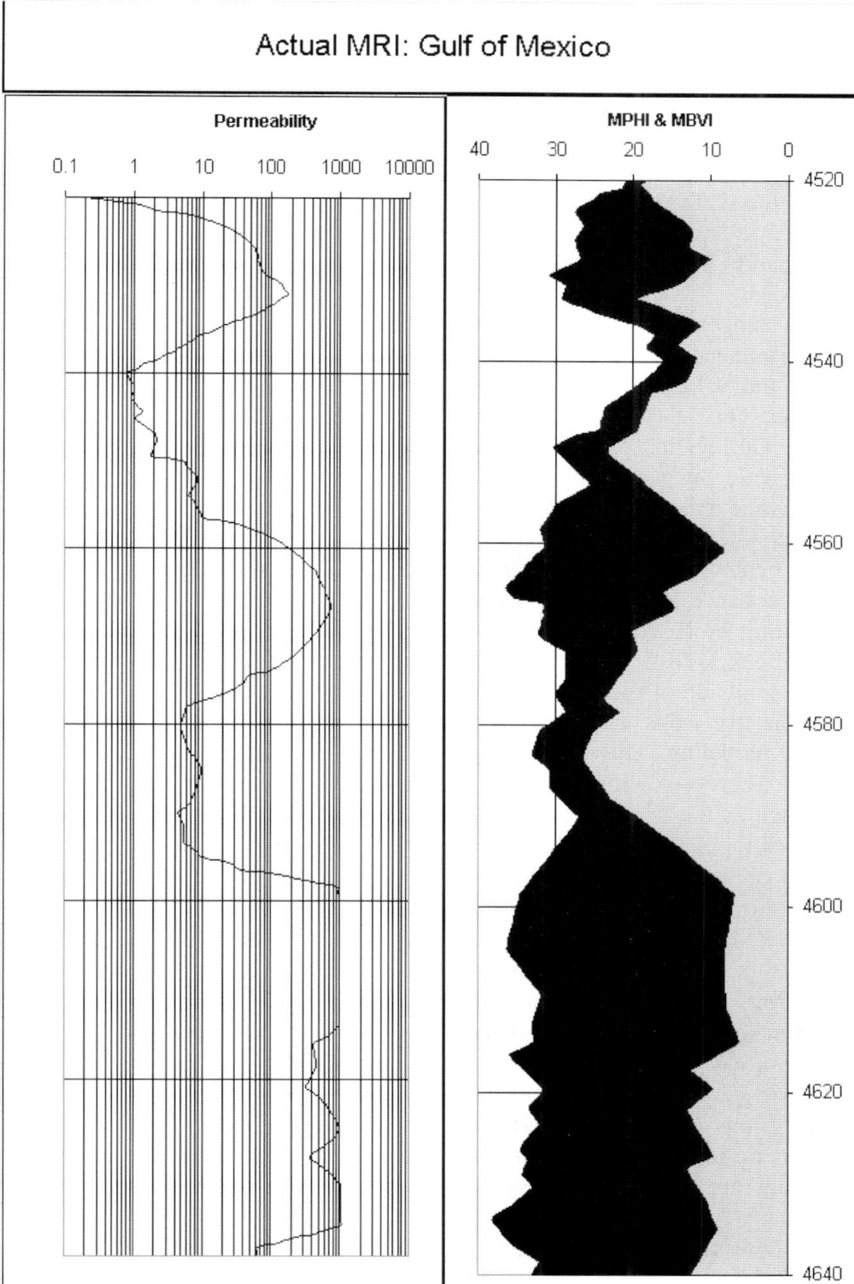

Fig. 17. Actual MR logs for the well in Gulf of Mexico.

field would be among the important consideration. In other cases (such as the one in this study) we have to do with the data that is available and make the best of it.

As seen in Fig. 2, there are six wells in this part of the field that have magnetic resonance logs. The goal is to use the magnetic resonance logs from these wells and develop a predictive, intelligent model that can generate virtual (synthetic) magnetic resonance logs from conventional logs such as gamma ray, SP, induction, and density logs for all the wells in the figure. As was mentioned in the prior section, in this field some of the wells did not have porosity indicator logs. Therefore synthetic version of these logs had to be constructed for these wells prior to generation of virtual magnetic resonance logs.

Prior to using all the six wells with magnetic resonance logs to generate virtual magnetic resonance logs, a test and verification process should be performed in order to confirm the validity of the approach for the specific field and formation under investigation. This test and verification process is the main subject of this portion of this article. During this process we demonstrate that the methodology of generating virtual magnetic resonance logs is a valid and useful process in a field-wide basis. We demonstrate this by using five of the wells, Christian Alice A5, Christian Alice 2, Christian Alice 6, Busby A5, and Busby 5, to develop an intelligent, predictive model and generate virtual magnetic resonance logs for well Beck Fred 5. Since the magnetic resonance logs for well Beck Fred 5 are available, but not used during the model building process, it would provide an excellent verification well. Furthermore, since well Beck Fred 5 is on the edge of the section of the field being studied, and is somewhat outside of the interpolation area, relative to wells Christian Alice A5 . . . Busby 5 (the five wells with magnetic resonance logs), it would stretch the envelope on accurate modeling. This is due to the fact that the verification is done outside of the domain where modeling has been performed. Therefore, one may claim that in a situation such as the one being demonstrated here, the intelligent, predictive model is capable of extrapolation as well as interpolation. Please note that here, extrapolation is mainly an areal extrapolation rather an extrapolation based on the log characteristics.

Fig. 18 shows the actual and virtual magnetic resonance logs (MPHI – effective porosity, and MBVI – irreducible water saturation) for well Beck Fred 5. This figure shows that this methodology is quite a promising one. Although one may argue that the virtual logs under-estimate both effective porosity and irreducible water saturation in many cases, the fact that they are capable of detecting the trend and identifying the peaks and valleys of the formation characteristics are very encouraging. It is believed that using virtual porosity indicator logs such as neutron porosity, density porosity and bulk density logs during the training process has contributed to the under-estimation of the magnetic resonance logs. Although it was demonstrated that the virtual porosity indicator logs are quite accurate, it is desirable to train the networks with the best possible data.

Fig. 19 shows the actual and virtual magnetic resonance permeability logs – MPERM – for the same well (Beck Fred 5). Since MPERM log is not a direct measurement log rather a calculated log (it is a function of effective porosity and irreducible water saturation logs), it is expected that the virtual logs under-estimate the permeability when compared to actual calculated MPERM log. Again, the virtual log is capable of

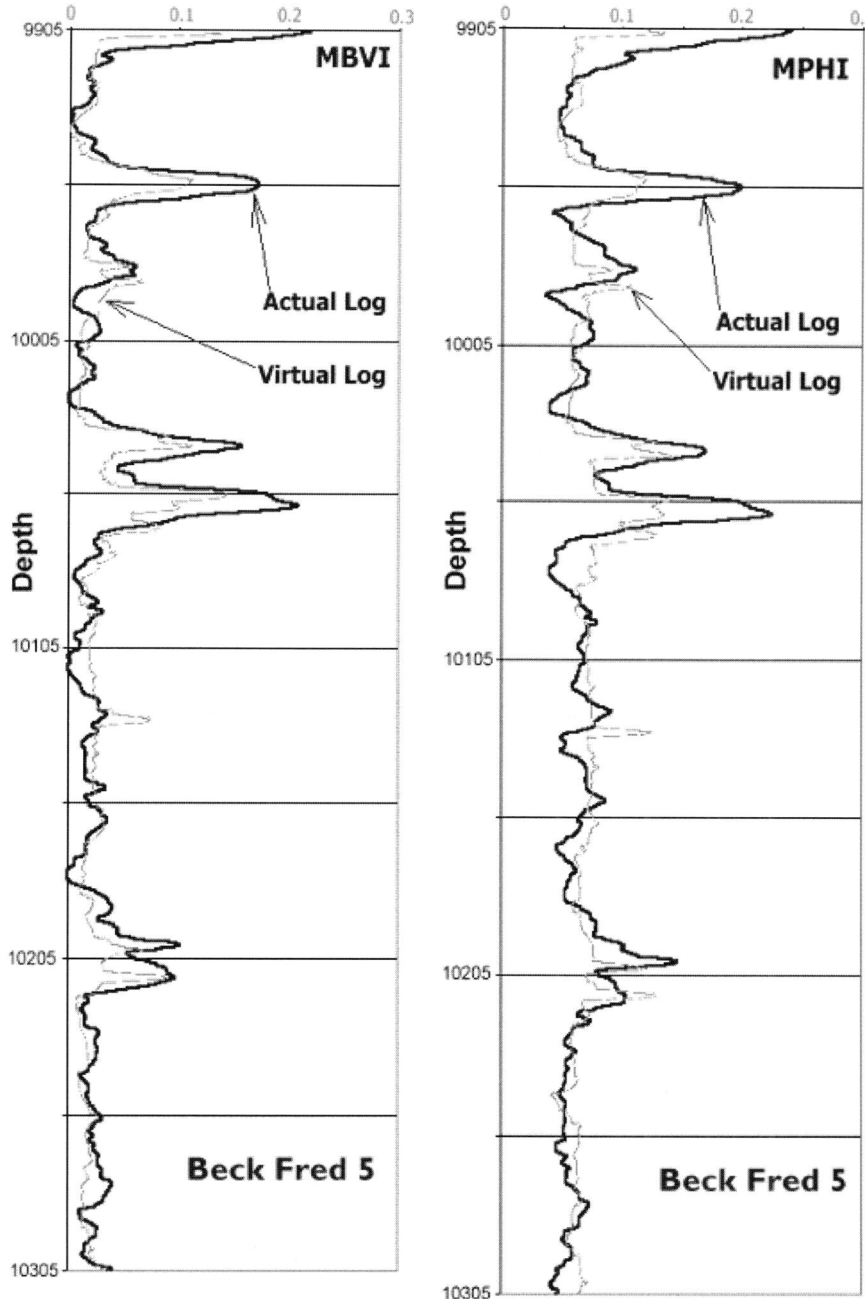

Fig. 18. Actual and virtual magnetic resonance logs for well Beck Fred 5.

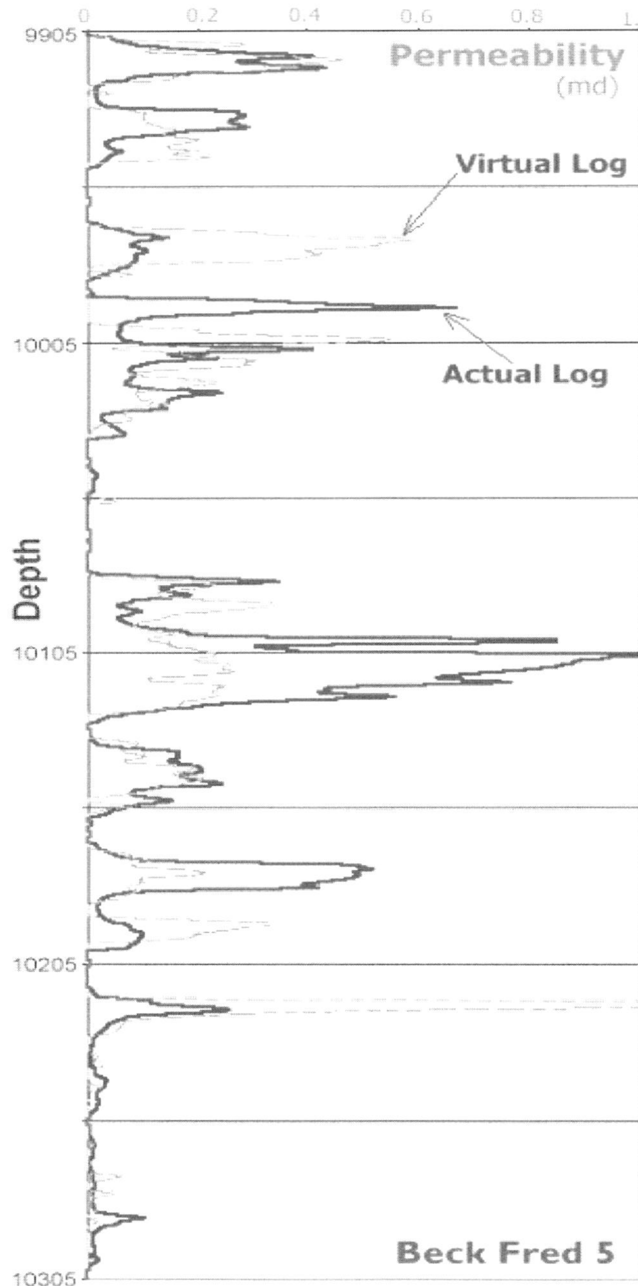

Fig. 19. Actual and virtual magnetic resonance permeability logs for well Beck Fred 5.

detecting most of the trends in permeability values in this formation. If the virtual log were used as a guide to identify perforation depth intervals in this formation, it would have done its job well.

In order to test and verify the effectiveness of the virtual magnetic resonance logs, as compared to its actual counterparts, they were used in a reserve estimation calculation. In this calculation all parameters were kept constant and the only difference between two sets of calculation were the use of virtual verses actual magnetic resonance logs. The logs shown in Fig. 18 are used to perform reserve estimate calculations. Using the virtual magnetic resonance logs the estimated reserves were calculated to be 138,630 MSCF/Acre while using the actual magnetic resonance logs the calculated reserve estimates were 139,324 MSCF/Acre for the 400 ft of pay in this well. The 0.5% difference in the calculated estimated reserves based on virtual and actual magnetic resonance logs demonstrates that operators can used this methodology effectively to reach at reserve estimates with much higher accuracy at a fraction of the cost. This will allow operators make better reserve management, and operational decisions.

4. CONCLUSIONS

A new methodology was introduced that has the potential to reduce the cost of reservoir characterization from well logs significantly. This methodology uses the conventional well logs and generates virtual or synthetic magnetic resonance logs for all the wells in a field. The development process requires that only a handful of wells in a field be logged using the magnetic resonance logging tools. Then the data generated from the magnetic resonance logging process is coupled with the conventional log data and used to develop an intelligent, predictive model. After testing and verifying the predictive model's accuracy, it can be applied to all the wells in the field that have only conventional logs. At the end of the process all the wells in the field will have magnetic resonance logs. This process will help engineers in the filed to acquire a much better handle on the reservoir characteristics at a fraction of the cost of running magnetic resonance logs on all the wells in the field. This is especially true and beneficial for fields that have many producing wells the already have been cased.

It was also demonstrated that virtual magnetic resonance logs could provide reserve estimates that are highly accurate when compared to the reserve estimates that can be acquired from actual magnetic resonance logs. The neural networks that are constructed and trained for a particular formation may not be used to generate virtual MR logs for other formations. This is similar to the case of virtual measurement of formation permeability the methodology is formation dependent (Mohaghegh et al., 1996a).

ACKNOWLEDGEMENTS

The author would like to express his appreciation to his undergraduate and graduate students that contributed significantly to this research program throughout the past several years. These students are Mark Richardson, Carrie Goddard, and Andrei Popa.

The author would also like to acknowledge the Consortium for Virtual Operations Research at West Virginia University and its member companies for supporting most of the above graduate students.

REFERENCES

Austin, J. and Faulkner, T., 1993. Magnetic resonance imaging log evaluates low-resistivity pay. *Am. Oil Gas Reporter*, August.

Mohaghegh, S., Arefi, R. and Ameri, S., 1995. Design and development of an artificial neural network for estimation of formation permeability.*SPE Comput. Appl. J.*, December, pp. 151–154.

Mohaghegh, S., Arefi, R. and Ameri, S., 1996a. Virtual measurement of heterogeneous formation permeability using geophysical well log responses. *Log Analyst*, March–April, pp. 32–39.

Mohaghegh, S., Arefi, R. and Ameri, S., 1996b. Reservoir characterization with the aid of artificial neural networks. *J. Pet. Sci. Eng.*, 16: 263–274.

Mohaghegh, S., Balan, B. and Ameri, S., 1997. Determination of permeability from well log data.*SPE Formation Eval. J.*, September, pp. 263–274.

Mohaghegh, S., Koperna, G., Popa, A.S. and Hill, D.G., 1999. Reducing the cost of field-scale log analysis using virtual intelligence techniques. *1999 SPE Eastern Regional Conference and Exhibition*, October 21–22, Charleston, WV, SPE Paper 57454,

Developments in Petroleum Science, 51
Editors: M. Nikravesh, F. Aminzadeh and L.A. Zadeh

Chapter 28

ARTIFICIAL NEURAL NETWORKS LINKED TO GIS

Y. YANG and M.S. ROSENBAUM [1]

Civil Engineering Division, The Nottingham Trent University, Newton Building, Burton Street, Nottingham NG1 4BU, UK

ABSTRACT

As a database and decision support tool, GIS has been applied to a wide range of social and engineering situations where the spatial relationships are of significance. Simultaneously, the application of artificial neural networks has developed, providing an ability to handle unknown relationships. A brief introduction to GIS and artificial neural networks is presented, paying particular attention to the GIS 'overlay' operation and the concept of 'relative strength of effect'. An integrated spatial analysis utilising these two systems is presented wherein an artificial neural network has been incorporated as a mapping tool within a raster-based GIS to provide a predictive capability. This method is capable of incorporating dynamic change encompassed within existing observations. The technique has been applied to sediment prediction in Gothenburg harbour.

1. INTRODUCTION

As a tool for storage, retrieval, analysis and display of spatial data, Geographical Information Systems (GIS) have been widely applied within earth science. Most applications to engineering, including harbours, have focused on simple overlay analysis. However, the actual relations tend to be dynamic and uncertain. Statistics have been explored as a basis for multiple parameter evaluation (Stein et al., 1988), but difficulties remain concerning the nature of interdependencies between the variables (Burrough, 1999).

The utility offered by artificial neural networks (ANN) has therefore been explored in order to assess their potential for assisting analysis of spatial data. As a tool for simulating the intuitive reasoning process of human brain, ANN is able to map complex mechanisms without any prior knowledge concerning them. The handling of spatial information has been enabled by developing a system combining ANN with a GIS.

The ANN algorithm has been programmed in C^{++} to create a neural network for relative strength of effect (NRSE) (Yang and Rosenbaum, 1999). The raster GIS 'Idrisi' (Eastman, 1999) was selected because of its extensive usage and the ability to readily interface with external software. A raster structure stores the information cell-by-cell, the cell dimensions determining the resolving capability of the system. The alternative

[1] Tel.: +44 (115) 848-2099, Fax: +44 (115) 848-6450,
E-mail: mike.rosenbaum@ntu.ac.uk

structure is the vector system whereby objects are defined by points, lines and polygons. The combined GIS and ANN system has been applied to sedimentology studies within Gothenburg harbour, Sweden, as part of the H-SENSE project (Stevens, 1999).

2. GEOGRAPHICAL INFORMATION SYSTEMS AND THE OVERLAY OPERATION

GIS appeared in the late 1960s (Coppock and Rhind, 1991) as a mapping, planning and management tool applicable to large areas of terrain. GIS can be regarded as "a set of tools for the input, storage and retrieval, manipulation and analysis, and output of spatial data" (Marble et al., 1984). It is essentially a database designed for handling large quantities of spatial data. With the aid of artificial intelligence, GIS is becoming a potent tool for decision support. In this sense, GIS could also be regarded as "a decision support system involving the integration of spatially referenced data in a problem solving environment" (Cowen, 1988).

The functionality of a GIS can be considered in four categories (Malczewski, 1999):

- Data input
- Data storage and management
- Data manipulation and analysis
- Data output

Data input and output represent the processes of collection and formatting to permit spatial representation by both the GIS and the user. One reason for the popularity of GIS arises from its ability to act as a vehicle for communicating information based on its graphical user interface, to which the spatial character of the information lends support. The data storage and management concerns structure, organisation and retrieval of information. The data manipulation and analysis is a unique contribution by a GIS, providing functions such as measurement (DeMers, 1999), classification (Davis, 1996), overlay (Berry, 1993), connectivity (Eastman, 1993), statistical modelling (Zhang and Griffith, 1997), multivariate analysis (Johnson, 1978), cluster analysis and discriminant analysis (Griffith and Amrhein, 1991), principle component analysis and factor analysis (Bailey and Gatrell, 1995), time-series analysis (Hepple, 1981), geostatistical analysis (Burrough and McDonnell, 1998), and mathematical modelling and simulation (Steyaert and Goodchild, 1994).

The overlay operation is perhaps the most useful: a process whereby a new map is produced as a function of two or more input maps covering the same area. Each layer (map) in the GIS contains the values for an attribute. The new layer can be considered to be the result of an exhaustive execution of a function to all the cells at the same spatial location (Fig. 1).

The overlay operation can be simple: addition, subtraction, multiplication, division, AND, OR, NOT, or it could consist of a complex sequence of operations, including fuzzy sets. However, all the functions in a GIS require specific knowledge of the values for each input image at the location, together with the relationships between the input layers, in order to establish the overlay function. A problem arises where data is missing or the knowledge of the interactions is incomplete. Such instances are often inherent in situations involving pollution, controlled by many parameters and where the sampling

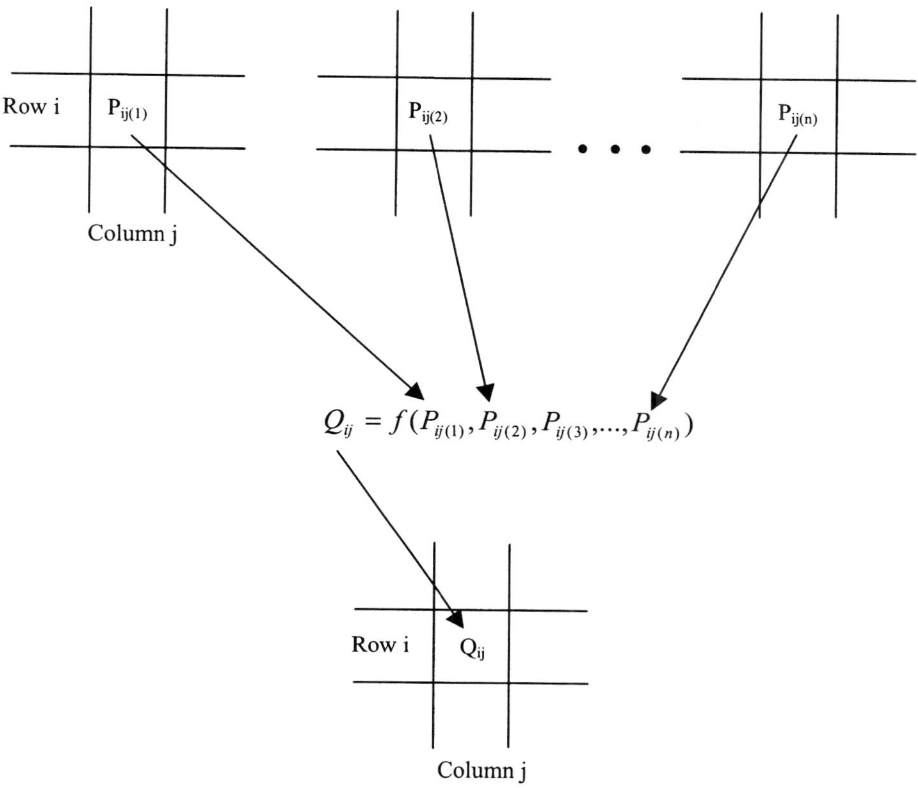

Fig. 1. The GIS 'overlay' operation for a specific location represented by a series of input images.

is limited by time and financial support. It may also be difficult to establish reasonable frequency distributions for a probabilistic basis to the analysis. Assistance is now to hand from a powerful tool that has emerged from artificial intelligence: the Artificial Neural Network (ANN).

3. ARTIFICIAL NEURAL NETWORKS

Artificial Neural Networks aim to simulate the mechanisms of the human brain when establishing inter-relations between a variety of information sources. This is realised as intuitive reasoning rather than the logical reasoning normally executed by machine. The capability offered by ANN to incorporate uncertainty as well as data which is dynamic in character has led to a number of studies to establish its applicability to civil engineering problems concerning structures (Ghaboussi et al., 1990), underground excavation (Zhang et al., 1991; Lee and Sterling, 1992), geotechnical engineering (Ellis et al., 1995; Yang and Zhang, 1997) and spatial analysis (Demyanov et al., 1998).

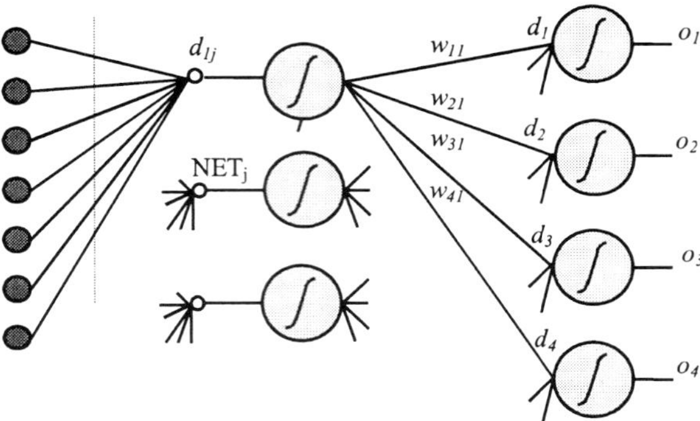

Fig. 2. Back-propagation (BP) algorithm. The weight changes in the hidden unit layer j.

One of the most promising mapping schemes is the back-propagation (BP) network (Rumelhart and McClelland, 1986). The back-propagation neural network architecture is a hierarchical design consisting of fully interconnected layers or rows of processing units (Fig. 2). The interconnections are called weights, and provide the means for ANN to save knowledge, the process of 'learning'. The function in the node is called the 'activation function', which transfers the inputs received by this node to its output and thus to the next layer (Fig. 3). This process modifies the weights by incorporating the errors in the mapped output. Based on the calculation of error gradients, such errors are then back-propagated from the output neurones to all the hidden neurones; subsequently all the weights are adjusted with respect to the errors.

The BP process is repeated until the error output has been reduced to a specified minimum value. The weights are then fixed and saved as a record of the knowledge pertaining to this system. Thus for a given input, an output is then associated with the fixed weight system.

The information processing operation facilitated by back-propagation performs an approximation of the bounded mapping function $f : \boldsymbol{A} \subset \boldsymbol{R}^n \rightarrow \boldsymbol{R}^m$. This function is from a compact subset $\boldsymbol{A}$ of $\boldsymbol{n}$-dimensional Euclidean space to a bounded subset $f[\boldsymbol{A}]$ of $\boldsymbol{m}$-dimensional Euclidean space, by means of training with examples (x_1, y_1), $(x_2, y_2), \ldots, (x_k, y_k), \ldots$ of the mapping, where $y_k = f(x_k)$. It is assumed that the mapping function f is generated by selecting x_k vectors randomly from $\boldsymbol{A}$ in accordance with a fixed probability density function $\rho(\boldsymbol{x})$.

The operational use to which this network is put once training has been performed (on a set of experimental or observed data) makes use of the random selection of input vectors $\boldsymbol{x}$ in accordance with $\rho(\boldsymbol{x})$. ANN then models the mapping by utilising simple neurones based on either a linear or a non-linear activation function. Because of the large number of neurone connections, model behaviour is characterised by co-operation between neurones. Thus an ill definition introduced by a few neurones does not influence the outcomes from its associated mapping.

ANN has a robust nature with respect to uncertain or deficient information, even though such information influences many aspects of a complex system, for example the

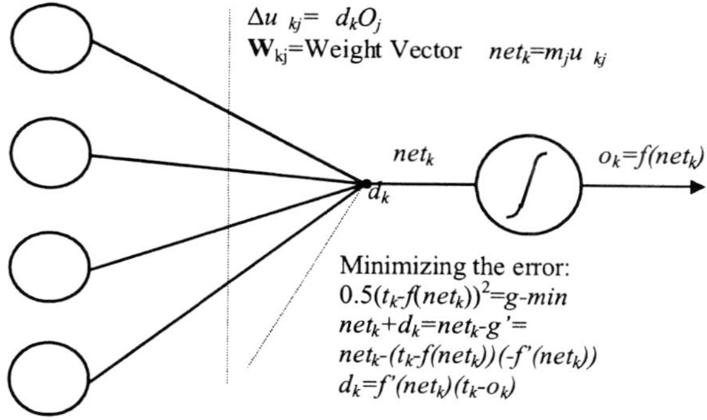

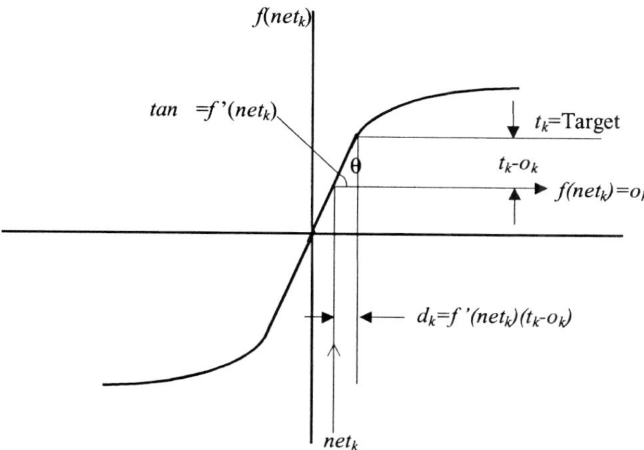

Fig. 3. The node and its activation function in the Back-propagation (BP) algorithm.

propagation of uncertainty. ANN can apply additional neurones and weights as required to take full account of such influences, and thus possesses an in-built capability for including any relation once ANN has been trained using a reference data set.

ANN may well have a large number of nodes, yet the activation function at each node is very simple. The complex knowledge contained within the training data is saved in the form of the connections between the various nodes. The connections and activation functions determine the behaviour of the neural network. Thus, no matter how complicated the mechanisms are, ANN has the capability of mapping them without having to incorporate a prior supposition or simplification.

The existence of large numbers of nodes needed to represent the knowledge provides the robust structure for uncertain or incomplete inputs. The limited connection weights have the advantage of dramatically reducing the requirement for computer memory. Kolmogorov's 'Mapping Neural Network Existence Theorem' (Hecht-Nielsen, 1990) has demonstrated that ANN is capable of implementing a mapping function to any desired degree of accuracy, and can thus enhance the decision support ability of GIS.

4. RELATIVE STRENGTH OF EFFECT

As already mentioned, ANN learns how to handle the contributory factors by use of training data sets. From these a 'Relative Strength of Effect' (RSE) can be computed along with a 'Global Relative Strength of Effect' (GRSE) (Yang and Zhang, 1998). These provide measures of the contribution of every parameter to the system.

Having established a functional input/output relation, expressed by the set of weights determined by application of learning rules within elements of the network, interest turns to searching for a method of identifying what role these different factors play on the total system mechanism.

When the training process of the neural network ceases, the output O_k can then be written as

$$O_k = 1/\big(1+\exp(-e_k)\big) \tag{1}$$

where

$$\begin{aligned} e_k &= \sum_j O_j W_{jk} + \theta_k \\ O_j &= 1/\big(1+\exp(-e_j)\big) \\ e_j &= \sum_i O_i W_{ij} + \theta_j \end{aligned} \tag{2}$$

where θ is a threshold, so

$$O_k = \frac{1}{1+\exp\left[-\left(\sum_j W_{jk}\left(1\Big/\left(1+\exp\left[-\left(\sum_i W_{ij} O_i + \theta_j\right)\right]\right)\right) + \theta_k\right)\right]} \tag{3}$$

The activation function is sigmoidal, as shown in Fig. 3, and therefore can be differentiated. The change of O_k with change of O_i can thus be calculated as follows

$$\partial O_k / \partial O_i = \sum_j (\partial O_k / \partial O_j)(\partial O_j / \partial O_i) \tag{4}$$

When the number of layers is more than 3, the above equation can be rewritten as

$$\begin{aligned} \partial O_k / \partial O_i = \sum_{j_n} \sum_{j_{n-1}} \cdots \sum_{j_1} & \big(\partial O_k / \partial O_{j_n}\big)\big(\partial O_{j_n} / \partial O_{j_{n-1}}\big)\big(\partial O_{j_{n-1}} / \partial O_{j_{n-2}}\big) \\ & \times \big(\partial O_{j_{n-2}} / \partial O_{j_{n-3}}\big) \ldots \big(\partial O_{j_2} / \partial O_{j_1}\big)\big(\partial O_{j_1} / \partial O_i\big) \end{aligned} \tag{5}$$

where, $O_{j_n}, O_{j_{n-1}}, O_{j_{n-2}}, \ldots, O_{j_1}$ denote the hidden units in the n, $n-1$, $n-2, \ldots, 1$ hidden layer.

If the sigmoidal function is expressed as

$$f(x) = 1/\big(1+\exp(-x)\big) \tag{6}$$

then, by differentiating, the sigmoidal function becomes

$$\partial f/\partial x = f'(x) = f(x)f(x)\exp(-x) \tag{7}$$

Noting that $x = e_k = \sum_j W_{jk}O_j + |E_k|$, substitution of x in Eq. (7) by e_k, where k belongs to one of the hidden layers or output layer, enables the differentiation at every unit in the layer (hidden or output) to be written as

$$\partial f/\partial e_k = \exp(-e_k)/\left(1+\exp(-e_k)\right)^2 \tag{8}$$

If O_j is the input value of unit k received from the unit j, then

$$\begin{aligned}\partial f/\partial O_j &= W_{jk}(\partial f/\partial e_k)\\ &= W_{jk}\exp(-e_k)/\left(1+\exp(-e_k)\right)^2\end{aligned} \tag{9}$$

Now let $G(e_k) = \exp(-e_k)/\left(1+\exp(-e_k)\right)^2$; then

$$\partial f/\partial O_j = W_{jk}G(e_k) \tag{10}$$

so

$$\begin{aligned}\partial O_k/\partial O_j &= W_{jk}G(e_k)\\ \partial O_{j_n}/\partial O_{j_{n-1}} &= W_{j_{n-1}j_n}G(e_{j_n})\end{aligned} \tag{11}$$

If Eq. (11) is substituted into Eq. (5), the following is obtained

$$\begin{aligned}\partial O_k/\partial O_i = &\sum_{j_n}\sum_{j_{n-1}}\cdots\sum_{j_1} W_{j_nk}G(e_k)W_{j_{n-1}j_n}G(e_{j_n})W_{j_{n-2}j_{n-1}}G(e_{j_{n-1}})\\ &\times W_{j_{n-3}j_{n-2}}G(e_{j_{n-2}})\ldots W_{ij_1}G(e_{j_1})\end{aligned} \tag{12}$$

No matter which function is approximated by the neural network, all the terms on the right-hand side of Eq. (12) will exist (Hecht-Nielsen, 1990). This process can be solved in a manner somewhat like the differentiation of a function. Here concern is with the influence of input on output. Considering Eq. (12), a new parameter, relative strength of effect, RSE_{ki}, can be defined as the influence of input unit i on output unit k. This RSE value is analogous to a component of the fully coupled interaction matrix of a rock engineering system (RES), except that the RSE value changes according to variations in the input. This is an important additional feature of the ANN system characterization.

Definition 1: For a given sample set $\boldsymbol{S} = s_1, s_2, s_3, \ldots, s_j, \ldots s_r$, where, $s_j = \{\boldsymbol{X}, \boldsymbol{Y}\}$, $\boldsymbol{X} = \{x_1, x_2, x_3, \ldots, x_p\}$, $\boldsymbol{Y} = \{y_1, y_2, y_3, \ldots, y_q\}$, if there is a neural network trained by a BP algorithm with this set of samples, the RSE_{ki} will exist as

$$\begin{aligned}\mathrm{RSE}_{ki} = C&\sum_{j_n}\sum_{j_{n-1}}\cdots\sum_{j_1} W_{j_nk}G(e_k)W_{j_{n-1}j_n}G(e_{j_n})W_{j_n-2j_{n-1}}G(e_{j_{n-1}})\\ &\times W_{j_{n-3}j_{n-2}}G(e_{j_{n-2}})\ldots W_{ij_1}G(e_{j_1})\end{aligned} \tag{13}$$

The function G denotes the differentiation of the activation function, and C is a normalized constant which regulates the maximum absolute value of RSE_{ki} so that it ranges between -1 and $+1$. The output would have no demonstrable relation to the input if $\mathrm{RSE}_{ki} = 0$. RSE_{ki} is a dynamic parameter which changes as the input factors vary. It should be noted that the magnitude of RSE is controlled by the corresponding output unit, i.e. all the RSE values for each input unit with respect to the corresponding output unit are scaled by the same coefficient. The larger the absolute value of RSE_{ki}, the greater the effect which the corresponding input unit will have on the output unit. The sign of RSE_{ki} indicates the nature of its influence; thus a positive action would apply to the output when $\mathrm{RSE}_{ki} > 0$, whereas a negative action would apply when $\mathrm{RSE}_{ki} < 0$. A positive action means that the output increases with each increment of the corresponding input, and decreases with reduction of the corresponding input.

The meaning of RSE_{ki} reflects the influence that the input has on the output, as opposed to the differentiation of the mapping function itself. The RSE properties draw attention to those factors which dominate the state of the output (approaching the value 1), and to those factors which have little influence (having an RSE near to zero).

The items in the right side of Eq. (13) can be considered in two groups: the first is a group of weights W and the second is a group related to the differentiation of the sigmoidal function: the activation function (see Fig. 3). The weights of a neural network are fixed at the point where the process for learning has been completed, whereas the values of the second group will vary with the input.

Once a neural network has been trained with site observations, the influence of the input on the output needs to be determined. Because the knowledge provided by the site observations is contained within the weights of the neural network, the consequence of the input on the output can be determined using these weights. The differentiation of the simulated function can be regarded as being constant within the area of concern on site ('piecewise linear') thus obtaining a 'global' relative strength of effect (GRSE); within the domain this will not vary with change of input position.

The trained neural network can then be considered as a linear network, and its activation function will be linear, facilitating calculation of the values for RSE_{ki}.

Suppose the activation function is

$$F(x) = x \tag{14}$$

then its differential can be obtained as

$$G(x) = 1 \tag{15}$$

and the RSE can be written as

$$\mathrm{RSE}_{ki} = C \sum_{j_n} \sum_{j_{n-1}} \cdots \sum_{j_1} W_{j_n k} W_{j_{n-1} j_n} W_{j_{n-2} j_{n-1}} W_{j_{n-3} j_{n-2}} \cdots W_{i j_1} \tag{16}$$

The RSE in Eq. (16) represents the relative importance of every input unit on one output unit in the neural network in the 'global' sense, similar to the linear components of an interaction matrix in RES.

The global relative strength of effect (GRSE) can then be defined:

Definition 2: For a given sample set $S = \{\boldsymbol{s}_1, \boldsymbol{s}_2, \boldsymbol{s}_3, \ldots, \boldsymbol{s}_j, \ldots \boldsymbol{s}_r\}$, where $\boldsymbol{s}_j = \{\boldsymbol{X}, \boldsymbol{Y}\}$, $\boldsymbol{X} = \{x_1, x_2, x_3, \ldots, x_p\}$, $\boldsymbol{Y} = \{y_1, y_2, y_3, \ldots, y_q\}$, if there is a neural network trained by a

BP algorithm with this set of samples, the GRSE_{ki} will exist as

$$\mathrm{GRSE}_{ki} = C \sum_{j_n} \sum_{j_{n-1}} \cdots \sum_{j_1} W_{j_n k} W_{j_{n-1} j_n} W_{j_{n-2} j_{n-1}} W_{j_{n-3} j_{n-2}} \cdots W_{i j_1} \tag{17}$$

where C is a normalized constant which regulates the maximum absolute value of GRSE_{ki} as 1.

The GRSE_{ki} reveals the general consequence of every input unit on a chosen output unit. This yields a parameter useful for measuring the overall importance of input units on output units rather than revealing a numerical value for the influence at a specific location. The GRSE_{ki} is a more general parameter than the RSE_{ki}, the latter being location-specific.

The value of GRSE_{ki} enables assessment of how much influence the input unit will have on an output unit: the larger the absolute values of the weights, the more the effect that the input unit will have on the output. The GRSE is, in effect, a special case of RSE wherein all activation functions are linear. Thus the GRSE relates only to the weights themselves, and has no relation to the activation function. The GRSE reflects the relative dominance of each input on the output within an interval (rather than at a point, as does RSE). GRSE may therefore be regarded as a 'global' parameter; it is stable within its discerning interval.

The RSE reflects the dynamic role of an input parameter compared with other inputs, hence its value is determined by the specified values of the current inputs. The GRSE reveals the 'global' trend, providing an overall measure for the system indicating the prevailing dominance.

According to Eq. (13), the values of RSE_{ki} can be calculated using the following steps:

(1) Introduce all the values to the input neurons.

(2) Calculate all values of e_j (the values received by the hidden layer) in the hidden neurons and similarly e_k (the values received by the output layer) in the output neurons. The calculation progresses forwards, just as for the process of prediction, where e_j represents $e_{j_n}, e_{j_{n-1}}, e_{j_{n-2}}, \ldots, e_{j_1}$.

(3) Calculate the values of the G function (with differentiation of the activation function) in the output neurons and hidden neurons as

$$\begin{aligned} G(e_k) &= \exp(-e_k)/(1+\exp(-e_k))^2 \\ G(e_j) &= \exp(-e_j)/(1+\exp(-e_j))^2 \end{aligned} \tag{18}$$

(4) Assign the RS (intermediate) values of every output neuron as

$$\mathrm{RS}(e_k) = G(e_k) \tag{19}$$

where the RS is the value of RSE when it has not been scaled.

(5) Calculate the RS values of neurons in the previous layer as follows

$$\mathrm{RS}(e_{j_n}) = G(e_{j_n}) W_{j_n k} \mathrm{RS}(e_k) \tag{20}$$

(6) Calculate the RS values of neurons in other hidden layers

$$\mathrm{RS}(e_{j_{n-1}}) = G(e_{j_{n-1}}) \sum_{j_n} W_{j_{n-1} j_n} \mathrm{RS}(e_{j_n}) \tag{21}$$

(7) Repeat the calculation for the first hidden layer.

(8) Calculate the RS_{ki} value as

$$\mathrm{RS}_{k_i} = \sum_{j_1} W_{i j_1} \mathrm{RS}(e_{j_1}) \tag{22}$$

(9) Establish the number of output units as n, $\mathrm{RS}_{kx} = \max\{|\mathrm{RS}_{k1}|, |\mathrm{RS}_{k2}|, \ldots, |\mathrm{RS}_{kn}|\}$, and then scale the value of RS_{ki} to

$$\mathrm{RSE}_{ki} = \mathrm{RS}_{ki}/\mathrm{RS}_{kx} \tag{23}$$

Following this procedure, the RSE_{ki} value can be calculated, its value indicating the relative influence of variable i on variable k. Thus a comparison may be carried out to find the key input variables from all the input variables available on the basis of a consideration of their RSE_{ki}.

Similarly the values of GRSE_{ki} can be calculated using Eqs. (18)–(23), for $G(x) = 1$.

The importance of objectively analysing the factors which influence the behaviour of a system makes the linking of ANN with GIS of considerable interest with regard to decision support. Factor Analysis based on statistical reasoning is an established technique for measuring the relative importance of the various factors on the behaviour of the system, but it is difficult to use this to reveal the dynamic influence of every factor. ANN thus provides an additional tool, with RSE and GRSE providing new measures with the capability of revealing insights into the degree to which the available information influences the behaviour of the system, and thus to facilitate a sensitivity analysis as the basis for decision support.

5. INTEGRATION OF ANN WITH GIS

To be effective when handling spatial information, a tool like ANN requires a compatible data storage, retrieval, analysis and display environment. ANN can be regarded as a mapping function between inputs and outputs, and hence it is natural to apply ANN as a function of the overlay operation within a GIS. This can be offered by interfacing ANN with GIS, as shown in Fig. 4. The ANN model is linked to the analysis component.

The first step in applying ANN is to train it with an existing data set held within the spatial database of the GIS. ANN then operates as an overlay function. The data required as input to ANN is thereby fed from the GIS operation, following which ANN maps out an intuitive solution, generating outputs as attributes and images as well as corresponding values for RSE and GRSE (Yang and Rosenbaum, 1999) (Fig. 5). Unlike other multivariate methods such as factor analysis, RSE supposes that the roles of the different factors are likely to change both spatially and temporally. Combined with GRSE, RSE analysis is able to reveal the dynamic role of such factors, so determining the weights of influence for the different variables.

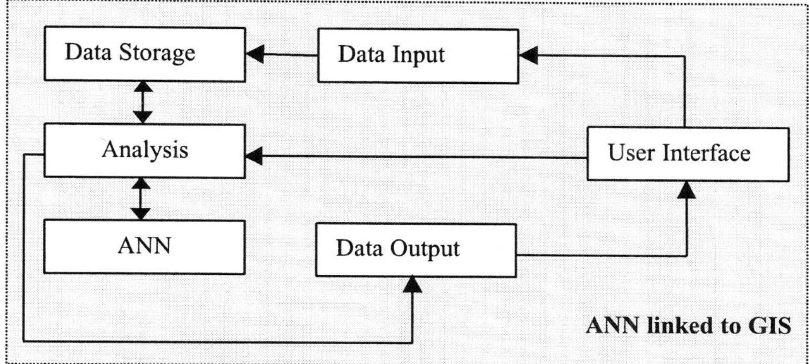

Fig. 4. A mechanism for linking ANN to a GIS.

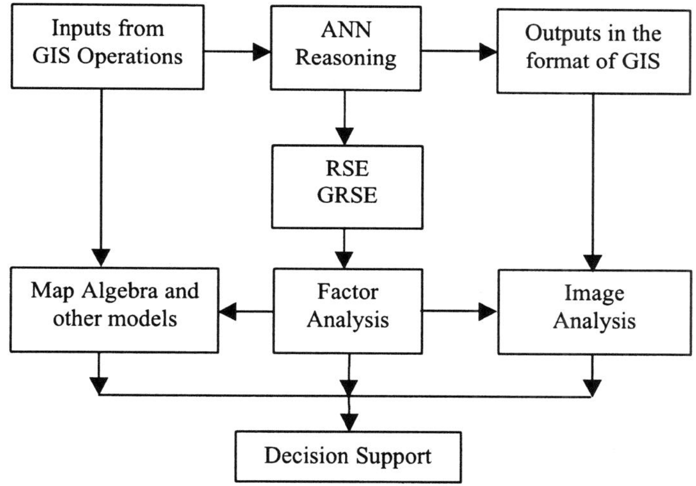

Fig. 5. The mechanisms whereby ANN can supplement GIS analysis.

The link between ANN and the GIS is effected using the newly developed neural network tool for relative strength of effect (NRSE), the structure of which is shown in Fig. 6. NRSE establishes the links between the factors within a data set provided for training. As a neural network tool, NRSE provides a flexible method for data input utilising ASCII files, Access databases or GIS (Idrisi) images. NRSE then establishes the neural network on the basis of manual selection of parameters from those which are available.

NRSE, in addition to enabling the standard BP method of training, can also permit the application of a dynamic BP method whereby the coefficients of learning and momentum are adapted dynamically according to the change of errors.

Four types of reasoning are offered by NRSE: single observation, multiple observations, RSE and GIS image. The single and multiple observation reasoning approaches

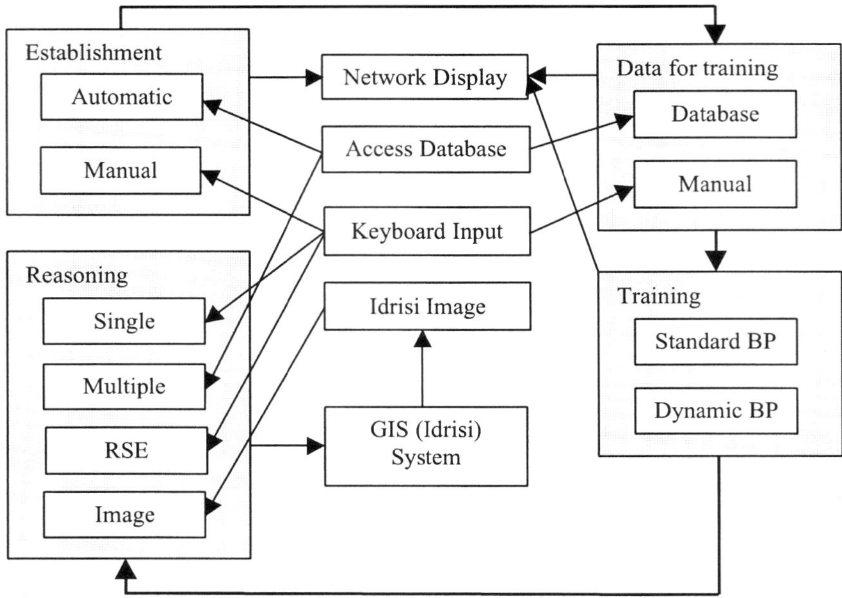

Fig. 6. The structure of NRSE.

are conducted using an Access table or SQL query. The RSE reasoning calculates corresponding values of RSE from the reasoning results. The GIS image reasoning executes a similar operation to single observation reasoning, but to each pixel of the raster image in turn.

Once reasoning has been completed, the results can be displayed on the screen or saved as an ASCII file. The training process can be displayed on the screen as a dynamic curve, together with the network structure and its associated parameters

6. APPLICATION OF NRSE TO ENVIRONMENTAL SEDIMENTOLOGY

NRSE has been applied to the prediction of sedimentology characteristics within the harbour area of Gothenburg, Sweden. Here sediment accumulation is the result of a number of environmental processes which are difficult to identify, yet their effects can be measured (Stevens, 1999). Such harbours, where the tidal influence is low, can create effective traps for sediment deposition, yet turbulence caused by river currents or shipping manoeuvres can cause renewed erosion, possibly re-introducing older, formerly buried and polluted, sediment into the harbour environment. Such behaviour is likely to be influenced by the seabed profile, water depth, distance from possible sources of pollution, the local biochemical environment, dredging, and climate, as well as a number of anthropogenic factors. These factors may act simultaneously and constitute a complicated interaction network, as shown in Fig. 7.

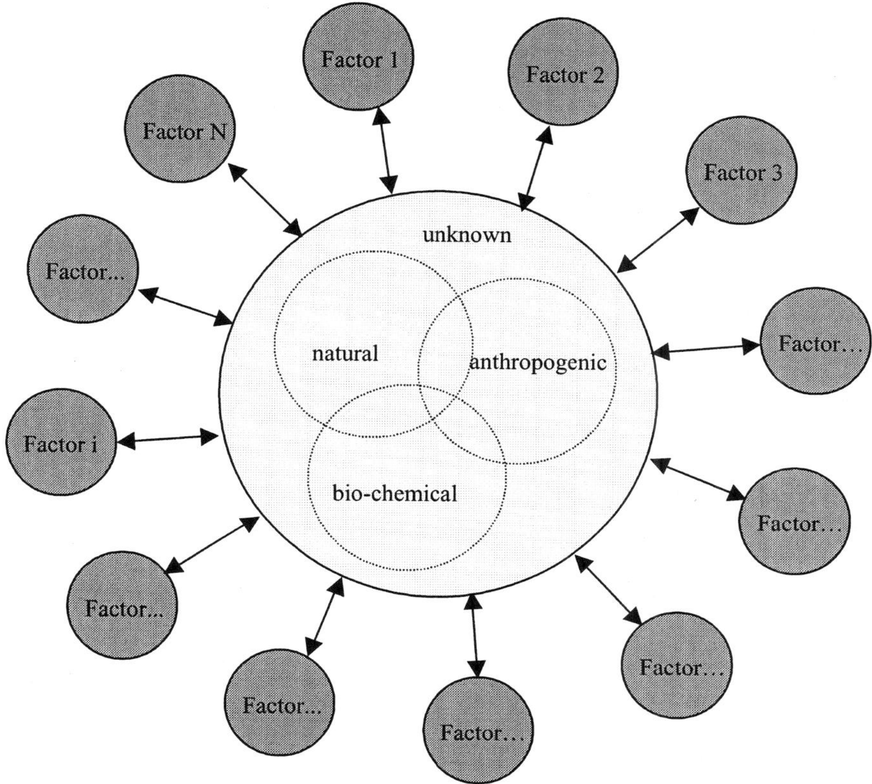

Fig. 7. The interaction network of factors influencing sedimentation.

Sediment accumulation may be thought of as the net effect of geoenvironmental processes in action, but these may be so complicated that representative measures may be difficult to attain, even more so their changes with time (Stevens, 1999). Factors may be related by consideration of natural (physical, chemical or biological) or anthropogenic mechanisms. Little is currently known about these interactions or their boundary constraints. However, it should be possible to measure the consequences of their interactions from the gathered data using the NRSE approach. NRSE has been employed to examine the sedimentology aspects of such a system in Gothenburg, Sweden, comparing natural and anthropogenic sediment sources, transport pathways, and geochemical changes.

The harbour investigation resulted in 139 sample stations, information for which was compiled using Access (Burton et al., 1999), which could subsequently be interfaced with NRSE. As an example, the influences of water depth and distance from the river mouth, bank and shipping lanes could be investigated on the distribution of sediment grain size on the harbour floor. The resulting network would have 4 input nodes and, say, 3 output nodes (one for each sediment grade: sand, silt and clay). The nature of

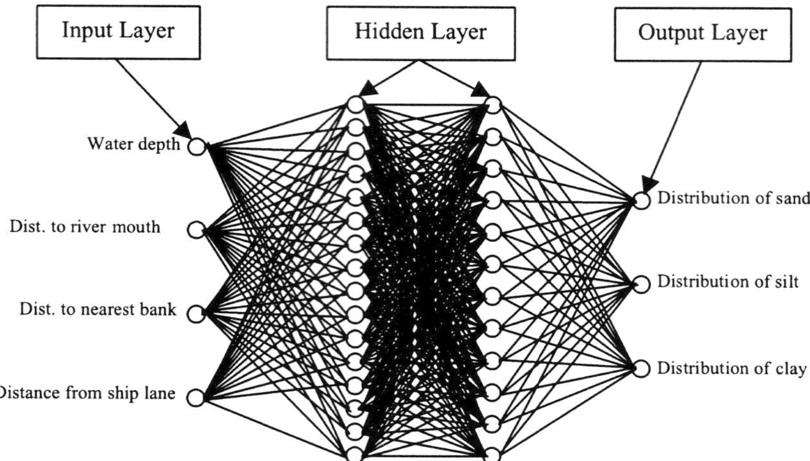

Fig. 8. The structure of the network established for Gothenburg harbour concerning sediment in the depth range 0–2 cm.

this system suggested adoption of two hidden layers, the first having 16 nodes and the second having 12 nodes (Fig. 8).

While NRSE was being trained in this example, the error reduced to less than 0.0005 after 1.2M iterations. Having trained the neural network, maps showing the predicted distribution of sand, silt or clay could be generated by reasoning. The relationship between the four input parameters and the output distributions are, by their nature, complex. No obvious relation between these parameters is apparent from a statistical standpoint, yet ANN has been able to establish a degree of interdependence.

Some errors remain (0.0005), yet ANN has effected a prediction of sediment characteristics which bears comparison with known field conditions on the basis of limited site investigation data and an incomplete set of input parameters. Of course, many factors would be expected to control the distribution of sediment grain size, but since it is impossible to include, or indeed measure, all of them, the neural network approach has provided a pragmatic basis for prediction, amenable to updating as new information becomes available.

The RSE and GRSE from the ANN model assist analysis of the model behaviour. The trained neural network yielded the GRSE values shown in Table 1, graphically portrayed in Fig. 9. This demonstrates that the distribution of silt is controlled most by water depth, and its negative value means that the deeper the water depth, the less likely will silt be encountered at that location. In way of contrast, the distribution of sand is seen to be controlled by the distance from the banks; again, its negative value denotes that a location near to the bank would favour sand deposition.

The value for GRSE reflects the dominant trends exhibited across the harbour environment, which may be illustrated by examining the values obtained from the consideration of the sediment conditions at the 3 sample sites listed in Table 2. The corresponding values for GRSE are shown in Table 3, which may be compared with the RSE values plotted in Fig. 10.

TABLE 1

GRSE from the neural network analysis of sediments within Gothenburg harbour (0–2 cm depth)

Outputs	Water depth	Distance from the mouth	Distance from the banks	Distance from ship lanes
Distribution of sand	+0.54	−0.08	−1.00	−0.45
Distribution of silt	−1.00	+0.52	+0.78	+0.04
Distribution of clay	+0.91	−0.96	+0.67	+1.00

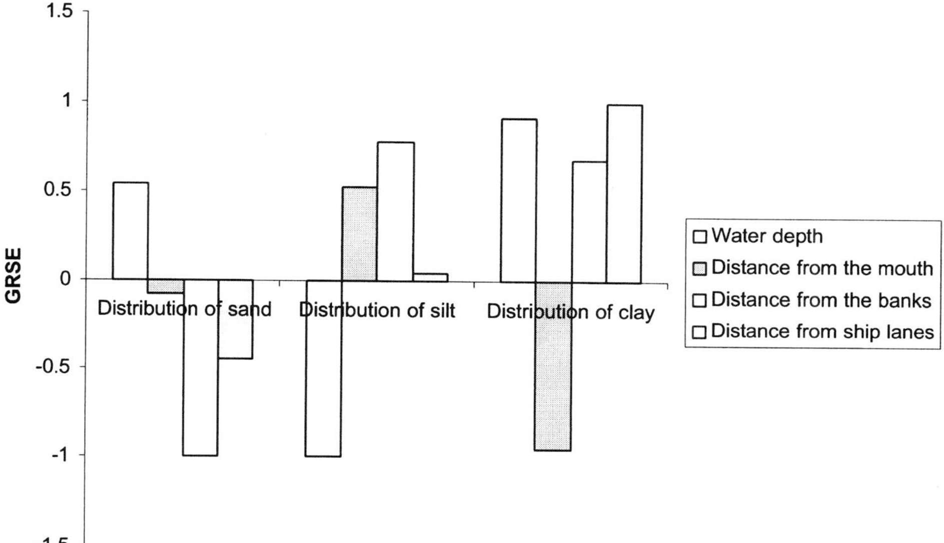

Fig. 9. The GRSE values for sediments in the depth range 0–2 cm for Gothenburg harbour.

Comparing the RSE values shown in Fig. 10, it is clear that they change from one place to another. For example, the RSE value for the influence of distance from the shipping lanes on the distribution of sand changes dramatically from negative at Sample

TABLE 2

The sediment conditions at 3 sample sites

Sample site no.	Water depth (m)	Distance from shipping lanes (m)	Distance from the mouth (m)	Distance from nearest bank (m)
1	12.5	200.0	14361.0	300.0
2	10.6	223.6	18448.0	360.5
3	2.0	100.0	200.0	141.4

TABLE 3

The RSE values for the 3 Sample Sites listed in Table 2

Sample site no.	Outputs	Water depth (m)	Distance from shipping lanes (m)	Distance from the mouth (m)	Distance from nearest bank (m)
1	Distribution of sand	−0.37	−0.63	+0.98	+1.00
	Distribution of silt	+0.28	+0.07	+1.00	−0.30
	Distribution of clay	+0.02	+0.04	−1.00	+0.33
2	Distribution of sand	+0.27	−0.41	+0.15	+1.00
	Distribution of silt	+0.04	−0.24	+0.01	−1.00
	Distribution of clay	−0.02	−0.07	−0.003	+1.00
3	Distribution of sand	−1.00	+0.99	+0.86	−0.28
	Distribution of silt	−1.00	−0.99	−0.89	−0.05
	Distribution of clay	+0.87	−0.80	−1.00	+0.01

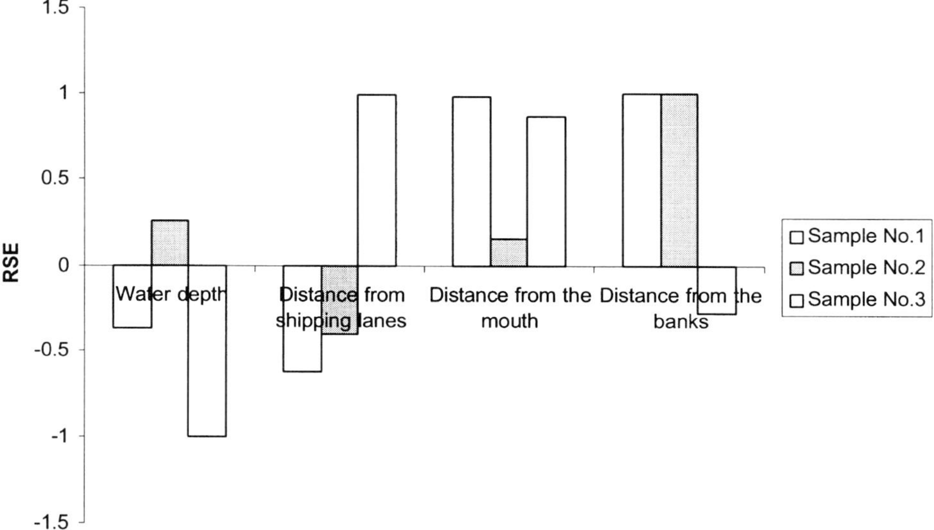

Fig. 10. The RSE values for the distribution of sand in Gothenburg harbour (equivalent distributions may be plotted for silt and for clay).

Site Nos.1 and 2 to positive at Sample Site No.3. This means that the distance from the shipping lane has a differing degree of influence on the distribution of sand dependant on the site location. ANN enables a series of such studies to be executed, GRSE and RSE being employed to reveal the nature of the interactions between the various parameters.

7. CONCLUSIONS

Both GIS and ANN provide powerful tools for investigating the problems encountered in geotechnical and geological engineering. GIS provides a practical basis for undertaking spatial analysis and modelling provided guidance is available from explicit field knowledge. ANN can supplement the available knowledge by establishing the relative importance of the factors, thus the combination and integration of ANN with GIS enhances the development and application of spatial modelling.

Using the sediments in Gothenburg harbour as a case history, ANN has been successfully linked to GIS using the NRSE algorithm, yielding a tool which can be used to assist decision support for engineering problems and harbour management. GIS provides an established platform for spatial data analysis, to which ANN can offer an effective tool, facilitating decision support. An example of its utility has been demonstrated based on sediment prediction within the harbour where the natural system is complex, making it otherwise difficult to establish a deterministic model, yet quantitative characterisation is desired.

The GRSE and RSE computed during the training of the neural network yield measures which are able to reveal the presence of trends within the system. Furthermore, the RSE yields a measure of the dynamic change as attention moves from one part of the system to another.

ACKNOWLEDGEMENTS

The authors would like to thank the EU for sponsoring the H-SENSE research project under the Transport RTD Programme of the 4th Framework Programme (Contract No. WA-97-SC.2050).

REFERENCES

Bailey, T.C. and Gatrell, A.C., 1995. *Interactive Spatial Data Analysis*. Longman, New York, NY, 413 pp.

Berry, J.K., 1993. Catographic modeling: the analytical capabilities of GIS. In: Goodchild, M., Parks, B. and Steyaert, L. (Eds.), *Environmental Modeling with GIS*. Oxford University Press, Oxford, pp. 58–74.

Burrough, P A., 1999. GIS and Geostatistics: essential partners for spatial analysis. In: Shi, W., Goodchild, M.F. and Fisher, P.F. (Eds.), *Proceedings of the International Symposium on Spatial Data Quality (ISSDQ '99)*, Hong Kong, 18–20 July, pp. 10–20.

Burrough, P.A. and McDonnell, R.A., 1998. *Principles of Geographical Information Systems*. Clarendon, Oxford, 346 pp.

Burton, C.L., Rosenbaum, M.S., Stevens, R.L. and Book, S., 1999. Creating a harbour sediment database. In: De Schutter, G. (Ed.), *Proceedings of the 4th International Congress for Characterisation and Treatment of Sediments (CATS IV)*, Antwerp, pp. 575–583.

Coppock, J.T. and Rhind, D.W., 1991. The history of GIS. In: D.J. Maguire, M.F. Goodchild and D.W. Rhind (Eds.), *Geographical Information Systems*. Longman Scientific and Technical, New York, NY, 1: pp. 21–43.

Cowen, D., 1988. GIS versus CAD versus DBMS: What are the differences? Photogrammetric Engineering and Remote Sensing, 54(2): 1551–1555.

Davis, B.E., 1996. GIS: a visual approach. On Word Press, Santa Fe, NM.

DeMers, M.N., 1999. *Fundamentals of Geographic Information Systems, 2nd edition*. Wiley, New York, NY, 512 pp.

Demyanov, V., Kanevski, M., Chernov, S., Savelieva, E. and Timonin, V., 1998. Neural network residual kriging application for climatic data. *J. Geogr. Inf. Decision Anal.*, (2)2: 234–252.

Eastman, J.R., 1993. *IDRISI: A Grid Based Geographic Analysis System, Version 4.1*. Graduate School of Geography, Clark University, Worcester, MA.

Eastman, J.R., 1999. Id*risi32: Guide to GIS and Image Processing*. Clark University, Worchester, MA, 2 volumes.

Ellis, G.W., Yao, C., Zhao, C. and Penumadu, D., 1995. Stress-strain modelling of sands using artificial neural networks. *J. Geotech. Eng.*, 121(5): 429–435.

Ghaboussi, J., Garrett, Jr., J.H. and Wu, X., 1990. Material modelling with neural networks. *Proceedings of the International Conference on Numerical Methods in Engineering: Theory and Applications*, Swansea, pp. 701–717.

Griffith, D.A. and Amrhein, C.G., 1991. *Statistical Analysis for Geographers*. Prentice Hall, Englewood Cliffs, NJ.

Hecht-Nielsen, R., 1990. *Neurocomputing*. Addison-Wesley, Reading, MA.

Hepple, L.W., 1981. Spatial and temporal analysis: time series analysis. In: Wrigley, N. and Bennett, R.J. (Eds.), *Quantitative Geography: A British View*. Routledge and Kegan Paul, pp. 92–96.

Lee, C. and Sterling, R., 1992. Identifying probable failure modes for underground openings using a neural network. *Int. J. Rock Mech. Mining Sci.*, 29(1): 46–67.

Johnson, R.J., 1978. *Multivariate Statistical Analysis in Geography: A Primer on the General Linear Model*. Longman.

Malczewski, J., 1999. *GIS and Multicriteria Decision Analysis*. Wiley, New York, NY.

Marble, D.F., Calkins, H.W. and Peuquet, D.J., 1984. *Basic Readings in Geographic Information Systems*. SPAD Systems Ltd., Williamsville, NY.

Rumelhart, D.E. and McClelland, J.L., 1986. *Parallel Distributed Processing: Explorations in the Microstructure of Cognition, Vol. 1*. MIT Press, Cambridge, MA, pp. 318–362.

Stein, A., van Dooremolen, W., Bouma, J. and Begt, A.K., 1988. Co-kriging point data on moisture deficit. *Soil Sci. Soc. Am. J.*, 52: 1418–1423.

Stevens, R.L., 1999. H-SENSE: Sediment perspectives upon harbour sustainability. In: De Schutter, G. (Ed.), *Proceedings of the 4th International Congress for Characterisation and Treatment of Sediments (CATS IV)*, Antwerp, pp. 617–624.

Steyaert, L.T. and Goodchild, M.F., 1994. Integrating geographic information systems and environmental simulation models: a status review. In: Michener, W.K. Brunt, J.W. and Stafford, S.G. (Eds.), *Environmental Information Management and Analysis*. Taylor and Francis, pp. 333–355.

Yang, Y. and Rosenbaum, M.S., 1999. Spatial data analysis with ANN: geoenvironmental modelling of harbour siltation. In: Shi, W., Goodchild, M.F. and Fisher, P.F. (Eds.), *Proceedings of the International Symposium on Spatial Data Quality (ISSDQ '99)*, Hong Kong, 18–20 July, pp. 534–541.

Yang, Y. and Zhang, Q., 1997. A hierarchical analysis for rock engineering using artificial neural networks. *Rock Mech. Rock Eng.*, 30(4): 207–222.

Yang, Y. and Zhang, Q., 1998. A new method for the application of artificial neural networks to rock engineering system. *Int. J. Rock Mech. Mining Sci.*, 35(6): 727–745.

Zhang, Z. and Griffith, D.A., 1997. Developing user-friendly spatial statistical analysis models for GIS: an example using ArcView. *Comput., Environ. Urban Systems*, 21(1): 5–29.

Zhang, Q., Song, J. and Nie, X., 1991. The application of neural network to rock mechanics and rock engineering. *Int. J. Rock Mech. Mining Sci.*, 28(6): 535–540.

Developments in Petroleum Science, 51
Editors: M. Nikravesh, F. Aminzadeh and L.A. Zadeh

Chapter 29

INTELLIGENT COMPUTING TECHNIQUES FOR COMPLEX SYSTEMS

MASOUD NIKRAVESH [1]

Berkeley Initiative in Soft Computing (BISC), Computer Science Division — Department of EECS, University of California, Berkeley, CA 94720, USA

ABSTRACT

One of the main objectives of this paper is to develop intelligent computing techniques for complex systems such as evaluating the mass and volume of contaminants in the heterogeneous soils and rocks using sparse borehole data. It will also focus on development of a robust and efficient technique to characterize and construct a static three-dimensional distribution of contaminant with the associated uncertainty factor (error bars at each point) and various possible realizations using minimum information. In addition, the technique will be used to optimize both the location and the orientation of each new well to be drilled based on data gathered from previous wells. This innovative research will not only reduce costs by focusing characterization on areas of greatest uncertainty but will also serve to help correlate and combine various types of data.

1. INTRODUCTION

At present, there are numerous contaminated sites around the world containing some unknown mass and volume of contaminant. In USA, There are numerous DOE contaminated sites (Hanford, INEEL, Savannah River, Oak Ridge, and others) containing contaminants that are threats to human health and the environment. In order to determine the extent of contamination, including the size, shape, and concentration distribution of the contaminated zone, a series of wells is usually drilled. The well data are then analyzed by geostatistical methods (De Marsily, 1986; Isaak and Srivastava, 1989; Cressie, 1993; Davis, 1986; Deutsch and Journal, 1982; Pannatier, 1996) in order to determine the spatial distribution of the contaminants. To use these techniques effectively, a large number of wells are usually drilled. Current methods of obtaining data are laborious and expensive. Therefore, minimizing sampling plans to reduce the number of wells to be drilled and samples to be taken are of great economic benefit and will have direct potential impact on the remediation plans of several DOE contaminated sites such as Hanford, INEEL, Savannah River, Oak Ridge, and others.

In this study, we propose neuro-statistical techniques as an alternative approach. *Neuro-statistical techniques, unlike regression analysis techniques, do not require spec-*

[1] Tel.: +1 (510) 643-4522, Fax: +1 (510) 642-5775, E-mail: nikravesh@cs.berkeley.edu,
URL: http://www.cs.berkeley.edu/~nikraves/

ification of structural relationships between the input and output data. These properties give the technique the ability to interpolate between typical patterns of data and generalize their learning in order to extrapolate to a region beyond their training domains.

Suppose there is a contaminant plume of unknown mass and volume. The approximate surface area covering the plume is known, and we have to determine how many wells one needs to drill in order to obtain a reasonable estimate of the contaminant mass in the plume. In this paper, we will consider two- and three-dimensional fields. In this paper, we will investigate three case studies: (1) a synthetic two-dimensional computer-generated concentration field, (2) a synthetic data field, which was created as an analogue to the fracture aperture of a core taken from the Yucca Mountain, and the data from the Alameda County sites (Nikravesh et al., 1996b). The benefits of using the numerical/simulated data sets are two fold: first, the data will be generated from complex geological settings that are difficult to correctly image and thus provide a rigorous test of imaging and predictive methods; and second unlike with real data, the correct results are known. Further, these data sets provide a standard for comparison among different methods and techniques used to estimate the mass or volume and distribution of contaminants.

The following is step by step description of the methodology and implementation.

2. NEURO-STATISTICAL METHOD (NIKRAVESH AND AMINZADEH, 1997)

Using conventional statistical methods such as ordinary least-squares (LS), partial least-squares (PLS), and non-linear quadratic partial least-squares (QPLS), only linear and simple non-linear information can be extracted from data sets. However, if a prior information regarding the non-linear input–output mapping is available, these methods become more useful.

Simple mathematical models may become inaccurate as several assumptions are made to further simplify the models. On the other hand, complex models may become inaccurate when additional equations involving a more or less approximate description of phenomena are included in the model. In some cases, these models require a number of parameters, which are not physically measurable.

In contrary, neural network methods can be used to generate models from non-linear, complex, and multi-dimensional data, which are used in analyzing experimental, industrial, and field data sets. *Neural networks, unlike regression analysis techniques, do not require specification of structural relationships between the input and output data.* Neural networks have the ability to infer general rules and extract typical patterns from specific examples and recognize input–output mapping parameters from complex multi-dimensional field data. These properties give the neural networks the ability to interpolate between typical patterns of data and generalize their learning in order to extrapolate to a region beyond their training domains.

An application of neural networks for identification purposes requires a large number of data. Unlike statistical methods, conventional neural network models cannot deal with probability.

In this paper, we attempted to use the advantages of the neural network method in conjunction with statistical methods. The model uses neural network techniques, since the functional structure of the data is unknown. In addition, the model uses statistical techniques because the data and our requirements are imperfect.

Using this concept the conventional Levenberge–Marquardt algorithm has been modified (Appendix A). In this case, the final global error in the output at each sampling time is related to the network parameters and a modified version of the learning coefficient is defined. The following equations will briefly show the difference between conventional and modified techniques. In the conventional technique weights can be calculated by

$$\Delta W = \left(\underline{J}^{\mathrm{T}} J + \mu^2 \underline{I}\right)^{-1} \underline{J}^{\mathrm{T}} e \tag{1}$$

However, in the modified technique the weights are given by

$$\Delta W = \left(\underline{J}^{\mathrm{T}} \underline{\Lambda}^{\mathrm{T}} \underline{\Lambda} J + \Gamma^{\mathrm{T}} \Gamma\right)^{-1} \underline{J}^{\mathrm{T}} \underline{\Lambda}^{\mathrm{T}} \underline{\Lambda} e \tag{2}$$

$$\underline{\Lambda}^{\mathrm{T}} \underline{\Lambda} = \underline{\hat{V}}^{-1} \tag{3}$$

$$\hat{V}_{ij} = \frac{1}{2m+1} \sum_{k=-m}^{m} \hat{e}_{i+k} \hat{e}_{j+k} \tag{4}$$

$$\underline{\hat{V}} = \hat{\sigma}^2 \underline{I} \tag{5}$$

$$\hat{W} = W \pm k\hat{\sigma} \tag{6}$$

where e is error, k is gain, σ is variance, Γ is tuning parameter, and J is Jacobian matrix.

Fig. 1 shows the performance of the new network model (Nikravesh and Aminzadeh, 1997). Fig. 1a shows the predictive performance of the network model. Circles represent actual data, crosses represent the mean of the neural network predictions, squares represent the upper limit (max) of the network prediction and triangles represent the lower limit (min) of the network prediction. Fig. 1a shows that the network model has an excellent performance and also the actual value always lies between the upper and the lower limit predicted by neural network. Fig. 1b, c, and d are a magnification of Fig. 1a, b, and c, respectively. As one can see the trend in Fig. 1b, c, and d is the same as the trend in Fig. 1a.

Fig. 1e shows the distribution of the predicted values, because the output from the network is a distribution rather than a crisp value. Fig. 1e and f show that the actual value is bounded between the one standard deviation from the most probable value. Fig. 1f, g, and h show the comparison between actual data, the most probable prediction based on the neural network, upper limit, lower limit, and one standard deviation from the most probable prediction. Using this technique the upper and lower bounds are tightened.

Even though the Levenberge–Marquardt algorithm is faster and more robust than the conventional algorithm, it requires more memory. In order to overcome this disadvantage, we need to reduce the complexity of the neural network model and/or reduce a number of data points in each step of training. In the former case, we will use the alternative conditional expectation (ACE) technique (Breiman and Friedman, 1985), a non-parametric statistical technique, in order to reduce the network structure. This will be done by extracting the patterns, which exist in the data (Fig. 2a, b, d,

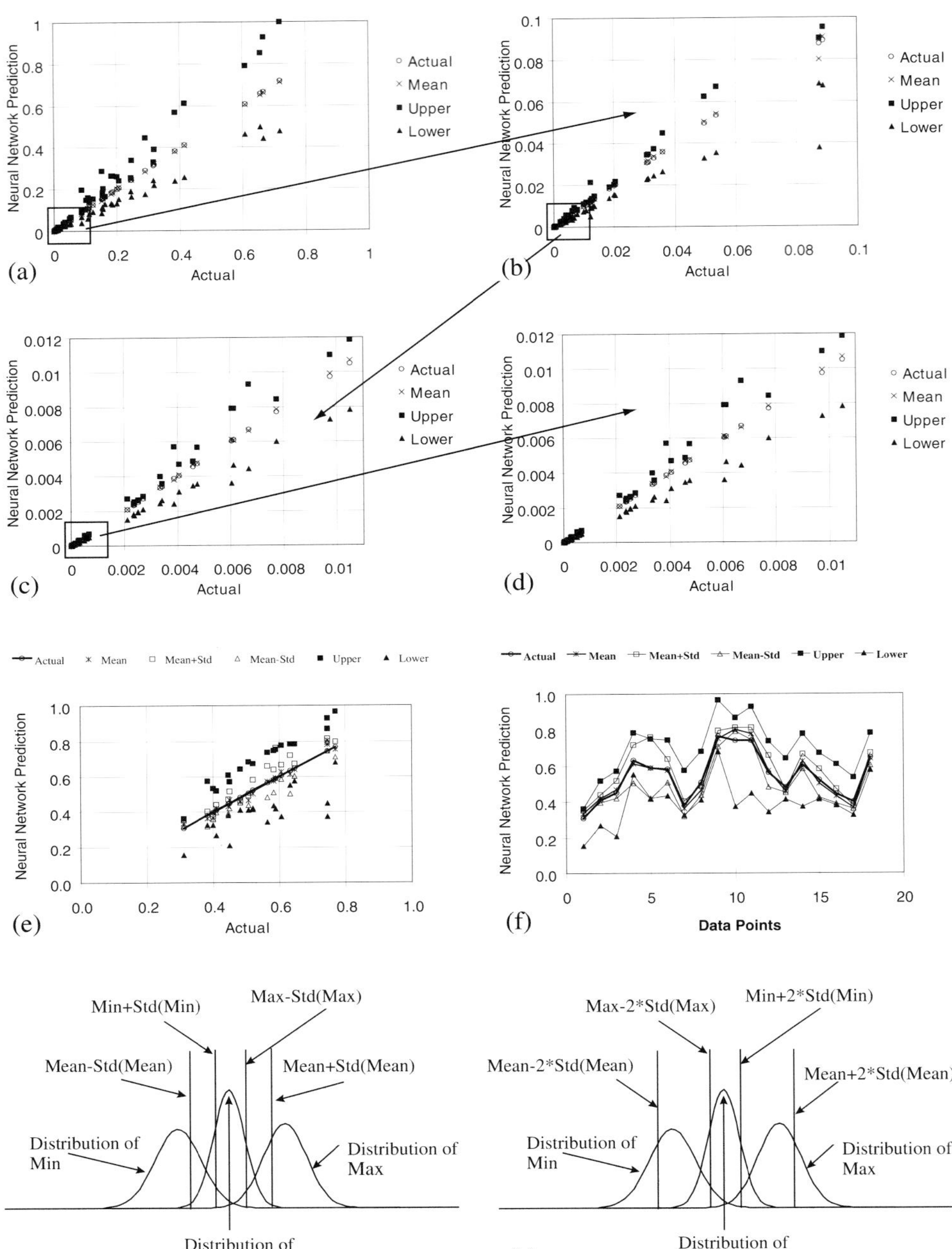

Fig. 1. Performance of the neuro-statistical model.

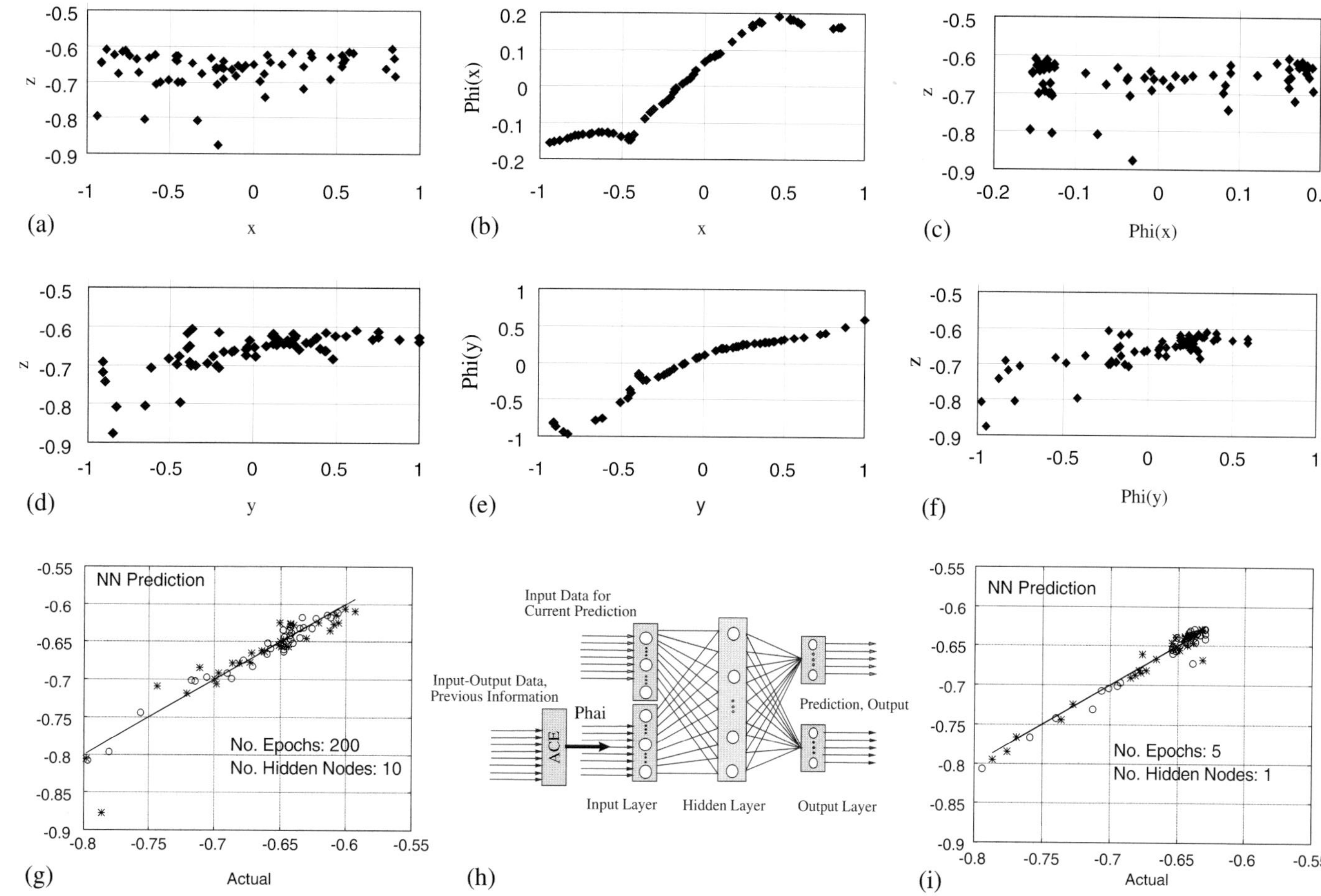

Fig. 2. ACE, neural network, and ACE–neural network models and performances. (a) Actual data. (b) Underlying pattern. (c) Transformed space. (d) Actual data. (e) Underlying pattern. (f) Transformed space. (g) Performance of conventional neural network. (h) Hybrid ACE model. (i) Typical performance of ACE–neural.

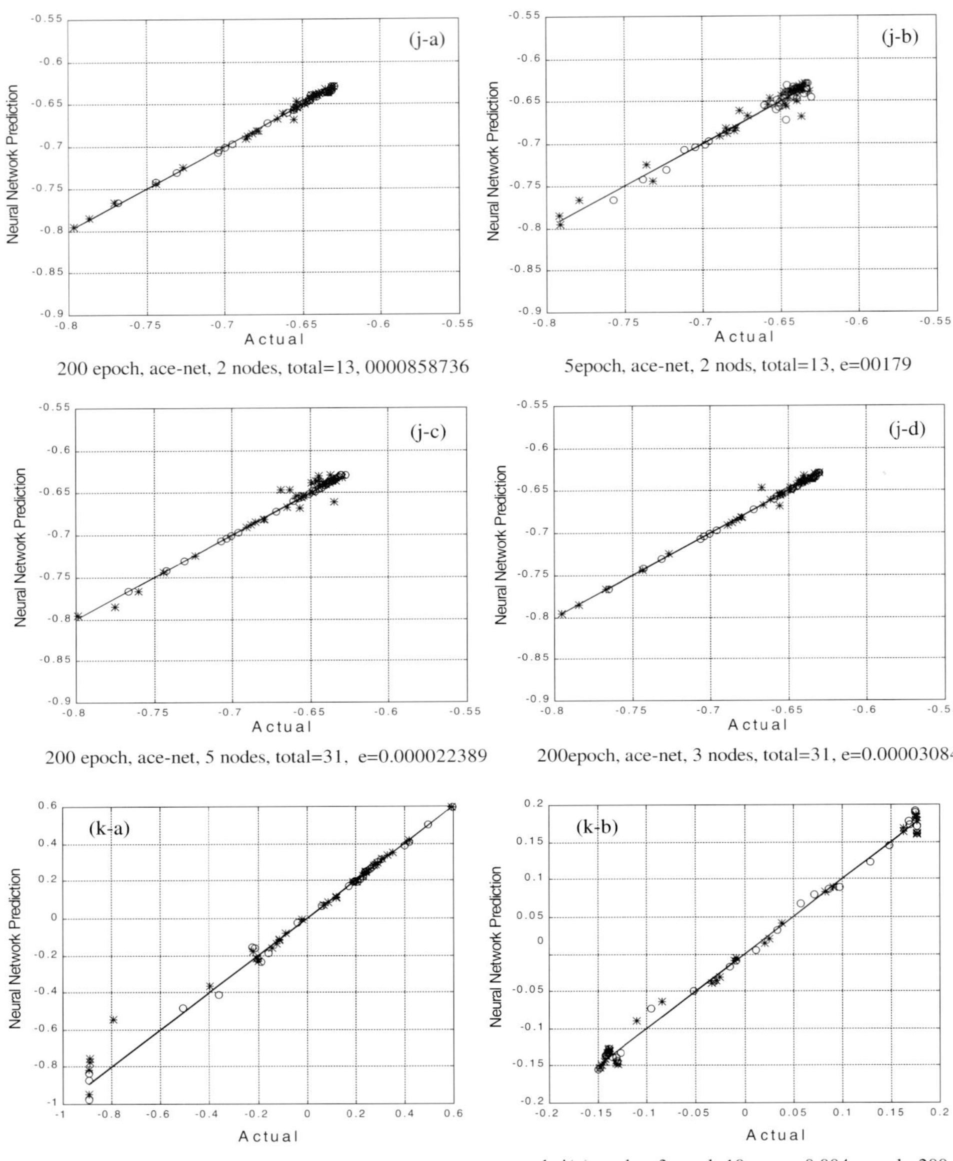

Fig. 2 (continued). (j) Performance of ACE–neural network. (k) Performance of neural network model for mapping x-phai(x) and y-phai(y) functions.

and e). In the next section, we will introduce this approach. Fig. 2k(a) and 2k(b) show the performance of the neural networks has been used to map x-phai(x) and y-phai(y) functions.

To reduce a number of data points used in each step of training, we will divide the data sets into several sub-data sets based on the pattern extracted (Fig. 2b and e) using the ACE technique. In this case, we will use the recursive technique (Azimi and Liou, 1992; Nikravesh et al., 1996a) for network training and then use the modified Levenberge–Marquardt algorithm. Therefore, the final global error in the output at each sampling time is related to the network parameters and a modified version of the learning coefficient is defined.

3. HYBRID NEURAL NETWORK–ALTERNATIVE CONDITIONAL EXPECTATION (HNACE/ACE NEURAL NETWORK) TECHNIQUE (NIKRAVESH AND AMINZADEH, 1997)

Recently, the application of the non-parametric statistical method such as the Alternative Conditional Expectation (ACE) scheme for modeling the complex multi-dimensional field data has greatly been increased. Statistical techniques such as ACE are considered to be appropriate to deal with the nature of uncertainty. In addition, the underlying patterns and structures recognized by ACE are more visible than neural network structure (Fig. 2b and e). The ACE method is a statistical technique which can be used to find an optimal transformation that maximizes the correlation between the transformed variables in a reduced and normally distributed space (Fig. 2c and f). Since the ACE technique transforms the variables into a scaled domain in an optimal fashion, it can be used for scaling the input–output data for a neural network structure (Fig. 2c and f). In addition, since this technique can find the correlation between the variables, it can be used to reduce the complexity of a network structure by eliminating the variables, which do not introduce any new information or knowledge into the network model (Fig. 2g, h and i). This can be done by examining the transformed variables (Fig. 2b, e and h).

4. APPLICATION OF A NEURO-STATISTICAL METHOD FOR SYNTHETIC DATA SETS

A Neuro-Statistical model was used to predict the distribution of data and the boundary of the contaminant as a function of x and y coordinates. In addition, we tried to optimize the location and orientation of each new well to be drilled based on data gathered from previous wells. The new well is drilled based on the information about the distribution of concentration at the surface. We explored a possibility of drilling vertical and slanted boreholes up to 30 Deg. from vertical regardless of orientation.

Fig. 3 shows the comparison between the conventional technique and a new technique for calculating total mass of contaminant. Data shown in Fig. 3a is generated using a fractal surface generator software (Russ, 1995) with a fractal dimension of 2.5. The fractal surface option was selected with the midpoint displacement algorithm (Peitgen and Saupe, 1988). Fig. 3b is generated based on simple statistical techniques. In this case, wells are drilled vertically and in the optimal locations. Fig. 3c shows the performance of the neuro-statistical model. In this case, the plume's boundary is predicted by the model at each step. Therefore, the location and the orientation of each new well to be drilled are optimized based on the data gathered from previous wells. As

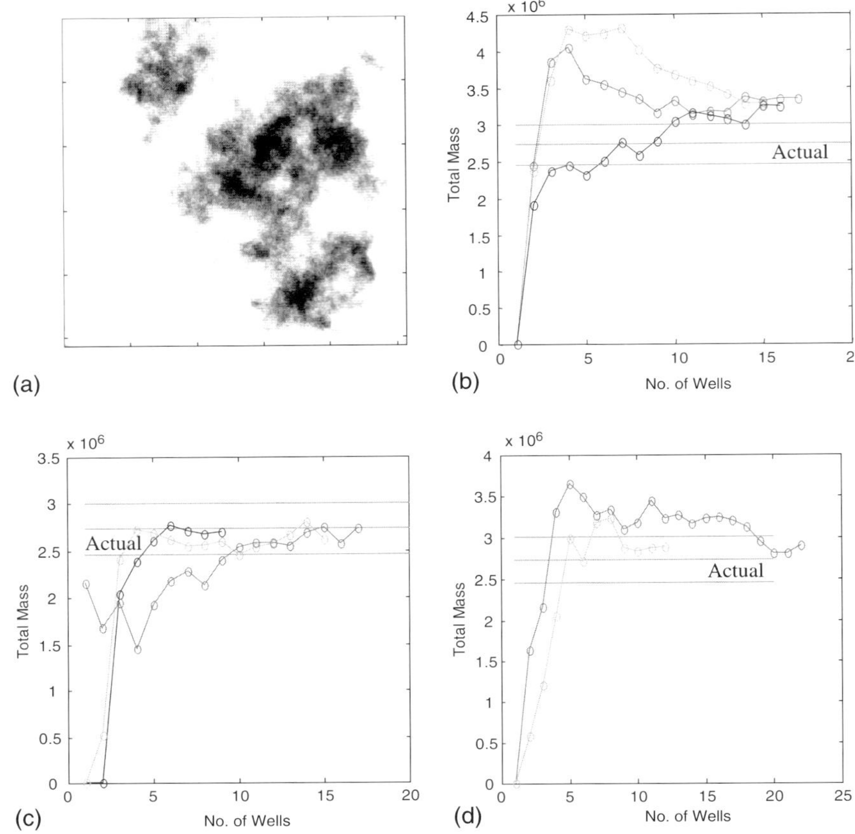

Fig. 3. Calculation of total mass; Example 1. (a) Actual image. (b) Conventional technique, different starting point. (c) Neuro-Statistical, different starting point. (d) Adaptation of conventional technique based on Neuro-Statistical model.

evident in Fig. 3c, the neural network model shows that a trend converges. The model converges to an actual total mass and is not sensitive to initial starting points.

Figs. 4 and 5 show the performance of the new technique for prediction of a total mass of contaminant for two other examples. Data for Example 2 is based on the fracture aperture distribution from the Yucca Mountain tuff core. Data for Example 3 is generated based on data from Examples 1 and 2. Here we can see the same trend as in Example 1. This trend can be used for extrapolation purposes. For all cases, the model converges to the certain value.

Fig. 6, which was generated based on the concept presented in Eqs. (2) to (6), shows that the model can not only predict the upper and lower limits, but also the most probable and mean values, which bound the total contaminant. The upper and lower bounds of the most probable prediction are tightened based on the concept shown in Fig. 1f to h. In this case, the mean value is simply equal to the upper limit plus lower

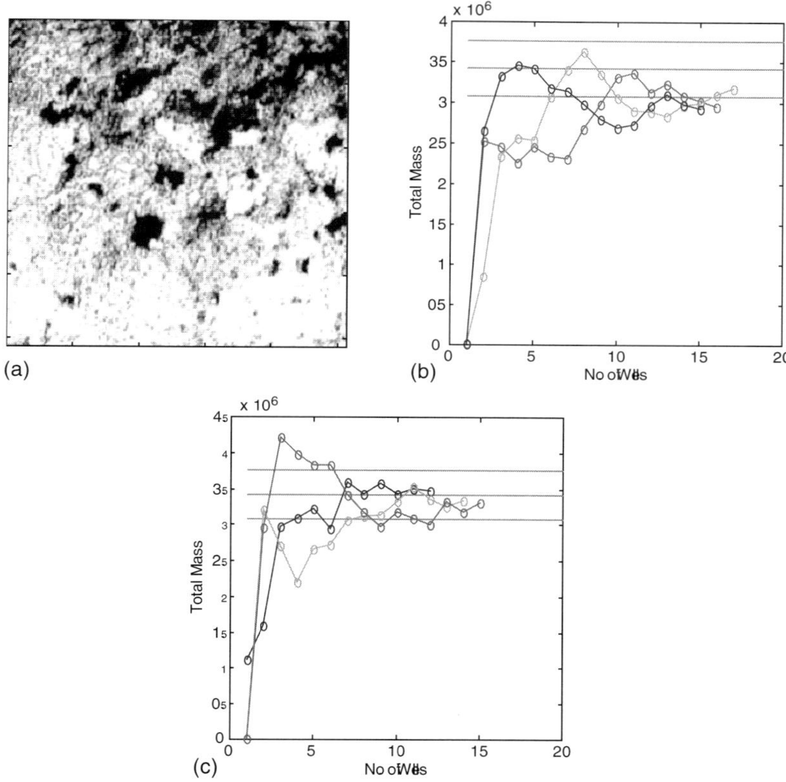

Fig. 4. Calculation of total mass; Example 2. (a) Actual image. (b) Conventional technique, different starting point. (c) Neuro-Statistical, different starting point.

limit divided by 2. Once the mean value converges to the most probable value, drilling can be stopped. Comparison of the described model with a conventional neural network models showed that the most probable parameter for the new model is closely related to the crisp value of the conventional models.

5. APPLICATION OF NEURO-STATISTICAL METHOD FOR A METAL-CONTAMINATED FILL AT ALAMEDA COUNTY (NIKRAVESH ET AL., 1996B)

In this study, we developed a series of neural network models to analyze the actual field data from a metal-contaminated site in Alameda County, California (Nikravesh et al., 1996b) and investigate the effectiveness of neuro-statistical techniques, which have been presented in our previous work (Nikravesh et al., 1996b).

Fig. 7 shows the topology of the earth. Table 1 shows the statistics of Soil Metal Concentrations and mass of the metal calculated based on simple statistical techniques. The

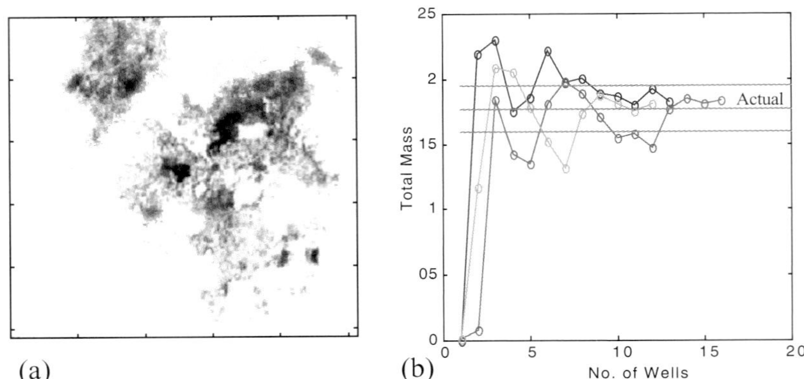

Fig. 5. Calculation of total mass; Example 3. (a) Actual image. (b) Neuro-Statistical, different starting point.

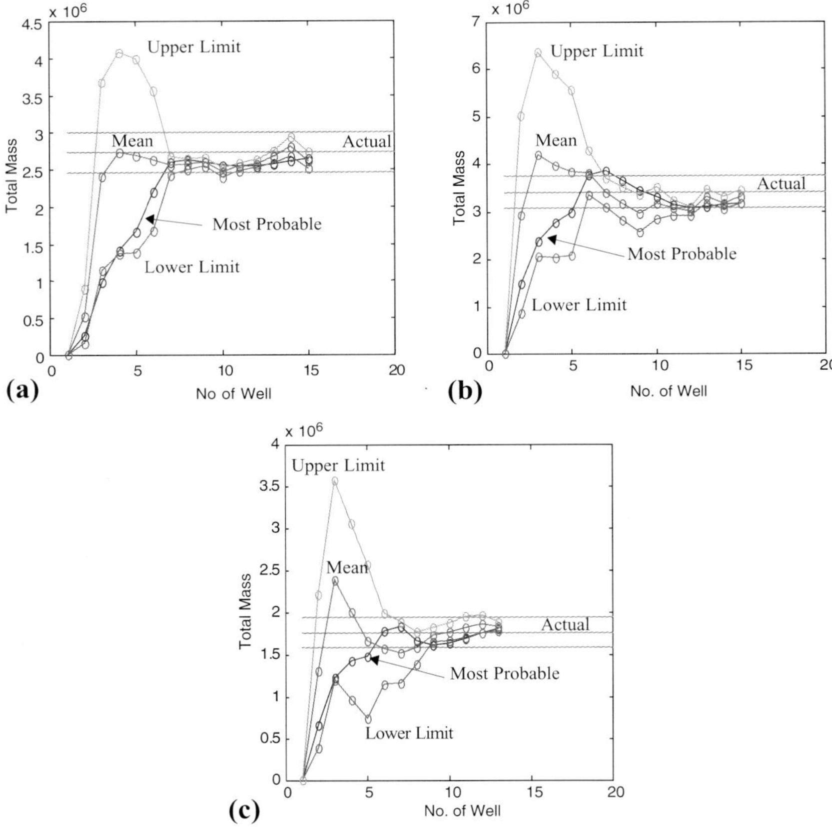

Fig. 6. Calculation of total mass; upper, lower, mean and the most probable. (a) Example 1; (b) Example 2; (c) Example 3.

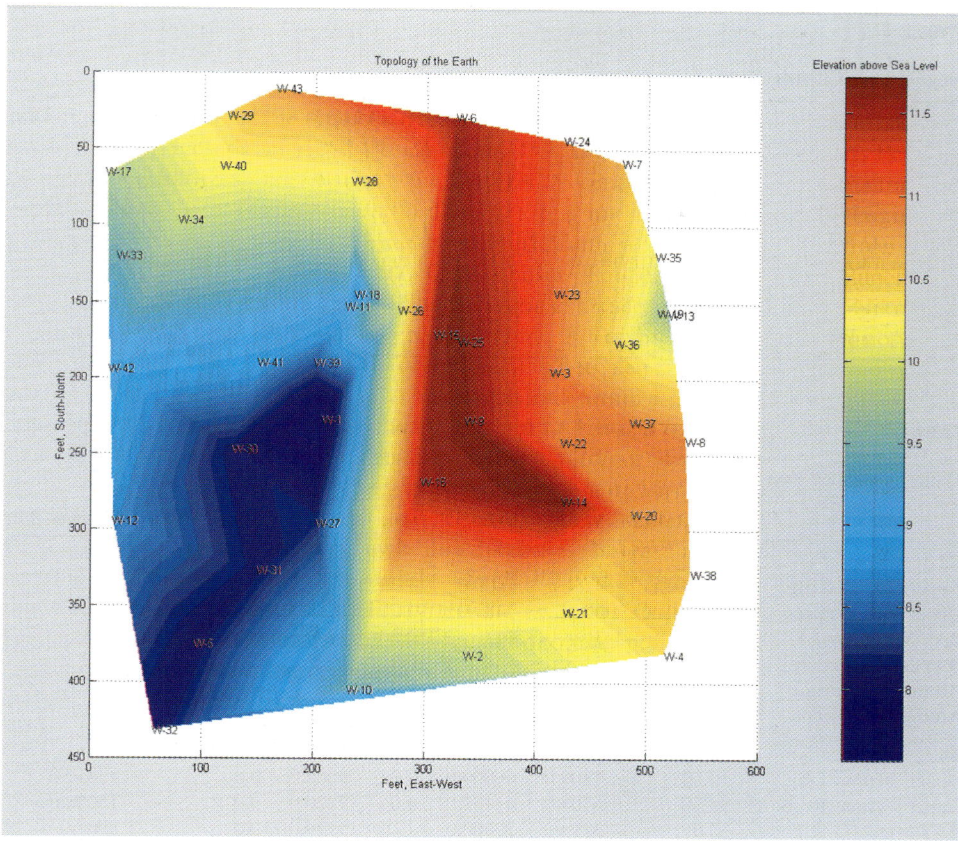

Fig. 7. Surface topology.

soil samples are log-normally distributed. Therefore, based on the assumption that soil metal samples are to be representative of the site, we were able to estimate the total metal masses underneath the site. In addition, using the statistical techniques, the data has been interpolated and extrapolated around the actual data points, along the z direction (Elevation). Using this technique, we were able to increase the number of data points by a factor of 10 (Table 2). The new data set and the actual data set have been checked based on conventional statistical methods to insure that the new data has not seriously changed the original statistics (Table 2). In addition, the new data has been tested for their relevance to the original data based on conventional statistical methods (Table 2).

In this study, an ACE-neural network model was developed to reconstruct the topology of the earth as a function of coordinates x and y. The network had 4 nodes in the input layer (dimensionless location, x and y, Phi(x) and Phi(y)), 5 nodes in the hidden layer with non-linear transfer function, and one node in the output layer (elevation above sea level, dimensionless) with a non-linear transfer function. Phi(x) and Phi(y) are non-linear and non-parametric statistical transformation of x and y

TABLE 1

Simple statistical analysis based on actual data

	Mean	Max.	Min.	St. Dev.	Mean − St. Dev.	Mean + St. Dev.
Sb	14	180	0	29	0	43
Zn	12999	60000	0	12116	883	25115
Pb	11864	43500	8	12364	0	24228
Cu	1304	3600	0	1062	242	2366
Cd	40	1400	0	123	0	163
As	200	18000	0	1307	0	1507
Hg	2	62	0	6.7	0	8.7
Ba	4077	92000	24	13518	0	17595

TABLE 2

Typical expanded data

	No. of data	Mean	Max.	Min.	Std.	Mean − Std.	Mean + Std.
Cd							
Upper section (Orig.)	104.0	47.4	1400.0	0.0	170.7	0.0	218.1
Upper section (Recon.)	968.0	60.7	1489.0	0.0	183.1	0.0	243.8
BAY-MUD (Orig.)	28,0	7.8	44.0	0.0	14.5	0.0	22.3
BAY-MUD (Recon.)	211,0	11.3	44.2	0.0	15.7	0.0	27.0
All the data (Orig.)	132.0	39.0	1400.0	0.0	152.5	0.0	191.5
All the data (Recon.)	1179.0	51.9	1489.0	0.0	167.0	0.0	218.9
Zn							
Upper section (Orig.)	102.0	4809.0	60000.0	0.0	9001.0	0.0	13810.0
Upper section (Recon.)	952.0	5708.0	61132.0	0.0	9320.0	0.0	15028.0
BAY-MUD (Orig.)	33.0	3178.0	19000.0	25.0	5729.0	0.0	8907.0
BAY-MUD (Recon.)	242.0	4554.0	19002.0	25.0	6319.0	0.0	10873.0
All the data (Orig.)	135.0	4410.4	60000.0	0.0	8330.4	0.0	12741.2
All the data (Recon.)	1194.0	5474.2	61132.0	0.0	8804.3	0.0	14278.5

recognized by ACE technique. To analyze the data based on the neural network model, two series of network models have been developed. The first model predicts the concentration of the metals as a function of the x and y direction (in the direction of the topology of the surface at certain elevation and Phi(x) and Phi(y). In this case, the elevation is referred to elevation from the original topology of the Earth as shown in Fig. 7. This model is called two-dimensional mapping. The second model predicts the concentration of a certain metal as a function of topology, elevation (elevation from the original earth surface as shown in Fig. 7, and Phi(x) and Phi(y)). This model is the three-dimensional model.

Even though the three-dimensional model has some advantages, it can be inaccurate for extrapolation purposes far beyond the range, in which the network (especially conventional neural network models) has been trained. The two-dimensional model is not appropriate when insufficient data are available for a specific surface and elevation.

Once enough data are available for a certain elevation in a certain surface, the two-dimensional model (neuro-statistical model) can be used. This model can map the input–output data locally, therefore representing the behavior of the site in this location more accurately. It is important to mention that the effect of other neighboring data has already been included in the data based on preprocessing. If the data for the two-dimensional model was insufficient, the three-dimensional model should be used. In this case, a higher weight has been assigned to the data of the surface under study and a lower weight has been assigned to the neighboring surfaces.

Table 3 shows the comparison between neural network and simple statistical techniques for prediction of total of mass of the metal contaminants. Fig. 8 shows the

TABLE 3

Comparison between neural network and simple statistical technique predictions (values for simple statistical technique are multiplied by factor two)

	Neural network prediction			Linear model (simple statistical method) multiplied by two		
	Mean	Upper	Lower	Mean	Upper	Lower
Sb						
Average ppm	73.2	75.7	55.0	51.6	58.7	44.6
Total Mass. kg	7292.7	7549.1	5479.7	5148.2	5850.2	4446.2
Zn						
Average ppm	18804.2	26388.1	12554.3	19674.9	25685.3	13664.6
Total Mass. kg	1874686.7	2630761.7	1251603.0	1961490.9	2560693.1	1362288.7
Ratio of Zn: Above bay-mud/(below bay-mud); based on same volume						
Average ppm (ratio)	2.9	3.8	2.7	3.0	3.0	3.0
Total Mass. kg	2.9	3.8	2.7	3.0	3.0	3.0
Pb						
Average ppm	10788.4	16170.0	6771.0	10777.3	14069.5	7485.0
Total Mass. kg	1075549.2	1612067.6	675035.9	1074438.3	1402660.9	746215.6
Cu						
Average ppm	1425.5	1774.2	934.7	1438.0	1877.2	998.7
Total Mass. kg	142117.6	176880.3	93184.2	143357.3	187150.5	99564.1
Cd						
Average ppm	281.4	312.5	206.5	255.7	333.9	177.6
Total Mass. kg	28054.6	31157.7	20585.8	25495.3	33283.7	17706.9
As						
Average ppm	887.7	986.4	641.7	761.3	993.8	528.7
Total Mass. kg	88501.2	98336.9	63971.1	75894.3	99078.7	52709.8
Hg						
Average ppm	7.6	9.3	5.4	7.1	9.3	4.9
Total Mass. kg	757.6	930.3	534.8	710.4	927.5	493.4
Ba						
Average ppm	17181.4	22447.8	13100.8	15228.2	19880.1	10576.2
Total Mass. kg	1712896.0	2237928.2	1306084.0	1518171.4	1981947.0	1054395.8

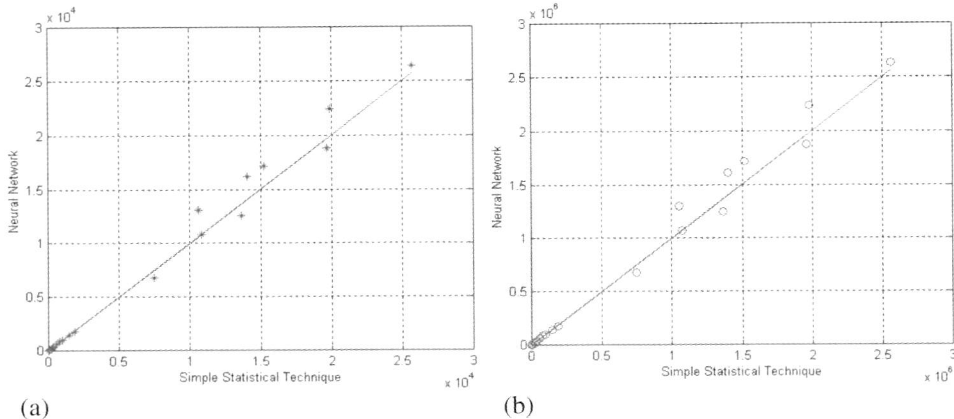

(a) (b)

Fig. 8. Comparison between neural network and simple statistical techniques predictions as shown in Table 3. (a) Prediction of average ppm of metals; (b) Prediction of total mass of contaminated ppm of metal. (Values for simple statistical method are multiplied by factor two.)

results as shown in Table 3. It is concluded that the performance of the simple statistical and ACE-neural network model for prediction of total mass of the metal contaminated is the same. ACE–neural network predicts roughly twice the mass calculated based on simple statistical techniques. For further comparison, an arbitrary surface has been chosen and the concentration of the Metals have been calculated at this surface using the neural network and table lookup technique in 3D. The results are shown in Fig. 9. It is clear that neural model predictions are quite different than the statistical technique. In addition, the surfaces generated by table lookup show quite different behavior. The table lookup technique shows a wavy and discontinuous surface whereas the neural network predicts a smooth, continuous and non-wavy surface. Therefore, it is concluded that the input–output mappings are quite different. In this study, based on the results shown in Fig. 9, it appears that no simple structure exists. This may confirm that soil samples are randomly distributed (log-normally as we mentioned earlier). Therefore, soil metal samples may be representative for the site. That may also be why the total mass estimate using simple statistical techniques is close to the ACE-Neural Network Model even though the 3D reconstruction of data is quite different.

6. CONCLUSION

In order to plan the remediation of contaminant sites, it is of great importance to estimate the mass and a three-dimensional extent of contaminants in soils and rocks. Estimation of the volume and mass of contaminants is usually based on sparse well data. However, data from contaminant sites are often difficult to analyze due to their uncertainty. In addition, methods of obtaining data are laborious and expensive. Therefore, minimizing sampling plans to reduce the number of wells to be drilled and samples to be taken are of great economic benefit. In this paper, we presented a robust

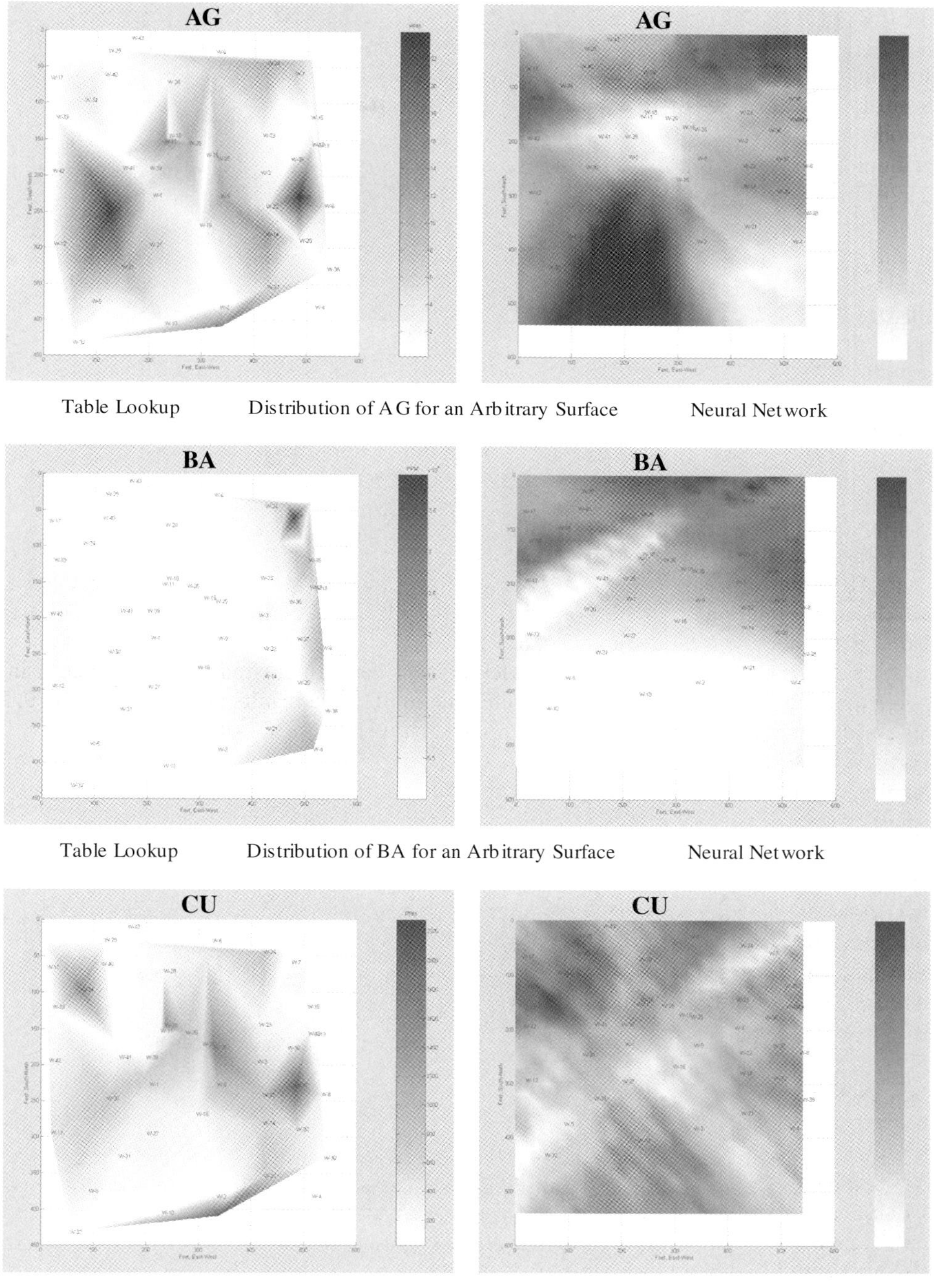

Fig. 9. Comparison between table lookup and neural network model for prediction of the distribution of metal contaminant for an arbitrary surface.

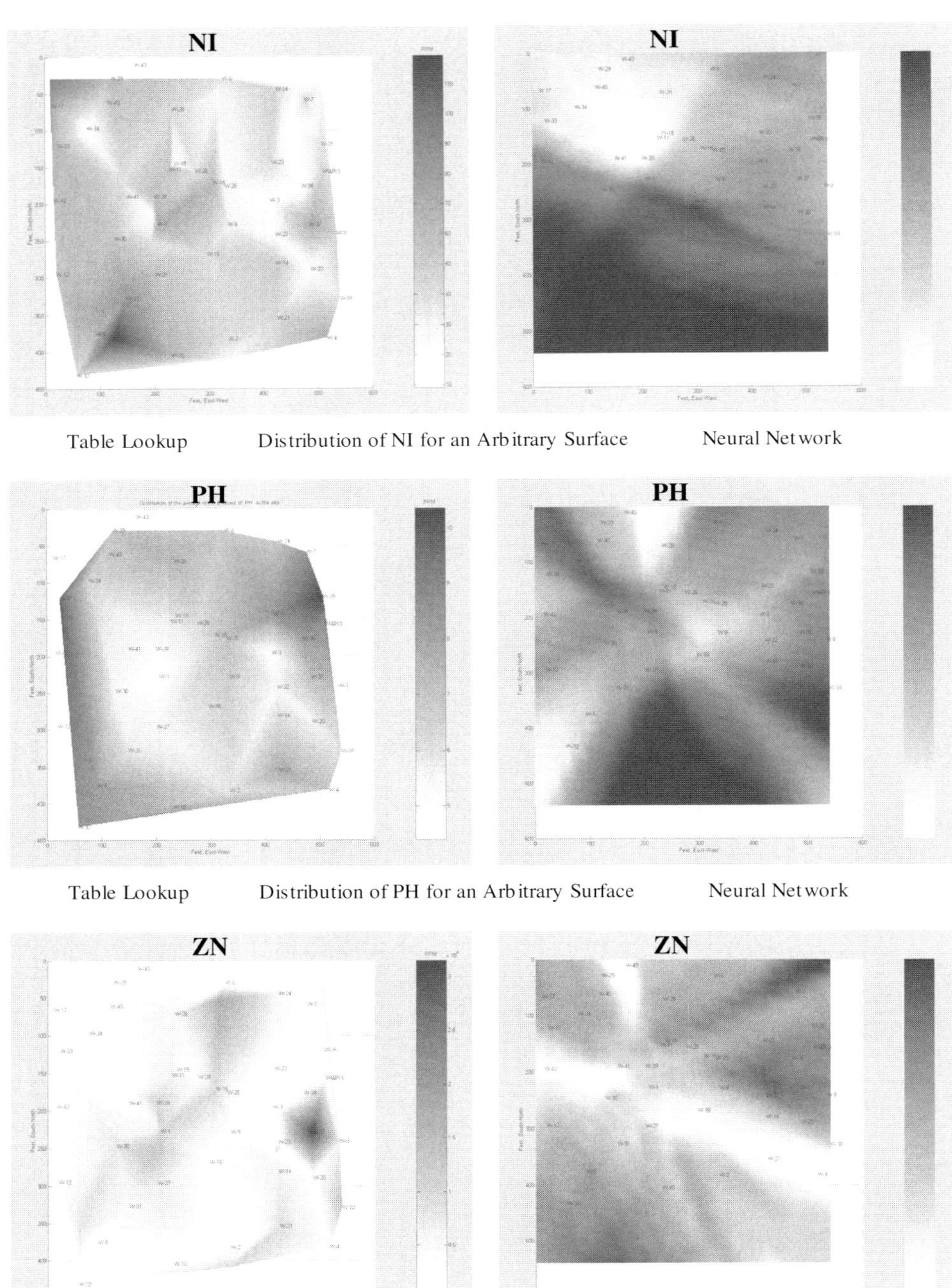

Table Lookup Distribution of NI for an Arbitrary Surface Neural Network

Table Lookup Distribution of PH for an Arbitrary Surface Neural Network

Table Lookup Distribution of ZN for an Arbitrary Surface Neural Network

Fig. 9 (continued).

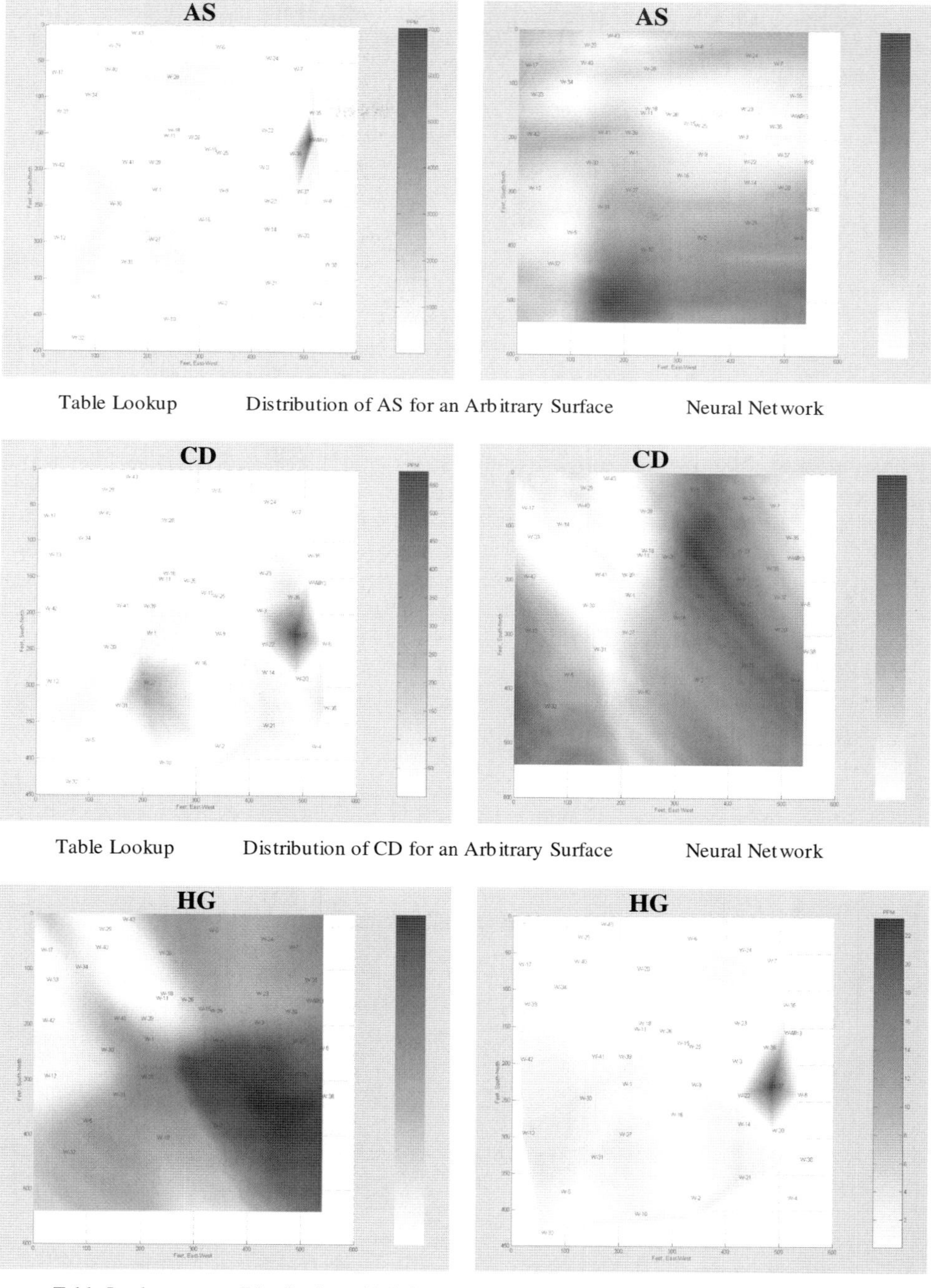

Table Lookup — Distribution of AS for an Arbitrary Surface — Neural Network

Table Lookup — Distribution of CD for an Arbitrary Surface — Neural Network

Table Lookup — Distribution of HG for an Arbitrary Surface — Neural Network

Fig. 9 (continued).

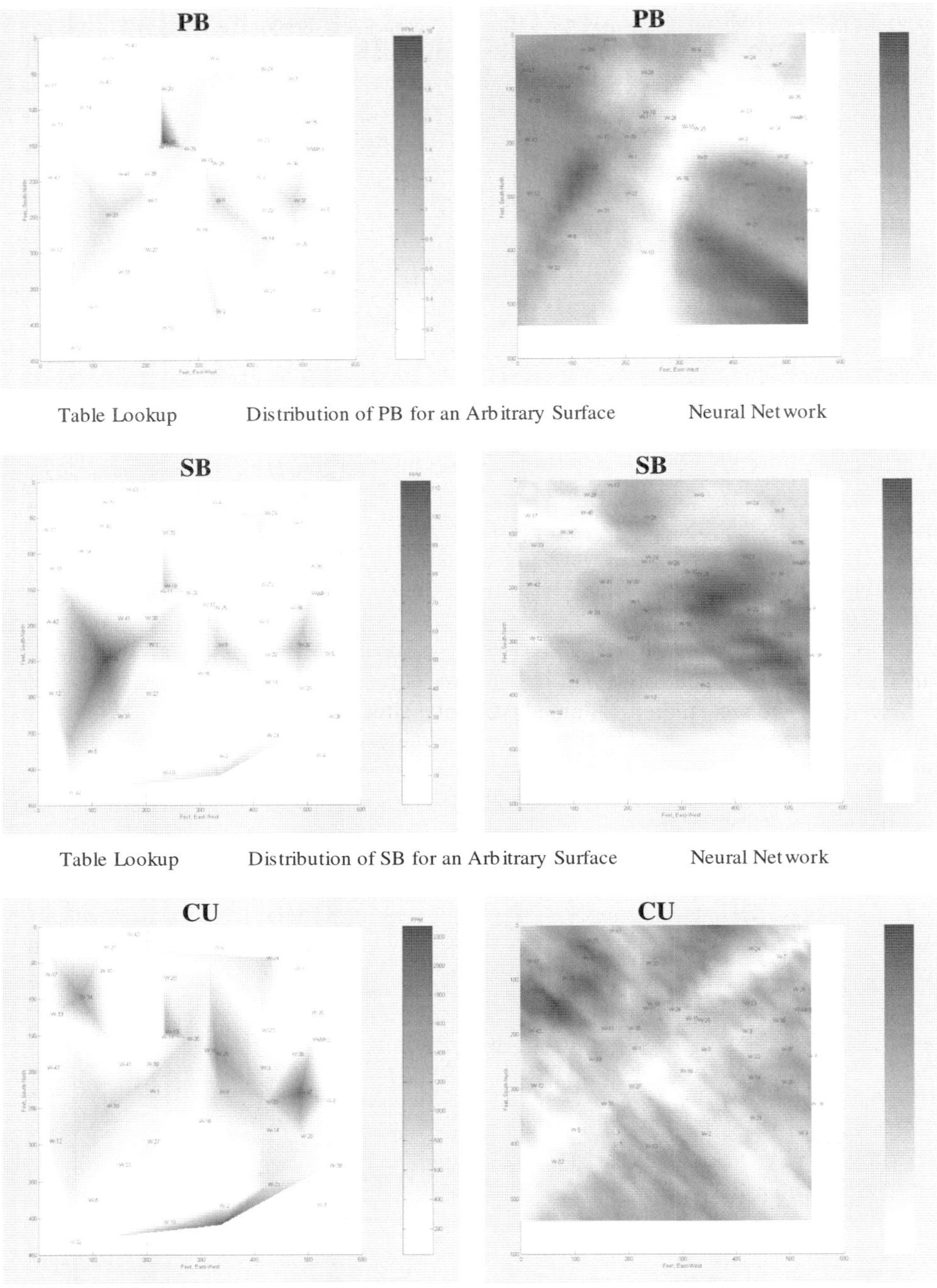

Fig. 9 (continued).

model that is based on a neural network technique. The model has the ability to infer general rules and extract typical patterns from the data available, and recognize input–output mapping parameters from complex multi-dimensional field data. This approach gives our model the ability to interpolate between typical patterns of data, and then extrapolate to a region beyond the training domains. The model requires minimum input information to estimate the total mass and/or volume, shape, and distribution of a contaminant in 2D and 3D spaces.

In this study, the neuro-statistical model has been used instead of conventional network models. MTALAB® (The Math Works, 1995) software has been used to develop the Neuro-Statistical Techniques. Neuro-Statistical models can predict the uncertainty, probability, degree of acceptance, and the confidence level for each prediction. In the first case study, the plume's boundary is predicted by the model at each step. The location and orientation of each new well to be drilled are optimized based on the information gathered from previous wells. The model shows that a trend of the total mass converges to the expected mass when the number of wells drilled increased to a range from 5 to 10 wells. It is important that this trend is not sensitive to initial starting points.

Both ACE-Neural network and a simple statistical technique showed the same performance for prediction of the total mass of the metal. This may due to the fact that data are normally distributed and soil metal samples may be representative of the site. The ACE-neural network technique smooths the data and eliminates the outliers whereas the simple table lookup technique generates an uneven surface and distorts the area near outliers, as well as creates a wavy, non-continuous surface with artificial peaks and valleys.

For all cases we studied (Alameda County sites, Yucca Mountain tuff core, and computer generated examples), predicted mass converged to the actual total mass (Cox et al., 1997; Nikravesh, 1997; Nikravesh and Aminzadeh, 1998; Nikravesh et al., 1999). This study showed that ACE-neural network approach is very promising. Therefore, it would be an attractive alternative to conventional neural network or conventional statistical techniques. However, additional work is needed to investigate the stability and the predictive capability of neuro-statistical techniques for sparse well data within a dynamic environment.

APPENDIX A. ROBUST ALGORITHM FOR TRAINING THE NEURAL NETWORK MODELS (NON-LINEAR MODEL FOR IMPRECISE DATA)

Let the data to fit:

$$Y_i, X_{1i}, X_{2i}, X_{mi} \qquad i = 1,2,\ldots,n$$

Let the model to be fitted:

$$\begin{aligned} E(y) &= f(x_1,x_2,\ldots,x_m,\beta_1\beta_2,\ldots,\beta_k) \\ &= f(X,\beta) \end{aligned}$$

where $x_1,x_2,\ldots,x_m$ = independent variables; $\beta_1,\beta_2,\ldots,\beta_k$ = population values of k parameters; and $E(y)$ = expected value of the dependent variable y.

A.1. Current methods:

A.1.1. Gauss method or the Gausee–Newton method

$$Y(X_i, \hat{\beta} + \delta_t) = f(X_i, \hat{\beta}) + \sum_{j=1}^{k} \left(\frac{\partial f_i}{\partial \beta_j}\right)(\delta_t)_j$$

$$\langle Y \rangle = f_0 + \underline{P}\delta_t$$

$$\langle \Phi \rangle = \sum_{i=1}^{n} \left[Y_i - \langle Y_i \rangle\right]^2 = \left\| Y - \langle Y \rangle \right\|^2$$

$$\frac{\partial \langle \Phi \rangle}{\partial \delta_j} = 0 \Rightarrow \underline{A}\delta_t = g$$

with,

$$\underline{A} = \underline{P}^{\mathrm{T}}\underline{P}; \qquad \underline{P} = \frac{\partial f_i}{\partial \hat{\beta}_j}; \qquad g = \underline{P}^{\mathrm{T}}(Y - f_0)$$

A.1.2. Gradient methods

Current value in the direction of the negative gradient of Φ:

$$\frac{\partial \langle \Phi \rangle}{\partial \delta_j} = 0 \Rightarrow \delta_g = \left(\frac{\partial \Phi}{\partial \hat{\beta}}\right)^{\mathrm{T}}$$

A.1.3. Levenberg–Marquardt/Marquardt–Levenberg

Let: $\left(\underline{A} + \lambda^2 \underline{I}\right)\delta_0 = g$

Then δ_0 minimizes $\langle \Phi \rangle$ on the sphere whose radius $\|\delta\|$ satisfies $\|\delta\|^2 = \|\delta_0\|^2$.

Minimize: $\langle \Phi(\delta) \rangle = \left\| Y - f_0 - \underline{P}\delta \right\|^2$

under $\|\delta\|^2 = \|\delta_0\|^2$ constraint.

Therefore:

Minimize: $\langle \hat{\Phi}(\delta) \rangle = \left\| Y - f_0 - \underline{P}\delta \right\|^2 + \lambda^2\left(\|\delta\|^2 - \|\delta_0\|^2\right)$

$$\frac{\partial \langle \hat{\Phi}(\delta) \rangle}{\partial \delta} = 0 \qquad \text{and} \qquad \frac{\partial \langle \hat{\Phi}(\delta) \rangle}{\partial (\lambda^2)} = 0$$

$$\|\delta\|^2 - \|\delta_0\|^2 = 0$$

$$\left(\underline{P}^{\mathrm{T}} P + \lambda^2 \underline{I}\right)\delta_0 = g = P^{\mathrm{T}}(Y - f_0)$$

A.2. Proposed method

$$(\underline{A}+\lambda^2\underline{I})\delta_0 = g \qquad \text{with } \|\delta\|^2 = \|\delta_0\|^2$$

Therefore;

$$\left[(\underline{A}+\lambda^2\underline{I})\delta_0\right] \times \underline{\Lambda}^{\mathrm{T}}\underline{\Lambda} = g \times \underline{\Lambda}^{\mathrm{T}}\underline{\Lambda} \qquad \text{with } \|\delta\|^2 = \|\delta_0\|^2$$

Therefore;

$$(\underline{P}^T\underline{\Lambda}^{\mathrm{T}}\underline{\Lambda P}+\lambda'^2\underline{I})\delta_0 = \underline{P}^{\mathrm{T}}\underline{\Lambda}^{\mathrm{T}}\underline{\Lambda}(Y-f_0)$$

$$(\underline{P}^{\mathrm{T}}\hat{\sigma}^2\underline{IP}+\lambda'^2\underline{I})\delta_0 = \underline{P}^{\mathrm{T}}\hat{\sigma}^2\underline{I}(Y-f_0)$$

$$\delta_0 = (\underline{P}^{\mathrm{T}}\hat{\sigma}^2\underline{IP}+\lambda'^2\underline{I})^{-1}\underline{P}^{\mathrm{T}}\hat{\sigma}^2\underline{I}(Y-f_0)$$

with

$$\delta_0 - \hat{\sigma} < \hat{\delta}_0 < \delta_0 + \hat{\sigma}$$

$$(\underline{P}^{\mathrm{T}}\underline{\Lambda}^{\mathrm{T}}\underline{\Lambda P}+\Gamma^{\mathrm{T}}\Gamma)\delta_0 = \underline{P}^{\mathrm{T}}\underline{\Lambda}^{\mathrm{T}}\underline{\Lambda}(Y-f_0)$$

A.3. Neural network models

Current:

$$\Delta W = (\underline{J}^{\mathrm{T}}J+\mu^2\underline{I})^{-1}\underline{J}^{\mathrm{T}}e$$

Proposed:

$$\Delta W = (\underline{J}^{\mathrm{T}}\underline{\Lambda}^{\mathrm{T}}\underline{\Lambda}J+\Gamma^{\mathrm{T}}\Gamma)^{-1}\underline{J}^{\mathrm{T}}\underline{\Lambda}^{\mathrm{T}}\underline{\Lambda}e$$

and

$$\hat{W} = W \pm k\hat{\sigma}$$

REFERENCES

Azimi, M.R. and Liou, R.J., 1992. Fast learning process of multilayer neural networks using recursive least squares method. *IEEE Trans. Signal Processing*, 40.

Breiman, L. and Friedman, J.H., 1985. Estimating optimal transformations for multiple regression and correlation. *J. Am. Stat. Assoc.*, 80(391).

Cox, B.L., Nikravesh, M. and Faybishenko, B., 1997. Estimating total mass of contaminant plumes from sparse well data. *Am. Geophys. Union Fall Meet.*, Dec. 8–12, San Francisco, CA, AGU No. AN: H22I-10.

Cressie, N.A.C., 1993. *Statistics for Spatial Data, Revised Edition*. John Wiley and Sons, Inc., New York, NY.

Davis, J.C., 1986. *Statistics and Data Analysis in Geology, 2nd Edition*. John Wiley and Sons, Inc., New York, NY.

De Marsily, G., 1986. *Quantitative Hydrogeology: Groundwater Hydrology for Engineers*. Academic Press, London, p. 440.

Deutsch, C.V. and Journal, A.G., 1982. *GSLIB, Geostatistical Software Library and User's Guide*. Oxford University Press, New York, NY.

Isaak, E.H. and Srivastava, R.M., 1989. *Applied Geostatistics*. Oxford University Press, New Yrok, NY, p. 561.

Nikravesh, M., 1997. Neuro-statistical Methods for contaminant site. In: Tsang, C.F., Mironenko, V. and Pozdniakov, S. (Eds.), *Proceedings of the Joint Russian–American Hydrogeology Seminar*. Russian–American Center for Contaminants Transport Studies, LBNL, Berkeley, CA, July 8–9, 1997, pp. 312–329.

Nikravesh, M. and Aminzadeh, F., 1997. *Knowledge Discovery from Data Bases: Intelligent Data Mining Technique*. FACT Inc. and LBNL Proposal, submitted to SBIR-NASA.

Nikravesh, M. and Aminzadeh, F., 1998. Neuro-statistical technique for characterization of contaminant sites using sparse well data. *JCIS'98, The Fourth Joint Conference on Information Sciences*, October 23–28, 1998, Research Triangle Park, NC.

Nikravesh, M., Farell, A.E. and Stanford, T.G., 1996a. Model identification of nonlinear time variant processes via artificial neural networks. *J. Comput. Chem. Eng.*, 20(11).

Nikravesh, M.M., Soroush, A.R. and Kovseck, T.W., 1996b. Patzek, identification and control of industrial-scale processes via neural networks, *CPC-V Conference,* Tahoe City, CA, January 1996.

Nikravesh, M., Cox, B.L., Faybishenko, B. and Aminzadeh, F., 1999. Characterization of contaminant sites using sparse well Data. *1998, SPE Annual Technical Conference and Exhibition*, New Orleans, LA, September 27–30, 1998 and *1999 Exploration and Production Environmental Conference*, Austin, TX, March 1999, SPE Paper #49330.

Pannatier, Y., 1996. *VARIOWIN, Software for Spatial Data Analysis in 2D*. Springer Verlag, New York, NY.

Peitgen, H.O. and Saupe, D. (Eds.) 1988. *The Science of Fractal Images*. Springer Verlag, New York, NY.

Russ, J.C., 1995. *Fractal Surfaces*. Plemun Press, New York, NY.

The Math Works, 1995. The Math Works, Natick.

Developments in Petroleum Science, 51
Editors: M. Nikravesh, F. Aminzadeh and L.A. Zadeh

Chapter 30

MULTIVARIATE STATISTICAL TECHNIQUES INCLUDING PCA AND RULE BASED SYSTEMS FOR WELL LOG CORRELATION

JONG-SE LIM [1]

Division of Ocean Development Engineering, Korea Maritime University, Dongsam-Dong, Yeongdo-Gu, Puasn, 606-791, Republic of Korea

ABSTRACT

This paper describes the effective techniques for automated well log correlation using both multivariate statistical techniques including principal component analysis (PCA) and rule based systems. The correlation of wireline logging data is on the basis of a large set of subjective rules for pattern recognition that are intended to represent human logical processes. The data processed are the characteristics of the shapes extracted along log traces by object-oriented programming. The correlation of zones between wells is made by rule-based inference program. This method has the advantage over the conventional methods considering the capability of handling the shifting, thickening, and thinning strata in well-to-well log correlation.

Use of statistical techniques can be helpful for obtaining more reliable correlation results for complex geologic formation. The efficient and reliable pattern recognition for well-to-well correlation can be established by using the first principal component log, since it has the largest common part of variance of all available well log data. In addition, the electrofacies derived by selecting, weighting, and combining multivariate well logs can be used as important information of lithology identification for more reliable correlation.

The correlated results with logging data in the Korea Continental Shelf and oversea fields show that this method can be used to make it more reliable and efficient to well-to-well log correlation rather than the traditional methods in which only one approach was adopted. It has been applied as an effective tool in planning the development and production strategy of the fields.

1. INTRODUCTION

Reservoir characterization is the process of describing various reservoir characteristics using all the available data to provide reliable reservoir models for accurate reservoir performance prediction. The reservoir characteristics include pore and grain size distributions, permeability, porosity, facies distribution, and depositional environment. The types of data to need for describing the characteristics are core data, well logs, well

[1] Tel: +82 (51) 410-4682, Fax: +82 (51) 404-3986, E-mail: jslim@kmaritime.ac.kr

test data, production data and seismic survey. Especially, well log data can provide valuable but indirect information about mineralogy, texture, sedimentary structure and fluid content of a reservoir. Generally well logs appear to be continuous information with intensive vertical resolutions.

One of the most common uses of well logs is to develop well correlation for reservoir studies. Interwell or well-to-well stratigraphic correlation can be defined as the correct determination of the spatial equivalence of rocks based on their petrophysical properties. Most correlation methods in the literatures depend on the curve matching procedures that determine the degree of similarity between two series in the space domain (Olea and Davis, 1986).

Cross-correlation is known to the widely used method to attempt to correlate well logs by the computer applications. However, a difference exists between comparing well-log traces by cross-correlation methods and making correlation by human experts. Traditional cross-correlation methods can only detect shifts, but have no provision for stretching or shrinking of stratigraphic intervals.

Human eyes can identify similar characteristics from log traces despite their different scaling in depth or amplitude. Because most human decisions are qualitative in nature rather than quantitative, it is possible to allow computers to make decisions based on logic and reasoning rather than on numbers alone. Artificial intelligence methods attempt to translate this human-like reasoning into computer code (Olea and Davis, 1986; Startzman and Kuo, 1987; Olea, 1994). The data processed are mainly the qualitative, not the quantitative information or characteristics of log traces. Moreover, the correlation is made by an artificial intelligence technique rather than by numerical computation procedures. However, most correlation methods using artificial intelligence has been excluded the statistical characteristics of data. The introduction of the statistical techniques can compensate for lack of statistical information in correlation (Lim et al., 1998). Using combination of artificial intelligence and statistical analysis can obtain more reliable correlation results for complex geologic formation. Traditionally well-to-well correlation has been performed using a couple of logs such as resistivity log or gamma ray log. The better correlation often can be obtained when all available wireline logs are used (Elek, 1988). In this study, an expert system is developed for correlation between wells using both multivariate statistical analysis and rule based systems.

2. MULTIVARIATE STATISTICAL ANALYSIS

Two multivariate statistical techniques, principal component analysis and cluster analysis, are applied to wireline log data for correlation.

2.1. Principal component analysis (PCA)

Principal component analysis is the most commonly used technique for allowing high-dimensional representations to be compressed into a lower dimensionality. Also it has many additional useful properties that can be applied in systematic log interpretation.

If m log response from a sequence of zones are plotted as points in a space with mutually orthogonal axes, they form cloud in m-dimensional space. Principal components are the eigenvectors of this cloud, computed to locate the major axes in order of importance. These axes provide a new framework of reference which is aligned with natural axes of the cloud, rather than the original log measurement axes. The orientation of the principal components are computed from either the covariance or correlation matrix of the zone log data. The correlation matrix is the more common choice, because most logs are recorded in radically different units. In order to avoid artificial and undue weighting by any of the logs, the original data should be standardized to dimensionless units by subtracting the mean and dividing by the standard deviation. The covariance matrix of standardized data is the correlation matrix (Davis, 1986; Doveton, 1994).

Conceptually, an ideal well-to-well log correlation should include all available well-log data. Because this can be a very demanding process, interpreters often use only a selectively reduced data set such as resistivity logs. In many situations data reduction may result in an erroneous or misleading correlation.

PCA is a multivariate statistical techniques often used successfully in various log analysis applications (Elek, 1988; Lim et al., 1998). This technique can establish the effective quantity of data set information for easier handling by the interpreter. In a simple example, Fig. 1a shows a hypothetical crossplot of density and neutron porosity readings from a sequence of logged zones. Because the logs are recorded in radically different units, it is appropriate to rescale them in a standard form. Standardization is achieved by relocating the cloud centroid to a new origin and making units of variation on each log axis equal to one standard deviation. The standardization converts the variance/covariance matrix to a matrix of correlation coefficients. The correlation matrix is described geometrically by an ellipse. The eigenvalue associated with each eigenvector gives the relative length associated with each axis (Fig. 1b). The eigenvectors are called the 'principal components' of the correlation matrix and define a rotated set of axes that are keyed with the principal sources of variation in the raw data (Fig. 1c). In this example data points projected onto the first principal component

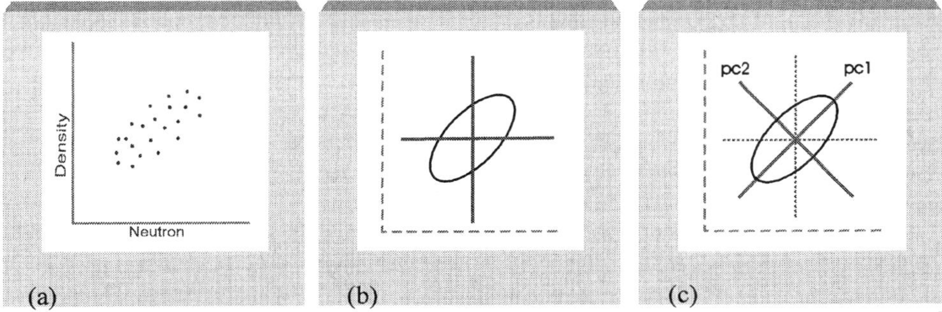

Fig. 1. Schematic diagram sequence of the geometry involved in principal component analysis. (a) Hypothetical neutron-density crossplot; (b) correlation matrix ellipse superimposed on standardized neutron-density axes; (c) location of ellipse principal axes by eigenvectors.

are scaled in terms of the major sources of common variation between the neutron and density logs (Lim et al., 1998).

When this operation is applied to a larger set of logs, the mathematics remains unchanged although it is difficult to visualize the geometry. The eigenvectors of the correlation matrix locate the axes of a hyper-ellipsoid that represents the data cloud, while the eigenvalues measure their relative elongation. By ordering the eigenvalues from the largest to the smallest, the associated eigenvectors are ordered in their importance in absorbing the total variation of the data cloud.

PCA technique calls for the interpreter to compute the first principal component for each well; this results in a dimensionless log containing the largest common part of variances of the input logs. While the interpreters had to assimilate and utilize a large number of logs conventionally, they need to perform pattern recognition with only one log for correlation.

2.2. Electrofacies determination

Subsurface lithology is traditionally determined from cores. Cores are generally not continuous and do not provide complete descriptions of formations crossed by a well. The lithology based on core data is not sufficiently accurate and precise for the quantitative use. On the other hand, well-logs have the advantage of providing a continuous record over the entire well and can be obtained in conditions where coring is impossible. Therefore well-log data can give a good lithologic descriptions of formations.

An electrofacies is defined as "the set of log responses which characterizes a bed and permits it to be distinguished from the others" (Serra and Abbott, 1982). The electrofacies derived by selecting, weighting, and combing well-log data can be used as an indicator of lithology. Once good correlation between electrofacies and core analysis are established on a local basis, significant geological information can be extracted from well-logs alone. Therefore, electrofacies can be used as an important information of lithology identification for more reliable correlation (Bucheb and Evans, 1994). The importance of electrofacies characterization in reservoir description and management has been widely recognized. This kind of data partitioning is to simplify a complex data set into some homogeneous and simple subgroups and to produce the better correlation between dependent and independents within distinct subgroups for further petrophysical properties regression(Lee and Datta-Gupta, 1999).

Hierarchical cluster analysis is used to determine the electrofacies in this study (Lim et al., 1997). This multivariate statistical technique is to group similar objects and to distinguish them from other dissimilar objects on the basis of their measured characteristics.

The basic steps are shown in Fig. 2 as an illustration of the clustering of hypothetical zones based on their log responses. First, a database of attribute measurements is compiled on the basis of the collective treatment of the attributes. The clustering algorithm is applied to the similarity matrix as an iterative process. The pairs of objects with highest similarities are merged, the matrix is re-computed, and the procedure repeats. Ultimately all the objects will be linked together as a hierarchy, which is most commonly shown

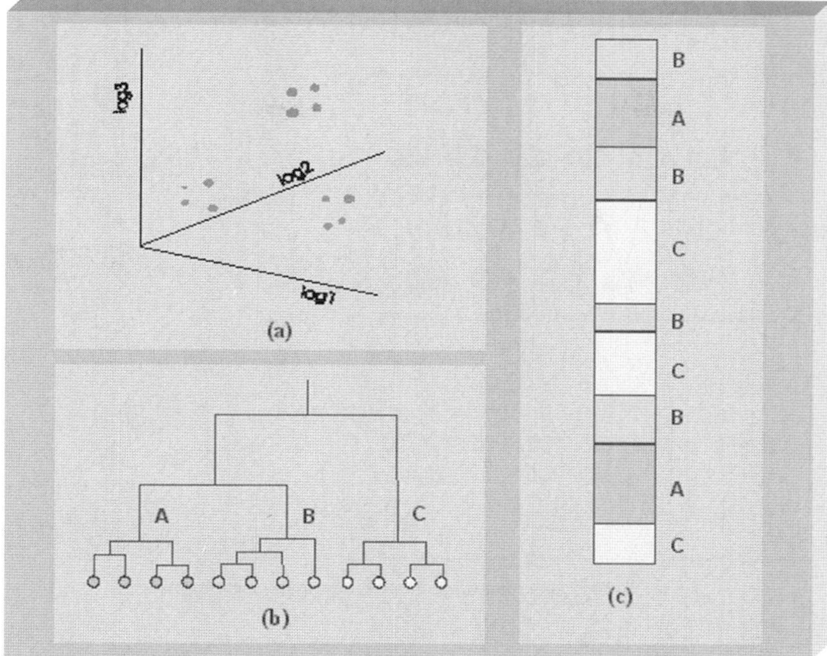

Fig. 2. Stages of cluster analysis of log data: (a) multivariate database of zones; (b) dendrogram of zones according to hierarchical clustering of the zones based on their similarities; (c) classification of zones.

as a dendrogram. At this point, the objects are in one giant cluster. A certain decision must now be made concerning where to cut the tree diagram into branches that coincide with distinctive groupings. The choice may be based either on visual inspection, a mathematical criterion that appears to reveal a natural breaking point, or some measure that can be used to check potential clusters against some external standard. Electrofacies identification can be used as lithology information in correlation procedures.

3. RULE-BASED CORRELATION SYSTEM

The rule-based correlation system used in this study is composed of database, rule base, and inference program (Startzman and Kuo, 1987; Lim et al., 1998).

3.1. Database

The digital wireline log data are translated into qualitative information such as the characteristics of the shapes extracted along the log traces. Recently object oriented programming has become very popular in the artificial intelligence applications. In object oriented programming, the domain is viewed as a collection of objects that have certain characteristics and objects communicated with each other through messages.

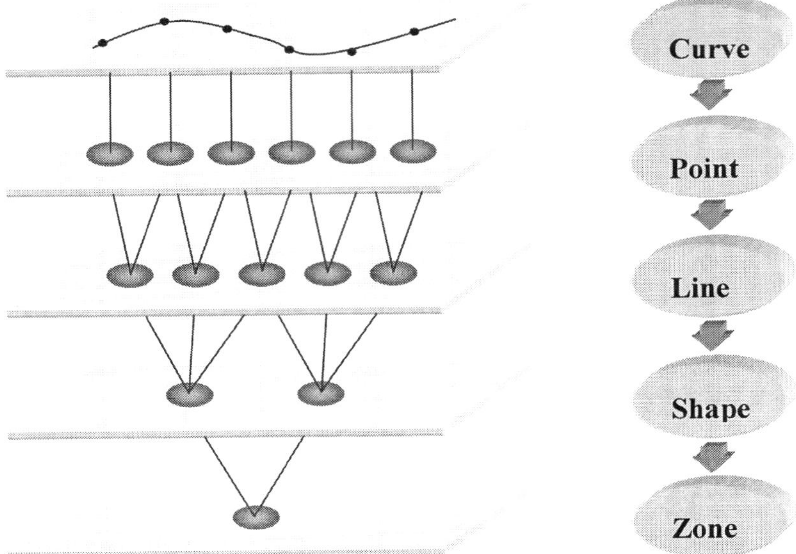

Fig. 3. Representation of qualitative information as objects.

Instead of naming a procedure to perform an operation on an object, a message is sent to the object, which responds using its own procedure (method) for performing operations.

The representation of objects can be explained in Fig. 3. Each log curve can be represented by four different objects sequentially; point, line, shape, and zone objects. The processes are performed by the object preparation program and result in a database. This database contains a collection of expressions that describe the characteristics of each zone; thickness, relative position, shape, average amplitude, and lithology.

3.2. Rule base

The rule base consists of a number of if–then rules. The rules work on the initial database contained zonal characteristics to produce similarity information on the zones. A typical if/then rule or production rule has two parts: an IF (or condition) part and a THEN (or action) part.

RULE

IF (Condition 1)

(Condition 2)

⋮

THEN (Action)

The IF part may contain more than one condition. The THEN part indicates the conclusion drawn if all conditions in the IF part are satisfied.

3.3. *Inference program*

The program that applies rules and provides a logical inference for deriving solutions is called the inference program. The degree of similarity in thickness, relative position, shape and lithology is expressed in terms of relative quality. Electrofacies can be used as lithologic information for correlation. The inference program conducts zonal correlation.

4. APPLICATIONS

4.1. *Verification*

For the validation of the export system developed in this study, the correlated results are compared with a previous work for resistivity logs of wells in Kurten Field (Startzman and Kuo, 1987). Dash line in Fig. 4 is the result from the previous work and solid line is the result from this study. As it can be seen in figure, the correlated results from the two systems are close matched.

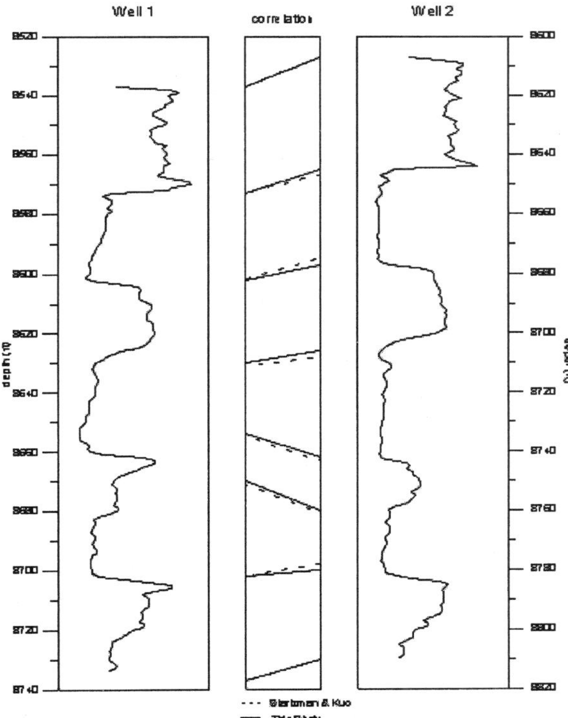

Fig. 4. Comparison of the correlation result with previous work

4.2. Comparison of methods

Well-logs in Louisiana are used for comparison of methods(Olea and Davis, 1986). Fig. 5 shows the well-log traces of a 600-ft interval centered at a depth of 10,130 ft taken from two wells. Each well-log consists of two log traces: the curve in the left track represents spontaneous potential (SP) and that in the right track is resistivity. The resistivity log usually is less sensitive to mud characteristics and has a higher resolution, Therefore, it is preferred when only one log is used for correlation.

Cross-correlation method can be selected as the measure of similarity between well-logs because of its desirable statistical properties and ease of computation. A short reference interval (correlation length) is matched vs. successive positions in a second well. In Fig. 5, the resistivity log curve from Well 5 within a 100-ft interval centered on the depth of 10,130 ft is compared with the resistivity log curve in all possible 100-ft intervals of the log segment from Well 7. Of all possible matches, the 100-ft interval centered at 10,040 ft in Well 7 is the most similar to the interval of the same length centered at 10,130 ft in the reference well (dash line).

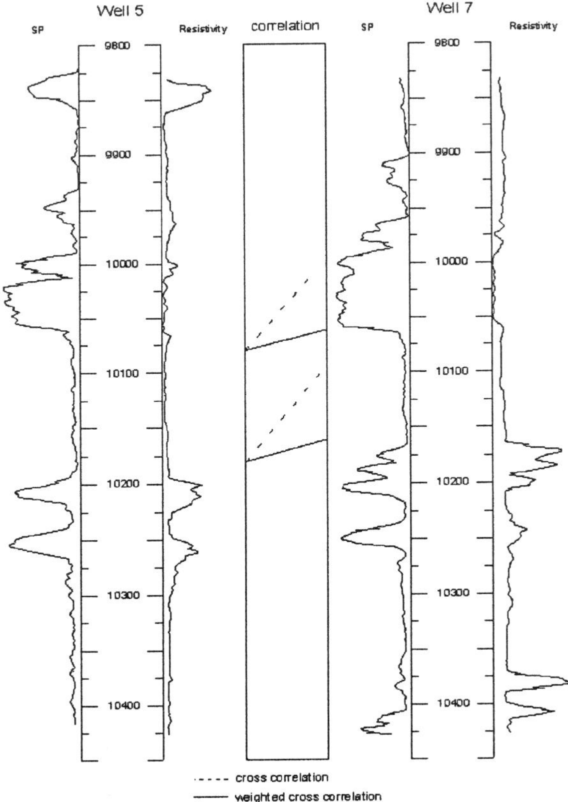

Fig. 5. Interwell correlation by cross-correlation methods

However, the SP logs indicate that reference interval in Well 5 is entirely with a shale, while the interval of maximum cross-correlation within Well 7 includes both shales and sandstone. The weighted cross-correlation is processed on pairs of wells that ideally have same pair of wireline logs: a shale log (gamma ray or SP) and a correlation log (typically a high-resolution resistivity log) (Olea, 1994). This method can modify the results obtained by simple cross-correlation method. The new correlation is found at a depth of 10,112 ft, which is the geologically correct correlation (solid line in Fig. 5).

Although this modified method may make a reasonable match, it cannot consider the stretching or shrinking of stratigraphic interval because of a fixed correlation length. However, pattern recognition can detect similar characteristics from log traces despite their different scaling in depth or amplitude. In addition, the use of principal component log can solve the problem of geological incorrect match and make correlation more efficient and accurate. Fig. 6 shows the first principal component logs computed for each well. Even if two well-logs are used in this example, all the available wireline logs can be included for PCA. Pattern recognition can handle the variation in strata thickness. The new correlation by pattern recognition is shown in Fig. 6, in which 124-ft interval

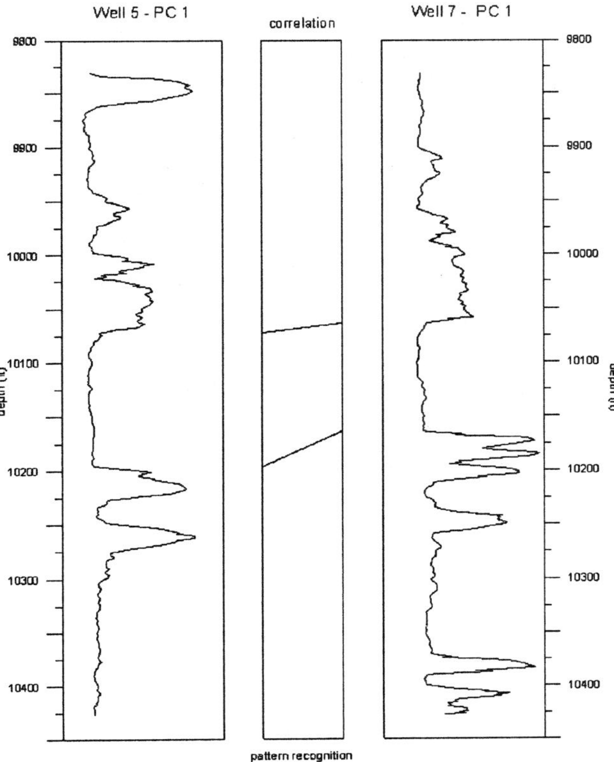

Fig. 6. Interwell correlation by pattern recognition method.

of Well 5 (10,072 ft to 10,196 ft) corresponds to 100-ft interval of Well 7 (10,064 ft to 10,164 ft). The new approach can obtain more reliable and efficient result rather than conventional methods.

4.3. Field examples

The expert system is applied to correlate between wells in the Korea Continental Shelf and oversea field.

For Well D-1 and Well D-2 in the Korea Continental Shelf, the logs selected for correlation were gamma-ray log (GR), induction log deep (ILD), spherically focused log (SFLU), and neutron log (NPHI) of a 200-m interval taken from two wells (Figs. 7 and 8). These logs were chosen based on the descriptive statistical results and the quality of data. The principal components were computed from the correlation matrix of log measurements. The first principal component collectively accounted for 80% of the total variability. Hierarchical clustering was used to group the segmented principal component log traces. The three groups of electrofacies were determined on the basis of the ratio of the within-group variance to the total variance. Fig. 9

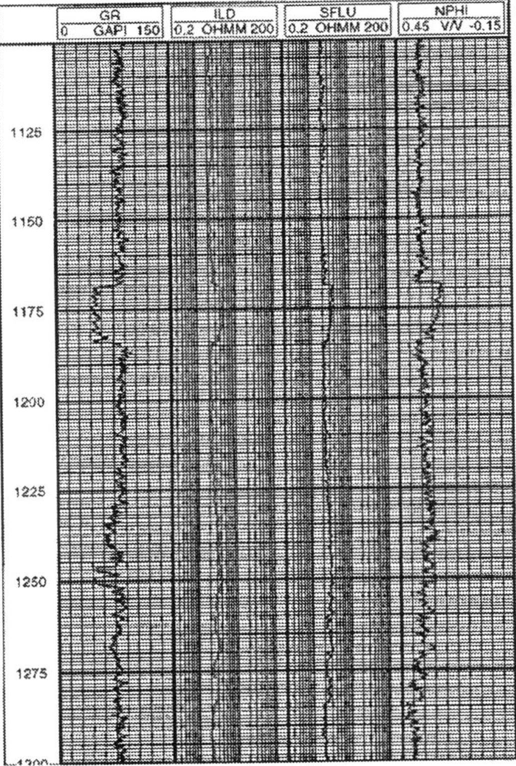

Fig. 7. Measured well logs of Well D-1.

shows lithologic description from electrofacies. Correlation results between Well D-1 and Well D-2 with the first principal component logs and electrofacies are shown in Fig. 10.

For Well L-1 and Well L-2 in oversea field, gamma-ray log (GR), induction log deep (ILD), neutron log (NPHI), and short normal (SN) of a 600-ft interval were selected for well-to-well correlation (Figs. 11 and 12). The first principal component collectively accounted for 89% of the total variability. This example is more complex than previous one. The new approach can be particularly useful for the formation with more complex geologic sequences. Lithologic description from electrofacies is shown on Fig. 13. Correlations between Well L-1 and Well L-2 were successfully made on the basis of the first principal component logs with electrofacies classification (Fig. 14).

5. CONCLUSIONS

This method can make more reliable and efficient well-to-well log correlation rather than the traditional methods in which only one approach was adopted.

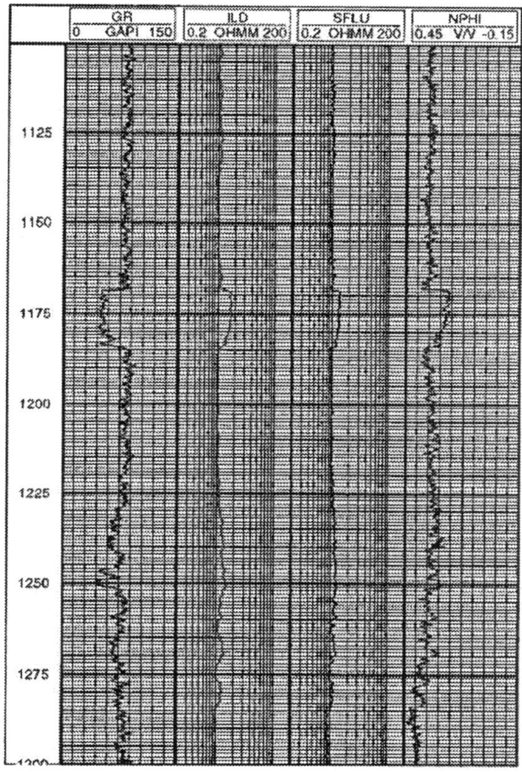

Fig. 8. Measured well logs of Well D-2.

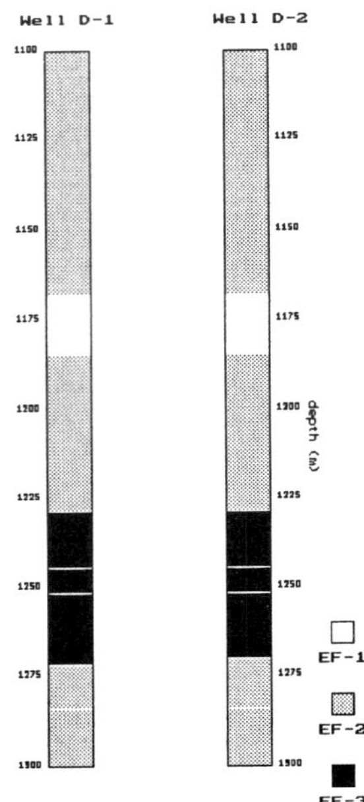

Fig. 9. Electrofacies classification for Well D-1 and Well D-2.

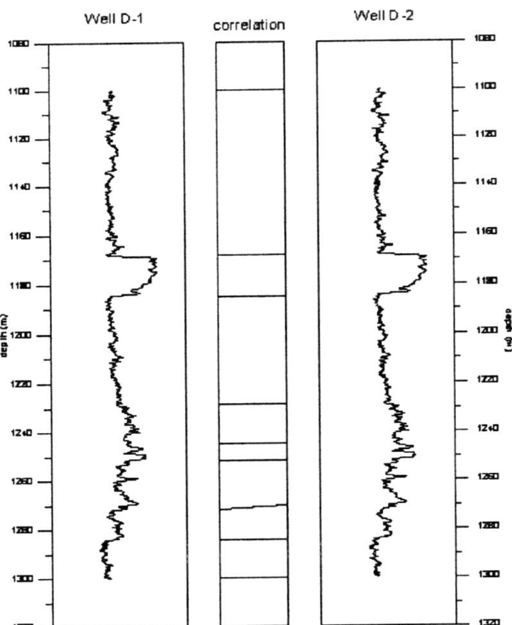

Fig. 10. Well-to-well correlation result between Well D-1 and Well D-2 with the first principal component logs.

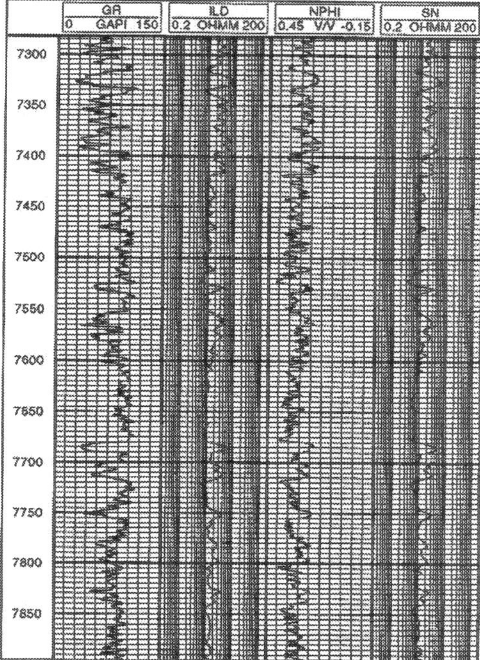

Fig. 11. Measured well logs of Well L-1.

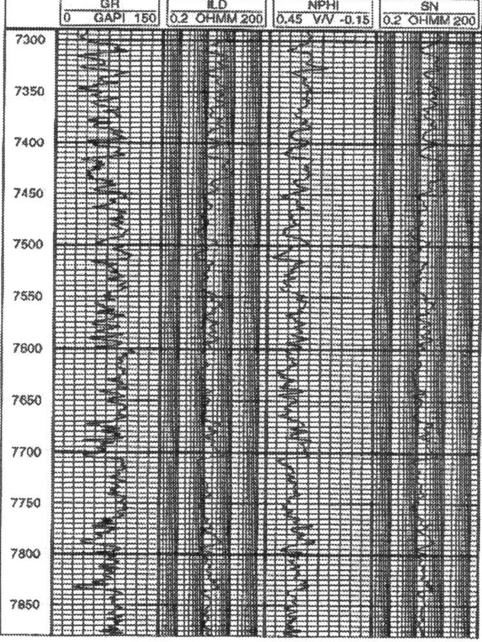

Fig. 12. Measured well logs of Well L-2.

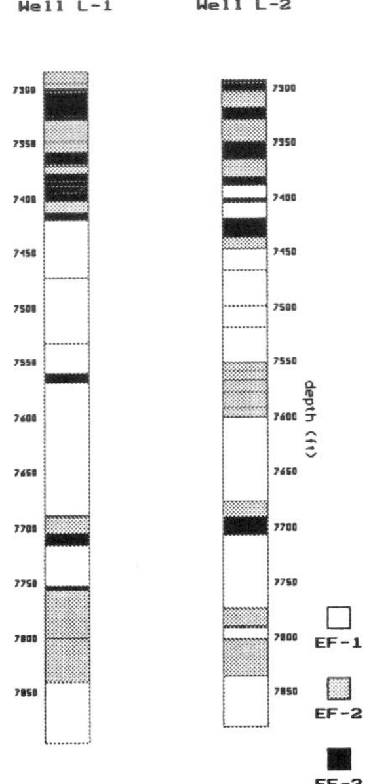

Fig. 13. Electrofacies classification for Well L-1 and Well L-2.

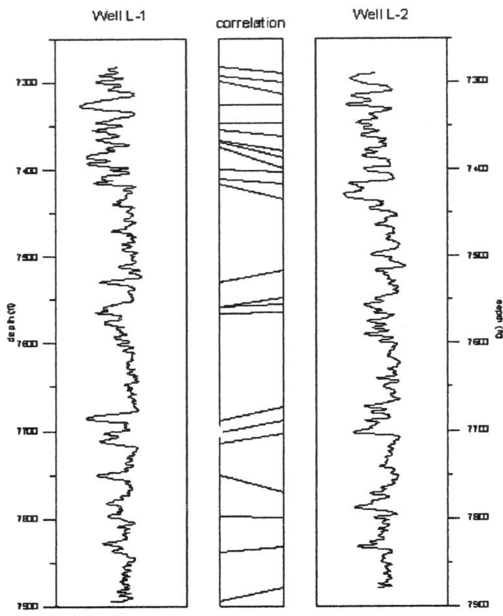

Fig. 14. Well-to-well correlation result between Well L-1 and Well L-2 with the first principal component logs.

The expert system is capable of handling the shifting, thickening, and thinning strata in well-to-well log correlation. Use of the principal components included all available well-log data can be obtained correlation results with high degree of reliability. Electrofacies identification can be applied to the complex geologic formation as lithology information for correlation. This procedure saves time and also minimized the subjectivity involved in the well-to-well log correlation process.

REFERENCES

Al-Sabti, H.M., 1991. Lithology determination of clastic reservoir facies from well logs, Saudi Arabia. *1991 SPE Middle East Oil Show*, Bahrain, Nov. 16–19, Paper SPE 21457.

Al-Sabti, H.M. and Al-Bassam, K.A., 1993. 3D electrofacies model, Safaniya Reservoir, Safaniya Field. *1993 SPE Middle East Oil Technical Conference and Exhibition*, Bahrain, Apr. 3–6, Paper SPE 25611.

Bucheb, J.A. and Evans, H.B., 1994, Some Applications of Methods Used in Electrofacies Identification, The Log Analyst (Jan.–Feb.) 14–24.

Busch, J.M., Fortney, W.G., and Berry, L.N., 1987. Determination of lithology from well logs by statistical analysis. *Soc. Pet Eng. Formation Eval.*, Dec., pp. 412–418.

Davis, J.C., 1986. *Statistics and Data Analysis in Geology, 2nd Edition*. Wiley, New York, NY.

Definer, P., Peyret, O. and Serra, O., 1987. Automatic determination of lithology from well logs. *SPE Formation Evalion*, pp. 303–314.

Descalzi, C., Rognoni, A. and Cigni, M., 1988. Synergetic log and core data treatment through cluster analysis: methodology to improve reservoir description. *1988 SPE International Meeting on Petroleum Engineering*, Tianjin, China, Nov. 1–4, Paper SPE 17637.

Doveton, J.H., 1994. *Geologic Log Analysis Using Computer Methods*. American Association of Petroleum Geologists, Tulsa, OK.

Elek, I., 1988. Some applications of principal component analysis: well-to-well correlation, zonation. *Geobyte*, pp. 46–55.

Elek, I., 1990. Fast porosity estimation by principal component analysis. *Geobyte*, pp. 25–34.

Gill, D., Shomrony, A. and Fligelman, H., 1993. Numerical zonation of log suites and logfacies recognition by multivariate clustering. *Am. Assoc. Pet. Geol. Bull.*, 77(10): 1781–1791.

Lang, W.H., Jr., 1986. Correlation with Multiple Logs. *Log Analyst*, Jan.–Feb., pp. 43–52.

Lee, S.H. and Datta-Gupta, A., 1999. Electrofacies characterization and permeability predictions in carbonate reservoirs: role of multivariate analysis and nonparametric regression. *SPE Annual Technical Conference and Exhibition*, Houston, TX, Oct. 3–6, Paper SPE 56658.

Lim, J.-S., Kang, J.M. and Kim, J., 1997. Multivariate statistical analysis for automatic electrofacies determination from well log measurements. *SPE Asia Pacific Oil and Gas Conference and Exhibition*, Kuala Lumpur, Malaysia, Apr. 14–16, Paper SPE 38028.

Lim, J.-S., Kang, J.M. and Kim, J., 1998,. Artificial intelligence approach for well-to-well log correlation. *SPE India Oil and Gas Conference and Exhibition*, New Delhi, India, Apr. 7–9, Paper SPE 39541.

Olea, R.A., 1994. Expert system for automated correlation and interpretation of wireline logs. *Math. Geol.*, 26(8): 879–897.

Olea, R.A. and Davis, J.C., 1986. An artificial intelligence approach to lithostratigraphic correlation using geophysical well logs. *SPE Annual Conference and Exhibition*, New Orleans, LA, Oct. 5–8, Paper SPE 15603.

Serra, O. and Abbott, H.T., 1982. The contribution of logging data to sedimentology and stratigraphy. *Soc. Pet. Eng. J.*, pp. 117–131.

Soto, B.R. and Soto, T.C., 1995. An object oriented expert system to enhance the log analysis of the Columbian basins. *1995 SPE Petroleum Computer Conference*, Houston, TX, Jun. 11–14, Paper SPE 30196.

Startzman, R.A. and Kuo, T.B., 1987. A rule-based system for well log correlation. *SPE Formation Evalion*, pp. 311–319.

Wolff, M. and Pelissier-Combescure, J., 1982. FACIOLOG – automatic electrofacies determination. *1982 SPWLA Annual Logging Symposium*, Corpus Cristi, TX, Jul. 6–9, Paper FF.

AUTHOR INDEX

(Page numbers in italics correspond to the reference list at the end of each chapter)

Aadland, A., 369, *394*
Aanonsen, S.I., 369, *394*
Aasen, J.O., *394*
Abbott, H.T., 676, *687*
Abdul Majeed, G.H., 415, 419, 438, *442*
Abdus Sattar, A., *442*
Accarain, P., 415, *442*
Acinas, J.R., 190, *215*
Acock, B., 329, *362*
Adams, R., *31*, *497*
Adams, R.D., 15, 19, *29*, 476, *496*
Agarwal, R.G., 445, *466*
Agrawal, R.B., 67, *75*
Agterberg, F.P., 15, *29*
Ahmed, T., 415, 423, *442*
Aki, K., *142*, *496*
Al-Bassam, K.A., *687*
Al-kaabi, A.W., 415, *442*
Al-Marhoun, M.A., 415, 418, 435, 438, *442, 443*
Al-Matter, D., *442*
Al-Sabti, H.M., *687*
Alabert, F., 94, *95*
Alcott, J.M., 191, *217*
Ali, J.K., 415, *442*
Alikhan, A.A., 418, 422, 423, 426, 435, 439–441, *442*
Altany, D.A., *32*, *497*
Ameri, S., *411*, *467*, *632*
Aminian, K., *467*
Aminiand, K., *411*
Aminzadeh, F., 6, 7, 9, 10, 12–15, 18–20, *29*, *31*, *32*, 120, *141*, *142*, 178, *184*, *185*, *319*, 471, 472, *496*, *497*, 652, 653, 657, 669, *672*
Amos, C.L., 190, *216*
Amrhein, C.G., 634, *650*
Ånes, H.M., *155*
Angevine, C.L., 190, 192, *215*
Anterion, F., 327, *361*
Archer, J.S., 250, *271*
Arefi, R., *467*, *632*
Arens, G., *245*
Areti, R., *411*
Armentrout, J.M., 536, *550*, *551*
Asgarpour, S., *442*
Ashby, S.F., *395*
Assad, Y., *443*
Assilian, S., 196, *216*
Atiya, A., 379, *394*
Attanucci, V., *395*
Aubry, M.-P., *551*
Austin, J., 605, *632*
Averill, R., *362*
Azimi, M.R., 657, *671*

Babuška, R., *48*
Bahorich, M., 160, *172*
Bahorich, M.S., *173*
Bailey, H.W., *411*
Bailey, T.C., 634, *649*
Baker, J.E., 71, *75*
Balan, B., *632*
Baldwin, J.L., 19, *29*, 119, *142*, 293, *307*, 472, *496*
Ball, L., *551*
Banik, A., *602*
Barbosa, H.J.C., *362*
Barhen, J., 13, *29*, *30*, *184*
Barker, J.W., *362*
Barlett, R.M., *319*
Barton, M.D., 274, *287*
Basinski, P., 596, *601*
Bateman, A.R.M., *29*, *142*, *496*
Baygun et al., 16, *30*
Beal, M., 475, *496*
Beale, M., 431, *442*
Beckett, D., *411*
Beckmann, J.P., *551*
Begg, S.H., 275, *287*
Beggs, H.D., 418, 419, 422, 423, 432, 438, 439, 441, *442*, *444*
Begt, A.K., *650*
Behrenbruch, P., 369, *394*
Behrens, R.A., 274, *287*
Bellis, J.K., *411*
Bemuth, H., 475, *496*
Benediktsson, J.A., 13, *30*
Benson, D.J., *216*

Berggren, W.A., 532, *551*
Berman, M., *245*
Bernath, A., *287*
Berry, J.K., 634, *649*
Berry, L.N., *687*
Beucher, H., *287*
Beyer, H.-G., 67, *75*
Bezdek, J.C., 9, 16, 23, 27, *30*, *48*, *142*, 474, 494, *496*
Bishop, C.M., 504, 505, 508, *525*
Bissell, R., 328, *361*
Bissell, R.C., 327, 328, *361*
Bittencourt, A.C., 370, *394*
Blanc, G., *287*, 328, *362*
Blanchet, C., *244*
Bliefnick, D., *411*
Bloch, R.B., *551*
Blockley, D., *551*
Boadu, F.K., 18, 19, *30*, 120, 129, 135, *142*, 471, 472, *496*
Boerner, S., *271*
Bois, P., 16, *30*
Bonissone, P.P., *49*
Book, S., *649*
Borden, G., *442*
Bos, C., 330, 337, *361*
Bosence, D., 190, 191, *215*, *217*
Bosence, D.W.J., 190, 191, *215*
Bosscher, H., 190, 195–198, *215*
Boucher, M., *362*
Bouma, J., *650*
Bowen, B.E., *552*
Braithaite, K., *411*
Brand, P.J., *95*
Brebbia, C.A., 190, *215*
Breiman, L., 21, *30*, 653, *671*
Breitenbach, E.A., 368, *394*
Brevik, I., 151, *155*
Bridge, J.S., 190, 191, *215*, *216*
Bril, A.H., *172*, *173*
Briones, 415, *442*
Broecker, W.A., 190, 191, 205, 206, *215*
Brooks, L.C., *601*
Brown, A.A., *411*
Brown, C., *216*
Brown, C.B., *216*
Brown, E., *216*
Bryant, I.D., 190, *215*
Bubb, J.N., *217*
Bucheb, J.A., 676, *687*
Buddemeier, R.W., 190, *215*
Bujak, J.P., 534, *552*
Burkhart, T.D., *155*
Burrough, P.A., 633, 634, *649*
Burton, C.L., 645, *649*
Burton, R., 190, *215*
Burtschy, B., 220, *244*
Busch, J.M., *687*
Bush, M.D., 330, *361*

Caers, J., 176, 181, *185*, *525*
Caixerio, E., *442*
Calkins, H.W., *650*
Camp, C.C., *32*, *142*, *497*
Cannon, R.L., 18, *30*
Capilla, J.E., *362*, *525*
Carter, J.N., 60, *75*, 330, 344, *361*
Carter, R.D., 326, *361*, *466*
Carter, R.R., *287*
Cartwright, H.M., *361*
Chalker, B.E., 195, *215*
Chang, H.-C., 310, *319*
Chapellier, D., *245*
Chappaz, R.J., 16, *30*
Charnock, M.A., 536, 539, *551*
Chatterjee, S., 10, 14, 15, 18, *29*, *141*, 471, *496*
Chavent, G., 326, *361*
Chawathé, A., *287*
Chawathe, A., 18, 19, *30*, 293, *307*, 471, 472, *496*
Chen, D.C., *30*
Chen, H., *216*
Chen, H.C., 16, *30*, *319*, 528, *551*
Chen, S., 431, *442*
Chen, W.H., 326, *361*
Chernov, S., *650*
Cheung, V., *442*
Childs, C., *363*
Chiles, J.-P., *115*
Cho, S., 178, *185*
Chopra, A.K., *287*
Chork, C.Y., *319*
Christakos, G., 384, *394*
Christie, M.A., 274, *287*
Chu, L., 275, *287*
Cichowicz, L., 445, *467*
Cieniawski, S.E., 382, *394*
Cigni, M., *687*
Cinco-Ley, H., 445, 464, *466*
Clement, J.F., 536, *550*
Coats, K.H., 325, *361*
Collinson, J.D., 193, *216*
Collister, J.W., *29*, *496*
Compton, P., *319*
Cook, M., *271*
Cook, W.J., 77, *83*
Copestake, P., 532, 534, *552*
Coppock, J.T., 634, *649*

Corbett, P., *32*, *362*
Coskuner, G., 369, *394*
Cowan, C.F.N., *442*
Cowen, D., 634, *649*
Cox, B.L., 669, *671, 672*
Craig, M., *245*
Craig, P.S., 328, *361*
Crawford, K.D., *31*
Cressie, N.A.C., 651, *671*
Cronquist, C., *442*
Cross, T.A., *215*
Culver, S.J., 539, *551*
Cunningham, W.H., *83*
Cybenko, G., 24, *30*, 138, *142*

Dagan, G., *525*
Dalley, R.M., 160, *172*
Darken, C.J., 432, *443*
Datta-Gupta, A., *363*, 676, *687*
Dave, J.V., *30*
Davies, J., *551*
Davies, L.D., *75*
Davis, B.E., 634, *649*
Davis, J.C., 178, *185*, 651, *671*, 674, 675, 680, *687*
Davis, M., 94, *95*
de Groot, P., 12, *29*
de Groot, P.F.M., *172, 173*, 399, *411*
de Jong, K.A., 65, *76*
de Marsily, G., 328, *362*, *525*, 651, *671*
Deb, K., 67, 71, *75*
Definer, P., *687*
Delfiner, P., *115*
Delzoppo, G., *319*
DeMers, M.N., 634, *650*
Demicco, R.V., 48, *48*, 191, 211, *215*
Dempsey, J.R., *361*
Demuth, H., 431, *442*
Demyanov, V., 635, *650*
den Boer, L.D., *287*
Déquirez, P.Y., *244*
Derain, J.F., 228, *244*
Desbrandes, R., 415, *442*
Descalzi, C., *687*
Deschamps, T., 327, *361*
Deutsch, C., 276, *287*
Deutsch, C.V., 86, 87, 93, *95*, *115*, 274, *287*, *361*, 500, 506, 510, *525*, 559, *581*, 651, *671*
Di Nola, A., 42, *48*
Diaz, B., *32*
Dickenson, B., 536, *551*
Ding, H., 13, *30*
Dokla, M.E., 418, 419, 435, 438, *442*
Dominguez, N., *466*
Dougherty, D.E., 367, 384, 385, *394, 395*
Doveton, J.H., 675, *687*
Dowla, F.U., 367, *395*
Dowsland, K.A., 384, 385, *394*
Doyen, P.M., 274, *287*
Dranfield, P., *287*
Driankov, D., 48, *48*
Dronen, O.M., *394*
Dubes, R.C., 313, *319*
Dubois, D., 40, 42, *48*
Dubrule, O., *361*
Dumay, J., 228, *244*
Dunham, R.J., 401, *411*
Dunn, P.A., 191, *215*
Dupuy, M., *361*
Durlofsky, L.J., 275, *287*
Durrans, S.R., *319*
Durrer, E.J., 326, *363*
Dvorkin, J., *525*
Dyrnes, O., *394*

Eastman, J.R., 633, 634, *650*
Eby, D., 361, *362*
Echols, R.C., *550*
Edington, D.H., 192, *215*
Egeland, T., *362*
Eheart, J.W., *394, 395*
Ehlers, J., *32*
Ehlig-Economides, C.A., *443*
Eide, A.L., *394*
Ekart, D., *31*, *497*
Ekart, D.D., *29*, *31*, *496*
Ekren, B.O., *155*
Elam, F.M., *442*
Elek, I., 674, 675, *687*
Elgibly, A.A., *442*
Ellis, G.W., 635, *650*
Elsharkawy, A.M., 415, 418, 422, 423, 426, 435, 439–441, *442*
Emery, D., 191, *215*
Ershaghi, I., 293, *307*, 415, *443*
Erson, O.K., *30*
Eschard, R., *287*
Eshelman, L.J., 67, *75*
Esogbue, A.O., 256, *271*
Evans, H.B., 676, *687*
Eymard, R., *361*

Falgout, R.D., *395*
Fang, J.H., *30*, *32*, *142*, 191, *215*, *319*, *497*, 528, *551*
Faraj, A., 220, 221, *244, 245*
Farell, A.E., *31*, *142*, *497*, *672*

Farmer, C.L., 94, *95*, 329, *362*
Farmer, S., 160, *172*
Farmer, S.L., *173*
Farshad, F.F., 415, 418, 419, 423, 435, 439, 441, *442–444*
Fasanino, G., *362*
Fassett, J.E., 593, *601*
Faulkner, T., 605, *632*
Faybishenko, B., *671, 672*
Fearn, L.B., *551*
Feuchtwanger, T., *244*
Filev, D.P., 48, *49*
Filewicz, M., 531, 532, *551*
Fligelman, H., *687*
Flint, S.S., 190, *215*
Floris, F., *361*
Fogwell, T.W., *395*
Foley, L., 528, *551*
Forster, A., 190, *215*
Fortney, W.G., *687*
Fossen, H., *155*
Fournier, F., 220, 228, *244*, *287*
Frais, F., 329, *362*
Franseen, E.K., 190, *215*
Freeman, B., *363*
Friedman, J.H., 21, *30*, *244*, 653, *671*
Friedman, M., 594, 601, *602*
Fritzke, B., 278, *287*
Fu, K.S., 313, *319*
Fujii, H., 370, *394*
Furlong, K., *216*
Furuhashi, T., *30, 31*, *496*
Furushashi, T., *496*

Gabriels, P.W., 144, *155*
Gallagher, L.T., *411*
Gamber, J.H., *552*
Garbeer, J.O., *442*
Garcia, A.F.D, Loula, E.L.M., *362*
Garrett Jr., J.H.*650*
Gary, A.C., 531, 532, *551*
Gastaldi, C., *271*
Gatens, J.M., 445, *466*
Gatrell, A.C., 634, *649*
Gaudiani, P., *245*
Gavalas, G.R., 327, *361–363*
Gawith, D.E., *155*
Geary, R.C., 223, *244*
Gedeon, T.D., *32*, *142*, *185*, *319*, *497*
Gelatt, C.D., *395*
Geldart, L.P., 15, *32*, *271*
Gelfand, B., *362*
Geroch, S., *551*
Ghaboussi, J., 635, *650*
Gharbi, R.B., 415, *442*
Ghetto, G., 415, *443*
Gibson, T.G., 529, *551*
Gill, D., *687*
Girosi, F., 432, *443*
Glangeaud, F., 228, *244*
Glasø, O., *443*
Glover, C.W., *29*, *184*
Goggin, D., *362*
Goldberg, D.E., 53, 71, 72, *75*, 368, 382, *395*
Goldstein, M., *361*
Gomez, S., 329, *362*
Gomez-Hernandez, J., *525*
Gómez-Hernández, J.J., *362*
Gomide, F., 48, *49*
Goodall, J., *551*
Goodchild, M.F., 634, *650*
Goodman, D.M., *395*
Goodman, E., *362*
Goovaerts, P., *115*, 220, *244*
Gosselin, O., *362*
Gou, Y., *245*
Gradstein, F.M., *551*
Graham, M., *271*
Grant, P.M., *442*
Grassle, J.F., *551*
Green, A., 220, *245*
Griffith, C.M., 312, *319*
Griffith, D.A., 634, *650*
Griffiths, C.M., 15, *29, 30*
Gringarten, A.C., 445, *466*
Grussaute, T., *361*
Guërillot, D., *287*
Guérillot, D., *287*
Guardiano, F., 503, *525*
Guerreiro, J.N.C., 330, *362*
Gulati, S., *30*
Gulley, N., 9, 27, *30*, 141, *142*, 474, 475, *496*
Gum, T., *601*
Gupta, M.M., 13, *30*, 474, *497*
Gutteridge, P., *155*

Habiballah, 415, *443*
Haga, O., *155*
Hague, Y., *217*
Haldorsen, H.H., 94, *95*
Hampton, M.J., *411*
Harbaugh, J.W., 190, 191, *216, 217*
Harff, J., 190, *216*
Harik, G.R., 72, *75*
Harris, S.P., *361*
Hassibi, M., 293, *307*
Hastings, B.S., *217*

Hatleid, W.G., *217*
Hawkins, D.J., *245*
He, N., *288*
Heath, A.E., *363*
Hecht-Nielsen, R., 19, 23, 24, *30*, 137, 138, *142*, 472, 473, *496*, 637, 639, *650*
Heggland, R., 158, 159, *172, 173*
Hegstad, B.K., 505, *525*
Hellendoorn, H., 48, *48*
Heller, P.L., *215*
Hellums, L.J., *363*
Henderson, J.H., *361*
Hepple, L.W., 634, *650*
Hernandez, P.M.A., *444*
Herrick, D.C., *551*
Hesthammer, J., *155*
Hibbert, B., *319*
Hilde, E., *155*
Hill, D.G., *632*
Hill, N.C., 426, 440, 441, *443*
Hillier, F.S., 77, *83*
Hinderaker, L., *362*
Hinton, G.E., *185*, *395*, *497*
Hirasaki, G.J., 326, *362*
Hirsche, K., 250, *271*
Holden, L., *361, 362*, *394*
Holland, J.H., 53, 71, *75*
Holloway, S., 532, *551*
Holmes, N., *551*
Holmes, N.A., 535–538, *551*
Hopkins, C.W., 445, *466*
Horikawa, S., *30*, 472, *496*
Horikawa, S.I., 474, *496*
Horne, R.N., 328, *362*, 370, *394*
Horvei, N.A., *155*
Hovland, M., 159, *172*
Hu, L-Y., 328, *362*
Huang, C., 329, *362*
Huang, M.D., 355, *362*
Huang, Z., 15, *30*, 313, *319*
Huijbregts, C., *115*
Huijbregts, C.J., 333, *362*
Huijbregts, Ch.J., 222, *245*
Hunt, P.E., *32*, *497*
Hurst, A., *551*

Instefjord, R., *155*
Ioannides, N.S., *552*
Isaak, E.H., 651, *672*
Isaaks, E., *115*
Isaaks, E.H., 85–87, 92, *95*, 222, 232, *245*
Ishiguro, A., *363*
Ishii, Y., *32*, *497*

Jacquard, P., 325, *362*
Jahn, F., 250, *271*
Jahns, H.O., 325, *362*
Jain, A.K., 313, *319*
Jain, C., 325, *362*
Jakobsen, K.G., *155*
Jamshidi, M., 7, 15, *29*
Jang, J.S.R., 9, 26, 27, *30*, 140, 141, *142*, 474, 475, 493, *496*
Jensen, J., 333, *362*
Ji, C., 379, *394*
Jian, F.X., 309, *319*, *411*
Jiang, F.X., *32*, *142*, *497*
Jizba, D., *155*
Johann, P., 274, *287*
Johansson, E.M., 379, *395*
Johnson, R.J., 634, *650*
Johnson, V.M., 13, *30*, *395*
Johnstad, S.E., 144, *155*
Johnston, D.H., *155*
Jones, R.C., *287*
Jones, R.W., 536, 539, *551*
Joslin, T.H., *395*
Journal, A.G., 651, *671*
Journel, A.C., *361*
Journel, A.G., 86, 87, 93, 94, *95*, *115*, 176, *185*, 222, *245*, 333, *362*, 503, 506, 510, *525*, 559, *581*
Judd, A.G., 159, *172*
Juniardi, I.R., 415, *443*
Justice, J.H., 229, *245*

Kacprzk, J., 256, *271*
Kalkomey, C., *271*
Kalogerakis, N., 327, *363*
Kaminski, M.A., 536, *551*
Kandel, A., 48, *48*
Kanevski, M., *185*, *650*
Kang, J.M., *687*
Karaboga, D., 77, *83*
Karatzas, G.P., 367, *395*
Karcher, B., *361*
Karprapu, S., *32*, *142*, *497*
Karr, C.L., *32*, *142*, *319*, *497*
Kartoatmodjo, T., 418, 419, 422, 423, 435, 439–441, *443*
Katz, D.L., 414, *443*
Katz, S., *142*, *496*
Kaufmann, A., 474, *497*
Keith, B.D., 402, *411*
Kemp, L.F., *361*
Kendall, C.G.St.C., *215*, *217*
Kent, D.V., *551*

Kerre, E.E., 47, *49*
Kikani, J., 369, *395*
Killam, B.R., 384, *394*
Killough, J.E., *361*
Kim, J., *687*
King, M.J., 275, *287*
King, P.R., 275, *287*
Kinsey, D.W., 190, *217*
Kirkpatrick, S., 368, *395*
Kirlin, R.L., 161, *173*
Kitanidis, P.K., 512, *525*
Klawonn, F., 204, 211, *216*
Klimentos, T., 19, *31*, 120, *142*, 472, *497*
Klir, G., 211, *215*
Klir, G.J., 36, 40, 46–48, *48, 49*, 192, 194, 204, 211, *216*, 256, 260, 263, *271*
Knauss, J.A., 193, *216*
Kneller, B., 543, *551*
Knopp, C.R., *443*
Knox, P.R., 274, *287*
Knox, R.W.O'B., 532, *551*
Kobayashi, S., 67, 68, *76*
Koehler, F., *32*
Kohonen, T., 9, 24, 28, *30*, 138, *142*, 277, *287*, 474, 495, *497*
Kopaska-Merkel, D.C., *319*
Koperna, G., *632*
Kortright, M.E., *30*
Kosko, B., *49*, 204, 211, *216*, 529, *551*
Koster, J.K., *155*
Kovscek, A.R., 328, *363*
Kovseck, T.W., *672*
Koza, J.R., 60, *75*
Kramer, M.A., *443*
Kravitz, J.H., 190, 191, *216*
Krebs, W.N., *552*
Krzyzak, A., 431, *443*
Ku, T.-L., *216*
Kumar, M., 369, *395*
Kumoluyi, 415, *443*
Kuo, T.B., 674, 677, 679, *688*
Kuromiya, A., *30, 31*, *496*

LA Coe, M.K., *552*
Labedi, R., 419, 422, 426, 438, 439, 441, *443*
Lake, L., *362*
Lake, L.W., *363*
Lamy, P., *361*
Landa, J.L., 328, *362*
Landrø, M., *155*
Lang Jr., W.H., *687*
Lankester, H.G., *215*
Lasater, J.A., *443*
Lashgari, B., 18, *31*
Lavedan, G., *362*
LaVenue, A.M., *525*
Lavergne, M., 228, *245*
Lebart, L., 220, 221, *244, 245*
LeBlance, J.L., *442*
Lee, C., 635, *650*
Lee, C.S.G., 24, *31*, 138, *142*
Lee, G., 211, *216*
Lee, J.W., 415, *442*
Lee, S.H., 676, *687*
Lee, T.-W., 184, *185*
Lee, T.C., 278, *287*
Lee, T.D., *550*
Leeder, M.P., 190, *216*
Leeder, M.R., 191, *215*
Lehr, M.A., 24, *32*, 472, *497*
Lehr, S.K., *142*
Lemke, W., *216*
Lemonier, P., *361*
Leonard, J.A., 431, *443*
Lepine, O., *361*
Lerche, I., *215*
Levenick, J.R., 344, *362*
Levey, R., 19, *31*, 476, *497*
Levey, R.A., *29*, *31*, *496*
Lewis, J.J.M., *288*
Li, J., 210, *216*
Li, M.Z., 190, *216*
Li, Y., 14, *31, 32*
Lia, O., *362*
Lieberman, G.J., 77, *83*
Lim, J.-S., 674–677, *687*
Lin, C.T., 24, *31*, 138, *142*, 211, *216*
Lin, M.D., 382, *395*
Lin, Y.K., 313, *319*
Liou, R.J., 657, *671*
Lippmann, R.P., 13, *31*
Lisle, J.L., 590, *601*
Liu, X., 14, *31*
Livnat, Y., *31*, *497*
Loosmore, G.A., *395*
Loucks, R.G., 191, *216*, 398, 399, 405, 408, *411*
Lowenstein, T.K., 210, 211, *216, 217*
Lumley, D.E., 151, *155*
Luo, S., *216*
Lutes, B., 369, *394*
Lyle, W.D., *551*

Macallister, D.J., *31*
Macaulay, C.I., 398, 402, 405, *411*
MacDonald, C.J., *95*
Mack, G.H., *216*
Mackey, S.D., 190, 191, *216*

MacLeod, M.K., *287*
MacQueen, J., 27, *31*, 141, *142*, 494, *497*
Macy, R.B., 293, *307*
Mahfoud, S.W., 72, *75*
Mahgoub, I.S., *443*
Malczewski, J., 634, *650*
Mallick, S., 17, *31*
Malta, S.M.C., *362*
Mamdani, E.H., 196, *216*
Mancini, E.A., *216*
Mantoglou, A., 94, *95*
Manzocchi, T., *362*
Marble, D.F., 634, *650*
Marfurt, K.J., 161, *173*
Mari, J.L., 228, *244*, *245*
Marietta, M.G., *525*
Marryott, R.A., 384, 385, *394*, *395*
Masoner, L.O., *395*
Matheron, G., 85, 94, *95*, 219–221, *245*
Mathews, O., *362*
Matsushita, S., *31*
Mavko, G., 508, *525*
Mayer, A.S., 329, *362*
Mayers, D., *361*
McCaffrey, W., 543, *551*
McCain, W.D., 426, 440, 441, *443*
McCann, C., 19, *31*, 120, *142*, 472, *497*
McClelland, J.L., 636, *650*
McCormack, M.D., 13, 14, 17, *31*
McDonnell, R.A., 634, *649*
McKay, D.M., *319*
McKenny, R.S., *155*
McKinney, D.C., 382, *395*
Mclauchlin, L., *442*
McPherson, J.G., *307*
McQuilken, *411*
McVay, D., 445, 455, *467*
McVey, D.S., *467*
Medsker, L.R., 22, 23, *31*, 137, *142*
Meehan, N., 276, *287*
Meldahl, P., 159, 172, *172*, *173*
Mendel, J.M., 48, *49*, 199, *216*
Mengshoel, O.J., 72, *75*
Merriam, D.F., 190, *215*
Michalewicz, Z., 382, *395*
Miller, R.E., 77, *83*
Millero, F.J., *216*
Millheim, K.K., 445, *467*
Mitchell, M., 53, *76*
Mitchum Jr., R.M., *217*
Mitchum, R.M., 536, *551*
Mo, Y., *287*
Mohaghegh, S., 399, *411*, 415, *443*, 447, *467*, 606, 631, *632*
Mohamood, M.A., 415, *443*
Moiola, R.J., *307*
Monson, G.D., 18, 19, *31*, 471, 472, *497*
Moody, J.E., 432, *443*
Moody, R.T.J., *411*
Moore, R.E., 39, *49*
Morineau, A., *245*
Morse, J.W., 190, 191, *216*
Mosher, C., *287*
Mudge, D.C., 532, 534, *552*
Mukerji, T., *525*
Mulholland, M., 312, *319*
Murray Jr., G.H., 590, *601*
Mutti, E., *307*
Myers, K.J., 191, *215*

Nauck, D., 204, 211, *216*
Neal, J.E., 535, *552*
Needham, D.T., *363*
Nell, P., *362*
Nelson, R., 586, *601*
Nepveu, M., *361*
Neumaier, A., 39, *49*
Newberry, J.D.H., *288*
Nguyen, H.T., 48, *49*
Niakan, M.R., *443*
Nie, X., *650*
Nikravesh, M., 5, 17–21, *29*, *31*, *32*, 119, 120, *142*, *185*, 471, 472, 491, *496*, *497*, 652, 653, 657, 669, *671*, *672*
Nikravesh, M.M., 652, 659, *672*
Nittrourer, C.A., 190, 191, *216*
Nordlund, U., 190, 191, 193, 199, *216*
Norland, U., 528, *552*
Normark, W.R., *307*
Novák, V., *49*
Novak, B., *31*, *497*
NPC (National Petroleum Council), 5, *31*
Nutter, R., *411*
Nwnakwo, B., *551*

O'Connell, S., *307*
O'Connor, S.J., 532, *552*
Obomanu, D.A., 418, 419, 435, 438, *443*
Okpobiri, G.A., 418, 419, 435, 438, *443*
Olea, R., *115*
Olea, R.A., 674, 680, 681, *687*
Oliver, D., *288*
Oliver, D.S., *363*
Olsen, S.R., *394*
Omre, H., 347, *361*, *362*, *525*
Ono, I., 67, 68, *76*

Onstein, A., *155*
Osborne, O.A., 415, *443*
Osman, M.E., 418, 419, 435, 438, *442*
Osorio, J.G., *442*
Ostermann, R.D., 415, *443*
Ostlund, H.G., *216*
Ostrander, W.J., 15, *31*
Otte, D.N., *29*, *142*, *496*
Ouenes, A., 329, *362*, 591, 595, 596, *601*, *602*
Ovchinnikov, S., *49*
Owolabi, O., *443*
Owolabi, O.O., *443*

Pachepsky, Y., 329, *362*
Pacht, J.A., 536, *552*
Pal, S.K., 23, *30*, 137, *142*, *496*
Pan, Y., 40, *48*
Panda, M.N., 275, *287*
Pandya, A.S., 293, *307*
Pannatier, Y., 651, *672*
Panone, F., *443*
Paola, C., *215*
Parcell, W.C., 192, *216*
Park, J., 431, *443*
Peaceman, D.W., 275, *287*
Peakall, J., *216*
Pedrycz, W., 48, *48*, *49*
Peitgen, H.O., 657, *672*
Pelissier-Combescure, J., 309, *319*, *688*
Peng, C.P., *395*
Penumadu, D., *650*
Peterson, A.M., 278, *287*
Petrosky, G.E., 418, 419, 423, 435, 439, 441, *443*
Peuquet, D.J., *650*
Peyret, O., *687*
Pezeshk, S., 19, *32*, 119, *142*, 472, *497*
Pham, D.T., 77, *83*
Philps, B., *411*
Pickering, K.T., *552*
Pierce, A.C., *361*
Pilkey, O.H., 190, *216*
Piltet, W.R., *287*
Pinder, G.F., 367, *395*
Piron, M., *245*
Pita, J.A., 18, 19, *31*, 471, 472, *497*
Pittman, E.D., 402, *411*
Poeter, E.P., *215*
Poggio, T., 432, *443*
Pollock, C.B., *466*
Polmar, L., *215*
Popa, A.S., *632*
Pope, G.A., *363*
Porras, J.M., *32*
Portella, R.C.M., 329, *362*
Posamentier, H.W., *217*
Potter, D., 17, *32*
Pottorf, B.R., *552*
Poumot, C., 528, *552*
Powell, M.J.D., 432, *443*
Prade, H., 40, 42, *48*
Preston, P., *319*
Pulleyblank, W.R., *83*
Punch, W., *362*
PUNQ, 338, *363*

Qi, Y., *362*
Quenes, A., *30*, *496*
Quirein, J.A., *32*, *497*

Racey, A., 398, 399, *411*
Raghavan, R., *466*
Rahon, 276, *287*
RamaRao, B.S., *525*
Ramey Jr., H.J., *466*
Ramsey, C., *288*
Ramsey, L.A., *443*
Ranjithan, S., *394*, *395*
Rankey, E.C., *216*
Reading, H.G., 193, *216*
Reeves, C.R., 77, 80, *83*, 368, *395*
Reheis, G.M., *363*
Rencher, A.C., 224, *245*
Reynolds, A.C., 274, *288*, *363*
Rhind, D.W., 634, *649*
Richardson, S., 591, *601*, *602*
Ritzel, B.J., 382, *395*
Rizzo, D.M., 384, *395*
Roberts, S.M., 210, *216*
Robinson, J.R., 422, 439, *442*
Rodgers, K., *551*
Rogers, L.L., 13, *30*, 367, *395*
Rogers, S.J., 19, *32*, 119, *142*, 309, *319*, 472, *497*
Roggero, F., *361*
Rognoni, A., *687*
Romeder, J.-M., 224, *245*
Romeo, F., *362*
Romero, C., 339, 349, *363*
Root, S., *551*
Rosenbaum, M.S., 633, 642, *649*, *650*
Rosenblatt, F., 13, *32*
Ross, C.A., *217*
Ross, W., *215*
Röth, G., *525*
Royer, J.-J., 220, 223, 224, *245*
Rumelhart, D.E., 177, *185*, 368, 378, *395*, 472, 473, *497*, 636, *650*
Ruspini, E.H., 48, *49*

Russ, J.C., 657, *672*
Rzasa, M.J., *442*

Sahuquillo, A., *362*, *525*
Saleh, A.M., 415, *443*
Salman, N.H., *442*
Samaniego-V., F., *466*
Sanches, E., *48*
Sanchez, E., 204, 211, *216*
Sandberg, I.W., 431, *443*
Sangiovanni-Vincentelli, A., *362*
Sangree, J.B., *217*, *551*
Santamaria, G.N.E., *444*
Sarg, J.F., *32*, 191, *216*, *497*
Saupe, D., 657, *672*
Savelieva, E., *650*
Schaffer, J.D., 67, *75*
Schatzinger, R.A., 274, *287*, *288*
Schlager, W., 190, 191, 195–198, *215*, *216*
Schmidt, Z., 418, 419, 422, 423, 435, 439–441, *443*
Schrijver, A., *83*
Schubert, G., 192, *217*
Schuelke, J.S., 18, 19, *32*, 471, 472, *497*
Seheult, A.H., *361*
Seifert, D., 274, *288*
Seinfeld, J.H., *361*
Seinfield, J.H., *362*, *363*
Selva, C., *32*
Sen, M.K., 329, *363*
Serra, O., 402, *411*, 676, *687*
Sessa, S., *48*
Seymour, R.H., *155*
Shaffer, B.L., 536, *552*
Shah, P.C., *362*
Shanmugam, G., 294, *307*
Sharma, Y., *361*
Shaw, A.B., 534, *552*
Sheriff, R.E., 15, *32*, *271*
Sherrif, R.E., *32*
Shibata, T., *216*
Shibli, S.A.R., 177, 179, *185*
Shimada, N., *32*, *497*
Shomrony, A., *687*
Shyllon, E.A., 250, *271*
Siegfried, R.W., *29*
Simpson, E.H., 537, *552*
Sivia, D.S., 347, 348, *363*
Slater, G.E., 326, *363*
Slingerland, R., 190–192, *216*
Smart, P., *217*
Smith, N.D., 191, *216*
Smith, P.J., *155*
Smith, S.G., *395*
Smith, S.V., 190, *215*, *217*
Smith, T.D., 369, *395*
Smoot, J.P., 211, *217*
Snedden, J.W., *551*
Solheim, O.A., *155*
Song, J., *650*
Soroush, A.R., *672*
Soto, B.R., *687*
Soto, T.C., *687*
Souza, O., *287*
Spears, W.M., 65, *76*, 345, *363*
Spencer, R.J., 210, *216*
Srinivasan, S., *287*, 508, *525*
Srivastava, M., 503, *525*
Srivastava, R.M., 85–87, 92, *95*, 222, 232, *245*, 651, *672*
Srivastava, S., *115*
St. Clair, U., *271*
St. Clair, U.H., *49*
Standing, M.B., 414, 418, 419, 432, 438, *444*
Stanford, T.G., *31*, *142*, *497*, *672*
Stanley, D.A., *32*, *142*, *319*, *497*
Staples, R., *155*
Startzman, R.A., 674, 677, 679, *688*
Stattegger, K., *216*
Stearn, D.W., 594, 601, *602*
Stein, A., 633, *650*
Stein, J.A., 535, *552*
Sterling, R., 635, *650*
Stevens, C.E., *395*
Stevens, R.L., 634, 644, 645, *649*, *650*
Stewart, I.J., 532, *552*
Steyaert, L.T., 634, *650*
Stoffa, P.L., 329, *363*
Stoisits, R.F., *31*
Stollar, R.L., *395*
Strebelle, S., *115*, 519, *525*
Strønen, L.K., *155*
Sugeno, H., 211, *217*
Sugeno, M., 26, *32*, *49*, 126, 140, *142*, 493, *497*
Sultan, A.J., *601*
Sung, 415, *444*
Suro-Perez, V., 94, *95*
Sutton, R.P., 415, *444*
Swaby, P., *361*
Swain, P.H., *30*
Swisher, C.C., *551*
Switzer, P., 220, *245*
Syswerda, G., 384, *395*
Syvitski, J.P.M., 191, *217*

Tagawa, T., *30*
Taggart, I.J., *32*, *142*, 293, *307*, *319*, *411*, *497*

Tahrani, H.O., *444*
Takagi, T., 211, *217*
Takahashi, T., 190, 191, 205, 206, *215*
Tamhane, D., 16, *32*, 177, *185*, *319*
Tan T.B., 327, *363*
Tan, T.B.S., *363*
Tanaka, Y., 329, *363*
Taner, M.T., 14, *32*
Tarantola, A., 507, *525*
Telford, W.M., 251, 252, *271*
Tetzlaff, D.L., 190, 191, *217*
Tham, M.K., *287*
The Math Works™, *32*, *142*, *497*
Thieler, E.R., 190, *216*
Thomas, G.W., 326, *363*
Thomas, L.K., 326, *363*
Thompson III, S., *217*
Thurmond, V., *216*
Timonin, V., *650*
Tingdahl, K.M., 163, 169, *173*
Tjelmeland, H., *362*, *525*
Todd, R.G., *217*
Tompson, A.F.B., 367, *395*
Tomutsa, L., 274, *288*
Tong, R.M., *49*
Toomanian, N.B., *29*
Toomarian, N.B., *184*
Torriero, D., *244*
Tran, T., 503, *525*
Tran, T.T., *287*
Trube, A.S., *444*
Tucker, M.E., 401, *411*
Tuckey, J.W., *244*
Tura, A., *32*, 151, *155*
Turcotte, D.L., 192, *217*
Tyler, K., *525*

U.S. Geological Survey, National Oil and Gas Resource Assessment Team, *32*
Uchikawa, Y., *30*, *31*, *363*, *496*
Uchkawa, Y., *496*
Ungar, L.H., *443*

Vail, P.R., 191, *217*, *551*
van Dooremolen, W., *650*
Van Leekwijck, W., 47, *49*
van Wagoner, J.C., *217*
Vasquez, M.E., 418, 419, 422, 423, 432, 438, 439, 441, *444*
Veatch, R.W., 326, *363*
Vecchi, M.P., *395*
Veezhinathan, J., 14, *32*
Verdière, S., *287*
Villa, M., *443*
Vining, B.A., 532, *552*
Vomvoris, E.G., 512, *525*
von Altrock, C., 22, 23, *32*, 137, *142*

Wackernagel, H., 220, 222, 232, *245*
Wackowski, R.K., 369, *395*
Wagner, B.J., 80, *83*, 367, *395*
Wagner, D., *32*
Walker, D., 532, *552*
Walker, E.A., 48, *49*
Walker, R.G., *307*
Wall, C.G., 250, *271*
Walsh, J.J., 336, *362*, *363*
Waltham, D., 190, 191, *215*, *217*
Waltham, D.A., *215*
Wang, L., 176, 181, *185*, 276, *287*, 500, 503, *525*
Wang, Y., 328, *363*
Wasserman, M.L., *361*
Waterhouse, M., *551*
Watney, L.W., *216*
Watney, W.L., *215*
Watson, A.T., 327, *363*
Watterson, J., *363*
Watts, G.F.T., 144, *155*
Weiss, W.W., *30*, *496*, *601*
Wendebourg, J., 190, 191, *217*
Wheatley, C.L., *29*, *142*, *496*
Whitaker, F., 190, *217*
Whitlatch, R.B., *551*
Whitley, D., 383, *395*
Widmier, J.M., *217*
Widrow, B., 24, *32*, *142*, 472, *497*
Wilgus, C.K., 191, *217*
Williams, D.L., *361*
Williams, R.J., *185*, *395*, *497*
Williamson, M.A., 15, *30*
Wilson, J.L., 94, *95*
Wloszczczowski, D., 220, *245*
Wolff, M., 309, *319*, *688*
Womg, D., *442*
Wong, G., *245*
Wong, P.M., 4, 19, *32*, 119, *142*, 176–179, *185*, 293, *307*, 309, 310, *319*, 399, *411*, 472, *497*
Wornardt, W.W., *551*
Wright, J., *32*
Wright, V.P., 401, *411*
Wu, X., *650*
Wu, Z., 328, *363*

Xue, P., *31*

Yager, R.R., *32*, 48, *49*
Yamaoka, M., *30*, *31*, *496*

Yang, P-H., 327, *363*
Yang, W., *216*
Yang, Y., 633, 635, 638, 642, *650*
Yao, C., *650*
Yasukawa, T., 26, *32*, 126, 140, *142*, 493, *497*
Yielding, G., *362, 363*
York, S.D., 369, *395*
Yoshioka, K., 18, 19, *32*, 471, 472, *497*
Yuan, B., 36, 46–48, *48, 49*, 192, 194, 204, 211, *216*, *271*

Zadeh, L.A., 13, 15, 19, 23, *32*, 45, 48, *49*, 137, *142*, 192, 194, *216, 217*, 256, *271*, 472, 473, *497*, 529, *552*
Zak, M., *30*
Zellou, A., *601*
Zellou, A.M., 592, 598, *602*
Zhang, Q., 635, 638, *650*
Zhang, X., 14, *32*
Zhang, Z., 634, *650*
Zhao, C., *650*
Zhou, 415, *444*
Ziegler, V.M., 369, *395*
Zimmermann, H.J., 48, *49*
Zurada, J.M., 293, *307*

SUBJECT INDEX

absolute error, 418, 430, 431
absolute minimum, 75
abundance, 113, 400, 402, 405, 506, 528, 531, 536, 537
abundance gradients, 528
acceptance probability, 506
accurately model subsurface distribution, 189, 214
ACE (alternative conditional expectation), 21, 653, 655, 657
ACE method, 657
ACE technique, 657, 662
ACE technique transforms, 657
ACE-neural network, 669
ACE-neural network approach, 669
ACE-neural network model, 661, 664
ACE-neural network technique, 669
adaptive neural network, 189
added noise, 350
agglutinated, 542
agglutinated foraminifera, 530, 535, 536, 538, 542
agglutinated foraminifera community, 536
agglutinated foraminiferal, 536, 537
agglutinated foraminiferal biofacies, 542
agglutinated foraminiferal populations, 527
AI (artificial intelligence), 584, 589
AI technology, 601
AI tools, 585
algorithm, 9, 13, 21, 25, 27, 29, 51–53, 59, 66, 67, 69, 72, 74, 82, 94, 108, 111, 151, 153, 158, 165, 140, 141, 158, 159, 161, 164, 177, 204, 211, 273, 276, 279, 282–284, 286, 293, 307, 313, 327–329, 335, 338, 342 347, 359, 366, 383–386, 473, 474, 493, 494, 496, 500, 508, 513, 514, 519, 565, 633, 636, 637, 649, 653, 657, 669
alternative conditional expectation (ACE), 21, 653, 657
alternative method sequential simulation method, 519
alternative selection strategies, 360
analysis component, 642
analysis framework, 244
analysis process, 146
analysis selection range, 244
analysis techniques, 7, 14, 147
analyzing hydrocarbon reservoir estimate, 267
ANN-based, 366, 393
annealing algorithm, 94, 385
ANN (artificial neural network), 365, 366, 377–380, 382, 388–390, 392, 393
antecedent membership functions, 211, 212
antecedent parameter, 196
antecedent propositions, 214
API, 416, 431
API gravity, 372
applied artificial neural network, 19, 120, 472
applied frac jobs, 449
approximate fuzzy model, 126
approximate gradients, 342
approximate linear relationship, 135
approximate methods, 256
approximate relationship, 377
approximate solution, 15, 473
aquifer, 85, 377
aquifer analysis, 360
aquifer characteristics, 360
aquifer properties, 325
aquifer system, 372
arithmetic scale, 194
artificial classification boundaries, 194
artificial facies, 568
artificial intelligence (AI), 3, 267, 292, 584, 634, 635, 674
artificial intelligence applications, 677
artificial intelligence methods, 674
artificial intelligence problems, 415
artificial intelligence technique, 674
artificial logs, 569
artificial neural network (ANN), 4, ,9, 13, 19, 51, 120, 176, 177, 309, 399, 431, 447, 472, 474, 606, 633, 635
artificial neuron, 24, 138
Ashtart field range, 405
Ashtart oilfield, 397–399
associated fuzzy propositions, 44
associated log magnitude, 297
associative networks, 13
associative neural network, 293
automated algorithm, 306
automated clustering technique, 123

automated history history matching, 325
automated pattern classification, 292
automated pattern classification technique, 293
automated pattern recognition, 295
automated processing system, 291
automated techniques, 6, 289
automatic clustering, 8
automatic history matching, 325, 326
automatic history matching process, 328
automatic lithological pattern classification process, 302
automatic reservoir characterisation algorithm, 360
averaged permeability, 513, 516–519
averaged permeability field, 513, 514, 516
averaged permeability model, 513
averaged quantization errors, 282

backpropagation (BP), 137, 607
backpropagation algorithm, 27, 121, 126, 139, 141
backpropagation approach, 25, 140
backpropagation artificial neural network, 159
backpropagation learning, 23, 137
backpropagation learning algorithm, 22, 137, 378
backpropagation method, 211
backpropagation neural, 177
backpropagation neural network, 14, 314
backpropagation training, 380
basin sedimentation rates, 535
basin showing regional orthogonal fractures, 583
Bayesian, 7, 347
Bayesian approaches, 328
Bayesian estimates dependent, 328
Bayesian history matching, 327
Bayesian linear strategy, 328
bioevent, 527, 533–535, 541, 543
biofacies model, 536
biofacies modelling, 527, 535
biostratigraphical correlation, 543
biostratigraphical model applied, 527
black-oil, 413, 417
black-oil model, 274
bottom hole, 328, 329, 341, 346, 348, 360, 413, 547, 548
bottom hole pressures, 346, 359
bottom markers, 556
BP (back propagation), 13, 370, 373, 550, 636, 637, 685
BP algorithm, 639, 641
BP method, 643
BP network, 416
BP network model, 417
BP process, 636
building spatially descriptive models, 274
built input patterns, 293

c-means, 9, 27
c-means algorithm, 9, 18, 27
c-means techniques, 9
calculated dip, 159, 163
calculated log, 628
calculated reserve estimates, 631
calculated reservoir properties, 325
calculated semivariogram, 564
calibration neural network, 508, 509
calibration process, 514, 539
capture knowledge, 43
capture relationships, 74
carbonate, 191, 193, 311, 398, 583, 586, 592
carbonate depositional systems, 190
carbonate oil reservoir, 590
carbonate production, 195, 205
carbonate reservoir, 398, 584
carbonate rock, 398, 402, 410
carbonate sediment, 191, 204
carbonate sediment production, 189, 205–207
carbonate sediment production pattern, 208
Carpinteria field, 553–556, 568, 579, 580
Carpinteria offshore field, 553
Carpinteria structure, 554
cells, 199, 211, 258, 273–275, 327, 333, 334, 546, 557, 589, 634
changing boundary conditions, 190
characterisation method, 332
characterisation process, 342
characterise core porosity, 309
characteristic function, 34, 36, 193, 194
characteristic log, 292
chimney, 157–160, 165, 167, 168, 170–172
chimney bodies, 159
chimney classification network, 168
chimney detection, 158, 159
chimney interpretation, 159
chimney node, 159
chimney prediction, 159
chromosome, 17, 54, 60, 64, 65, 70, 74, 330, 343–346, 360
chromosome length, 330
class, 12, 15, 23, 36–38, 46, 100, 111, 114, 137, 159, 221, 229, 235, 237–239, 294, 295, 309, 312–314, 316–318, 389, 397–399, 401, 402, 410, 431, 475, 524
class boundaries, 318
class distribution, 313
classical arithmetic, 40
classical c-means algorithm, 474, 494
classical factorial discriminant analysis, 221
classical gaussian assumptions, 507

classical interval, 262
classical logic, 46
classical neural network, 23, 137
classical pattern recognition, 471
classical probability distribution function, 45
classical property, 35
classical regression techniques, 415
classical set, 33, 34, 194
classical set theory, 34–37, 40
classification, 8, 12, 13, 168, 171, 229, 292, 295, 309, 312–315, 318, 397, 400–402, 405, 415, 460, 471, 601, 634, 677
classification algorithms, 9, 474
classification characteristics, 286
classification model, 313, 314
classification model generation, 313
classification neural network, 169
classification problems, 309
classification process, 12, 295–297
classification tools, 504
classifier, 16, 273, 285
classify fractures, 601
classify porosity, 309, 310
clastic reservoirs, 309
clastic sedimentation, 533
clay, 21, 135, 136, 194, 201, 606, 645–648
clay content, 129, 130, 132, 135, 136
cluster, 9, 20, 27, 28, 126, 136, 141, 205, 252, 273, 276, 278–282, 284, 295, 296, 313, 314, 319, 330, 360, 471, 474–476, 478, 483, 485–491, 494, 495, 677
cluster analysis, 9, 474, 634, 674, 677
cluster changes, 27, 141, 494
cluster delineation, 282, 283
cluster distributions, 483
cluster identification, 281
cluster identification process, 280
cluster node, 313, 314
cluster patterns, 483
clustering algorithm, 276, 312, 313, 676
clustering analysis, 313
clustering classifier, 313
clustering information, 295
clustering methods, 9, 27, 474, 494
clustering result, 314
clustering technique, 10, 295, 313, 318, 471, 483, 490
co-kriging, 107, 114, 175, 232, 559
co-kriging approach, 559
co-kriging estimate, 107
co-kriging technique, 559
combination GA, 358
combinatorial optimization, 77–79
combinatorial problems, 78, 79, 82
combined crossover probability, 330
combined GIS, 634
combining computation, 475
compartment boundaries, 300, 303
compartmentalization, 298, 299
complementary techniques, 145
complex anisotropic structures, 181, 503
complex community structure, 536
complex depositional history, 541
complex genome structure, 61
complex geologic formation, 673, 674, 687
complex geologic sequences, 683
complex geology, 100
complex integration, 368
complex model, 18, 339, 472, 652
complex multi-dimensional field, 652, 657, 669
complex neural networks, 22, 137
complex nonlinear system, 140, 493
complex representation techniques, 360
complex reservoir, 326, 331, 583
complex sedimentary structure, 289, 306
complex sedimentary systems, 295
complex trace signal, 477
component analysis, 6, 18, 178, 184, 219, 224, 226, 610, 634, 673–675
component extraction technique, 220
component lithofacies, 309
component log, 673, 681, 683, 684, 686
component log traces, 682
comprehensive software, 466
compression methods, 6
compression techniques, 6
computational advantage, 211
computational benefits, 585
computational intelligence, 447
computational methods, 256
computational problems, 79
computational resources, 81, 324
computational stratigraphy, 539, 550
computational tractability, 365
computational units, 22, 137
computer algorithm, 255, 267
computer models, 13, 177
computer simulation, 445
computer systems, 9, 474
computing methodologies, 4
computing systems composed, 415
computing techniques, 189, 214
computing technology, 274
conceptual geological model, 501
conceptual model, 553
conceptual reasoning, 473

conditional distribution, 95, 104, 105, 109, 111, 176, 510, 524
conditional expectation, 505, 509, 657
conditional fuzzy propositions, 46
conditional Gaussian distribution, 105
conditional probability, 95, 111, 504, 505, 510
conditional probability distribution, 505
conditional sequential simulation, 104
conditional simulation, 333, 505–507, 510, 511
conditional simulation needs, 506
conditional simulation technique, 519
conditioned cells, 333
conditioning porosity, 574–576
conditioning shale fraction, 575, 576
confidence interval, 4, 11, 22
confidence range, 11
connected cells, 280
considering inference results, 531
consistent match, 548
consistent model, 589
constrained fuzzy arithmetic, 40
constrained models, 274
constrained system, 92
constructing membership functions, 256
constructing reservoir models, 107
constructing suitable input patterns, 293
contaminant, 80, 651, 652, 663, 664, 367, 368, 651, 652, 657, 658, 664, 665, 669
contiguity analysis, 221
contiguity coefficient, 220, 223, 224, 231
contiguity relation, 220
continental shelf, 673, 682
continuity correlation, 298
conventional algorithm, 25, 140, 473, 653
conventional analysis methods, 3
conventional core, 528
conventional engineering, 454
conventional gas reservoir, 598
conventional geological models, 302
conventional geostatistics, 231
conventional hydraulic fracture simulation possible, 466
conventional log, 21, 605, 606, 609, 611, 628, 631
conventional log analysis, 623
conventional mathematical representations, 192
conventional means, 360
conventional modeling tools, 251
conventional models, 659
conventional network models, 669
conventional neural network, 21, 655, 669
conventional neural network models, 120, 473, 652, 662
conventional porosity indicator logs, 611
conventional porosity logs, 609
conventional probability, 11
conventional regression techniques, 429
conventional reservoirs, 598
conventional simulators, 275
conventional statistical approaches, 21
conventional statistical methods, 652, 661
conventional statistical techniques, 669
conventional training set, 179, 180
conventional wireline logs, 606
Cook formation, 144, 151, 153
coral reef, 42, 195, 197, 198
core analysis, 176, 247, 253, 255, 310, 318, 597, 599, 676
core analysis report, 310
core descriptions, 310, 312
core information, 398
core interpretations, 397, 399, 400
core measurement, 144, 405, 409
core parameters, 401
core permeability, 406–408
core plug, 310, 409
core plug analyses, 398
core plug measurements, 397, 402, 409
core plug permeability, 410
core plug porosity measurements, 404
core plug preparation, 409
core porosity, 402, 403, 405–408, 410
core porosity log, 406
core porosity measurements, 408
correlate fracture density, 21
correlated field, 554
correlated probability field model, 570
correlated results, 673, 679
correlation analysis, 541
correlation approach, 534
correlation coefficient, 100, 175, 179, 184, 225, 404, 410, 418, 463, 509, 559, 562, 589, 607, 611, 616, 675
correlation length, 275, 579, 680, 681
correlation log, 681
correlation measures, 111
correlation methods, 674
correlation model, 550
correlation range, 333
correlation result, 679, 683
correlation structure, 273, 274
coupled sets, 189, 215
covariance analysis, 161
covariance modeling, 107
covariograms, 225
critical component, 376
critical parameter, 165

critical parameters, 513
critical research, 393
cross-correlation, 112, 289, 292, 512, 674, 680
cross-correlation method, 674, 680, 681
cross-correlation process, 299, 307
cross-section, 213, 289, 301, 302, 305, 370
crossed covariograms, 241
crossover operator, 55, 57, 60, 64–66, 70, 71, 74, 330, 331, 342–344, 361
crossover probability, 67, 330
crude oil, 414–416, 419, 422, 423, 429, 432
crude oil systems, 413, 416–418, 429
crude oil viscosity, 418, 430
curvature, 311, 583, 585, 591, 599
curvature method, 591
curvilinear, 175
curvilinear fluvial, 176
curvilinear porosity, 180

decision analysis techniques, 369
deep core, 189
deep fault, 586, 588, 589
deep resistivity, 607
defuzzification, 47, 248, 531, 539
defuzzification method, 47
defuzzification process, 539
delta functions, 68
delta simulation, 189
dense outcrop sampling, 111
dense training information, 184
density logs, 95, 609, 628, 676
density porosity, 559, 562, 609–611, 628
density porosity log, 559
density turbidite flows, 532
depletion, 371
deposition, 189, 191, 193, 200, 211, 213, 214, 311, 369, 532, 536, 543, 554, 557
depositional model, 214, 527, 532
depositional pattern, 554, 579
depositional process, 93
depositional system, 190, 532
depositional texture, 401
design frac jobs, 466
design issues involved, 51
design optimization, 458
design optimization problems, 368
design optimum stimulation, 466
design options, 60
design parameters, 455
design principles, 368
detected matching, 306
deterministic coefficients, 16
deterministic estimates, 93
deterministic function, 232
deterministic model, 649
deterministic optimization techniques, 367
deterministic parameters, 16
deterministic processing, 16
deterministic techniques, 553
diagenetic, 274, 310
diagenetic features, 402
diagenetic processes, 159, 398
differential wave, 16
dimensional chromosomes, 344, 345
dimensional cross-section, 524
dimensional crossover, 345
dimensional operator, 344
dimensional semivariogram, 563
dimensional structure, 343, 344
dimensionality, 9, 28, 175, 182, 332, 367, 474, 495, 674
dip, 157–167, 171, 532
dip estimation algorithm, 158
dip information, 157, 171
dip variance calculated, 170
dip-steered similarity, 167, 169, 170
dip-steering, 165, 166, 168, 171
dip-variance, 168
dip-variance attribute, 171
directional variogram, 219, 233, 234, 557
discriminant analysis, 221, 309, 634
discriminant analysis generalization, 221
discriminant factorial analysis, 224
discrimination function, 7
displacement, 335, 336, 338, 513, 554
displacement algorithm, 657
distributed input, 278
distributed pattern, 415
distributed representation, 415
distribution function, 86, 97–99, 559
distribution model, 102, 104, 109
distribution quantifies spatial continuity, 100
dominant class, 314, 401
dominant lithofacies, 310
dominant reservoir compartments, 298
drainage, 144, 213, 584
drainage pattern, 154
drainage simulation, 147
Driankov, 48
drilled injectors, 374
drilling, 15, 153, 154, 337, 365, 374–377, 392, 491, 528, 543, 545, 555, 600, 657, 659
drilling phase, 340
drilling success rates, 5
driver, 202, 204, 584, 589, 595, 601
DT (travel time), 20, 120–123, 126–128, 136

DT prediction, 123
dual permeability, 586
dynamic BP method, 643
dynamic information, 583
dynamic parameter, 639
dynamic process, 275, 311
dynamic reservoir, 144
dynamic reservoir modeling, 491
dynamic response, 415
dynamic sedimentary process simulators, 190

eastern fault, 147
eastern fault segment, 149
economic benefits, 451
economic decision, 609
economic optimization genetic algorithm, 449
economic parameters, 393, 446, 449
economic performance, 374
economical modeling approach, 247
economical potential, 247
editing noise, 14
effective permeability, 275
effective porosity, 249, 605, 606, 609, 616, 628
effective recovery strategies, 369
efficient geobody detection, 285
efficient processing, 6
efficient production, 5
efficient result, 682
efficient well-to-well log correlation, 683
El Garia formation, 397–402, 405, 407, 408, 410
El Garia reservoir, 399
electrofacies, 673, 676, 679, 682, 683
electrofacies characterization, 676
electrofacies classification, 683, 684, 686
electrofacies identification, 677, 687
encoding method, 65
engineering analysis, 595
engineering application, 387
engineering constraints, 80, 377
environmental conditions, 530
environmental gradients, 528
environmental planning objectives, 366
environmental processes, 644
error bars, 651
error function, 326, 509
error function trained, 505
error gradients, 636
error method, 607
error variance, 101
estimate, 22, 85, 86, 90, 91, 93, 97–101, 107, 109, 110, 164, 247, 250–253, 262–267, 332–334, 349, 375, 413, 418, 419, 422, 455, 460, 503, 504, 510, 511, 535, 574, 607, 623, 652, 661, 664, 669
estimate crude oil viscosity, 422
estimate dip, 158
estimate distribution, 85
estimate fracture, 591
estimate parameter, 85
estimate undersaturated oil viscosity, 422
estimated compressibility, 423
estimated dip, 164
estimated maximum, 164
estimated models, 21
estimated net, 154
estimated parameter, 86
estimated saturation changes, 153
estimates, 4, 91, 93, 94, 98, 101, 104, 106, 110, 114, 146, 159, 267, 365, 374, 387, 389, 391, 392, 510
estimating oil gravity, 423, 426
estimating oil viscosity, 423
estimating permeability, 110, 398
estimating porosity, 175, 184
estimating rock properties, 176
estimating undersaturated oil compressibility, 423, 426
estimation, 61, 85, 87, 88, 90, 93, 97, 98, 109, 110, 163, 278, 325, 326, 329, 330, 393, 413, 414, 419, 471, 664
estimation errors, 290
estimation errors introduced, 388
estimation process, 94, 368
estimation process attempts, 85
estimation process results, 85
estimation processes, 86
estimation technique, 333
estimation variance, 85
evolutionary computing, 5
expected geological statistics, 325
expected length, 65
expected performance, 388
expected production, 338
expected soft computing techniques, 22
experimentally measured properties, 413
experimentally measured solution, 429
expert geologists, 499
expert knowledge, 476
expert system, 3, 213, 215, 311–313, 315–318, 674, 682, 687
expert system approach, 309, 310
expert system knowledge acquisition, 312
expert system result, 317
expert system technique, 315–318
exploration, 3–5, 14, 55, 59, 69, 72, 73, 214, 251, 314, 598
exploration applications, 4, 18

exploration approach, 233
exploration phase, 73
exploration problems, 17, 18
exploration risk, 5
exploration seismology, 251
extrapolate mapping, 19, 472
extrapolate production-log, 471, 475
extremely sparse distributions, 284

facies, 111–113, 211, 252, 292–294, 311, 402, 503, 504, 506–509, 554, 557, 568–570, 572, 573, 580, 584, 596, 597
facies approach, 572, 574–576
facies categories, 505
facies changes, 598
facies classification, 293
facies distribution, 572, 580, 673
facies information, 508, 509
facies models, 572
facies simulation, 182
factor analysis, 634, 642
factorial, 220, 226, 227, 235, 237
factorial component, 219, , 220, 224–229, 235, 244
factorial methods, 229
factorial techniques, 220
fang, 191, 528
far offset traces, 12
Faraj, 220–222, 224, 226, 228, 230, 232, 234, 236, 238, 240, 242, 244
fault, 148, 149, 153, 157, 210, 281, 290, 303, 327, 330, 331, 335–340, 371, 415, 553–555, 583, 584, 586, 589
fault density, 586
fault displacement, 335, 336
fault network, 586
fault outcrop, 335
fault parameters, 339, 343
fault pattern, 153, 583, 586
fault permeability, 335
fault permeability scaling, 337
fault properties, 334, 336
fault recognition, 555
fault rock thickness, 336
fault thickness, 336
fault traces, 339
faulted limestone, 583
faulted limestone reservoir, 584, 586
faulted oil reservoir, 584
field averaged, 341
field cases, 329, 338
field characteristic, 247
field fracture, 601
field fracture modeling, 589
field oil, 348
field oil rates, 341
field operator, 553, 554
field performance, 548
field production, 545, 547
field properties, 331, 332, 334
field property chromosome, 344
field results, 451–454, 458
field simulation scenario, 514
field structure, 586
field-wide, 398, 542, 628
field-wide mudstone, 527, 535, 550
field-wide reservoir characterization, 623
field-wide reservoir interpretation integrating, 553
fit individual, 71
fit individuals, 55, 360
fit observations, 207
flood basin, 203
flood project, 365
flows, 22, 137, 294, 543
fluvial reservoir, 175, 184, 369, 500
focused log, 682
foraminifera, 531, 539, 542
foraminiferal, 529, 533, 539, 541, 554
foraminiferal community, 542
foraminiferal community structure, 535
foraminiferal model, 536
foraminiferal population, 541
forecast gas storage, 466
forecasting tools, 366
formation characteristics, 606, 628
formation dependent, 631
formation estimate, 247, 248
formation interval, 543
formation permeability, 631
formation sandstones, 532
formation thickness, 252, 259
frac design, 455
frac design parameters, 455, 458
frac jobs, 449, 450, 455
frac parameters, 457
frac parameters combination, 457
fracture, 22, 401, 445, 450, 583–586, 589–594, 596, 598, 600, 601, 652, 658
fracture analysis, 599
fracture density, 595
fracture design, 445
fracture design process, 466
fracture identification, 583
fracture network, 583, 590, 592, 593
fracture parameters, 594
fracture prediction, 596

fracture set interpreted, 589
fracture system, 592
fractured cluster, 281
fractured models, 585
fractured reservoir, 21, 583, 586, 601
fractured reservoir characterization, 21, 583, 601
fractured reservoir model, 584
fractured reservoir modeling, 589
fractured reservoirs, 5, 7, 22, 601
fractured sandstone, 598
fractured shale, 5, 303
fracturing driver, 589
fracturing factor, 595, 596
fracturing process, 455
free gas, 158, 341
fully pelagic conditions, 541, 542
function approximation, 431
function evaluation, 52, 73, 331, 342, 345, 350, 358, 359, 368
function evaluations before selection, 73
function measures, 348
function model predicts, 426
function network, 417, 418, 431
function network model, 418, 429
function neural network model, 413
functional evaluation, 7
functional relationship, 89, 93, 610
functional structure, 119, 129, 653
fusion, 4–7, 17, 18, 119
fusion concepts, 6
fuzziness, 15, 16, 23, 120, 137, 250, 473
fuzzy analysis, 43
fuzzy approach, 262
fuzzy arithmetic, 40
fuzzy boundaries, 483
fuzzy c-means, 474, 475, 490, 494
fuzzy c-means clustering, 471, 483, 494
fuzzy c-means clustering technique, 489
fuzzy c-means distribution, 487
fuzzy c-means techniques, 474, 475
fuzzy cluster, 9, 27, 28, 474, 483, 494, 495
fuzzy clustering, 9, 27, 474, 494
fuzzy clustering technique, 9, 27, 474, 494
fuzzy coefficients, 16, 22
fuzzy composition, 42
fuzzy descriptions, 247
fuzzy encoding means, 318
fuzzy event, 45
fuzzy geometry, 43
fuzzy inference, 43, 47, 194
fuzzy inference rules, 26, 140, 201, 493
fuzzy inference system, 27, 141, 197, 198, 200, 204, 211, 493
fuzzy interval, 38–40, 42, 43, 46, 48
fuzzy logic, 3, 4, 6, 7, 9, 11, 12, 15, 16, 18, 19, 21–23, 27, 33, 43, 48, 119, 120, 137, 189, 191, 192, 194, 200, 204, 211, 214, 266, 471–473, 494, 527–531, 538, 541, 548, 583, 584, 589, 590, 598, 601
fuzzy logic analysis, 19, 472
fuzzy logic approach, 527
fuzzy logic inference result, 531
fuzzy logic methods, 309
fuzzy logic model, 9, 189, 202, 203, 214, 472, 474, 527, 550
fuzzy logic model results, 548
fuzzy logic modelling, 532, 539, 542, 546, 550
fuzzy logic ranking, 598
fuzzy logic rules, 120, 530
fuzzy logic system, 189, 195, 198, 200–202, 204, 205, 207–209, 211, 213, 215, 530, 531
fuzzy logic system approach, 196
fuzzy logic techniques, 119, 120, 129
fuzzy logic toolbox, 211
fuzzy membership, 17
fuzzy membership function, 27, 141, 256–259, 493
fuzzy model, 548
fuzzy modeling, 16
fuzzy modelling results, 548
fuzzy pattern recognition, 15
fuzzy propositions, 43–46, 194
fuzzy quantifiers, 46
fuzzy reasoning, 5
fuzzy regression analysis, 16
fuzzy relation, 40–42, 45, 47
fuzzy rules, 16, 17, 26, 46, 140, 266
fuzzy set, 15, 23, 26, 33–38, 40, 42–44, 45–48, 137, 138, 141, 189, 192–194, 196, 199, 200, 204, 211, 256, 260, 262, 265, 473, 493, 528, 529, 634
fuzzy set characteristic, 473
fuzzy set representation, 192
fuzzy set theory, 15, 16, 18, 33–36, 38, 43, 48, 189, 193, 473
fuzzy sets theory, 16
fuzzy system, 27, 42, 43, 48, 141, 204, 211, 256, 267, 493, 530
fuzzy system approach, 256
fuzzy system modeling, 247, 248, 256
fuzzy system modeling approach, 248, 257
fuzzy system technique, 267
fuzzy systems association, 48
fuzzy technique, 9, 27, 474, 483, 494
fuzzy theory, 473
fuzzy truth, 44

GA (genetic algorithm), 4, 8, 17, 51–60, 64, 71, 73–75, 77, 80, 82, 329–332, 344, 349, 351, 355, 357–361, 365, 366, 370, 382–386, 389–392
GA applications, 382
GA approach, 329
GA generation, 355, 358
GA generations, 386
GA literature, 60
GA methodology, 323
GA optimisation, 350
GA research, 53, 65
GA strategies, 359
gamma ray, 18, 119, 120, 136, 252, 400, 402, 403, 407, 471, 542, 556, 563, 605, 607, 628, 681
gamma ray log, 293, 406, 539, 674, 682, 683
gas chimneys, 7
gas distribution encountered, 545
gas exploration, 190
gas fields, 446
gas gravity, 413, 414, 416–419, 426, 428, 429, 432, 441
gas oil, 432
gas price, 375, 449
gas production, 5, 365, 373, 375, 380, 381, 387, 388
gas production rates, 348
gas research, 494
gas reservoir, 5, 13, 583, 593, 599
gas resources, 5
gas sandstone, 445
gas storage, 445, 464
gas storage field, 445, 446, 466
gas systems, 413, 415, 419, 432
gas-oil, 329, 348
Gaussian, 105, 109, 163, 333, 524
Gaussian distributed, 347
Gaussian distribution, 284, 333, 334
Gaussian errors, 341
Gaussian field, 333, 361
Gaussian function, 432
Gaussian membership functions, 126, 207, 209
Gaussian model, 90, 94
Gaussian random field, 500
Gaussian random function model, 94
Gaussian simulations, 102
gene-based crossover, 70
gene-based operator, 70, 71
generalization accuracy, 379, 381, 393
generalization capabilities, 451
generalization capabilities correlation coefficients, 463
generalization performance, 378
generalization property, 123, 126
generated clusters, 313, 314
generated combination, 385
generated event, 14
generated individuals, 59, 74
generated logs, 605
generated per set, 74
generates stochastic realizations, 104
generation process, 350
genetic algorithm (GA), 3, 4, 7, 17, 51–53, 58, 75, 77, 204, 211, 323, 329, 341, 342, 358, 360, 365, 366, 368, 382, 447–449, 450, 455, 460, 466
genetic algorithm method, 52
genetic approach, 293
genetic computing, 4, 471
genetic optimization, 455, 456
genetic reservoir characterization, 293
genetic variability, 67
genome, 54–57, 60, 62–65, 67, 71, 73, 342, 343, 349
genome design, 57, 60, 343
genome information, 54
genome structure, 55, 343
geobody, 276, 278, 286
geobody detection, 273
geobody encountered, 276
geobody identification, 286
geographic structures, 220
geographical information systems, 633
geologic characterization, 95
geologic depositional, 554
geologic history, 197, 198
geologic information, 87, 592
geologic interpretation, 228, 585
geologic layer, 284
geologic model, 289, 553, 554, 598, 599
geologic modeling, 297
geologic parameters, 598
geologic property distributions, 94
geologic structure, 219, 228, 554, 555
geologic structure model, 556
geological association, 336
geological bodies, 16
geological component, 328
geological continuity, 99–101, 110, 112, 113, 499, 524
geological depositional, 579
geological engineering, 649
geological facies, 599
geological fact, 299
geological faults, 324, 335
geological features, 251, 328, 329

geological formations, 251
geological information, 16, 182, 325, 499
geological interpretation, 111, 157, 499
geological knowledge, 332
geological layer, 61
geological model, 290, 291, 306, 332, 553, 598
geological modeling, 289, 292, 307
geological parameter, 476, 583, 586
geological pattern, 102, 104
geological perspective, 325
geological problems, 4
geological processes, 11, 324, 325
geological property field, 336
geological representations, 21, 491
geological statistics, 332
geological structure, 13, 111, 115, 500, 553
geological uncertainty, 528
geologically complex oil reservoir, 553
geologically complex reservoir, 359, 553
geologically correct correlation, 681
geologically reasonable deposition, 338
geophysical analysis, 143, 146
geophysical analysis methods, 10
geophysical analysis techniques, 16
geophysical exploration, 251
geophysical log signals, 295
geophysical logs, 556
geophysical modeling, 3
geophysical wave propagation, 22
geophysics, 3, 16, 276, 528
geoscience, 13, 17
geoscience applications, 13, 14
geostatistical algorithms, 512, 598
geostatistical analysis, 16, 231, 555, 580, 634
geostatistical applications, 87, 175
geostatistical characterization, 553
geostatistical distributions, 555
geostatistical estimation, 98
geostatistical mapping, 250
geostatistical method, 85, 332, 333, 500, 584, 651
geostatistical model, 512, 513, 562
geostatistical modeling, 111, 274, 499
geostatistical modeling approach, 275
geostatistical parameters, 332, 338, 339, 343
geostatistical simulation, 110, 111, 328, 349, 351
geostatistical simulation techniques, 104
geostatistical technique, 16, 85, 93–95, 98, 102, 175, 220, 274, 553, 554
geostatistical tools, 85, 93
geostatistics, 85–87, 97–100, 106, 107, 110, 112–114, 220–222, 241, 274, 332, 499, 510, 521, 524, 598, 601
GIS (geographical information systems), 253, 633–635, 638, 642–644, 649
Gothenburg, 633, 634, 644–649
gradient information, 342
gradient methods, 327, 360
gradient optimization method, 379
gradient-based methods, 328, 329
gradient-based optimisation, 359
grain density, 293
gravity, 4, 16, 311, 430, 531
gray, 61, 63, 73, 110, 159, 310, 381
gray encoding, 61
Groot, 12, 399, 411
groundwater, 79, 80, 190, 377
groundwater optimization, 367
groundwater remediation, 190, 377, 380, 391
groundwater remediation literature, 367
groundwater remediation problems, 367
groundwater remediation strategies, 366
GRSE (global relative strength of effect), 638, 640–642, 646–649
Gullfaks, 143, 144, 146, 149, 153–155
Gullfaks field, 144, 150, 151, 154
Gullfaks oil field, 143
Gullfaks time lapse, 146

hedge, 265–267
heterogeneity filtering, 235
heterogeneous input, 282
heterogeneous reservoir, 81, 148, 300, 310, 330, 605
heterogeneous structure, 301
heuristic approaches, 77
heuristic methods, 77, 368
heuristic relationship, 10
heuristic technique, 80, 82
hierarchical cluster analysis, 676
hierarchical clustering, 677, 682
hierarchical design, 636
hierarchical fuzzy model, 17
high resolution reservoir heterogeneity characterization, 289
high-resolution correlation, 527, 535, 548
high-resolution resistivity log, 681
hill-climber, 75, 357, 359
history match, 327, 338, 373, 513, 517, 522, 523, 545, 547, 548, 553, 554
history match, 548, 550, 586
history matching, 325, 327–329, 340, 341, 359, 512, 513, 527, 545–548, 550, 601
history matching problems, 327, 360
history matching process, 147, 359, 550
history-matching algorithms, 274

history-matching techniques, 491
horizontal borehole, 486
horizontal distribution, 229
horizontal permeability, 324
horizontal resolution, 228
horizontal structures, 237
horizontal variograms, 499
human decisions, 674
human error, 15, 16, 120, 473
human expert, 7, 43
human expert input, 289
human expert needs, 313
human input, 14
human logical processes, 673
hybrid, 7, 8, 43, 279, 327, 474, 493, 546, 548
hybrid ace model, 655
hybrid fuzzy systems, 43
hybrid method, 10, 333, 360
hybrid model, 547, 548, 550
hybrid system developed, 466
hybrid techniques, 310
hydraulic fracture, 446, 451
hydraulic fracture design, 456
hydraulic fracture simulation, 445
hydraulic fractures, 445, 446, 464
hydraulic fracturing, 445, 453, 455
hydraulic fracturing jobs, 450
hydraulic fracturing process, 445
hydraulic fracturing simulators, 445
hydraulic properties, 309
hydrocarbon, 15, 153, 159, 247, 252, 257, 258
hydrocarbon asset, 323
hydrocarbon estimate, 250
hydrocarbon formation, 247, 250, 255, 265, 266
hydrocarbon history, 158
hydrocarbon pore, 250
hydrocarbon production, 323
hydrocarbon reservoir, 143, 247, 248, 250, 252, 256, 262, 323, 324, 328, 335, 347
hydrocarbon reservoir characterization, 250
hydrocarbon reservoir estimate, 247, 248, 256, 257, 260
hydrocarbon saturation, 597
hydrocarbon system, 413
hypothetical delta simulation, 199
hypothetical markers, 290

identify laterally persistent mudstones, 527
ill-defined problems, 256
implicit heuristic relationship, 11
improve production, 373
improved characterization, 554
improved chimney, 170–172
improved history matching, 527
improved probability modeling, 283
improved reservoir management, 5
improved reservoir simulation, 583
in-fill drilling, 143, 146, 153, 154
in-fill drilling pattern, 366
includes wireline technology, 249
increasingly complex set, 189
indicator, 94, 101, 108, 109, 302, 303, 411, 451, 500, 505, 585, 586, 596, 676
indicator kriging, 108, 109, 503
indicator kriging estimate, 109
indicator methods, 109
indicator model, 94
indicator pattern, 297, 298, 303
indicator process, 303
indicator random functions, 108
indicator signals, 293
indicator simulation, 109
indicator simulation realization, 109
indicator variogram, 101, 108, 110, 559–562, 565–568
induction log deep, 682, 683
induction logs, 403, 406
induction resistivity, 120, 136
infill drilling, 20, 490, 583, 590, 601
information content, 18, 119, 455, 471, 500
injection ANNs, 389
injection island gas, 361
input attributes, 160, 171, 400, 401
input degrees, 201
input factors, 639
input functions, 201, 205
input layer, 23, 138, 159, 400, 417, 432, 473, 607, 634, 661
input log, 293, 676
input node, 160, 171, 406
input parameter, 192, 248, 256, 410, 610, 641, 646
input pattern, 292, 293, 295, 297, 307
input permeability, 571, 572
input porosity, 559–562, 571
input shale fraction, 566–568, 570
input signal, 165
input statistics, 329
input structure, 16
input units, 14, 641
integrate linguistic descriptions, 16
integrated clustering techniques, 472
integrated reservoir characterization, 475
integrated spatial analysis, 633
intelligent computing techniques, 651
intelligent model, 623, 628

intelligent reservoir characterization, 3, 4, 19, 471, 475, 493
intelligent systems, 4, 7, 22, 445
intelligent techniques, 5, 6, 311
intelligent technology, 6
interaction network, 644, 645
interbedded mudstones, 548
interbedded shale intervals, 290
interpretation aid tools, 241
interval analysis, 262
interval arithmetic, 39
interwell correlation, 680, 681
intuitive reasoning, 635
intuitive reasoning process, 633
intuitive solution, 642
intuitive understanding, 77
inverse functions, 35, 39
inverse model, 126
inverse problems, 329, 507
inverse relationship, 335
inverse rule, 384
inverse thickness, 598, 599
inversion, 328, 329, 360, 514, 516
inversion operators, 458
inversion process, 514
inversion techniques, 17
IRESC (intelligent reservoir characterization), 19, 475, 493

Juan basin, 593, 596, 599, 600

k-mean, 18, 483
k-mean clustering, 471
k-mean technique, 126
k-means, 9, 27, 141, 292, 474, 475, 483, 490, 494
k-means algorithm, 18
k-means cluster, 483
k-means cluster analysis, 309
k-means clustering, 27, 141, 483, 494
k-means clustering technique, 488
k-means distribution, 485, 486
k-means technique, 9, 27, 141, 474, 494
Kalman filtering, 18
KFM (Kohonen feature map), 277, 278, 281–285
KFM clustering algorithm, 286
KFM clusters, 282
KFM error, 283
KFM model, 282–284
KFM network, 284
knowledge acquisition, 120
knowledge acquisition techniques, 37
knowledge extraction, 17, 126, 129
knowledge management systems, 120
knowledge representation, 23, 137
knowledge-based approach, 123, 393
knowledge-based fuzzy system, 43, 45, 47, 48
knowledge-based model, 19, 120, 472
known alternative optimization methods, 77
known constraints, 40
known field conditions, 646
known patterns, 23, 137
known porosity cases, 312
known porosity classes, 310
known properties, 57, 335
known rock properties, 4, 177
Kohonen self-organizing, 495
kriging, 85, 86, 90, 93, 101, 102, 104–107, 109, 110, 114, 180, 232, 333, 334, 503, 504, 514, 559
kriging estimate, 102, 108
kriging factorial analysis, 220
kriging system, 101, 107
kriging variance, 102, 105, 175

laminated systems, 289
lamination characteristics, 291
lamination distribution, 306
lamination markers, 289
lamination pattern, 291
large-scale spatial structures, 231, 235
large-scale structures, 229, 235, 237–239, 243
lateral connectivity, 291
lateral continuity, 289, 299, 527
lateral continuity patterns, 302
lateral correlation, 294, 295, 299, 300, 305, 306
lateral correlation algorithm, 301
lateral correlation process, 294, 299, 302, 307
lateral correlation results, 301
lateral validity, 159
laterally persistent mudstone, 532, 542, 543
layer neural network, 504, 607
learning algorithm, 9, 368, 378, 474
learning capabilities, 21, 204, 211
learning coefficient, 25, 140, 653, 657
learning method, 27, 141
learning model, 9, 29, 474, 495
learning phase, 14
learning process, 25, 140, 295, 400, 473, 610
learning rule, 29, 496, 638
learning techniques, 27, 141, 177, 493
limestone, 399, 541, 586
limestone depositional texture, 397
limestone texture, 397
linear averaging, 500
linear convergence properties, 326
linear estimate, 91, 101
linear method, 81

linear models, 207
linear network, 640
linear optimization techniques, 79
linear programming, 79
linear programming methods, 326
linear reduction, 184
linear regression, 101, 570
linear regression techniques, 85
linear relationship, 175, 207, 325
linear sediment production functions, 208
linear system, 93, 102, 325, 326
linear technique, 79
linguistic descriptions, 310, 311
linguistic encoding technique, 314
linguistic hedge, 265–267
linguistic information, 123, 318, 319
linguistic petrographical descriptions, 309
linguistic quantifiers, 247, 248
linguistic statements, 310
litho-seismic, 247, 248, 250, 252, 253, 256, 267
lithofacies, 292, 293, 309, 310, 475
lithofacies classification, 292
lithofacies distribution results, 309
lithofacies information, 293
lithofacies recognition, 309
lithofacies sets, 310
lithological composition, 290
lithological descriptions, 310, 311
lithological indicator, 597
lithological layer, 121
lithological logs, 289
lithological models, 300, 304
lithological pattern classification, 291, 306
lithological patterns, 289, 293, 297, 303, 307
lithological structure, 289, 299
lithology, 12, 18, 119, 121, 123, 249, 252, 253, 255, 275, 400, 401, 471, 507, 508, 510, 583, 584, 589, 596, 676, 678, 679
lithology identification, 673, 676
lithology information, 677, 687
lithology input, 293
lobes, 294, 527, 532
log, 4, 5, 8, 10, 15, 18–22, 95, 119-121, 123, 126, 135, 179, 250, 255, 259, 289, 290, 292–294, 296, 303, 309, 319, 335, 337, 347, 400–403, 406–410, 464, 471, 472, 476, 491, 512, 513, 556, 557, 559, 583, 584, 591, 592, 605–609, 611, 616, 623, 628, 631, 673–678, 680, 682, 683, 685
log analysis, 258, 554
log analysis applications, 675
log characteristics, 628
log correlation, 555, 673
log databases, 19, 120, 472
log information, 143, 248
log interpretations, 297
log measurement, 144, 675, 682
log response, 295, 297, 397
log segment, 680
log signals, 176, 291, 294, 295, 297, 307
log traces, 673, 674, 677, 680
logging, 143, 176, 249, 251, 252, 293, 297, 340, 673
logging suites, 397, 400
logging techniques, 250
long range structure, 506
lookup technique, 664, 669

magnetic field, 329
magnetic resonance, 605
magnetic resonance log, 606
magnetic resonance logging, 623
magnetic resonance logging process, 631
magnetic resonance logging technique, 605
magnetic resonance logging tools, 623, 631
magnetic resonance logs, 605, 606, 609, 611, 623, 628, 631
magnetic resonance tools, 623
mapping algorithm, 500
mapping applications, 250
mapping function, 636, 638, 640, 642
mapping method, 591, 598
mapping neural network, 637
mapping parameters, 652, 669
mapping process, 417
mapping technique, 599
Marsily, 328, 329, 651
matching filter, 145
matching parameters, 327, 329, 330
matching process, 585
matching reservoir properties, 327
material balance, 250, 267, 360, 413
mathematical algorithms, 399
mathematical function, 605
mathematical literature, 80
mathematical model, 524
mathematical modeling, 15, 26, 140, 474, 493
mathematical modelling, 634
mathematical models, 18, 472, 500, 652
mathematical optimization, 77
mathematical results, 105
mathematical techniques, 3
mathematical theory, 325
mathematical tools, 4, 219, 220, 366
matheron, 85, 94, 219–221, 229
mature field, 553
mature hydrocarbon reservoir, 16

mature oil field, 553
maximize compression, 279
maximize deliverability, 445
maximum approach, 538
maximum bottom hole, 341
maximum continuity, 89
maximum correlation, 332, 333, 337
maximum cross-correlation, 681
maximum defuzzification, 539
maximum dip, 163
maximum hydrocarbon, 372
maximum injection rates, 341
maximum noise fraction, 220
maximum production, 207
maximum structural curvature, 589
maximum variance, 224
MBVI, 605–609, 611, 616, 623, 628
MBVI logs, 607
MD, 335, 397–399, 405, 407–409, 515, 580, 586, 594
mean permeability, 513
mean population, 386
mean results, 231
mean thickness, 233
mean upscaling, 276
mean variance, 333
measured bottom hole, 548
measured characteristics, 676
measured core plug, 398, 405
measured core plug permeabilities, 408
measured core porosity, 397, 406, 407
measured directly, 402
measured logs, 397, 398
measured porosity, 409
measured pressures, 326
measurement log, 628
membership function, 16, 17, 23, 34, 37, 120, 130, 137, 193, 194, 196, 197, 199–201, 204, 205, 207–209, 211, 256–259, 260, 473, 529, 530, 537, 538
membership function parameters, 27, 141, 493
membership grade, 16, 260, 266
membership grade functions, 247, 256
minimal adaptation, 391
minimal clay content, 597
minimized quantization error standpoint, 286
minimizing sampling plans, 651, 664
minimum bottom hole pressures, 341
minimum contiguity, 223
minimum correlation, 333
minimum deliverability, 448
minimum economic production, 341
minimum energy, 355
minimum error, 180, 277
minimum error variance, 102
minimum information, 651
minimum input information, 669
minimum log, 293
minimum operator, 41
minimum possible error, 317
minimum quantization error, 277
minimum spatial continuity, 89
minimum spatial variability, 224
mining, 5–8, 17, 22, 85, 110, 114, 120
mining applications, 313, 333
mining approaches, 3
mining methods, 18
mining problems, 110
mining system, 314
mining techniques, 6
mining wireline logs, 120
minor geologic markers, 290
miocene, 370–373
miocene complex, 371
miocene oil, 372
miocene reservoir, 371, 372
miocene sand, 371
mlp (multi-layer perceptrons network), 397, 400, 402
mlp neural network, 403, 405, 406, 408
model behavior, 429
model building process, 446, 609, 611, 628
model fit, 564
model fracture, 584
model free function, 610
model identification methods, 9, 27, 474, 494
model layer, 547
model membership sets, 529
model multipoint statistics, 177
model porosity, 106
model representation, 347
model reproducing facies, 189
model rock properties, 19, 120, 472
model sediment dispersal systems, 189
modeling process, 585, 623
modeling results, 211
modeling sedimentation, 204
modeling tools, 248
modelling capability, 499
modelling framework, 342
modelling parameters, 347
modelling techniques, 360
modern computational resources, 79
modern geostatistics, 98
modified GA, 329, 330, 351
modified objective function, 328

modified rank selection, 330
module design optimization, 449
MPHI, 605–608, 611, 616, 623, 628
MPHI logs, 623
MR (magnetic resonance) log, 606, 616, 623
MR log measures effective porosity, 606
MR logs, 617, 623, 625, 627
MRI (magnetic resonance imaging), 605
MRI logs, 606, 607, 614–616, 618–623
mudstone, 311, 401, 527, 532, 535–539, 541–543, 550
mudstone continuity, 542, 546
mudstone deposition, 535
mudstone interval, 535, 541–543
mudstones interbedded, 527, 535, 538
multi-attribute analysis, 20, 471, 491
multi-dimensional chromosomes, 70
multi-dimensional field, 472
multi-disciplinary analysis methods, 3
multi-disciplinary approach, 146
multi-Gaussian, 105
multi-Gaussian approach, 99
multi-Gaussian distribution model, 108
multi-Gaussian model, 105, 108
multiple linear regression model, 19, 120, 472
multiple stochastic realizations, 115
multiple-point geostatistics, 100, 110, 111, 114
multiple-point information, 111, 499
multiple-point relation, 106, 113
multiple-point statistics, 111, 499–501
multipoint relationships, 176
multipoint statistics, 175–177, 181, 184
multipoint training, 177
multivariate analysis, 634
multivariate database, 677
multivariate Gaussian behavior, 565
multivariate methods, 642
multivariate statistical analysis, 674
multivariate statistical technique, 673–676
multivariate upscaling, 273, 285
mutation operator, 57, 59, 61, 74, 75, 343, 345, 346, 355, 358
mutation probability, 330

natural evolution, 17
natural evolution success, 71
natural fracture, 593, 596, 597
natural fracture systems, 583
natural fracturing, 593, 595, 596, 600
natural gas storage field, 445
naturally fractured reservoir, 583, 584
near outliers, 669
net pay, 598
net profit, 365, 374, 375, 387–392
net thickness, 250, 255, 258, 263
net-to-gross distribution, 527, 545–547
net-to-gross parameter, 547
network applications, 79
network cluster, 279, 280
network connection, 381
network design, 415
network model, 126, 415, 416, 423, 426, 653, 655, 657, 662
network parameters, 25, 140, 505, 509, 653, 657
network performance, 415
network prediction, 473, 653, 463
network processes, 14
network structure, 29, 60, 280, 496, 644, 653, 657
network target, 318
network training, 403, 406–408, 510, 657
networks structures, 280
neural computing, 5
neural model, 450, 455, 466, 611
neural model predictions, 664
neural module, 455–458
neural net, 113, 455, 461, 462
neural net approach, 519
neural net calibration, 113
neural net function, 106
neural net input, 506
neural net technique, 519
neural net technology, 23, 137
neural network algorithm, 273
neural network analysis, 598, 600, 647
neural network applications, 416
neural network approach, 309, 310, 405, 646
neural network clustering, 28, 471, 483, 495
neural network clustering technique, 488
neural network clusters, 483
neural network distribution, 487
neural network input layer, 171
neural network method, 411, 524, 652, 653
neural network model, 16, 19, 25, 26, 120, 121, 123, 126, 130, 136, 138, 140, 179, 278, 315, 415, 466, 472, 473, 474, 493, 504, 599, 611, 653, 656, 658, 659, 662, 665, 669
neural network modeling process, 585
neural network parameters, 610
neural network performance, 122, 124, 125, 127, 128
neural network predicted results, 169
neural network prediction, 397, 410, 460, 653
neural network preliminary experiments, 316
neural network structure, 26, 140, 493, 610, 657
neural network system, 23, 137
neural network technique, 119, 129, 318, 653, 669

neural network trained, 175, 184, 505, 506, 639, 640
neural network training, 106, 184, 380, 401, 403, 406–408, 505
neural network training performance, 410
neural network yield measures, 649
neural pattern recognition methods, 229
neuro-fuzzy, 17, 19, 23, 120, 137, 472
neuro-fuzzy approach, 21
neuro-fuzzy inference, 211
neuro-fuzzy inference system, 27, 141, 204, 493
neuro-fuzzy logic model, 213
neuro-fuzzy logic model generated, 212
neuro-fuzzy model, 17, 26, 120, 126, 129, 130, 136, 140, 212, 493
neuro-fuzzy modeling, 26, 140, 493
neuro-fuzzy system, 23, 137, 211
neuro-fuzzy technique, 19, 126, 472
neuro-statistical method, 657, 659
neuro-statistical model, 654, 657, 658, 663, 669
neuro-statistical techniques, 651, 659, 669
neurocomputing, 601
neuron, 22, 24, 137–139, 277–282, 317, 431, 432, 641
neutron log, 95, 556, 557, 559, 562, 682, 683
neutron porosity, 293, 400, 402, 403, 406, 407, 410, 559, 609–611, 628, 675
neutron porosity logs, 580, 611
noise filtering, 289, 293
noise reduction, 18
non-linear information, 18, 652
non-linear mapping, 504
non-linear methods, 9
non-linear model, 669
non-linear modeling, 18
non-linear multipoint relationship, 500
non-linear problems, 326, 447
non-linear processing, 7, 13
non-linear relationship, 19, 20, 397, 401, 583
non-linear signal, 22
non-linear spatial relationships, 176
non-linear statistical techniques, 25
non-linear system, 26
non-linear transfer function, 661
nonlinear information, 471
nonlinear mapping, 120, 126, 129, 132, 136, 472
nonlinear methods, 80
nonlinear optimisation, 325, 380
nonlinear optimization methods, 80
nonlinear regression methods, 326
nonlinear relationship, 119, 135, 136, 472, 490
nonlinear statistical techniques, 140
nuclear magnetic resonance (NMR), 605
numerical analysis, 475
numerical computation, 674
numerical experiments, 177
numerical model, 197, 323–325, 335, 337, 344, 349, 378, 367
numerical modeling, 10, 192
numerical results, 325
numerical significance, 34, 193
numerical simulation model, 331
numerical simulations, 190
numerical solution approaches, 204, 211

object-based algorithm, 500
object-based modeling approach, 276
object-based modeling perspective, 286
object-based modeling strategy, 276
object-based simulation, 501
object-based upscaling, 276
objective function, 25, 28, 52, 55, 72, 77–80, 94, 139, 327–329, 331, 348–350, 355, 357, 358, 360, 368, 375, 377, 382, 394, 495
offset analysis, 15
offshore, 191, 200, 397, 398, 531
offshore australia, 310
offshore carbonate, 191
oil API gravity, 422
oil compressibility, 418, 429, 441
oil density, 413, 416–418, 422, 423, 429, 432, 440
oil exploration, 190, 251
oil field, 85, 143, 247, 248, 250, 253, 258, 261, 325, 413, 416
oil formation, 414, 416–419, 429, 432
oil gravity, 413, 414, 416–419, 425, 429, 432
oil price, 143, 154, 375
oil production, 307, 326, 348, 350, 369, 372, 379–381, 388, 592
oil reserve, 5
oil reservoir, 583
oil reservoir characterization, 248
oil sand, 265, 266
oil simulators, 366
oil system, 413, 416–418, 429
oil viscosity, 413, 416–418, 422, 424, 429, 432
optimal crossover operators, 361
optimal estimate, 247, 265, 266
optimal estimation, 255
optimal exploration, 594
optimal fracture design, 445
optimal model, 179
optimal mutation operators, 361
optimal network, 181
optimal processing, 476, 478
optimal solution, 78, 80, 81, 182, 365, 384, 385
optimisation algorithm, 57, 328, 348, 357

optimisation method, 329, 351, 355
optimisation process, 61
optimisation technique, 325
optimization algorithm, 25, 140, 513
optimization analysis, 393
optimization methodology, 458
optimization methods, 366
optimization process, 394, 448, 449, 457
optimization strategies, 365
optimization technique, 17, 27, 79, 81, 121, 126, 141, 365, 366, 370, 493
optimized frac design, 455
optimum combination, 449
optimum design, 457
optimum estimation, 165
optimum list, 449
optimum mutation rates, 361
optimum stimulation, 449
optimum strategy, 22
outcrop measurements, 184
outcrop modelling, 524
overall deliverability potential, 445
overall model uncertainty, 335
overall permeability statistics, 572
overlay analysis, 633
overlay function, 634, 642

p-wave, 13, 129, 130, 132, 135, 136, 144, 145, 159
p-wave attenuation, 129, 130, 132, 135, 136
palaeogene, 527, 532, 535
palaeogene turbidite reservoir, 537
paleocave system, 476
parallel algorithms, 367
parallel processing, 473
parallel processing techniques, 3
parallel structures, 13
pattern class, 295
pattern classification, 291, 293, 295, 303
pattern classification algorithm, 303
pattern classification process, 293
pattern recognition, 4, 5, 7, 9, 14–16, 26, 27, 97, 110, 140, 289, 293, 295, 399, 415, 474, 493, 494, 500, 501, 503, 673, 676, 681
pattern recognition applications, 9, 474
pattern recognition method, 681
pattern recognition problems, 431
pattern recognition process, 295, 297, 300
pattern recognition techniques, 9, 298, 309, 474
pattern recognition tools, 500
pattern reproduction, 97, 110, 500, 501
PCA (principal component analysis), 178, 181, 184, 219, 228, 229, 231–233, 673–675, 681
PCA components, 233
PCA technique, 676
pelagic biofacies, 536
pelagic conditions, 536, 541
pelagic content, 542
pelagic mudstones, 527, 532
performance characteristics, 388
performance estimates, 445
performance measures, 368, 376
performance predictions, 382
permeability analyses, 400
permeability connectivity, 110
permeability correlations, 568
permeability distribution, 545, 568, 572, 580
permeability estimation, 327
permeability features, 516
permeability field, 512–514, 516–519, 523
permeability interval, 623
permeability logs, 572
permeability model, 98, 512, 572, 585
permeability prediction, 398, 408, 410
permeability prediction results, 409
permeability scaling, 337
permeability simulation realizations, 575, 576
persistent mudstone units, 542
petrographical characteristics, 309, 310
petroleum, 3, 110, 119, 155, 325, 366, 368, 583, 584
petroleum engineering, 3, 85, 175, 276, 279, 413, 415
petroleum engineering applications, 273, 285
petroleum engineering field, 370
petroleum geoscience, 145, 146, 151
petroleum geostatistics, 499
petroleum literature, 416
petroleum recovery research, 273
petroleum reservoir, 16, 17, 309
petroleum reservoir characterisation, 309
petroleum reservoir simulation, 80
petroleum seismology, 13
petrophysical measurements, 11
petrophysical properties regression, 676
petrophysical property, 113, 190, 250, 293, 333, 335, 339, 674
petrophysical property fields, 329, 351
petrophysics, 528, 599
phase models, 326
phase problems, 327
phase-locking, 165, 166, 170, 171
physical behavior, 414, 473
physical characteristics, 78
physical experiments, 10
physical models, 6, 250

physical models solved, 5
physical performance, 415
physical principles, 324, 382
physical process, 355, 500
physical properties, 15, 251, 273, 279, 324, 377, 394
physical relationship, 119, 377
physical representation, 54
physical response, 500
physical significance, 11
physical transfer function, 507
plug permeability, 586
poor convergence characteristics, 282
poor correlation, 402, 406
poor crossover operator, 55, 57
poor estimates, 325
poor GA, 57
poor predictive tools, 413
poor training performance, 402
population composition, 528, 537
population generation, 59
population members, 82
population strategies, 360
porosity class, 311, 312
porosity classification, 310
porosity distribution, 177, 562
porosity estimation, 179
porosity indicator logs, 609–611, 628
porosity limestone, 586
porosity limestone layers, 586
porosity log, 403, 406, 410
porosity models, 572
porosity prediction, 406
porosity prediction results, 404
porosity range, 404, 410
porosity realizations, 559, 568
predicted chimney bodies, 169
predicted core permeability, 409
predicted core porosity, 404
predicted oil gravity, 430
predicted oil viscosity, 430
predicted permeability, 406–410
predicted porosity, 403–405, 409
predicted porosity classes, 312
predicted undersaturated oil compressibility, 430
predicting carbonate sedimentation production patterns, 206
predicting depositional texture, 401
prediction accuracy, 453
predictive methods, 652
predictive model, 628, 631
probabilistic approach, 73
probabilistic chromosomes, 360
probabilistic classification, 503
probabilistic reasoning, 3, 4, 471
probability density, 281–283
probability density function, 45, 274, 636
probability distribution function, 68, 71, 349
probability model, 98, 104, 109, 113, 114, 504
probability theory, 251
process information, 415
process modeling, 447
production characteristics, 347
production criteria, 377
production history, 328, 330, 338, 346, 348, 350, 361, 455, 456, 464, 518, 592, 599
production information, 144, 516, 521
production log, 471, 476, 483, 584
production optimization, 5, 21, 493
production plans, 323
production rates, 329, 341, 348, 372
production response, 513, 516
production rule, 678
production scenario, 515
production strategies, 527, 532
production strategy, 22, 673
production system, 370
production-log, 20, 471, 475, 483, 486, 489–491
productive sandstone units, 596
profit function, 449
propagation approach, 473
propagation network, 417, 431
propagation network model, 415
propagation neural network, 295, 416
property distribution, 93, 180, 184, 553, 557, 580
property field, 274, 332
proximity analysis, 223

qualitative analysis, 483, 484, 623
qualitative information, 16, 191, 677, 678
quantitative biostratigraphical correlation techniques, 536
quantitative characterisation, 649
quantitative estimate, 250
quantitative geologic models, 598
quantization technique, 293

random changes, 57
random combinations, 457
random errors, 248, 249
random field, 329, 333
random function, 86–88, 93, 94, 221
random function model, 91, 94, 524
random functions, 221
random methods, 378
random noise, 227, 229–231, 235, 237, 238
random process, 91, 93

random sampling, 81, 82, 248, 388
random selection, 397, 400, 636
random structures, 224
rank selection, 71
RBF (radial basis function), 415–417, 431, 432
RBF model, 422
RBF network, 413, 418
RBF network model, 417, 423
RBF networks, 431
realistic geostatistical models, 98
realistic reserve estimation, 605
realistic sedimentation dispersal patterns, 189, 213
recognition technology, 289
regional components, 224, 227, 236
regional dip, 161
regional mudstones, 527
regional orthogonal fractures, 583
regression analysis, 399
regression function, 106
regression model, 417, 432, 508, 585
regression network, 416, 417
regression neural network, 15
regression relation, 106
relational database, 447, 448, 466
remote file system, 368
remote sensing, 4, 13, 113
researchers tackle uncertainty, 473
reserve estimate, 623, 631
reserve management, 631
reservoir analysis, 16
reservoir characterisation, 309, 323–325, 328, 329, 331, 339, 347, 358, 360
reservoir characterisation process, 324, 325, 331, 335, 338, 339, 360
reservoir characteristics, 605, 606, 631, 673
reservoir characterization, 16, 19, 21, 98, 107, 110, 248–251, 293, 392, 493, 499, 524, 607, 631, 673
reservoir classification, 302
reservoir compartment boundaries, 307
reservoir compartmentalization, 292, 298, 299, 300, 303
reservoir compartmentalization process, 306
reservoir compartments, 289, 291, 293, 295, 302
reservoir complex, 371
reservoir connectivity, 20, 471, 491
reservoir depletion, 372
reservoir descriptions, 334, 342, 347, 359
reservoir engineering, 18, 77, 119, 324, 325, 329, 344, 359, 415, 471
reservoir engineering application, 22, 78, 79
reservoir engineering optimization, 77, 82
reservoir engineering relationships, 585
reservoir facies model, 328
reservoir formation, 154, 251
reservoir fracture, 598, 599
reservoir heterogeneity, 107, 309
reservoir impact, 144
reservoir interval, 12, 398, 405
reservoir length scale, 275
reservoir log, 595
reservoir management, 143, 154, 251, 323, 360, 368, 369, 391, 554, 606
reservoir management decisions, 393
reservoir management goal, 366
reservoir management parameters, 369
reservoir model, 3, 21, 22, 108, 111, 323, 335, 339, 344, 348, 359, 370, 373, 328, 331, 336, 370, 377, 491, 512, 513, 519, 520, 521, 524, 545, 548
reservoir model prediction, 21
reservoir modeling, 21, 175, 176, 184, 274, 276, 290, 378, 393, 493, 584
reservoir modeling problems, 177
reservoir modeling techniques, 175
reservoir modelling, 323, 527, 532, 537
reservoir parameter, 248, 249, 326
reservoir performance, 274, 323, 512, 595
reservoir performance prediction, 673
reservoir permeability, 513
reservoir permeability field, 510
reservoir permeability models, 512
reservoir pressures, 417, 432
reservoir properties, 3, 10–12, 14, 22, 248, 250, 252, 273, 274, 276, 285, 292, 325–327, 333, 598
reservoir properties simulated, 333
reservoir property estimation, 22
reservoir property fields, 337, 344, 349
reservoir quality, 144, 397–399, 402, 411, 580
reservoir quality prediction, 400
reservoir response, 369, 512
reservoir rock, 252, 397, 398
reservoir rock deposition, 586
reservoir rock properties, 553, 554, 556
reservoir sand, 289
reservoir sand distribution, 545
reservoir sandstones, 548
reservoir scale, 228
reservoir simulation, 16, 22, 280, 286, 348, 366, 368, 376, 413, 429, 491, 513, 527, 550, 586, 601, 606
reservoir simulation applications, 279
reservoir simulation literature, 274
reservoir simulation model, 335
reservoir simulation performance, 276

reservoir simulation problems, 275
reservoir simulator, 79, 80, 275, 276, 284, 326, 327, 339, 347, 365, 366, 370, 392–394
reservoir units, 176
resistivity log, 15, 136, 401, 555, 563, 607, 674, 675, 679, 680
resistivity logging, 252
resistivity reservoirs, 605, 606
resources management literature, 80
resources optimization applications, 384
restimulation, 445, 446, 459, 466
restimulation jobs, 446
result interpretation aid tools, 224
Rhob, 120, 123, 125, 136, 609
Rhob logs, 121
rigorous fuzzy logic systems, 214
robust model, 669
robust optimization, 17
robust structure, 637
robust technique, 411
rock characteristics, 310
rock core, 95
rock engineering system, 639
rock fracturing, 598
rock parameter, 129, 135
rock pore, 579
rock properties, 4, 13, 18, 19, 107, 119, 120, 175, 176, 472, 500
rock properties fields, 328
rock property distributions, 557
rock property measurements, 10
rock sequences, 293
rock texture, 402
rule system, 205
rule trace, 318
rule-based correlation system, 677
rule-based inference, 673
running magnetic resonance logs, 631
running multiple simulations, 342

salt, 5, 371, 532
salt bodies, 371
salt core, 213
sand bodies, 94, 294
sand deposition, 646
sand facies, 500, 506
sand lobes, 294
sand thickness, 450
sandstone, 19, 94, 120, 130, 144, 252, 310, 311, 335, 445, 446, 455, 472, 514, 527, 536, 538, 542, 543, 550, 554, 583, 593, 681
sandstone deposition, 537
sandstone formations, 593
sandstone gas reservoir, 593
sandstone lobes, 532
sandstone reservoir, 584, 593
sandstones rocks, 120
saturation changes, 151, 153
saturation determination, 275
saturation functions, 326, 516
saturation hydrocarbon, 247, 250, 255, 260, 265, 266
saturation logs, 628
SBX, 69, 74
SBX crossover, 68
SBX operator, 67, 69
scalar wavelet classification, 12
scale information, 107, 302
scale variability, 88
scaling functions, 275
scaling techniques, 473
secondary attributes, 106
secondary conditioning, 562
secondary density, 562
secondary faults, 555
sediment accumulation, 190, 644, 645
sediment characteristics, 646
sediment conditions, 646, 647
sediment core, 210
sediment deposition, 200, 644
sediment dispersal, 191
sediment grade, 645
sediment grain, 200, 645, 646
sediment production, 191, 204–208
sediment production modeled, 207, 208
sediment thickness, 211
sedimentary accumulation, 190
sedimentary facies, 189, 190, 214
sedimentary geologist, 214
sedimentary layers, 299, 302
sedimentary patterns, 294
sedimentary portions, 189
sedimentary process, 190, 191, 203
sedimentary process simulator, 191, 192
sedimentary rock, 311, 586
sedimentary rock properties, 309
sedimentary sequences, 191
sedimentary structure, 297–299, 302, 674
sedimentary systems, 194, 204
sedimentation, 201, 202, 204, 294, 535, 645
sedimentation models, 189
sedimentation pattern, 295
sedimentation process, 290
sedimentation rates, 535
sedimentologic simulations, 189
sedimentology, 189, 536, 590, 634, 645

sedimentology characteristics, 644
seismic acquisition, 5
seismic analysis, 145
seismic attenuation, 129, 136
seismic attribute, 8, 10, 11, 15, 18, 19, 106, 119, 120, 135, 175, 177, 178, 181, 182, 184, 471, 472, 475–478, 598
seismic calibration, 248
seismic chimney, 158, 160, 167, 171
seismic chimney detection, 157
seismic classification method, 171
seismic clustering analysis, 15
seismic differences, 150, 152
seismic event, 158, 165, 167, 171, 508, 510
seismic features, 144
seismic information, 18, 119, 120, 229, 471, 508
seismic interpretation, 16, 372
seismic inversion, 17, 508
seismic mapping, 17, 148
seismic measurements, 10, 176
seismic modeling, 148
seismic monitoring, 143
seismic parameters, 143–145, 150
seismic pattern, 12, 180
seismic pattern analysis, 399
seismic processing, 159, 160
seismic reservoir characterization, 399
seismic resolution, 9, 107, 373
seismic response, 12, 15, 146, 148, 157, 158, 167, 507
seismic signal, 14, 153, 157, 161
seismic signal processing, 7
seismic simulation model, 507
seismic time lapse analysis, 154
seismic trace, 166, 228, 229, 476, 477
seismic travel, 145, 598
seismic wave propagation, 16
selection criteria, 71
selection pressures, 71
self-organizing networks, 9, 474
semivariogram, 557, 558, 564, 581
semivariogram model, 333
sequential conditional simulation, 110
sequential Gaussian simulation, 105, 109, 332, 333, 563
sequential Gaussian simulation algorithm, 105, 559, 565
sequential Gaussian simulation approach, 565
sequential Gaussian simulation method, 559
sequential Gaussian simulation technique, 108
sequential indicator simulation, 94, 109
sequential simulation, 104, 105, 107, 111, 182, 601
sequential simulation algorithm, 104
sequential simulation approach, 562
sequential simulation methodology, 111
sequential simulation technique, 105, 557
set accuracy, 380
set boundaries, 204, 211
set concept, 193
set information, 675
set representation, 194
shale bodies, 113, 514, 579, 580
shale boundaries, 297
shale content, 179, 335, 568
shale continuity, 302
shale fraction, 324, 338, 556, 563, 564, 568–575, 577–580
shale fraction distribution, 565
shale fraction models, 572
shale fraction realizations, 574
shale fraction reproduced, 580
shale intervals, 293
shale laminations, 289
shale layers, 289
shale log, 681
shale patterns, 111
shale permeability, 330
shale sequences, 335
shallow carbonate, 205
shallow membership function, 201
shallow resistivity logs, 410
shift mutation operator, 347, 351
sigmoidal function, 638–640
signal processing, 228
significant error propagation, 573
significant permeability, 580
significant seismic changes, 150
siliciclastic sedimentary process simulators, 191
siliciclastic sediments, 189
siliciclastic shelf, 193
similarity analysis, 291, 293
similarity analysis process, 291, 294
similarity association, 278
similarity characterization, 289, 307
similarity characterization process, 298
similarity match, 29, 496
similarity method, 160
simplified function, 449
simplified input patterns, 296
simplified network, 380
simplified reservoir model, 545
simulated annealing, 77, 94, 329, 351, 355, 357, 365, 366, 368, 382
simulated bottom hole pressures, 346
simulated field conditions, 446

simulated permeability, 574–576, 580
simulated permeability realizations, 573
simulated porosity, 577, 579
simulated porosity realizations, 562, 563
simulated pressures, 325
simulated realizations, 505, 510, 524, 562, 565, 580
simulated reservoir models, 111, 519
simulated reservoir rock properties, 574
simulated shale fraction, 578, 579
simulated shale fraction realizations, 569
simulation algorithm, 94, 106
simulation approach, 93
simulation cells, 274
simulation methods, 98
simulation model, 275, 284, 332, 337, 373, 546
simulator predictions, 370, 379, 387, 389, 393
six-layer model, 527, 545–548
six-layer reservoir model, 543
skin, 337
skin chromosome, 345
skin factor, 324, 331, 337, 339, 343, 345, 346, 351, 378
skin factor parameters, 346
skin parameters, 337
smallest depositional units, 307
smallest geobody, 285
soft computing, 3, 4, 7, 22, 33, 447, 471
soft computing methods, 22, 365
soft computing techniques, 3–5, 129
soft computing technology, 4
soft conditioning, 95
soft information, 191, 510, 524
soft training, 500, 508
software application, 447, 463–466
SP, 120–126, 136, 250, 252, 293, 605, 607, 609, 628, 680, 681
SP logs, 120, 681
sparse borehole, 651
spatial analysis, 635, 649
spatial behavior, 222
spatial component, 219, 223, 228, 235–237, 239, 240, 242
spatial connectivity, 512
spatial contiguity analysis, 219, 241
spatial contiguity constraints, 244
spatial contiguity relations, 219
spatial continuity, 89, 99, 100, 237, 514, 557, 559
spatial continuity parameters, 563, 579
spatial correlation, 99–101, 107, 598
spatial correlation structure, 274
spatial covariance, 108
spatial database, 642
spatial distribution, 85, 89, 98, 309, 325, 513, 594, 651
spatial estimation, 99, 114
spatial estimation technique, 101, 114
spatial filtering, 219
spatial filtering method, 220
spatial information, 224, 226, 377, 633, 642
spatial pattern recognition, 98
spatial pattern reproduction, 102
spatial proximity analysis, 220, 223
spatial scale structure, 219
spatial sense, 176
spatial variability, 87, 89, 93, 106, 219, 223, 226, 227, 231, 235–238, 240, 241, 244, 328, 512, 513, 559
spatial variability exhibited, 521
spatial variability reproduced, 226
SPE, 447
spontaneous potential, 252, 563, 605, 607, 680
spontaneous potential logs, 556
stacked reservoirs, 4, 371
stacked turbidite sandstones, 527, 532
statistical analysis, 11, 662, 674
statistical characteristics, 219, 674
statistical distribution, 12, 97, 332, 565
statistical pattern classification, 292
statistical pattern recognition, 504, 524
statistical sciences, 97, 521
statistical technique, 528, 653, 657, 659, 661, 663, 664, 669, 673, 674
statistical variance, 168
stimulation jobs, 446, 449, 460
stimulation optimization process, 449
stimulation parameters, 448, 460
stochastic aspect, 585
stochastic conditional simulation, 333
stochastic distributions, 553
stochastic framework, 585
stochastic geologic modeling, 289
stochastic methods, 16
stochastic model, 111, 274, 328
stochastic models techniques, 21
stochastic optimization methods, 17
stochastic parameters, 16
stochastic permeability, 512
stochastic permeability fields, 329
stochastic principles, 86
stochastic realization, 102, 104, 580
stochastic reservoir model, 501, 506, 508, 521
stochastic reservoir simulation, 499, 500
stochastic simulation, 102, 110, 113, 182
stochastic simulation generates multiple reservoir models, 102

stochastic simulation method, 102
stochastic simulation techniques, 93
stochastic techniques, 554
stochastic theory, 86, 91
stochastic universal sampling, 71
stochastically generated realizations, 93
stochastically generated reservoir models, 102
storage field, 445, 466
stratigraphic analysis, 17
stratigraphic facies, 8
stratigraphic features, 289, 297
stratigraphic interpretation, 398
stratigraphic interval, 402, 674, 681
stratigraphic markers, 556
stratigraphic model, 20, 189, 190, 204, 475, 493
stratigraphic units, 404, 409
stratigraphical modelling, 528
stratigraphy model, 553
structural changes, 598, 599
structural curvature, 594
structural dip, 159, 171
structural dip information, 159
structural features, 596
structural history, 586
structural information, 524, 598
structural mapping, 4
structural model, 3, 113, 339, 500, 524, 555
structural parameters, 473
structural properties, 342, 598
structural relationships, 119, 473, 652
structural uncertainty, 338
structure-related fractures, 583
subsurface distribution, 190
subsurface fracture network, 592
subsurface geologic structures, 228, 242, 251
subsurface geology, 189, 215, 219, 228
subsurface lithology, 676
subsurface pattern, 593
subsurface problems, 79
subsurface properties, 4, 21
subsurface reservoir, 499
subsurface reservoir model, 111
subsurface structure, 251, 258, 259
supervised classification, 314
supervised clustering, 315, 316
supervised clustering algorithm, 315, 317
supervised clustering approach, 309, 310
supervised clustering models, 315
supervised clustering technique, 313, 318
supervised gradient, 607
supervised learning, 177, 295
supervised learning approach, 293, 400
supervised neural network, 177, 398, 400, 401
supervised technique, 295
supervised training model, 8
synthetic conventional logs, 611
synthetic field, 299–301
synthetic magnetic resonance, 605
synthetic magnetic resonance logs, 605, 631
synthetic model, 325, 327, 328, 338, 347
synthetic MR logs, 607
synthetic reservoir model, 327, 338
synthetic stratigraphic models, 190, 214
systematic errors, 248, 249
systematic log interpretation, 674
systematic optimization techniques, 366

tectonic fractures, 583
tectonic history, 590
tectonic phase, 586
ten-layer model, 527, 545, 547, 548, 550
ten-layer numerical model, 337
ten-layer reservoir model, 545
tessellation, 273, 276, 279
three-dimensional chromosomes, 361
three-dimensional computer simulators, 445
three-dimensional distribution, 575, 651
three-dimensional fields, 652
three-dimensional hypothetical, 189
three-dimensional model, 211, 662, 663
three-dimensional realization, 559
three-dimensional relations, 40
three-dimensional reservoir model, 329
three-dimensional seismic subsurface information, 215
three-dimensional semivariogram, 557
three-dimensional stratigraphic models, 191
three-phase reservoir models, 327
time lapse seismic analysis, 143
total field, 388
total hydrocarbon production, 377, 387
total variance, 178, 181, 682
traditional cross-correlation methods, 674
traditional geostatistical algorithms, 512
traditional geostatistics, 110, 111
traditional indicator method, 506
traditional indicator model variogram, 507
traditional mutation operators, 345
traditional numerical optimisation algorithms, 342
traditional spontaneous potential, 556
traditional systems modelling, 43
traditional variogram model, 111
train ANNs, 380, 388
train artificial neural networks, 365, 366
trained calibration neural network, 508, 510

trained network, 177, 180, 181, 366, 368, 398
trained neural network, 106, 159, 401, 404, 408, 505, 509, 611, 640, 646
trained neural network regression techniques, 415
training accuracy, 379
training algorithm, 25, 139, 211, 400, 610
training cells, 589
training neural network models, 473
training neural networks, 182, 367
training patterns, 177
training performance, 402, 407
training phase, 15, 121, 123, 130, 278, 400, 476
training process, 279, 280, 413, 417, 607, 628, 638, 644
training progress, 283
training reservoir model, 111
training set, 176, 179, 295, 317, 379, 380, 397, 400, 407, 408, 417, 585, 589, 599
transfer function, 24, 25, 114, 138, 139, 379, 513
tuning parameter, 211, 653
tuning process, 414
tuning sedimentary models, 215
turbidite, 371, 541, 543
turbidite flows, 542, 543, 547, 550
turbidite reservoir, 294, 373, 536, 554
turbidite sand, 303
turbidite sandstone, 527, 535, 537, 542
turbidite sediments, 294, 295
turbidite sequences, 289, 294
two-dimensional fuzzy set, 211
two-dimensional input, 284
two-dimensional mapping, 662
two-dimensional model, 189, 327, 662, 663
two-dimensional projection, 41
two-phase model, 327

uncertainty analysis, 5, 394, 574
uncertainty assessment, 97
uncertainty indicator, 175
uncertainty model, 98
undersaturated oil compressibility, 413, 416, 418, 427, 429
undrained oil, 146
upscaled reservoir property, 284, 286
upscaled simulation, 273
upscaling algorithms, 273
upscaling technique, 275

variogram, 8, 21, 100–102, 104, 110, 111, 115, 176, 219–222, 224, 225, 231–233, 235, 241, 274, 328, 332, 499, 500, 506, 524, 557, 565
variogram function, 500
variogram information, 106
variogram model, 108, 110, 113, 328, 512, 517, 558
virtual effective porosity log, 611
virtual intelligence techniques, 447
virtual log, 616, 628, 631
virtual magnetic resonance logs, 605–607, 610, 611, 616, 623, 624, 626, 628, 629, 631
virtual magnetic resonance permeability logs, 628, 630
virtual MBVI log, 611
virtual MPHI, 607, 608, 611
virtual MPHI log, 616
virtual MR log methodology, 616
virtual MR methodology, 607
virtual MRI logs, 607, 616, 623
virtual neutron porosity, 611
virtual neutron porosity logs, 613
virtual permeability log, 616
virtual porosity indicator logs, 611, 612, 628
viscosity correlations, 422
Voronoi tessellation, 273, 276, 278, 279

wavelet classification, 12
wavelet response, 12
wavelet transforms, 273
weighted absolute differences, 330
weighted cross-correlation, 681
weighted linear combination, 91
well-log, 107, 476, 675, 676, 680, 687
well-log attributes, 106
well-log traces, 674, 680
well-to-well correlation, 673, 674, 683
well-to-well correlation result, 684, 686
well-to-well log correlation, 673, 675, 687
well-to-well log correlation process, 687
well-to-well stratigraphic correlation, 674
wireline, 398, 401–403, 405, 406, 410
wireline log, 408, 674, 677
wireline logging, 249, 673
wireline logs, 16, 17, 120, 136, 397, 398, 400, 536, 674, 681
worst function evaluation, 73